Current Topics in Behavioral Neurosciences

Volume 27

Series editors

Mark A. Geyer, La Jolla, CA, USA
Bart A. Ellenbroek, Wellington, New Zealand
Charles A. Marsden, Nottingham, UK
Thomas R.E. Barnes, London, UK

About this Series

Current Topics in Behavioral Neurosciences provides critical and comprehensive discussions of the most significant areas of behavioral neuroscience research, written by leading international authorities. Each volume offers an informative and contemporary account of its subject, making it an unrivalled reference source. Titles in this series are available in both print and electronic formats.

With the development of new methodologies for brain imaging, genetic and genomic analyses, molecular engineering of mutant animals, novel routes for drug delivery, and sophisticated cross-species behavioral assessments, it is now possible to study behavior relevant to psychiatric and neurological diseases and disorders on the physiological level. The *Behavioral Neurosciences* series focuses on "translational medicine" and cutting-edge technologies. Preclinical and clinical trials for the development of new diagnostics and therapeutics as well as prevention efforts are covered whenever possible.

More information about this series at http://www.springer.com/series/7854

Eleanor H. Simpson · Peter D. Balsam
Editors

Behavioral Neuroscience of Motivation

Springer

Editors
Eleanor H. Simpson
New York State Psychiatric Institute
Columbia University Medical Center
New York, NY
USA

Peter D. Balsam
Department of Psychology
Barnard College, Columbia University
New York, NY
USA

ISSN 1866-3370   ISSN 1866-3389 (electronic)
Current Topics in Behavioral Neurosciences
ISBN 978-3-319-80044-8   ISBN 978-3-319-26935-1 (eBook)
DOI 10.1007/978-3-319-26935-1

© Springer International Publishing Switzerland 2016
Softcover reprint of the hardcover 1st edition 2016
This work is subject to copyright. All rights are reserved by the Publisher, whether the whole or part of the material is concerned, specifically the rights of translation, reprinting, reuse of illustrations, recitation, broadcasting, reproduction on microfilms or in any other physical way, and transmission or information storage and retrieval, electronic adaptation, computer software, or by similar or dissimilar methodology now known or hereafter developed.
The use of general descriptive names, registered names, trademarks, service marks, etc. in this publication does not imply, even in the absence of a specific statement, that such names are exempt from the relevant protective laws and regulations and therefore free for general use.
The publisher, the authors and the editors are safe to assume that the advice and information in this book are believed to be true and accurate at the date of publication. Neither the publisher nor the authors or the editors give a warranty, express or implied, with respect to the material contained herein or for any errors or omissions that may have been made.

Printed on acid-free paper

This Springer imprint is published by Springer Nature
The registered company is Springer International Publishing AG Switzerland

Preface

Motivation, defined as the energizing of behavior in pursuit of a goal, is a fundamental element of our interaction with the world and with each other. All animals share motivation to obtain their basic needs, including food, water, sex and social interaction. Meeting these needs is a requirement for survival, but in all cases the goals must be met in appropriate quantities and at appropriate times. Therefore motivational drive must be modulated as a function of both internal states as well as external environmental conditions. The regulation of motivated behaviors is achieved by the coordinated action of molecules (peptides, hormones, neurotransmitters etc), acting within specific circuits that integrate multiple signals in order for complex decisions to be made.

In the past few decades, there has been a great deal of research on the biology and psychology of motivation which is reviewed in this volume. Much of the work reviewed involves the investigation of specific aspects of motived behavior using multiple levels of analyses. In this way, the underpinning neurobiological mechanisms that support relevant psychological processes can be identified. In this volume, the first part considers the neurobiology of components of healthy motivational drive, and includes chapters that are focused on specific motivational goals e.g. food, sex, social interaction, escape. The second part is concerned with neural measures and correlates of motivation in humans and other animals. The next three parts of the book deal with disorders in which abnormal motivation plays a major role. Much space is devoted to this aspect of motivation because deficits in motivation occur in a number of psychiatric disorders, affecting a large population, and severe disturbance of motivation can be devastating. Deficits in motivation fall into two distinct categories: Apathy and pathological deficits in motivation which are commonly seen in patients with schizophrenia and affective disorders. The other category involves problematic excesses in behavior including addictions, the pathological misdirection of motivation. Each of these categories is addressed in separate parts, which are followed by a part on the development of treatments for disorders of motivation. The first chapter in this volume provides a more detailed roadmap of the content and also discussion of the themes that cut across chapters and parts of the book. It is hoped that the collection of reviews in the volume will expose scientists to a breadth of ideas from several different sub-disciplines, thereby inspiring new directions of research that may increase our understanding of motivational regulation and bring us closer to effective treatments for disorders of motivation.

Contents

The Behavioral Neuroscience of Motivation: An Overview
of Concepts, Measures, and Translational Applications   1
Eleanor H. Simpson and Peter D. Balsam

Part I   The Neurobiology of Components of Motivational Drive

Regulation of the Motivation to Eat   15
Stephen C. Woods and Denovan P. Begg

Sexual Motivation in the Female and Its Opposition by Stress   35
Ana Maria Magariños and Donald Pfaff

Oxytocin, Vasopressin, and the Motivational Forces that Drive
Social Behaviors   51
Heather K. Caldwell and H. Elliott Albers

Roles of "Wanting" and "Liking" in Motivating Behavior:
Gambling, Food, and Drug Addictions   105
M.J.F. Robinson, A.M. Fischer, A. Ahuja, E.N. Lesser and H. Maniates

Circadian Insights into Motivated Behavior   137
Michael C. Antle and Rae Silver

The Neural Foundations of Reaction and Action in Aversive
Motivation   171
Vincent D. Campese, Robert M. Sears, Justin M. Moscarello,
Lorenzo Diaz-Mataix, Christopher K. Cain and Joseph E. LeDoux

**Part II   Neural Measures and Correlates of Motivation Signals
and Computations**

Neurophysiology of Reward-Guided Behavior: Correlates Related
to Predictions, Value, Motivation, Errors, Attention, and Action   199
Gregory B. Bissonette and Matthew R. Roesch

Mesolimbic Dopamine and the Regulation of Motivated Behavior 231
John D. Salamone, Marta Pardo, Samantha E. Yohn, Laura López-Cruz,
Noemí SanMiguel and Mercè Correa

**Learning and Motivational Processes Contributing to
Pavlovian–Instrumental Transfer and Their Neural
Bases: Dopamine and Beyond** 259
Laura H. Corbit and Bernard W. Balleine

**Multiple Systems for the Motivational Control of Behavior
and Associated Neural Substrates in Humans** 291
John P. O'Doherty

**The Computational Complexity of Valuation and Motivational
Forces in Decision-Making Processes** 313
A. David Redish, Nathan W. Schultheiss and Evan C. Carter

Part III   Apathy and Pathological Deficits in Motivation

**The Neurobiology of Motivational Deficits in Depression—An
Update on Candidate Pathomechanisms** 337
Michael T. Treadway

**Motivational Deficits and Negative Symptoms in Schizophrenia:
Concepts and Assessments** 357
L. Felice Reddy, William P. Horan and Michael F. Green

**Motivational Deficits in Schizophrenia and the Representation
of Expected Value** 375
James A. Waltz and James M. Gold

**Mechanisms Underlying Motivational Deficits in Psychopathology:
Similarities and Differences in Depression and Schizophrenia** 411
Deanna M. Barch, David Pagliaccio and Katherine Luking

**Methods for Dissecting Motivation and Related Psychological
Processes in Rodents** 451
Ryan D. Ward

**Part IV   Addiction and the Pathological Misdirection of
Motivated Behaviour**

Motivational Processes Underlying Substance Abuse Disorder 473
Paul J. Meyer, Christopher P. King and Carrie R. Ferrario

**Skewed by Cues? The Motivational Role of Audiovisual Stimuli
in Modelling Substance Use and Gambling Disorders** 507
Michael M. Barrus, Mariya Cherkasova and Catharine A. Winstanley

Part V Developments in Treatments for Motivation Pathologies

The Role of Motivation in Cognitive Remediation for People with Schizophrenia 533
Alice M. Saperstein and Alice Medalia

Distress from Motivational *Dis*-integration: When Fundamental Motives Are *Too* Weak or *Too* Strong 547
James F.M. Cornwell, Becca Franks and E. Tory Higgins

Motivation and Contingency Management Treatments for Substance Use Disorders 569
Kimberly N. Walter and Nancy M. Petry

Index 583

Contributors

A. Ahuja Department of Psychology, Wesleyan University, Middletown, CT, USA

H. Elliott Albers Center for Behavioral Neuroscience, Neuroscience Institute, Georgia State University, Atlanta, GA, USA

Bernard W. Balleine Brain and Mind Research Institute, University of Sydney, Camperdown, NSW, Australia

Peter D. Balsam Psychology Departments, Barnard College, Columbia University, New York, NY, USA; Department of Psychiatry, New York State Psychiatric Institute, New York, NY, USA

Deanna M. Barch Departments of Psychology, Psychiatry and Radiology, Washington University, St. Louis, MO, USA

Michael M. Barrus Djavad Mowafaghian Centre for Brain Health, Department of Psychology, University of British Columbia, Vancouver, BC, Canada

Denovan P. Begg School of Psychology, University of New South Wales, Sydney, Australia

Gregory B. Bissonette Program in Neuroscience and Cognitive Science, Department of Psychology, University of Maryland, College Park, MD, USA

Christopher K. Cain Emotional Brain Institute at NYU and Nathan Kline Institute, New York, USA

Heather K. Caldwell Laboratory of Neuroendocrinology and Behavior, Department of Biological Sciences, School of Biomedical Sciences, Kent State University, Kent, OH, USA

Vincent D. Campese Center for Neural Science, NYU, New York, USA

Evan C. Carter Department of Ecology, Evolution, and Behavior, University of Minnesota, St. Paul, USA

Mariya Cherkasova Department of Medicine, Division of Neurology, Djavad Mowafaghian Centre for Brain Health, University of British Columbia, Vancouver, BC, Canada

Laura H. Corbit Department of Psychology, University of Sydney, Sydney, NSW, Australia

James F.M. Cornwell Department of Behavioral Sciences and Leadership, United States Military Academy, New York, NY, USA

Mercè Correa Department of Psychology, University of Connecticut, Storrs, CT, USA; Àrea de Psicobiologia, Universitat Jaume I, Castelló, Spain

Lorenzo Diaz-Mataix Center for Neural Science, NYU, New York, USA

Carrie R. Ferrario Department of Pharmacology, University of Michigan Medical School, Ann Arbor, MI, USA

A.M. Fischer Department of Psychology, Wesleyan University, Middletown, CT, USA

Becca Franks Animal Welfare Program, University of British Columbia, Vancouver, BC, Canada

James M. Gold Department of Psychiatry, Maryland Psychiatric Research Center, University of Maryland School of Medicine, Baltimore, MD, USA

Michael F. Green VA Greater Los Angeles Healthcare System, University of California, Los Angeles, CA, USA

E. Tory Higgins Department of Psychology, Columbia University, New York, NY, USA

William P. Horan VA Greater Los Angeles Healthcare System, University of California, Los Angeles, CA, USA

Christopher P. King Behavioral Neuroscience Program, Department of Psychology, University at Buffalo, Buffalo, NY, USA

Joseph E. LeDoux Center for Neural Science, Emotional Brain Institute at NYU and Nathan Kline Institute, New York, USA

E.N. Lesser Department of Psychology, Wesleyan University, Middletown, CT, USA

Katherine Luking Neurosciences Program, Washington University, St. Louis, MO, USA

Laura López-Cruz Àrea de Psicobiologia, Universitat Jaume I, Castelló, Spain

Ana Maria Magariños Laboratory of Neurobiology and Behavior, The Rockefeller University, New York, NY, USA

H. Maniates Department of Psychology, Wesleyan University, Middletown, CT, USA

Alice Medalia Department of Psychiatry, Columbia University Medical Center, New York, USA

Paul J. Meyer Behavioral Neuroscience Program, Department of Psychology, University at Buffalo, Buffalo, NY, USA

Justin M. Moscarello Center for Neural Science, NYU, New York, USA

John P. O'Doherty California Institute of Technology, Pasadena, USA

David Pagliaccio Neurosciences Program, Washington University, St. Louis, MO, USA

Marta Pardo Àrea de Psicobiologia, Universitat Jaume I, Castelló, Spain

Nancy M. Petry Department of Medicine, Calhoun Cardiology Center, University of Connecticut School of Medicine (MC 3944), Farmington, CT, USA

Donald Pfaff Laboratory of Neurobiology and Behavior, The Rockefeller University, New York, NY, USA

L. Felice Reddy VA Greater Los Angeles Healthcare System, University of California, Los Angeles, CA, USA

A. David Redish Department of Neuroscience, University of Minnesota, Minneapolis, USA

M.J.F. Robinson Department of Psychology, Wesleyan University, Middletown, CT, USA

Matthew R. Roesch Program in Neuroscience and Cognitive Science, Department of Psychology, University of Maryland, College Park, MD, USA

John D. Salamone Department of Psychology, University of Connecticut, Storrs, CT, USA

Noemí SanMiguel Àrea de Psicobiologia, Universitat Jaume I, Castelló, Spain

Alice M. Saperstein Department of Psychiatry, Columbia University Medical Center, New York, USA

Nathan W. Schultheiss Department of Neuroscience, University of Minnesota, Minneapolis, USA

Robert M. Sears Emotional Brain Institute at NYU and Nathan Kline Institute, New York, USA

Eleanor H. Simpson Department of Psychiatry, New York State Psychiatric Institute, Columbia University, New York, NY, USA

Michael T. Treadway Department of Psychology, Emory University, Atlanta, USA

Kimberly N. Walter Department of Medicine and Calhoun Cardiology Center, University of Connecticut School of Medicine (MC 3944), Farmington, CT, USA

James A. Waltz Department of Psychiatry, Maryland Psychiatric Research Center, University of Maryland School of Medicine, Baltimore, MD, USA

Ryan D. Ward Department of Psychology, University of Otago, Dunedin, New Zealand

Catharine A. Winstanley Department of Psychology, Djavad Mowafaghian Centre for Brain Health, University of British Columbia, Vancouver, BC, Canada

Stephen C. Woods Department of Psychiatry and Behavioral Neuroscience, University of Cincinnati, Cincinnati, OH, USA

Samantha E. Yohn Department of Psychology, University of Connecticut, Storrs, CT, USA

The Behavioral Neuroscience of Motivation: An Overview of Concepts, Measures, and Translational Applications

Eleanor H. Simpson and Peter D. Balsam

Abstract Motivation, defined as the energizing of behavior in pursuit of a goal, is a fundamental element of our interaction with the world and with each other. All animals share motivation to obtain their basic needs, including food, water, sex and social interaction. Meeting these needs is a requirement for survival, but in all cases the goals must be met in appropriate quantities and at appropriate times. Therefore motivational drive must be modulated as a function of both internal states as well as external environmental conditions. The regulation of motivated behaviors is achieved by the coordinated action of molecules (peptides, hormones, neurotransmitters etc), acting within specific circuits that integrate multiple signals in order for complex decisions to be made. In the past few decades, there has been a great deal of research on the biology and psychology of motivation. This work includes the investigation of specific aspects of motived behavior using multiple levels of analyses, which allows for the identification of the underpinning neurobiological mechanisms that support relevant psychological processes. In this chapter we provide an overview to the volume "The Behavioural Neuroscience of Motivation". The volume includes succinct summaries of; The neurobiology of components of healthy motivational drive, neural measures and correlates of motivation in humans and other animals as well as information on disorders in which abnormal motivation plays a major role. Deficits in motivation occur in a number of psychiatric disorders, affecting a large population, and severe disturbance of motivation can be devastating. Therefore, we also include a section on the development of treatments for disorders of motivation. It is hoped that the collection of reviews in the volume will expose scientists to a breadth of ideas from several different subdisciplines,

---

E.H. Simpson (✉)
Department of Psychiatry, New York State Psychiatric Institute,
Columbia University, New York, NY, USA
e-mail: es534@cumc.columbia.edu

P.D. Balsam
Department of Psychiatry, New York State Psychiatric Institute,
Psychology Departments of Barnard College and Columbia University,
New York, NY, USA
e-mail: Balsam@columbia.edu

thereby inspiring new directions of research that may increase our understanding of motivational regulation and bring us closer to effective treatments for disorders of motivation.

Keyword Motivation · Cost-benefit analysis · Addiction · Apathy · Translational research

Contents

1	Why Motivation Is Important to Understand..	2
2	What We Mean by the Word Motivation	3
3	A Simplified Overview of How Motivation Might Work in the Brain	4
4	Cost–Benefit Computation as the Arbiter of Motivated Behavior.....................................	5
5	Research Approaches to Understanding Motivation...	7
6	Organismal Level Biology Is Critical to Understanding Motivation	8
7	Motivation Gone Wrong..	8
8	Treatments...	10
References		11

1 Why Motivation Is Important to Understand

Understanding what drives motivated behavior in humans is a truly fascinating endeavor. But as important as our curiosity for knowing what drives us as individuals, and what supports individual differences in levels of motivation among our friends and colleagues, is the critical question; why do motivational processes get disrupted when the clinical and personal consequences can be so devastating? As we will see across this volume, motivated behaviors involve biological and psychological processes that have undergone evolution at numerous levels, from individual molecules all the way to species-specific social organization. While motivational processes represent heritable traits of fitness, humans suffer from a number of disorders of motivation that can be organized into two distinct categories. The first category is composed of the apathy and pathological deficits in motivation commonly seen in patients with schizophrenia and affective disorders. The second category involves problematic excesses in behavior including addictions, the pathological misdirection of motivation. Developing treatments for disorders of motivation requires a detailed understanding of how motivated behavior occurs, how it is dynamically regulated under normal conditions, and how it is disrupted in disease. This volume provides reviews of recent research in each of these areas.

2 What We Mean by the Word Motivation

The concept of motivation is a useful summary concept for how an individual's past history and current state interact to modulate goal-directed activity. In this book, the authors examine the motivation to pursue many different goals. One general aspect of motivated behaviors is that they lead to a goal and obtaining the goal is rewarding. Thus, motivation, defined as the energizing of behavior in pursuit of a goal, is a fundamental property of all deliberative behaviors. One of the earliest psychological theories of motivation, Hull's drive theory, posited that behaviors occur to reduce biological needs, thereby optimizing the organism's potential for survival (Hull 1943). However in Hull's theory, motivational drive functioned solely to energize responding, drive was not responsible for initiating, or maintaining the direction of action. Later, motivation was conceptualized to consist of both a goal-directed, directional component and an arousal, activational component (Duffy 1957; Hebb 1955). This is the framework of motivation still in use, such that if motivation were a vector—its length would represent the amplitude, or intensity of pursuit, and the angle of the vector would represent its focus on a specific goal. In this analogy, a motivation vector affected by apathy might have a reduced length in all directions and a motivation vector affected by addiction might have an increased length and a less flexible direction. The chapters in this volume explicitly acknowledge that motivation affects which responses occur as well as the vigor of those responses. It appears that we are just beginning to understand that these two aspects of motivation have both common and distinctive neural underpinnings. For example, circadian factors may energize the general motive of seeking food or mate (Antle and Silver, this volume), but the specific actions that occur in pursuit of these goals are regulated by different substrates (see Caldwell and Alders, Woods and Begg, Magarinos and Pfaff, all in this volume). Similarly, local cues that signal food availability may energize many food seeking actions, signals for specific foods differentially energize actions associated with obtaining the specific outcome. Again, the neural substrates of the general and specific effects are somewhat distinctive. At each level, whole classes of specific actions are made more or less likely by these factors (Neuringer and Jensen 2010). We suggest that there is generally a hierarchical structure to motivation in the sense that general arousal factors such as sleep–wake cycles will affect many different motives, that activation of specific motives (e.g., hunger, thirst, social motives) can activate many specific actions that could lead to many specific outcomes within a general class of goals, and that more temporally and situationally specific factors determine the specific actions that occur in pursuit of that goal (Timberlake 2001). With this in mind, it is clear that disruptions in motivation can occur at multiple levels of control which suggests there may be multiple interacting ways to attempt to treat disruptions.

3 A Simplified Overview of How Motivation Might Work in the Brain

Many different factors influence motivation, including the organism's internal physiological states, the current environmental conditions, as well as the organism's past history and experiences. In order for all these factors to influence motivation, information about them must be processed in a number of ways; it must be evaluated and encoded, and unless the motives are novel, the valuation and encoding will be affected by learning and retrieval processes. A simplified overview of how such diversity of information must be processed and integrated to result in motivation (both response selection and action vigor) is shown in Fig. 1. Here, we organize the problem into a single, highest order concept that motivated behaviors represent the actions associated with the highest net value that results from a cost–benefit analysis that encompasses all of the potential influencing factors and processes.

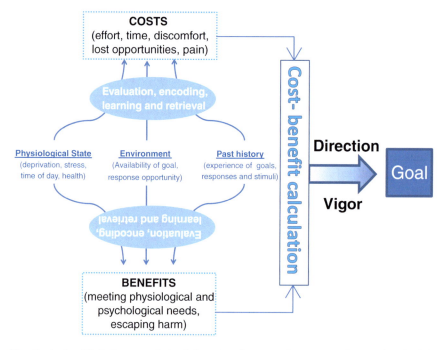

Fig. 1 A simplified diagram of the influencing factors and processes that are involved in motivation. This framework of motivation places cost–benefit analysis central to the concept of motivation. Three major categories of factors are known to influence motivation: the individual's physiological state, the environment, and the individual's past history. Information about all 3 categories of factors will be subject to a number of processes (represented inside the blue oval), including evaluation and encoding. In almost all circumstances, the motive, environment, and physiological state will not be novel; therefore, information will also undergo learning and retrieval processes. All of the combined processes result in weighting of all the costs and benefits related to the motive, and the output of the cost–benefit calculation will impact upon the direction and vigor of action that the individual takes toward the motive goal

4 Cost–Benefit Computation as the Arbiter of Motivated Behavior

The *costs* associated with behavioral action may include physical effort, mental effort, time, loss of potential opportunities, discomfort, and danger (the risk of pain and potential death). The *benefits* associated with behavioral action might include fulfilling physiological and psychological needs, obtaining reinforcement secondary to those needs, escaping from harm, or avoidance of some of the costs listed above. As mentioned above, information entering the cost–benefit computation for any specific motive will be processed in several ways. The value of every cost and every benefit must be calculated and encoded. The concept of encoding value and experimental methods for measuring encoded value are discussed in detail by Redish et al., in this volume. It is important to consider that value must be encoded when a goal is obtained and then stored for future retrieval when obtaining that goal again becomes relevant. When that happens in the current moment, the assessment of value must be conditioned both on this past experience as well as the current state and environmental conditions. Goals are likely to be obtained with some temporal distance from the initiation, or even conclusion of behavioral output. Single neuron activity in several brain regions including orbitofrontal cortex, anterior cingulate, and basolateral amygdala has been shown to correlate with reward prediction and this work is reviewed by Bissonette and Roesch in this volume. The encoded values of costs and benefits do not belong in absolute scales because the values of all costs and benefits are rendered relative to the animal's current physiological state as well as the current conditions of the surrounding environment.

Much has been learned about the role of dopamine in reinforcement learning, and its impact on motivated behavior from experimental manipulations of the dopamine system in rodents. This work is comprehensively reviewed by Salamone et al. in this volume. In addition to learning about the costs and benefits of a particular action, subjects also learn about specific signals that are associated with obtaining particular goals. Such signals can have an enormous influence on motivated behavior, and several chapters in this volume provide details of when and how environmental cues can influence response selection and response activation. These include Corbit and Balleine's chapter on learning and motivational processes contributing to Pavlovian–instrumental transfer and John O'Doherty's chapter on the neural substrates of motivational control in humans. Cue learning is also discussed in the context of motivational disorder, in the chapters by Meyer et al. and Barrus et al., that deal with substance abuse disorders and gambling.

Another aspect of the computation that energizes specific action has to do with signals that a particular goal is currently available. These signals occur on multiple timescales. Specific times of day can become associated with the opportunity to obtain specific goals, and discrete cues can signal the opportunity to achieve a goal as well as what specific ways there are to achieve it. For example, when meals occur at a regular time of day, there are behavioral, hormonal, and neural changes that occur in anticipation of a meal time that give rise to the motivation to seek food (Antle and Silver, in this volume). Encountering a restaurant can activate the specific behavioral sequences that lead to the ordering of food and the specific foods themselves can activate specific consummatory responses. All along this sequence of temporally organized behavior, there are concomitant changes in hormonal and neural states that energize and guide action (Woods and Begg in this volume).

After effective encoding of all the relevant costs and benefits, a computational process (the cost–benefit computation) is required to resolve the appropriate direction and vigor of action to be taken. This complex interplay of factors and processes is schematized in Fig. 1.

How conceptually the cost–benefit computation is made is currently unclear. It is still unknown whether the value of costs and benefits are calculated on the same scale or not, whether their weights are integrated or subtracted such that, for example, the amount of predicted effort reduces the value of the predicted reward. Or perhaps, there is a circuit component that acts as a comparator of these two component values. An additional complication is that for any given motive, there are often multiple types of costs and potentially multiple types of benefits involved because many different types of control systems and circuits are at play (e.g., neuroendocrine, circadian, Pavlovian). This leads to the question of how so much diversity of information can all be used to make an appropriate response selection and determine action vigor. Do all factors enter into a singular, highly complex equation, as our simplified diagram (Fig. 1) may seem to imply? Or do some systems continually run in parallel, with behavioral output as the result of a hierarchical switching from one system to another? Or perhaps there is fluctuation in the degree to which different factors influence the computation, e.g., the relative weights of physical and mental costs depend on the energy state of the organism.

Furthermore, it is possible that these alternative regulatory schemes are not mutually exclusive. For a detailed discussion on the potential mechanisms by which multiple deliberative processes that are running in parallel may each influence motivation (see Redish et al. in this volume). In the case of appetitive conditioning, there is evidence to suggest that animals can rapidly switch between responding that is driven by two different control systems, goal directed or habitual (Gremel and Costa 2013). Audiovisual cues can trigger the rapid switching, implying that these two alternate circuits are constantly online and available in parallel. On the other hand, instead of a multi-tiered, hierarchical, or switching system, other work suggests that all information enters a singular computation process, and the output of this meta-computation is what drives motivation. This concept is favored by Magarinos and Pfaff (this volume) whose work on the sexual motivation of female rodents may suggest that for this specific motive, at least some factors may be integrated into a single decision-making process.

Above we have described the complex situation of many different factors influencing a single motive. It must also be recognized that at any given time, there may be competition for multiple goals and that imbalances in the strength of the motivations for each goal can cause conflict and dysfunctional behavior. The chapter by Cornwell et al. describes how human well-being depends not only on satisfying specific motives, but also on ensuring that motives work together such that no individual motive is too weak or too strong. It is becoming clear that different motivational systems have control elements that are unique to each system but that there may also be common substrates, perhaps close to the final steps that determine behavioral output. This is well illustrated in the chapter on defensive motivation by Campese et al. and in the chapter on social motivation by Caldwell and Alders. Again, the neurobiological mechanism whereby different motive systems interact is an important but not yet well-understood problem. Of particular interest will be to understand how defensive motivations interact with appetitive ones. The vast majority of modern work on motivation concerns itself with the mechanisms of appetitive motivation. Campese et al. show how to leverage what is now known about fear learning to understand the neurobiological mechanisms of defensive motivations. In a similar vein, Cornwell et al. argue for the importance of understanding how promotion/prevention motives in humans is an important modulator of other motives. Hopefully, the future will include a greater focus on understanding defensive motivations.

5 Research Approaches to Understanding Motivation

To increase our understanding of motivation in the brain, there are numerous approaches that can be taken. In this volume, many different academic approaches are represented as the research reviewed includes clinical, experimental, and comparative psychology; and several neuroscience subfields including, cognitive, molecular, cellular, behavioral, and systems neuroscience. This means that specific questions or single hypotheses can be, and often are being, approached at multiple levels of analysis. Indeed, it is when research programs combine a number of techniques, or use information derived from a few different techniques to propose (and test) new hypotheses that the most compelling results are obtained. For example, the work described by O'Doherty in this volume includes the use of human fMRI studies to investigate potential action-value signals that have been proposed from rat electrophysiological recordings. The research described by Redish et al. considers computational models of decision making and tests these models by measuring neuronal activity during deliberative behavior. In the chapter by Ward, the approach to testing motivational deficits in mice has very much been informed by the data from molecular and clinical studies in humans. By phenocopying in mice the molecular changes that have been detected in patients using PET imaging techniques, the behavioral consequences can be probed under well-controlled conditions. In a similarly translational manner, the research described by Robinson et al. in this volume applies electrophysiological and optogenetic techniques in rodents to probe behaviors that are altered in people with addictions. In the chapter by Barrus et al., the authors discuss the development of rodent paradigms designed to test various psychological theories of substance use and gambling disorders.

6 Organismal Level Biology Is Critical to Understanding Motivation

When multiple levels of analyses are used to investigate motivational processes, a critically important concept becomes apparent. While the evolution of traits that support motivation occurs at the level of molecules, proteins, cells, and circuits, it is the entire organism, and its interaction with the environment that is selected. An example of this concept is easily seen in the research on circadian modulation of motivation (Antle and Silver) and in the work on motivation for eating (Woods and Begg). For example, in the case of feeding we know many of the molecules and circuits involved in both the intrinsic, homeostatic factors which drive the motivational to eat, such as hormones and peptides, and we also know the neuromodulators and circuits that are responsible for some of the extrinsic/environmental influences on eating such as predictive cues. We are beginning to understand how these signals are integrated in order for decisions to be made and behavioral responses to occur; though as described above, understanding the mechanism of integration is currently a critical area of research.

7 Motivation Gone Wrong

Patients with many different psychiatric diagnoses may experience deficits in motivation, including depression, schizophrenia, bipolar disorder, PTSD, and anxiety disorders. In this volume, we focus on the neurobiology of motivational deficits in depression and schizophrenia primarily because these are the two illnesses in which pathological deficits in motivation play a major role in patient functioning and clinical outcome (Barch et al. 2014; Strauss et al. 2013). As such, far more research has been done on motivation in depression and schizophrenia than any other illness. In the last few decades, it has been recognized that the motivational deficits in Schizophrenia and depression share similarities, but also distinct differences. These differences occur because as mentioned above, there are many components involved in motivated behavior and each of them represent potential vulnerabilities that may be involved in different pathophysiological mechanisms. An excellent review of the similarities and differences in mechanisms underlying motivational deficits in depression and schizophrenia is provided by Barch et al. in this volume. The central difference in these types of pathologies is that many depressed patients suffer from impairments of in-the-moment hedonic reaction. Such anhedonia can diminish an individual's capacity for anticipation, learning, and effort. In contrast, patients with schizophrenia demonstrate relatively intact in-the-moment hedonic processing. Instead, patients suffer impairments in other components involved in translating reward experience to anticipation and action selection.

There are also separate chapters that go into more specific detail for each of these pathophysiological conditions. An update on candidate pathomechanisms for motivational deficits in depression is provided by Treadway's chapter. This volume devotes two chapters to the topic of motivation deficit in schizophrenia because this area of research has been more active than it has been in depression. This is likely because antipsychotic medications that successfully ameliorate the positive symptoms of schizophrenia (delusions, hallucinations, etc.) have been available for some time, leaving patients with the residual negative symptoms, of which amotivation is the primary driver of poor outcome and low quality of life (Kiang et al. 2003).

Current concepts of motivation deficits and how motivation is assessed in patients with schizophrenia is reviewed by Reddy et al. Waltz and Gold extend these concepts into the exploration of the relationship between amotivation and the representation of expected value. The clinical research reviewed in the chapters that deal with apathy and motivation in humans is complimented by a chapter on methods for dissecting motivation and related psychological processes in rodents (Ward). Research using animal models is critical for several obvious reasons, including the availability of genetic manipulations, molecular modifications as well as invasive in vivo monitoring procedures that are not possible in human subjects. What hasn't previously been obvious is how well we can use such animal models to investigate the various components of motivation that are particularly relevant to human disease. Ward describes such procedures and explains how best to leverage our current clinical knowledge using state-of-the-art mouse models.

On the flip side of apathy may be when motivation for a specific goal can come to dominate action in maladaptive ways as appears to be the case in addictions. Excessive behavior for many types of rewards including drugs, food, gambling, and sex can be problematic. In addiction, rapid and strong learning about what leads to reward, excesses in experiencing the hedonic value of rewards, exaggeration in representing those values, and dominance in being guided by those representations can all lead to significant narrowing in the diversity of motives. Several theories exist that attempt to explain the process of addiction in terms of disruption of motivational processes. Each theory differs in the emphasis on which specific aspects of motivation are primarily affected. The chapters by both Meyer et al. and Robinson et al. describe the motivational processes underlying substance abuse disorder. The chapter by Barrus et al. extends this discussion into the field of gambling. Barrus et al. suggest that many of the processes affected in gambling are the same as those affected in drug addiction, and therefore, the paradigms that have been successfully used to study drug addiction in animal models can be successfully modified to identify neurobiological mechanisms related to gambling. The central hypothesis in these analyses of addictions of drugs and gambling (as well as addiction to food and other things) is that an aberration in reward processing and/or in the control by cues associated with these rewards underlies the problematic nature of addictive behavior and its resistance to change.

8 Treatments

Given the modern emphasis on reward processes as a fundamental component of motivation, it is encouraging that modern cognitive/behavioral approaches to treating motivational disturbances focus on creating reward contingencies that modify deficits or excesses in behavior. Saperstein and Medalia describe how in schizophrenia patients motivation enhancing techniques are critical to treatment-related improvements within cognitive remediation therapy. In the case of addictions, Walter and Petry provide an overview of research indicating that contingency management is a demonstrably effective psychosocial treatment for substance use disorders. The central concept of contingency management is that extrinsic motivators are used to change patients' behaviors. Specifically, reinforcement is provided when patients demonstrate abstinence. In the descriptions of both treatment approaches, the chapters consider the important role that intrinsic motivation may play in clinical success.

We are hopeful that the great progress in understanding the neurobiology of motivation described in this book will influence new ideas that will lead to novel pharmacological, physiological and psychological/psychosocial approaches to treatments for disorders of motivation. The identification of novel pharmacological treatments is dependent on the ability of preclinical researchers to investigate potential targets and screen potential candidate compounds using truly meaningful endophenotypic assays. The chapters in this volume that describe clinical studies of patient with disorders of motivation describe how motivation has been dissected into a number of component processes and the specific processes that are selectively disrupted in disease have been identified (Reddy et al., Barch et al., Waltz and Gold). To identify drugs that will be effective for disorders of motivation, preclinical assays must focus on the same specific processes affected in humans (see Ward in this volume). A recent example of the development of the kind of research tools that are needed for the purpose of investigating potential treatment targets is the strategy of dissecting goal-directed action from arousal by modifying previously existing rodent behavioral tasks (Bailey et al. 2015a). These new tools can then be used to assay specific effects of drugs that affect novel treatment targets (Simpson et al. 2011) and Bailey et al. (2015b). This novel approach was directly inspired by the literature on the selectivity of processes disrupted in humans with disorders of motivation.

In addition to pharmacological treatments, there is also the possibility that electrophysiological treatments for disorders of motivation may be developed. Deep brain stimulation (DBS) is currently used to treat a number of neurological and psychiatric conditions (Kocabicak et al. 2015; Kringelbach et al. 2007; Udupa and Chen 2015). DBS has been used to treat essential tremor, Parkinson's disease, treatment refractory major depression, severe obsessive–compulsive disorder, and chronic pain for several years. Several other applications are in experimental stages, including clinical trials for pervasive addiction and symptoms of schizophrenia. Such an invasive procedure requires many successive small-scale clinical trials before optimal procedures can be successfully developed. A significantly less invasive procedure used to modulate brain activity is transcranial magnetic stimulation (TMS). While the mechanism(s) by which TMS alters neuronal function and network activity is not understood, due to its noninvasiveness, hundreds of clinical trials have been conducted for a long list of neuropsychiatric conditions, including schizophrenia, and craving/addiction. A comprehensive review of studies of repetitive TMS has recently been published (Lefaucheur et al. 2014).

Lastly, we are also hopeful that the emerging understanding that there are multiple systems driving motivation on an organismal level will lead to the development of treatment schemes that are more comprehensive than those that have been developed in the past. It may be that subtle adjustments in several of the factors that are involved in disorders of motivation (the endocrine system, circadian system, neurotransmitter function, etc.) can result in greater improvements and less side effects than treatments that focus on a single system.

References

Bailey MR, Jensen G, Taylor K, Mezias C, Williamson C, Silver R, Simpson EH, Balsam PD (2015a) A novel strategy for dissecting goal-directed action and arousal components of motivated behavior with a progressive hold-down task. Behav Neurosci 129:269–280

Bailey MR, Williamson C, Mezias C, Winiger V, Silver R, Balsam PD, Simpson EH (2015b) The effects of pharmacological modulation of the serotonin 2C receptor on goal-directed behavior in mice. Psychopharmacology. doi: 10.1007/s00213-015-4135-3

Barch DM, Treadway MT, Schoen N (2014) Effort, anhedonia, and function in schizophrenia: reduced effort allocation predicts amotivation and functional impairment. J Abnorm Psychol 123:387–397

Duffy E (1957) The psychological significance of the concept of arousal or activation. Psychol Rev 64:265–275

Gremel CM, Costa RM (2013) Orbitofrontal and striatal circuits dynamically encode the shift between goal-directed and habitual actions. Nat Commun 4:2264

Hebb DO (1955) Drives and the C.N.S. (conceptual nervous system). Psychol Rev 62:243–254

Hull CL (1943) Principles of behavior: an introduction to behavior theory. D. Appleton-century company, Oxford

Kiang M, Christensen BK, Remington G, Kapur S (2003) Apathy in schizophrenia: clinical correlates and association with functional outcome. Schizophr Res 63:79–88

Kocabicak E, Temel Y, Hollig A, Falkenburger B, Tan S (2015) Current perspectives on deep brain stimulation for severe neurological and psychiatric disorders. Neuropsychiatr Dis Treat 11:1051–1066

Kringelbach ML, Jenkinson N, Owen SL, Aziz TZ (2007) Translational principles of deep brain stimulation. Nat Rev Neurosci 8:623–635

Lefaucheur JP, Andre-Obadia N, Antal A, Ayache SS, Baeken C, Benninger DH, Cantello RM, Cincotta M, de Carvalho M, de Ridder D, Devanne H, di Lazzaro V, Filipovic SR, Hummel FC, Jaaskelainen SK, Kimiskidis VK, Koch G, Langguth B, Nyffeler T, Oliviero A, Padberg F, Poulet E, Rossi S, Rossini PM, Rothwell JC, Schonfeldt-Lecuona C, Siebner HR, Slotema CW, Stagg CJ, Valls-Sole J, Ziemann U, Paulus W, Garcia-Larrea L (2014) Evidence-based guidelines on the therapeutic use of repetitive transcranial magnetic stimulation (rTMS). Clin Neurophysiol 125:2150–2206

Neuringer A, Jensen G (2010) Operant variability and voluntary action. Psychol Rev 117:972–993

Simpson EH, Kellendonk C, Ward RD, Richards V, Lipatova O, Fairhurst S, Kandel ER, Balsam PD (2011) Pharmacologic rescue of motivational deficit in an animal model of the negative symptoms of schizophrenia. Biol Psychiatry 69:928–935

Strauss GP, Horan WP, Kirkpatrick B, Fischer BA, Keller WR, Miski P, Buchanan RW, Green MF, Carpenter WT Jr (2013) Deconstructing negative symptoms of schizophrenia: avolition-apathy and diminished expression clusters predict clinical presentation and functional outcome. J Psychiatr Res 47:783–790

Timberlake W (2001) Motivational modes in behavior systems. In: Mowrer RR, Klein SB (eds) Handbook of contemporary learning theories. Erlbaum Associates, Hillsdale, pp 155–209

Udupa K, Chen R (2015) The mechanisms of action of deep brain stimulation and ideas for the future development. Prog Neurobiol

Part I
The Neurobiology of Components of Motivational Drive

Regulation of the Motivation to Eat

Stephen C. Woods and Denovan P. Begg

Abstract Although food intake is necessary to provide energy for all bodily activities, considering food intake as a motivated behavior is complex. Rather than being a simple unconditioned reflex to energy need, eating is mediated by diverse factors. These include homeostatic signals such as those related to body fat stores, to food available and being eaten, and to circulating energy-rich compounds like glucose and fatty acids. Eating is also greatly influenced by non-homeostatic signals that convey information related to learning and experience, hedonics, stress, the social situation, opportunity, and many other factors. Recent developments identifying the intricate nature of the relationships between homeostatic and non-homeostatic influences significantly add to the complexity underlying the neural basis of the motivation to eat. The future of research in the field of food intake would seem to lie in the identification of the neural circuitry and interactions between homeostatic and non-homeostatic influences.

Keywords Adipose tissue · Adiposity signal · Agouti-related protein · Allostasis · Amygdala · Anorexia nervosa

Contents

1 Motivation	16
2 Eating	18
3 Homeostatic Versus Non-homeostatic Eating	21
4 Homeostatic Influences	23
5 Non-homeostatic Influences over the Motivation to Eat	26
6 Integration of Homeostatic and Non-homeostatic Influences over Food Intake	27
7 Summary	28
References	29

S.C. Woods (✉)
Department of Psychiatry and Behavioral Neuroscience, University of Cincinnati,
2170 East Galbraith Road, Cincinnati, OH 45237, USA
e-mail: steve.woods@uc.edu

D.P. Begg
School of Psychology, University of New South Wales, Sydney, Australia

The concept of motivation has taken many forms. It appears in our folklore, in our traditions and customs, in our great philosophical systems, and in our more recent science of behavior. Sometimes it is made explicit so that it may be scrutinized, but more often it is implicit, unanalyzed, and unquestioned. The concept of motivation has been variously identified as an unquestionable fact of human experience, as an indisputable fact of behavior, and as a mere explanatory fiction. (Bolles 1967)

Is the motivation to eat excess food, when living in an environment enriched with an overabundance of palatable and hedonically pleasing foods, so compelling that the so-called epidemic of obesity is inevitable? Stated another way, are homeostatic circuits simply overwhelmed by the reward value of overeating?

It is axiomatic that organisms must acquire energy in the form of nutrients to power growth and all other physiological functions including behavior. Most adult animals living in an environment with ample available food ingest nutrients at regular intervals, roughly matching energy intake to energy expenditure over an extended period of time and maintaining body weight/adiposity/energy stores within a narrow range. Individuals who have not ingested sufficient nutrients to maintain weight for some time will seek food and work harder to obtain it and, over time and with food available, will take in sufficient nutrients to reverse any energy deficits that have occurred. This is generally considered to be due to a need-based or homeostatic motivation intended to restore stored energy to optimal levels within the body. The strength of the motivation to obtain food by deprived animals underlies the foundations of many influential theories of learning and other behaviors of the last century (Hull 1931; Miller et al. 1950; Skinner 1930); i.e., organisms easily learn operant tasks that enable them to acquire needed food, and as they consume the food, the motivation apparently diminishes.

1 Motivation

In considering the motivation to eat, an important question is whether eating elicited by a critical deficit of energy is 'motivated?' And is such eating a common event in our everyday lives? The answer lies in the extent to which the act of eating is an unconditioned as opposed to an operant response. Simple reflexes such as the patellar response or salivation in response to food on the tongue have easily identifiable unconditioned stimuli and are consequently considered to be unconditioned responses. The concept of motivation does not enter into a description of the behavior; i.e., one is not considered to be motivated to knee-jerk, or to reflexively salivate. If eating that occurs in response to a deficit of stored or available energy is likewise a simple reflex, should it be considered motivated?

A perhaps more informative question to ask is whether any instances of food intake are truly unconditioned responses. The answer is that unconditioned eating is extremely rare and may never naturally occur in normal situations. It is undeniable that when the energy available to receptor cells in neural circuits linked to food intake is acutely lowered, eating is initiated. This occurs, for example, when blood glucose and consequently glucose entering the brain (where it is a necessary source of energy) is precipitously lowered by systemic insulin administration (Lotter and Woods 1977; MacKay et al. 1940) or when the capability of brain cells to derive energy from available glucose is blocked by the administration of 2-deoxyglucose (Smith and Epstein 1969). In these instances, if food is available, the onset of eating is quick and robust across species including humans (Grossman 1986; Langhans 1996). Analogously, when an individual has become acclimated to deriving most of its energy from lipids as opposed to carbohydrates, compounds that block the conversion of fatty acids to cellular energy can also elicit acute eating (Langhans and Scharrer 1987). In all of these instances, the rapidly precipitated lack of utilizable energy is recognized by dedicated receptor cells and triggers a number of protective reflexes, including increasing glucose and fatty acid secretion into the circulation from storage organs and decreasing energy expenditure by non-critical tissues, and if food is at hand, it elicits eating behavior as well. Such behavior is innately predetermined and necessary for survival. It is not clear, however, whether such a rapid series of events ever happens other than in laboratories.

In contrast to what occurs when available energy is instantaneously lowered by pharmacological means, when an individual is chronically deprived of food, reductions of fuels (glucose or fatty acids) occur gradually as the body slowly consumes any available stores to keep vital tissues alive. When this occurs, the motivation to eat is strong and continuous, but the eating is not an acutely elicited unconditioned response. This can be readily demonstrated by requiring calorically deprived and underweight individuals to overcome hardships in order to obtain food. In such instances, they eat less if at all. As a common example, the availability of only unpalatable food to an animal results in reduced consumption over time (Sclafani et al. 1996) and maintenance of a body weight that is lower than that of control animals eating regular chow (Ferguson and Keesey 1975; Keesey and Boyle 1973; Naim et al. 1980); the individual is below optimal weight and in principle motivated to eat, but it constrains its behavior. The point is that eating is conditional as opposed to unconditional in physiologically relevant chronic food deprivation situations. We consequently contend that the concept of motivation to eat, rather than applying directly to energy-deficit or homeostatic-need-based behavior, applies principally (if not entirely) to situations that cannot be considered as unconditioned and rather are based on experience as discussed below. That said, there is no doubt that a food-deprived state amplifies the effect of other, non-homeostatic factors, on the motivation to eat.

Another potentially ambiguous situation with regard to motivation, homeostasis, and food taking concerns the concept of satiation. During a meal, ingested food interacts with receptors on the lining of the tongue and digestive tract to generate neural and hormonal signals that facilitate the digestive process, and some of the same signals also influence the brain to elicit a perception of satiation/fullness, and as these satiation signals accumulate, eating eventually ends (Moran 2004; Smith and Gibbs 1984; Woods 2009). Administration of the most-studied such satiation signal, cholecystokinin (CCK), to humans or animals, causes them to reduce meal size, and the administration of compounds that block the action of endogenous CCK results in an increase in meal size (Moran and Kinzig 2004; Smith and Gibbs 1985).

Analogous findings have been found with numerous putative endogenous satiation signals including glucagon-like peptide-1 (GLP-1), peptide YY (PYY), bombesin family peptides, amylin, and apolipoprotein A-IV (apoAIV) (see reviews in Begg and Woods 2013; Woods and D'Alessio 2008), and such phenomena are thought by some to reflect a natural, genetically determined or hardwired, braking system on the motivation to eat. It could be argued that this type of influence over meal size is unconditional, but as detailed below, there is considerable evidence that the ability of endogenous satiation signals, as well as of compounds that influence their availability or potency, is not only conditional but even conditionable.

So, how is the issue of the motivation to eat to be conceptualized? We contend, like many before us (see comprehensive review in Bolles 1967), that the concept of motivation to eat implies some sort of incentive, whether homeostatically or non-homeostatically based. One is motivated to obtain what one knows is, or might be, available. This implies some sort of experience or learning with the situation. Caloric deprivation interacts with this in basic ways. A deprived individual works harder, or moves faster, or suffers more hardships, or more readily acquires new skills, in order to obtain food. Greater deprivation leads to increased activity and exploration, activities likely to bring one in contact with food. All of these fit well with the contemporary concept of motivation (Bouton 2011).

2 Eating

Eating is a complex behavior. In addition to providing both micro- (vitamins, minerals) and macronutrients (calories), ingesting food interacts with the reflexive control of many critical parameters including the circulating levels of plasma fuels (glucose, fatty acids, and others) and plasma osmolality (sodium, water, and other constituents), with blood pressure, with body temperature, with the maintenance of body fat, and others. Eating itself is not a regulated variable, but rather is a behavior that can be recruited in the maintenance of any of several key physiological parameters (Woods 2009). Given ample available food with no constraints, rats adopt a pattern of eating a large number of small meals each day [i.e., they become 'nibblers' (Collier et al. 1986, 1992)], and by so doing, they maintain a relatively consistent amount of body fat without having to deal with the metabolic consequences of large meals (Woods 1991, 2002). However, an individual will readily abandon this habitual and presumably preferred eating pattern when conditions are changed. These include situations such as having food available only at limited times each day, having to work hard to gain access to food, receiving treatments that limit the amount that can be consumed in one meal, having access to other activities (exercise; socializing), and the presence of nearby predators. When such constraints are imposed, animals eat at different times, or less often but with larger individual meals, or adopt any strategy that allows them to acquire sufficient calories to maintain body fat at its customary level and in the environment in which they find themselves (see reviews in Collier and Johnson 2004; Schneider et al. 2013;

Woods 2002). Likewise, eating can occur when unanticipated opportunities arise, or as the social situation dictates. All of these kinds of laboratory influences also impact human food taking. Babies, like most animals, prefer a large number of small meals each day, and as they age and social constraints are imposed (e.g., for the sanity of the parents), adapt to a schedule of less frequent and larger meals, and this persists throughout life as meals and snacking become integrated with life schedules and events.

The point is that eating itself (timing, meal size, etc.) is adaptable to a broad range of environmental conditions. As discussed above, living in a stable environment with ample and consistent food, such as occurs in most laboratory situations, provides the luxury of being able to establish regular eating patterns and optimally integrate caloric intake with other behaviors (Schneider et al. 2013; Strubbe and Woods 2004). The amount eaten in individual meals at specific and clearly predictable times of the day becomes habitual; i.e., because the food supply in most experiments is reliable, and the food itself is the same from meal to meal and day to day, the individual readily learns to make responses to take in and process energy most efficiently (Woods 1991). Highly predictable temporal cues (such as the timing of lights-on and lights-off) become optimal times to eat because the individual learns to make meal-anticipatory (also known as cephalic) responses at those times, responses that circumvent the negative impact of meals, such as extreme hyperglycemia, hyperthermia, and hyperlipidemia (Woods 2009). Stimuli reliably associated with consuming the customary food, such as odors, tastes, and mouth feel, therefore become bellwethers of food quality and quantity; such stimuli become distal controllers, and presumably motivators, of eating because when consistent associations with food and its caloric content occur, individuals can acquire the ability to use these distal cues to guide food-taking behavior. They are distal cues in the sense that while their presence has been reliably associated with energy content in the past, they do not inherently denote energy content themselves. In a less predictable environment, the association between distal cues and actual energy content can break down, requiring the individual to rely to a greater extent upon more proximal cues; i.e., cues more closely tied to actual caloric content. Satiation signals such as CCK are intermediate cues, being secreted in response to a physicochemical analysis of ingested food by intestinal cells. To summarize, when the food ingested is consistent for every bout of eating, diverse sensory signals become accurate indicators of the number of calories that can be anticipated to enter the blood from the intestines over the subsequent hours. These include sensations related to the amount of food passing through the mouth, or to the level of CCK once eating has begun, or to the degree of gastric distension as the meal progresses, or to any other stimuli that reliably correlate with ingested caloric content. Nonetheless, none of these surrogate 'satiation' signals is necessarily hardwired to a specific calorie content. The ultimate proximal cues that directly signal energy status to the neural circuits influencing eating are glucose, fatty acids, amino acids or other energy-rich molecules or their metabolites. The point is that there is an energy-reliability gradient that maps onto stimuli an individual can use to guide how much is eaten during a meal. In a *highly predictable* world, distal cues increase the motivation to eat and a meal's initiation, and once the meal is underway, distal and perhaps intermediate cues signal how many calories are being eaten and consequently when the meal should end.

Therefore, in a consistent environment, distal cues such as the time of day collectively help prepare the body to anticipate and hence cope with the impending nutrients. An important principle is that it is advantageous to accurately gauge the best time to start eating as well as when to stop the meal so that the meal's adverse metabolic consequences are minimized (Woods 1991). Hence, signals that herald a meal is imminent, such as the timing of lights going off, elicit cephalic responses such as increased insulin and ghrelin prior to when the meal actually begins (Drazen et al. 2006; Teff 2000), and these cephalic responses render the digestive process more efficient. While time of day is a key determinant of mealtime in most laboratory settings, animals can also learn to make meal-anticipatory behaviors in response to arbitrary stimuli that reliably indicate food availability, and those same stimuli can elicit eating, even in sated animals (Sclafani 1997; Weingarten 1983). Therefore, the most reliable indicators of food availability acquire the ability to control meal onset, presumably by increasing the motivation for food. As an example, secretion of cephalic insulin renders the individual more glucose tolerant; in the absence of cephalic insulin individuals either must consume very small meals or else appear diabetic (Teff 2011).

The continuum of cues from distal to intermediate to proximal also correlates with time until ingested nutrients become available to the body via the blood. An individual using taste or other oral cues can stop eating long before most nutrients are digested and absorbed, with the assurance that sufficient but not excessive calories have been secured. As the individual shifts to using more proximal cues, the lag between ingestion and the physiological consequences lessens, increasing the probability of overconsuming and having to deal with too many calories entering the blood at once (leading to symptoms of diabetes). Controlling intake via distal cues additionally bestows the advantage of being able to eat relatively large meals because appropriate anticipatory responses can be made sufficiently far in advance to lessen the meal's metabolic impact. As discussed above, anticipatory responses made prior to the actual absorption and even ingestion of food, such as the secretion of cephalic insulin, reduce the prandial increase of glucose and other nutrients that would otherwise occur (Teff 2011; Woods 1991). An individual faced with a novel food (novel in the sense that its odor, taste, and other attributes have not been encountered previously, or else have not been associated with any particular food) eats very little when the food is first made available, even if it is food-deprived. Because the individual has no distal cues on which to rely, it appears neophobic (Armelagos 2014; Birch and Fisher 1998). Small meals continue to be consumed until the consequences of early ingestion prove to be safe, and reliable distal cues can be established (Rozin 1990). In such an instance, the absence of functional distal cues caused by the lack of experience with the food elicits little or no motivation to eat.

3 Homeostatic Versus Non-homeostatic Eating

Separation of influences over the motivation to eat into those that are need- or deprivation-based (homeostatic) and those that are not (hedonic, opportunistic, social, based on learning, stress or emotional-related, cognitive) has considerable face validity. Until the last ten or fifteen years, the brain was considered to have distinct and for the most part anatomically separated circuits collating these two types of influences, with integration occurring mainly very close to the behavioral act of eating itself in hindbrain motor control areas as depicted in Fig. 1. However, the ability to identify and molecularly and genetically map specific neuronal types, neurotransmitters and their receptors, as well as entire neural circuits has revealed that in fact, there is considerable cross talk and overlap at almost every circuit previously thought to be a distinct homeostatic or non-homeostatic pathway, as depicted in Fig. 2. There are many reviews of these phenomena and the change in thinking about the neurocircuitry of food intake (Berthoud 2004; Figlewicz and Sipols 2010; Petrovich 2013; Richard et al. 2013; Rinaman 2010; Zheng and Berthoud 2007) such that only a brief outline is presented here. An important principle is that homeostatic signals, in addition to triggering activity in the

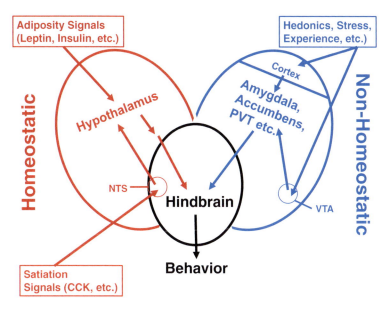

Fig. 1 Historic depiction of how homeostatic and non-homeostatic influences over the motivation to eat are considered. Large sections of the brain were considered distinct for the two types of influences and for the most part not in much contact with one another until the final decision to behave, i.e., to initiate or stop eating at the level of the hindbrain. NTS—nucleus of the solitary tract, which receives satiation signals during meals. VTA in the midbrain which is the source of the mesolimbic and mesocortical dopamine tracts. PVT—paraventricular nucleus of the thalamus is a region that can integrate information on cognitive, emotional, and anxious states with the reward value of food

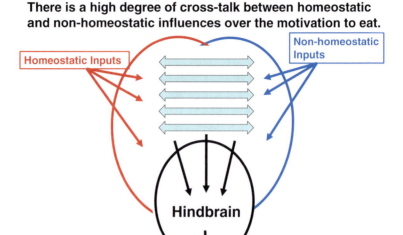

Fig. 2 More nuanced depiction of how homeostatic and non-homeostatic influences over the motivation to eat are considered. In this conception, integration of the diverse types of influence over the motivation to eat occurs throughout the brain. For example, traditional non-homeostatic factors such as stress and learned responses interact with adiposity signals to modify food intake

hypothalamus, also influence activity in essentially every non-homeostatic relay area. As an example, in addition to dense populations in the hypothalamus (see below), receptors for the adiposity-signaling molecules leptin and insulin are also located in neurons in the ventral tegmental area (VTA) of the midbrain (Figlewicz et al. 2003; Pardini et al. 2006), an area where the hedonic value of stimuli is determined and relayed to other areas via axons that release dopamine. Local administration of either insulin or leptin into the vicinity of the VTA dampens positive hedonic signals being relayed to other brain areas (Davis et al. 2010; Davis et al. 2011; Figlewicz and Benoit 2009; Konner et al. 2011). Insulin receptors are also expressed in areas of the neocortex (Corp et al. 1986; Zahniser et al. 1984) where experiential factors are integrated and in the hippocampus which is necessary for many forms of learning (Corp et al. 1986; Harvey et al. 2006; Zahniser et al. 1984).

Historically, subareas within the hypothalamus were considered to be the major controllers of specific behaviors (e.g., drinking, eating, satiety, sex). The roots of this thinking go back to the pioneering work of Hetherington and Ranson (1940, 1942), Hess (1956) and particularly to the influential treatise of Eliot Stellar (1954). For even though Stellar cautioned against considering behavior-controlling centers as being isolated and argued that each is under the influence of numerous types of inputs, he is generally regarded as popularizing the notion of eating and satiety centers. One consequence was that a generation of scholars worked at identifying inputs, outputs, and integrative capacities of specific hypothalamic areas such as the ventromedial nuclei (VMN, considered a satiety center), the lateral hypothalamus (LH, considered a hunger and thirst center), the paraventricular nuclei (PVN, considered an integrative center that influences autonomic nervous system activity and the stress axis), and many more. Individual 'centers' such as these were considered the mediators of specific homeostatic behaviors.

More recently, scholars began dissecting what was considered the more primitive limbic brain and soon identified important pathways passing from the midbrain VTA to the nucleus accumbens and to areas of the cerebral cortex that signal reward and hedonics (Berridge 1996; Kelley 1999, 2004; Meredith et al. 2008). While different groups typically stuck with one or the other type of neural circuit (homeostatic vs. non-homeostatic), over the ensuing years, it has become obvious that a better conception considers the brain to have one complex circuit that integrates myriad and diverse signals that influence the initiation and consumption of meals (see Fig. 2). All that differs is the origin of specific information that influences the decision to eat or not, and where that information gets integrated in order to make the most informed decisions.

4 Homeostatic Influences

As a short aside, we have argued elsewhere (Ramsay and Woods 2014) that terms such as homeostasis (and allostasis) that are often invoked to explain the apparent regulation of parameters such as core body temperature, plasma glucose, blood pressure, body fat, and others using terms such as set points, central controllers, error signals, and the like are misleading at best. Rather, there is sound evidence to suggest that there is in fact little actual basis for considering such terms and that the levels of body temperature or body fat represent a balance of all of the influences present (Ramsay and Woods 2014; Romanovsky 2007; Woods and Ramsay 2011). Nonetheless, we use the terms homeostasis/homeostatic throughout this article to denote influences on the motivation to eat based upon caloric deficits or surfeits as the terms are in common usage.

To review, the so-called homeostatic brain circuits that are thought to control eating that is based upon need or deprivation have been well described. Numerous signals related to all aspects of metabolism converge on circuits in the mediobasal hypothalamus and elsewhere in the brain. Included are signals whose levels in the circulation are directly proportional to the amount of fat stored in adipose tissue; i.e., the so-called adiposity signals such as insulin and leptin. Insulin, which is secreted by pancreatic B cells, and leptin, which is secreted mainly by adipocytes, are both able to penetrate the blood–brain barrier and gain access to brain cells expressing insulin and leptin receptors, respectively. Many such cells (neurons and glial cells) are found in the hypothalamic arcuate nuclei (ARC) and nearby areas. If body weight decreases due to dieting or being starved, circulating insulin and leptin levels decline and a diminished 'adiposity' signal reaches these hypothalamic cells. One consequence is that circuits that enhance food taking are stimulated, whereas activity in those that enhance energy expenditure is reduced, and there is an increased probability of eating more food at mealtime until insulin and leptin levels return to baseline. Such a mechanism explains, at least in part, the tendency to regain weight after dieting. Conversely, if one overeats and gains weight, the increased insulin and leptin signals favor eating less and losing body fat. There are numerous reviews of these phenomena (Begg and Woods 2012; Belgardt and Bruning 2010; Guyenet and Schwartz 2012; Morton et al. 2014; Myers and Olson 2012; Schwartz et al. 2000; Sohn et al. 2013).

The ARC contains neurons that have a mainly catabolic action, favoring decreased food intake and increased energy expenditure and consequently loss of body fat (see Fig. 3). The best known of these are neurons expressing the large peptide proopiomelanocortin (POMC). POMC-synthesizing cells in the hypothalamus cleave the larger POMC molecule into smaller bioactive peptides, especially α-melanocyte concentrating hormone (αMSH) (Elmquist et al. 2005; Schwartz et al. 2000; Woods 2009). Administration of insulin or leptin into the vicinity of the ARC causes reduced food intake and loss of body weight/fat, a response that is mediated by, as well as mimicked by, the local administration of αMSH or related compounds (Benoit et al. 2002; Halaas et al. 1995; Niswender and Schwartz 2003; Seeley and Woods 2003; Seeley et al. 1997). People or animals with a genetic

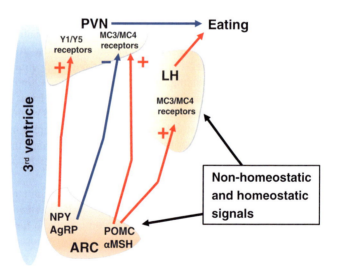

Fig. 3 Proopiomelanocortin (POMC) neurons in the arcuate nucleus of the hypothalamus (ARC) project to both the LH and the PVN (as well as many other areas). These neurons release α-melanocyte-stimulating hormone (αMSH) onto neurons expressing MC3/MC4 receptors, resulting in suppression of food intake. Also within the ARC, neuropeptide Y (NPY)/agouti-related peptide (AgRP) neurons also project to the PVN; NPY acts on Y1/Y5 receptors in the PVN to stimulate food intake, while AgRP acts to antagonize MC3/MC4 receptor signaling, further increasing food intake

disruption of this system (e.g., lacking leptin or its receptor, or lacking αMSH-3 or αMSH-4 receptors; or else are lacking insulin receptors uniquely in neurons) are hyperphagic and obese (Benoit et al. 2000; Farooqi and O'Rahilly 2006; Schwartz et al. 2000; Tao 2010). The ARC also has neurons opposing the action of αMSH. These neurons generate two anabolic peptides, neuropeptide Y (NPY) and agouti-related protein (AgRP). NPY acts on its receptors to activate eating, and with prolonged NPY signaling, body weight and adiposity are increased (Schwartz et al. 2000). AgRP functions as an endogenous antagonist of αMSH receptors and so blocks the action of leptin and insulin (Morton and Schwartz 2001; Schwartz et al. 2003). Both αMSH-secreting POMC neurons and NPY/AgRP neurons project from the ARC to several other hypothalamic areas and especially to the PVN where neurons express receptors for NPY and for αMSH/AgRP (Morton et al. 2006; Schwartz et al. 2000; Seeley and Woods 2003). The PVN in turn coordinates several activities related to metabolism. It projects to hindbrain areas that control the act of eating, it influences activity of the sympathetic and parasympathetic nervous systems, and it influences the secretions of the pituitary gland.

While this is a first-order and greatly simplified view of the role of homeostatic adiposity signals in influencing hypothalamic activity, it is nonetheless instructive. The importance of this adiposity-signaling circuit can be inferred from what happens when its activity is chronically altered. As discussed above, inhibiting the activity of insulin or leptin within the brain, and especially within the hypothalamus, results in individuals (including humans) with a greatly increased motivation to eat; when this condition occurs chronically, they maintain an elevated body weight (Begg and Woods 2013; Bruning et al. 2000; Farooqi and O'Rahilly 2006; Obici et al. 2002; Schwartz et al. 2000). Their exaggerated motivation to eat can be normalized by increasing the leptin or insulin signal locally within the brain, and if this occurs chronically, body weight returns to normal as well. If the insulin/leptin signal within the region of the ARC is experimentally or pathologically elevated above normal, individuals have reduced motivation to eat and they lose weight (Benoit et al. 2002; Schwartz et al. 2000; Woods et al. 1979, 1998). Thus, the determination by the brain as to how much body fat to carry is greatly influenced by this hypothalamic circuit that is sensitive to leptin and insulin. Motivation to eat or not to eat comes into play when an individual's body fat has been displaced from its customary level.

Satiation signals such as CCK that are secreted during meals interact with the adiposity-signal system in that their potency is enhanced when leptin and/or insulin is higher than normal. When the insulin or leptin signal locally in the brain is increased, the ability of satiation signals to reduce meal size is enhanced, and a lower insulin or leptin signal reduces the efficacy of satiation signals (Emond et al. 1999, 2001; Matson et al. 1997, 2000; Riedy et al. 1995). Thus, an overweight individual, with an increased insulin and/or leptin signal in the brain, tends to eat smaller meals because the effects of meal-generated satiation signals acting in the hypothalamus, hindbrain, and other brain areas are potentiated. Conversely, a starved individual will have a tendency to eat larger-than-normal meals. This homeostatic explanation accounts for maintenance of body weight in most individuals within relatively strict limits, but does not take into account motivation and consequent increased food intake due to non-homeostatic factors. Of note is that when living in an environment rich in palatable and hedonically pleasing foods, the motivation to eat is strong and can overpower the homeostatic influences, and this is abetted by the body's becoming relatively resistant to the actions of leptin and insulin as obesity progresses.

As noted above, CCK and other satiation signals are considered intermediate cues with regard to influencing meal size. When the macronutrient profile of available food is consistent, a certain amount of CCK secretion (as well, of course, of other satiation signals) indicates a certain number and type of calories entering the intestine from the stomach. Over hundreds of meals, the CCK–calorie content association becomes strong such that a certain level of CCK (and other satiation signals) triggering its receptors reliably causes a perception of fullness and the meal ends (Woods 2009). This association can be weakened, however, by frequently administering exogenous CCK during meals, thus weakening the CCK–calorie bond and rendering CCK a poor predictor of calories already consumed. When this is done, rats learn to ignore the CCK signal in the specific environment where the bond has been weakened (i.e., they no longer reduce their meal size when exogenous CCK is administered), but they do respond to the same CCK signal by reducing meal size normally in other environments (Duncan et al. 2005; Goodison and Siegel 1995). This suggests that the ability of so-called satiation signals to reduce eating is based on learned associations rather than being a hardwired reflex.

The motivation to eat based on the homeostatic maintenance of body weight was historically touted by ourselves and many others to be the primary determinant of eating (e.g., Elmquist 2001; Schwartz et al. 2000; Woods et al. 1998). However, research over the last decade or two has led to a major reconsideration of this model. Perhaps most importantly, the crisis of steadily increasing obesity rates around the world belies the importance of hardwired homeostatic controllers. Further, as detailed by (Woods and Langhans 2012), the ability of insulin, leptin, CCK, and other so-called homeostatic signals to influence food intake varies widely depending upon environmental, experiential, and other factors. Finally, the realization that circuits conveying learning, stress, hedonics, cognitive activity, etc., interact so extensively with the homeostatic circuits suggests that there is considerable adaptability and flexibility in the controls over the motivation to eat.

5 Non-homeostatic Influences over the Motivation to Eat

Detailed research on the non-homeostatic side of the brain circuits influencing eating has traditionally lagged that of the homeostatic circuits discussed above. Most researchers today base non-homeostatic eating upon the reward properties of specific foods or food-associated stimuli, and the so-called reward circuitry has been well established in the brain (see reviews in Berthoud 2004; Figlewicz and Sipols 2010; Petrovich 2013; Richard et al. 2013; Rinaman 2010; Zheng and

Berthoud 2007). A primary focus has been the neurons that synthesize dopamine in the midbrain and project anteriorly to several brain regions including the nucleus accumbens (Baik 2013). Stimulation of these dopaminergic fibers is highly rewarding as animals will learn responses that activate them. Likewise, neural circuits activated by palatable foods, by many drugs of abuse, by sex, and by other rewarding activities converge in the midbrain to activate those same dopaminergic neurons (Bjorklund and Dunnett 2007; see Salamone et al. in this volume). While there are several dopaminergic circuits emanating from the midbrain, some are concerned mainly with motor control and not discussed here. However, those originating in the VTA that project to cortical regions (mesocortical tract), and those that project to limbic areas such as the nucleus accumbens and parts of the amygdala (mesolimbic tract), convey aspects of reward and enhance motivation (Hegarty et al. 2013). Considerable research has identified specific properties of the circuits activated by the nucleus accumbens and other dopamine-sensitive regions, circuits that in turn project to the cortex, to the hypothalamus and elsewhere. Because the rewarding and consequently motivation-enhancing properties of these circuits relate to diverse behaviors in addition to eating, they are detailed in several other chapters in this volume and need not be reiterated here. The model that is generally accepted, and which is in common use today, was originated by Berridge and Robinson (1998). It posits that distinct components within the nucleus accumbens mediate what they term 'liking' and 'wanting,' with liking referring to the perception of pleasure that has become associated with stimuli typical of a particular food. It is based on experience, requires cortical inputs to the nucleus accumbens, and adds a positive hedonic aspect to consuming that food. Wanting, on the other hand, is considered as the motivational or driving component of eating. It is mediated by dopaminergic fibers from the VTA and is enhanced by food deprivation.

6 Integration of Homeostatic and Non-homeostatic Influences over Food Intake

An important example of how knowledge, and thinking, about the complexity of the interactions between homeostatic and non-homeostatic influences over food intake involves the LH. As discussed above, the LH was historically considered the quintessential 'hunger' or 'feeding center' of the brain because pharmacological or direct neural stimulation there elicits eating, even in sated individuals, and because lesions of the LH result in hypophagia and weight loss (Nicolaidis 1981; Teitelbaum and Epstein 1962). Further, because the LH receives diverse signals related to adipose stores, food available and being eaten, and current circulating levels of glucose and other energy-rich compounds, it is well positioned to integrate homeostatic factors and directly influence food intake (Bernardis and Bellinger 1996). The LH has also been historically recognized as an area where localized stimulation, in addition to eliciting eating, also elicits other motivated behaviors including drinking and sex, and more recently, the LH has come to be viewed as an important site that integrates arousal with the reward value of specific stimuli related to food, water, sex, and drugs of abuse (Aston-Jones et al. 2009; Berridge 2009). A key neurotransmitter conveying this integrated information from the LH to other brain areas is the neuropeptide, orexin-A (Borgland et al. 2009; Cason et al. 2010). In particular, orexinergic neurons convey information from the LH to the midbrain VTA, and the message is then forwarded to the nucleus accumbens by dopaminergic neurons, and this hypothalamic–midbrain–accumbens system has come to be recognized as an important circuit indicating food reward since the accumbens in turn also innervates the LH (Aston-Jones et al. 2009). More recently, a second LH-originating circuit utilizing orexinergic fibers has been identified that innervates the paraventricular nucleus of the thalamus (PVT) (Kirouac et al. 2005). The PVT is an important site where information on cognition, emotion, and anxiety is integrated with the reward value of food and drugs (Li et al. 2010a, b), and fibers from the PVT have now been identified that proceed to the nucleus accumbens where they modulate dopaminergic activity (Choi et al. 2012). Thus, the LH, rather than being considered a 'hunger center,' might more appropriately be considered as one hub in a complex neural circuit that integrates homeostatic with diverse non-homeostatic information to inform all motivated behaviors. Consistent with this, a recent review concluded that LH orexinergic neurons function as integrators of information from the internal and external environments with the level of vigilance and arousal to inform decisions on multiple motivated behaviors (Sakurai 2014).

To provide a more specific example of how homeostatic and non-homeostatic influences interact in the motivation to eat, consider the phenomenon of stress-induced eating. There is compelling evidence that some stressors, and especially those eliciting psychological stress, increase the motivation to consume so-called comfort foods (Adam and Epel 2007; Dallman 2010). The reason seems to be that consuming these foods dampens the response to stressors and is consequently, by definition, rewarding. Recent research using a rat model has identified circuits in the amygdala and cortex that are stimulated when sweet and otherwise hedonically pleasing foods are consumed (Ulrich-Lai et al. 2010; Ulrich-Lai and Ryan 2014). Further, there is evidence that consuming such comfort foods on a regular basis strengthens synaptic connections in the amygdala, connections that tie consumption of the food to the stress axis dampening phenomenon (Christiansen et al. 2011). Note that this is tied to overeating certain foods as opposed to being based on a change of body weight, although it may be that if the behavior becomes habitual, body weight will follow.

7 Summary

Examining food intake as a motivated behavior is complex, given that eating is regulated by diverse factors including 'homeostatic' brain circuits influenced by adiposity and satiation signals, as well as by 'non-homeostatic' circuits that convey information related to learning and experience, hedonics, stress, and many other factors. Recent developments identifying the intricate nature of the relationships between homeostatic and non-homeostatic influences significantly add to the complexity underlying the neural basis of the motivation to eat. The future of research in the field of food intake would seem to lie in the examination of the interactions between homeostatic and non-homeostatic influences.

Living in a world with a seemingly endless supply of highly palatable food makes it appear as though the homeostatic circuits that have evolved to regulate eating in historic environments are overwhelmed by the motivation to eat excess food for its rewarding value. This situation is exacerbated by a novel type of distal cue in the form of omnipresent advertisements that appeal to hedonic and other non-homeostatic motivations to eat. At the same time, the reliability of customary distal cues of caloric content, such as taste, smell, and mouthfeel, is often degraded by the use of artificial sweeteners and fats, and this in turn dictates that more proximal cues be used to help determine meal size and runs the risk of overeating (Woods 1991, 2009). The reliability of distal cues is also compromised by the use of super-sweet compounds such as those containing added high-fructose corn syrup or artificial sweeteners, a practice that has been associated with the development of metabolic disorders and obesity (DiNicolantonio et al. 2015; Swithers 2013). Finally, since taking in food is itself a biological stressor (Woods 1991), the complexity and interactions of so many homeostatic, hedonic, cognitive, and emotional factors that can become associated with meals and which may also therefore become associated with stress responses may predispose some individuals to eating disorders such as anorexia nervosa. Therapies that treat both the biological/homeostatic (i.e., starvation-based) and the learned and other non-homeostatic factors associated with food are likely to be most efficacious.

References

Adam TC, Epel ES (2007) Stress, eating and the reward system. Physiol Behav 91(4):449–458
Armelagos GJ (2014) Brain evolution, the determinates of food choice, and the omnivore's dilemma. Crit Rev Food Sci Nutr 54(10):1330–1341
Aston-Jones G, Smith RJ, Moorman DE, Richardson KA (2009) Role of lateral hypothalamic orexin neurons in reward processing and addiction. Neuropharmacology 56(Suppl 1):112–121
Baik JH (2013) Dopamine signaling in reward-related behaviors. Front Neural Circ 7:152
Begg DP, Woods SC (2012) The central insulin system and energy balance. Handb Exp Pharmacol 209:111–129
Begg DP, Woods SC (2013) The endocrinology of food intake. Nat Rev Endocrinol 9(10):584–597
Belgardt BF, Bruning JC (2010) CNS leptin and insulin action in the control of energy homeostasis. Ann N Y Acad Sci 1212:97–113
Benoit S, Schwartz M, Baskin D, Woods SC, Seeley RJ (2000) CNS melanocortin system involvement in the regulation of food intake. Horm Behav 37(4):299–305
Benoit SC, Air EL, Coolen LM, Strauss R, Jackman A, Clegg DJ, Woods SC (2002) The catabolic action of insulin in the brain is mediated by melanocortins. J Neurosci 22(20):9048–9052
Bernardis LL, Bellinger LL (1996) The lateral hypothalamic area revisited: ingestive behavior. Neurosci Biobehav Rev 20(2):189–287

Berridge K (1996) Food reward: brain substrates of wanting and liking. Neurosci Biobehav Rev 20:1–25

Berridge KC (2009) 'Liking' and 'wanting' food rewards: brain substrates and roles in eating disorders. Physiol Behav 97(5):537–550

Berridge KC, Robinson TE (1998) What is the role of dopamine in reward: hedonic impact, reward learning, or incentive salience? Brain Res Brain Res Rev 28(3):309–369

Berthoud HR (2004) Mind versus metabolism in the control of food intake and energy balance. Physiol Behav 81(5):781–793

Birch LL, Fisher JO (1998) Development of eating behaviors among children and adolescents. Pediatrics 101:539–549

Bjorklund A, Dunnett SB (2007) Dopamine neuron systems in the brain: an update. Trends Neurosci 30(5):194–202

Bolles RC (1967) Theory of motivation. Harper & Row, New York

Borgland SL, Chang SJ, Bowers MS, Thompson JL, Vittoz N, Floresco SB, Bonci A (2009) Orexin A/hypocretin-1 selectively promotes motivation for positive reinforcers. J Neurosci 29(36):11215–11225

Bouton ME (2011) Learning and the persistence of appetite: extinction and the motivation to eat and overeat. Physiol Behav 103(1):51–58

Bruning JC, Gautam D, Burks DJ, Gillette J, Schubert M, Orban PC, Kahn CR (2000) Role of brain insulin receptor in control of body weight and reproduction. Science 289(5487):2122–2125

Cason AM, Smith RJ, Tahsili-Fahadan P, Moorman DE, Sartor GC, Aston-Jones G (2010) Role of orexin/hypocretin in reward-seeking and addiction: implications for obesity. Physiol Behav 100(5):419–428

Choi DL, Davis JF, Magrisso IJ, Fitzgerald ME, Lipton JW, Benoit SC (2012) Orexin signaling in the paraventricular thalamic nucleus modulates mesolimbic dopamine and hedonic feeding in the rat. Neuroscience 210:243–248

Christiansen AM, Dekloet AD, Ulrich-Lai YM, Herman JP (2011) "Snacking" causes long term attenuation of HPA axis stress responses and enhancement of brain FosB/deltaFosB expression in rats. Physiol Behav 103(1):111–116

Collier G, Johnson DF (2004) The paradox of satiation. Physiol Behav 82(1):149–153

Collier GH, Johnson DF, Hill WL, Kaufman LW (1986) The economics of the law of effect. J Exp Anal Behav 48:113–136

Collier GH, Johnson DF, Morgan C (1992) The magnitude-of-reinforcement function in closed and open economies. J Exp Anal Behav 57:81–89

Corp ES, Woods SC, Porte D Jr, Dorsa DM, Figlewicz DP, Baskin DG (1986) Localization of ^{125}I-insulin binding sites in the rat hypothalamus by quantitative autoradiography. Neurosci Lett 70:17–22

Dallman MF (2010) Stress-induced obesity and the emotional nervous system. Trends Endocrinol Metab 21(3):159–165

Davis JF, Choi DL, Benoit SC (2010) Insulin, leptin and reward. Trends Endocrinol Metab 21(2):68–74

Davis JF, Choi DL, Schurdak JD, Fitzgerald MF, Clegg DJ, Lipton JW, Benoit SC (2011) Leptin regulates energy balance and motivation through action at distinct neural circuits. Biol Psychiatry 69(7):668

DiNicolantonio JJ, O'Keefe JH, Lucan SC (2015) Added fructose: a principal driver of type 2 diabetes mellitus and its consequences. Mayo Clin Proc 90(3):372–381

Drazen DL, Vahl TP, D'Alessio DA, Seeley RJ, Woods SC (2006) Effects of a fixed meal pattern on ghrelin secretion: evidence for a learned response independent of nutrient status. Endocrinology 147(1):23–30

Duncan EA, Davita G, Woods SC (2005) Changes in the satiating effect of cholecystokinin over repeated trials. Physiol Behav 85:387–393

Elmquist JK (2001) Hypothalamic pathways underlying the endocrine, autonomic, and behavioral effects of leptin. Physiol Behav 74(4–5):703–708

Elmquist JK, Coppari R, Balthasar N, Ichinose M, Lowell BB (2005) Identifying hypothalamic pathways controlling food intake, body weight, and glucose homeostasis. J Comp Neurol 493 (1):63–71

Emond M, Ladenheim EE, Schwartz GJ, Moran TH (2001) Leptin amplifies the feeding inhibition and neural activation arising from a gastric nutrient preload. Physiol Behav 72(1–2):123–128

Emond M, Schwartz GJ, Ladenheim EE, Moran TH (1999) Central leptin modulates behavioral and neural responsivity to CCK. Am J Physiol 276:R1545–R1549

Farooqi S, O'Rahilly S (2006) Genetics of obesity in humans. Endocr Rev 27(7):710–718

Ferguson NB, Keesey RE (1975) Effect of a quinine-adulterated diet upon body weight maintenance in male rats with ventromedial hypothalamic lesions. J Comp Physiol Psychol 89(5):478–488

Figlewicz DP, Benoit SC (2009) Insulin, leptin, and food reward: update 2008. Am J Physiol Regul Integ Comp Physiol 296(1):R9–R19

Figlewicz DP, Evans SB, Murphy J, Hoen M, Baskin DG (2003) Expression of receptors for insulin and leptin in the ventral tegmental area/substantia nigra (VTA/SN) of the rat. Brain Res 964(1):107–115

Figlewicz DP, Sipols AJ (2010) Energy regulatory signals and food reward. Pharmacol Biochem Behav 97(1):15–24

Goodison T, Siegel S (1995) Learning and tolerance to the intake suppressive effect of cholecystokinin in rats. Behav Neurosci 109:62–70

Grossman SP (1986) The role of glucose, insulin and glucagon in the regulation of food intake and body weight. Neurosci Biobehav Rev 10:295–315

Guyenet SJ, Schwartz MW (2012) Clinical review: Regulation of food intake, energy balance, and body fat mass: implications for the pathogenesis and treatment of obesity. J Clin Endocrinol Metab 97(3):745–755

Halaas JL, Gajiwala KS, Maffei M, Cohen SL, Chait BT, Rabinowitz D, Friedman JM (1995) Weight-reducing effects of the plasma protein encoded by the obese gene. Science 269(5223):543–546

Harvey J, Solovyova N, Irving A (2006) Leptin and its role in hippocampal synaptic plasticity. Prog Lipid Res 45(5):369–378

Hegarty SV, Sullivan AM, O'Keeffe GW (2013) Midbrain dopaminergic neurons: a review of the molecular circuitry that regulates their development. Dev Biol 379(2):123–138

Hess WR (1956) Hypothalamus und thalamus: experimental-dokumente. Thieme, Stuttgart, Germany

Hetherington AW, Ranson SW (1940) Hypothalamic lesions and adiposity in the rat. Anat Rec 78(2):149–172

Hetherington AW, Ranson SW (1942) The spontaneous activity and food intake of rats with hypothalmic lesions. Am J Physiol 136:609–617

Hull CL (1931) Goal attraction and directing ideas conceived as habit phenomena. Psychol Rev 38(6):487–506

Keesey RE, Boyle PC (1973) Effects of quinine adulteration upon body weight of LH-lesioned and intact male rats. J Comp Physiol Psychol 84(1):38–46

Kelley AE (1999) Functional specificity of ventral striatal compartments in appetitive behaviors. Ann N Y Acad Sci 877:71–90

Kelley AE (2004) Ventral striatal control of appetitive motivation: role in ingestive behavior and reward-related learning. Neurosci Biobehav Rev 27(8):765–776

Kirouac GJ, Parsons MP, Li S (2005) Orexin (hypocretin) innervation of the paraventricular nucleus of the thalamus. Brain Res 1059(2):179–188

Konner AC, Hess S, Tovar S, Mesaros A, Sanchez-Lasheras C, Evers N, Bruning JC (2011) Role for insulin signaling in catecholaminergic neurons in control of energy homeostasis. Cell Metab 13(6):720–728

Langhans W (1996) Metabolic and glucostatic control of feeding. Proc Nutr Soc 55:497–515

Langhans W, Scharrer E (1987) Role of fatty acid oxidation in control of meal pattern. Behav Neural Biol 47:7–16

Li Y, Li S, Wei C, Wang H, Sui N, Kirouac GJ (2010a) Changes in emotional behavior produced by orexin microinjections in the paraventricular nucleus of the thalamus. Pharmacol Biochem Behav 95(1):121–128

Li Y, Li S, Wei C, Wang H, Sui N, Kirouac GJ (2010b) Orexins in the paraventricular nucleus of the thalamus mediate anxiety-like responses in rats. Psychopharmacology 212(2):251–265

Lotter EC, Woods SC (1977) Injections of insulin and changes of body weight. Physiol Behav 18(2):293–297

MacKay EM, Calloway JW, Barnes RH (1940) Hyperalimentation in normal animals produced by protamine insulin. J Nutr 20:59–66

Matson CA, Reid DF, Cannon TA, Ritter RC (2000) Cholecystokinin and leptin act synergistically to reduce body weight. Am J Physiol Regul Integr Comp Physiol 278(4):R882–890

Matson CA, Wiater MF, Kuijper JL, Weigle DS (1997) Synergy between leptin and cholecystokinin (CCK) to control daily caloric intake. Peptides 18:1275–1278

Meredith GE, Baldo BA, Andrezjewski ME, Kelley AE (2008) The structural basis for mapping behavior onto the ventral striatum and its subdivisions. Brain Struct Funct 213(1–2):17–27

Miller NE, Bailey CJ, Stevenson JA (1950) Decreased "hunger" but increased food intake resulting from hypothalamic lesions. Science 112(2905):256–259

Moran TH (2004) Gut peptides in the control of food intake: 30 years of ideas. Physiol Behav 82(1):175–180

Moran TH, Kinzig KP (2004) Gastrointestinal satiety signals II, Cholecystokinin. Am J Physiol 286:G183–G188

Morton GJ, Cummings DE, Baskin DG, Barsh GS, Schwartz MW (2006) Central nervous system control of food intake and body weight. Nature 443(7109):289–295

Morton GJ, Meek TH, Schwartz MW (2014) Neurobiology of food intake in health and disease. Nat Rev Neurosci 15(6):367–378

Morton GJ, Schwartz MW (2001) The NPY/AgRP neuron and energy homeostasis. Int J Obes Relat Metab Disord 25:S56–62

Myers MG Jr, Olson DP (2012) Central nervous system control of metabolism. Nature 491 (7424):357–363

Naim M, Brand JG, Kare MR, Kaufmann NA, Kratz CM (1980) Effects of unpalatable diets and food restriction on feed efficiency in growing rats. Physiol Behav 25(5):609–614

Nicolaidis S (1981) Lateral hypothalamic control of metabolic factors related to feeding. Diabetologia 20(Suppl):426–434

Niswender KD, Schwartz MW (2003) Insulin and leptin revisited: adiposity signals with overlapping physiological and intracellular signaling capabilities. Front Neuroendocrinol 24:1–10

Obici S, Feng Z, Karkanias G, Baskin DG, Rossetti L (2002) Decreasing hypothalamic insulin receptors causes hyperphagia and insulin resistance in rats. Nat Neurosci 5(6):566–572

Pardini AW, Nguyen HT, Figlewicz DP, Baskin DG, Williams DL, Kim F, Schwartz MW (2006) Distribution of insulin receptor substrate-2 in brain areas involved in energy homeostasis. Brain Res 1112(1):169–178

Petrovich GD (2013) Forebrain networks and the control of feeding by environmental learned cues. Physiol Behav 121:10–18

Ramsay DS, Woods SC (2014) Clarifying the roles of homeostasis and allostasis in physiological regulation. Psychol Rev 121(2):225–247

Richard JM, Castro DC, Difeliceantonio AG, Robinson MJ, Berridge KC (2013) Mapping brain circuits of reward and motivation: in the footsteps of Ann Kelley. Neurosci Biobehav Rev 37(9 Pt A):1919–1931

Riedy CA, Chavez M, Figlewicz DP, Woods SC (1995) Central insulin enhances sensitivity to cholecystokinin. Physiol Behav 58:755–760

Rinaman L (2010) Ascending projections from the caudal visceral nucleus of the solitary tract to brain regions involved in food intake and energy expenditure. Brain Res 1350:18–34

Romanovsky AA (2007) Thermoregulation: some concepts have changed. Functional architecture of the thermoregulatory system. Am J Physiol Regul Integr Comp Physiol 292(1):R37–46

Rozin P (1990) Acquisition of stable food preferences. Nutr Rev 48(2):106–113 (discussion 114–131)

Sakurai T (2014) The role of orexin in motivated behaviours. Nat Rev Neurosci 15(11):719–731
Schneider JE, Wise JD, Benton NA, Brozek JM, Keen-Rhinehart E (2013) When do we eat? Ingestive behavior, survival, and reproductive success. Horm Behav 64(4):702–728
Schwartz MW, Woods SC, Porte D Jr, Seeley RJ, Baskin DG (2000) Central nervous system control of food intake. Nature 404(6778):661–671
Schwartz MW, Woods SC, Seeley RJ, Barsh GS, Baskin DG, Leibel RL (2003) Is the energy homeostasis system inherently biased toward weight gain? Diabetes 52:232–238
Sclafani A (1997) Learned controls of ingestive behavior. APPETITE 29:153–158
Sclafani A, Lucas F, Ackroff K (1996) The importance of taste and palatability in carbohydrate-induced overeating in rats. Am J Physiol 270:R1197–R1202
Seeley RJ, Woods SC (2003) Monitoring of stored and available fuel by the CNS: implications for obesity. Nat Rev Neurosci 4(11):901–909
Seeley RJ, Yagaloff KA, Fisher SL, Burn P, Thiele TE, van Dijk G, Schwartz MW (1997) Melanocortin receptors in leptin effects. NATURE 390(6658):349
Skinner BF (1930) On the conditions of elicitation of certain eating reflexes. Proc Natl Acad Sci U S A 16(6):433–438
Smith GP, Epstein AN (1969) Increased feeding in response to decreased glucose utilization in rat and monkey. Am J Physiol 217:1083–1087
Smith GP, Gibbs J (1984) Gut peptides and postprandial satiety. Fed Proc 43(14):2889–2892
Smith GP, Gibbs J (1985) The satiety effect of cholecystokinin. Recent progress and current problems. Ann N Y Acad Sci 448:417–423
Sohn JW, Elmquist JK, Williams KW (2013) Neuronal circuits that regulate feeding behavior and metabolism. Trends Neurosci 36(9):504–512
Stellar E (1954) The physiology of motivation. Psychol Rev 61:5–22
Strubbe JH, Woods SC (2004) The timing of meals. Psychol Rev 111:128–141
Swithers SE (2013) Artificial sweeteners produce the counterintuitive effect of inducing metabolic derangements. Trends Endocrinol Metab 24(9):431–441
Tao YX (2010) The melanocortin-4 receptor: physiology, pharmacology, and pathophysiology. Endocr Rev 31(4):506–543
Teff K (2000) Nutritional implications of the cephalic-phase reflexes: endocrine responses. APPETITE 34(2):206–213
Teff KL (2011) How neural mediation of anticipatory and compensatory insulin release helps us tolerate food. Physiol Behav 103(1):44–50
Teitelbaum P, Epstein AN (1962) The lateral hypothalamic syndrome: recovery of feeding and drinking after lateral hypothalamic lesions. Psychol Rev 69:74–90
Ulrich-Lai YM, Christiansen AM, Ostrander MM, Jones AA, Jones KR, Choi DC, Herman JP (2010) Pleasurable behaviors reduce stress via brain reward pathways. Proc Natl Acad Sci 107 (47):20529–20534
Ulrich-Lai YM, Ryan KK (2014) Neuroendocrine circuits governing energy balance and stress regulation: functional overlap and therapeutic implications. Cell Metab 19(6):910–925
Weingarten HP (1983) Conditioned cues elicit feeding in sated rats: a role for learning in meal initiation. Science 220:431–433
Woods SC (1991) The eating paradox: how we tolerate food. Psychol Rev 98(4):488–505
Woods SC (2002) The house economist and the eating paradox. APPETITE 38:161–165
Woods SC (2009) The control of food intake: behavioral versus molecular perspectives. Cell Metab 9(6):489–498
Woods SC, D'Alessio DA (2008) Central control of body weight and appetite. J Clin Endocrinol Metab 93(11 Suppl 1):S37–50
Woods SC, Langhans W (2012) Inconsistencies in the assessment of food intake. Am J Physiol Endocrinol Metab 303(12):E1408–E1418
Woods SC, Lotter EC, McKay LD, Porte D Jr (1979) Chronic intracerebroventricular infusion of insulin reduces food intake and body weight of baboons. Nature 282:503–505
Woods SC, Ramsay DS (2011) Food intake, metabolism and homeostasis. Physiol Behav 104 (1):4–7

Woods SC, Seeley RJ, Porte D Jr, Schwartz MW (1998) Signals that regulate food intake and energy homeostasis. Science 280(5368):1378–1383

Zahniser NR, Goens MB, Hanaway PJ, Vinych JV (1984) Characterization and regulation of insulin receptors in rat brain. J Neurochem 42(5):1354–1362

Zheng H, Berthoud HR (2007) Eating for pleasure or calories. Curr Opin Pharmacol 7(6):607–612

Sexual Motivation in the Female and Its Opposition by Stress

Ana Maria Magariños and Donald Pfaff

Abstract A well worked-out motivational system in laboratory animals produces estrogen-dependent female sex behavior. Here, we review (a) the logical definition of sexual motivation and (b) the basic neuronal and molecular mechanisms that allow the behavior to occur. Importantly, reproductive mechanisms in the female can be inhibited by stress. This is interesting because, in terms of the specificity of neuroendocrine dynamics in space and time, the two families of phenomena, sex and stress, are the opposite of each other. We cover papers that document stress effects on the underlying processes of reproductive endocrinology in the female. Not all of the mechanisms for such inhibition have been clearly laid out. Finally, as a current topic of investigation, this system offers several avenues for new investigation which we briefly characterize.

Keywords Corticosterone · CRF · Estrogen · Lordosis · Oxytocin · Stress

Contents

1	Logic of Motivation	36
2	Sexual Mechanisms	36
3	Comparing Sex and Stress	38
4	Stress Mechanisms	39
5	Stress Impact upon Reproductive Behaviors in Females	40
6	Open Questions from the Works Referred To	44
7	Useful Research Topics for the Future	44
	References	46

A.M. Magariños (✉) · D. Pfaff
Laboratory of Neurobiology and Behavior, The Rockefeller University, New York, NY, USA
e-mail: amagarinos@rockefeller.edu

D. Pfaff
e-mail: pfaff@mail.rockefeller.edu

1 Logic of Motivation

How do we justify defining a specific behavior as "motivated"? Suppose a human or a laboratory animal, studied under well-controlled conditions, is presented with a precisely defined stimulus and no detectable response occurs. Then, some kind of experimental manipulation (say, nutritional or hormonal), is imposed. A short time later (too short a time for maturational changes), the same human or laboratory animal is presented with the same precisely defined stimulus and the expected behavioral response *does* occur. What explains the appearance of the response on this second trial? We can infer that the manipulation we applied (say, nutritional or hormonal) had a "motivational" effect. That is, as the intervening variable, the concept of motivation identifies the logically required cause of behavioral change (Pfaff 1982).

In the case of female sex behavior, the crucial intervening variable, the motivational force is supplied by a relatively simple chemical, the steroid hormone estradiol. Because of the non-homeostatic nature of this hormonally driven set of mechanisms, we are dealing with a simpler motivational system than faced, for example, by Woods and Begg (this volume).

2 Sexual Mechanisms

When the male laboratory rat approaches the female, often by anogenital investigation and subsequently by mounting, if the female's hypothalamus has not been adequately exposed hormonally, then she will reject the male by fighting or by running away. If instead, her ventromedial hypothalamic neurons have been exposed for at least 18 h of high levels of estrogens supplemented by at least an hour of progesterone, then when mounted she will perform lordosis behavior (Pfaff and Schwartz-Giblin 1988). This vertebral dorsiflexion is absolutely required for reproduction for, without this behavior, it is impossible for the male to supply sperm for fertilization.

In all cases, when the male mounts, somatosensory signals are sent from the flanks and rump of the female's skin to the spinal cord and then, through well-studied ascending pathways, to the medulla and the midbrain. At that point, for lordosis to occur, an estrogen-dependent signal from the ventromedial hypothalamus must have arrived at the midbrain. If indeed it has, the midbrain effects on medullary reticular neurons and their subsequent facilitation of activity in axial muscle motoneurons in the spinal cord will permit lordosis, the vertebral dorsiflexion, to occur.

For estrogens to facilitate lordosis, new mRNA and new protein synthesis must occur. This requirement follows naturally from the fact that the estrogen receptor (ER)-alpha, the zinc finger protein discovered in the cell nuclei of neurons in the ventromedial nucleus of the hypothalamus (Pfaff 1968), is a ligand-activated transcription factor. Following ER-alpha activation by estrogenic binding to its ligand binding domain, ER-alpha in turn binds to specific portions of DNA associated with estrogen-responsive genes. Several of these discovered in hypothalamic and limbic brain tissues have been shown to foster lordosis behavior (Pfaff and Schwartz-Giblin 1988).

For estrogen-facilitated behaviorally important genes to be turned on (Fig. 1), not only must ER-alpha bind to estrogen-response elements of the DNA, but also certain changes in the chromatin structure must occur. That is, chromatin, the protein structure that envelopes DNA in the cell, must be unwound so that DNA response elements are physically accessible to ligand-bearing estrogen receptors. Some of these chromatin changes comprise chemical alterations in the N-termini of specific histones and have been shown to follow estrogen treatment in ventromedial hypothalamic neurons and preoptic area neurons. In only one case, pan acetylation of histone H4 in ventromedial hypothalamus, has such a histone modification been linked causally to lordosis behavior (Gagnidze et al. 2013).

Sex hormone effects are not limited to mating behaviors themselves. In several cases, it has been demonstrated that estrogens in females and androgens in males enhance communications between the sexes through the distance senses and then cutaneous senses. We have pictured this series of courtship communications as "hormone-dependent behavioral funnels" (Pfaff 2006) that have the effect of bringing reproductively competent conspecifics into the same place at the same time, for mating with minimal exposure to predators.

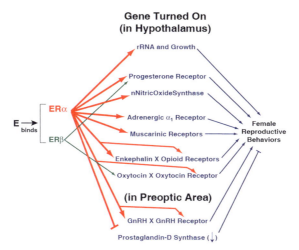

Fig. 1 Partial list of genes activated by estrogens. Estrogens (E), having bound to estrogen receptors (ERα and β), elevate the expression levels of numerous genes that participate in some aspects of female reproductive behavior. Note the exception for the bottom example, which works by disinhibition (adapted from Pfaff and Schwartz-Giblin 1988)

3 Comparing Sex and Stress

Although similar in having strong hormonal components, reproductive physiology and stress physiology have opposite characteristics in the following ways. Reproductive events in the female focus from whole body adaptations to a specific behavioral event, whereas stress can be initiated from a specific behavioral event to eventually affect the whole body. That is over the duration of the entire estrus cycle of the female, across the body from the ovaries to the pituitary and brain, cellular events are occurring such that on the late afternoon of the day of proestrus, during this brief period, (a) the ovulatory surge of luteinizing hormone will be released, coupled with (b) accepting fertilization by the performance of lordosis behavior. Thus, events throughout the body for a long period of time determine a specific pair of coupled events during a brief period of time.

Stress is just the opposite. An individual behavioral event at a specific time can trigger changes throughout the body that last a long time (Fig. 2).

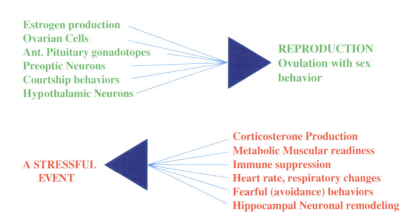

Fig. 2 Comparing sex and stress. Reproductive events in the female focus from whole body adaptations over a long time frame to a specific behavioral/neuroendocrine event. Instead, stress can be initiated from a specific behavioral event to eventually affect the whole body and generate lasting consequences

4 Stress Mechanisms

The hallmark of the stress response is the activation of the hypothalamic–pituitary–adrenal (HPA) axis. A constellation of coordinated endocrine, autonomic, immune, and behavioral responses are set in motion to secure a highly regulated reaction to a real or perceived stressful challenge (Selye 1936; Chrousos and Gold 1992; Sapolsky et al. 2000). Energy mobilization, increased cardiovascular activity, better cognitive acuity, and suppression of energetically expensive anabolic processes are well-recognized consequences of stress. Prolonged upregulation of the HPA axis due to sustained and intense stress challenges or unsuccessful stress coping mechanisms can trigger metabolic, vascular, immune, and psychiatric disorders. It only follows that reproductive function is adversely affected by stress, and given the omnipresent cross talk between the HPA and the hypothalamic–pituitary–gonadal (HPG) axes (Selye 1939; reviewed, Whirledge and Cidlowski 2013), this disruption can potentially impose the outcome of the reproductive success and survival trade-off (Wingfield and Sapolsky 2003). Behavioral details are below.

In response to a stressful event, the neuropeptides corticotropin-releasing hormone (CRF) and arginine vasopressin (AVP) are synthesized in the parvocellular portion of the hypothalamic paraventricular nucleus (PVN). CRF is also synthesized in neurons of the central and medial nuclei of the amygdala and the bed nucleus of the stria terminalis (Swanson et al. 1983). CRF is then transported to the adenohypophysis via the hypothalamo-hypophyseal portal system, and after binding to specific receptors, it triggers the synthesis and breakdown of pro-opiomelanocortin (POMC), resulting in the release of beta-endorphin, alpha-melanocyte-stimulating hormone (α-MSH), and adrenocorticotropin hormone (ACTH) to the systemic circulation. AVP potentiates CRH-induced secretion of ACTH which in turn stimulates the cortex and the medulla of the adrenal gland to synthesize glucocorticoids and adrenaline, respectively (Chrousos and Gold 1992).

Corticosterone, being the main glucocorticoid detected in the circulation of rodents, represents the end-product of the stress-induced activation of the HPA axis and, upon arrival in the brain, exerts feedback mechanisms at extra hypothalamic, hypothalamic, and hypophyseal levels, thus preventing the over-activation of the system (Chrousos and Gold 1992).

As with sex hormones introduced above, steroid stress hormonal actions lead us to the use of the techniques of molecular biology and high-resolution neuroanatomy. The effects of adrenal steroids mentioned above, and other effects not covered here, are mediated by glucocorticoid receptors (GRs) and mineralocorticoid receptors (MRs), both of them ligand-activated transcription factors (de Kloet et al. 2005). Their molecular endocrine dynamics are under investigation (reviewed, DeFranco and Guerrero 2000). Among those molecular properties, two are useful to summarize here: (a) MR affinity for corticosterone is ten times higher that of GRs, thus becoming occupied at very low circulating levels of corticosteroids, i.e., during basal conditions. MR density is higher in limbic brain regions such as the hippocampus, and they mediate the cognitive processing and the onset of the stress response (de Kloet et al. 2005); and (b) in contrast, GRs, although ubiquitously distributed in the brain, are enriched in the hypothalamic PVN and they become activated during the stress-induced surge of corticosteroids. One important consequence of MR and GR occupation by glucocorticoids is to promote the recovery phase and the termination of the stress response, by interfering with transcription factors and repressing the signaling by neuropeptides such as CRF and AVP (Joëls and Baram 2009). Indeed, restriction of the magnitude and duration of stress responses is important, because initially adaptive responses yielding reversible remodeling of brain areas, associated with learning and memory, can give way to deleterious effects on the cell biology of hippocampal neurons (Magariños et al. 1997; reviewed, McEwen and Magariños 2001; McEwen 2007).

5 Stress Impact upon Reproductive Behaviors in Females

Stresses of various sorts can interfere with female reproductive behaviors by directly blocking mechanisms for lordosis behavior and by indirectly interfering, namely, by blocking natural estrus cycles that are permissive for lordosis in cycling females (Fig. 3).

Indirect interference by stress The activation of the HPA axis by a variety of stress paradigms such as immune and inflammatory challenges, psychogenic stress, and insulin-induced hypoglycemia interferes with the precise timing of reproductive hormone release and, therefore, disrupts ovulation (Matteri et al. 1984; Ferin 2007). In particular, gonadotropin-releasing hormone (GnRH) is the central regulator of the proper attainment and maintenance of reproductive function and is produced by a few specialized neurons of the paraventricular nucleus and the preoptic area of the hypothalamus. Stress has been shown to delay the pubertal onset in female rodents and CRF signaling, via its cognate receptors, has also been implicated in the impairment of both the GnRH pulse frequency and the GnRH pulse generator's ability to stimulate pituitary gonadotrophs to synthesize luteinizing (LH) and

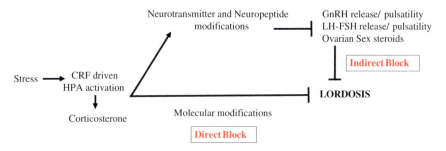

Fig. 3 Stress can interfere with female reproductive behavior through direct or indirect mechanisms

follicular stimulating hormone (FSH) in rodents and monkeys (Li et al. 2006; Rivier and Rivest 1991; Chand and Lovejoy 2011; Ono et al. 1984). In turn, the responsiveness of the specialized anterior pituitary cells to GnRH is also affected by CRF administration, thus decreasing the amplitude of the gonadotropin response as well as its tonic secretion (Kinsey-Jones et al. 2009; Castellano et al. 2010; Cates et al. 2004).

The activation of the HPA axis in female rodents and sheep ahead of the estrogen-induced preovulatory LH surge impairs the proestrus LH peak, while the concomitant administration of a GnRH agonist restores the LH surge (reviewed in Ferin 2007).

Further support for CRF's mediating role in the stress effect on the HPG axis activity comes from experimental evidence of CRF antagonists advancing not only the onset of puberty (Engelbregt et al. 2001; Iwasa et al. 2010; Knox et al. 2009) but also abolishing the stress suppression of LH release and reinstating its pulsatility (Reviewed Tellam et al. 2000).

Both CRF-induced suppression and stress-induced suppression of pulsatile LH secretion are enhanced by the gonadal steroid estrogen. In this regard, in vitro studies showing that CRF induced, and a selective CRF-R2 antagonist blocked, the reduction in GnRH mRNA expression in a GnRH cell line, also demonstrated that estrogen acted synergistically with CRF in suppressing GnRH expression. Furthermore, estrogen alone upregulated CRF-R2 expression, suggesting a possible mechanism for the sensitizing influence of estrogen on CRF-induced suppression and stress-induced suppression of the GnRH pulse generator (Kinsey-Jones et al. 2006).

While the effects of CRF, just described, seem to be indirectly mediated by monoamines, opioids, the sympathetic nervous system, glucocorticoids as well as limbic activation, a direct regulation of the HPG axis by CRF and CRF-like peptides has been described by Tellam et al. (1998). After transfecting a neuronal GnRH-expressing cell line with mouse or chicken, GnRH promoter/luciferase constructs, CRF, and CRF-like peptides decreased luciferase reporter activity indicating a direct suppressive effect on GnRH-expressing neurons.

The numbers of routes by which stress can alter female reproductive processes are legion. GnRH release into the anterior pituitary can also be repressed by a deficient cellular energy metabolism (Chen et al. 1992). Indeed, metabolic stress such as insulin-induced hypoglycemia, and the consequent energy deficiency, suppresses pituitary LH secretion in several species, including the rat (Cates et al. 2004). The connectivity of GnRH neurons with extra-preoptic, ERα and β expressing, energy sensors such as the caudal hindbrain explains in part why the induction of glucopenia within the noradrenergic neurons of the caudal hindbrain inhibits estrogen positive-feedback activation and the LH surge (Briski and Sylvester 1998). In vitro studies using hypothalamic slices showed that a drop in glucose can be detected by GnRH cells resulting in the decrease of their firing activity (Zhang et al. 2007). This response can be abolished by inhibition of the energy sensor adenosine 5′-monophosphate-activated protein kinase (AMPK) (Roland and Moenter 2011). A role for the associated lactate deficit of the caudal hindbrain has been recently suggested as a possible mechanism to explain hypoglycemia-induced inhibition of LH release (Shrestha and Briski 2015).

The suppression of LH pulsatile release by insulin-induced hypoglycemic stress is mediated by calcitonin gene-related peptide (Li et al. 2004) and can be completely prevented by icv administration of a CRH antagonist (Cates et al. 2004).

Infectious stress and consequent immune activation suppress reproductive behavior. This is partially mediated by glucocorticoid secretion, secondary to CRF activation of brain interleukin (IL-1) (Berkenbosch et al. 1987). Still another route of interference: The hypothalamic neurokinin B (NKB)/NK3R signaling is required for acute systemic stress (as in LPS challenge)-induced suppression of the GnRH pulse generator (Grachev et al. 2014).

Kisspeptin, a hypothalamic peptide encoded by the Kiss 1 gene and initially characterized by loss-of-function mutations inducing hypogonadotropic hypogonadism, stimulates GnRH secretion and LH release (Hujibregts et al. 2015; Seminara et al. 2003). Stress inhibits the expression of kisspeptin in part through a mechanism that involves glucocorticoid receptor signaling (Kinsey-Jones et al. 2009).

While the greatest number of studies has concentrated on demonstrating the interference of reproductive processes by stress, we must note that some work makes the opposite point. Studies by Matsuwaki et al. (2003, 2004) claim that glucocorticoids (GCs) are protective to both pulsatile and surge secretion of gonadotropin during a specific type of stress, infectious stress, induced by tumor necrosis factor alpha (TNF-α) injection, through a mechanism that involves inhibition of prostaglandin synthesis in the brain. Important for the present argument, they indicate that GCs counteract the inhibitory effect of infectious stress on LH secretion (Matsuwaki et al. 2003). Follow-up studies by the same investigators revealed similar protective effects by GC for other types of stress such as an acute restraint and 2-deoxyglucose-induced hypoglycemia (Matsuwaki et al. 2006).

Direct effect of stress on lordosis-facilitating brain mechanisms In addition to blocking naturally cycling hormonal mechanisms that regulate female reproductive behaviors in the normal case, some work has indicated the ability of stress directly to block molecular neurobiological steps proximal to lordosis.

Linda Uphouse and her laboratory addressed the question of whether stress could interfere with the performance of lordosis behavior under well-defined endocrine conditions. In this type of work, the female is placed into the cage of a sexually experienced "stud" male rat and receives many mounts by the male. The percentage of those mounts that elicit the type of behavior required for fertilization, the vertebral dorsiflexion of lordosis, constitutes the "lordosis quotient." When ovariectomized female rats were hormonally primed with estradiol, but not progesterone, they showed lordosis behavior, but if a mild challenge such as five minutes of restraint stress was added to the paradigm, lordosis performance was reduced (White and Uphouse 2004). In addition, priming doses of estradiol between 4 and 10 micrograms followed by 250 micrograms of progesterone reduced the lordosis-inhibiting effects of stress in a dose-dependent fashion. In later studies, the lordosis-protective effect of progesterone was itself blocked by a progesterone receptor antagonist CDB-4124. Similarly, the antiprogestin RU486, a classical progesterone receptor blocker, exaggerated the decline in lordosis behavior after restraint stress (Uphouse and Hiegel 2013; Uphouse et al. 2015). A probable mechanism of action involving the regulation of the serotonergic system has been suggested. In fact, the priming of both estrogen and progesterone dose dependently reduced the effectiveness of the bilateral infusion into the hypothalamic ventromedial nucleus of the 5HT-1A receptor agonist (±)-8-hydroxy-2-(di-n-propylamino)tetralin (8OHDPAT) in inhibiting lordosis behavior (Truitt et al. 2003). Uphouse et al. have begun to investigate the involvements of other types of serotonin receptors as they interact with the oppositional effects of stress on lordosis. This must be considered a preliminary investigation because of the complexity of the neuroanatomy of ascending serotonergic projections and the neurochemical complexity of serotonin receptors expressed from 14 separate genes. Nevertheless, after replicating the lordosis-inhibiting effectors of stress, Uphouse et al. (2011) reported that after an intraperitoneal injection of the 5-HT3 receptor antagonist tropisetron, in combination with restraint stress, lordosis behavior was inhibited further. That is, 5-HT3 receptor antagonism enhanced the lordosis-inhibiting effect of stress. Further study with bilateral cannulation indicated that the drug effect worked through the ventromedial nucleus of the hypothalamus, at the top of the lordosis behavior circuit.

As well, the same brief, mild restraint stress lasting 5 min interfered with female reproductive behavior at a different point in the chain of behaviors that lead to successful fertilization. That is, such stress reduces the amount of time that females spend in the vicinity of a stud male (Uphouse et al. 2005). Normally, the so-called courtship behaviors of females hormonally prepared for reproduction include locomotion that seeks out the male and encourages him to mount in the right position.

What are the mechanisms by which stress could reduce female reproductive behavior directly, that is, independent of pituitary and ovarian cycles? Evidence for one example is clear. It is well established that oxytocin, applied systemically or intracerebroventricular (ICV), working through the oxytocin receptor, can facilitate lordosis behavior in female rats (Arletti and Bertolini 1985; Gorzalka and Lesterm 1987; Witt and Insel 1991; McCarthy et al. 1994; Benelli et al. 1994; Pedersen and Boccia 2002). For this behavioral action, a permissive effect of progesterone in the ventromedial hypothalamus has been reported (Schulze and Gorzalka 1991; Schumacher et al. 1990). Thus, the chain of molecular steps is as follows: estrogens increase gene expression for oxytocin (Chung et al. 1991; Dellovade et al. 1999) and the oxytocin receptor (Schumacher et al. 1992; Quiñones-Jenab et al. 1997), and, in turn, oxytocin increases lordosis. As part of this behavioral effect, oxytocin increases the firing rates of ventromedial hypothalamic neurons (Kow et al. 1991). Importantly, for the present argument, Pournajafi-Nazarloo et al. (2013) have reported that social isolation stress reduces oxytocin receptor gene expression. Thus, this type of stress, at least, would cut into the chain of molecular steps by which estrogens facilitate lordosis behavior.

In fact, an early theoretical article (McCarthy et al. 1991) hypothesized that a unifying principle of oxytocin's actions on reproductive behaviors would be that oxytocin protects these behaviors against disruptive effects of stress. The behavioral and molecular data cited above support that theory.

In addition to regulation of lordosis behavior itself, some workers have addressed social–motivational behaviors related to preferences for sexual partners. Gonadal hormone signaling prepares for and promotes crucial aspects of the female rodent sexual behavior associated with sexual motivation, proceptivity, and receptivity (Pfaff and Schwartz-Giblin 1988). The partner preference paradigm (PPP) is an experimental setup commonly used to evaluate approach and avoidance motivation in female rats. Regarding receptivity, while estrogen priming of ovariectomized rats is necessary and sufficient for lordosis behavior, progesterone facilitates it, possibly through the activation of membrane receptors of the progesterone–adipoQ receptor superfamily (Frye et al. 2013).

6 Open Questions from the Works Referred To

While a large literature, reviewed above, treats the "indirect" inhibition of female reproductive behavior by interfering with normal hormonal rhythms, "direct" brain mechanism alterations comprise a newer and more difficult subject.

The clearest conclusions are based on estrogenic actions in brain. But one line of research, nascent, delves into the roles of progesterone and its metabolites in the brain. In the normal case, progesterone massively amplifies the facilitatory actions of estrogens on lordosis. But we note that progesterone metabolites not only are present in the brain, but are even produced in the brain as so-called neurosteroids, and that they act in such a manner as to reduce anxiety by allosteric modulation of the GABA-a receptor. What is unknown is whether they, by themselves, would dampen the stress effect that reduces lordosis as documented by Uphouse et al. (2010).

Another set of unresolved issues deal with the stress itself. Sometimes, the behavioral effects of acute restraint stress are short-lived, but other times, if the stress is severe enough, the effects may last for a long time. In addition, the implications of chronic stress might be much different than the effects of acute stress on reproduction. Even the definition of stress can be elusive. Thus, the nature of stress (vague, non-specific, and difficult, but of great consequence to public health) can be contrasted to the nature of reproductive behavior by laboratory animals (specific, concrete, but of no consequence to public health).

7 Useful Research Topics for the Future

During this era of great concentration on molecular mechanisms in the brain, synaptic physiology and biophysics, etc., global concepts such as motivation are just beginning to come back into play. Looking forward we can think of four aspects of the topic we have covered that could provide fertile intellectual ground for launching work in the future.

Evolution What might be the biological value of the stressful inhibitory influences on female reproductive behavior that we have referred to above? An evolutionary biologist might say that it is advantageous for the female to ensure that her mating will result in the raising of young in a stress-free environment. Food present. Predators absent. After all, the female will have exposed herself to predation when she leaves her burrow to mate. This is only worth it, biologically, if stress will be low enough for her to go through a normal gestation and raise the young. On the other hand, Wingfield and Sapolsky (2003) have given examples of situations in nature by which the effects of stress are circumvented and reproduction can proceed even in the presence of stress (successful breeding of semelparous fish despite severe activation of adrenal function; resistance of the gonadal axis of seasonal breeders to stress; and in some avian species characterized by biparental care, females that lose their mate maintain the ability to successfully raise their young in spite of maintaining a high reactivity to stressful challenges).

Our work and references relate to experimentation under controlled conditions with laboratory rodents. Certainly, evolutionary biologists would want to apply motivational concepts, as introduced at the beginning of this chapter, to a variety of species studied in the wild.

Sex differences The references in this chapter concentrate on female sex behavior. The same dynamics do not necessarily apply to males of the same species. For example, Waldherr and Neumann (2007) have studied males mating under stress and have detailed some of the mechanisms by which such mating reduces anxiety.

Arousal Both sex and stress can be discussed in the context of arousal (Pfaff et al. 2007). Generalized arousal of the CNS has been portrayed as very high during sexual episodes (Pfaff 2006). As well, it is difficult to think of high levels of stress in the absence of CNS arousal. How these systems intersect requires further experimentation.

Molecular Mechanisms Surely, the relations between sexual motivation and stress can now be explored using techniques of modern molecular biology: (a) measurement techniques such as microarrays, polymerase chain reactions, and RNAseq, and (b) genomic manipulations such as small interfering RNA (siRNA), gene knockouts, viral vectors, and optogenetics. For the analysis of hippocampal reactions to stress, a start has already been made in which the suppression of transposon expression was observed following acute restraint stress, in hippocampal neurons (Hunter et al. 2012, 2015). This suppression was likely accomplished by transcriptionally repressive lysine trimethylation, at least on histone H3 (Hunter et al. 2009).

As a side point, some students of brain and behavior, thinking about "explanations" of behavior, will tend to contrast evolutionary explanations with mechanistic explanations. That is not correct. Evolutionary biology treats the modifications and development of a given type of behavior as it changes during long epochs of time. Mechanistic work simply examines how the behavior is produced by a defined neural circuit right now, at "time t." In fact, what is actually changing during those long epochs of time are, indeed, the mechanisms that influence that type of behavior. Thus, evolutionary approaches and mechanistic approaches converge.

References

Arletti R, Bertolini A (1985) Oxytocin stimulates lordosis behavior in female rats. Neuropeptides 6:247–253

Benelli A, Poggioli R, Luppi P, Ruini L, Bertolini A, Arletti R (1994) Oxytocin enhances, and oxytocin antagonism decreases, sexual receptivity in intact female rats. Neuropeptides 27: 245–250

Berkenbosch F, van Oers J, del Rey A, Tilders F, Besedovsky H (1987) Corticotropin-releasing factor-producing neurons in the rat activated by interleukin-1. Science 238:524–530

Briski KP, Sylvester PW (1998) Role of endogenous opiates in glucoprivic inhibition of the luteinizing hormone surge and Fos expression by preoptic gonadotropin-releasing hormone neurons in ovariectomized steroid-primed female rats. J Neuroendocrinol 10:769–776

Castellano JM, Bentsen AH, Mikkelsen JD, Tena-Sempere M (2010) Kisspeptins: bridging energy homeostasis and reproduction. Brain Res 1364:129–138

Cates PS, Li XF, O'Byrne KT (2004) The influence of 17β-oestradiol on corticotrophin-releasing hormone induced suppression of luteinizing hormone pulses and the role of CRH in hypoglycaemic stress-induced suppression of pulsatile LH secretion in the female rat. Stress 7:113–118

Chand D, Lovejoy DA (2011) Stress and reproduction: controversies and challenges. Gen Comp Endocrinol 171:253–257

Chen GA, Feng Q, Zhang LZ, Liu YX (1992) Hypoglycemic stress and gonadotropin-releasing hormone pulse generator activity in the rhesus monkey: role of the ovary. Neuroendocrinology 56:666–673

Chrousos GP, Gold PW (1992) The concepts of stress and stress system disorders: overview of physical and behavioral homeostasis. JAMA 267:1244–1252

Chung SR, McCabe JT, Pfaff DW (1991) Estrogen influences on oxytocin mRNA expression in preoptic and anterior hypothalamic regions studied by in situ hybridization. J Comp Neurol 307:281–295

De Kloet ER, Joëls M, Holsboer F (2005) Stress and the brain: from adaptation to disease. Nat Rev Neurosci 6:463–475

DeFranco DB, Guerrero J (2000) Nuclear matrix targeting of steroid receptors: specific signal sequences and acceptor proteins. Crit Rev Eukaryotic Gene Expr 10:39–44

Dellovade T, Zhu Y, Pfaff D (1999) Thyroid hormones and estrogen affect oxytocin gene expression in hypothalamic neurons. J Neuroendocrinol 11:1–10

Engelbregt MJT, Van Weissenbruch MM, Popp-Snijders C, Lips P, Delemarre-Van de Waal HA (2001) Body mass index, body composition, and leptin at onset of puberty in male and female rats after intrauterine growth retardation and after early postnatal food restriction. Pediatr Res 50:474–478

Ferin M (2007) Effects of stress on Gonadotropin secretion. In: Fink G (ed) Encyclopedia of stress. Elsevier, Amsterdam, vol 2, pp 283–288

Frye CA, Walf AA, Kohtz AS, Zhu Y (2013) Horm Behav 64:539–545

Gagnidze K, Weil ZM, Faustino LC, Schaafsma SM (2013) Pfaff DW early histone modifications in the ventromedial hypothalamus and preoptic area following oestradiol administration. J Neuroendoc 25:939–955

Gorzalka BB, Lesterm GLL (1987) Oxytocin-induced facilitation of lordosis behavior in rats is progesterone-dependent. Neuropeptides 10:55–65

Grachev P, Li XF, Hu MH, Li SY, Millar RP, Lightman SL, O'Byrne KT (2014) Neurokinin B signaling in the female rat: a novel link between stress and reproduction. Endocrinology 155:2589–2601

Hujibregts L, Tata B, deRoux N (2015) Gonadotropic axis deficiency: a neurodevelopmental disorder. Res Persp Endoc Interact 13:155–162

Hunter RG, McCarthy KJ, Milne TA, Pfaff DW, McEwen BS (2009) Regulation of hippocampal H3 histone methylation by acute and chronic stress. Proc Natl Acad Sci 109:17657–17662

Hunter RG, Murakami G, Dewell S, Seligsohn M, Baker ME, Datson NA, McEwen BS, Pfaff DW (2012) Acute stress and hippocampal histone H3 lysine 9 trimethylation, a retrotransposon silencing response. Proc Nat Acad Sci 109(43):17657–17662

Hunter R, Gagnidze K, McEwen B, Pfaff D (2015) Stress and the dynamic genome: steroids, epigenetics and the transposome. Proc Natl Acad Sci 112:6828–6833

Iwasa T, Matsuzaki T, Murakami M, Fujisawa S, Kinouchi R, Gereltsetseg G, Kuwahara A, Yasui T, Irahara M (2010) J Physiol 588:821–829

Joëls M, Baram TZ (2009) The neuro-symphony of stress. Nature Rev Neurosci 10:459–466

Kinsey-Jones JS, Li XF, Bowe JE, Lightman SL, O'Byrne KT (2006) Corticotrophin-releasing factor type 2receptor-mediated suppression of gonadotrophin-releasing hormone mRNA expression in GT1-7 cells. Stress 9:215–222

Kinsey-Jones JS, Li XF, Knox AMI, Wilkinson ES, Zhu XL, Chaudhary AA, Milligan SR, Lightman SL, O'Byrne KT (2009) Down-regulation of hypothalamic kisspeptin and its receptor, Kiss1r, mRNA expression is associated with stress-induced suppression of luteinizing hormone secretion in the female rat. J Neuroendocrinol 21:20–29

Knox AMI, Li XF, Kinsey-Jones JS, Wilkinson ES, Wu XQ, Cheng YS, Milligan SR, Lightman SL, O'Byrne KT (2009) Neonatal lipopolysaccharide exposure delays puberty and alters hypothalamic Kiss1 and Kiss 1r mRNA expression in the female rat. J Neuroendoc 21:683–689

Kow L-M, Johnson AE, Ogawa S, Pfaff DW (1991) Electrophysiological actions of oxytocin on hypothalamic neurons, in vitro: neuropharmacological characterization and effects of ovarian steroids. Neuroendocrinology 54:526–535

Li XF, Bowe JE, Kinsey-Jones JS, Brain SD, Lightman SL, O'Byrne KT (2006) Differential role of corticotropin-releasing factor receptor types 1 and 2 in stress-induced suppression of pulsatile luteinizing hormone secretion in the female rat. J Neuroendoc 18:602–610

Li XF, Bowe JE, Mitchell JC, Brain SD, Lightman SL, O'Byrne KT (2004) Stress-induced suppression of the gonadotropin-releasing hormone pulse generator in the female rat: a novel neural action for calcitonin gene-related peptide. Endocrinology 145:1556–1563

Magariños AM, Garcia Verdugo JM, McEwen BS (1997) Chronic stress alters synaptic terminal structure in hippocampus. Proc Nat Acad Sci 94:14002–14008

Matsuwaki T, Watanabe E, Suzuki M, Yamanouchi KN, Nishihara M (2003) Glucocorticoid maintains pulsatile secretion of luteinizing hormone under infectious stress condition. Endocrinology 144:347–3482

Matsukawi T, Suzuki M, Yamanouchi K, Nishihara M (2004) Glucocorticoid counteracts the suppressive effect of tumor necrosis factor-α on the surge of luteinizing hormone secretion in rats. J Endocrinol 181:509–513

Matsuwaki T, Kayasuga Y, Yamanouchi K, Nishihara M (2006) Maintenance of gonadotropin secretion by glucocorticoids under stress conditions through the inhibition of prostaglandin synthesis in the brain. Endocrinology 147:1087–1093

Matteri RL, Watson JG, Moberg GP (1984) Stress or acute adrenocorticotrophin treatment suppresses LHRH-induced LH release in the ram. J Reprod Fertil 72:385–393

McCarthy MM, Chung SR, Ogawa S, Kow L-M, Pfaff DW (1991) Behavioral effects of oxytocin: is there a unifying principle? In: Jard S, Jamison R (eds) Vasopressin. Colloque INSERM/John Libbey Eurotext Ltd. vol 208, pp. 195–212

McCarthy MM, Kleopoulos SP, Mobbs CV, Pfaff DW (1994) Infusion of antisense oligodeoxynucleotides to the oxytocin receptor in the ventromedial hypothalamus reduces estrogen–induced sexual receptivity and oxytocin receptor binding in the female rat. Neuroendocrinology 59:432–440

McEwen BS, Magariños AM (2001) Stress and hippocampal plasticity: implications for the pathophysiology of affective disorders. Hum Psychopharmacol 16:S7–S19

McEwen BS (2007) Physiology and neurobiology of stress and adaptation: Central role of the brain. Physiol Rev 87:873–904

Ono N, Lumpkin MD, Samson WK, McDonald JK, McCann SM (1984) Intrahypothalamic action of corticotrophin-releasing factor (CRF) to inhibit growth hormone and LH release in the rat. Life Sci 35:1117–1123

Pedersen CA, Boccia ML (2002) Oxytocin maintains as well as initiates female sexual behavior: effects of a highly selective oxytocin antagonist. Horm Behav 41:170–177

Pfaff DW (1968) Autoradiographic localization of radioactivity in rat brain after injection of tritiated sex hormones. Science 161:1355–1356

Pfaff DW (ed) (1982) The physiological mechanisms of motivation. Springer, Heidelberg, New York

Pfaff DW, Schwartz-Giblin S (1988) Cellular mechanisms of female reproductive behaviors. In: Knobil E, Neill J (eds) The physiology of reproduction. Raven Press, New York, pp 1487–1568 (Chapter 35)

Pfaff DW (2006) Brain Arousal and information theory: neural and genetic mechanisms. Harvard University Press, Cambridge

Pfaff DW, Martin EM, Ribeiro AC (2007) Relations between mechanisms of CNS arousal and mechanisms of stress. Stress 10:316–325

Pournajafi-Nazarloo H, Kenkel W, Mohsenpour SR, Sanzenbacher L, Saadat H, Partoo L, Yee J, Azizi F, Carter CS (2013) Exposure to chronic isolation modulates receptors mENAs for oxytocin and vasopressin in the hypothalamus and heart. Peptides 43:20–26

Quiñones-Jenab V, Jenab S, Ogawa S, Adan RAM, Burbach PH, Pfaff DW (1997) Effects of estrogen on oxytocin receptor messenger ribonucleic acid expression in the uterus, pituitary and forebrain of the female rat. Neuroendocrinology 65:9–17

Rivier C, Rivest S (1991) Effect of stress on the activity of the hypothalamic-pituitary-gonadal axis: peripheral and central mechanisms. Biol Reprod 45:523–532

Roland AV, Moenter SM (2011) Regulation of gonadotropin-releasing hormone neurons by glucose. Trends Endocrinol Metab 22:443–449

Sapolsky RM, Romero LM, Munck AU (2000) How do glucocorticoids influence stress responses? Integrating permissive, suppressive, stimulatory, and preparative actions. Endocr Rev 21:55–89

Schulze HG, Gorzalka BB (1991) Oxytocin effects on lordosis frequency and lordosis duration following infusion into the medial preoptic area and ventromedial hypothalamus of female rats. Neuropeptides 18:99–106

Schumacher M, Coirini H, Pfaff DW, McEwen BS (1990) Behavioral effects of progesterone associated with rapid modulation of oxytocin receptors. Science 250:691–694

Schumacher M, Coirini H, Flanagan LM, Frankfurt M, Pfaff DW, McEwen BS (1992) Ovarian steroid modulation of oxytocin receptor binding in the ventromedial hypothalamus. In: Pedersen CA et al (eds) Oxytocin in maternal, sexual, and social behaviors. Annals, New York Academy of Sciences, vol 652, pp 374–386

Selye H (1936) Syndrome produced by diverse nocuous agents. Nature 138:32

Selye H (1939) Effects of adaptation to various damaging agents on the female sex organs in the rat. Endocrinology 25:615–624

Seminara SB, Messager S, Chtzidaki EE, …, Aparicio SAJR, Colledge WH (2003) The GPR54 gene as a regulator of Puberty. New England J Med 349:1614–1627

Shrestha PK, Briski KP (2015) Hindbrain lactate regulates preoptic gonadotropin-releasing hormone (GnRH) neuron GnRH-I protein but not AMPK responses to hypoglycemia in the steroid-primed ovariectomized female rat. Neuroscience 298:1–8

Swanson LW, Sawchenko PE, Rivier J, Vale WW (1983) Organization of ovine corticotropin-releasing factor immunoreactive cells and fibers in the rat brain: an immunohistochemical study. Neuroendocrinology 36:165–186

Tellam DJ, Perone M, Dunn IC, Radovick S, Brennand J, Castro MG, Rivier JE, Lovejoy DA (1998) Direct regulation of GnRH transcription by CRF-like peptides in an immortalized neuronal cell line. NeuroReport 9:3135–3140

Tellam DJ, Mohammad DA, Lovejoy DA (2000) Molecular integration of hypothalamo-pituitary-adrenal axis-related neurohormones on the GnRH neuron. Biochem Cell Biol 78:216–305

Truitt W, Harrison L, Guptarak J, White S, Hiegel C, Uphouse L (2003) Progesterone attenuates the effect of the 5-HT1A receptor agonist, 8-OH-DPAT, and of mild restraint on lordosis behavior. Brain Res 974:202–211

Uphouse L, Hiegel C (2013) An antiprogestin, CDB4124, blocks progesterone's attenuation of the negative effects of a mild stress on sexual behavior. Behav Brain Res 240:21–25

Uphouse L, Selvamani A, Lincoln C, Morales L, Comeaux D (2005) Mild restraint reduces the time hormonally primed rats spend with sexually active males. Behav Brain Res 157:343–350

Uphouse L, Guptarak J, Hiegel C (2010) Progesterone reduces the inhibitory effect of a serotonin 1B receptor agonist on lordosis behavior. Pharmacol Biochem Behav 97:317–324

Uphouse L, Heckard D, Hiegel C, Guptarak J, Maswood S (2011) Tropisetron increases the inhibitory effect of mild restraint on lordosis behavior of hormonally primed, ovariectomized rats. Behav Brain Res 219:221–226

Uphouse L, Hiegel C, Martinez G, Solano C, Gusick W (2015) Repeated estradiol benzoate treatment protects against the lordosis-inhibitory effects of restraint and prevents effects of the antiprogestin, RU486. Pharmacol Biochem Behav 137:1–6

Waldherr M, Neumann ID (2007) Centrally released oxytocin mediates mating-induced anxiolysis in male rats. PNAS 104:16681–16684

Whirledge S, Cidlowski JA (2013) A role for glucocorticoids in stress-impaired reproduction: Beyond the hypothalamus and pituitary. Endocrinology 154:4450–4468

White S, Uphouse L (2004) Estrogen and progesterone dose-dependently reduce disruptive effects of restrain on lordosis behavior. Horm Behav 45:201–208

Wingfield JC, Sapolsky RM (2003) Reproduction and resistance to stress: when and how. J Neuroendocrinol 15:711–724

Witt DM, Insel TR (1991) A selective oxytocin antagonist attenuates progesterone facilitation of female sexual behavior. Endocrinology 128:3269–3276

Zhang C, Bosch MA, Levine JE, Ronnekleiv OK, Kelly MJ (2007) Gonadotropin-releasing hormone neurons express KATP channels that are regulated by estrogen and responsive to glucose and metabolic inhibition. J Neurosci 27:10153–10164

Oxytocin, Vasopressin, and the Motivational Forces that Drive Social Behaviors

Heather K. Caldwell and H. Elliott Albers

Abstract The motivation to engage in social behaviors is influenced by past experience and internal state, but also depends on the behavior of other animals. Across species, the oxytocin (Oxt) and vasopressin (Avp) systems have consistently been linked to the modulation of motivated social behaviors. However, how they interact with other systems, such as the mesolimbic dopamine system, remains understudied. Further, while the neurobiological mechanisms that regulate prosocial/cooperative behaviors have been extensively examined, far less is understood about competitive behaviors, particularly in females. In this chapter, we highlight the specific contributions of Oxt and Avp to several cooperative and competitive behaviors and discuss their relevance to the concept of social motivation across species, including humans. Further, we discuss the implications for neuropsychiatric diseases and suggest future areas of investigation.

Keywords Aggression · Competitive behavior · Cooperative behavior · Dopamine · Epigenetics · Neuropsychiatric disorders · Oxytocin receptor · Pair bonding · Social behavior network · Social communication · Social recognition memory · Vasopressin 1a receptor · Vasopressin 1b receptor

---

H.K. Caldwell
Laboratory of Neuroendocrinology and Behavior, Department of Biological Sciences, Kent State University, Kent, OH, USA
e-mail: hcaldwel@kent.edu

H.K. Caldwell
School of Biomedical Sciences, Kent State University, Kent, OH 44242, USA

H.E. Albers (✉)
Center for Behavioral Neuroscience, Neuroscience Institute, Georgia State University, Atlanta, GA 30302, USA
e-mail: biohea@gsu.edu

Contents

1	Overview...	52
2	Origins and Mechanisms of Motivated Social Behaviors	53
	2.1 Cooperative and Competitive Behaviors...	54
	2.2 Brain Areas that Regulate Cooperative and Competitive Behaviors.......	56
3	Neuroendocrine Modulation of Social Behaviors..	58
	3.1 The Oxytocin System	58
	3.2 The Vasopressin System	59
	3.3 Signaling by Oxytocin and Vasopressin in the Brain..............................	60
	3.4 Epigenetics	62
4	Oxytocin/Vasopressin and Cooperative and Competitive Behaviors: Social Memory, Social Interactions, and Aggression	63
	4.1 Social Recognition Memory	63
	4.2 Cooperative Behavior...	65
	4.3 Competitive Behavior	67
	4.4 Social Communication	73
	4.5 Interactions Between Oxytocin, Vasopressin, and Dopamine in the Regulation of Cooperation/Competition	74
5	Cooperativity and Competitiveness in Humans...	75
	5.1 Nonapeptides and Social Cognition in Healthy Humans.........................	76
	5.2 Oxytocin, Vasopressin, and the Mesolimbic Dopamine System.............	78
6	Implications for Neuropsychiatric Disorders...	79
	6.1 Autism Spectrum Disorder	79
	6.2 Personality Disorder...	80
	6.3 Schizophrenia	81
	6.4 Posttraumatic Stress Disorder	82
7	Conclusions and Future Directions	83
References...		84

1 Overview

Motivation is a dominant construct in psychology, psychiatry, and neuroscience, as trying to understand why animals, including humans, do what they do is at the core of these disciplines. Although motivation can be defined in a variety of ways, a key component is that motivated behaviors are directed toward (approach) or away (avoidance) from a stimulus. Motivation also contains emotional elements with approach linked to positive hedonic valence and avoidance linked to negative valence. This review focuses on social motivation, which, like other forms of motivation, is influenced by past experience and an individual's internal state. Social motivation is, however, intrinsically more dynamic and less predictable because the drive to approach or avoid another individual(s) depends in large measure on how that individual behaves.

Recent studies of the neurobiology of social behavior have often characterized social behavior as having a positive valence, described as prosocial or affiliative interactions (e.g.,pair bonding, maternal behavior), or as having a negative valence, described as negative social interactions (e.g., aggression, territoriality). Although such a dichotomy is convenient and can have descriptive value, a closer look at these behaviors suggests that social motivation is more complex. For example, while the formation of a pair bond in a species like prairie voles has positive behavioral elements, such as highly affiliative behaviors directed toward a partner, it is also associated with mate guarding, in which males display selective aggression toward other voles. Thus, in the context of a pair bond, simply ascribing positive valence to the affiliative behaviors and negative valence to aggression is an oversimplification. Further, all aggressive behaviors are not the same, nor are the effects on the players. The fact is, winning is rewarding (Martinez et al. 1995; Meisel and Joppa 1994), and there is even the possibility that losing can be rewarding as long as the defeat is not too severe (Gil et al. 2013). Therefore, assigning hedonic valence to social behaviors (e.g., aggressive behavior) or to mating strategies (e.g., pair bonding) must be done with great care, particularly when linking approach or avoidance with the neural mechanisms underlying motivation.

Historically, investigations into the neurobiology of motivation have primarily focused on the mesolimbic dopamine (DA) system where DA neurons were thought of as "reward" neurons. It has now been recognized that the role of the mesolimbic DA system in hedonic mechanisms is far more complex. Understanding **social** motivation requires us to expand our studies of the neural mechanisms of motivation beyond this system into the networks that control the expression of social behavior in response to social stimuli. Of the myriad of neurochemical signals that are known to be involved in the modulation of social behaviors, two neuropeptides, oxytocin (Oxt) and vasopressin (Avp), stand out as being critical across species. Because of the vast literature on the role of these two nine-amino acid neuropeptides, or nonapeptides, in regulating social behavior, this chapter will focus on mammalian social behavior and provide examples of the powerful contributions of these two neuropeptide systems to cooperative and competitive behaviors.

2 Origins and Mechanisms of Motivated Social Behaviors

When considering the evolutionary origins of motivated behaviors, most simply put, it all comes down to fitness. Animals engage in species-specific behaviors because over evolutionary time these behaviors were either selected for via natural or sexual selection, or occurred through some other mechanism of evolution. In the context of sexual behavior, males are described as "ardent" and females as "choosy," which is reflected in their physiology. Males make a lot of sperm and often display behaviors that that will result in the fertilization of as many eggs as possible over their reproductive lifetime. Female mammals on the other hand typically have to invest in the gestation and the care of offspring, so they tend to be more selective about their mates. These differences in selective pressures result in vastly different behavioral displays between males and females. However, these behavioral differences are not limited to sexual behaviors. For instance, in mammals, biparental behavior is scarce, occurring in fewer than 6 % of rodent species and 5–10 % of all mammals (Kleiman 1977). Thus, a male that engages in parental care does so because the "cost" of not being paternal is too high; for instance, while reproductive success may be compromised, his proximity to the female increases the likelihood that he is the sire of the offspring. There are also sex differences in displays of cooperative and competitive behaviors, with females often displaying more cooperative behaviors across the lifetime and males displaying more competitive behaviors, particularly during the breeding season.

In nature, the diversity in cooperativity and competitiveness observed across animal species is striking. Some animals have social structures that are characterized by high levels of cooperativity, such as that observed in species that form long-term social bonds like pair bonds. In other species, high levels of competitive behaviors serve to establish and maintain social dominance relationships. Overlaid on the complexity of social life for a given species is a lack of stability, as social behaviors often change over the seasons and over the lifetime. Further, the different behavioral strategies employed by an animal have their particular costs and benefits. To explore these costs and benefits we can take a closer look at cooperative and competitive behaviors.

2.1 Cooperative and Competitive Behaviors

Cooperative behaviors, associated with affiliative behaviors, are thought to have evolved from reproductive and parental behaviors, in turn being permissive for the development of longer-term social bonds (Crews 1997). Competitive behaviors too are important in the formation of social bonds, as intraspecific interactions are universal and often governed by dominance relationships. Some evolutionary advantages to forming social bonds include localization of resources, lower predation due to group aggression, and increased reproductive opportunities (Alexander 1974). Social bonds have been extensively studied in primates and in some instances have been shown to increase evolutionary fitness (Silk 2007). In free-ranging baboons, females that have strong social bonds with one another live longer than those who have weaker social bonds (Silk et al. 2009). Even in humans social relationships can have profound effects on an individual's health, including improved mood and a longer life (House et al. 1988; Rodriguez-Laso et al. 2007; Baumeister and Leary 1995).

However, being social does have its cost, such as increased susceptibility to disease, parasites, or injury (Alexander 1974; Crews 1997). In order for animals to live in groups, they must be able to tolerate close proximity; thus, keeping levels of aggression in check becomes particularly important. It also requires a memory of the members of the social group, as this allows animals to identify familiar stimuli, which in turn is permissive for adaptive behavioral responses. While many mammalian species live in groups, some species show "shifts" in the nature of their social interactions depending on where they are in their breeding cycle. Some species display high levels of affiliative behaviors in the non-breeding season, while others increase their intraspecific aggression when resources are scarce (Anacker and Beery 2013).

Over the last several decades, investigation of the neural mechanisms underlying social behavior has focused primarily on "prosocial" behaviors, rather than competitive behaviors. The tremendous progress that has been made in understanding the neural mechanisms underlying phenomena such as maternal behavior and pair bonding has likely contributed to this imbalance. Unfortunately, investigation of more competitive behaviors, such as aggression, has been on the decline for a variety of reasons (see Blanchard et al. 2003). Within studies of competitive behavior, males have been the main experimental subjects, perhaps because of Darwin's emphasis on male–male competition and female mate choice in the context of sexual selection (Darwin 1871). More recently, however, the importance of competitive behaviors in females has been recognized. Not only do female mammals compete for resources and mates to achieve reproductive benefits, but female competition is widespread in the animal kingdom (Rosvall 2011; Stockley and Bro-Jorgensen 2011; Huchard and Cowlishaw 2011). Females compete for resources such as food, nest sites, and protection using a variety of strategies including intergroup aggression, dominance relationships, and territoriality, as well as through the inhibition of the reproductive capacity of other females. In many primate species female aggression is associated with rank and ultimately reproductive goals (for review, see Stanyon and Bigoni 2014). As a result, investigation of female competitive behavior is essential to understanding social behavior and its translational implications.

In rodents, one reason that there are few data on female competitive behavior is that in commonly studied laboratory species, i.e., rats and mice, females display little or no competitive behavior (Blanchard and Blanchard 2003). This contrasts with what is observed in another laboratory rodent, whose utility as an experimental model continues to increase, Syrian hamsters (*Mesocricetus auratus*). Female Syrian hamsters display a range of competitive strategies including the expression of high levels of spontaneous offensive aggression, the rapid formation of robust dominance relationships, and the ability to inhibit the reproductive capacity of other females (Albers et al. 2002; Huck et al. 1988). While there is very little known about the neural mechanisms controlling female offensive aggression in any mammalian species, studies in hamsters have provided a good deal of information about how gonadal hormones influence this form of aggression (Albers et al. 2002).

In males, high levels of aggressive behaviors, intermale and territorial in particular, are often at their peak with the onset of breeding. Other forms of aggression are closely linked to parental behaviors, thus allowing for the defense of young, mates, food, or territories. Seasonal shifts from high cooperativity/affiliative behaviors to competitive behaviors are mediated primarily by changes in gonadal steroids, which are often linked to changes in photoperiod, though numerous neurotransmitter/neuropeptides are also involved. In many mammalian species, androgen concentrations are very high during the breeding season, as they are needed to support reproductive behaviors as well as the physiology of the gonads. During the non-breeding season, seasonal breeders will often undergo a period of gonadal quiescence, whereby the testes shrink in size and levels of circulating androgens plummet. However, it should be noted that in some species, such as hamsters (also seen in birds), androgens, specifically dehydroepiandrosterone (DHEA), produced by the adrenal glands may help to support aggressive behavior during the non-breeding season by serving as a prohormone or neurosteroid for the brain when gonadally derived androgen levels are low (for review, see Soma et al. 2015).

2.2 Brain Areas that Regulate Cooperative and Competitive Behaviors

The social behavior neural network (SBNN) hypothesis by Newman (1999) proposes that a network composed of neural groups or "nodes" including, but not limited to, the extended amygdala, the bed nucleus of the stria terminalis (BNST), lateral septum (LS), periaqueductal gray (PAG), medial preoptic area (MPOA), ventromedial hypothalamus (VMH), and anterior hypothalamus (AH) controls social behavior. Each node within the SBNN meets several criteria: reciprocal connectivity, neurons with gonadal steroid hormone receptors, and having been identified as being important to more than one social behavior. The SBNN hypothesis has gained traction in the field in recent years (reviewed by Albers 2012, 2015; Crews 1997; Goodson and Kingsbury 2013). It represents a more nuanced and complicated approach to the understanding of social behavior, as it takes the regulation of these behaviors beyond the examination of just a single neuroanatomical area and supposes that the output of the network is an emergent property. The identification of SBNN for different species is an important next step in understanding the complexity of behavior. While previous approaches have been more simplistic, examining specific neural anatomical areas, single neurotransmitter/neuropeptides, and behaviors, the foundation has now been laid for impactful studies focused on how social behavior emerges from complex neural networks. A large number of different types of motivated social behaviors are thought to be controlled by the SBNN, including offensive and defensive aggression, social recognition memory, parental behavior, and social communication. Importantly, Oxt/Avp and their receptors are found throughout the SBNN and are ideally suited to regulate the expression of social behavior because of their plasticity in response to factors that influence social behavior (Fig. 1) (reviewed in Kelly and Goodson 2014; Goodson and Kingsbury 2013; Albers 2012, 2015; Caldwell et al. 2008a; Adkins-Regan 2009; Bosch and Neumann 2012).

Motivated behaviors also arise from a network of reciprocally connected brain regions that determine the salience of stimuli, assign motivational value, and initiate appropriate action (reviewed by Love 2014). The ventral tegmental area (VTA) is a key region in this network in that VTA neurons producing DA project to a large number of cortical and limbic structures, forming the foundation underlying the

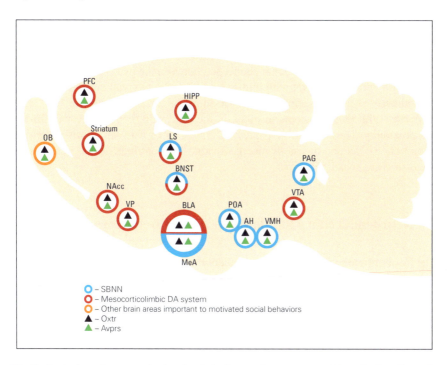

Fig. 1 Oxytocin and vasopressin signaling in brain areas important to social motivation. Oxytocin (Oxtr) and vasopressin receptors (Avprs) are found throughout the structures of the social behavior neural network (SBNN) and the mesocorticolimbic dopamine (DA) system. Their localization in these nuclei is critical for oxytocin's and vasopressin's modulation of socially motivated behaviors and may serve as the functional connection between the SBNN and DA systems, particularly by their action in the lateral septum (LS) and extended amygdala, including the bed nucleus of the stria terminalis (BNST). Other abbreviations: *AH* anterior hypothalamus; *BLA* basolateral amygdala; *HIPP* hippocampus; *MeA* medial amygdala; *NAcc* nucleus accumbens; *OB* olfactory bulb; *PAG* periaqueductal gray; *PFC* prefrontal cortex; *POA* preoptic area; *VP* ventral pallidum; *VTA* ventral tegmental area; *VMH* ventromedial hypothalamus

motivational circuitry. This network plays a critical role in social as well as nonsocial behavior and appears to provide an alerting signal for unexpected stimuli. Within this network, there are distinct groups of DA neurons that determine motivational value, being excited by appetitive stimuli and inhibited by aversive stimuli. Other groups of DA neurons appear to encode motivational salience, but not valence, in that they are excited by the intensity of the stimulus, regardless of whether it is appetitive or aversive (reviewed by Love 2014). See chapters by Bissonette and Roesch, Robinson et al., Redish et al., and Salamone et al. for detailed discussions of these issues.

While the SBNN and the mesolimbic DA system are distinct from one another, they are thought to dynamically interact and support decision making in the context of motivated social behaviors (O'Connell and Hofmann 2011a, b). O'Connell and Hofmann (2011b) have proposed that these two systems should be considered as part of a larger social decision-making network (SDM) that is relatively conserved across species. Across these two systems, there are two neuroanatomical areas, or nodes, of overlap—the LS and the extended amygdala—including the BNST. These areas provide the functional connection between the two systems by acting as relays, providing the SBNN with information from the motivational network about the salience of a social stimulus in turn allowing for an appropriate behavioral response. With these concepts in mind, in the following sections, we will discuss the possibility that Oxt and Avp provide critical links between specific elements within the SBNN and the motivational network that contribute to the motivational forces driving social behaviors.

3 Neuroendocrine Modulation of Social Behaviors

The first hormones implicated in the regulation of social behaviors were the gonadal steroids (Berthold 1849), since changes in social behaviors are observed following gonadectomy. While the gonadal steroids are important, so too are the evolutionarily ancient Oxt and Avp neuropeptide systems, as well as their non-mammalian homologues. Oxt and Avp are both primarily synthesized in the paraventricular nucleus (PVN) and supraoptic nucleus (SON) of the hypothalamus. Their genes sit in opposite transcriptional direction on the chromosome as the result of the duplication of an ancestral vasotocin gene (Acher and Chauvet 1995; Acher et al. 1995) and they are synthesized as part of a larger precursor preprohormone (Hara et al. 1990). Since they are so structurally similar, Oxt and Avp are considered "sister" hormones, though their actions both peripherally and centrally can differ significantly from one another. Interestingly, across mammalian species, the roles of Oxt and Avp with regard to cooperative and competitive behaviors tend to be fairly conserved (Caldwell et al. 2008a; Caldwell and Young 3rd 2006; Lee et al. 2009a; Adkins-Regan 2009; Neumann 2008; Veenema and Neumann 2008; Carter et al. 2008; Albers 2012, 2015).

3.1 The Oxytocin System

Oxytocin literally meaning "sharp childbirth" is known for its peripheral actions on the regulation of uterine contraction as well as the facilitation of milk ejection (Dale 1906; Ott and Scott 1910). In rats, Oxt is synthesized in larger, magnocellular neurons, of the PVN and SON that project to the posterior pituitary and mediate the aforementioned actions. However, it is Oxt that is synthesized in the smaller, parvocellular neurons of the PVN that project centrally and mediate many of the central actions of Oxt. It should be noted, however, that this compartmentalization of function by magnocellular versus parvocellular neurons is not found in all species. For example, in Syrian hamsters, magnocellular neurons do not exclusively project to the posterior pituitary and seem to project centrally (Ferris et al. 1992b). Also, in mice and several vole species, there are reports of parvocellular Oxt neurons outside of the PVN (Castel and Morris 1988; Jirikowski et al. 1990; Wang et al. 1996) as well as reports of Oxt collaterals from the SON/PVN to the nucleus accumbens (NAcc) (Ross et al. 2009a).

Thus far, only a single Oxt receptor (Oxtr) has been identified, and it is thought to be the primary mechanism for the transduction of the Oxt signal (Kimura et al. 1992; Kubota et al. 1996); however, see the section titled "**Signaling by Oxytocin and Vasopressin in the Brain**." The Oxtr is a member of the seven-transmembrane G-protein-coupled receptor family and signals through $G_{\alpha q/11}$ GTP-binding proteins and $G_{\beta\lambda}$ (Ku et al. 1995; Gimpl and Fahrenholz 2001; Zingg and Laporte 2003), which results in the hydrolysis of phosphatidylinositol. The structure and sequence of the Oxtr is similar to the Avp receptors (Gimpl and Fahrenholz 2001). In rats and mice, the Oxtr is most often visualized with receptor autoradiography through the use of a potent and specific ^{125}I-labeled antagonist (Kremarik et al. 1993; Veinante and Freund-Mercier 1997). The Oxtr is observed in several areas of the brain, including the hippocampal formation, LS, central amygdala (CeA), olfactory tubercle, NAcc shell, dorsal caudate–putamen, BNST, medial amygdala (MeA), and VMH (Kremarik et al. 1993; Veinante and Freund-Mercier 1997; Insel et al. 1991), but there are seasonal as well as species- and sex-specific differences.

3.2 The Vasopressin System

Avp is named for its involvement in the constriction of blood vessels, but is also important to salt and water balance. The peripheral actions of Avp are primarily mediated by the magnocellular neurons of the PVN and SON, which result in Avp release from the posterior pituitary. Centrally, Avp is more widely expressed than Oxt, and its distribution can vary substantially between species. Avp immunoreactive (ir) cell bodies are consistently found in several hypothalamic nuclei including the suprachiasmatic nucleus (SCN), PVN, SON as well as in groups of accessary nuclei (Sofroniew 1983). Outside of the hypothalamus, Avp-ir neuronal cell bodies can be observed in the BNST and MeA in most rodent species examined to date (Sofroniew 1985). Interestingly, in Syrian hamsters, neuronal cell bodies containing Avp are absent in the BNST and MeA (Albers et al. 1991). Projections from Avp-producing neurons form a dense vasopressinergic network throughout the brain (Buijs et al. 1983, 1987; De Vries and Buijs 1983; Sawchenko and Swanson 1982).

Avp receptors can be divided into two classes: Avp1 and Avp2 receptors (Avpr1 and Avpr2, respectively), both of which are seven-transmembrane G-protein-coupled receptors that are similar in structure to the Oxtr. There are two subtypes of the Avpr1: the Avpr1a and the Avpr1b. Peripherally, the Avpr1a mediates the effects of Avp on vasoconstriction and can be found in the liver, kidney, platelets, and smooth muscle (Ostrowski et al. 1992; Watters et al. 1998). Centrally, the Avpr1a is found in a variety of brain nuclei (Johnson et al. 1995; Tribollet et al. 1997; Ostrowski et al. 1994; Szot et al. 1994). The Avpr1a is modulated by gonadal hormones and photoperiod in some brain regions, but not others (Johnson et al. 1995; Young et al. 2000; Caldwell and Albers 2004b; Caldwell et al. 2008b). The Avpr1b was originally described in the anterior pituitary, where it is prominent on the corticotrophs; however, it can also be found in the brain (Antoni 1984; Lolait et al. 1995). In rats, there is a lack of consensus about the central distribution of the Avpr1b, with some groups reporting Avpr1b in the olfactory bulb, piriform cortical layer II, LS, cerebral cortex, hippocampus, PVN, SCN, cerebellum, and red nucleus (Lolait et al. 1995; Saito et al. 1995; Vaccari et al. 1998; Hernando et al. 2001; Stemmelin et al. 2005). However, a later study by Young and colleagues, which used more stringent conditions for in situ hybridization histochemistry (ISHH), determined that the Avpr1b in rats and mice is more discretely localized with prominent expression in hippocampal field CA2 pyramidal neurons (Young et al. 2006). The Avpr2 is found in the periphery and is primarily expressed in the kidney; it has not been localized to the brain. Its role in the kidney is to transduce the antidiuretic effects of Avp within the renal collecting ducts (Bankir 2001).

3.3 Signaling by Oxytocin and Vasopressin in the Brain

Neuropeptides can act in a highly localized manner, similar to classic neurotransmitter release at the synapse. However, neuropeptides can also be released in a much more diffuse manner, potentially impacting large numbers of neurons at multiple sites (Engelmann et al. 2000; Landgraf and Neumann 2004; Ludwig 1998; Ludwig and Leng 2006). This diversity of action has long been recognized, although the role of these different types of signaling mechanisms in the control of social behavior is not well understood. Neuropeptides such as Oxt and Avp are usually packaged in large dense-core vesicles (LDCV) that can be found in all areas of neurons including the presynaptic terminal (Jakab et al. 1991; Buijs and Swaab 1979; van Leeuwen et al. 1978). Because of the broad distribution of LDCV throughout the cell, neuropeptides can act locally at the synapse or much more broadly when released from non-synaptic regions (e.g., dendrites) in what is called volume transmission. Many factors affect the profiles of neuropeptide release such as the size of the neurons from which they are released, the spread of peptide after release, and the timing and intensity of its degradation by peptidases. While the spatial and temporal profiles of peptide release via volume transmission are not well understood (Leng and Ludwig 2008), estimates suggest that they may travel as far as 4–5 mm from their site of release (Engelmann et al. 2000). Magnocellular neurons in the hypothalamus represent the largest pool of nonapeptides in the brain, and there is evidence that they are activated by a variety of stimuli related to social behavior (Delville et al. 2000). As a result, it seems likely that volume transmission of nonapeptides from these neurons plays an important role in regulating social behavior by acting on nonapeptide receptors throughout the SBNN.

Another consideration is that neuropeptides are commonly found in neurons that also produce small molecular weight "classical neurotransmitters" (e.g., amino acids) (for review, see van den Pol 2012; Albers 2015), thus allowing for their co-release. In most cases, classical neurotransmitters are packaged in small synaptic vesicles (SSVs) in presynaptic terminals. The exocytosis of both SSVs and LDCVs at synapses is Ca^{2+} dependent. Because SSVs are usually in closer proximity to the membrane than LDCV, less activity is required for classic neurotransmitter release than for neuropeptide release. Therefore, synaptic release of neuropeptides is thought to lag behind that of neurotransmitter release and to require more electrical activity. The functional significance of synaptic co-release of classical neurotransmitters and neuropeptides is not known, but dynamic interactions of these signals will likely be important in understanding neuropeptide regulation of behavior (Bamshad et al. 1996). In summary, the ways that Oxt and Avp contribute to neurochemical signaling within the brain are varied and complex. As a result, researchers are left to untangle very intricate pharmacological interactions when they consider the effects of these neuropeptides on behavior.

In addition to the diversity seen in the ways that Oxt and Avp can signal, Oxt and Avp also have a high degree of similarity in their structure as well as in the structure of their canonical receptors (Maybauer et al. 2008; Gimpl and Fahrenholz 2001; Manning et al. 2012; Song et al. 2014b). Due to these structural similarities, there is a substantial amount of cross talk between these systems (Schorscher-Petcu et al. 2010; Sala et al. 2011), with Oxt and Avp having similar affinities for the Oxtr, Avpr1a, and Avpr1b in rats and mice (Manning et al. 2008, 2012). For example, both Oxt and Avp can induce communicative behavior in hamsters when injected into the AH by activating Avpr1a rather than the Oxtr (Song et al. 2014b). Similarly, both Oxt and Avp can enhance social recognition and social reward by activating the Oxtr and not the Avpr1a (Song et al. 2014a, b, 2015; Ragnauth et al. 2004). These data indicate that the effects of Oxt and Avp on different social behaviors can result from their activation of the Avpr1a, the Oxtr, or both.

Oxt and Avp receptors are distributed within structures of the mesolimbic DA system, providing a means by which these systems may interact. These areas include the amygdala, the hippocampus, the VTA, the prefrontal cortex (PFC), the NAcc, and the ventral pallidum (VP) (Vaccari et al. 1998; Baskerville and Douglas 2010; Curtis et al. 2008). It is also important to remember that there is a great deal of interspecies and interindividual variability in the distribution of Oxt and Avp receptors and that these differences play a major role in producing differences in the expression of social behavior.

Recent work in primates has shed some light on the complexity of these systems in a broader evolutionary context, as there are some significant evolutionary changes in the Oxt and Avp systems. As discussed above, in rats and mice, the Oxtr and Avprs are relatively non-selective for Oxt and Avp; however, in humans, the

OXTR has significant selectivity for Oxt over Avp. The functional significance of these differences in receptor selectivity is not known but is likely to be important. Aside from cross talk, there are also sequence differences in Oxt across primate species. Specifically, work by Lee et al. (2011) challenged the idea that Oxt is an invariant nine-amino acid sequence by determining that some New World primates have an amino acid substitution in the 8th position, a leucine rather than a proline. The implications of these findings are still being explored (Cavanaugh et al. 2014), but this coupled with variations in the *Avpr1a* gene suggests that there has been, and perhaps continues to be, an evolutionary shift in these systems and reinforces the importance of studying these systems together rather than in isolation (for review, see Ragen and Bales 2013). These findings are likely to impact work in preclinical models and may help to resolve some of the conflicting findings in the literature between primates and other mammalian species.

3.4 Epigenetics

The advent of epigenetics has also challenged our understanding of how these systems are regulated and potentially how they regulate other systems. Epigenetics refers to changes in gene transcription that are not due to changes in nucleotide sequence, but rather are "above" the genome. Epigenetic mechanisms include histone modification and gene methylation, both of which alter the ability of the transcriptional machinery to access promoter regions. While much of this work originally centered on Oxt and maternal behavior in animal models, it has now been expanded to human studies.

Elegant work from Dr. Michael Meaney's laboratory found that in rats, the quality of mother–infant interactions affected DNA methylation and histone acetylation patterns in the offspring (e.g., Champagne et al. 2001, 2004; Fish et al. 2004; Francis et al. 1999; Weaver et al. 2004). During the first postpartum week, low licking-and-grooming (LG) dams have reduced Oxtr binding in the MPOA compared to high LG dams (Champagne et al. 2001; Francis et al. 2000). Further, microinjection of an Oxtr antagonist reduces licking and grooming in high LG dams, with essentially no effect in low LG dams (Champagne et al. 2001). For more complete reviews of this work, see (Champagne 2008; Bridges 2015).

More recently, work in humans suggests that changes in the methylation of the *OXTR* are associated with a variety of diseases/disorders, including anorexia nervosa (Kim et al. 2014), the perception of fear and anxiety (Puglia et al. 2015; Ziegler et al. 2015), psychopathy (Dadds et al. 2014), and autism (Gregory et al. 2009). However, a direct causal link between methylation patterns and behavior has yet to be made; however, given the animal literature, it seems likely that this is an important means by which early life experience may be able to directly impact behavior.

4 Oxytocin/Vasopressin and Cooperative and Competitive Behaviors: Social Memory, Social Interactions, and Aggression

Within mammals, social structures can vary widely. Take for instance naked mole rats and their large insect-like colony structure, prairie voles and their pair bonding that results in a lifetime social dyad, or more solitary species such as Syrian hamsters. With this diversity, it might be assumed that there are vast differences in the types of behaviors that these species are capable of displaying and in the neurotransmitters and/or brain areas that are important to their regulation. This, however, does not seem to be the case. While the details may differ, there are a limited number of social behaviors common to most species and they include the ability to remember others of the same species (social memory) as well as the social behaviors that determine the nature of their relationships with conspecifics (e.g., affiliation/dominance). Furthermore, the roles of Oxt and Avp in the modulation of social behaviors have been evolutionarily conserved across species. For example, aggression is modulated by social experience-induced changes in the expression of Avpr1a in the hypothalamus in both male prairie voles and hamsters.

Since 90 % of mammals are not biparental, it is common for them to live in large social groups and display both cooperativity and competitiveness, depending on the context. Even in species that are more solitary and highly competitive, individuals have the capacity for social recognition, social communication, as well as the potential to form stable long-standing social relationships. There are also sex differences as well as environmental effects, such as photoperiod, which can cause shifts in social behaviors from the breeding to the non-breeding season. In this section, we will explore the role of Oxt and Avp in selected forms of cooperative and competitive behavior in these differing and complex social contexts.

4.1 Social Recognition Memory

Displays of social behaviors often depend on whether the interaction is with an individual that is familiar or unknown. Thus, the ability to recognize individuals and remember them, i.e., social recognition memory, plays an important role in the decision to approach or avoid. There are considerable data that both Oxt and Avp contribute to social recognition memory. Oxt is thought to facilitate social memory by altering the processing of socially salient olfactory information (for review, see Lee et al. 2009a; Gabor et al. 2012; Wacker and Ludwig 2012). In males, Oxt infused into the olfactory bulb (OB), lateral ventricles, and MPOA facilitates social recognition memory (Dluzen et al. 1998; Benelli et al. 1995; Popik and Van Ree 1991). Some of the particulars of the circuit have been worked out with the assistance of genetic knockouts of the Oxt system, with Oxt knockout (Oxtr −/−) mice and forebrain-specific Oxtr knockout (Oxtr FB/FB, where CRE recombinase is driven by a CaMKIIα promoter) mice having deficits in social recognition memory (Lee et al. 2008a, b; Takayanagi et al. 2005; Macbeth et al. 2009; Ferguson et al. 2000; Hattori et al. 2015). Based on this work, a four-gene micronet involving Oxt, the Oxtr, estrogen receptor α, and estrogen receptor β has been proposed as being critical to the regulation of social recognition memory in both males and females (Choleris et al. 2003). In particular, estrogen-dependent Oxt signaling in the MeA appears to be key for normal social recognition memory. Infusion of Oxt into the MeA of Oxt −/− mice can rescue deficits in social recognition memory (Ferguson et al. 2001), and infusion of an Oxtr antisense DNA or an Oxtr antagonist into control mice can block social recognition memory (Choleris et al. 2007; Ferguson et al. 2001). Female Oxt −/− mice also have a disrupted Bruce effect (Bruce 1959), whereby they terminate their pregnancy when exposed to their mate, which suggests that they do not remember him (Wersinger et al. 2008).

Recent work suggests that Oxt may also play a role in human social recognition. Generally speaking, exogenous Oxt enhances the memory for faces (Savaskan et al. 2008; Guastella et al. 2008; Rimmele et al. 2009). Further, a common single-nucleotide polymorphism (SNP) (rs237887) in the *OXTR* is moderately associated with face recognition memory in families from the UK and Finland that have a child with an autism spectrum disorder (ASD) (Skuse et al. 2014). Taken together, these data suggest that Oxt and the Oxtr have a conserved role in the modulation of social memory across species.

Avp also appears to be important for social memory. Androgen-dependent Avp projections from the MeA and BNST to the LS, all parts of the SBNN, are important for individual recognition (De Vries et al. 1984; Mayes et al. 1988; Bluthe et al. 1990, 1993). Microinjections of Avp into the LS of control or AVP-deficient Brattleboro rats facilitates social memory, whereas microinjections of Avpr1a antagonists or infusions of antisense Avpr1a oligonucleotides into the LS of control rats impairs social recognition memory (Engelmann and Landgraf 1994; Landgraf et al. 1995). The use of Avpr1a −/− and Avpr1b −/− mice has also provided some insight into the contributions of Avp to social memory. However, in the NIMH line of Avpr1a −/− mice, the findings have been mixed (Hu et al. 2003), with one group reporting that males have impaired social recognition that can be rescued by the overexpression of Avpr1a in the LS (Bielsky et al. 2003, 2005; Bielsky and Young 2004) and another group reporting no deficits in social recognition, but rather in olfaction (Wersinger et al. 2007b). While the reason for the discrepancy remains unknown, it is obvious from previous reports that Avpr1a in the LS is important for normal social recognition memory. Interestingly, Avpr1b −/− mice have mild impairments in social recognition memory (Wersinger et al. 2002). Further, lesions and genetic silencing of the CA2 region of the hippocampus, where the Avpr1b is prominently expressed, also results in impaired social recognition memory (Stevenson and Caldwell 2012; Hitti and Siegelbaum 2014).

While the aforementioned data provide strong support for a role of Avp in social recognition, how or if Avp modulates social recognition has really only been studied in a small number of species. Further, the data are limited by the fact that the vast majority of studies have only examined social recognition for very short intervals (<2 h). However, there are some studies that suggest that Avp may be important for long-term social recognition. Specifically, injection of an Avpr1a antagonist into the LS can block the formation of a mating-induced pair bond in male prairie voles, while injection of Avp into the LS institutes a pair bond in the absence of mating (Liu et al. 2001). Since social recognition is a necessary part of pair bonding and pair bonds can last a lifetime, these data suggest that Avp signaling via the Avpr1a can induce long-term changes in social recognition. There are also data to suggest that Avpr1a antagonists administered into the MeA can alter maternal memories in rats. Normally, after a ten-day separation from their pups, mothers display full maternal behavior within about 12 h of re-exposure. However, in peripartum mothers, in which an Avpr1a antagonist is infused into the MeA, the latency to display full maternal behavior does not occur for approximately 60 h; the antagonist had no effect on the initial expression of maternal behavior on the day of parturition (Nephew and Bridges 2008).

More recently, we have found that male Syrian hamsters can recognize social odors of other male Syrian hamsters for 24 h. Injection of Oxt or Avp intracerebroventricular (i.c.v) extends their social memory to 48 h. Interestingly, these effects are mediated by the Oxtr and not the Avpr1a (Song et al. 2015). In the broader context of animal behavior, it will also be important to determine whether Avp mediates some of the more complex forms of social recognition found by rodents such as kin recognition and the true recognition of specific individuals (e.g., Mateo and Johnston 2003; Johnston and Peng 2008; Petrulis 2009).

4.2 Cooperative Behavior

Although cooperative behavior is certainly not limited to species that pair bond, pair bonding species do provide an important model system in which to investigate the neurobiology of cooperation. That said, it remains important to recognize that these bonds are formed by mating and are limited to the cooperation of a male and a female. Pair bonding, formed by mating, represents one form of cooperative behavior. Pair bonds are somewhat unique among mammals in that they are only seen in 3–5 % of mammalian species (Kleiman 1977). Defined as a preference for contact with a familiar sexual partner, selective aggression toward unfamiliar conspecifics, biparental care, socially regulated reproduction, and incest avoidance (Carter et al. 1995; Carter and Getz 1993), pair bonding can be found in titi monkeys (*Callicebus cupreus*), marmosets (*Callithrix penicillata* and *Callithrix jacchus jacchus*), California mice (*Peromyscus californicus*), and prairie voles (*Microtus ochrogaster*). Humans too can have strong selective bonds between mates, but also show cooperative behaviors in many other contexts as well. The extent to which human pair bonds and other forms of human cooperative behavior are regulated by Oxt and Avp remains to be determined.

Our understanding of the mechanisms by which Oxt and Avp contribute to pair bonding comes primarily from work in prairie voles. Prairie voles live in extended family groups and are considered a socially monogamous species (Carter et al. 1995) (social monogamy is distinct from sexual monogamy as most individuals have extra-pair copulations). The formation of a pair bond is experimentally tested in the laboratory using a partner preference test (Williams et al. 1992), whereby the preference of a male or female for an animal in which they have previously cohabitated, i.e., the partner, versus a "stranger," is assessed. If the experimental subject spends twice as much time with the "partner" animal, then it is said to have formed a pair bond with that individual (Insel and Hulihan 1995; Carter and Getz 1993; Williams et al. 1994; Carter et al. 1995).

Due to the diversity in social structures within the genus *Microtus*, comparative studies between vole species have provided significant insight into the neural regulation of social bonding. By comparing the neurochemistry of monogamous vole species, such as the prairie or pine vole (*Microtus pinetorum*), to non-monogamous voles, such as the montane (*Microtus montanus*) or meadow (*Microtus pennsylvanicus*) voles, scientists have explored how variations in Oxt and Avp neurochemistry between highly related species can result in significant differences in social behavior (Young et al. 2008, 2011; Adkins-Regan 2009). While there are differences in Oxt- and Avp-ir cells or their projections between species, most profound are the changes in the neuroanatomical distribution of the Oxtr and the Avpr1a. Relative to non-monogamous voles, monogamous voles have higher densities of Oxtr, as measured using Oxtr autoradiography and ISHH, in the NAcc, the PFC, and the BNST. In contrast, promiscuous voles have higher Oxtr density in the LS, VMH, and the cortical nucleus of the amygdala (Insel and Shapiro 1992; Young et al. 1996; Smeltzer et al. 2006). Evidence that the differences in the distribution of the Oxtr between species might be behaviorally meaningful comes primarily from pharmacological studies.

In female prairie voles, central infusion of an Oxtr antagonist blocks the formation of the pair bond, but has no effect on sexual behavior, whereas central infusion of Oxt facilitates the pair bond in the absence of mating (Insel et al. 1995; Williams et al. 1994; Cho et al. 1999) and can decrease male-directed aggression (Bales and Carter 2003). In the aforementioned studies, the infusions were i.c.v.; however, manipulation of Oxtr signaling, using Oxtr antagonists and RNAi knockdown of the Oxtr within the NAcc, inhibits formation of a partner preference (Liu and Wang 2003; Young et al. 2001; Keebaugh et al. 2015), and overexpression of the Oxtr in the NAcc of adult female prairie voles accelerates the formation of partner preference (Ross et al. 2009b). However, overexpression of the Oxtr in the NAcc of non-monogamous meadow voles is not sufficient to promote pair bond formation (Ross et al. 2009b), which suggests that all of the required neurocircuitry is not in place for this species.

There are also differences in the distribution of the Avpr1a between vole species. Prairie voles have a higher density of Avpr1a, as measured using receptor autoradiography and ISHH, within the MeA, accessory olfactory bulb, diagonal band, thalamus, VP, and BNST compared to montane voles (Young et al. 1997; Insel et al. 1994). Montane voles, on the other hand, have a higher density of Avpr1a in the medial PFC and the LS (Smeltzer et al. 2006; Insel et al. 1994). These differences in

Avpr1a distribution are thought to contribute to differences in social organization between monogamous and non-monogamous vole species. This hypothesis is supported by data in pine voles and meadow voles, which suggest similar social structure-specific distributions of Avpr1a between these species (Insel et al. 1994). Further support for this hypothesis comes from pharmacological manipulations of the Avpr1a in prairie voles. When an Avp antagonist is injected i.c.v. prior to mating, the formation of a partner preference is inhibited. Conversely, Avp infusion facilitates the formation of the partner preference (Winslow et al. 1993; Cho et al. 1999). Some of the more interesting data that support a role for the differential distribution of the Avpr1a in the formation of social bonds come from a study in which the prairie vole Avpr1a gene was overexpressed in the ventral forebrain of meadow voles, resulting in increases in the amount of time meadow voles spent huddled with their partners compared to controls (Lim et al. 2004).

4.3 Competitive Behavior

The most conspicuous form of competitive behavior is aggression. Offensive and defensive aggressions have been studied intensely and almost exclusively in male mammals. Further, neural circuits that overlap much of the SBNN have been proposed for each of these forms of aggression (Delville et al. 2000; Choi et al. 2005). Although frequently characterized as a negative social interaction, aggression plays a highly constructive role in the formation of social relationships. In the vast majority of mammals that do not form pair bonds, dominance relationships provide social bonds that serve many adaptive functions (e.g., resource distribution) and that ultimately reduce social conflict. Typically, dominance relationships are rapidly determined through aggressive interactions but are primarily maintained through social communication (e.g., scent marking, vocalization), thereby reducing the dangers associated with intense fighting (Albers et al. 2002; Fernald 2014).

4.3.1 Oxytocin and Competitive Behavior

Some of the earliest evidence that Oxt is involved in both aggression and social communication came from studies in squirrel monkeys. In male squirrel monkeys, with established dominant–subordinate relationships, i.c.v. administration of Oxt increases aggression in dominant males, while having no effect in subordinate males (Winslow and Insel 1991b). In contrast, Oxt stimulates scent marking in subordinate males but does not alter scent marking in dominant males (Winslow and Insel 1991b), thus demonstrating that social experience can determine the behavioral response to Oxt acting in the brain. Studies in rats suggest one mechanism that might contribute to the effects of social experience on the behavioral response to Oxt. Dominant male rats have significantly higher levels of Oxtr mRNA in the MeA 3 h after the social encounter that defined their relationship. Further, infusion of an Oxtr antagonist immediately after the establishment of subordinate status increases the duration of the dominant–subordinate relationship (Timmer et al. 2011). Social status also significantly impacts the circulating levels of serum Oxt in primates. In rhesus monkeys, dominant females have significantly higher levels of serum Oxt than subordinates (Michopoulos et al. 2011, 2012). The importance of social experience in determining the behavioral response to Oxt/Avp is a theme seen repeatedly in this chapter.

Historically, there have been little data supporting a role for Oxt in the regulation of intermale aggression in laboratory species of rodents. However, some recent work suggests that pharmacological treatment with Oxt may have antiaggressive effects in adult males. Work from Jaap Koolhaas' laboratory has found that pharmacological enhancement of Oxt in rats, by intranasal treatment or i.c.v. infusion, reduces offensive aggression and promotes prosocial exploratory behaviors (Calcagnoli et al. 2013, 2014, 2015a). Furthermore, these inhibitory effects seem to be mediated by Oxt acting via the Oxtr in the CeA; however, it should be noted that blocking endogenous Oxt signaling in the CeA has no effect (Calcagnoli et al. 2015b). Thus, it is proposed that these findings should be considered in the context of pharmacological effects rather than being directly regulated by the local endogenous Oxt system (Calcagnoli et al. 2015b).

The role of Oxt in the neural regulation of female offensive aggression is sparse; as mentioned previously, females of the most commonly studied laboratory rodents, unlike many other mammalian species, rarely display aggressive behaviors outside of the peripartum period. However, there is evidence in female Syrian hamsters that Oxt can influence competitive behaviors. Specifically, microinjection of Oxt into the MPOA-AH reduces offensive aggression, and injection of an Oxtr antagonist increases offensive aggression directed toward a female intruder (Harmon et al. 2002a). Traditionally, studies of Oxt in female rodents have focused on maternal aggression. This is a unique, and transient, hormonal/physiological time in a female's life and is characterized by high levels of nurturing behaviors directed toward pups and aggressive behaviors directed toward intruders. During the peripartum period, Oxt has anxiolytic effects (Bosch and Neumann 2008), specifically via its actions in the PVN and CeA (Blume et al. 2008; Jurek et al. 2012; Windle et al. 1997; Huber et al. 2005; Viviani et al. 2011; Knobloch et al. 2012). These decreases in anxiety permit females to attend to their pups. But even with lower levels of anxiety and strong bonds with her offspring, dams can display high levels of maternal aggression. Depending on the brain area and the behavioral state of the animal, Oxt and/or Oxtr antagonists can either increase or decrease maternal aggression (for review, please see Bosch 2013). For instance, in rats bred for low-anxiety rats, Oxt in the PVN increases maternal aggression (Bosch et al. 2005), but when microinjected into the BNST of Wistar rats decreases maternal aggression (Consiglio et al. 2005). In hamsters, injection of Oxt into the amygdala facilitates maternal aggression (Ferris et al. 1992a). So, while central Oxt is important to aggression in females, its effects are context and site specific. There are also numerous studies that support the assertion that developmental exposure to Oxt is important for the proper development of motivated social behaviors, including aggression, but those studies will not be reviewed in this chapter (recently reviewed in Miller and Caldwell 2015; Hammock 2015).

4.3.2 Vasopressin and Competitive Behavior

Most of what we know about the role of nonapeptides in competitive behavior comes from studies of Avp and aggressive behavior. Further, much of the work has focused on the Avpr1a, as this was the first centrally identified receptor, and as such, there are a number of pharmacological tools available. However, the Avpr1b appears also to be important for displays of aggressive behavior. Because of the differences in the distribution of these two receptor subtypes, this section will be divided by subtype.

4.3.3 The Vasopressin 1a Receptor and Competitive Behavior

The first evidence for a role for Avp in aggression came from studies in male hamsters, in which the injection of Avpr1a antagonists into the AH significantly inhibited offensive aggression (Ferris and Potegal 1988; Potegal and Ferris 1990). Subsequent studies have replicated these results and shown that Avp injected into the AH stimulates offensive aggression (Caldwell and Albers 2004a; Ferris et al. 1997). However, the ability of Avp to stimulate aggression by its action in the AH appears to depend on an individual's prior social experience. Avp injected into the AH increases aggression in male hamsters previously trained to fight other hamsters and in hamsters socially isolated for at least four weeks, but not in hamsters housed in social groups. The ability of social experience to enhance the response of the AH to Avp appears to be mediated by experience-dependent increases in the number of Avpr1a in the AH (Albers et al. 2006; Cooper et al. 2005). In summary, Avp has powerful effects on offensive aggression in male hamsters, but only if social experience has upregulated Avpr1a receptors in the AH.

Avp can also have potent effects on aggression in male prairie voles through a similar mechanism. Sexually naïve male voles are essentially non-aggressive, choosing to explore intruder males as opposed to attacking them (Winslow et al. 1993). Following mating-induced pair bonding, males display high levels of aggression toward conspecifics other than their mate and have increases in the density of Avpr1a receptors in the AH (Gobrogge et al. 2009). Further, overexpression of the Avpr1a in the AH using viral vector gene transfer increases aggression in non-pair-bonded males. Thus, in male hamsters and prairie voles, species with very different types of social organization, an individual's social experience can modulate the number of Avpr1a in the AH and thereby regulate the intensity of aggression that is expressed.

Another hypothalamic region where Avp influences aggression is the ventrolateral hypothalamus (VLH). Avp injected into the VLH facilitates aggression in gonadally intact males and castrated males given testosterone but does not facilitate aggression in castrated controls (Delville et al. 1996). Avpr binding is also reduced in the VLH following castration and precastration levels of binding can be restored by testosterone. In contrast, it is not known whether testosterone influences the ability of Avp to alter aggression when injected into the AH. However, since castration reduces Avpr binding within this region (Delville et al. 1996), it seems possible that castration could reduce aggression stimulated by Avp. It should be noted that the relationship between gonadal hormones and aggression is complex; however, it is clear that testosterone does not simply induce aggression in males (for review, see Johnson et al. 1995; Young et al. 2000; Demas et al. 2007). In females, while there is evidence that gonadal hormones can affect Avpr1a binding within the VLH (Delville and Ferris 1995), the specific effects on aggression are unknown.

There are also extra-hypothalamic regions where aggression and Avp activity have been linked. In both male rats and mice selected for varying levels of aggression, a negative correlation has been observed between Avp fiber density in the LS and the amount of intermale aggression (Compaan et al. 1993; Everts et al. 1997). Interestingly, these differences in Avp and aggression are not related to differences in circulating levels of testosterone (Elkabir et al. 1990). The role of septal Avp in aggression has also been studied in male rats bred for low or high anxiety. Release of Avp into the LS is significantly lower in the much more aggressive low-anxiety rats than in the high-anxiety rats that exhibit lower levels of aggression. In addition, septal administration of Avp to the highly aggressive group and administration of the Avpr1a antagonist to the low aggressive group did not alter the levels of aggression expressed (Beiderbeck et al. 2007). The level of aggressiveness and the pattern of Avp expression in the LS and BNST are also correlated in mice. The monogamous California mice (*Peromyscus californicus*) have shorter attack latencies and increased Avp-ir in the BNST and LS compared to the polygamous, white-footed mice (*Peromyscus leuopus*) (Bester-Meredith et al. 1999). Interestingly, when California mice pups are cross-fostered to white-footed mice dams, they are less aggressive in adulthood than those reared by the same species, and they have less Avp-ir in the BNST and SON compared to controls (Bester-Meredith and Marler 2001). These data in mice, like those from hamsters, suggest that changes in the social environment are able to alter Avp neurocircuitry and the behavior driven by that circuitry. In addition, although relationships between aggressiveness and Avp expression and release within the LS have been found, Avp may not have any direct effects on male aggression by its actions in the LS.

Despite the powerful effects of Avp on aggression in the hypothalamus, not all central manipulations of Avp have been found to influence offensive aggression. Although i.c.v. injections of an Avpr1a antagonist increase the latency to the onset of aggression in highly aggressive California mice, i.c.v. injections of Avp or Avpr1a antagonists have no effect on aggression in white-footed mice (Bester-Meredith and Marler 2001). In non-pair-bonded male prairie voles with extensive experience with aggression, i.c.v. administration of an Avpr1a antagonist does not inhibit aggression (Winslow et al. 1993). There is other evidence from rats that i.c.v. administration of Avp does not alter the expression of aggression nor does deletion of the Avpr1a receptor in mice (Elkabir et al. 1990; Wersinger et al. 2007b). While it is clear that Avp can stimulate Aggression by its action in some brain sites, it remains possible that Avp might act to reduce aggression by its action in other brain regions or by its action on other receptors (e.g., the Oxtr) to reduce/inhibit aggression. It is also important to recognize that aggression is a complex behavior and may be facilitated by neurochemical systems other than Avp, at least in some cases.

In females, Avp has very different effects on offensive aggression than it does in males. As described above, in male hamsters, injection of Avpr1a antagonists into the AH inhibits offensive aggression and injection of Avp stimulates aggression (Ferris and Potegal 1988; Potegal and Ferris 1990; Caldwell and Albers 2004a; Ferris et al. 1997). In contrast, in female hamsters, an Avpr1a antagonist injected into the AH stimulates offensive aggression and injection of Avp inhibits aggression in the resident–intruder test (Gutzler et al. 2010). More recently, similar sex differences have been seen in the effects of Avp and Avpr1a antagonists on social play behavior in rats (Bredewold et al. 2014; Veenema et al. 2013). For example, injection of Avpr1a antagonists in the LS increases social play in juvenile males and reduces social play in juvenile females. In a related study, a negative correlation was found between Avp mRNA levels in the BNST and social play in male juvenile rats (Paul et al. 2014). As social play is hypothesized to be a precursor to aggressive behavior, these data are consistent with Avp-related sex differences in the regulation of competitive behaviors.

4.3.4 The Vasopressin 1A Receptor and Maternal Aggression

Maternal aggression is an intense form of aggression displayed by lactating mothers confronted by intruders (Lonstein and Gammie 2002). Studies using Avpr1a −/− mice found that maternal aggression does not differ from that seen in wild-type mice (Wersinger et al. 2007b). In contrast, i.c.v. administration of Avp in lactating rats reduces and an Avpr1a antagonist increases aggression toward male intruders (Nephew and Bridges 2008; Nephew et al. 2010). The effects of i.c.v. administration of Avp and an Avpr1a antagonist on maternal aggression were also examined in rat strains that had been selectively bred for high (HAB) or low (LAB) anxiety (Bosch and Neumann 2010). In both strains, Avp was found to increase and an Avpr1a antagonist to decrease maternal aggression, respectively. Other studies have used microdialysis to examine the role of Avp in specific brain regions in the regulation of maternal aggression in HAB and LAB rats (Bosch and Neumann 2010). In HAB, but not LAB, rats, Avp is positively correlated with maternal aggression in the CeA but not the PVN. In addition, administration of an Avpr1a antagonist into the CeA reduces aggression in HAB rats, while administration of Avp into the CeA increases aggression in LAB rats. The ability of Avp to promote maternal aggression by acting in the CeA does not appear to be restricted to HAB rat strains since similar results have been reported in Sprague-Dawley rats (Meddle and Bosch, unpublished; cited in, Bosch 2011). In the future, it will be important to clarify the effects of Avp on maternal aggression, its site(s) of action, and whether the effects of Avp on aggression are related to anxiety levels.

4.3.5 The Vasopressin 1b Receptor and Competitive Behavior

There is compelling evidence that the Avpr1b is essential for displays of aggressive behavior directed toward a conspecific (for review, see Caldwell et al. 2008a, c; Stevenson and Caldwell 2012). In resident–intruder and neutral cage aggression tests, Avpr1b −/− mice display fewer attacks and have longer attack latencies than Avpr1b +/+ controls (Wersinger et al. 2002, 2007a). Further, in a reversal of a resident–intruder test where the experimental mice are intruders rather than residents, Avpr1b −/− mice display normal defensive avoidance behaviors when attacked by a stimulus animal, but are less likely to initiate retaliatory attacks (Wersinger et al. 2007a). Pharmacological studies using the Avpr1b antagonist, SSR149415, support the findings of work in Avpr1b −/− mice. Syrian hamsters orally administered SSR149415 have reductions in the frequency and duration of offensive attacks, in chase behaviors, in flank marking, and in the olfactory investigation that often precedes and accompanies an offensive attack (Blanchard et al. 2005). Mice given SSR149415 display fewer defensive bites when forced to encounter a threatening predator and reductions in the duration of offensive aggression in a resident–intruder test (Griebel et al. 2002).

The deficits in aggressive behaviors observed in Avpr1b −/− mice are not limited to males. Following parturition, female Avpr1b −/− mice have reductions in maternal aggressive behaviors, compared to control mice, as measured by longer attack latencies and fewer attacks, directed toward a male intruder (Wersinger et al. 2007a). Interestingly, the disruption of the Avpr1b does not affect all forms of aggressive behavior. In a nonsocial context, such as the predation of a cricket, Avpr1b −/− and Avpr1b +/+ mice have similar attack latencies (Wersinger et al. 2007a). Based on the genetic and pharmacological data, it has been hypothesized that the disruption of the *Avpr1b* does not specifically disrupt aggressive behavior, but rather the ability to have the appropriate behavioral response within a given social context (Caldwell et al. 2008c; Young et al. 2006; Stevenson and Caldwell 2012).

With prominent expression within the pyramidal neurons of the CA2 region of the hippocampus, recent work has focused on what the Avpr1b may be doing here. To this end, Pagani et al. 2015 have shown that replacement of the Avpr1b in the dorsal CA2 region of Avpr1b −/− mice restores socially mediated attack behaviors. Further, selective Avpr1b antagonists result in the production of N-methyl-D-aspartic-acid-dependent excitatory postsynaptic responses specifically within the CA2 region of control mice, but not Avpr1b −/− mice (Pagani et al. 2015). While the hippocampus is not currently a recognized node in the SBNN, it is a part of the motivational pathway. The CA2 region is structurally unique, as it does not receive rich mossy fiber input from the dentate gyrus (Tamamaki et al. 1988), and is the only part of the hippocampus to receive input from the posterior hypothalamus and the perforant pathway (Bartesaghi et al. 2006; Borhegyi and Leranth 1997; Vertes and McKenna 2000), which connects the entorhinal cortex to the hippocampal formation (Bartesaghi and Gessi 2004). Further, there is a vasopressinergic projection from the PVN to the CA2 region (Cui et al. 2013). Based on the findings described here, we, and others, have hypothesized that the CA2 region of the hippocampus may aid in the formation and/or recall of accessory olfactory-based memories (Caldwell et al. 2008c; Young et al. 2006). Thus, it seems likely that this is a region that will need to be included in future discussions of the SBNN and its interactions with the mesolimbic DA system.

4.4 Social Communication

While Avp plays a key role in the regulation of social communication in hamsters (see below), Oxt also contributes to its regulation. Hamsters communicate using a form of scent marking called flank marking, and the expression of flank marking is essential for the maintenance of dominant/subordinate relationships (Johnston 1985). After the rapid establishment of dominance, aggressive behavior declines and flank marking increases in dominant hamsters and to a lesser extent in subordinate hamsters (Ferris et al. 1987). In the absence of flank marking, aggression remains high and a stable relationship is not formed.

Oxt injected into the areas extending from the MPOA to the posterior medial and lateral aspects of the AH (referred to from here on as the MPOA-AH) of dominant female hamsters induces flank marking in a dose-dependent manner but only when the dominant hamsters are tested with their subordinate partners (Harmon et al. 2002b). Oxt does not induce flank marking when injected into the MPOA-AH of socially naive female hamsters tested with an opponent or alone. In males, by contrast, Oxt induces flank marking in dominant hamsters when they are tested with their subordinate partner or alone (Harmon et al. 2002b). These data indicate that social experience, social context, and sex interact to regulate the ability of Oxt to stimulate flank marking by its actions in the MPOA-AH in hamsters.

Although Oxt can stimulate flank marking, Avp plays the predominate role in regulating its expression. Avp stimulates high levels of flank marking in male and female hamsters by acting on the Avpr1a in the rostral hypothalamus (Ferris et al. 1984, 1985; Albers et al. 1986). It is of note that the MPOA-AH is substantially larger than the site where Avp can induce aggression (Ferris et al. 1986a). Avp can also induce flank marking following its injection into the LS, BNST, and PAG (Irvin et al. 1990; Hennessey et al. 1992). Gonadal hormones modulate the ability of Avp to stimulate flank marking by regulating the number of hypothalamic Avpr1a (Huhman and Albers 1993; Albers et al. 1988). In the LS, BNST, and PAG, gonadal hormones have only small effects on the ability of Avp to stimulate flank marking, suggesting that the MPOA-AH may be the primary site where gonadal hormones influence the ability of Avp to stimulate flank marking (Albers and Cooper 1995). Interestingly, Avp seems to induce flank marking regardless of the social context.

For example, in hamsters with an established dominant/subordinate relationship, injection of Avp produces high levels of flank marking in either the dominant or subordinate hamster during social interactions (Ferris et al. 1986b). Similar results have been obtained in squirrel monkeys where Avp injected i.c.v. induces scent marking in both dominant and subordinate males (Winslow and Insel 1991a).

4.5 Interactions Between Oxytocin, Vasopressin, and Dopamine in the Regulation of Cooperation/Competition

As discussed in detail above, Oxt and Avp act within brain regions considered to be components of the mesolimbic DA system to influence cooperative and competitive behaviors. There is also evidence that DA in the mesolimbic DA system plays important roles in the regulation of both cooperative and competitive behaviors. Non-selective DA antagonists block mating-induced partner preferences in both male and female prairie voles, and treatment with the non-selective DA agonist apomorphine facilitates partner preference in the absence of mating (Aragona et al. 2003; Wang et al. 1999). The NAcc shell, and not the core, appears to be the site of action for these drugs since local administration into the NAcc shell, but not the core, has the same effects as systemic administration. Increases in DA activity within the NAcc that occurs following mating is necessary for the formation of pair bonds in male prairie voles (Aragona et al. 2003). Activation of DA D2 receptors in the NAcc can induce a pair bond in cohabiting voles in the absence of mating, while activation of D1 receptors can block pair bonding (Aragona et al. 2006). In addition, D1 receptors are increased in the NAcc in male prairie voles following the formation of a pair bond. Mate guarding aggression in males can also be inhibited by D1 antagonists injected into the NAcc (Aragona and Wang 2009). Other regions in the network also play a key role in the formation of cooperative behavior. Pair-bonded voles have lower concentrations of D1 receptors and higher concentrations of D2 receptors in the medial prefrontal cortex than non-pair-bonded voles (Smeltzer et al. 2006). Enhancement of DA release in the VTA via administration of GABA or glutamate antagonists can also induce pair bonding in male voles (Curtis and Wang 2005).

There is also considerable evidence that the mesolimbic DA system plays a critical role in the regulation of competitive behavior and in particular aggression (for review, see de Almeida et al. 2005). For instance, the non-selective DA receptor agonist apomorphine stimulates aggression and flank marking in male hamsters (Hyer et al. 2012). Social experience can regulate the expression of DA in several nodes of the SBBN. The amount of the rate-limiting synthetic enzyme for DA, tyrosine hydroxylase, increases significantly in several regions of the SBNN including the LS and BNST as well as within the shell of the NAcc in males trained to fight as compared to controls (Schwartzer et al. 2013). In addition, activation of D1 and D2 receptors in the NAcc influences the expression of aggression and its rewarding properties (Couppis and Kennedy 2008; Miczek et al. 2002); although selective aggression in male voles appears to require only the D1 receptor activation (Aragona et al. 2006). While DA is involved in various aspects of the preparation, expression, and consequences of aggression, its precise role remains elusive.

The importance of interactions between Oxt, Avp, and DA in social motivation has been widely discussed, and yet there are few studies demonstrating a direct link between these neuropeptides and DA in controlling social behavior. Partner preferences induced by the activation of D2 receptors in the NAcc can be prevented by administration of an Oxtr antagonist and partner preference induced by i.c.v. Oxt administration can be blocked by a D2 antagonist administered in the NAcc (Liu and Wang 2003). In addition, overexpression of Avpr1a in the VP of the non-pair-bonded meadow voles results in mating-induced partner preferences that can be blocked by a D2 antagonist given prior to mating (Lim et al. 2004).

Despite the mounting empirical evidence suggesting that Oxt and Avp can directly interact with DA to influence social behavior, the majority of evidence for this interaction comes from studies showing either that Oxt and/or Avp act within structures that comprise the mesolimbic DA system, or that manipulations of DA can influence the same social behaviors that are influenced by Oxt and Avp. Interestingly, there are very little data on whether Oxt or Avp might contribute the rewarding properties of social behavior. A limited amount of data suggest that Oxt can induce a conditioned place preference (CPP) when given peripherally to male rats or centrally to female mice (Liberzon et al. 1997; Kent et al. 2013). Recently, we investigated whether injection of Oxt or Avp into the VTA could produce a conditioned place preference in male hamsters (Song et al. 2014a). Both Oxt and Avp increase CPP when injected into the VTA. The administration of selective Oxt and Avpr1a agonists and antagonists revealed that the rewarding properties of both Oxt and Avp in the VTA are mediated by the Oxtr and not the Avpr1a.

5 Cooperativity and Competitiveness in Humans

As we all know, cooperation and competition are a hallmark of nearly all human endeavors. Successful cooperation and competition require a set of social skills collectively termed social cognition, which allow an individual to engage in social behaviors that are appropriate for a particular context. Since the emergence of the social cognition field in the late 1960s and early 1970s, there has been a concerted scientific effort to understand the complicated cognitive processes that underlie human social interactions. While social cognition and many disciplines that are brought to bear in this field are interesting, there is accumulating evidence that Oxt and Avp in humans are important to social cognition. This section will highlight what we know about Oxt and Avp in humans, in particular how exogenous administration of these nonapeptides impacts measures of social cognition.

As described throughout this chapter, there is considerable compelling evidence that Oxt promotes social behaviors, at least in specific contexts. In humans, the study of social behaviors includes testing procedures designed to measure trust, the ability to read facial expressions, the memory for socially salient information, such as faces, and more recently measures of empathy. In most human studies focused on the therapeutic effects of Oxt and Avp, they are exogenously administered intranasally. This delivery system is preferable as it is considered noninvasive and some assert that Oxt and Avp are able to cross the blood–brain barrier (Born et al. 2002); however, this latter assertion is questionable given their size, hydrophilic nature, as well as numerous other issues.

5.1 Nonapeptides and Social Cognition in Healthy Humans

5.1.1 Oxytocin and Social Cognition in Humans

Investigation into the role of Oxt in human cognition has dramatically increased over the last decade. More recently, attempts have been made to reconcile the existing data on the role of Oxt into a broader theory and understanding of human cognition. For example, De Dreu proposes that Oxt plays a critical role in the motivation of cooperation and competition in humans (De Dreu and Kret 2015; De Dreu 2012). This hypothesis stems from the idea that humans are social animals and are likely to cooperate with others, even with those genetically unrelated. He suggests that Oxt motivates humans to like and empathize with others in their group, to comply with group norms, and to reciprocate trust with other group members, while competing with out-group members. While De Dreu's hypothesis is in reference to the endogenous Oxt system, this idea is supported by some recent work using intranasal Oxt. Essentially, intranasal Oxt increases "in-group" favoritism when individuals are asked to use intuitive decision making; however, Oxt has an opposite effect if individuals are asked to use reflective decision making (Ma et al. 2015). These data suggest that an individual's "cognitive style" may contribute to the effects of Oxt, which has implications for both endogenous and exogenous Oxt.

Given the literature in animal models, and the proposed prosocial effects of Oxt, in recent years, there has been a surge in the number of clinical studies that have administered exogenous Oxt to improve social interactions in healthy individuals. While the data do suggest that intranasal Oxt as a therapeutic agent has some efficacy, long-term and dose–response studies have yet to be completed. In humans, intranasal Oxt may influence the processing of social information in several ways, including selective attention, enhancement of the memory, and/or the appraisal of socially relevant information (Guastella and MacLeod 2012). Intranasal Oxt also increases trust, as measured by an individual's willingness to accept social risk during a social interaction (Zak et al. 2005; Kosfeld et al. 2005). However, similar to what is observed in other species, the effects of Oxt on trust are nuanced and often sex specific. For instance, if subjects are provided with information that suggests that a trustee is untrustworthy, then intranasal Oxt does not facilitate trust, or as the authors state, "Oxt makes people trusting, not gullible" (Mikolajczak et al. 2010). In females that have had their trust betrayed, intranasal Oxt results in less restoration of trust behavior compared to controls (Yao et al. 2014). However, in males, intranasal Oxt does not alter trust behavior in individuals that have had their trust betrayed, whereas placebo controls decrease their trust in response to betrayal (Baumgartner et al. 2008). Further, when intranasal Oxt treatment in males is coupled with functional magnetic resonance imaging (fMRI), there is a reduction in activity in areas of the brain associated with processing fearful stimuli, such as the amygdala and some areas of the midbrain, and reward feedback, such as the striatum (Baumgartner et al. 2008).

Oxt also has sex-specific effects in the context of reciprocal altruism, as measured using the prisoner's dilemma game, which examines cooperative exchange (Chen et al. 2015b; Feng et al. 2014; Rilling et al. 2014). Specifically, intranasal Oxt administered to females causes them to treat computer partners more like human partners; this same effect is not observed in males (Rilling et al. 2014). In this same context, there are also striking sex differences in neural activation, as measured by fMRI. In males, Oxt increases activation in the caudate/putamen while decreasing activation in females. The authors suggest that in this context, Oxt may increase the reward or salience of positive social interactions among males, while having the opposite effect in females (Feng et al. 2014; Rilling et al. 2014). Recent work also indicates that Oxt selectively improves kinship recognition in women but not men and that Oxt improves men's performance in competition recognition (Fischer-Shofty et al. 2013). Taken together, these data suggest that intranasal Oxt does not universally promote social interactions, but rather that the effects are subtle and dependent upon the social context and sex of the individuals. As mentioned previously, there has been a flood of clinical studies using intranasal Oxt, and with those studies, there is evidence that it can improve the ability to infer another individual's mental state, improves facial recognition memory, alters the processing of faces, and can facilitate empathy (Heinrichs et al. 2009; Rimmele et al. 2009; Domes et al. 2007; Guastella et al. 2008, 2009; Zak et al. 2007; Hurlemann et al. 2010; Unkelbach et al. 2008; Bos et al. 2015).

While the intranasal administration of Oxt can definitely influence aspects of social cognition, it is the individual differences in endogenous Oxt and their association with social cognition that is equally, if not more, important. Not only are individual differences in Oxt genetically heritable (Rubin et al. 2014), but also they appear to be stable (Feldman et al. 2007) and can be modulated by social interactions (Feldman et al. 2010). Further, individual differences in Oxt and Oxt signaling may increase an individual's vulnerability to certain neuropsychiatric disorders characterized by altered social cognition; see the section titled "**Implications for Neuropsychiatric Disorders**." Some recent work suggests that endogenous Oxt may underlie natural variations in social perception (Lancaster et al. 2015). Further, in a more recent study, plasma Oxt is associated with increases in activity in brain areas that are important to social cognition such as the superior temporal sulcus, inferior frontal gyrus, and medial prefrontal cortex (Lancaster et al. 2015).

Interpretation of some of the aforementioned data is muddled by our lack of understanding about the functional relationship between peripheral and central Oxt. But, it does follow that peripheral physiology and central physiology (behavioral output) are often going to be coupled, even if the systems are separate. However, while there is some evidence that they can be associated with one another (Landgraf and Neumann 2004), this does not always appear to be the case (Amico et al. 1990; Rosenblum et al. 2002; Winslow et al. 2003; Leng and Ludwig 2015).

5.1.2 Vasopressin and Social Cognition in Humans

The role of Avp in the regulation of social behavior in humans has not been studied as extensively as Oxt, though it is often associated with antisocial rather than prosocial behaviors. In males, intranasal Avp increases electromyogram (EMG) activity to socially neutral facial expressions (Thompson et al. 2004), as well as the memory for happy and angry faces (Guastella et al. 2010b). This suggests that Avp acts to bias an individual to perceive a neutral stimulus as an aggressive or threatening stimulus or enhances the encoding of negative social cues (Thompson et al. 2004; Guastella et al. 2010b). In females, Avp decreases EMG responses to happy and angry faces, suggesting that Avp acts to increase the perception of friendliness (Thompson et al. 2006). These sex-specific responses to Avp are supported by other studies as well (Feng et al. 2014; Rilling et al. 2014; Thompson et al. 2006) and may reflect sex differences in Avp neurochemistry and in the types of behavioral strategies employed during social interactions. Other effects include decreases in emotional recognition in men (Uzefovsky et al. 2012) and increases in empathy concern in individuals who received more "parental warmth" in their early family life (Tabak et al. 2015). It should be noted that the same issues that plague intranasal Oxt studies, such as the separation of central and peripheral Avp, are issues with the aforementioned studies as well.

5.2 *Oxytocin, Vasopressin, and the Mesolimbic Dopamine System*

Studies employing nasally administered Oxt and Avp have found numerous examples where these peptides alter the activity in various structures of the mesolimbic DA system. For example, Oxt augments the ventral striatum response to viewing the faces of romantic partners and to reciprocated cooperation from human partners (Rilling et al. 2012; Scheele et al. 2013). In women, Oxt increases activity in the VTA in response to both positive and negative facial expressions, but reduces the VTA response to reciprocated cooperation (Groppe et al. 2013; Chen et al. 2015a). While it is clear that Oxt and Avp can significantly alter activity in several structures in the mesolimbic DA system, the functional significance of these responses is not known. It is also possible that these neuropeptides alter social motivation without engaging DA neurons. For example, recent data suggests that Oxt enhances the attractiveness of female faces and correlates with increased activity in the striatum, while having no measureable impact on DA signaling (Striepens et al. 2014). However, more sensitive methods may be needed to detect subtle changes in behaviorally evoked DA release.

6 Implications for Neuropsychiatric Disorders

Oxt and Avp have also been implicated in a variety of neuropsychiatric disorders, particularly those characterized by aberrant social interactions and/or heightened aggression, such as ASD, personality disorder, schizophrenia, and posttraumatic stress disorder (PTSD). Further, many of these neuropsychiatric disorders are also characterized by dysfunctions of the mesolimbic DA system (for review, see Dichter et al. 2012). As many of these disorders are complex and have multiple etiologies, it becomes even more important not only to understand the potential contributions of Oxt and Avp, as well as their therapeutic effects, but also to determine how these systems interact with the DA system.

6.1 Autism Spectrum Disorder

ASD is characterized by repetitive behaviors, communication difficulties, and abnormal sociability (Matson and Nebel-Schwalm 2007a, b). In preclinical models of ASD, specifically Oxt −/− and Oxtr −/− mice, behavioral deficits are observed that are consistent with some of the symptoms of ASD (Crawley et al. 2007; Lee et al. 2008a, b; Macbeth et al. 2009; Wersinger et al. 2008; Winslow and Insel 2002; Ferguson et al. 2000, 2001). Evidence that Oxt may have a role in ASD comes from several sources. There are reports of reduced concentrations of Oxt in the cerebral spinal fluid (CSF) of autistic children, and reduced CSF Oxt is correlated with impairments in social functioning (Modahl et al. 1998). There are also increases in the amount of the Oxt prohormone in the blood of autistic children, which is indicative of incomplete processing of Oxt into its biologically active form (Green et al. 2001). Oxt treatment in adults with ASD results in the reduction of repetitive behaviors and improvements in emotional recognition (Hollander et al. 2003, 2007) and can increase eye gazing, a behavior important to social communication (Auyeung et al. 2015). In youth and adults with ASD, intranasal Oxt enhances emotion recognition (Domes et al. 2013, 2014; Guastella et al. 2010a) and increases social interactions (Andari et al. 2010). Further, in adults with Asperger's syndrome, Oxt-mediated increases in emotional recognition are associated with an increase in activity in the left amygdala (as measured in fMRI) (Domes et al. 2014).

There are also some genetic and epigenetic links between the Oxt system and ASD. There are data in the Chinese Han population, in Finnish families, in Caucasian children, and in individuals with "high-functioning" ASD, suggesting that portions of the *OXTR* gene may contain susceptibility loci for ASD (Wu et al. 2005; Ylisaukko-oja et al. 2006; Jacob et al. 2007; Wermter et al. 2010; Nyffeler et al. 2014). Epigenetic modifications of the *OXTR* gene have also been reported, with hypermethylation of the *OXTR* promoter found in ASD subjects and subsequent reductions in *OXTR* mRNA (Gregory et al. 2009). A more recent study focused on whether the levels of DNA methylation of the *OXTR* could predict individual variability in social perception found that there are significant associations between the degree of *OXTR* methylation and social perception (Jack et al. 2012). Though the sample sizes in these latter two epigenetic studies are small, the data are provocative and will likely facilitate more research in this area.

Data implicating Avp in the etiology of ASD are sparse, but there have been studies suggesting that polymorphisms of the *AVPR1A* may contribute to ASD (Kim et al. 2002; Wassink et al. 2004; Yirmiya et al. 2006; Kantojarvi et al. 2015; Tansey et al. 2011). Further, two of the polymorphisms, RS3 and RS1, have been linked to differential activation of the amygdala (Meyer-Lindenberg et al. 2009), providing a possible neural substrate with which the Avp system may interact to mediate a genetic risk for ASD.

6.2 Personality Disorder

Personality disorder is characterized by disconnect between an individual's behavior and cultural norms. Those diagnosed with personality disorder have impairments in at least two of the following areas: (1) cognition, (2) affectivity, (3) interpersonal functioning, and (4) impulse control (American Psychiatric Association 2013). To date, only a handful of studies have examined how exogenous Oxt treatment affects individuals diagnosed with personality disorder, and they have been performed primarily in individuals diagnosed with borderline personality disorder (BPD). Females diagnosed with BPD and treated with intranasal Oxt have reductions in stress reactivity (Simeon et al. 2011), decreases in hypersensitivity to threat, and decreases in amygdala activation in response to angry faces compared to controls; suggesting that Oxt in these individuals may decrease their reaction to a perceived social threat (Bertsch et al. 2013; Brune et al. 2013). This same research group also reports that females with BPD have reduced plasma Oxt concentrations (Bertsch et al. 2012). There are also data that suggest that intranasal Oxt can worsen trust in BPD and that this worsening is correlated with the patients' history of childhood trauma (Ebert et al. 2013). One issue with all of the aforementioned studies is that BPD patients were not used as controls, only healthy individuals, making it more difficult to parse out what is going on specifically within BPD patients. However, these data do reinforce the importance of past history and social context.

One study has measured Oxt in the CSF of individuals with personality disorder, which includes individuals diagnosed with intermittent explosive disorder, as well as BPD. This study found that while CSF Oxt concentrations are not correlated with having a personality disorder, a life history of suicidal behavior is inversely correlated with Oxt (Lee et al. 2009b). The authors suggest that these data are consistent with the previous work in animal models demonstrating that Oxt can reduce aggressive behaviors (Consiglio and Lucion 1996; Giovenardi et al. 1998; Harmon et al. 2002a; Bales and Carter 2003).

Since individuals diagnosed with a personality disorder often have more impulsive behaviors, which can result in increased aggression, it is not surprising that Avp has been examined in these individuals. However, the data appear to be contradictory. A study by Coccaro et al. (1998) found a positive correlation between Avp in the CSF of personality-disordered individuals that have a life history of aggressive behavior. However, an earlier study found no differences in CSF Avp between violent offenders and controls (Virkkunen et al. 1994). It may be that differences in the populations studied account for the inconsistency in the findings, but it seems that more work in this area is warranted.

6.3 Schizophrenia

There are three broad categories of symptoms that characterize schizophrenia: (1) positive (e.g., hallucinations and delusion), (2) negative (e.g., anhedonia, impaired social behavior), and (3) cognitive/attentional (e.g., impaired memory and executive function) (American Psychiatric Association 2013). However, in humans, while its role has remained controversial, Oxt has been linked to schizophrenia since the 1970s when it was used as an antipsychotic (Bujanow 1974, 1972; for a recent review, see Rich and Caldwell 2015). Altered CSF concentrations of Oxt are reported in patients diagnosed with schizophrenia (Beckmann et al. 1985; Linkowski et al. 1984). However, the data are conflicting with some studies reporting an increase in Oxt and the Oxt carrier protein neurophysin I (Linkowski et al. 1984; Beckmann et al. 1985) and another reporting no change in CSF Oxt concentrations (Glovinsky et al. 1994). However, patients with higher plasma levels of Oxt have less severe positive symptoms and exhibit fewer social deficits (Rubin et al. 2010, 2011).

There are also reports of SNPs in the promotor regions of the *OXT* and *OXTR* genes that may contribute to symptom severity and treatment efficacy in schizophrenic patients (Teltsh et al. 2012; Watanabe et al. 2012; Montag et al. 2013). SNPs of the *OXTR* gene are associated with the severity of symptoms and the improvement of the positive symptoms of schizophrenia following treatment with antipsychotics (Souza et al. 2010a, b). Additionally, postmortem analysis of brain tissue from unmedicated schizophrenia patients has altered immunoreactivity of the

Oxt carrier protein, neurophysin I, in the PVN, internal palladium, and substantia nigra (Mai et al. 1993). Most recently, in patients with schizophrenia and polydipsia, decreases in plasma Oxt were found to correlate with the ability to correctly identify facial emotions (Goldman et al. 2008). Thus, it appears that alterations in the Oxt system underlie all three symptom domains. Given the dysregulation of the Oxt system in patients with schizophrenia, Oxt has been studied as a candidate for use as a therapeutic.

Some studies suggest that Oxt may have antipsychotic properties (for review, see Macdonald and Feifel 2012; Bakermans-Kranenburg and van Jzendoorn 2013). Previous work found that injections of Oxt can reduce the symptoms of psychosis and anhedonia in patients with schizophrenia (McEwen 2004; Churchland and Winkielman 2012). In healthy patients, intranasal Oxt increases holistic processing, divergent thinking, and creative cognition (De Dreu et al. 2013), and studies in schizophrenic patients report that intranasal Oxt can be beneficial. Specifically, intranasal Oxt can facilitate social cognition (Davis et al. 2013; Feifel et al. 2010; Pedersen et al. 2011; Averbeck et al. 2012) and alleviate some of the cognitive deficits and positive symptoms (Pedersen et al. 2011). Yet, intranasal Oxt may be most effective as an adjunctive therapy to already prescribed antipsychotics, where chronic Oxt treatment is able to ameliorate some of the negative symptoms and the cognitive deficits, as well as the positive symptoms (Feifel et al. 2010, 2012; Modabbernia et al. 2013). While this research suggests that Oxt treatment has the potential to improve symptoms in all three domains, where in the brain and how these effects are mediated remains unknown.

Support for a potential role for Avp in schizophrenia comes from studies indicating that treatment with neuroleptics improves psychiatric symptoms and reduces (or normalizes) Avp in blood plasma (Peskind et al. 1987; Raskind et al. 1987). In studies using an animal model that lacks Avp, the Brattleboro rat, there are reports of deficits in behaviors associated with schizophrenia, specifically social discrimination and prepulse inhibition of the startle reflex; these deficits can be rescued following treatment with antipsychotics (Feifel and Priebe 2001, 2007; Feifel et al. 2004, 2007, 2009).

Given that schizophrenia is a complicated multiple etiology neuropsychiatric disorder, it may be that Oxt and Avp only contribute to certain types of schizophrenia. Further, it is likely that their action within specific parts of the brain is particularly important. Thus, the use of preclinical models continues to be critical to help improve our understanding of the specific neural substrates where these neuropeptides may act to contribute to the symptoms associated with schizophrenia as well as to determine their therapeutic efficacy (for review, see Rich and Caldwell 2015; Feifel 2011, 2012; Rosenfeld et al. 2010).

6.4 Posttraumatic Stress Disorder

Work examining Oxt and Avp in the context of PTSD has only recently begun, but PTSD is characterized by some symptoms related to Oxt and Avp function.

Specifically dysregulation of stress reactivity and problems with intimate relationship interactions, which relies on normal social cognition (American Psychiatric Association 2013; Monson et al. 2009). While there have been no clinical studies looking at Oxt, intranasal Oxt has been proposed to be used as an adjunctive treatment to help enhance psychotherapy, as it may help ultimately reduce the fear response at the level of the amygdala and diminish the hormonal stress response (Koch et al. 2014). In regard to Avp, there has been one clinical study. Intranasal Avp administered to men, but not women, diagnosed with PTSD results in improved social cognition with their heterosexual partner. Further, men's urinary Avp is negatively correlated with the severity of their PTSD (Marshall 2013). While these data are limited in scope, they do suggest that understanding the role of Avp in modulating social cognition may have clinical relevance in some psychiatric conditions.

7 Conclusions and Future Directions

Oxt and Avp are key neurochemical signals that act throughout the brain to influence socially motivated behaviors, having powerful effects on both cooperative and competitive behaviors mediated by their actions within the SBNN. There is also a substantial body of evidence that the mesolimbic DA system is involved in regulating cooperative and competitive behaviors. More recently, these two networks have been combined into a larger social decision-making network. However, there is a lack of understanding about the neurochemical linkages between the "social" and "motivational" elements within these networks. We propose that the Oxt and Avp systems represent a bridge between the social and motivational elements. More specifically, we propose that Oxt and Avp provide the critical links between specific elements of the SBNN and DA within the motivational network that produce the forces that drive social behaviors. While there is little direct support for this hypothesis, there is ample evidence that Oxt and Avp can act within structures comprising the mesolimbic DA system and that manipulations of DA can influence the same social behaviors as Oxt and Avp.

These interactions also have important implications for neuropsychiatric disorders characterized by aberrant social behaviors, particularly those with known disruptions in Oxt and Avp signaling. Rigorous testing of preclinical models continues to be the best way to uncover the specific mechanisms (i.e., neurochemistry, substrates, and circuits) that are important to socially motivated behaviors. Thus, it is crucial that a variety of species be studied to help determine both the conserved and unique mechanisms of Oxt and Avp actions across development, sexes, and behaviors. Overall, this is an exciting time in the field of socially motivated behaviors, with the advent of new experimental tools and an improvement in our ability to examine entire circuits we are now on the brink of making significant leaps in our understanding of these systems.

References

Acher R, Chauvet J (1995) The neurohypophysial endocrine regulatory cascade: precursors, mediators, receptors, and effectors. Front Neuroendocrinol 16(3):237–289

Acher R, Chauvet J, Chauvet MT (1995) Man and chimera: selective versus neutral oxytocin evolution. Adv Exp Med Biol 395:615–627

Adkins-Regan E (2009) Neuroendocrinology of social behavior. ILAR J 50(1):5–14

Albers HE (2012) The regulation of social recognition, social communication and aggression: vasopressin in the social behavior neural network. Horm Behav 61(3):283–292. doi:10.1016/j.yhbeh.2011.10.007

Albers HE (2015) Species, sex and individual differences in the vasotocin/vasopressin system: relationship to neurochemical signaling in the social behavior neural network. Front Neuroendocrinol 36:49–71. doi:10.1016/j.yfrne.2014.07.001

Albers HE, Cooper TT (1995) Effects of testosterone on the behavioral response to arginine vasopressin microinjected into the central gray and septum. Peptides 16(2):269–273

Albers HE, Pollock J, Simmons WH, Ferris CF (1986) A V1-like receptor mediates vasopressin-induced flank marking behavior in hamster hypothalamus. J Neurosci 6(7):2085–2089

Albers HE, Liou SY, Ferris CF (1988) Testosterone alters the behavioral response of the medial preoptic-anterior hypothalamus to microinjection of arginine vasopressin in the hamster. Brain Res 456:382–386

Albers HE, Rowland CM, Ferris CF (1991) Arginine-vasopressin immunoreactivity is not altered by photoperiod or gonadal hormones in the Syrian hamster (*Mesocricetus auratus*). Brain Res 539:137–142

Albers HE, Huhman KL, Meisel RL (2002) Hormonal basis of social conflict and communication. In: Pfaff DW, Arnold AP, Etgen AM, Fahrbach SE, Rubin RT (eds) Hormones, brain and behavior. Academic Press, Amsterdam, pp 393–433

Albers HE, Dean A, Karom MC, Smith D, Huhman KL (2006) Role of V1a vasopressin receptors in the control of aggression in Syrian hamsters. Brain Res 1073–1074:425–430

Alexander RD (1974) The evolution of social behavior. Annu Rev Ecol Syst 5:325–383

American Psychiatric Association (2013) Diagnostic and statistical manual of mental disorders, 5th edn. American Psychiatric Association, Washington, D.C.

Amico JA, Challinor SM, Cameron JL (1990) Pattern of oxytocin concentrations in the plasma and cerebrospinal fluid of lactating rhesus monkeys (*Macaca mulatta*): evidence for functionally independent oxytocinergic pathways in primates. J Clin Endocrinol Metab 71(6):1531–1535. doi:10.1210/jcem-71-6-1531

Anacker AM, Beery AK (2013) Life in groups: the roles of oxytocin in mammalian sociality. Front Behav Neurosci 7:185. doi:10.3389/fnbeh.2013.00185

Andari E, Duhamel JR, Zalla T, Herbrecht E, Leboyer M, Sirigu A (2010) Promoting social behavior with oxytocin in high-functioning autism spectrum disorders. Proc Natl Acad Sci USA 107(9):4389–4394. doi:10.1073/pnas.0910249107 0910249107 [pii]

Antoni FA (1984) Novel ligand specificity of pituitary vasopressin receptors in the rat. Neuroendocrinology 39:186–188

Aragona BJ, Wang Z (2009) Dopamine regulation of social choice in a monogamous rodent species. Front Behav Neurosci 3:15. doi:10.3389/neuro.08.015.2009

Aragona BJ, Liu Y, Curtis JT, Stephan FK, Wang ZX (2003) A critical role for nucleus accumbens dopamine in partner preference formation of male prairie voles (*Microtus ochrogaster*). J Neurosci 23:3483–3490

Aragona BJ, Liu Y, Yu YJ, Curtis JT, Detwiler JM, Insel TR, Wang Z (2006) Nucleus accumbens dopamine differentially mediates the formation and maintenance of monogamous pair bonds. Nat Neurosci 9(1):133–139. doi:10.1038/nn1613 nn1613 [pii]

Auyeung B, Lombardo MV, Heinrichs M, Chakrabarti B, Sule A, Deakin JB, Bethlehem RA, Dickens L, Mooney N, Sipple JA, Thiemann P, Baron-Cohen S (2015) Oxytocin increases eye contact during a real-time, naturalistic social interaction in males with and without autism. Transl Psychiatry 5:e507. doi:10.1038/tp.2014.146

Averbeck BB, Bobin T, Evans S, Shergill SS (2012) Emotion recognition and oxytocin in patients with schizophrenia. Psychol Med 42(02):259–266. doi:10.1017/S0033291711001413

Bakermans-Kranenburg MJ, van Jzendoorn MH (2013) Sniffing around oxytocin: review and meta-analyses of trials in healthy and clinical groups with implications for pharmacotherapy. Transl Psychiatry 3:e258. doi:10.1038/tp.2013.34

Bales KL, Carter CS (2003) Sex differences and developmental effects of oxytocin on aggression and social behavior in prairie voles (*Microtus ochrogaster*). Horm Behav 44(3):178–184

Bamshad M, Cooper TT, Karom M, Albers HE (1996) Glutamate and vasopressin interact to control scent marking in Syrian hamsters (*Mesocricetus auratus*). Brain Res 731:213–216

Bankir L (2001) Antidiuretic action of vasopressin: quantitative aspects and interaction between V1a and V2 receptor-mediated effects. CardiovascRes 51(3):372–390

Bartesaghi R, Gessi T (2004) Parallel activation of field CA2 and dentate gyrus by synaptically elicited perforant path volleys. Hippocampus 14(8):948–963

Bartesaghi R, Migliore M, Gessi T (2006) Input-output relations in the entorhinal cortex-dentate-hippocampal system: evidence for a non-linear transfer of signals. Neuroscience 142(1):247–265

Baskerville TA, Douglas AJ (2010) Dopamine and oxytocin interactions underlying behaviors: potential contributions to behavioral disorders. CNS Neurosci Ther 16(3):e92–123. doi:10.1111/j.1755-5949.2010.00154.x

Baumeister RF, Leary MR (1995) The need to belong: desire for interpersonal attachments as a fundamental human motivation. Psychol Bull 117(3):497–529

Baumgartner T, Heinrichs M, Vonlanthen A, Fischbacher U, Fehr E (2008) Oxytocin shapes the neural circuitry of trust and trust adaptation in humans. Neuron 58(4):639–650. doi:10.1016/j.neuron.2008.04.009 S0896-6273(08)00327-9 [pii]

Beckmann H, Lang RE, Gattaz WF (1985) Vasopressin–oxytocin in cerebrospinal fluid of schizophrenic patients and normal controls. Psychoneuroendocrinology 10(2):187–191

Beiderbeck DI, Neumann ID, Veenema AH (2007) Differences in intermale aggression are accompanied by opposite vasopressin release patterns within the septum in rats bred for low and high anxiety. Eur J Neurosci 26(12):3597–3605. doi:10.1111/j.1460-9568.2007.05974.x

Benelli A, Bertolini A, Poggioli R, Menozzi B, Basaglia R, Arletti R (1995) Polymodal dose-response curve for oxytocin in the social recognition test. Neuropeptides 28(4):251–255

Berthold AA (1849) Transplantation der Hoden (*Transplantation of testes*). Arch Ant Physiol Wissenschr Med 42–46

Bertsch K, Schmidinger I, Neumann ID, Herpertz SC (2012) Reduced plasma oxytocin levels in female patients with borderline personality disorder. Horm Behav. doi:10.1016/j.yhbeh.2012.11.013

Bertsch K, Gamer M, Schmidt B, Schmidinger I, Walther S, Kastel T, Schnell K, Buchel C, Domes G, Herpertz SC (2013) Oxytocin and reduction of social threat hypersensitivity in women with borderline personality disorder. Am J Psychiatry 170(10):1169–1177. doi:10.1176/appi.ajp.2013.13020263

Bester-Meredith JK, Marler CA (2001) Vasopressin and aggression in cross-fostered California mice (*Peromyscus californicus*) and white-footed mice (*Peromyscus leucopus*). Horm Behav 40(1):51–64

Bester-Meredith JK, Young LJ, Marler CA (1999) Species differences in paternal behavior and aggression in *Peromyscus* and their associations with vasopressin immunoreactivity and receptors. Horm Behav 36:25–38

Bielsky IF, Young LJ (2004) Oxytocin, vasopressin, and social recognition in mammals. Peptides 25:1565–1574

Bielsky IF, Hu SB, Szegda KL, Westphal H, Young LJ (2003) Profound impairment in social recognition and reduction in anxiety-like behavior in vasopressin V1a receptor knockout mice. Neuropsychopharmacology 29(3):483–493

Bielsky IF, Hu SB, Ren X, Terwilliger EF, Young LJ (2005) The V1a vasopressin receptor is necessary and sufficient for normal social recognition: a gene replacement study. Neuron 47 (4):503–513

Blanchard DC, Blanchard RJ (2003) What can animal aggression research tell us about human aggression? Horm Behav 44(3):171–177

Blanchard RJ, Wall PM, Blanchard DC (2003) Problems in the study of rodent aggression. Horm Behav 44(3):161–170

Blanchard RJ, Griebel G, Farrokhi C, Markham C, Yang M, Blanchard DC (2005) AVP V1b selective antagonist SSR149415 blocks aggressive behaviors in hamsters. Pharmacol Biochem Behav 80(1):189–194. doi:10.1016/j.pbb.2004.10.024 S0091-3057(04)00352-1 [pii]

Blume A, Bosch OJ, Miklos S, Torner L, Wales L, Waldherr M, Neumann ID (2008) Oxytocin reduces anxiety via ERK1/2 activation: local effect within the rat hypothalamic paraventricular nucleus. Eur J Neurosci 27(8):1947–1956. doi:10.1111/j.1460-9568.2008.06184.x

Bluthe RM, Schoenen J, Dantzer R (1990) Androgen-dependent vasopressinergic neurons are involved in social recognition in rats. Brain Res 519(1–2):150–157

Bluthe RM, Gheusi G, Dantzer R (1993) Gonadal steroids influence the involvement of arginine vasopressin in social recognition in mice. Psychoneuroendocrinology 18(4):323–335

Borhegyi Z, Leranth C (1997) Distinct substance P- and calretinin-containing projections from the supramammillary area to the hippocampus in rats; a species difference between rats and monkeys. Exp Brain Res Experimentelle Hirnforschung Experimentation Cerebrale 115 (2):369–374

Born J, Lange T, Kern W, McGregor GP, Bickel U, Fehm HL (2002) Sniffing neuropeptides: a transnasal approach to the human brain. Nat Neurosci 5(6):514–516. doi:10.1038/nn849

Bos PA, Montoya ER, Hermans EJ, Keysers C, van Honk J (2015) Oxytocin reduces neural activity in the pain circuitry when seeing pain in others. NeuroImage 113:217–224. doi:10.1016/j.neuroimage.2015.03.049

Bosch OJ (2011) Maternal nurturing is dependent on her innate anxiety: the behavioral roles of brain oxytocin and vasopressin. Horm Behav 59(2):202–212. doi:10.1016/j.yhbeh.2010.11.012

Bosch OJ (2013) Maternal aggression in rodents: brain oxytocin and vasopressin mediate pup defence. Philos Trans R Soc Lond B Biol Sci 368(1631):20130085. doi:10.1098/rstb.2013.0085

Bosch OJ, Neumann ID (2008) Brain vasopressin is an important regulator of maternal behavior independent of dams' trait anxiety. Proc Natl Acad Sci USA 105(44):17139–17144. doi:10.1073/pnas.0807412105

Bosch OJ, Neumann ID (2010) Vasopressin released within the central amygdala promotes maternal aggression. Eur J Neurosci 31(5):883–891. doi:10.1111/j.1460-9568.2010.07115.x

Bosch OJ, Neumann ID (2012) Both oxytocin and vasopressin are mediators of maternal care and aggression in rodents: from central release to sites of action. Horm Behav 61(3):293–303. doi:10.1016/j.yhbeh.2011.11.002

Bosch OJ, Meddle SL, Beiderbeck DI, Douglas AJ, Neumann ID (2005) Brain oxytocin correlates with maternal aggression: link to anxiety. J Neurosci 25(29):6807–6815. doi:10.1523/JNEUROSCI.1342-05.2005 25/29/6807 [pii]

Bredewold R, Smith CJ, Dumais KM, Veenema AH (2014) Sex-specific modulation of juvenile social play behavior by vasopressin and oxytocin depends on social context. Front Behav Neurosci 8:216. doi:10.3389/fnbeh.2014.00216

Bridges RS (2015) Neuroendocrine regulation of maternal behavior. Front Neuroendocrinol 36:178–196. doi:10.1016/j.yfrne.2014.11.007

Bruce HM (1959) An exteroceptive block to pregnancy in the mouse. Nature 184:105

Brune M, Ebert A, Kolb M, Tas C, Edel MA, Roser P (2013) Oxytocin influences avoidant reactions to social threat in adults with borderline personality disorder. Human Psychopharmacol 28(6):552–561. doi:10.1002/hup.2343

Buijs RM, Swaab DF (1979) Immuno-electron microscopical demonstration of vasopressin and oxytocin synapses in the limbic system of the rat. Cell Tissue Res 204(3):355–365

Buijs RM, De Vries GJ, Van Leeuwen FW, Swaab DF (1983) Vasopressin and oxytocin: distribution and putative functions in the brain. Prog Brain Res 60:115–122

Buijs RM, Gash DM, Boer GJ (1987) Vasopressin localization and putative functions in the brain. Vasopressin: principles and properties. Plenum Press, New York, pp 91–115

Bujanow W (1972) Hormones in the treatment of psychoses. Br Med J 4(5835):298

Bujanow W (1974) Letter: is oxytocin an anti-schizophrenic hormone? Can Psychiatr Assoc J 19 (3):323

Calcagnoli F, de Boer SF, Althaus M, den Boer JA, Koolhaas JM (2013) Antiaggressive activity of central oxytocin in male rats. Psychopharmacology (Berl) 229(4):639–651. doi:10.1007/s00213-013-3124-7

Calcagnoli F, Meyer N, de Boer SF, Althaus M, Koolhaas JM (2014) Chronic enhancement of brain oxytocin levels causes enduring anti-aggressive and pro-social explorative behavioral effects in male rats. Horm Behav 65(4):427–433. doi:10.1016/j.yhbeh.2014.03.008

Calcagnoli F, Kreutzmann JC, de Boer SF, Althaus M, Koolhaas JM (2015a) Acute and repeated intranasal oxytocin administration exerts anti-aggressive and pro-affiliative effects in male rats. Psychoneuroendocrinology 51:112–121. doi:10.1016/j.psyneuen.2014.09.019

Calcagnoli F, Stubbendorff C, Meyer N, de Boer SF, Althaus M, Koolhaas JM (2015b) Oxytocin microinjected into the central amygdaloid nuclei exerts anti-aggressive effects in male rats. Neuropharmacology 90:74–81. doi:10.1016/j.neuropharm.2014.11.012

Caldwell HK, Albers HE (2004a) Effect of photoperiod on vasopressin-induced aggression in Syrian hamsters. Horm Behav 46(4):444–449

Caldwell HK, Albers HE (2004b) Photoperiodic regulation of vasopressin receptor binding in female Syrian hamsters. Brain Res 1002(1–2):136–141

Caldwell HK, Young 3rd WS (2006) Oxytocin and vasopressin: genetics and behavioral implications. In: Lim R (ed) Neuroactive proteins and peptides, vol 3rd. Handbook of Neurochemistry and Molecular Neurobiology, 3rd edn. Springer, New York, pp 573–607

Caldwell HK, Lee HJ, Macbeth AH, Young WS 3rd (2008a) Vasopressin: behavioral roles of an "original" neuropeptide. Prog Neurobiol 84(1):1–24. doi:10.1016/j.pneurobio.2007.10.007 S0301-0082(07)00192-X [pii]

Caldwell HK, Smith DA, Albers HE (2008b) Photoperiodic mechanisms controlling scent marking: interactions of vasopressin and gonadal steroids. Eur J Neurosci 27(5):1189–1196. doi:10.1111/j.1460-9568.2008.06071.x EJN6071 [pii]

Caldwell HK, Wersinger SR, Young WS 3rd (2008c) The role of the vasopressin 1b receptor in aggression and other social behaviours. Prog Brain Res 170:65–72. doi:10.1016/S0079-6123 (08)00406-8 S0079-6123(08)00406-8 [pii]

Carter CS, Getz LL (1993) Monogamy and the prairie vole. Sci Am 268(6):100–106

Carter CS, DeVries AC, Getz LL (1995) Physiological substrates of mammalian monogamy: the prairie vole model. Neurosci Biobehav Rev 19(2):303–314

Carter CS, Grippo AJ, Pournajafi-Nazarloo H, Ruscio MG, Porges SW (2008) Oxytocin, vasopressin and sociality. Prog Brain Res 170:331–336. doi:10.1016/S0079-6123(08)00427-5 S0079-6123(08)00427-5 [pii]

Castel M, Morris JF (1988) The neurophysin-containing innervation of the forebrain of the mouse. Neuroscience 24(3):937–966

Cavanaugh J, Mustoe AC, Taylor JH, French JA (2014) Oxytocin facilitates fidelity in well-established marmoset pairs by reducing sociosexual behavior toward opposite-sex strangers. Psychoneuroendocrinology 49:1–10. doi:10.1016/j.psyneuen.2014.06.020

Champagne FA (2008) Epigenetic mechanisms and the transgenerational effects of maternal care. Front Neuroendocrinol 29(3):386–397. doi:10.1016/j.yfrne.2008.03.003

Champagne F, Diorio J, Sharma S, Meaney MJ (2001) Naturally occurring variations in maternal behavior in the rat are associated with differences in estrogen-inducible central oxytocin receptors. Proc Natl Acad Sci USA 98(22):12736–12741

Champagne FA, Chretien P, Stevenson CW, Zhang TY, Gratton A, Meaney MJ (2004) Variations in nucleus accumbens dopamine associated with individual differences in maternal behavior in the rat. J Neurosci 24(17):4113–4123. doi:10.1523/JNEUROSCI.5322-03.2004

Chen X, Hackett P, DeMarco A, Gautam P, Feng C, Haroon E, Rilling J (2015a) Oxytocin and vasopressin effects on the neural response to social cooperation among women: a within-subject study. Abstract presented at the Organization for Human Brain Mapping

Chen X, Hackett PD, DeMarco AC, Feng C, Stair S, Haroon E, Ditzen B, Pagnoni G, Rilling JK (2015b) Effects of oxytocin and vasopressin on the neural response to unreciprocated cooperation within brain regions involved in stress and anxiety in men and women. Brain Imaging Behav. doi:10.1007/s11682-015-9411-7

Cho MM, DeVries AC, Williams JR, Carter CS (1999) The effects of oxytocin and vasopressin on partner preferences in male and female prairie voles (*Microtus ochrogaster*). Behav Neurosci 113(1071):1079

Choi GB, Dong HW, Murphy AJ, Valenzuela DM, Yancopoulos GD, Swanson LW, Anderson DJ (2005) Lhx6 delineates a pathway mediating innate reproductive behaviors from the amygdala to the hypothalamus. Neuron 46(4):647–660. doi:10.1016/j.neuron.2005.04.011

Choleris E, Gustafsson JA, Korach KS, Muglia LJ, Pfaff DW, Ogawa S (2003) An estrogen-dependent four-gene micronet regulating social recognition: a study with oxytocin and estrogen receptor-alpha and -beta knockout mice. Proc Natl Acad Sci USA 100:6192–6197

Choleris E, Little SR, Mong JA, Puram SV, Langer R, Pfaff DW (2007) Microparticle-based delivery of oxytocin receptor antisense DNA in the medial amygdala blocks social recognition in female mice. Proc Natl Acad Sci USA 104(11):4670–4675. doi:10.1073/pnas.0700670104 0700670104 [pii]

Churchland PS, Winkielman P (2012) Modulating social behavior with oxytocin: how does it work? What does it mean? Horm Behav 61(3):392–399. doi:10.1016/j.yhbeh.2011.12.003

Coccaro EF, Kavoussi RJ, Hauger RL, Cooper TB, Ferris CF (1998) Cerebrospinal fluid vasopressin levels: correlates with aggression and serotonin function in personality-disordered subjects. Arch Gen Psychiatry 55(8):708–714

Compaan JC, Buijs RM, Pool CW, de Ruiter AJ, Koolhaas JM (1993) Differential lateral septal vasopressin innervation in aggressive and nonaggressive male mice. Brain Res Bull 30 (1–2):1–6

Consiglio AR, Lucion AB (1996) Lesion of hypothalamic paraventricular nucleus and maternal aggressive behavior in female rats. Physiol Behav 59(4–5):591–596

Consiglio AR, Borsoi A, Pereira GA, Lucion AB (2005) Effects of oxytocin microinjected into the central amygdaloid nucleus and bed nucleus of stria terminalis on maternal aggressive behavior in rats. Physiol Behav 85(3):354–362. doi:10.1016/j.physbeh.2005.05.002 S0031-9384(05) 00161-7 [pii]

Cooper MA, Karom M, Huhman KL, Albers HE (2005) Repeated agonistic encounters in hamsters modulate AVP V1a receptor binding. Horm Behav 48(5):545–551. doi:10.1016/j.yhbeh.2005. 04.012

Couppis MH, Kennedy CH (2008) The rewarding effect of aggression is reduced by nucleus accumbens dopamine receptor antagonism in mice. Psychopharmacol (Berl) 197(3):449–456. doi:10.1007/s00213-007-1054-y

Crawley JN, Chen T, Puri A, Washburn R, Sullivan TL, Hill JM, Young NB, Nadler JJ, Moy SS, Young LJ, Caldwell HK, Young WS (2007) Social approach behaviors in oxytocin knockout mice: comparison of two independent lines tested in different laboratory environments. Neuropeptides 41(3):145–163

Crews D (1997) Species diversity and the evolution of behavioral controlling mechanisms. Ann NY Acad Sci 807:1–21

Cui Z, Gerfen CR, Young WS 3rd (2013) Hypothalamic and other connections with dorsal CA2 area of the mouse hippocampus. J Comp Neurol 521(8):1844–1866. doi:10.1002/cne.23263

Curtis JT, Wang Z (2005) Glucocorticoid receptor involvement in pair bonding in female prairie voles: the effects of acute blockade and interactions with central dopamine reward systems. Neuroscience 134(2):369–376. doi:10.1016/j.neuroscience.2005.04.012

Curtis JT, Liu Y, Aragona BJ, Wang Z (2008) Neural regulation for social behavior in rodents. In: Wolff JO, Sherman PW (eds) Rodent societies: an ecological and evolutionary perspective. University of Chicago Press, Chicago, pp 185–194

Dadds MR, Moul C, Cauchi A, Dobson-Stone C, Hawes DJ, Brennan J, Ebstein RE (2014) Methylation of the oxytocin receptor gene and oxytocin blood levels in the development of psychopathy. Dev Psychopathol 26(1):33–40. doi:10.1017/S0954579413000497

Dale HH (1906) On some physiological actions of ergot. J Physiol (Lond) 34:163–206

Darwin C (1871) The descent of man, and selection in relation to sex. John Murray, London

Davis MC, Lee J, Horan WP, Clarke AD, McGee MR, Green MF, Marder SR (2013) Effects of single dose intranasal oxytocin on social cognition in schizophrenia. Schizophr Res 147(2–3):393–397. doi:10.1016/j.schres.2013.04.023

de Almeida RM, Ferrari PF, Parmigiani S, Miczek KA (2005) Escalated aggressive behavior: dopamine, serotonin and GABA. Eur J Pharmacol 526(1–3):51–64. doi:10.1016/j.ejphar.2005.10.004

De Dreu CK (2012) Oxytocin modulates cooperation within and competition between groups: an integrative review and research agenda. Horm Behav 61(3):419–428. doi:10.1016/j.yhbeh.2011.12.009

De Dreu CK, Kret ME (2015) Oxytocin conditions intergroup relations through upregulated in-group empathy, cooperation, conformity, and defense. Biol Psychiatry. doi:10.1016/j.biopsych.2015.03.020

De Dreu CK, Baas M, Roskes M, Sligte DJ, Ebstein RP, Chew SH, Tong T, Jiang Y, Mayseless N, Shamay-Tsoory SG (2013) Oxytonergic circuitry sustains and enables creative cognition in humans. Soc Cogn Affect Neurosci. doi:10.1093/scan/nst094

De Vries GJ, Buijs RM (1983) The origin of the vasopressinergic and oxytocinergic innervation of the rat brain with special reference to the lateral septum. Brain Res 273(2):307–317

De Vries GJ, Buijs RM, Sluiter AR (1984) Gonadal hormone actions on the morphology of the vasopressinergic innervation of the adult rat brain. Brain Res 298:141–145

Delville Y, Ferris CF (1995) Sexual differences in vasopressin receptor binding within the ventrolateral hypothalamus in golden hamsters. Brain Res 681:91–96

Delville Y, Mansour KM, Ferris CF (1996) Testosterone facilitates aggression by modulating vasopressin receptors in the hypothalamus. Physiol Behav 60(1):25–29

Delville Y, De Vries GJ, Ferris CF (2000) Neural connections of the anterior hypothalamus and agonistic behavior in golden hamsters. Brain Behav Evol 55(2):53–76

Demas GE, Albers HE, Cooper MA, Soma KK (2007) Novel mechanisms underlying neuroendocrine regulation of aggression: a synthesis of bird, rodent and primate studies. In: Blaustein JD (ed) Behavioral neurochemistry, neuroendocrinology and molecular neurobiology. Kluwer Press, Heidelberg, pp 337–372

Dichter GS, Damiano CA, Allen JA (2012) Reward circuitry dysfunction in psychiatric and neurodevelopmental disorders and genetic syndromes: animal models and clinical findings. J Neurodevelopmental Dis 4(1):19. doi:10.1186/1866-1955-4-19

Dluzen DE, Muraoka S, Engelmann M, Landgraf R (1998) The effects of infusion of arginine vasopressin, oxytocin, or their antagonists into the olfactory bulb upon social recognition responses in male rats. Peptides 19(6):999–1005

Domes G, Heinrichs M, Michel A, Berger C, Herpertz SC (2007) Oxytocin improves "mind-reading" in humans. Biol Psychiatry 61(6):731–733. doi:10.1016/j.biopsych.2006.07.015 S0006-3223(06)00939-5 [pii]

Domes G, Heinrichs M, Kumbier E, Grossmann A, Hauenstein K, Herpertz SC (2013) Effects of intranasal oxytocin on the neural basis of face processing in autism spectrum disorder. Biol Psychiatry 74(3):164–171. doi:10.1016/j.biopsych.2013.02.007

Domes G, Kumbier E, Heinrichs M, Herpertz SC (2014) Oxytocin promotes facial emotion recognition and amygdala reactivity in adults with asperger syndrome. Neuropsychopharmacology 39(3):698–706. doi:10.1038/npp.2013.254

Ebert A, Kolb M, Heller J, Edel MA, Roser P, Brune M (2013) Modulation of interpersonal trust in borderline personality disorder by intranasal oxytocin and childhood trauma. Soc Neurosci 8 (4):305–313. doi:10.1080/17470919.2013.807301

Elkabir DR, Wyatt ME, Vellucci SV, Herbert J (1990) The effects of separate or combined infusions of corticotrophin-releasing factor and vasopressin either intraventricularly or into the amygdala on aggressive and investigative behaviour in the rat. Regul Pept 28(2):199–214

Engelmann M, Landgraf R (1994) Microdialysis administration of vasopressin into the septum improves social recognition in Brattleboro rats. Physiol Behav 55(1):145–149

Engelmann M, Wotjak CT, Ebner K, Landgraf R (2000) Behavioural impact of intraseptally released vasopressin and oxytocin in rats. Exp Physiol 85(Spec No):125S–130S

Everts HGJ, De Ruiter AJH, Koolhaas JM (1997) Differential lateral septal vasopressin in wild-type rats: correlation with aggression. Horm Behav 31:136–144

Feifel D (2011) Is oxytocin a promising treatment for schizophrenia? Expert Rev Neurother 11 (2):157–159. doi:10.1586/ern.10.199

Feifel D (2012) Oxytocin as a potential therapeutic target for schizophrenia and other neuropsychiatric conditions. Neuropsychopharmacology 37(1):304–305. doi:10.1038/npp. 2011.184

Feifel D, Priebe K (2001) Vasopressin-deficient rats exhibit sensorimotor gating deficits that are reversed by subchronic haloperidol. Biol Psychiatry 50:425–433

Feifel D, Priebe K (2007) The effects of cross-fostering on inherent sensorimotor gating deficits exhibited by Brattleboro rats. J Gen Psychol 134(2):173–182

Feifel D, Melendez G, Shilling PD (2004) Reversal of sensorimotor gating deficits in Brattleboro rats by acute administration of clozapine and a neurotensin agonist, but not haloperidol: a potential predictive model for novel antipsychotic effects. Neuropsychopharmacology 29 (4):731–738 10.1038/sj.npp.13003781300378 [pii]

Feifel D, Melendez G, Priebe K, Shilling PD (2007) The effects of chronic administration of established and putative antipsychotics on natural prepulse inhibition deficits in Brattleboro rats. Behav Brain Res 181(2):278–286. doi:10.1016/j.bbr.2007.04.020 S0166-4328(07) 00230-6 [pii]

Feifel D, Mexal S, Melendez G, Liu PY, Goldenberg JR, Shilling PD (2009) The brattleboro rat displays a natural deficit in social discrimination that is restored by clozapine and a neurotensin analog. Neuropsychopharmacology 34(8):2011–2018. doi:10.1038/npp.2009.15 npp200915 [pii]

Feifel D, Macdonald K, Nguyen A, Cobb P, Warlan H, Galangue B, Minassian A, Becker O, Cooper J, Perry W, Lefebvre M, Gonzales J, Hadley A (2010) Adjunctive intranasal oxytocin reduces symptoms in schizophrenia patients. Biol Psychiatry 68(7):678–680. doi:10.1016/j. biopsych.2010.04.039 S0006-3223(10)00479-8 [pii]

Feifel D, Macdonald K, Cobb P, Minassian A (2012) Adjunctive intranasal oxytocin improves verbal memory in people with schizophrenia. Schizophr Res 139(1–3):207–210. doi:10.1016/j. schres.2012.05.018

Feldman R, Weller A, Zagoory-Sharon O, Levine A (2007) Evidence for a neuroendocrinological foundation of human affiliation: plasma oxytocin levels across pregnancy and the postpartum period predict mother-infant bonding. Psychol Sci 18(11):965–970. doi:10.1111/j.1467-9280. 2007.02010.x

Feldman R, Gordon I, Schneiderman I, Weisman O, Zagoory-Sharon O (2010) Natural variations in maternal and paternal care are associated with systematic changes in oxytocin following parent-infant contact. Psychoneuroendocrinology 35(8):1133–1141. doi:10.1016/j.psyneuen. 2010.01.013

Feng C, Hackett PD, DeMarco AC, Chen X, Stair S, Haroon E, Ditzen B, Pagnoni G, Rilling JK (2014) Oxytocin and vasopressin effects on the neural response to social cooperation are modulated by sex in humans. Brain Imaging Behav. doi:10.1007/s11682-014-9333-9

Ferguson JN, Young LJ, Hearn EF, Matzuk MM, Insel TR, Winslow JT (2000) Social amnesia in mice lacking the oxytocin gene. Nat Genet 25:284–288

Ferguson JN, Aldag JM, Insel TR, Young LJ (2001) Oxytocin in the medial amygdala is essential for social recognition in the mouse. J Neurosci 21(20):8278–8285
Fernald RD (2014) Communication about social status. Curr Opin Neurobiol 28:1–4. doi:10.1016/j.conb.2014.04.004
Ferris CF, Potegal M (1988) Vasopressin receptor blockade in the anterior hypothalamus suppresses aggression in hamsters. Physiol Behav 44:235–239
Ferris CF, Albers HE, Wesolowski SM, Goldman BD, Luman SE (1984) Vasopressin injected into the hypothalamus triggers a complex stereotypic behavior in golden hamsters. Science 224:521–523
Ferris CF, Pollock J, Albers HE, Leeman SE (1985) Inhibition of flank-marking behavior in golden hamsters by microinjection of a vasopressin antagonist into the hypothalamus. Neurosci Lett 55:239–243
Ferris CF, Meenan DM, Albers HE (1986a) Microinjection of kainic acid into the hypothalamus of golden hamsters prevents vasopressin-dependent flank-marking behavior. Neuroendocrinology 44:112–116
Ferris CF, Meenan DM, Axelson JF, Albers HE (1986b) A vasopressin antagonist can reverse dominant/subordinate behavior in hamsters. Physiol Behav 38:135–138
Ferris CF, Axelson JF, Shinto LH, Albers HE (1987) Scent marking and the maintenance of dominant/subordinate status in male golden hamsters. Physiol Behav 40:661–664
Ferris CF, Foote KB, Meltser HM, Plenby MG, Smith KL, Insel TR (1992a) Oxytocin in the amygdala facilitates maternal aggression. Ann NY Acad Sci 652:456–457
Ferris CF, Pilapil CG, Hayden-Hixson D, Wiley RG, Koh ET (1992b) Functionally and anatomically distinct populations of vasopressinergic magnocellular neurons in the female golden hamster. J Neuroendocrinol 4(2):193–205. doi:10.1111/j.1365-2826.1992.tb00159.x
Ferris CF, Melloni RH Jr, Koppel G, Perry KW, Fuller RW, Delville Y (1997) Vasopressin/serotonin interactions in the anterior hypothalamus control aggressive behavior in golden hamsters. J Neurosci 17(11):4331–4340
Fischer-Shofty M, Levkovitz Y, Shamay-Tsoory SG (2013) Oxytocin facilitates accurate perception of competition in men and kinship in women. Social Cogn Affective Neurosci 8 (3):313–317. doi:10.1093/scan/nsr100
Fish EW, Shahrokh D, Bagot R, Caldji C, Bredy T, Szyf M, Meaney MJ (2004) Epigenetic programming of stress responses through variations in maternal care. Ann NY Acad Sci 1036:167–180. doi:10.1196/annals.1330.011
Francis D, Diorio J, Liu D, Meaney MJ (1999) Nongenomic transmission across generations of maternal behavior and stress responses in the rat. Science 286(5442):1155–1158
Francis DD, Champagne FC, Meaney MJ (2000) Variations in maternal behaviour are associated with differences in oxytocin receptor levels in the rat. J Neuroendocrinol 12(12):1145–1148
Gabor CS, Phan A, Clipperton-Allen AE, Kavaliers M, Choleris E (2012) Interplay of oxytocin, vasopressin, and sex hormones in the regulation of social recognition. Behav Neurosci 126 (1):97–109. doi:10.1037/a0026464
Gil M, Nguyen NT, McDonald M, Albers HE (2013) Social reward: interactions with social status, social communication, aggression, and associated neural activation in the ventral tegmental area. Eur J Neurosci 38(2):2308–2318. doi:10.1111/ejn.12216
Gimpl G, Fahrenholz F (2001) The oxytocin receptor system: structure, function, and regulation. Physiol Rev 81(2):629–683
Giovenardi M, Padoin MJ, Cadore LP, Lucion AB (1998) Hypothalamic paraventricular nucleus modulates maternal aggression in rats: effects of ibotenic acid lesion and oxytocin antisense. Physiol Behav 63(3):351–359
Glovinsky D, Kalogeras KT, Kirch DG, Suddath R, Wyatt RJ (1994) Cerebrospinal fluid oxytocin concentration in schizophrenic patients does not differ from control subjects and is not changed by neuroleptic medication. Schizophr Res 11(3):273–276
Gobrogge KL, Liu Y, Young LJ, Wang Z (2009) Anterior hypothalamic vasopressin regulates pair-bonding and drug-induced aggression in a monogamous rodent. Proc Natl Acad Sci USA 106(45):19144–19149. doi:10.1073/pnas.0908620106 0908620106 [pii]

Goldman M, Marlow-O'Connor M, Torres I, Carter CS (2008) Diminished plasma oxytocin in schizophrenic patients with neuroendocrine dysfunction and emotional deficits. Schizophr Res 98(1–3):247–255. doi:10.1016/j.schres.2007.09.019 S0920-9964(07)00427-6 [pii]

Goodson JL, Kingsbury MA (2013) What's in a name? Considerations of homologies and nomenclature for vertebrate social behavior networks. Horm Behav 64(1):103–112. doi:10.1016/j.yhbeh.2013.05.006

Green L, Fein D, Modahl C, Feinstein C, Waterhouse L, Morris M (2001) Oxytocin and autistic disorder: alterations in peptide forms. Biol Psychiatry 50(8):609–613 S0006322301011398 [pii]

Gregory SG, Connelly JJ, Towers AJ, Johnson J, Biscocho D, Markunas CA, Lintas C, Abramson RK, Wright HH, Ellis P, Langford CF, Worley G, Delong GR, Murphy SK, Cuccaro ML, Persico A, Pericak-Vance MA (2009) Genomic and epigenetic evidence for oxytocin receptor deficiency in autism. BMC Med 7(1):62. doi:10.1186/1741-7015-7-62 1741-7015-7-62 [pii]

Griebel G, Simiand J, Serradeil-Le Gal C, Wagnon J, Pascal M, Scatton B, Maffrand JP, Soubrie P (2002) Anxiolytic- and antidepressant-like effects of the non-peptide vasopressin V1b receptor antagonist, SSR149415, suggest an innovative approach for the treatment of stress-related disorders. Proc Natl Acad Sci USA 99(9):6370–6375

Groppe SE, Gossen A, Rademacher L, Hahn A, Westphal L, Grunder G, Spreckelmeyer KN (2013) Oxytocin influences processing of socially relevant cues in the ventral tegmental area of the human brain. Biol Psychiatry 74(3):172–179. doi:10.1016/j.biopsych.2012.12.023

Guastella AJ, MacLeod C (2012) A critical review of the influence of oxytocin nasal spray on social cognition in humans: evidence and future directions. Horm Behav 61(3):410–418. doi:10.1016/j.yhbeh.2012.01.002

Guastella AJ, Mitchell PB, Dadds MR (2008) Oxytocin increases gaze to the eye region of human faces. Biol Psychiatry 63(1):3–5. doi:10.1016/j.biopsych.2007.06.026 S0006-3223(07)00617-8 [pii]

Guastella AJ, Carson DS, Dadds MR, Mitchell PB, Cox RE (2009) Does oxytocin influence the early detection of angry and happy faces? Psychoneuroendocrinology 34(2):220–225. doi:10.1016/j.psyneuen.2008.09.001 S0306-4530(08)00231-X [pii]

Guastella AJ, Einfeld SL, Gray KM, Rinehart NJ, Tonge BJ, Lambert TJ, Hickie IB (2010a) Intranasal oxytocin improves emotion recognition for youth with autism spectrum disorders. Biol Psychiatry 67(7):692–694. doi:10.1016/j.biopsych.2009.09.020 S0006-3223(09)01122-6 [pii]

Guastella AJ, Kenyon AR, Alvares GA, Carson DS, Hickie IB (2010b) Intranasal arginine vasopressin enhances the encoding of happy and angry faces in humans. Biol Psychiatry 67(12):1220–1222. doi:10.1016/j.biopsych.2010.03.014

Gutzler SJ, Karom M, Erwin WD, Albers HE (2010) Arginine-vasopressin and the regulation of aggression in female Syrian hamsters (*Mesocricetus auratus*). Eur J Neurosci 31(9):1655–1663. doi:10.1111/j.1460-9568.2010.07190.x

Hammock EA (2015) Developmental perspectives on oxytocin and vasopressin. Neuropsychopharmacology 40(1):24–42. doi:10.1038/npp.2014.120

Hara Y, Battey J, Gainer H (1990) Structure of mouse vasopressin and oxytocin genes. Brain Res Mol Brain Res 8:319–324

Harmon AC, Huhman KL, Moore TO, Albers HE (2002a) Oxytocin inhibits aggression in female Syrian hamsters. J Neuroendocrinol 14(12):963–969

Harmon AC, Moore TO, Huhman KL, Albers HE (2002b) Social experience and social context alter the behavioral response to centrally administered oxytocin in female Syrian hamsters. Neuroscience 109(4):767–772

Hattori T, Kanno K, Nagasawa M, Nishimori K, Mogi K, Kikusui T (2015) Impairment of interstrain social recognition during territorial aggressive behavior in oxytocin receptor-null mice. Neurosci Res 90:90–94. doi:10.1016/j.neures.2014.05.003

Heinrichs M, von Dawans B, Domes G (2009) Oxytocin, vasopressin, and human social behavior. Front Neuroendocrinol 30(4):548–557. doi:10.1016/j.yfrne.2009.05.005 S0091-3022(09) 00029-6 [pii]

Hennessey AC, Whitman DC, Albers HE (1992) Microinjection of arginine-vasopressin into the periaqueductal gray stimulates flank marking in Syrian hamsters (*Mesocricetus auratus*). Brain Res 569:136–140

Hernando F, Schoots O, Lolait SJ, Burbach JP (2001) Immunohistochemical localization of the vasopressin V1b receptor in the rat brain and pituitary gland: anatomical support for its involvement in the central effects of vasopressin. Endocrinology 142(4):1659–1668

Hitti FL, Siegelbaum SA (2014) The hippocampal CA2 region is essential for social memory. Nature 508(7494):88–92. doi:10.1038/nature13028

Hollander E, Novotny S, Hanratty M, Yaffe R, DeCaria CM, Aronowitz BR, Mosovich S (2003) Oxytocin infusion reduces repetitive behaviors in adults with autistic and Asperger's disorders. Neuropsychopharmacology 28(1):193–198. doi:10.1038/sj.npp.1300021 1300021 [pii]

Hollander E, Bartz J, Chaplin W, Phillips A, Sumner J, Soorya L, Anagnostou E, Wasserman S (2007) Oxytocin increases retention of social cognition in autism. Biol Psychiatry 61(4):498–503. doi:10.1016/j.biopsych.2006.05.030 S0006-3223(06)00729-3 [pii]

House JS, Landis KR, Umberson D (1988) Social relationships and health. Science 241 (4865):540–545

Hu SB, Zhao ZS, Yhap C, Grinberg A, Huang SP, Westphal H, Gold P (2003) Vasopressin receptor 1a-mediated negative regulation of B cell receptor signaling. J Neuroimmunol 135(1–2):72–81

Huber D, Veinante P, Stoop R (2005) Vasopressin and oxytocin excite distinct neuronal populations in the central amygdala. Science 308(5719):245–248. doi:10.1126/science.1105636

Huchard EC, Cowlishaw G (2011) Female-female aggression around mating: an extra cost of sociality in a multimale primate society. Behav Ecol 22(5):1003–1011. doi:10.1093/beheco/arr083

Huck UW, Lisk RD, McKay MV (1988) Social dominance and reproductive success in pregnant and lactating golden hamsters (*Mesocricetus auratus*) under seminatural conditions. Physiol Behav 44(3):313–319

Huhman KL, Albers HE (1993) Estradiol increases the behavioral response to arginine vasopressin (AVP) in the medial preoptic-anterior hypothalamus. Peptides 14:1049–1054

Hurlemann R, Patin A, Onur OA, Cohen MX, Baumgartner T, Metzler S, Dziobek I, Gallinat J, Wagner M, Maier W, Kendrick KM (2010) Oxytocin enhances amygdala-dependent, socially reinforced learning and emotional empathy in humans. J Neurosci 30(14):4999–5007. doi:10.1523/JNEUROSCI.5538-09.2010 30/14/4999 [pii]

Hyer MM, Rycek LM, Floody OR (2012) Effects of apomorphine on mating behavior, flank marking and aggression in male hamsters. Pharmacol Biochem Behav 101(4):520–527. doi:10.1016/j.pbb.2012.02.019

Insel TR, Hulihan TA (1995) A gender-specific mechanism for pair bonding: oxytocin and partner preference formation in monogamous voles. Behav Neurosci 109:782–789

Insel TR, Shapiro LE (1992) Oxytocin receptor distribution relects social organization in monogamous and polygamous voles. Proc Natl Acad Sci USA 89:5981–5985

Insel TR, Gelhard R, Shapiro LE (1991) The comparative distribution of forebrain receptors for neurohypophyseal peptides in monogamous and polygamous mice. Neuroscience 43(2–3):623–630

Insel TR, Wang ZX, Ferris CF (1994) Patterns of brain vasopressin receptor distribution associated with social organization in microtine rodents. J Neurosci 14:5381–5392

Insel TR, Winslow JT, Wang ZX, Young L, Hulihan TJ (1995) Oxytocin and the molecular basis of monogamy. Adv Exp Med Biol 395:227–234

Irvin RW, Szot P, Dorsa DM, Potegal M, Ferris CF (1990) Vasopressin in the septal area of the golden hamster controls scent marking and grooming. Physiol Behav 48:693–699

Jack A, Connelly JJ, Morris JP (2012) DNA methylation of the oxytocin receptor gene predicts neural response to ambiguous social stimuli. Front Hum Neurosci 6:280. doi:10.3389/fnhum. 2012.00280

Jacob S, Brune CW, Carter CS, Leventhal BL, Lord C, Cook EH Jr (2007) Association of the oxytocin receptor gene (OXTR) in Caucasian children and adolescents with autism. Neurosci Lett 417(1):6–9. doi:10.1016/j.neulet.2007.02.001 S0304-3940(07)00135-8 [pii]

Jakab RL, Naftolin F, Leranth C (1991) Convergent vasopressinergic and hippocampal input onto somatospiny neurons of the rat lateral septal area. Neuroscience 40(2):413–421

Jirikowski GF, Caldwell JD, Stumpf WE, Pedersen CA (1990) Topography of oxytocinergic estradiol target neurons in the mouse hypothalamus. Folia Histochem Cytobiol 28:3–9

Johnson AE, Barberis C, Albers HE (1995) Castration reduces vasopressin receptor binding in the hamster hypothalamus. Brain Res 674:153–158

Johnston RE (1985) Communication. In: Seigel HI (ed) The hamster: reproduction and behavior. Plenum Press, New York, pp 121–149

Johnston RE, Peng A (2008) Memory for individuals: hamsters (*Mesocricetus auratus*) require contact to develop multicomponent representations (concepts) of others. J Comp Psychol 122 (2):121–131. doi:10.1037/0735-7036.122.2.121

Jurek B, Slattery DA, Maloumby R, Hillerer K, Koszinowski S, Neumann ID, van den Burg EH (2012) Differential contribution of hypothalamic MAPK activity to anxiety-like behaviour in virgin and lactating rats. PLoS One 7(5):e37060. doi:10.1371/journal.pone.0037060

Kantojarvi K, Oikkonen J, Kotala I, Kallela J, Vanhala R, Onkamo P, Jarvela I (2015) Association and promoter analysis of AVPR1A in finnish autism families. Autism Res Official J Int Soc Autism Res. doi:10.1002/aur.1473

Keebaugh AC, Barrett CE, Laprairie JL, Jenkins JJ, Young LJ (2015) RNAi knockdown of oxytocin receptor in the nucleus accumbens inhibits social attachment and parental care in monogamous female prairie voles. Soc Neurosci 1–10. doi:10.1080/17470919.2015.1040893

Kelly AM, Goodson JL (2014) Social functions of individual vasopressin-oxytocin cell groups in vertebrates: what do we really know? Front Neuroendocrinol. doi:10.1016/j.yfrne.2014.04.005

Kent K, Arientyl V, Khachatryan MM, Wood RI (2013) Oxytocin induces a conditioned social preference in female mice. J Neuroendocrinol 25(9):803–810. doi:10.1111/jne.12075

Kim SJ, Young LJ, Gonen D, Veenstra-VanderWeele J, Courchesne R, Courchesne E, Lord C, Leventhal BL, Cook EH Jr, Insel TR (2002) Transmission disequilibrium testing of arginine vasopressin receptor 1A (AVPR1A) polymorphisms in autism. Mol Psychiatry 7(5):503–507

Kim YR, Kim JH, Kim MJ, Treasure J (2014) Differential methylation of the oxytocin receptor gene in patients with anorexia nervosa: a pilot study. PLoS One 9(2):e88673. doi:10.1371/journal.pone.0088673

Kimura T, Tanizawa O, Mori K, Brownstein MJ, Okayama H (1992) Structure and expression of a human oxytocin receptor. Nature 356:526–529

Kleiman DG (1977) Monogamy in mammals. Q Rev Biol 52:39–69

Knobloch HS, Charlet A, Hoffmann LC, Eliava M, Khrulev S, Cetin AH, Osten P, Schwarz MK, Seeburg PH, Stoop R, Grinevich V (2012) Evoked axonal oxytocin release in the central amygdala attenuates fear response. Neuron 73(3):553–566. doi:10.1016/j.neuron.2011.11.030

Koch SB, van Zuiden M, Nawijn L, Frijling JL, Veltman DJ, Olff M (2014) Intranasal oxytocin as strategy for medication-enhanced psychotherapy of PTSD: salience processing and fear inhibition processes. Psychoneuroendocrinology 40:242–256. doi:10.1016/j.psyneuen.2013.11.018

Kosfeld M, Heinrichs M, Zak PJ, Fischbacher U, Fehr E (2005) Oxytocin increases trust in humans. Nature 435(7042):673–676. doi:10.1038/nature03701 nature03701 [pii]

Kremarik P, Freund-Mercier MJ, Stoeckel ME (1993) Histoautoradiographic detection of oxytocin- and vasopressin-binding sites in the telencephalon of the rat. J Comp Neurol 333 (3):343–359

Ku CY, Qian A, Wen Y, Anwer K, Sanborn BM (1995) Oxytocin stimulates myometrial guanosine triphosphatase and phospholipase-C activities via coupling to G alpha q/11. Endocrinology 136(4):1509–1515

Kubota Y, Kimura T, Hashimoto K, Tokugawa Y, Nobunaga K, Azuma C, Saji F, Murata Y (1996) Structure and expression of the mouse oxytocin receptor gene. Mol Cell Endocrinol 124 (1–2):25–32
Lancaster K, Carter CS, Pournajafi-Nazarloo H, Karaoli T, Lillard TS, Jack A, Davis JM, Morris JP, Connelly JJ (2015) Plasma oxytocin explains individual differences in neural substrates of social perception. Front Hum Neurosci 9:132. doi:10.3389/fnhum.2015.00132
Landgraf R, Neumann ID (2004) Vasopressin and oxytocin release within the brain: a dynamic concept of multiple and variable modes of neuropeptide communication. Front Neuroendocrinol 25(2–4):150–176
Landgraf R, Gerstberger R, Montkowski A, Probst JC, Wotjak CT, Holsboer F, Engelmann M (1995) V1 vasopressin receptor antisense oligodeoxynucleotide into septum reduces vasopressin binding, social discrimination abilities, and anxiety-related behavior in rats. J Neurosci 15(6):4250–4258
Lee HJ, Caldwell HK, Macbeth AH, Tolu SG, Young WS 3rd (2008a) A conditional knockout mouse line of the oxytocin receptor. Endocrinology 149(7):3256–3263. doi:10.1210/en.2007-1710 en.2007-1710 [pii]
Lee HJ, Caldwell HK, Macbeth AH, Young WS 3rd (2008b) Behavioural studies using temporal and spatial inactivation of the oxytocin receptor. Prog Brain Res 170:73–77. doi:10.1016/S0079-6123(08)00407-X S0079-6123(08)00407-X [pii]
Lee HJ, Macbeth AH, Pagani JH, Young WS 3rd (2009a) Oxytocin: the great facilitator of life. Prog Neurobiol 88(2):127–151. doi:10.1016/j.pneurobio.2009.04.001 S0301-0082(09) 00046-X [pii]
Lee R, Ferris C, Van de Kar LD, Coccaro EF (2009b) Cerebrospinal fluid oxytocin, life history of aggression, and personality disorder. Psychoneuroendocrinology 34(10):1567–1573. doi:10.1016/j.psyneuen.2009.06.002 S0306-4530(09)00192-9 [pii]
Lee AG, Cool DR, Grunwald WC Jr, Neal DE, Buckmaster CL, Cheng MY, Hyde SA, Lyons DM, Parker KJ (2011) A novel form of oxytocin in New World monkeys. Biology letters 7(4):584–587. doi:10.1098/rsbl.2011.0107
Leng G, Ludwig M (2008) Neurotransmitters and peptides: whispered secrets and public announcements. J Physiol 586(Pt 23):5625–5632. doi:10.1113/jphysiol.2008.159103
Leng G, Ludwig M (2015) Intranasal Oxytocin: myths and delusions. Biol Psychiatry. doi:10.1016/j.biopsych.2015.05.003
Liberzon I, Trujillo KA, Akil H, Young EA (1997) Motivational properties of oxytocin in the conditioned place preference paradigm. Neuropsychopharmacology 17(6):353–359. doi:10.1016/S0893-133X(97)00070-5
Lim MM, Wang Z, Olazabal DE, Ren X, Terwilliger EF, Young LJ (2004) Enhanced partner preference in a promiscuous species by manipulating the expression of a single gene. Nature 429(6993):754–757
Linkowski P, Geenen V, Kerkhofs M, Mendlewicz J, Legros JJ (1984) Cerebrospinal fluid neurophysins in affective illness and in schizophrenia. Eur Arch Psychiatry Neurol Sci 234 (3):162–165
Liu Y, Wang ZX (2003) Nucleus accumbens oxytocin and dopamine interact to regulate pair bond formation in female prairie voles. Neuroscience 121:537–544
Liu Y, Curtis JT, Wang ZX (2001) Vasopressin in the lateral septum regulates pair bond formation in male prairie voles (*Microtus ochrogaster*). Behav Neurosci 115:910–919
Lolait SJ, O'Carroll AM, Mahan LC, Felder CC, Button DC, Young WS III, Mezey E, Brownstein MJ (1995) Extrapituitary expression of the rat V1b vasopressin receptor gene. Proc Natl Acad Sci USA 92(15):6783–6787
Lonstein JS, Gammie SC (2002) Sensory, hormonal, and neural control of maternal aggression in laboratory rodents. Neurosci Biobehav Rev 26(8):869–888
Love TM (2014) Oxytocin, motivation and the role of dopamine. Pharmacol Biochem Behav 119:49–60. doi:10.1016/j.pbb.2013.06.011
Ludwig M (1998) Dendritic release of vasopressin and oxytocin. J Neuroendocrinol 10 (12):881–895

Ludwig M, Leng G (2006) Dendritic peptide release and peptide-dependent behaviours. Nat Rev Neurosci 7(2):126–136. doi:10.1038/nrn1845

Ma Y, Liu Y, Rand DG, Heatherton TF, Han S (2015) Opposing oxytocin effects on intergroup cooperative behavior in intuitive and reflective minds. Neuropsychopharmacology. doi:10.1038/npp.2015.87

Macbeth AH, Lee HJ, Edds J, Young WS 3rd (2009) Oxytocin and the oxytocin receptor underlie intrastrain, but not interstrain, social recognition. Genes Brain Behav 8(5):558–567. doi:10.1111/j.1601-183X.2009.00506.x GBB506 [pii]

Macdonald K, Feifel D (2012) Oxytocin in schizophrenia: a review of evidence for its therapeutic effects. Acta Neuropsychiatrica 24(3):130–146. doi:10.1111/j.1601-5215.2011.00634.x

Mai JK, Berger K, Sofroniew MV (1993) Morphometric evaluation of neurophysin-immunoreactivity in the human brain: pronounced inter-individual variability and evidence for altered staining patterns in schizophrenia. J Hirnforsch 34(2):133–154

Manning M, Stoev S, Chini B, Durroux T, Mouillac B, Guillon G (2008) Peptide and non-peptide agonists and antagonists for the vasopressin and oxytocin V1a, V1b, V2 and OT receptors: research tools and potential therapeutic agents. Prog Brain Res 170:473–512. doi:10.1016/S0079-6123(08)00437-8

Manning M, Misicka A, Olma A, Bankowski K, Stoev S, Chini B, Durroux T, Mouillac B, Corbani M, Guillon G (2012) Oxytocin and vasopressin agonists and antagonists as research tools and potential therapeutics. J Neuroendocrinol 24(4):609–628. doi:10.1111/j.1365-2826.2012.02303.x

Marshall AD (2013) Posttraumatic stress disorder and partner-specific social cognition: a pilot study of sex differences in the impact of arginine vasopressin. Biol Psychol 93(2):296–303. doi:10.1016/j.biopsycho.2013.02.014

Martinez M, Guillen-Salazar F, Salvador A, Simon VM (1995) Successful intermale aggression and conditioned place preference in mice. Physiol Behav 58(2):323–328

Mateo JM, Johnston RE (2003) Kin recognition by self-referent phenotype matching: weighing the evidence. Animal Cogn 6(1):73–76. doi:10.1007/s10071-003-0165-z

Matson JL, Nebel-Schwalm M (2007a) Assessing challenging behaviors in children with autism spectrum disorders: a review. Res Dev Disabil 28(6):567–579. doi:10.1016/j.ridd.2006.08.001 S0891-4222(06)00073-4 [pii]

Matson JL, Nebel-Schwalm MS (2007b) Comorbid psychopathology with autism spectrum disorder in children: an overview. Res Dev Disabil 28(4):341–352. doi:10.1016/j.ridd.2005.12.004 S0891-4222(06)00049-7 [pii]

Maybauer MO, Maybauer DM, Enkhbaatar P, Traber DL (2008) Physiology of the vasopressin receptors. Best Pract Res Clin Anaesthesiol 22(2):253–263

Mayes CR, Watts AG, McQueen JK, Fink G, Charlton HM (1988) Gonadal steroids influence neurophysin II distribution in the forebrain of normal and mutant mice. Neuroscience 25(3):1013–1022

McEwen BB (2004) Brain-fluid barriers: relevance for theoretical controversies regarding vasopressin and oxytocin memory research. Adv Pharmacol 50(531–592):655–708. doi:10.1016/S1054-3589(04)50014-5

Meisel RL, Joppa MA (1994) Conditioned place preference in female hamsters following aggressive or sexual encounters. Physiol Behav 56(5):1115–1118

Meyer-Lindenberg A, Kolachana B, Gold B, Olsh A, Nicodemus KK, Mattay V, Dean M, Weinberger DR (2009) Genetic variants in AVPR1A linked to autism predict amygdala activation and personality traits in healthy humans. Mol Psychiatry 14(10):968–975. doi:10.1038/mp.2008.54 mp200854 [pii]

Michopoulos V, Checchi M, Sharpe D, Wilson ME (2011) Estradiol effects on behavior and serum oxytocin are modified by social status and polymorphisms in the serotonin transporter gene in female rhesus monkeys. Horm Behav 59(4):528–535. doi:10.1016/j.yhbeh.2011.02.002

Michopoulos V, Higgins M, Toufexis D, Wilson ME (2012) Social subordination produces distinct stress-related phenotypes in female rhesus monkeys. Psychoneuroendocrinology 37(7):1071–1085. doi:10.1016/j.psyneuen.2011.12.004

Miczek KA, Fish EW, De Bold JF, De Almeida RM (2002) Social and neural determinants of aggressive behavior: pharmacotherapeutic targets at serotonin, dopamine and gamma-aminobutyric acid systems. Psychopharmacology (Berl) 163(3–4):434–458. doi:10.1007/s00213-002-1139-6

Mikolajczak M, Gross JJ, Lane A, Corneille O, de Timary P, Luminet O (2010) Oxytocin makes people trusting, not gullible. Psychol Sci 21(8):1072–1074. doi:10.1177/0956797610377343

Miller TV, Caldwell HK (2015) Oxytocin during development: possible organizational effects on behavior. Front Endocrinol 6:76. doi:10.3389/fendo.2015.00076

Modabbernia A, Rezaei F, Salehi B, Jafarinia M, Ashrafi M, Tabrizi M, Hosseini SM, Tajdini M, Ghaleiha A, Akhondzadeh S (2013) Intranasal oxytocin as an adjunct to risperidone in patients with schizophrenia : an 8-week, randomized, double-blind, placebo-controlled study. CNS Drugs 27(1):57–65. doi:10.1007/s40263-012-0022-1

Modahl C, Green L, Fein D, Morris M, Waterhouse L, Feinstein C, Levin H (1998) Plasma oxytocin levels in autistic children. Biol Psychiatry 43(4):270–277 S0006322397004393 [pii]

Monson CM, Taft CT, Fredman SJ (2009) Military-related PTSD and intimate relationships: from description to theory-driven research and intervention development. Clin Psychol Rev 29 (8):707–714. doi:10.1016/j.cpr.2009.09.002

Montag C, Brockmann EM, Bayerl M, Rujescu D, Muller DJ, Gallinat J (2013) Oxytocin and oxytocin receptor gene polymorphisms and risk for schizophrenia: a case-control study. World J Biol Psychiatry Official J World Fed Soc Biol Psychiatry 14(7):500–508. doi:10.3109/15622975.2012.677547

Nephew BC, Bridges RS (2008) Arginine vasopressin V1a receptor antagonist impairs maternal memory in rats. Physiol Behav 95(1–2):182–186. doi:10.1016/j.physbeh.2008.05.016

Nephew BC, Byrnes EM, Bridges RS (2010) Vasopressin mediates enhanced offspring protection in multiparous rats. Neuropharmacology 58(1):102–106. doi:10.1016/j.neuropharm.2009.06.032

Neumann ID (2008) Brain oxytocin: a key regulator of emotional and social behaviours in both females and males. J Neuroendocrinol 20(6):858–865. doi:10.1111/j.1365-2826.2008.01726.x JNE1726 [pii]

Newman SW (1999) The medial extended amygdala in male reproductive behavior. A node in the mammalian social behavior network. Ann NY Acad Sci 877:242–257

Nyffeler J, Walitza S, Bobrowski E, Gundelfinger R, Grunblatt E (2014) Association study in siblings and case-controls of serotonin- and oxytocin-related genes with high functioning autism. J Molecul Psychiatry 2(1):1. doi:10.1186/2049-9256-2-1

O'Connell LA, Hofmann HA (2011a) Genes, hormones, and circuits: an integrative approach to study the evolution of social behavior. Front Neuroendocrinol 32(3):320–335. doi:10.1016/j.yfrne.2010.12.004

O'Connell LA, Hofmann HA (2011b) The vertebrate mesolimbic reward system and social behavior network: a comparative synthesis. J Comp Neurol 519(18):3599–3639. doi:10.1002/cne.22735

Ostrowski NL, Lolait SJ, Bradley DJ, O'Carroll A, Brownstein MJ, Young WS 3rd (1992) Distribution of V1a and V2 vasopressin receptor messenger ribonucleic acids in rat liver, kidney, pituitary and brain. Endocrinology 131(1):533–535

Ostrowski NL, Lolait SJ, Young Iii WS (1994) Cellular localization of vasopressin V1a receptor messenger ribonucleic acid in adult male rat brain, pineal, and brain vasculature. Endocrinology 135(4):1511–1528

Ott I, Scott JC (1910) The action of infundibulin upon the mammary secretion. Proc Soc Exp Biol (NY) 8:48–49

Pagani JH, Zhao M, Cui Z, Avram SK, Caruana DA, Dudek SM, Young WS (2015) Role of the vasopressin 1b receptor in rodent aggressive behavior and synaptic plasticity in hippocampal area CA2. Mol Psychiatry 20(4):490–499. doi:10.1038/mp.2014.47

Paul MJ, Terranova JI, Probst CK, Murray EK, Ismail NI, de Vries GJ (2014) Sexually dimorphic role for vasopressin in the development of social play. Front Behav Neurosci 8:58. doi:10.3389/fnbeh.2014.00058

Pedersen CA, Gibson CM, Rau SW, Salimi K, Smedley KL, Casey RL, Leserman J, Jarskog LF, Penn DL (2011) Intranasal oxytocin reduces psychotic symptoms and improves theory of mind and social perception in schizophrenia. Schizophr Res 132(1):50–53. doi:10.1016/j.schres. 2011.07.027

Peskind ER, Raskind MA, Leake RD, Ervin MG, Ross MG, Dorsa DM (1987) Clonidine decreases plasma and cerebrospinal fluid arginine vasopressin but not oxytocin in humans. Neuroendocrinology 46(5):395–400

Petrulis A (2009) Neural mechanisms of individual and sexual recognition in Syrian hamsters (*Mesocricetus auratus*). Behav Brain Res 200(2):260–267. doi:10.1016/j.bbr.2008.10.027 S0166-4328(08)00594-9 [pii]

Popik P, Van Ree JM (1991) Oxytocin but not vasopressin facilitates social recognition following injection into the medial preoptic area of the rat brain. Eur Neuropsychopharmacol 1:555–560

Potegal M, Ferris CF (1990) Intraspecific aggression in male hamsters is inhibited by intrahypothalamic vasopressin-receptor antagonist. Aggressive Behav 15:311–320

Puglia MH, Lillard TS, Morris JP, Connelly JJ (2015) Epigenetic modification of the oxytocin receptor gene influences the perception of anger and fear in the human brain. Proc Natl Acad Sci USA 112(11):3308–3313. doi:10.1073/pnas.1422096112

Ragen BJ, Bales KL (2013) Oxytocin and vasopressin in non-human primates. In: Choleris E, Pfaff DW, Kavaliers M (eds) Oxytocin, vasopressin and related peptides in the regulation of behavior. Cambridge University Press, Cambridge, pp 288–308

Ragnauth AK, Goodwillie A, Brewer C, Muglia LJ, Pfaff DW, Kow LM (2004) Vasopressin stimulates ventromedial hypothalamic neurons via oxytocin receptors in oxytocin gene knockout male and female mice. Neuroendocrinology 80:92–99

Raskind MA, Courtney N, Murburg MM, Backus FI, Bokan JA, Ries RK, Dorsa DM, Weitzman RE (1987) Antipsychotic drugs and plasma vasopressin in normals and acute schizophrenic patients. Biol Psychiatry 22(4):453–462 0006-3223(87)90167-3 [pii]

Rich ME, Caldwell HK (2015) A role for oxytocin in the etiology and treatment of schizophrenia. Front Endocrinol 6:90. doi:10.3389/fendo.2015.00090

Rilling JK, DeMarco AC, Hackett PD, Thompson R, Ditzen B, Patel R, Pagnoni G (2012) Effects of intranasal oxytocin and vasopressin on cooperative behavior and associated brain activity in men. Psychoneuroendocrinology 37(4):447–461. doi:10.1016/j.psyneuen.2011.07.013

Rilling JK, Demarco AC, Hackett PD, Chen X, Gautam P, Stair S, Haroon E, Thompson R, Ditzen B, Patel R, Pagnoni G (2014) Sex differences in the neural and behavioral response to intranasal oxytocin and vasopressin during human social interaction. Psychoneuroendocrinology 39:237–248. doi:10.1016/j.psyneuen.2013.09.022

Rimmele U, Hediger K, Heinrichs M, Klaver P (2009) Oxytocin makes a face in memory familiar. J Neurosci 29(1):38–42. doi:10.1523/JNEUROSCI.4260-08.2009 29/1/38 [pii]

Rodriguez-Laso A, Zunzunegui MV, Otero A (2007) The effect of social relationships on survival in elderly residents of a Southern European community: a cohort study. BMC Geriatr 7:19. doi:10.1186/1471-2318-7-19

Rosenblum LA, Smith EL, Altemus M, Scharf BA, Owens MJ, Nemeroff CB, Gorman JM, Coplan JD (2002) Differing concentrations of corticotropin-releasing factor and oxytocin in the cerebrospinal fluid of bonnet and pigtail macaques. Psychoneuroendocrinology 27(6):651–660

Rosenfeld AJ, Lieberman JA, Jarskog LF (2010) Oxytocin, dopamine, and the amygdala: a neurofunctional model of social cognitive deficits in schizophrenia. Schizophr Bull. doi:10.1093/schbul/sbq015 sbq015 [pii]

Ross HE, Cole CD, Smith Y, Neumann ID, Landgraf R, Murphy AZ, Young LJ (2009a) Characterization of the oxytocin system regulating affiliative behavior in female prairie voles. Neuroscience 162(4):892–903. doi:10.1016/j.neuroscience.2009.05.055 S0306-4522(09) 00965-8 [pii]

Ross HE, Freeman SM, Spiegel LL, Ren X, Terwilliger EF, Young LJ (2009b) Variation in oxytocin receptor density in the nucleus accumbens has differential effects on affiliative behaviors in monogamous and polygamous voles. J Neurosci 29(5):1312–1318. doi:10.1523/JNEUROSCI.5039-08.2009 29/5/1312 [pii]

Rosvall KA (2011) Intrasexual competition in females: evidence for sexual selection? Behav Ecol Official J Int Soc Behav Ecol 22(6):1131–1140. doi:10.1093/beheco/arr106

Rubin LH, Carter CS, Drogos L, Pournajafi-Nazarloo H, Sweeney JA, Maki PM (2010) Peripheral oxytocin is associated with reduced symptom severity in schizophrenia. Schizophr Res 124(1–3):13–21. doi:10.1016/j.schres.2010.09.014 S0920-9964(10)01543-4 [pii]

Rubin LH, Carter CS, Drogos L, Jamadar R, Pournajafi-Nazarloo H, Sweeney JA, Maki PM (2011) Sex-specific associations between peripheral oxytocin and emotion perception in schizophrenia. Schizophr Res 130(1–3):266–270. doi:10.1016/j.schres.2011.06.002

Rubin LH, Carter CS, Bishop JR, Pournajafi-Nazarloo H, Drogos LL, Hill SK, Ruocco AC, Keedy SK, Reilly JL, Keshavan MS, Pearlson GD, Tamminga CA, Gershon ES, Sweeney JA (2014) Reduced levels of vasopressin and reduced behavioral modulation of oxytocin in psychotic disorders. Schizophr Bull 40(6):1374–1384. doi:10.1093/schbul/sbu027

Saito M, Sugimoto T, Tahara A, Kawashima H (1995) Molecular cloning and characterization of rat V1b vasopressin receptor: evidence for its expression in extra-pituitary tissues. Biochem Biophys Res Commun 212:751–757

Sala M, Braida D, Lentini D, Busnelli M, Bulgheroni E, Capurro V, Finardi A, Donzelli A, Pattini L, Rubino T, Parolaro D, Nishimori K, Parenti M, Chini B (2011) Pharmacologic rescue of impaired cognitive flexibility, social deficits, increased aggression, and seizure susceptibility in oxytocin receptor null mice: a neurobehavioral model of autism. Biol Psychiatry 69(9):875–882. doi:10.1016/j.biopsych.2010.12.022 S0006-3223(10)01314-4 [pii]

Savaskan E, Ehrhardt R, Schulz A, Walter M, Schachinger H (2008) Post-learning intranasal oxytocin modulates human memory for facial identity. Psychoneuroendocrinology 33(3):368–374. doi:10.1016/j.psyneuen.2007.12.004

Sawchenko PE, Swanson LW (1982) Immunohistochemical identification of neurons in the paraventricular nucleus of the hypothalamus that project to the medulla or to the spinal cord in the rat. J Comp Neurol 205(3):260–272

Scheele D, Wille A, Kendrick KM, Stoffel-Wagner B, Becker B, Gunturkun O, Maier W, Hurlemann R (2013) Oxytocin enhances brain reward system responses in men viewing the face of their female partner. Proc Natl Acad Sci USA 110(50):20308–20313. doi:10.1073/pnas.1314190110

Schorscher-Petcu A, Sotocinal S, Ciura S, Dupre A, Ritchie J, Sorge RE, Crawley JN, Hu SB, Nishimori K, Young LJ, Tribollet E, Quirion R, Mogil JS (2010) Oxytocin-induced analgesia and scratching are mediated by the vasopressin-1A receptor in the mouse. J Neurosci 30 (24):8274–8284. doi:10.1523/JNEUROSCI.1594-10.2010

Schwartzer JJ, Ricci LA, Melloni RH Jr (2013) Prior fighting experience increases aggression in Syrian hamsters: implications for a role of dopamine in the winner effect. Aggress Behav 39 (4):290–300. doi:10.1002/ab.21476

Silk JB (2007) The adaptive value of sociality in mammalian groups. Philos Trans R Soc Lond B Biol Sci 362(1480):539–559. doi:10.1098/rstb.2006.1994 Q05478288G021600 [pii]

Silk JB, Beehner JC, Bergman TJ, Crockford C, Engh AL, Moscovice LR, Wittig RM, Seyfarth RM, Cheney DL (2009) The benefits of social capital: close social bonds among female baboons enhance offspring survival. Proc Biol Sci 276(1670):3099–3104. doi:10.1098/rspb.2009.0681 rspb.2009.0681 [pii]

Simeon D, Bartz J, Hamilton H, Crystal S, Braun A, Ketay S, Hollander E (2011) Oxytocin administration attenuates stress reactivity in borderline personality disorder: a pilot study. Psychoneuroendocrinology 36(9):1418–1421. doi:10.1016/j.psyneuen.2011.03.013

Skuse DH, Lori A, Cubells JF, Lee I, Conneely KN, Puura K, Lehtimaki T, Binder EB, Young LJ (2014) Common polymorphism in the oxytocin receptor gene (OXTR) is associated with human social recognition skills. Proc Natl Acad Sci USA 111(5):1987–1992. doi:10.1073/pnas.1302985111

Smeltzer MD, Curtis JT, Aragona BJ, Wang Z (2006) Dopamine, oxytocin, and vasopressin receptor binding in the medial prefrontal cortex of monogamous and promiscuous voles. Neurosci Lett 394(2):146–151. doi:10.1016/j.neulet.2005.10.019 S0304-3940(05)01183-3 [pii]

Sofroniew MV (1983) Morphology of vasopressin and oxytocin neurones and their central and vascular projections. Prog Brain Res 60:101–114

Sofroniew MV (1985) Vasopressin- and neurophysin-immunoreactive neurons in the septal region, medial amygdala, and locus coeruleus in colchicine-treated rats. Neuroscience 15 (2):347–358

Soma KK, Rendon NM, Boonstra R, Albers HE, Demas GE (2015) DHEA effects on brain and behavior: insights from comparative studies of aggression. J Steroid Biochem Mol Biol 145:261–272. doi:10.1016/j.jsbmb.2014.05.011

Song Z, Larkin T, Albers HE (2014a) Microinjection of arginine-vasopressin (AVP) in the ventral tegmental area (VTA) enhances conditioned place preference and social interaction. Abstract presented at the Society for Neuroscience

Song Z, McCann KE, McNeill JKt, Larkin TE, 2nd, Huhman KL, Albers HE (2014b) Oxytocin induces social communication by activating arginine-vasopressin V1a receptors and not oxytocin receptors. Psychoneuroendocrinology 50C:14–19. doi:10.1016/j.psyneuen.2014.08.005

Song Z, Larkin T, O'Malley M, Albers HE (2015) Both oxytocin and arginine vasopressin enhance social recognition by acting on oxytocin receptors. Abstract presented at the Society for Behavioral Neuroendocrinology

Souza RP, de Luca V, Meltzer HY, Lieberman JA, Kennedy JL (2010a) Schizophrenia severity and clozapine treatment outcome association with oxytocinergic genes. Int J Neuropsychopharmacol 13(6):793–798. doi:10.1017/S1461145710000167

Souza RP, Ismail P, Meltzer HY, Kennedy JL (2010b) Variants in the oxytocin gene and risk for schizophrenia. Schizophr Res 121(1–3):279–280. doi:10.1016/j.schres.2010.04.019 S0920-9964(10)01295-8 [pii]

Stanyon R, Bigoni F (2014) Sexual selection and the evolution of behavior, morphology, neuroanatomy and genes in humans and other primates. Neurosci Biobehav Rev 46P4:579–590. doi:10.1016/j.neubiorev.2014.10.001

Stemmelin J, Lukovic L, Salome N, Griebel G (2005) Evidence that the lateral septum is involved in the antidepressant-like effects of the vasopressin V(1b) receptor antagonist SSR149415. Neuropsychopharmacology 30:35–42

Stevenson EL, Caldwell HK (2012) The vasopressin 1b receptor and the neural regulation of social behavior. Horm Behav 61(3):277–282. doi:10.1016/j.yhbeh.2011.11.009

Stockley P, Bro-Jorgensen J (2011) Female competition and its evolutionary consequences in mammals. Biol Rev Cambridge Philos Soc 86(2):341–366. doi:10.1111/j.1469-185X.2010.00149.x

Striepens N, Matusch A, Kendrick KM, Mihov Y, Elmenhorst D, Becker B, Lang M, Coenen HH, Maier W, Hurlemann R, Bauer A (2014) Oxytocin enhances attractiveness of unfamiliar female faces independent of the dopamine reward system. Psychoneuroendocrinology 39:74–87. doi:10.1016/j.psyneuen.2013.09.026

Szot P, Bale TL, Dorsa DM (1994) Distribution of messenger RNA for the vasopressin V1a receptor in the CNS of male and female rats. Brain Res Mol Brain Res 24(1–4):1–10

Tabak BA, Meyer ML, Castle E, Dutcher JM, Irwin MR, Han JH, Lieberman MD, Eisenberger NI (2015) Vasopressin, but not oxytocin, increases empathic concern among individuals who received higher levels of paternal warmth: a randomized controlled trial. Psychoneuroendocrinology 51:253–261. doi:10.1016/j.psyneuen.2014.10.006

Takayanagi Y, Yoshida M, Bielsky IF, Ross HE, Kawamata M, Onaka T, Yanagisawa T, Kimura T, Matzuk MM, Young LJ, Nishimori K (2005) Pervasive social deficits, but normal parturition, in oxytocin receptor-deficient mice. Proc Natl Acad Sci USA 102(44):16096–16101

Tamamaki N, Abe K, Nojyo Y (1988) Three-dimensional analysis of the whole axonal arbors originating from single CA2 pyramidal neurons in the rat hippocampus with the aid of a computer graphic technique. Brain Res 452(1–2):255–272

Tansey KE, Hill MJ, Cochrane LE, Gill M, Anney RJ, Gallagher L (2011) Functionality of promoter microsatellites of arginine vasopressin receptor 1A (AVPR1A): implications for autism. Mol Autism 2(1):3. doi:10.1186/2040-2392-2-3

Teltsh O, Kanyas-Sarner K, Rigbi A, Greenbaum L, Lerer B, Kohn Y (2012) Oxytocin and vasopressin genes are significantly associated with schizophrenia in a large Arab-Israeli pedigree. Int J Neuropsychopharmacol/Official Sci J Collegium Int Neuropsychopharmacologicum 15(3):309–319. doi:10.1017/S1461145711001374

Thompson R, Gupta S, Miller K, Mills S, Orr S (2004) The effects of vasopressin on human facial responses related to social communication. Psychoneuroendocrinology 29(1):35–48

Thompson RR, George K, Walton JC, Orr SP, Benson J (2006) Sex-specific influences of vasopressin on human social communication. Proc Natl Acad Sci USA 103(20):7889–7894

Timmer M, Cordero MI, Sevelinges Y, Sandi C (2011) Evidence for a role of oxytocin receptors in the long-term establishment of dominance hierarchies. Neuropsychopharmacology 36 (11):2349–2356. doi:10.1038/npp.2011.125

Tribollet E, Barberis C, Arsenijevic Y (1997) Distribution of vasopressin and oxytocin receptors in the rat spinal cord: sex-related differences and effect of castration in pudendal motor nuclei. Neuroscience 78(2):499–509

Unkelbach C, Guastella AJ, Forgas JP (2008) Oxytocin selectively facilitates recognition of positive sex and relationship words. Psychol Sci 19(11):1092–1094. doi:10.1111/j.1467-9280. 2008.02206.x

Uzefovsky F, Shalev I, Israel S, Knafo A, Ebstein RP (2012) Vasopressin selectively impairs emotion recognition in men. Psychoneuroendocrinology 37(4):576–580. doi:10.1016/j. psyneuen.2011.07.018

Vaccari C, Lolait SJ, Ostrowski NL (1998) Comparative distribution of vasopressin V1b and oxytocin receptor messenger ribonucleic acids in brain. Endocrinology 139:5015–5033

van den Pol AN (2012) Neuropeptide transmission in brain circuits. Neuron 76(1):98–115. doi:10. 1016/j.neuron.2012.09.014

van Leeuwen FW, Swaab DF, de Raay C (1978) Immunoelectronmicroscopic localization of vasopressin in the rat suprachiasmatic nucleus. Cell Tissue Res 193(1):1–10

Veenema AH, Neumann ID (2008) Central vasopressin and oxytocin release: regulation of complex social behaviours. Prog Brain Res 170:261–276. doi:10.1016/S0079-6123(08)00422-6 S0079-6123(08)00422-6 [pii]

Veenema AH, Bredewold R, De Vries GJ (2013) Sex-specific modulation of juvenile social play by vasopressin. Psychoneuroendocrinology 38(11):2554–2561. doi:10.1016/j.psyneuen.2013. 06.002

Veinante P, Freund-Mercier MJ (1997) Distribution of oxytocin- and vasopressin-binding sites in the rat extended amygdala: a histoautoradiographic study. J Comp Neurol 383(3):305–325

Vertes RP, McKenna JT (2000) Collateral projections from the supramammillary nucleus to the medial septum and hippocampus. Synapse 38(3):281–293

Virkkunen M, Kallio E, Rawlings R, Tokola R, Poland RE, Guidotti A, Nemeroff C, Bissette G, Kalogeras K, Karonen SL (1994) Personality profiles and state aggressiveness in finnish alcoholic, violent offenders, fire setters, and healthy volunteers. Arch Gen Psychiatry 51(1):28–33

Viviani D, Charlet A, van den Burg E, Robinet C, Hurni N, Abatis M, Magara F, Stoop R (2011) Oxytocin selectively gates fear responses through distinct outputs from the central amygdala. Science 333(6038):104–107. doi:10.1126/science.1201043

Wacker DW, Ludwig M (2012) Vasopressin, oxytocin, and social odor recognition. Horm Behav 61(3):259–265. doi:10.1016/j.yhbeh.2011.08.014

Wang Z, Zhou L, Hulihan TJ, Insel TR (1996) Immunoreactivity of central vasopressin and oxytocin pathways in microtine rodents: a quantitative comparative study. J Comp Neurol 366 (4):726–737

Wang Z, Yu G, Cascio C, Liu Y, Gingrich B, Insel TR (1999) Dopamine D2 receptor-mediated regulation of partner preferences in female prairie voles (*Microtus ochrogaster*): a mechanism for pair bonding? Behav Neurosci 113(3):602–611

Wassink TH, Piven J, Vieland VJ, Pietila J, Goedken RJ, Folstein SE, Sheffield VC (2004) Examination of AVPR1a as an autism susceptibility gene. Mol Psychiatry 9(10):968–972

Watanabe Y, Kaneko N, Nunokawa A, Shibuya M, Egawa J, Someya T (2012) Oxytocin receptor (OXTR) gene and risk of schizophrenia: case-control and family-based analyses and meta-analysis in a Japanese population. Psychiatry Clin Neurosci 66(7):622. doi:10.1111/j.1440-1819.2012.02396.x

Watters JJ, Poulin P, Dorsa DM (1998) Steroid homone regulation of vasopressinergic neurotransmission in the central nervous system. Prog Brain Res 119:247–261

Weaver IC, Cervoni N, Champagne FA, D'Alessio AC, Sharma S, Seckl JR, Dymov S, Szyf M, Meaney MJ (2004) Epigenetic programming by maternal behavior. Nat Neurosci 7(8):847–854. doi:10.1038/nn1276

Wermter AK, Kamp-Becker I, Hesse P, Schulte-Korne G, Strauch K, Remschmidt H (2010) Evidence for the involvement of genetic variation in the oxytocin receptor gene (OXTR) in the etiology of autistic disorders on high-functioning level. Am J Med Genet B Neuropsychiatr Genet 153B(3):629–639. doi:10.1002/ajmg.b.31032

Wersinger SR, Ginns EI, O'Carroll AM, Lolait SJ, Young WS III (2002) Vasopressin V1b receptor knockout reduces aggressive behavior in male mice. Mol Psychiatry 7(9):975–984

Wersinger SR, Caldwell HK, Christiansen M, Young WS 3rd (2007a) Disruption of the vasopressin 1b receptor gene impairs the attack component of aggressive behavior in mice. Genes Brain Behav 6(7):653–660. doi:10.1111/j.1601-183X.2006.00294.x

Wersinger SR, Caldwell HK, Martinez L, Gold P, Hu SB, Young WS 3rd (2007b) Vasopressin 1a receptor knockout mice have a subtle olfactory deficit but normal aggression. Genes Brain Behav 6(6):540–551

Wersinger SR, Temple JL, Caldwell HK, Young WS 3rd (2008) Inactivation of the oxytocin and the vasopressin (Avp) 1b receptor genes, but not the Avp 1a receptor gene, differentially impairs the Bruce effect in laboratory mice (*Mus musculus*). Endocrinology 149(1):116–121. doi:10.1210/en.2007-1056 en.2007-1056 [pii]

Williams JR, Catania KC, Carter CS (1992) Development of partner preferences in female prairie voles (*Microtus ochrogaster*): the role of social and sexual experience. Horm Behav 26 (3):339–349 0018-506X(92)90004-F [pii]

Williams JR, Insel TR, Harbaugh CR, Carter CS (1994) Oxytocin administered centrally facilitates formation of a partner preference in prairie voles (*Microtus ochrogaster*). J Neuroendocrinol 6:247–250

Windle RJ, Shanks N, Lightman SL, Ingram CD (1997) Central oxytocin administration reduces stress-induced corticosterone release and anxiety behavior in rats. Endocrinology 138: 2829–2834

Winslow J, Insel TR (1991a) Vasopressin modulates male squirrel monkeys' behavior during social separation. Eur J Pharmacol 200(1):95–101

Winslow JT, Insel TR (1991b) Social status in pairs of male squirrel monkeys determines the behavioral response to central oxytocin administration. J Neurosci 11(7):2032–2038

Winslow JT, Insel TR (2002) The social deficits of the oxytocin knockout mouse. Neuropeptides 26(2–3):221–229

Winslow JT, Hastings N, Carter CS, Harbaugh CR, Insel TR (1993) A role for central vasopressin in pair bonding in monogamous prairie voles. Nature 365:545–548

Winslow JT, Noble PL, Lyons CK, Sterk SM, Insel TR (2003) Rearing effects on cerebrospinal fluid oxytocin concentration and social buffering in rhesus monkeys. Neuropsychopharmacology 28(5):910–918. doi:10.1038/sj.npp.1300128

Wu S, Jia M, Ruan Y, Liu J, Guo Y, Shuang M, Gong X, Zhang Y, Yang X, Zhang D (2005) Positive association of the oxytocin receptor gene (OXTR) with autism in the Chinese Han population. Biol Psychiatry 58(1):74–77. doi:10.1016/j.biopsych.2005.03.013 S0006-3223(05) 00310-0 [pii]

Yao S, Zhao W, Cheng R, Geng Y, Luo L, Kendrick KM (2014) Oxytocin makes females, but not males, less forgiving following betrayal of trust. Int J Neuropsychopharmacol 17(11):1785–1792. doi:10.1017/S146114571400090X

Yirmiya N, Rosenberg C, Levi S, Salomon S, Shulman C, Nemanov L, Dina C, Ebstein RP (2006) Association between the arginine vasopressin 1a receptor (AVPR1a) gene and autism in a family-based study: mediation by socialization skills. Mol Psychiatry 11(5):488–494. doi:10.1038/sj.mp.4001812 4001812 [pii]

Ylisaukko-oja T, Alarcon M, Cantor RM, Auranen M, Vanhala R, Kempas E, von Wendt L, Jarvela I, Geschwind DH, Peltonen L (2006) Search for autism loci by combined analysis of autism genetic resource exchange and finnish families. Ann Neurol 59(1):145–155. doi:10.1002/ana.20722

Young LJ, Huot B, Nilsen R, Wang Z, Insel TR (1996) Species differences in central oxytocin receptor gene expression: comparative analysis of promoter sequences. J Neuroendocrinol 8(10):777–783

Young LJ, Winslow JT, Nilsen R, Insel TR (1997) Species differences in V1a receptor gene expression in monogamous and nonmonogamous voles: behavioral consequences. Behav Neurosci 111(3):599–605

Young LJ, Wang Z, Cooper TT, Albers HE (2000) Vasopressin (V1a) receptor binding, mRNA expression and transcriptional regulation by androgen in the Syrian hamster brain. J Neuroendocrinol 12(12):1179–1185

Young LJ, Lim MM, Gingrich B, Insel TR (2001) Cellular mechanisms of social attachment. Horm Behav 40(2):133–138. doi:10.1006/hbeh.2001.1691 S0018-506X(01)91691-5 [pii]

Young WS, Li J, Wersinger SR, Palkovits M (2006) The vasopressin 1b receptor is prominent in the hippocampal area CA2 where it is unaffected by restraint stress or adrenalectomy. Neuroscience 143(4):1031–1039

Young KA, Liu Y, Wang Z (2008) The neurobiology of social attachment: a comparative approach to behavioral, neuroanatomical, and neurochemical studies. Comp Biochem Physiol C Toxicol Pharmacol 148(4):401–410. doi:10.1016/j.cbpc.2008.02.004 S1532-0456(08)00032-X [pii]

Young KA, Gobrogge KL, Liu Y, Wang Z (2011) The neurobiology of pair bonding: insights from a socially monogamous rodent. Front Neuroendocrinol 32(1):53–69. doi:10.1016/j.yfrne.2010.07.006 S0091-3022(10)00055-5 [pii]

Zak PJ, Kurzban R, Matzner WT (2005) Oxytocin is associated with human trustworthiness. Horm Behav 48(5):522–527. doi:10.1016/j.yhbeh.2005.07.009 S0018-506X(05)00175-3 [pii]

Zak PJ, Stanton AA, Ahmadi S (2007) Oxytocin increases generosity in humans. PLoS One 2(11): e1128. doi:10.1371/journal.pone.0001128

Ziegler C, Dannlowski U, Brauer D, Stevens S, Laeger I, Wittmann H, Kugel H, Dobel C, Hurlemann R, Reif A, Lesch KP, Heindel W, Kirschbaum C, Arolt V, Gerlach AL, Hoyer J, Deckert J, Zwanzger P, Domschke K (2015) Oxytocin receptor gene methylation: converging multilevel evidence for a role in social anxiety. Neuropsychopharmacology 40(6):1528–1538. doi:10.1038/npp.2015.2

Zingg HH, Laporte SA (2003) The oxytocin receptor. Trends Endocrinol Metab 14(5):222–227

Roles of "Wanting" and "Liking" in Motivating Behavior: Gambling, Food, and Drug Addictions

M.J.F. Robinson, A.M. Fischer, A. Ahuja, E.N. Lesser and H. Maniates

Abstract The motivation to seek out and consume rewards has evolutionarily been driven by the urge to fulfill physiological needs. However in a modern society dominated more by plenty than scarcity, we tend to think of motivation as fueled by the search for pleasure. Here, we argue that two separate but interconnected subcortical and unconscious processes direct motivation: "wanting" and "liking." These two psychological and neuronal processes and their related brain structures typically work together, but can become dissociated, particularly in cases of addiction. In drug addiction, for example, repeated consumption of addictive drugs sensitizes the mesolimbic dopamine system, the primary component of the "wanting" system, resulting in excessive "wanting" for drugs and their cues. This sensitizing process is long-lasting and occurs independently of the "liking" system, which typically remains unchanged or may develop a blunted pleasure response to the drug. The result is excessive drug-taking despite minimal pleasure and intense cue-triggered craving that may promote relapse long after detoxification. Here, we describe the roles of "liking" and "wanting" in general motivation and review recent evidence for a dissociation of "liking" and "wanting" in drug addiction, known as the incentive sensitization theory (Robinson and Berridge 1993). We also make the case that sensitization of the "wanting" system and the resulting dissociation of "liking" and "wanting" occurs in both gambling disorder and food addiction.

Keywords "Wanting" · "Liking" · Motivation · Incentive salience · Sensitization · Addiction · Gambling · Obesity · Overconsumption

---

M.J.F. Robinson (✉) · A.M. Fischer · A. Ahuja · E.N. Lesser · H. Maniates
Department of Psychology, Wesleyan University, 207 High Street, Judd Hall, Middletown, CT 06459, USA
e-mail: mjrobinson@wesleyan.edu
URL: http://robinsonlab.research.wesleyan.edu/

© Springer International Publishing Switzerland 2015
Curr Topics Behav Neurosci (2016) 27: 105–136
DOI 10.1007/7854_2015_387
Published Online: 27 September 2015

Contents

1	Introduction	106
2	"Liking" and "Wanting"	106
3	Drug Addiction	112
	3.1 Evidence for the Incentive Sensitization Theory	115
	3.2 The Role of "Liking," and Alternate Hypotheses of Addiction	117
4	Gambling	119
5	Food Addiction	123
6	Concluding Remarks	127
References		127

1 Introduction

Most people enjoy eating fatty, salty, and sugary foods. Many of us happily indulge in the occasional opportunity to gamble or get a thrill from visiting a casino. Others manage to recreationally use psychoactive drugs (such as alcohol), even in large quantities, without allowing them to consume their lives. Yet for each of these activities you can find countless examples of people who overindulge, even to the point of serious adverse consequences detrimental to their work, family, or health. In some instances, the enjoyment provided by these activities steadily declines, and all that is left is the unrelenting desire to carry on. Motivational structures of the brain provide the initial spark to seek out and consume the resources needed to survive, yet these systems can be hijacked by stimuli that surpass what is typically encountered in nature and may lead people astray, often with devastating consequences. It is in these moments, when the dichotomy between our survival needs and our wants is greatest, that the complexity of the system is exposed, and we can gain insights into its function.

2 "Liking" and "Wanting"

While we typically want the things that we like and like the things that we want, these concepts are not synonymous. The intuitive nature of these words helps nurture understanding of relatively complex motivational concepts. The fact that our language developed to have these two separate words shows how important the distinction between these ideas is to motivation. In 1993, Terry Robinson and Kent Berridge at the University of Michigan refined the use of the words in the context of motivational research (Robinson and Berridge 1993). They posited that the brain contains two distinct systems; one system responsible for hedonic pleasure, or "liking," and another separate yet interconnected system responsible for "wanting," or what Robinson and Berridge termed incentive salience. To ease discussion of these concepts, we will refer to "wanting" (with quotation marks) as a specific subcomponent of the colloquial understanding of the word wanting. Wanting (without quotation marks) typically refers to conscious, cognitive desire, while we will use "wanting" to refer to the visceral feeling of desire. Similarly, "liking" refers to the core process of hedonic pleasure and is deemed separate from the subjective experience of conscious pleasure.

When you take a bite into your favorite food, the look, taste, texture, and smell of the food come together to comprise the pleasure experienced from the affective hedonic impact or "liking" aspects of the food. This "liking" component associated with attaining a reward goes beyond mere sensory properties. Rewards such as food possess clear sensory components of taste and smell. However, separate brain circuits account for how much a food is "wanted" or "liked." For example, the once-liked sweet taste of chocolate ice cream can become strongly disliked if paired with violent sickness, despite retaining its sweet sensory properties (Garcia et al. 1985; Rozin 2000; Reilly and Schachtman 2009; Berridge et al. 2010). The converse seems anecdotally true of the bitter taste of beer or coffee, as these tastes often become desired and pleasant with repeated exposure and cultural norms. Further, both "wanting" and "liking" can be strongly modulated by internal physiological states. Hunger will make foods more desired and pleasurable (Cabanac 1971), whereas satiation can dampen the pleasure elicited by chocolate in a self-proclaimed chocoholic (Small et al. 2001; Lemmens et al. 2009), a dynamic shift in hedonic tone referred to as "alliesthesia" (Cabanac 1971).

"Wanting," or incentive salience, is the acquisition of a visceral and unconscious desire for a reward. The motivational value given to that reward can be conferred to cues and objects related to the reward or its retrieval (Bindra 1978) transforming them into "wanted" incentives. Noticing cues that predict food (or monetary gains) can help one accrue more rewards and, therefore, evolutionary fitness (Hollis 1984). In turn, these cues are imbued with incentive salience and become capable of triggering motivation and bursts of reward-seeking (Holmes et al. 2010; Peciña and Berridge 2013). For example, the smell of freshly brewed coffee or the distinct lights and layout of a casino may prompt the need for a pick-me-up or create the urge to play. Three fundamental characteristics apply to cues or conditioned stimuli (CSs) that have been imbued with incentive salience. First, these cues become "motivational magnets." Attention and behavior is "drawn" to them, like a magnet, making such cues difficult to ignore. Experimentally, we can see this demonstrated during Pavlovian conditioned approach (PCA) or autoshaping, where an animal will sniff, lick, and even bite inedible objects such as a protruding metal lever because it has repeatedly predicted delivery of a tasty edible reward (Brown and Jenkins 1968; Hearst and Jenkins 1974; Boakes et al. 1978; Robinson et al. 2014c). This irrational behavior, referred to as sign-tracking, is often evolutionarily adaptive behavior similar to that prompted by the nature of the reward, but appears irrational due to the arbitrary nature of the stimulus (such as an inedible metal lever).

Secondly, beyond simply attracting attention, reward-related cues can become the focus of motivation and themselves act as reinforcers. They may even foster new behaviors that increase interaction and contact with these cues. In a laboratory setting, animals will display this type of behavior, known as conditioned reinforcement (Robbins et al. 1983), when they are first trained to associate a cue with a reward and then given the opportunity in a novel task to work for a presentation of that cue alone. The animal is no longer receiving a reward that has any innate value like food (unconditioned stimulus, or UCS); all that generates and sustains their behavior is the cue that was once associated with the UCS reward. In our daily lives, this phenomenon can be seen in routine behaviors like walking by our favorite bakery simply to experience the aroma of freshly baked goods or when a recent ex-smoker might linger around other smokers for the opportunity to experience the smell of second-hand smoke.

Finally, a reward-associated cue attributed with incentive salience can trigger sudden surges in effort to obtain a reward. Experimentally, we see this in a task called Pavlovian-to-instrumental transfer (PIT) that measures cue-induced peaks in "wanting," seen as surges in effort to obtain a previously available reward. For example, many people report needing a cup of coffee to start their day. But on days when they do not have time to go and buy coffee, the simple sight or smell of someone else's coffee nearby can trigger a powerful enough urge, on top of what might be already strong motivation, to go and get coffee. Here, cues become powerful enough to direct and sometimes dictate behavior.

Although cortical influences are now responsible for cognitive processes we use to consciously determine our behavior, these cortical inputs are an evolutionarily more recent addition to motivated behavior (Swanson 2000; Cardinal et al. 2002; Bernard et al. 2005; Swanson 2005). It is likely that incentive salience developed in living organisms that lacked higher level cortical functioning to guide them toward activities like feeding, drinking, and procreating. For example, animals lacking any clear cortical structures, such as the Atlantic cod and the cuttlefish, show evidence of incentive salience attribution in the form of sign-tracking (Purdy et al. 1999; Nilsson et al. 2008). Nevertheless, not all behavior is determined by subcortical systems. There are likely earlier-evolved reflexes that are now suppressed in order to let cortical mechanisms guide behavior. Despite this, it appears that we sometimes still rely on these primitive brain systems to provide a motivational spark toward fulfilling our biological needs (Robinson and Berridge 1993). Thus, many of our conscious wants arise from the subcortical "wanting" system, often despite any cognitive awareness of their subcortical origin.

Beyond cognitive intentions to seek out reward, most people believe that rewards are "wanted" and desired because they produce a conscious experience of pleasure. Although pleasure is a fundamental component of human existence, our ability to accurately discern its inner workings is limited (James 1884). Pleasure is generally described as a purely subjective experience, but evidence suggests that subjective pleasure is only one of the components of reward that is experienced (Berridge et al. 2009; Dai et al. 2010; Litt et al. 2010). Instead rewards can influence behavior even in the absence of conscious awareness (Fischman and Foltin 1992; Winkielman et al. 2005). For example, Fischman and Foltin showed that recovering addicts would consistently choose a very low dose of cocaine over an injection of saline, despite reporting no more subjective feelings of pleasure than with saline, no cardiovascular responses, and indicating that they thought they were sampling both options equally (Fischman and Foltin 1992). Similarly, a study by Winkielman and colleagues showed that subliminal presentations of happy faces made thirsty participants rate, pour and drink more of a sweet beverage despite reports of their conscious feelings remaining unaffected (Winkielman and Berridge 2003; Winkielman et al. 2005). Instead, these rewards may provoke unconscious pleasure or "liking" reactions that may be more objective as they do not rely on self-report and are typically contiguous with the experience of the hedonic stimulus. In contrast to conscious subjective pleasure, "liking" is an implicit response to hedonic stimuli that can be measured in behavior and physiology even in the absence of conscious liking. It is measured using a technique called taste reactivity that exploits objective "liking" reactions to sweet tastes (Grill and Norgren 1978). This method examines orofacial affective expressions, which are homologous across species, including humans, rats, and apes (Steiner et al. 2001; Berridge and Kringelbach 2008). Specifically, sweet tastes elicit positive "liking" responses such as lip licking and rhythmic tongue protrusions, whereas bitter tastes such as quinine produce negative "disliking" expressions such as gapes and headshakes.

Beyond a psychological dissociation, pleasure "liking" also appears to possess a more restrictive limbic brain circuit, both anatomically and neurochemically, which may predispose us more to states of desire than to states of pleasure (see Fig. 1). The "liking" and "wanting" systems in the brain have some structural and neurochemical overlap, but also separate substrates (Berridge et al. 2009, 2010; Castro et al. 2015). While both systems are contained within certain common mesolimbic structures, the "liking" components, or hedonic "hot spots," are only small subregions of these greater mesolimbic structures, including the nucleus accumbens and ventral pallidum (Berridge and Robinson 2003; Berridge et al. 2010). These hedonic "hot spots" were so named (Berridge 2003; Smith et al. 2007) for their ability to elicit positive facial hedonic reactions ("liking") to a sweet solution when neurochemically stimulated (such as by opioid and endocannabinoid neurotransmitters, but not dopamine stimulation). These increases in hedonic pleasure are restricted to stimulation of the small hedonic "hot spots" and cannot be readily elicited by stimulation of neighboring areas in the remaining mesolimbic system (Smith and Berridge 2007; Smith et al. 2007). These hot spots seem to function as a cooperative network that requires a unanimous vote to engender a "liking" response. While stimulation of just one hot spot will typically recruit others, pharmacologically inhibiting activity in one hot spot will prevent an enhancement of "liking" from opioid stimulation in one of the other hot spots (Smith and Berridge 2007; Castro and Berridge 2014). In contrast, "wanting" can be increased by raising dopamine levels in any part of the mesolimbic system (including those hot spots) and does not seem to require simultaneous activity from other motivation centers. "Wanting" can also be evoked by opioid stimulation (and certain other neurotransmitters) within the hot spots (in addition to the aforementioned effects of opioids on "liking"). For example, the opioid agonist DAMGO will enhance "liking" in the cubic millimeter hot spot of the accumbens medial shell, which makes up only 10 % of the entire nucleus accumbens (Peciña 2005; Smith and

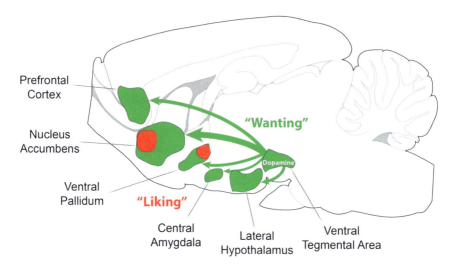

Fig. 1 Mesocorticolimbic circuitry of "liking" and "wanting." This sagittal view of a rodent brain depicts structures and circuitry underlying "liking" (*red*) and "wanting" (*green* and *red*). The nucleus accumbens medial shell contains a hedonic hot spot in the rostral half, where opioid and related stimulation increases "liking" reactions to sucrose taste. The caudal half of the ventral pallidum contains a similar opioid hedonic hot spot. The ventral tegmental area projects dopaminergic afferents to the above labeled areas, which when stimulated increase "wanting" and the attribution of incentive salience, including the areas that contain hedonic hot spots. Sagittal section adapted from Paxinos and Watson (2007)

Berridge 2007; Berridge et al. 2010). In contrast, the same DAMGO microinjection will potently enhance "wanting" in the entire nucleus accumbens (Peciña 2005). Some of the neural structures and pathways involved in "liking" and "wanting" are illustrated in Fig. 1.

Evidence for the existence of distinct neural pathways governing "liking" and "wanting" suggests that in some instances it might be possible to experience "wanting" without "liking" or vice versa. The earliest example of "wanting" without "liking" came from laboratory studies examining the impact of electrical stimulation of the lateral hypothalamus, a part of the brain that activates mesolimbic pathways and dopamine release (Berridge and Valenstein 1991). When electrically stimulated in such a fashion, animals eat voraciously but show no increase in their "liking" responses. Instead, they display a moderate increase in "disliking" to a sucrose solution, as if it became slightly unpleasant. Similar results have been found in mutant mice that have their dopamine transporter knocked down, which leads to excessive synaptic dopamine (Peciña et al. 2003), or in rats following amphetamine or drug sensitization-induced elevation of dopamine release (Wyvell and Berridge 2000; Tindell et al. 2005). More recently, studies have shown that stimulation of areas such as the central nucleus of the amygdala either pharmacologically using DAMGO or optogenetically will increase "wanting" for specific rewards and their cues independently of any changes in "liking" (Mahler and Berridge 2009;

DiFeliceantonio and Berridge 2012; Robinson et al. 2014b). In humans, studies show that dopamine levels are more highly correlated with subjective ratings of "wanting" a reward than with pleasure ratings of that same reward (Volkow et al. 2002; Leyton et al. 2002). In fact, certain highly addictive drugs such as nicotine are exceedingly "wanted" despite producing little to no feelings of pleasure or euphoria (Benowitz 1996; West 2009; Isomura et al. 2014).

Conversely, "liking" without "wanting" can also occur, specifically when dopaminergic transmission is disrupted. For example, in mutant mice that lack the ability to produce any dopamine in their brains, sweet solutions or food rewards will still be liked and preferred over water due to their hedonic impact (Cannon and Palmiter 2003; Robinson et al. 2005). Similarly, drugs that block dopamine transmission, such as the dopamine antagonist pimozide, or treatments (6-OHDA) that destroy over 99 % of mesolimbic and neostriatal dopamine afferents do not disrupt positive "liking" facial reactions to the taste of sucrose (Peciña et al. 1997; Berridge and Robinson 1998). However, these drugs do disrupt "wanting," in that the animals lack the motivation to feed themselves and display life-threatening aphagia and adipsia. In humans, drugs that block dopamine function completely fail to reduce the subjective ratings of pleasure people give to an addictive drug, such as amphetamine or methamphetamine (Brauer and De Wit 1997; Wachtel et al. 2002; Leyton 2010), yet diminish craving and cue-induced craving (Berger et al. 1996). Similarly, studies in which dopamine transmission was decreased by interfering with dopamine synthesis (acute phenylalanine/tyrosine depletion; APTD) show that the pleasurable and mood altering effects of a wide range of abused substances, such as alcohol (Leyton et al. 2000; Barrett et al. 2008), tobacco (Casey et al. 2006; Munafò et al. 2007), amphetamine (Leyton 2007), and cocaine (Leyton et al. 2005), remain intact, while traits related to "wanting" such as cocaine-induced confidence and drug craving are dramatically reduced (Leyton et al. 2005). These results demonstrate that both "wanting" and "liking" can occur independently, and since "liking" is controlled by a smaller portion of the brain and requires collective activation of different hot spots (making it easier to disrupt), it may be a more fragile and less critical component of motivated behavior than "wanting" (Fig. 1). That survival may be almost impossible in the total absence of "wanting," but not in the absence of "liking" may be evidence for this claim.

Natural rewards such as food, water, and sex all generate pleasure, while also triggering the release of mesolimbic dopamine and activating our "wanting" system (Hernandez and Hoebel 1988; Pfaus et al. 1990). In drug, food, and gambling addictions, we see evidence of hypersensitive "wanting" systems taking salience attribution to maladaptive levels, often with very little change in pleasure responding (Robinson and Berridge 2008; Rømer Thomsen et al. 2014; Robinson et al. 2015b). As such we often talk about how drugs of abuse hijack our natural "wanting" system and send it into overdrive. In the three following sections, we will examine the evidence and the insight that the incentive salience theory can provide in three types of addictive behavior: drug abuse, gambling disorder, and overeating and obesity. We will pay close attention to explaining what roles both "liking" and "wanting" may play in the development and maintenance of each addiction. We begin with drug addiction, as it has been the most extensively studied and provides the groundwork for explaining some of the changes seen in gambling and food addiction.

3 Drug Addiction

The symptoms and behaviors that characterize drug addiction can vary greatly from person to person, depending on the drug of choice, the circumstances of use and individual differences between users (Robinson and Berridge 1993; Cadet et al. 2014). Nevertheless, all cases of drug addiction possess three common, significant characteristics that a complete theory of drug addiction must explain (Hollis 1984; Robinson and Berridge 1993). These characteristics highlight the fractioning of the natural bond between "liking" and "wanting," which we believe occurs in multiple forms of addiction. They are as follows:

1. An increased intake and desire for the drug over time, often to the point of intense cravings.
2. Persistent and recurring bouts of craving, frequently triggered by drug-paired cues, that posses the ability to promote relapse, even long after drug-taking has ceased.
3. The dissociation of the pleasure generated by the drug, which tends to decrease or remain unchanged over time, from the desire for the drug, which increases over time and becomes hyper-responsive to drugs and drug stimuli.

The incentive sensitization theory of addiction aims to incorporate and explain these three tenets (Robinson and Berridge 1993; Holmes et al. 2010; Peciña and Berridge 2013). In this theory, Terry Robinson and Kent Berridge posit that repeated drug use causes the mesolimbic dopamine system of the brain responsible for the generation of "wanting" to experience **incentive sensitization,** which in turn leads to the symptoms of drug addiction. Incentive sensitization is defined as an increase in the sensitivity of the neural circuits responsible for the attribution of **incentive salience** ("wanting") to a drug—a process that occurs as a consequence of gradual and progressive neurological changes induced by repeated drug use. The attribution of incentive salience is mediated by dopamine projections to the nucleus accumbens and striatum from the ventral tegmental area and substantia nigra (Robinson and Berridge 2003). Sensitization of this neural circuitry over time, as a consequence of repeated drug consumption, elicits a greater dopaminergic response, and as a result, the incentive salience for the drug and its cues steadily increases. This means that the desire for a particular drug and the ability of its associated cues to trigger craving escalate with repeated drug consumption.

According to the incentive sensitization theory, an individual would first have to consume a potentially addictive drug. This initial behavior would likely be prompted by a desire to experience the expected pleasure ("liking") associated with being under the influence of the drug or by societal pressure to use the substance.

This sensation of euphoria or drug "liking" will likely initially prompt sporadic use that may evolve into repeated use over time. Each drug experience will incur surges in dopaminergic activity in the mesolimbic reward system. Over time, these recurring surges in dopamine release from repeated consumption will cause sensitization of mesolimbic dopamine pathways. The result is an enhanced dopaminergic response to the same initial dose of the drug (Robinson et al. 1988; Kalivas and Duffy 1990; Vezina 1993, 2004), in the form of enhanced dopamine overflow (Kalivas and Duffy 1993; Vanderschuren and Kalivas 2000; Vanderschuren et al. 2001), dopamine D1 receptor supersensitivity (Henry and White 1991; Hu et al. 2002), and enhanced intracellular mechanisms such as induction of immediate early genes (Hiroi et al. 1997). Greater dopamine signaling will result in an increase in the incentive salience assigned to the drug and its cues, which in turn will cause the drug to be "wanted" more. This process occurs independently of the pleasure produced by the drug and is not necessarily tied to "liking" (which may sometimes decrease with increased consumption) (Wyvell and Berridge 2000; Tindell et al. 2005). Due to the increasing incentive salience of the drug and its cues, the user is motivated to approach and consume the drug even more, which will only sensitize the brain further. Thus, a progressive increase in drug "wanting" and consumption occurs, without any paralleled increase in drug "liking," sometimes even despite "liking" the drug less. As a result, the drug becomes compulsively "wanted," in that the urge to consume may contradict cognitive wants to abstain (see Berridge and Robinson 2011), and drug-associated cues are able to trigger intense cravings, that may result in bouts of drug-seeking. In many instances, these urges to seek out and take drugs are appeased by top-down cognitive control, meaning that more often than not, cue-triggered impulses to take drugs are quashed by the knowledge of the undesired consequences. Yet in this war between subcortical impulses and cognitive intentions, it only takes the loss of a single battle in favor of subcortical "wanting" for relapse to occur, and the war to be lost (Berridge and Robinson 2011).

It has been suggested that compulsive drug-seeking might originate from a physical need for the drug or a powerful motivation to avoid the unpleasant symptoms of drug withdrawal (Wikler 1973; Khantzian 1985; Koob et al. 1989; Koob 1996). According to this view, drug-taking and the resulting dopamine release would satisfy a need, thereby satiating the user and reducing motivation to take more drug. This is counter-intuitive, as it implies that an addict would only take drugs to satisfy a need (and not beyond), when some of the hallmarks of addiction are a tendency to escalate the amount of drug taken and to regularly take more drug than intended (American Psychiatric Association 2013). Instead, the incentive sensitization theory suggests that dopaminergic activity produces incentive motivation for a reward and that a heightened/sensitized dopaminergic response to drug-taking events explains why a small hit of the drug triggers a greater urge for more drug, rather than producing any form of satiation or reduction in motivation (Robinson and Berridge 1993). The process of incentive sensitization is illustrated in Fig. 2.

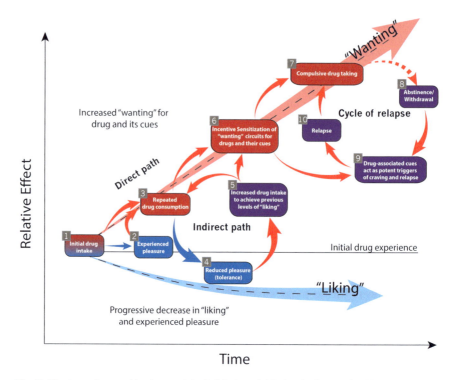

Fig. 2 The incentive sensitization model of addiction. Addiction is characterized by a progressive dissociation of drug "wanting" and "liking" with increasing incentive salience being attributed to drugs and their cues. This schematic model suggests a "direct path" to addiction that develops with repeated and escalating drug consumption leading to incentive sensitization of "wanting" and compulsive drug-taking [Steps: *1-3-6-7*]. A separate contributing factor to this phenomenon is highlighted by the "indirect path" loop, which suggests that with repeated drug-taking the experienced pleasure and euphoria caused by the drug fails to increase and may sometimes even diminish, which prompts intake of larger and larger doses of drug, thus contributing to the sensitization of mesolimbic dopamine circuits [Steps: *3-4-5*]. Finally, compulsive drug-taking is often punctuated by periods of abstinence and withdrawal, which all too frequently result in relapse, often triggered by cue-induced craving. This "cycle of relapse" characterizes drug addiction as a chronic relapsing disorder [Steps: *7-8-9-10*]. Adapted with permission from Robinson and Berridge (1993), Berridge et al. (2009)

The neural changes involved in the process of incentive sensitization are long-lasting (Shuster et al. 1975; Paulson et al. 1991; Castner and Goldman-Rakic 1999). More importantly, these changes persist beyond the cessation of drug-taking and beyond withdrawal, which often only lasts 1–2 weeks (Khavari et al. 1975; Stinus et al. 1998; Gekht et al. 2003). Withdrawal is typically described as an intense negative emotional state accompanied by dysphoria, anxiety, and irritability. Withdrawal avoidance-based theories of addiction suggest that these unpleasant symptoms are the primary motivator for unrelenting drug-taking and relapse (Koob et al. 1989). Although withdrawal may be a potent reason why many addicts relapse, it fails to explain why relapse frequently occurs even after withdrawal symptoms have subsided (Hunt et al. 1971). The persistence of incentive sensitization accounts for why drugs and their cues retain the ability to trigger craving and relapse for many years, even in "detoxified" addicts long after "recovery."

Finally, this entire process takes place independently of the pleasure induced by the drug, since the aforementioned sensitization affects only the motivational effects of the drug, but not to the euphoria it generates. Thus, the three facets of addiction outlined at the beginning of this section are addressed by the incentive sensitization theory.

The incentive sensitization theory also addresses the role of drug-related cues in addiction. Cues related to drug abuse (paraphernalia, contexts, etc.) can themselves take on added incentive salience through the process of sensitization. As a result, they are transformed into powerful "motivational magnets," able to induce cravings upon exposure and bring individuals into the proximity of drugs. Specifically, drug-related cues become increasingly capable of triggering increases in dopaminergic activity in the mesolimbic reward pathway resulting in excessive "wanting," even when presented in the absence of the drug (Leyton 2007; Vezina and Leyton 2009). A striking example of this can be seen with crack cocaine addicts, who when experiencing intense cravings will inspect the floor for any small, white specks and often try and smoke them, even when the specks are most likely dust or ordinary pebbles—a phenomenon known as "chasing ghosts" (Rosse et al. 1993).

In order to account for drug addiction, the incentive sensitization theory makes several verifiable assumptions. **Firstly**, the consumption of drugs must be able to affect areas of the brain involved in regulating "wanting," independent of influence from the brain's pleasure or "liking" systems. **Secondly**, excessive consumption of drugs should gradually render this neural circuitry hypersensitive to the motivational effects of said drugs. **Thirdly**, for this theory to apply to drug addiction more broadly, the neurological mechanism responsible for the attribution of incentive salience must be common to all addictive drugs. **Finally**, the neurobiological changes produced by excessive drug consumption must be long-lasting in order to account for instances of relapse occurring long after withdrawal symptoms have subsided. The next section will address all of these criteria and provide supporting experimental evidence.

3.1 Evidence for the Incentive Sensitization Theory

Let us begin by addressing the assumption that the neural system sensitized by drug consumption is one that only regulates "wanting," and acts independently of "liking." Since dopamine is the primary neural substrate that controls "wanting," an increase in the release of dopamine in response to a particular stimuli can be interpreted directly as an increase in "wanting" for those stimuli. Supporting evidence for the diverging role of these two systems comes from manipulations of the dopamine system in humans and animals that influences "wanting" while leaving "liking" intact. Drugs such as amphetamine, cocaine, methamphetamine, morphine, nicotine, alcohol, and even THC have been shown to increase transmission of dopamine in the nucleus accumbens and dorsal striatum (Robinson and Berridge 1993). A study by Di Chiara et al. further illustrated that while in humans, drugs of abuse such as cocaine, amphetamine, nicotine increase the concentration of dopamine in both the nucleus accumbens and the dorsal caudate nucleus, drugs that are typically not abused, such as antihistamines or antimuscarinic drugs, fail to show similar results (Di Chiara and Imperato 1988). Furthermore, drugs such as nicotine produce increases in dopamine transmission but fail to produce any reported "liking" or euphoria, suggesting the absence of any clear relationship between "liking" and the excessive "wanting" that leads to addiction (Rose et al. 2000; Caggiula et al. 2009; Balfour and Munafò 2015). The lack of any significant correlation between "liking" and "wanting" has also been shown for alcohol in humans (Hobbs et al. 2005; Ostafin et al. 2010). We can therefore suggest that a critical aspect of drug addiction is the sensitization of the mesolimbic dopaminergic system which results in an increase in dopaminergic response to drugs and their cues, and that these increases in incentive salience/"wanting" are independent of changes in "liking" (Ferrario et al. 2005; Ferrario and Robinson 2007; Robinson and Berridge 2008).

Our second assumption states that repeatedly consuming drugs that trigger activity in the "wanting" system gradually increases their incentive properties, which helps explain excessive drug use and the development of addiction. A study by Woolverton et al. reported that when rhesus monkeys were pretreated with methamphetamine injections, they became more likely to self-administer amphetamine if given the opportunity at a later time (Woolverton et al. 1984; Leyton 2010). This finding demonstrates that the initial exposure resulted in an "increased sensitivity to the reinforcing properties of the drug," because of which subjects were more motivated to consume the drug during subsequent trials as compared to controls. More recently, a study by Boileau et al. reported that when treated with three doses of amphetamine within the span of five days, healthy adult men demonstrated an increased release of dopamine in response to the third dose relative to the first (Berger et al. 1996; Boileau et al. 2006). When participants were re-tested a year later, they continued to demonstrate dopaminergic sensitization in brain areas such as the ventral striatum, which is involved in the regulation of "wanting." A similar sensitization of dopamine release has also been reported in Parkinson's patients who compulsively use dopaminergic drugs and exhibit dopamine dysfunction syndrome (DDS) (Leyton et al. 2000; Evans et al. 2006; Barrett et al. 2008). Studies like these establish that addictive drugs such as cocaine and amphetamine have the ability to produce sensitizing effects in the brain, especially in regard to how much they are "wanted" by the user.

Incentive sensitization theory also hinges on the idea that drug-induced sensitization occurs in a common neural network that is responsible for the attribution of incentive salience to all addictive drugs. One significant piece of evidence for this is a phenomenon known as **cross-sensitization.** Cross-sensitization, in the context of drug use, refers to when sensitization to one drug will produce a sensitized response to other drugs (such as between heroin and cocaine). In cases of cross-sensitization of "wanting," an individual, as a result of excessively consuming one drug, is rendered hyper-responsive to the motivational effects of other drugs, including ones that may have never been previously consumed. A study by Horger et al. found that when rats were given nine daily injections of amphetamine or nicotine, they acquired a cocaine self-administration habit much quicker than control animals, thus demonstrating that the pretreated rats were more susceptible to the reinforcing effects of the cocaine (Horger et al. 1992; Casey et al. 2006; Munafò et al. 2007). Similarly, a study by Cunningham et al. found that rats who were given intra-accumbens treatment of certain opiates (such as morphine) later proved to be sensitized to the behavioral effects of amphetamine (Cunningham and Kelley 1992; Leyton 2007). Cross-sensitization and resulting hyper-responsivity of dopaminergic systems also occurs between drugs of abuse and natural rewards (Avena and Hoebel 2003a) and drugs of abuse and stress (at both behavioral and physiological levels) (Piazza et al. 1990; Cruz et al. 2011; Garcia-Keller et al. 2013). This latter finding highlights the important role that stress may play in relapse, whereby stressful life events can act as powerful triggers of drug cravings and a history of stressful life events may even predispose a person to drug addiction. These examples of cross-sensitization support the idea of an underlying neural circuitry common to all addictive drugs.

Finally, the neural changes that underlie sensitization appear to be long-lasting. A study by Paulson and colleagues showed that when rats were pretreated with amphetamine, they exhibited sensitization an entire year after the pretreatment was discontinued (Paulson et al. 1991). Likewise, other studies have reported that mice demonstrate behavioral or psychomotor sensitization, in the form of increased locomotor activity, up to 3 months after cocaine exposure (Shuster et al. 1977) and up to 8 months after morphine exposure (Shuster et al. 1975), while monkeys still display a sensitized response to amphetamine even 2 years post-treatment (Castner and Goldman-Rakic 1999). Studies like these confirm that the sensitizing effects seen in the brain as a result of repeated drug consumption are long-lasting, which in turn explains the constant temptation as well as the tendency to relapse seen in many recovering addicts.

3.2 The Role of "Liking," and Alternate Hypotheses of Addiction

Initial drug consumption is often fueled by feelings of euphoria generated by taking the drug. In contrast to the sensitized response of the "wanting" system that develops in addicts toward drugs and their cues, the euphoria produced by drugs does not undergo the same transformation. There is no sensitization of "liking" systems in the brain. In fact, unlike "wanting," "liking" often undergoes a phenomenon known as **tolerance**, which is the opposite of sensitization. In other words, repeated drug consumption causes "liking" to decrease, and with time, the same dose of drug is no longer able to generate as much pleasure as it once could. As a consequence, an addict may be driven to chase that initial high by progressively escalating the amount of drug consumed, which further causes greater and more rapid incentive sensitization (represented by the indirect path [3-4-5] in Fig. 2). This progression may especially be the case for drugs such as opiates (e.g., heroin and some prescription painkillers) that trigger a strong "liking" response and produce rapid tolerance (Cochin and Kornetsky 1964; Lamb et al. 1991). The implication of this pattern is particularly striking, as it means that an addict can reach a point where a drug causes very little pleasure, and yet he/she may go to great lengths to fulfill an inexplicable craving for it. Supporting evidence comes from a study showing tolerance to the euphoric effects of psychostimulant drugs in cocaine-dependent abusers despite enhanced drug-seeking (Volkow et al. 1997; Mendelson et al. 1998). Several studies have also demonstrated that drug self-administration can be maintained in the absence of any subjective pleasure (Lamb et al. 1991; Fischman and Foltin 1992; Hart et al. 2001) and that drugs such as morphine concomitantly generate both positive reinforcing and negative aversive effects (Stolerman 1985; Bechara et al. 1993). These results highlight the limited role of "liking" in drug addiction and shift the explanation toward "wanting." "Wanting" is thought to be to blame rather than cognitive wanting, as awareness of desire does not seem to play a large role in drug-taking (Lamb et al. 1991; Fischman and Foltin 1992). Such lack of cognitive awareness would explain why addicts often have little insight into their hunger for drugs and drug-associated cues (Childress et al. 2008; Goldstein et al. 2009) and why drug-taking persists despite adverse consequences and a cognitive intent to remain abstinent.

The incentive sensitization theory is not the only explanation that has been put forward to account for drug addiction. There are three other main reasons frequently suggested to explain addiction and relapse. The first has to do with drug euphoria or pleasure and suggests that addicts resume drug-taking to experience intense pleasure (Wise 1982). While drug pleasure or "liking" certainly accounts for initial patterns of drug use, as previously mentioned, tolerance frequently develops with repeated drug use (although not equally for all drugs) and addicts often report knowing that relapse will fail to lead to intense pleasure but rather to more misery. The second explanation has to do with drug habits and the belief that drug-taking distorts learning and creates such robust habits that relapse is inevitable (Hyman et al. 2006; Everitt et al. 2008; Koob and Volkow 2010). This approach fails to incorporate the dimension of excessive "wanting" and compulsion that accompanies drug-taking, which otherwise distinguishes it from regular habits like brushing one's teeth and tying one's shoelaces. Certainly habits facilitate the repeated drug use that is characteristic of drug addiction and contributes to incentive sensitization, but they are unable to explain the flexibility and resourcefulness that addicts display when procuring drugs, and thus account better for drug-taking than for the craving-driven drug-seeking that typifies drug addiction primarily as a relapsing disorder. Finally, the intense negative emotional state of withdrawal produced as a result of drug abstinence is often suggested as the primary cause for relapse (Robinson and Berridge 1993; Koob and Volkow 2010). While withdrawal may be a potent reason many addicts do resume drug-taking, withdrawal is relatively short-lived and decays within days to weeks, depending on the drug (Wikler 1973; Khantzian 1985; Koob et al. 1989; Koob 1996). By contrast, relapse frequently occurs long after withdrawal has subsided, even many years later in fully detoxified individuals (Hunt et al. 1971). In fact, addicts often voluntarily undergo withdrawal in detoxification clinics to reduce tolerance and the monetary cost of their addiction (Kleber 2007; Robinson et al. 2013). In addition, certain drugs such as cocaine may produce relatively mild signs of physical withdrawal despite still being highly addictive, whereas certain pharmaceutical drugs such as sleeping pills induce high levels of tolerance and consequently withdrawal, and although they induce physical dependence, fail to produce some of the compulsive behavior seen in drug addiction (Graham and Vidal-Zeballos 1998; Wilkinson 1998). While all three of the aforementioned elements (pleasure, habit, withdrawal) are certainly present in most instances of drug abuse, they alone fall short of a full explanation that encompasses all aspects of addiction. Instead, incentive sensitization of "wanting" circuitry explains the escalation and compulsive pattern of drug-taking that occurs as addiction develops. It also accounts for the frequent incidence of relapse common to all addicts, which can often occur beyond withdrawal and in some cases for a lifetime. As an explanation for addiction, incentive sensitization theory is not limited to drugs of abuse. This divergence of "liking" and "wanting" can also be explored in the realm of gambling addiction.

4 Gambling

In the first four editions of the Diagnostic Statistical Manual of Mental Disorders (DSM), gambling disorder was classified as an impulse control disorder like kleptomania or pyromania. The 2013 release of the DSM-V, however, reclassified gambling disorder as a behavioral addiction (American Psychiatric Association 2013). Gambling disorder shares many characteristics with substance disorders, including the inability to cut down on gambling, continued gambling despite adverse consequences such as loss of money or job, and cravings for gambling (Potenza 2008). In this section, we will explore why gambling is attractive and potentially addictive, and if the transition from casual gambling to compulsive gambling can be explained by the same mechanisms that cause substance addiction.

Although few studies have specifically examined "wanting" and "liking" in gambling disorder, there is support for the idea that the incentive sensitization theory may apply to gambling disorder. The incentive sensitization theory posits that substance addictions cause drugs and their cues to take on increased salience and generate excessive motivation to consume more drug. In gambling addiction, gambling-related cues also seem to take on increased incentive salience, becoming motivational stimuli that drive behavior. One of the hallmarks of gambling, and indeed of most games, is the presence of uncertainty (Costikyan 2013). Studies in rats suggest that uncertainty pertaining to the probability and magnitude of the reward outcome can cause attribution of additional incentive salience to reward-related cues. A recent study by Anselme, Robinson, and Berridge showed that rats exposed to an uncertain reward schedule (where both the chances of receiving a reward and the magnitude of this reward vary) direct significantly more of their attention and behavior to the reward cue than rats exposed to a certain reward schedule (Anselme et al. 2013). In other words, uncertain reward-related cues appear to become stronger "motivational magnets." This finding is paradoxical since it contradicts the idea that the motivational value of a cue should be monotonically related to its predictive value. It is consistent with the incentive salience theory, however, and highlights the dissociation that can occur between the predictive value of a cue, driven by cue learning (CS-UCS association), and the attribution of cue "wanting" (CS attraction) (Zhang et al. 2009). Furthermore, cues that predict reward with a large degree of uncertainty are also more likely to acquire incentive salience. For example, distal cues that are on the periphery of our attention are typically ignored under certain and predictable reward conditions, but when reward conditions are unpredictable, these cues attract more attention (Robinson et al. 2014a). In fact the degree of incentive enhancement that uncertainty imparts to reward-related cues is similar to that produced by psychomotor sensitization through repeated amphetamine administration (Robinson et al. 2015a). This may not come as a surprise considering that cues that predict an uncertain reward (50 % probability) produce a greater dopamine signal, originating from the ventral midbrain, during the anticipation of the uncertain outcome (Fiorillo et al. 2003), and that this dopaminergic signal appears to promote risk-seeking behavior, as evidenced in gambling (Fiorillo 2011).

The role of uncertainty in attributing excessive incentive value can also be seen in humans. A set of studies by Brevers indicate that problem gamblers exhibit attentional bias toward gambling-related cues as compared to healthy controls, suggesting that these stimuli also take on increased salience in human gamblers and may possess "motivational magnet" properties (Brevers et al. 2014a, b). Thus, cues related to uncertain reward seem to acquire incentive salience, just as drug-related cues take on increased salience in substance addictions. Casinos are full of both uncertain reward and potentially salient reward-related cues, like sounds and flashing lights, which likely increase the potential for gambling to become addictive and are reported by problem gamblers as a crucial part of the gambling experience (Dow Schüll 2012).

There is also direct evidence for (cross-)sensitization of the dopaminergic system under gambling-like conditions. Uncertainty causes cross-sensitization of the dopaminergic system, as seen by increased reactivity to a single dose of amphetamine, in the same way that repeated exposure to drugs of abuse sensitizes this system. Zack and colleagues found that rats exposed to maximally uncertain conditions showed the greatest locomotor response to an amphetamine challenge (Zack et al. 2014). In a similar study, Singer and his collaborators found that rats trained to press a lever for reward on a variable schedule showed a greater locomotor response to amphetamine than those who were rewarded on a fixed schedule (Singer et al. 2012). As mentioned previously, heightened amphetamine-induced dopamine release in rats is associated with increased "wanting" but not increased "liking" (Wyvell and Berridge 2001). This implies that the escalating "wanting" that drives substance addictions may also be present in gambling disorder and is independent from "liking."

Cross-sensitization of dopaminergic systems from gambling has also been seen in humans. Boileau and colleagues found that problem gamblers have increased dopamine release in their dorsal striatum in response to amphetamine in comparison with healthy controls (Boileau et al. 2013). These results suggest that the escalating, sensitized "wanting" seen in rats exposed to uncertain reward is also present in human gamblers and possibly drives the transition from casual recreational gambling to compulsive gambling. Additionally, studies have found that problem gamblers have a sensitized dopaminergic response to gambling-related cues. Studies have correlated striatal dopamine release in problem gamblers with severity of problem gambling (Joutsa et al. 2012) and with self-reported levels of excitement during a gambling task (Linnet et al. 2011). However, certain studies instead report a blunted striatal dopamine response to cues in pathological gamblers (Miedl et al. 2012; Balodis et al. 2012). It has been suggested that such contradictory reports can be explained by the absence of familiar or relevant gambling cues during laboratory testing (Leyton and Vezina 2012), which when present instead produce an exaggerated striatal dopamine response (van Holst et al. 2012). This finding implies that while gambling-related cues take on increased incentive salience, other non-related or unfamiliar cues may become less important or even inhibit motivation. Similar arguments have been put forward to explain certain findings that suggest a role for dopamine deficiency across a variety of forms of addiction (Leyton 2007, 2014; Leyton and Vezina 2012, 2014).

Another key characteristic of addiction present in problem gamblers is their willingness to persist in gambling despite the negative consequences such as losing large amounts of money. A study by Linnet and colleagues found that problem gamblers have increased dopamine release in their ventral striatum as compared to healthy controls when they lost money in a gambling task, implying that loss still generates motivation in problem gamblers (Linnet et al. 2010). Additionally, a study by Clark and colleagues found that near misses (or almost winning) in a slot machine gambling task recruited areas of the brain that respond to wins. Participants in this study reported that near misses were significantly less pleasant than full misses, but triggered their urge to play more (Clark et al. 2009). These studies illustrate that although problem gamblers do not enjoy losses, they do find losses motivating, providing further evidence for a dissociation of "liking" and "wanting."

Lesion studies have implicated a number of brain regions involved in "liking" and "wanting" in gambling behavior. As previously mentioned, the nucleus accumbens is a component of the mesolimbic system with connections to prefrontal cortices and the dopaminergic neurons of the ventral tegmental area. Cardinal and Howes lesioned the nucleus accumbens core of rats and found that rats with these lesions were less likely to choose large uncertain rewards than controls when small certain rewards were also presented (Cardinal and Howes 2005). These results suggest that the nucleus accumbens core, a key component of the mesolimbic dopaminergic pathway, plays a role in mediating the desirability of uncertain reward.

Although further research is needed to fully understand the role of "liking" in human gambling, studies have supported the idea that "liking" is decreased in pathological gamblers. In a recent PET neuroimaging study, Mick and colleagues found reduced endogenous opioid release in pathological gamblers following an amphetamine challenge as compared to healthy controls. The problem gamblers also reported lower feelings of euphoria in response to the amphetamine challenge as compared to healthy controls (Mick et al. 2014). These results suggest that problem gamblers may experience a down-regulation in their "liking" system consistent with the incentive sensitization theory of addiction. Interestingly, opioid antagonists such as naltrexone (which is used to manage opioid and alcohol dependence) can help relieve gambling cravings and reduce problem gambling behaviors in some individuals. Although these results may at first seem controversial, as opioid-mediated "liking" seems to play less of a role in compulsive behavior than dopamine-mediated "wanting," there is increasing evidence that the opioid system is involved in regulating both "liking" and/or "wanting" in different regions of the brain (DiFeliceantonio et al. 2012; Castro and Berridge 2014).

Problem gambling, like substance addiction, seems to be rooted in the dysfunction or hijacking of the brain's natural reward system. This system drives animals to seek food, water, sex, and other rewards necessary for survival and propagating the species. It also likely evolved to make exploration and uncertainty motivating. Anselme posits that the motivational qualities of uncertainty are designed to compensate for the high rates of failure organisms experience when seeking resources (Anselme 2013). Resources are rarely fully predicted by external cues meaning that the appeal of uncertain cues may be a necessary requirement to overcome the unpleasantness of failure. If unpredictability were not motivating, the inevitable repeated failure experienced when seeking reward would extinguish behavior. The motivating qualities of uncertainty may therefore not be driven by pleasure or "liking," as could be argued in the case of food or sex, but instead by more primitive subcortical "wanting" systems. When purposefully programmed or designed as the outcome of a game or slot machine, uncertainty could drive the excessive "wanting" that arises below our conscious awareness and promote unhealthy gambling behavior.

5 Food Addiction

Here, we focus on the impact of highly palatable foods on the DA system (Genn et al. 2004; Avena et al. 2008; Tang et al. 2012) and examine how the incentive sensitization theory may explain food addiction and its associated health risks: obesity and binge eating.

Overeating is one of the primary causes of obesity. Excessive "wanting" and "liking" for food, especially refined hyper-palatable food, may play a role in overeating. The recent rise in hyper-palatable foods that often combine high levels of sugar, sodium and fat may result in exacerbated hedonic reactivity, leading to a magnification of both "liking" and "wanting" and consequently overconsumption (Berridge et al. 2009; Davis and Carter 2009). Alternatively, overconsumption of highly palatable foods could be triggered by an amplification of "wanting" resulting from the progressive sensitization of mesolimbic dopamine circuits due to repeated exposure to sweet rewarding foods. Such a phenomenon has been demonstrated in animals following exposure to 12-h cycles of bingeing and overconsumption of sugar interspersed with cycles of dieting (Avena and Hoebel 2003a). After 21 days of this regimen, animals showed a sensitized locomotor response to amphetamine, suggesting an underlying sensitization of the mesolimbic dopaminergic system. Conversely, sensitization of this system in rats by daily amphetamine treatment results in hyperphagia and overconsumption of a sugar solution (Avena and Hoebel 2003b).

Overeating may not have one single explanation. Evidence from genetic studies suggests that for obese individuals with a BMI above 30, the presence or absence of binge eating disorder (BED) may be an important factor in determining the relative role of "liking" and "wanting." Specifically, obese individuals without BED were found to be more likely to possess certain polymorphisms of the dopamine D_2 receptor that suggest excessive "wanting." Yet obese individuals who also display BED might constitute a specific population subtype that is prone to binge eating due to an additional hyperactivity of their "liking" response to food. This enhanced hedonic response to food, linked to particular polymorphisms in their mu-opioid receptor gene, combined with excessive "wanting," may give rise to particularly intense addiction-like tendencies toward food (Davis et al. 2009; Davis and Carter 2009). Research also shows that individuals with a genetic leptin deficiency develop obesity at an early age and show both intense cravings for food and high levels of nucleus accumbens activity in response to food stimuli, even following a meal. When treated with medication to restore leptin levels, however, urges and pleasure reports for food are greatly reduced, as is activity in the accumbens (Farooqi et al. 2007; Farooqi and O'Rahilly 2009). Leptin may therefore regulate the suppression of "liking" and "wanting" following satiety. The development of leptin insensitivity with repeated exposure to a junk food diet may promote obesogenic behaviors through its interactions with the dopaminergic system (Pandit et al. 2011; Sáinz et al. 2015). In contrast, during states of hunger it appears that the pleasurable component of food may be enhanced by changes in activity in both opioid and endocannabinoid systems (Kirkham 2005, 2008). However, endocannabinoids also facilitate VTA dopamine which may trigger enhanced "wanting" for palatable foods independent of "liking" (Kirkham 2005; Cota et al. 2006).

The evidence of excessive "wanting" triggered by food in obese individuals suggests that food may act as an intensely potent reward, similar to certain drugs of abuse. This was examined in an experiment in which rats were given access to sugar water as well as to intravenous injections of cocaine. Results showed that over 90 % of the 132 rats in the experiment preferred to press the lever that allowed them access to the sugar solution instead of the lever that administered cocaine (Lenoir et al. 2007). This finding suggests that a commonly available and frequently ingested substance like sugar is strongly preferred over a "wanted" addictive substance like cocaine (which triggers a supranormal dopamine response), and may therefore be attributed with excessive incentive salience.

In addition, much like in drug addiction, the attractive and rewarding properties of hyper-palatable foods like sugar do not stay confined to the reward itself. Reward-related cues, in this case food cues, can be attributed with excessive incentive salience and become beacons that attract attention and trigger overconsumption. For example, overweight and obese individuals appear to direct greater attention to food-related cues than individuals of a normal weight, especially when food deprived (Nijs et al. 2010). In adolescents, it has been shown that the speed at which food stimuli attract attention is correlated with BMI (Yokum et al. 2011). Another study suggests that despite reduced hunger, obese individuals maintained increased attention to food images over non-food images, as compared to controls (Castellanos et al. 2009). In fact, many of the structures of the mesocorticolimbic dopamine system are also activated in people who have a healthy BMI and/or weight when confronted with food imagery (Tang et al. 2012)—much like how they are activated in drug addicts' brains when confronted with drug cues. A recent study suggests that food cues are excessively attractive only to a subpopulation of rats fed a junk food diet (Robinson et al. 2015b). In this model, rats were given free access to a human junk food diet, made of peanut butter, chocolate chip cookies, potato chips, and chocolate milk powder. Surprisingly only some of these animals (approximately 33–50 %) gained excessive amounts of weight, while the remaining animals maintained a steady weight gain, similar to that of animals provided with regular lab chow. The rats that over consumed junk food and displayed large amounts of weight gain initially displayed greater attraction and "wanting" for food-related cues, as seen by greater levels of cue-driven conditioned approach (e.g., sign-tracking), even before they were ever exposed to the junk food. Following extended access to the junk food diet, the animals that gained large amounts of weight perceived food cues themselves as a reward and were more willing to work simply for their presentation (conditioned reinforcement). This observation provides further evidence of excessive "wanting." This tendency for certain rats to over consume a palatable diet was not the result of a prior heightened pleasure response to sweet tastes, nor was it driven by increases in "liking" with repeated exposure to the junk food. If anything, chronic consumption of a palatable junk food diet led to an overall dampening of the "liking" reaction to increasingly sweet tastes (Robinson et al. 2015b), a phenomenon reminiscent of the sometimes blunted "liking" response seen in drug addicts, and that may similarly be related to tolerance. Further, these results echo previous animal findings suggesting a decoupling of "liking" and "wanting" in obesity (Shin et al. 2011).

Similar evidence for a neural dissociation of "liking" and "wanting" for food and food-related cues has also been demonstrated in humans using neuroimaging evidence (Jiang et al. 2014), with particular emphasis of the role of the striatum in "wanting" for food and its cues. One study also found that fMRI reactivity in obese participants in response to images of high-calorie food in those regions associated with motivation (the insula, ventral tegmental area, putamen, and fusiform gyrus) was inversely predictive of long-term efficacy of a weight loss program, although all participants reported liking the food in the pictures (Murdaugh and Cook 2012). Specifically, if these areas were more active when the participant was shown a picture of high-calorie food than when he or she was shown a control picture, that participant would likely have little success with a 9-month weight loss program. These findings suggest that it is the cues for food, as opposed to the food itself, which play a key role in weight maintenance and also further highlight the importance of individual differences. Specifically, the degree of mesolimbic brain reactivity may differ among individuals and may support a certain predisposition to food and its cues, where excessive attraction to food cues in certain individuals may promote weight gain and its maintenance. In a recent study, Yokum and colleagues found that food commercials caused striatal activation, whereas commercials that did not prominently feature food did not elicit activity in the same neural structures. More importantly, the degree of striatal activation in response to food commercials was predictive of adolescent weight gain one year later (Yokum et al. 2014). These results are further supported by findings that suggest that over time, cues may actually become the dominant driver of food overconsumption. In a recent fMRI study, Burger and Stice demonstrated that with repeated exposure, activity in the caudate progressively increased in response to cues that predicted delivery of a milk shake, while activity in the putamen and ventral pallidum showed a simultaneous decrease following receipt of the milk shake reward (Burger and Stice 2014). Crucially, in a 2-year follow-up, those who showed the greatest ventral pallidum increase to cues and the greatest decrease in caudate response to the milk shake also showed the largest increase in BMI. This finding suggests that the ability of food advertisements and food cues in general to be attributed with incentive salience and trigger surges of "wanting" might be the driving force behind our increasing waistlines. In spite of this support for individual differences, there is evidence to suggest that extended access to a palatable junk food diet sensitizes the mesolimbic dopamine system and renders it hyper-responsive to injections of amphetamine independent of whether animals gained excessive amounts of weight or were able to control their intake on that diet (Robinson et al. 2015b). Therefore, regular intake of palatable junk food, even in the absence of any overt weight gain, can sensitize and increase reactivity of the systems associated with "wanting" and the attribution to incentive salience—potentially leading to progressive susceptibility to overconsumption. An additional factor of particular importance for binge eating is the role of stress. Stress, and more specifically corticotropin-releasing factor (CRF) release, may produce cue-triggered peaks in "wanting," as it has been shown to induce surges in motivation for sugar-paired cues in the same manner as amphetamine microinjections into the nucleus accumbens shell (Peciña et al. 2006). This finding may explain how stress can provoke cue-triggered bursts of binge eating, and in this particular case, powerful sugar-seeking.

Historically, our strong drive for sugar was rooted in its scarcity and its importance in providing energy and nutrition for the brain. However, sugar as a reward has changed, both in its form and availability, and so has the environment in which we live. Today's food is designed, packaged, and presented in a manner that is a far cry from how it was when our distant ancestors expended energy foraging and competing with other animals for resources. Supermarkets and cafeterias have negated the need to forage, yet the neural systems responsible for motivation and "wanting" continue to reward consumption. Food today, especially that containing highly rewarding ingredients such as sugar, fat, and salt, is readily available. These ingredients are refined and modified to enhance their rewarding and sensory properties. Sugar, for example, is now omnipresent in our food (Gearhardt et al. 2011), with a 30 % increase in intake over the past four decades (Elliott et al. 2002; Johnson et al. 2007), and is increasingly present in the absence of fiber, which usually slows down its absorption and dampens any possible spike in blood sugar (Gearhardt et al. 2013; Schulte et al. 2015). In addition, advertisement campaigns now generate a slew of food-related cues that may trigger intense motivation to seek food, driving people to consume more food than may be dictated by physiological needs (Kelly et al. 2008; Harris et al. 2009). Food advertisements are tailored to attract our attention with increasingly tempting visuals of food. Neural systems that direct our incentive motivation cannot evolve rapidly enough to temper the temptation provided by the ever-increasing amount of persuasive cues for food we are bombarded with daily, whether we are physiologically hungry or not.

While there is still an ongoing debate as to whether food addiction can be considered a legitimate concept, there is little doubt that the overconsumption of palatable foods is a growing problem in Western society. Part of this problem results from the refinement and engineering of hyper-palatable foods that contain large quantities of often both sugar and fat and trigger strong initial hedonic "liking" responses. A more prominent role in the obesity epidemic seems to be played by the exacerbated "wanting" reactions elicited by these foods and their cues. Hyper-palatable foods activate mesolimbic dopamine reward pathways, spurring on motivation and attributing the food and its cues with incentive salience. The incessant bombardment of our brain by food advertisements triggers powerful urges to consume these foods beyond our caloric needs and often in spite of reduced pleasure.

6 Concluding Remarks

The incentive sensitization theory helps explain excessive drug-taking, gambling, and eating by allowing for a psychological and neural differentiation between "liking" and "wanting." According to this theory, the psychological and biological process responsible for the attribution of "wanting" to a reward may become dissociated from the hedonic "liking" experience generated by that same reward. The incentive sensitization theory states that in a non-addicted person, these "liking" and "wanting" systems may function in tandem so that a person may "like" what he or she "wants" and "want" what is "liked." In an addict's brain, however, these two systems become decoupled, so that a person feels excessive motivation for the reward and its cues, often despite a decrease in enjoyment. This theory of motivation was created to explain the progressive and incremental development of drug addiction and its persistence. Many of its principle tenets such as the dissociation of "liking" and "wanting," the sensitization of the mesolimbic dopamine system, and the incentive sensitization of the reward and its cues may possibly help provide an explanation for other addictive behaviors such as gambling, sex, Internet, shopping, and food addiction.

References

American Psychiatric Association (2013) American Psychiatric Association: Diagnostic and Statistical Manual of Mental Disorders, Fifth Edition (DSM-5®). American Psychiatric Association, Arlington

Anselme P (2013) Dopamine, motivation, and the evolutionary significance of gambling-like behaviour. Behav Brain Res 256C:1–4. doi:10.1016/j.bbr.2013.07.039

Anselme P, Robinson MJF, Berridge KC (2013) Reward uncertainty enhances incentive salience attribution as sign-tracking. Behav Brain Res 238:53–61. doi:10.1016/j.bbr.2012.10.006

Avena NM, Hoebel BG (2003a) A diet promoting sugar dependency causes behavioral cross-sensitization to a low dose of amphetamine. Neuroscience 122:17–20

Avena NM, Hoebel BG (2003b) Amphetamine-sensitized rats show sugar-induced hyperactivity (cross-sensitization) and sugar hyperphagia. Pharmacol Biochem Behav 74:635–639

Avena NM, Rada P, Hoebel BG (2008) Evidence for sugar addiction: behavioral and neurochemical effects of intermittent, excessive sugar intake. Neurosci Biobehav Rev 32:20–39. doi:10.1016/j.neubiorev.2007.04.019

Balfour DJK, Munafò MR (2015) The role of mesoaccumbens dopamine in nicotine dependence. 24:1–172. doi:10.1007/978-3-319-13482-6_3

Balodis IM, Kober H, Worhunsky PD et al (2012) Diminished frontostriatal activity during processing of monetary rewards and losses in pathological gambling. Biol Psychiatry 71:749–757. doi:10.1016/j.biopsych.2012.01.006

Barrett SP, Pihl RO, Benkelfat C et al (2008) The role of dopamine in alcohol self-administration in humans: individual differences. Europ Neuropsychopharmacol 18:439–447. doi:10.1016/j.euroneuro.2008.01.008

Bechara A, Martin GM, Pridgar A, van der Kooy D (1993) The parabrachial nucleus: a brain stem substrate critical for mediating the aversive motivational effects of morphine. Behav Neurosci 107:147–160

Benowitz NL (1996) Pharmacology of nicotine: addiction and therapeutics. Annu Rev Pharmacol Toxicol 36:597–613. doi:10.1146/annurev.pa.36.040196.003121

Berger SP, Hall S, Mickalian JD et al (1996) Haloperidol antagonism of cue-elicited cocaine craving. Lancet 347:504–508

Bernard LC, Mills M, Swenson L, Walsh RP (2005) An Evolutionary Theory of Human Motivation. Genet Soc Gen Psychol Monogr 131:129–184. doi:10.3200/MONO.131.2.129-184

Berridge KC (2003) Pleasures of the brain. Brain Cognit

Berridge KC, Ho C-Y, Richard JM, DiFeliceantonio AG (2010) The tempted brain eats: pleasure and desire circuits in obesity and eating disorders. Brain Res 1350:43–64. doi:10.1016/j.brainres.2010.04.003

Berridge KC, Kringelbach ML (2008) Affective neuroscience of pleasure: reward in humans and animals. Psychopharmacology 199:457–480. doi:10.1007/s00213-008-1099-6

Berridge KC, Robinson TE (2003) Parsing reward. Trends Neurosci 26:507–513

Berridge KC, Robinson TE (1998) What is the role of dopamine in reward: hedonic impact, reward learning, or incentive salience? Brain Res Rev 28:309–369

Berridge KC, Robinson TE (2011) Drug Addiction as Incentive Sensitization. In: Poland J, Graham G (eds) Addict & Responsibility. MIT Press, Cambridge, pp 21–54

Berridge KC, Robinson TE, Aldridge JW (2009) Dissecting components of reward: 'liking', "wanting", and learning. Curr Opin Pharmacol 9:65–73. doi:10.1016/j.coph.2008.12.014

Berridge KC, Valenstein ES (1991) What psychological process mediates feeding evoked by electrical stimulation of the lateral hypothalamus? Behav Neurosci 105:3–14

Bindra D (1978) How adaptive behavior is produced: a perceptual- motivational alternative to response-reinforcement. Behav Brain Sci 1:41–91

Boakes RA, Poli M, Lockwood MJ, Goodall G (1978) A study of misbehavior: token reinforcement in the rat. J Exp Anal Behav 29:115–134

Boileau I, Dagher A, Leyton M et al (2006) Modeling sensitization to stimulants in humans: an [11C]raclopride/positron emission tomography study in healthy men. Arch Gen Psychiatry 63:1386–1395. doi:10.1001/archpsyc.63.12.1386

Boileau I, Payer D, Chugani B, et al (2013) In vivo evidence for greater amphetamine-induced dopamine release in pathological gambling: a positron emission tomography study with [^{11}C]-(+)-PHNO. Mol Psychiatry 19:1305–1313. doi:10.1038/mp.2013.163

Brauer LH, De Wit H (1997) High dose pimozide does not block amphetamine-induced euphoria in normal volunteers. Pharmacol Biochem Behav 56:265–272

Brevers D, Bechara A, Hermoye L et al (2014a) Comfort for uncertainty in pathological gamblers: a fMRI study. Behav Brain Res 278C:262–270. doi:10.1016/j.bbr.2014.09.026

Brevers D, Koritzky G, Bechara A, Noël X (2014b) Cognitive processes underlying impaired decision-making under uncertainty in gambling disorder. Addict Behav 39:1533–1536. doi:10.1016/j.addbeh.2014.06.004

Brown PL, Jenkins HM (1968) Auto-shaping of the pigeon's key-peck. J Exp Anal Behav 11:1–8. doi:10.1901/jeab.1968.11-1

Burger KS, Stice E (2014) Greater striatopallidal adaptive coding during cue–reward learning and food reward habituation predict future weight gain. Neuroimage 99:122–128. doi:10.1016/j.neuroimage.2014.05.066

Cabanac M (1971) Physiological role of pleasure. Science 173:1103–1107

Cadet JL, Bisagno V, Milroy CM (2014) Neuropathology of substance use disorders. Acta Neuropathol 127:91–107. doi:10.1007/s00401-013-1221-7

Caggiula AR, Donny EC, Palmatier MI et al (2009) The role of nicotine in smoking: a dual-reinforcement model. Nebr Symp Motiv 55:91–109

Cannon CM, Palmiter RD (2003) Reward without dopamine. J Neurosci 23:10827–10831

Cardinal RN, Howes NJ (2005) Effects of lesions of the nucleus accumbens core on choice between small certain rewards and large uncertain rewards in rats. BMC Neurosci 6:37. doi:10.1186/1471-2202-6-37

Cardinal RN, Parkinson JA, Hall J, Everitt BJ (2002) Emotion and motivation: the role of the amygdala, ventral striatum, and prefrontal cortex. Neurosci Biobehav Rev 26:321–352

Casey KF, Benkelfat C, Young SN, Leyton M (2006) Lack of effect of acute dopamine precursor depletion in nicotine-dependent smokers. Europ Neuropsychopharmacol 16:512–520. doi:10.1016/j.euroneuro.2006.02.002

Castellanos EH, Charboneau E, Dietrich MS et al (2009) Obese adults have visual attention bias for food cue images: evidence for altered reward system function. Int J Obes (Lond) 33:1063–1073. doi:10.1038/ijo.2009.138

Castner SA, Goldman-Rakic PS (1999) Long-lasting psychotomimetic consequences of repeated low-dose amphetamine exposure in rhesus monkeys. Neuropsychopharmacol 20:10–28. doi:10.1016/S0893-133X(98)00050-5

Castro DC, Berridge KC (2014) Opioid Hedonic Hotspot in Nucleus Accumbens Shell: Mu, Delta, and Kappa Maps for Enhancement of Sweetness "Liking" and "Wanting". J Neurosci 34:4239–4250. doi:10.1523/JNEUROSCI.4458-13.2014

Castro DC, Cole SL, Berridge KC (2015) Lateral hypothalamus, nucleus accumbens, and ventral pallidum roles in eating and hunger: interactions between homeostatic and reward circuitry. Front Syst Neurosci 9:1–17. doi:10.3389/fnsys.2015.00090

Childress AR, Ehrman RN, Wang Z et al (2008) Prelude to passion: limbic activation by "unseen" drug and sexual cues. PLoS ONE 3:e1506. doi:10.1371/journal.pone.0001506

Clark L, Lawrence AJ, Astley-Jones F, Gray N (2009) Gambling near-misses enhance motivation to gamble and recruit win-related brain circuitry. Neuron 61:481–490. doi:10.1016/j.neuron.2008.12.031

Cochin J, Kornetsky C (1964) Development and loss of tolerance to morphine in the rat after single and multiple injections. J Pharmacol Exp Ther 145:1–10

Costikyan G (2013) Uncertainty in Games. MIT Press, Cambridge

Cota D, Barrera JG, Seeley RJ (2006) Leptin in energy balance and reward: two faces of the same coin? Neuron 51:678–680. doi:10.1016/j.neuron.2006.09.009

Cruz FC, Marin MT, Leão RM, Planeta CS (2011) Stress-induced cross-sensitization to amphetamine is related to changes in the dopaminergic system. J Neural Transm 119:415–424. doi:10.1007/s00702-011-0720-8

Cunningham ST, Kelley AE (1992) Evidence for opiate-dopamine cross-sensitization in nucleus accumbens: studies of conditioned reward. Brain Res Bull 29:675–680

Dai X, Brendl CM, Ariely D (2010) Wanting, liking, and preference construction. Emotion 10:324–334. doi:10.1037/a0017987

Davis C, Carter JC (2009) Compulsive overeating as an addiction disorder. A review of theory and evidence. Appetite 53:1–8. doi:10.1016/j.appet.2009.05.018

Davis CA, Levitan RD, Reid C et al (2009) Dopamine for "wanting" and opioids for "liking": a comparison of obese adults with and without binge eating. Obesity 17:1220–1225. doi:10.1038/oby.2009.52

Di Chiara G, Imperato A (1988) Drugs abused by humans preferentially increase synaptic dopamine concentrations in the mesolimbic system of freely moving rats. Proc Natl Acad Sci USA 85:5274–5278

DiFeliceantonio AG, Berridge KC (2012) Which cue to "want?" Opioid stimulation of central amygdala makes goal-trackers show stronger goal-tracking, just as sign-trackers show stronger sign-tracking. Behav Brain Res 230:399–408. doi:10.1016/j.bbr.2012.02.032

DiFeliceantonio AG, Mabrouk OS, Kennedy RT, Berridge KC (2012) Enkephalin surges in dorsal neostriatum as a signal to eat. Curr Biol 22:1918–1924. doi:10.1016/j.cub.2012.08.014

Dow Schüll N (2012) Addiction by Design: Machine Gambling in Las Vegas, 1st edn. Princeton University Press, Princeton

Elliott SS, Keim NL, Stern JS et al (2002) Fructose, weight gain, and the insulin resistance syndrome. Am J Clin Nutr 76:911–922

Evans AH, Pavese N, Lawrence AD et al (2006) Compulsive drug use linked to sensitized ventral striatal dopamine transmission. Ann Neurol 59:852–858. doi:10.1002/ana.20822

Everitt BJ, Belin D, Economidou D et al (2008) Neural mechanisms underlying the vulnerability to develop compulsive drug-seeking habits and addiction. Philos Trans R Soc, Biol Sci 363:3125–3135

Farooqi IS, Bullmore ET, Keogh J et al (2007) Leptin regulates striatal regions and human eating behavior. Science 317:1355. doi:10.1126/science.1144599

Farooqi IS, O'Rahilly S (2009) Leptin: a pivotal regulator of human energy homeostasis. Am J Clin Nutr 89:980S–984S. doi:10.3945/ajcn.2008.26788C

Ferrario CR, Gorny G, Crombag HS et al (2005) Neural and behavioral plasticity associated with the transition from controlled to escalated cocaine use. Biol Psychiatry 58:751–759. doi:10.1016/j.biopsych.2005.04.046

Ferrario CR, Robinson TE (2007) Amphetamine pretreatment accelerates the subsequent escalation of cocaine self-administration behavior. Eur Neuropsychopharmacol 17:352–357. doi:10.1016/j.euroneuro.2006.08.005

Fiorillo CD (2011) Transient activation of midbrain dopamine neurons by reward risk. Neuroscience 197:162–171. doi:10.1016/j.neuroscience.2011.09.037

Fiorillo CD, Tobler PN, Schultz W (2003) Discrete coding of reward probability and uncertainty by dopamine neurons. Science 299:1898–1902. doi:10.1126/science.1077349

Fischman MW, Foltin RW (1992) Self-administration of cocaine by humans: a laboratory perspective. Ciba Found Symp 166:165–180

Garcia J, Lasiter PS, Bermudez-Rattoni F, Deems DA (1985) A general theory of aversion learning. Ann N Y Acad Sci 443:8–21

Garcia-Keller C, Martinez SA, Esparza MA et al (2013) Cross-sensitization between cocaine and acute restraint stress is associated with sensitized dopamine but not glutamate release in the nucleus accumbens. Eur J Neurosci 37:982–995. doi:10.1111/ejn.12121

Gearhardt A, Roberts M, Ashe M (2013) If sugar is addictive…what does it mean for the law? J Law Med Ethics 41(Suppl 1):46–49. doi:10.1111/jlme.12038

Gearhardt AN, Davis C, Kuschner R, Brownell KD (2011) The addiction potential of hyperpalatable foods. CDAR 4:140–145. doi:10.2174/1874473711104030140

Gekht AB, Polunina AG, Briun EA, Gusev EI (2003) Neurological disturbances in heroin addicts in acute withdrawal and early post-abstinence periods. Zh Nevrol Psikhiatr Im S S Korsakova 103:9–15

Genn RF, Ahn S, Phillips AG (2004) Attenuated dopamine efflux in the rat nucleus accumbens during successive negative contrast. Behav Neurosci 118:869–873. doi:10.1037/0735-7044.118.4.869

Goldstein RZ, Craig ADB, Bechara A et al (2009) The neurocircuitry of impaired insight in drug addiction. Trends Cognit Sci 13:372–380. doi:10.1016/j.tics.2009.06.004

Graham K, Vidal-Zeballos D (1998) Analyses of use of tranquilizers and sleeping pills across five surveys of the same population (1985–1991): the relationship with gender, age and use of other substances. Soc Sci Med 46:381–395

Grill HJ, Norgren R (1978) The taste reactivity test. II. Mimetic responses to gustatory stimuli in chronic thalamic and chronic decerebrate rats. Brain Res 143:281–297

Harris JL, Bargh JA, Brownell KD (2009) Priming effects of television food advertising on eating behavior. Health Psychol 28:404–413. doi:10.1037/a0014399

Hart C, Ward A, Haney M et al (2001) Methamphetamine self-administration by humans. Psychopharmacology 157:75–81. doi:10.1007/s002130100738

Hearst ES, Jenkins HM (1974) Sign tracking: the stimulus-reinforcer relation and directed action. Psychonmic Soc, Austin

Henry DJ, White FJ (1991) Repeated cocaine administration causes persistent enhancement of D1 dopamine receptor sensitivity within the rat nucleus accumbens. J Pharmacol Exp 258:882–890

Hernandez L, Hoebel BG (1988) Feeding and hypothalamic stimulation increase dopamine turnover in the accumbens. Physiol Behav 44:599–606

Hiroi N, Brown JR, Haile CN et al (1997) FosB mutant mice: loss of chronic cocaine induction of fos-related proteins and heightened sensitivity to cocaine's psychomotor and rewarding effects.

Proc Natl Acad Sci USA 94:10397–10402. doi:10.2307/43218?ref=no-x-route: 2c8ba5cf31f36df291c5a2097b853938

Hobbs M, Remington B, Glautier S (2005) Dissociation of wanting and liking for alcohol in humans: a test of the incentive-sensitisation theory. Psychopharmacology 178:493–499. doi:10.1007/s00213-004-2026-0

Hollis KL (1984) The biological function of Pavlovian conditioning: the best defense is a good offense. J Exp Psychol Anim Behav Process 10:413–425

Holmes NM, Marchand AR, Coutureau E (2010) Pavlovian to instrumental transfer: a neurobehavioural perspective. Neurosci Biobehav Rev 34:1277–1295. doi:10.1016/j.neubiorev.2010.03.007

Horger BA, Giles MK, Schenk S (1992) Preexposure to amphetamine and nicotine predisposes rats to self-administer a low dose of cocaine. Psychopharmacology 107:271–276

Hu X-T, Koeltzow TE, Cooper DC et al (2002) Repeated ventral tegmental area amphetamine administration alters dopamine D1 receptor signaling in the nucleus accumbens. Synapse 45:159–170. doi:10.1002/syn.10095

Hunt WA, Barnett LW, Branch LG (1971) Relapse rates in addiction programs. J Clin Psychol 27:455–456

Hyman SE, Malenka RC, Nestler EJ (2006) Neural mechanisms of addiction: the role of reward-related learning and memory. Annu Rev Neurosci 29:565–598. doi:10.1146/annurev.neuro.29.051605.113009

Isomura T, Suzuki J, Murai T (2014) Paradise Lost: The relationships between neurological and psychological changes in nicotine-dependent patients. Addict Res Theor 22:158–165. doi:10.3109/16066359.2013.793312

James W (1884) What is an Emotion? Mind 9:188–205. doi:10.2307/2246769?ref=no-x-route: 661c887760fcf4a1f23afb46f8f75b0a

Jiang T, Soussignan R, Schaal B, Royet J-P (2014) Reward for food odors: an fMRI study of liking and wanting as a function of metabolic state and BMI. Soc Cogn Affect Neurosci. doi:10.1093/scan/nsu086

Johnson RJ, Segal MS, Sautin Y et al (2007) Potential role of sugar (fructose) in the epidemic of hypertension, obesity and the metabolic syndrome, diabetes, kidney disease, and cardiovascular disease. Am J Clin Nutr 86:899–906

Joutsa J, Johansson J, Niemelä S et al (2012) Mesolimbic dopamine release is linked to symptom severity in pathological gambling. Neuroimage 60:1992–1999. doi:10.1016/j.neuroimage.2012.02.006

Kalivas PW, Duffy P (1990) Effect of acute and daily cocaine treatment on extracellular dopamine in the nucleus accumbens. Synapse 5:48–58. doi:10.1002/syn.890050104

Kalivas PW, Duffy P (1993) Time course of extracellular dopamine and behavioral sensitization to cocaine. I. Dopamine axon terminals. J Neurosci off J Soc Neurosci 13:266–275

Kelly B, Hattersley L, King L, Flood V (2008) Persuasive food marketing to children: use of cartoons and competitions in Australian commercial television advertisements. Health Promot Int 23:337–344. doi:10.1093/heapro/dan023

Khantzian EJ (1985) The self-medication hypothesis of addictive disorders: focus on heroin and cocaine dependence. Am J Psychiatry 142:1259–1264

Khavari KA, Peters TC, Baity PL, Wilson AS (1975) Voluntary morphine ingestion, morphine dependence, and recovery from withdrawal signs. Pharmacol Biochem Behav 3:1093–1096

Kirkham T (2008) Endocannabinoids and the neurochemistry of gluttony. J Neuroendocrinol 20:1099–1100. doi:10.1111/j.1365-2826.2008.01762.x

Kirkham TC (2005) Endocannabinoids in the regulation of appetite and body weight. Behav Pharmacol 16:297–313

Kleber HD (2007) Pharmacologic treatments for opioid dependence: detoxification and maintenance options. Dialogues Clin Neurosci 9:455–470

Koob GF (1996) Drug addiction: the yin and yang of hedonic homeostasis. Neuron 16:893–896

Koob GF, Stinus L, le Moal M, Bloom FE (1989) Opponent process theory of motivation: neurobiological evidence from studies of opiate dependence. Neurosci Biobehav 13:135–140

Koob GF, Volkow ND (2010) Neurocircuitry of addiction. Neuropsychopharmacology 35:217–238. doi:10.1038/npp.2009.110

Lamb RJ, Preston KL, Schindler CW et al (1991) The reinforcing and subjective effects of morphine in post-addicts: a dose-response study. J Pharmacol Exp Ther 259:1165–1173

Lemmens SGT, Schoffelen PFM, Wouters L et al (2009) Eating what you like induces a stronger decrease of "wanting" to eat. Physiol Behav 98:318–325. doi:10.1016/j.physbeh.2009.06.008

Lenoir M, Serre F, Cantin L, Ahmed SH (2007) Intense sweetness surpasses cocaine reward. PLoS ONE 2:e698. doi:10.1371/journal.pone.0000698

Leyton M (2010) The neurobiology of desire: dopamine and the regulation of mood and motivational states in humans. In: Kringelbach ML, Berridge KC (eds) Pleasures of the Brain. Oxford University Press, New York, pp 222–243

Leyton M (2007) Conditioned and sensitized responses to stimulant drugs in humans. Prog Neuropsychopharmacol Biol Psychiatry 31:1601–1613. doi:10.1016/j.pnpbp.2007.08.027

Leyton M (2014) What's deficient in reward deficiency? J Psychiatry Neurosci 39:291–293

Leyton M, Boileau I, Benkelfat C et al (2002) Amphetamine-induced increases in extracellular dopamine, drug wanting, and novelty seeking: a PET/[11C]raclopride study in healthy men. Neuropsychopharmacology 27:1027–1035. doi:10.1016/S0893-133X(02)00366-4

Leyton M, Casey KF, Delaney JS et al (2005) Cocaine craving, euphoria, and self-administration: a preliminary study of the effect of catecholamine precursor depletion. Behav Neurosci 119:1619–1627. doi:10.1037/0735-7044.119.6.1619

Leyton M, Vezina P (2012) On cue: striatal ups and downs in addictions. Biol Psychiatry 72:e21–e22. doi:10.1016/j.biopsych.2012.04.036

Leyton M, Vezina P (2014) Dopamine ups and downs in vulnerability to addictions: a neurodevelopmental model. Trends Pharmacol Sci 35:268–276. doi:10.1016/j.tips.2014.04.002

Leyton M, Young SN, Blier P et al (2000) Acute tyrosine depletion and alcohol ingestion in healthy women. Alcohol Clin Exp Res 24:459–464

Linnet J, Møller A, Peterson E et al (2011) Dopamine release in ventral striatum during Iowa Gambling Task performance is associated with increased excitement levels in pathological gambling. Addiction 106:383–390. doi:10.1111/j.1360-0443.2010.03126.x

Linnet J, Peterson E, Doudet DJ et al (2010) Dopamine release in ventral striatum of pathological gamblers losing money. Acta Psychiatr Scand 122:326–333. doi:10.1111/j.1600-0447.2010.01591.x

Litt A, Khan U, Shiv B (2010) Lusting while loathing: parallel counterdriving of wanting and liking. Psychol Sci 21:118–125. doi:10.1177/0956797609355633

Mahler SV, Berridge KC (2009) Which cue to "want?" Central amygdala opioid activation enhances and focuses incentive salience on a prepotent reward cue. J Neurosci 29:6500–6513. doi:10.1523/JNEUROSCI.3875-08.2009

Mendelson JH, Sholar M, Mello NK et al (1998) Cocaine tolerance: behavioral, cardiovascular, and neuroendocrine function in men. Neuropsychopharmacology 18:263–271. doi:10.1016/S0893-133X(97)00146-2

Mick I, Myers J, Stokes PRA, et al. (2014) Endogenous opioid release in pathological gamblers after an oral amphetamine challenge: a [^{11}C] carfentanil pet study. Eur Neuropsychopharmacol 1–4

Miedl SF, Peters J, Büchel C (2012) Altered neural reward representations in pathological gamblers revealed by delay and probability discounting. Arch Gen Psychiatry 69:177–186. doi:10.1001/archgenpsychiatry.2011.1552

Munafò MR, Mannie ZN, Cowen PJ et al (2007) Effects of acute tyrosine depletion on subjective craving and selective processing of smoking-related cues in abstinent cigarette smokers. J Psychopharmacol (Oxford) 21:805–814. doi:10.1177/0269881107077216

Murdaugh D, Cook E (2012) fMRI reactivity to high-calorie food pictures predicts short- and long-term outcome in a weight-loss program. Neuroimage

Nijs IMT, Muris P, Euser AS, Franken IHA (2010) Differences in attention to food and food intake between overweight/obese and normal-weight females under conditions of hunger and satiety. Appetite 54:243–254. doi:10.1016/j.appet.2009.11.004

Nilsson J, Kristiansen TS, Fosseidengen JE et al (2008) Sign- and goal-tracking in Atlantic cod (Gadus morhua). Animal Behavior 11:651–659. doi:10.1007/s10071-008-0155-2

Ostafin BD, Marlatt GA, Troop-Gordon W (2010) Testing the incentive-sensitization theory with at-risk drinkers: wanting, liking, and alcohol consumption. Psychol Addict Behav 24:157–162. doi:10.1037/a0017897

Pandit R, de Jong JW, Vanderschuren LJMJ, Adan RAH (2011) Neurobiology of overeating and obesity: the role of melanocortins and beyond. Eur J Pharmacol 660:28–42. doi:10.1016/j.ejphar.2011.01.034

Paulson PE, Camp DM, Robinson TE (1991) Time course of transient behavioral depression and persistent behavioral sensitization in relation to regional brain monoamine concentrations during amphetamine withdrawal in rats. Psychopharmacology 103:480–492. doi:10.1007/BF02244248

Paxinos G, Watson C (2007) The rat brain in stereotaxic coordinates, 6 edn. Elsevier, Amsterdam

Peciña S (2005) Hedonic hot spot in nucleus accumbens shell: where do μ-opioids cause increased hedonic impact of sweetness? J Neurosci 25:11777–11786. doi:10.1523/JNEUROSCI.2329-05.2005

Peciña S, Berridge KC (2013) Dopamine or opioid stimulation of nucleus accumbens similarly amplify cue-triggered "wanting" for reward: entire core and medial shell mapped as substrates for PIT enhancement. Eur J Neurosci. doi:10.1111/ejn.12174

Peciña S, Berridge KC, Parker LA (1997) Pimozide does not shift palatability: separation of anhedonia from sensorimotor suppression by taste reactivity. Pharmacol Biochem Behav 58:801–811

Peciña S, Cagniard B, Berridge KC et al (2003) Hyperdopaminergic mutant mice have higher "wanting" but not 'liking' for sweet rewards. J Neurosci 23:9395–9402

Peciña S, Schulkin J, Berridge KC (2006) Nucleus accumbens corticotropin-releasing factor increases cue-triggered motivation for sucrose reward: paradoxical positive incentive effects in stress? BMC Biol 4:8. doi:10.1186/1741-7007-4-8

Pfaus JG, Damsma G, Nomikos GG et al (1990) Sexual behavior enhances central dopamine transmission in the male rat. Brain Res 530:345–348

Piazza PV, Deminiere JM, le Moal M, Simon H (1990) Stress- and pharmacologically-induced behavioral sensitization increases vulnerability to acquisition of amphetamine self-administration. Brain Res 514:22–26

Potenza MN (2008) The neurobiology of pathological gambling and drug addiction: an overview and new findings. Philos Trans Biol Sci 363:3181–3189

Purdy JE, Roberts AC, Garcia CA (1999) Sign tracking in cuttlefish (Sepia officinalis). J Comp Psychol 113:443–449

Reilly S, Schachtman TR (2009) Conditioned taste aversion: behavioral and neural processes. Oxford University Press, New York

Robbins TW, Watson BA, Gaskin M, Ennis C (1983) Contrasting interactions of pipradrol, d-amphetamine, cocaine, cocaine analogues, apomorphine and other drugs with conditioned reinforcement. Psychopharmacology 80:113–119

Robinson TE, Berridge KC (1993) The neural basis of drug craving: an incentive-sensitization theory of addiction. Brain Res Brain Res Rev 18:247–291

Robinson TE, Berridge KC (2003) Addiction. Annu Rev Psychol 54:25–53. doi:10.1146/annurev.psych.54.101601.145237

Robinson TE, Berridge KC (2008) The incentive sensitization theory of addiction: some current issues. Philos Trans R Soc, Biol Sci 363:3137–3146. doi:10.1098/rstb.2008.0093

Robinson TE, Jurson PA, Bennett JA, Bentgen KM (1988) Persistent sensitization of dopamine neurotransmission in ventral striatum (nucleus accumbens) produced by prior experience with (+)-amphetamine: a microdialysis study in freely moving rats. Brain Res 462:211–222

Robinson S, Sandstrom SM, Denenberg VH, Palmiter RD (2005) Distinguishing whether dopamine regulates liking, wanting, and/or learning about rewards. Behav Neurosci 119:5–15. doi:10.1037/0735-7044.119.1.5

Robinson MJF, Robinson TE, Berridge KC (2013) Incentive salience and the transition to addiction. Elsevier, Amsterdam, pp 391–399

Robinson MJF, Anselme P, Fischer AM, Berridge KC (2014a) Initial uncertainty in Pavlovian reward prediction persistently elevates incentive salience and extends sign-tracking to normally unattractive cues. Behav Brain Res 266:119–130. doi:10.1016/j.bbr.2014.03.004

Robinson MJF, Warlow SM, Berridge KC (2014b) Optogenetic excitation of central amygdala amplifies and narrows incentive motivation to pursue one reward above another. J Neurosci 34:16567–16580. doi:10.1523/JNEUROSCI.2013-14.2014

Robinson MJF, Anselme P, Suchomel K, Berridge KC (2015a) Amphetamine-induced sensitization and reward uncertainty similarly enhance incentive salience for conditioned cues. Behav Neurosci. doi:10.1037/bne0000064

Robinson MJF, Burghardt PR, Patterson CM et al (2015b) Individual differences in cue-induced motivation and striatal systems in rats susceptible to diet-induced obesity. Neuropsychopharmacology, epub ahead of print:1–11. doi:10.1038/npp.2015.71

Robinson TE, Yager LM, Cogan ES, Saunders BT (2014c) On the motivational properties of reward cues: individual differences. Neuropharmacology 76 Pt B:450–459. doi:10.1016/j.neuropharm.2013.05.040

Rose JE, Behm FM, Westman EC, Johnson M (2000) Dissociating nicotine and nonnicotine components of cigarette smoking. Pharmacol Biochem Behav 67:71–81

Rosse RB, Fay-McCarthy M, Collins JP et al (1993) Transient compulsive foraging behavior associated with crack cocaine use. Am J Psychiatry 150:155–156

Rozin P (2000) Disgust. In: Lewis M, Haviland-Jones JM (eds) Handbook of emotions. Guilford, New York, pp 637–653

Rømer Thomsen K, Fjorback LO, Møller A, Lou HC (2014) Applying incentive sensitization models to behavioral addiction. Neurosci Biobehav Rev 45C:343–349. doi:10.1016/j.neubiorev.2014.07.009

Sáinz N, Barrenetxe J, Moreno-Aliaga MJ, Martínez JA (2015) Leptin resistance and diet-induced obesity: central and peripheral actions of leptin. Metab, Clin Exp 64:35–46. doi:10.1016/j.metabol.2014.10.015

Schulte EM, Avena NM, Gearhardt AN (2015) Which foods may be addictive? The roles of processing, fat content, and glycemic load. PLoS ONE 10:e0117959. doi:10.1371/journal.pone.0117959

Shin AC, Townsend RL, Patterson LM, Berthoud H-R (2011) "Liking" and "wanting" of sweet and oily food stimuli as affected by high-fat diet-induced obesity, weight loss, leptin, and genetic predisposition. AJP: Regulatory. Integr Comp Physiol 301:R1267–R1280. doi:10.1152/ajpregu.00314.2011

Shuster L, Webster GW, Yu G (1975) Increased running response to morphine in morphine-pretreated mice. J Pharmacol Exp Ther 192:64–67

Shuster L, Yu G, Bates A (1977) Sensitization to cocaine stimulation in mice. Psychopharmacology 52:185–190

Singer BF, Scott-Railton J, Vezina P (2012) Unpredictable saccharin reinforcement enhances locomotor responding to amphetamine. Behav Brain Res 226:340–344. doi:10.1016/j.bbr.2011.09.003

Small DM, Zatorre RJ, Dagher A et al (2001) Changes in brain activity related to eating chocolate: from pleasure to aversion. Brain 124:1720–1733

Smith KS, Berridge KC (2007) Opioid limbic circuit for reward: interaction between hedonic hotspots of nucleus accumbens and ventral pallidum. J Neurosci 27:1594–1605. doi:10.1523/JNEUROSCI.4205-06.2007

Smith KS, Mahler SV, Peciña S, Berridge KC (2007) Hedonic hotspots: Generating sensory pleasure in the brain. In: Kringelbach ML, Berridge KC (eds) Pleasures of the brain. Oxford University Press, Oxford, pp 1–35

Steiner JE, Glaser D, Hawilo ME, Berridge KC (2001) Comparative expression of hedonic impact: affective reactions to taste by human infants and other primates. Neurosci Biobehav Rev 25:53–74

Stinus L, Robert C, Karasinski P, Limoge A (1998) Continuous quantitative monitoring of spontaneous opiate withdrawal: locomotor activity and sleep disorders. Pharmacol Biochem Behav 59:83–89

Stolerman IP (1985) Motivational effects of opioids: evidence on the role of endorphins in mediating reward or aversion. Pharmacol Biochem Behav 23:877–881

Swanson LW (2000) Cerebral hemisphere regulation of motivated behavior. Brain Res 886:113–164

Swanson LW (2005) Anatomy of the soul as reflected in the cerebral hemispheres: neural circuits underlying voluntary control of basic motivated behaviors. J Comp Neurol 493:122–131. doi:10.1002/cne.20733

Tang DW, Fellows LK, Small DM, Dagher A (2012) Food and drug cues activate similar brain regions: a meta-analysis of functional MRI studies. Physiol Behav 106:317–324. doi:10.1016/j.physbeh.2012.03.009

Tindell AJ, Berridge KC, Zhang J et al (2005) Ventral pallidal neurons code incentive motivation: amplification by mesolimbic sensitization and amphetamine. Eur J Neurosci 22:2617–2634. doi:10.1111/j.1460-9568.2005.04411.x

van Holst RJ, Veltman DJ, Büchel C et al (2012) Distorted expectancy coding in problem gambling: is the addictive in the anticipation? Biol Psychiatry 71:741–748. doi:10.1016/j.biopsych.2011.12.030

Vanderschuren LJ, De Vries TJ, Wardeh G et al (2001) A single exposure to morphine induces long-lasting behavioural and neurochemical sensitization in rats. Eur J Neurosci 14:1533–1538

Vanderschuren LJ, Kalivas PW (2000) Alterations in dopaminergic and glutamatergic transmission in the induction and expression of behavioral sensitization: a critical review of preclinical studies. Psychopharmacology 151:99–120

Vezina P (1993) Amphetamine injected into the ventral tegmental area sensitizes the nucleus accumbens dopaminergic response to systemic amphetamine: an in vivo microdialysis study in the rat. Brain Res 605:332–337

Vezina P (2004) Sensitization of midbrain dopamine neuron reactivity and the self-administration of psychomotor stimulant drugs. Neurosci Biobehav Rev 27:827–839. doi:10.1016/j.neubiorev.2003.11.001

Vezina P, Leyton M (2009) Conditioned cues and the expression of stimulant sensitization in animals and humans. Neuropharmacology 56(Suppl 1):160–168. doi:10.1016/j.neuropharm.2008.06.070

Volkow ND, Wang G-J, Fowler JS et al (2002) "Nonhedonic" food motivation in humans involves dopamine in the dorsal striatum and methylphenidate amplifies this effect. Synapse 44:175–180. doi:10.1002/syn.10075

Volkow ND, Wang GJ, Fowler JS et al (1997) Decreased striatal dopaminergic responsiveness in detoxified cocaine-dependent subjects. Nature 386:830–833. doi:10.1038/386830a0

Wachtel SR, Ortengren A, de Wit H (2002) The effects of acute haloperidol or risperidone on subjective responses to methamphetamine in healthy volunteers. Drug Alcohol Depend 68:23–33

West R (2009) The multiple facets of cigarette addiction and what they mean for encouraging and helping smokers to stop. COPD 6:277–283. doi:10.1080/15412550903049181

Wikler A (1973) Dynamics of drug dependence: Implications of a conditioning theory for research and treatment. Arch Gen Psychiatry 28:611–616

Wilkinson CJ (1998) The abuse potential of zolpidem administered alone and with alcohol. Pharmacol Biochem Behav 60:193–202

Winkielman P, Berridge KC (2003) Irrational wanting and subrational liking: how rudimentary motivational and affective processes shape preferences and choices. Polit Psychol 24:657–680. doi:10.2307/3792260?ref=no-x-route:d779aaa6e949d52d4e89ed4f70b1996e

Winkielman P, Berridge KC, Wilbarger JL (2005) Unconscious affective reactions to masked happy versus angry faces influence consumption behavior and judgments of value. Pers Soc Psychol Bull 31:121–135. doi:10.1177/0146167204271309

Wise RA (1982) Neuroleptics and operant behavior: The anhedonia hypothesis. Behav Brain Sci 5:39–53

Woolverton WL, Cervo L, Johanson CE (1984) Repeated methamphetamine administration on methamphetamine self-administration in rhesus monkeys. Pharmacol Biochem Behav 21:737–741. doi:10.1016/S0091-3057(84)80012-X

Wyvell CL, Berridge KC (2000) Intra-accumbens amphetamine increases the conditioned incentive salience of sucrose reward: enhancement of reward "wanting" without enhanced "liking" or response reinforcement. J Neurosci 20:8122–8130

Wyvell CL, Berridge KC (2001) Incentive sensitization by previous amphetamine exposure: increased cue-triggered "wanting" for sucrose reward. J Neurosci 21:7831–7840

Yokum S, Gearhardt AN, Harris JL, et al. (2014) Individual differences in striatum activity to food commercials predict weight gain in adolescents. Obesity doi:10.1002/oby.20882

Yokum S, Ng J, Stice E (2011) Attentional bias to food images associated with elevated weight and future weight gain: an fMRI study. Obesity 19:1775–1783. doi:10.1038/oby.2011.168

Zack M, Featherstone RE, Mathewson S, Fletcher PJ (2014) Chronic exposure to a gambling-like schedule of reward predictive stimuli can promote sensitization to amphetamine in rats. Front Behav Neurosci 8:36. doi:10.3389/fnbeh.2014.00036

Zhang J, Berridge KC, Tindell AJ et al (2009) A neural computational model of incentive salience. PLoS Comput Biol 5:e1000437. doi:10.1371/journal.pcbi.1000437

Circadian Insights into Motivated Behavior

Michael C. Antle and Rae Silver

> 'Time' he said, 'is what keeps everything from happening at once'.
>
> Ray Cummings, *The Girl in the Golden Atom*, 1922.

Abstract For an organism to be successful in an evolutionary sense, it and its offspring must survive. Such survival depends on satisfying a number of needs that are driven by motivated behaviors, such as eating, sleeping, and mating. An individual can usually only pursue one motivated behavior at a time. The circadian system provides temporal structure to the organism's 24 hour day, partitioning specific behaviors to particular times of the day. The circadian system also allows anticipation of opportunities to engage in motivated behaviors that occur at predictable times of the day. Such anticipation enhances fitness by ensuring that the organism is physiologically ready to make use of a time-limited resource as soon as it becomes available. This could include activation of the sympathetic nervous system to transition from sleep to wake, or to engage in mating, or to activate of the parasympathetic nervous system to facilitate transitions to sleep, or to prepare the body to digest a meal. In addition to enabling temporal partitioning of motivated behaviors, the circadian system may also regulate the amplitude of the drive state

---

M.C. Antle (✉)
Department of Psychology, University of Calgary, Calgary, AB, Canada
e-mail: antlem@ucalgary.ca

M.C. Antle
Department of Physiology and Pharmacology, University of Calgary, Calgary, AB, Canada

M.C. Antle
Hotchkiss Brain Institute, University of Calgary, Calgary, AB, Canada

R. Silver
Department of Psychology, Barnard College, New York, NY, USA
e-mail: qr@columbia.edu

R. Silver
Department of Psychology, Columbia University, New York, NY, USA

R. Silver
Department of Pathology and Cell Biology, Columbia University, New York, NY, USA

© Springer International Publishing Switzerland 2015

motivating the behavior. For example, the circadian clock modulates not only when it is time to eat, but also how hungry we are. In this chapter we explore the physiology of our circadian clock and its involvement in a number of motivated behaviors such as sleeping, eating, exercise, sexual behavior, and maternal behavior. We also examine ways in which dysfunction of circadian timing can contribute to disease states, particularly in psychiatric conditions that include adherent motivational states.

Keywords Rhythms · Suprachiasmatic nucleus · Anticipation · Eating · Exercise · Sleep · Sexual behavior · Maternal behavior · Amplitude

Contents

1 Circadian Considerations for Motivated Behaviors	139
1.1 Goal-Directed and Arousal Aspects of Motivation	139
1.2 Circadian Rhythms Impact Goal-Directed Motivation	140
1.3 Circadian Rhythms Impact Arousal Components of Motivation	141
1.4 Distinguishing Circadian Arousal and Goal-Directed Components of Behavior	142
2 Cellular, Molecular, and Network Basis of SCN Circadian Timing	143
2.1 Cellular Oscillation	144
2.2 Amplitude of SCN Oscillation	144
2.3 SCN Afferents and Efferents	146
2.4 Circadian Organization of Motivation	146
3 Circadian Regulation of Sleep–Wake/Arousal Cycles	146
3.1 Homeostatic Regulation of Sleep: Adenosine	147
3.2 Circadian Modulation of Sleep	147
3.3 Feedback of Sleep to Circadian System	148
4 Exercise/Activity	148
4.1 Locomotor Activity is Rewarding	148
4.2 Feedback of Exercise to Circadian System	149
4.3 Activity Influences Other Motivated Behaviors	150
5 Eating	150
5.1 Feeding Duration	151
5.2 Peripheral Factors Determining Eating	151
5.3 Anticipatory Behavior Entails Circadian Timing	151
5.4 Extra-SCN Circadian Oscillators Support Anticipatory Behaviors	152
5.5 Oscillators Mediating Anticipation of Feeding	152
5.6 Brain Oscillators	152
5.7 Peripheral Oscillators	153
5.8 Disrupted Circadian Control: Night Eating Syndrome	153
6 Mating/Sex	154
6.1 Sexual Behavior is Highly Motivated and is Under Temporal Control	154
6.2 Interaction Between Sex and Other Motivated Behaviors	155
6.3 Anticipatory Activity	156
6.4 Feedback of Sexual Behavior to the Circadian System	157
7 Maternal Behavior	157
7.1 Feedback of Maternal Behavior on the Circadian Clock	157

7.2	Neural Basis of Anticipation of Maternal Care	158
7.3	Cellular Clock and Maternal Care	158
8	Circadian Basis of Motivation Disorders	158
8.1	Depression and Schizophrenia	159
8.2	Attention Deficit Hyperactivity Disorder	159
8.3	Circadian Considerations for Treatment of Psychiatric Conditions	160
9	Conclusion	160
References		161

1 Circadian Considerations for Motivated Behaviors

One might reasonably ask why circadian rhythms are important for understanding motivated behaviors. In the context of evolution, the most important consideration is that motivated behaviors must satisfy biological needs that promote the survival of individuals and their offspring. These include eating and drinking, sleeping, mating, and parenting. Such motivated behaviors cannot be performed simultaneously. The circadian timing system provides temporal organization to motivated behaviors. A major function of the circadian clock is to regulate when during the day an animal will engage in specific behaviors, and when various biological functions are more or less likely to be expressed. When to satisfy these biological needs depends on a number of factors, both intrinsic and extrinsic to the organism. An important intrinsic factor is when the body, or a specific organ within the body, is ready to make use of the resource. Key among the extrinsic factors is the availability of the resource required to satisfy the need. Resources may only be available (or safely available) during a small temporal window. In such cases, the organism must structure its behavior to anticipate this availability. With respect to motivated behaviors, we suggest that the circadian timing system plays three roles. (1) Circadian rhythms provide the temporal structure that partition different goals to different parts of the day. (2) The circadian system orchestrates physiology and metabolism so that the body can most effectively satisfy the need underlying the goal. (3) The circadian system allows for anticipation of the availability of a temporally restricted goal.

1.1 Goal-Directed and Arousal Aspects of Motivation

There are two distinct, long-recognized components of motivated behavior, namely (1) a goal-directed, directional component and (2) an arousal, activational component (Hebb 1955; Duffy 1957; Salamone 1988)—and both are impacted by an internal circadian timing system. More specifically, Hebb (1955), in his beautifully worded paper on the "Conceptual Nervous System," written in an era when "neurologizing" was verboten, argues that "Motivation" refers in a rather general sense to the energizing of behavior, and especially to the sources of energy in a particular set of responses that keep them temporarily dominant over others and account for continuity and direction in behavior. Duffy (1957) remains on the behavioral level of analysis, and argues that "all variations in behavior may be described as variations in either the direction of behavior or the intensity of behavior" and that "confusion of the direction of behavior with the intensity of behavior, resulting in their fortuitous combination in certain psychological concepts..." (p. 256, Duffy 1957). Restated: Duffy suggests that response amplitude and goal direction are confounded in behavioral analyses of motivation. Today, "neurologizing" has become mandatory, and there is an increased emphasis on the neurobiological basis of motivation in studies of both non-human and human mammals. While much research effort has advanced the understanding of the neural and neurochemical basis of motivated behaviors (Chap. Oxytocin, Vasopressin, and the Motivational Forces that Drive Social Behaviors of this volume) (Gore and Zweifel 2013; Trifilieff et al. 2013), there has been limited progress in dissociating the activational and directional components (Bailey et al. 2015). Studies of goal-directed action selection and general arousal tend to examine each of these aspects in isolation, as separate central nervous system entities, with the former focusing on reward circuits (e.g., Richard et al. 2013) and the latter on arousal pathways (e.g., Pfaff et al. 2012). Behavioral paradigms that are used to assess models, drugs, environmental conditions, etc., generally do not distinguish between directional and arousal components. Experiments designed to examine the neural circuitry of goal-directed action selection and general arousal tend to consider each behavior and circuit in isolation, without consideration for their interrelationships on either a neural or behavioral level. Understanding the neural basis of motivation is considered especially important in understanding the symptomatology of depression, schizophrenia, and other emotional and affective disorders that have among their symptoms alterations in motivation and in circadian timing. As is discussed below, the circadian timing system impacts both arousal and goal-directed components of behavior.

1.2 *Circadian Rhythms Impact Goal-Directed Motivation*

The circadian system regulates goal-directed components of motivation such that motivated behaviors are normally expressed in a coordinated, temporally appropriate fashion. This coordination of goal seeking means that specific motivated behaviors occur at species-characteristic times of day. For instance, sleeping and feeding each occur at specific times of day in most species, including humans (as in Fig. 1). Other motivated behaviors, such as mating and exercise, are partitioned around these behaviors.

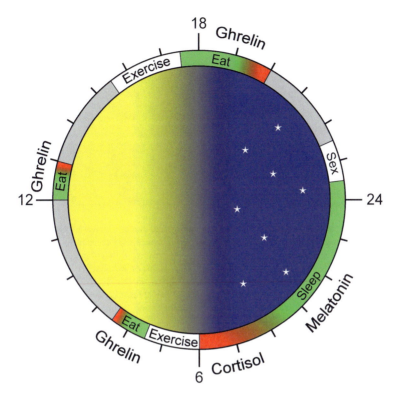

Fig. 1 The circadian clock gives temporal structure to our 24-h day. A typical North American might wake around dawn. They will engage in a number of motivated activities throughout their day, and the timing of these motivated behaviors tends to be consistent day-to-day. For instance, they will consume multiple meals at regular times, and hunger will increase over the regular mealtime until the meal is initiated (*green*-to-*red* gradient within the Eat box). The person may also engage in other motivated behaviors such as exercise or sex, and the timing of these behaviors may become habitual. Most people will have a particular phase angle of activity relative to dawn/dusk, with an average adult not initiating their major sleep bout until many hours after dusk, although this may be much earlier in "morning larks" and much later in "night-owls." Some people may also have a minor sleep bout (nap) mid day. The motivation to sleep will increase near the person's normal sleep time and continue to increase during their normal sleep period until sleep is initiated (*green*-to-*red* gradient in the sleep box). Many of these motivated behaviors are linked closely to neuroendocrine signals such as ghrelin which triggers hunger, cortisol which facilitates waking, and melatonin which is secreted during the night and facilitates sleep in diurnal species

1.3 Circadian Rhythms Impact Arousal Components of Motivation

Arousal level is related to motivation in that it impacts how vigorously a goal is pursued (amplitude of goal seeking). The circadian clock can also influence the amplitude of the motivated behavior (Fig. 2). As an example, the duration of sleep depends on circadian phase, not just on sleepiness (Czeisler et al. 1980). Thus, in a

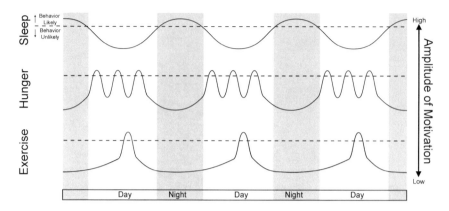

Fig. 2 An example of three behaviors (sleeping, eating, and exercise) from a hypothetical human. The desire to engage in different motivated behaviors changes over the course of the day (*solid lines*), and the timing of these behaviors can vary widely between different people. These rhythmic oscillations make the individual's motivated behaviors more or less likely. Typically, when the amplitude climbs above a certain threshold (*dashed lines*), the motivated behavior is likely to occur, and when the motivation amplitude falls below this threshold, the behavior is less likely, even in cases when the person did not engage in that behavior

classic study, monkeys were sleep deprived for various lengths of time, and recovery sleep was monitored (Klerman et al. 1999). Deprivation ended at the normal wake time for the groups with the longest durations of sleep deprivation (and thus the most tired), yet the monkeys didn't sleep! While the motivation to sleep should have been high due to homeostatic mechanisms, the circadian clock was able to override these homeostatic sleep mechanisms that were occurring at the wrong time of day, thus motivating the animal to be awake despite severe sleep loss. This illustrates the potency of the circadian clock in modifying motivation for a particular behavior. Circadian modulatory effects can also be seen in studies of cognitive performance following sleep deprivation. In a test of vigilance, subjects watch a red light and have to push a button when it turns on. Sleep deprivation beyond 17 h results in a dramatic increase in not only the latency to push the button, but also in complete misses of the light going on. Amazingly, these lapses in vigilance were under circadian control: After 8 days of sleep restricted to 4 h/day, subjects averaged about 8 more performance lapses at 0800 h than they had 10 h later at 1800 h (Mollicone et al. 2010).

1.4 Distinguishing Circadian Arousal and Goal-Directed Components of Behavior

Goal-directed and amplitude effects of the circadian system can be experimentally distinguished. This is commonly attempted in studies of desynchrony, such as occur in studies of shift work or jet lag. As an example, in an early study of individual time series (Reinberg et al. 1988), circadian period and amplitude were evaluated in order to understand tolerance to shift work, effect of age, duration of shift work, speed of rotation, and type of industry. The measures used included sleeping, working, oral temperature, grip strength of each hand, peak expiratory flow, and heart rate. Here, the timing of these responses could be analyzed separately from their amplitude. The results indicate that intolerance to shift work is associated with <u>both</u> reduced circadian amplitude and internal desynchronization of responses, such that various responses no longer occurred at appropriate/typical times of day.

Circadian rhythmicity is a clock-like process which modulates sleep propensity and many other motivated physiological and behavioral responses. The circadian system modulates virtually all physiological and behavioral responses by generating an oscillatory signal every day. Sleep (and other goal-directed, motivated responses) is normally expressed at species characteristic, specific times of day. They remain consolidated and appropriately timed with respect to each other, if the amplitude of the circadian signal is sufficiently strong. In this sense, the circadian system determines both the timing of goal-directed motivation and the amplitude or strength of responses.

2 Cellular, Molecular, and Network Basis of SCN Circadian Timing

In mammals, the suprachiasmatic nucleus (SCN) of the hypothalamus functions as the master circadian clock produces daily rhythms in physiology and behavior and synchronizes them to the environmental day/night cycle (Antle and Silver 2005). In environments lacking all external cues to time, these daily oscillations in behavior and physiology persist, albeit with a period slight different than 24 h. These endogenously generated rhythms are said to be free-running rhythms, as the internal biological clock is running free of any external time cues. When the SCN is lesioned, all circadian rhythms in physiology and behavior are lost (Moore and Eichler 1972; Stephan and Zucker 1972). Remarkably, these rhythms can be restored by transplanting a fetal SCN graft (Lehman et al. 1987), and the period of the restored rhythm is that of the donor rather the host (Ralph et al. 1990). Such studies prove that the SCN is necessary and sufficient for daily oscillations in behavior and physiology.

The SCN tissue is a multicellular oscillator in which many of its $\sim 20{,}000$ individual neurons function as cell-autonomous oscillators. These cells are networked together to produce a coherent tissue-level oscillation (Welsh et al. 2010). While numerous brain regions and body organs can exhibit circadian rhythmicity derived from the same molecular underpinnings (see below), they lack the network properties to maintain tissue-level oscillations in the long-term without organizing signals from the master circadian clock in the SCN, or from other exogenous rhythmic cues such as scheduled feeding. In contrast, the SCN continues to oscillate in the absence of phase-setting cues.

2.1 Cellular Oscillation

Circadian oscillation at the level of the individual cell emerges from interlocking positive and negative feedback loops in the transcription and translation of a number of circadian genes (Robinson and Reddy 2014). Briefly, the expression of clock genes is under control of E-box elements in their promoter regions. The E-boxes are activated by a dimer of CLOCK and BMAL1 proteins, leading to the translation of *period* and *cryptochrome* genes during the day. Translation into their corresponding proteins occurs with a time lag. The PERIOD and CRYPTOCHROME protein products of these genes then dimerize and return to the nucleus where they inhibit the activity of CLOCK and BMAL1, thus turning off their own expression. Levels of PERIOD and CRYPTOCHROME proteins fall steadily over the night, and when depleted, CLOCK and BMAL1 activity at E-boxes can resume, thus starting the cycle over again.

2.2 Amplitude of SCN Oscillation

The amplitude of SCN oscillation is key to achieving robust rhythmicity in physiology and behavior. A growing body of evidence suggests that coherent daily rhythms contribute to health, well-being, cognitive performance, and alertness (Ramkisoensing and Meijer 2015). In claiming that the circadian clock controls level of arousal, the suggestion is made that behavioral and physiological rhythms can have either high or low amplitude. A high-amplitude rhythm has a clear optimal peak time of expression and an equally clear amplitude trough (Fig. 3a). In contrast,

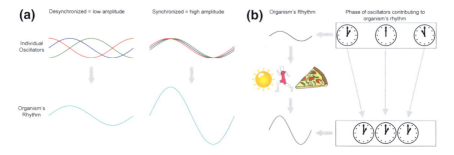

Fig. 3 **a** Circadian systems are composed of multiple oscillatory units. This could be thought of as the cells that make up an oscillating tissue, or even the various oscillating organs within an organism. Oscillators must be synchronized to yield a coherent high-amplitude output. When they are desynchronized, the overall rhythmic output exhibits diminished amplitude. **b** A number of stimuli can improve synchrony among oscillators within and between tissues. Light is the dominant cue for setting the phase of the master oscillator, but both feeding and exercise also affect oscillator function. Engaging in these motivated behaviors at regularly scheduled times may improve overall clock function and health

the absence of a rhythm points to a behavior that is not optimally expressed at its peak. As an example, both sleep disruption and aging reduce the amplitude of circadian rhythms, while exercise increases amplitude (Fig. 3b). Disruptions can also be caused by social or environmental factors, such as shift work or jet lag, or by disease states that involve circadian disruption including Parkinson's and Alzheimer's disease (Karatsoreos 2014; Lim et al. 2014; Ondo 2014; Sterniczuk et al. 2014). As such, circadian problems are often a marker of a number of diseases. Furthermore, deterioration of the 24-h rhythm increases the risk for the development or exacerbation not only of neurodegenerative disease, but also some cancers, depression, obesity, cardiovascular disease, and sleep disorders (Baron and Reid 2014; Uth and Sleigh 2014; Alibhai et al. 2015; Saini et al. 2015). Altogether, it is clear that daily oscillations in physiology, behavior, and motivation are regulated by the circadian system. In a healthy organism, the amplitude and timing of behavioral and physiological responses are both regulated by the circadian system. Circadian disruptions are a component of many compromised states.

The SCN is unique among the tissues of the body in that its clock cells form a network that is capable of sustained oscillation, even in vitro. Many of the factors determining amplitude of circadian oscillation at the cellular and system level of organization are understood. Basically, the amplitude of the SCN's electrical rhythm is high when the individual neurons in the SCN are appropriately synchronized and low when the neurons are poorly synchronized (Pauls et al. 2014; Ramkisoensing and Meijer 2015). Among the external signals that support synchrony is photic input from the eye, delivered to the SCN via a specialized retinohypothalamic tract (Schmidt et al. 2011). Greater amplitude oscillation is seen in SCN slices harvested from animals that have been housed in a light:dark cycle compared to those harvested from animals housed in constant darkness. As noted above, circadian rhythm amplitude is reduced in aging. Remarkably, transplanting young SCN tissue into the ventricle of intact aged rats or hamsters improves behavioral circadian rhythmicity (Van Reeth et al. 1994; Hurd et al. 1995). Such transplants also improve the amplitude of the circadian clock itself. Older rats have low-amplitude behavioral rhythms, and these are mirrored by low-amplitude expression rhythms of the immediate early gene FOS in the SCN. When given a fetal SCN graft, the rhythms of FOS expression in the SCN are augmented in these older animals (Cai et al. 1997). These studies provide powerful evidence of the sufficiency of SCN outputs to support robust rhythmic responses even in an aged body. In summary, the individual SCN cells form networks that sustain oscillation. Synchronization of the phases among individual oscillator cells determines the amplitude of SCN oscillation. Amplitude of SCN oscillation determines coherence of behavioral activity/rest rhythms, and age-related disruption can be corrected by providing a young SCN.

2.3 SCN Afferents and Efferents

The SCN is well positioned to send and receive information about the internal and external environment, so as to optimally coordinate timing of physiological and behavioral responses. The major inputs to the SCN arise from the eye, the raphe nuclei in the brain stem, and the intergeniculate leaflet of the thalamus, although numerous other regions also send direct projections to the SCN (Krout et al. 2002). Information regarding the internal state of the body, such as those provided by testicular and ovarian hormones, acts directly on hormone receptors within the SCN (Vida et al. 2010; Model et al. 2015) which are themselves under circadian control. In turn, the SCN sends projections to a number of areas in the brain (Kriegsfeld et al. 2004), some of which may serve as nodes to distribute circadian signals widely (Vujovic et al. in press) or integrate circadian signals with other homeostatic and sensory signals (Saper et al. 2005a).

2.4 Circadian Organization of Motivation

We review evidence that the coordination of intrinsic and extrinsic factors, and the anticipation of each of these events, is achieved by the circadian timing system. Behaviors that are critical to the survival of the organism and the species (e.g., feeding, drinking, sleeping, mating, and parenting) are highly motivated and highly motivating. In anticipation of performing these behaviors and associated physiological responses, the body prepares for their occurrence with a host of internal adjustments, coordinated by the circadian timing system. While these adjustments may not reach conscious awareness in humans, they nevertheless coordinate the expression of motivated behaviors and determine arousal levels. We therefore examine circadian modulation of a number of goal-directed, motivated behaviors, namely sleep/wake, exercise, feeding, mating, and maternal behavior. Disruption of circadian timing can produce motivational deficits and exacerbate or provoke emotional and affective disorders, and while the internal cues associated with these disruptions may not reach consciousness, their consequences do so.

3 Circadian Regulation of Sleep–Wake/Arousal Cycles

The window of time during which an organism is active, and thus the occurrence of motivated behaviors, is governed by the circadian clock, which appears to drive both sleep and wake times (Mistlberger 2005). Sleep itself is a motivated behavior that is homeostatically regulated. The drive to sleep increases with the duration of wakefulness, and animals that lose sleep will attempt to recover the loss when they are next able to sleep (Friedman et al. 1979). The interaction between the circadian and homeostatic drives to sleep has been described in the two-process model of sleep regulation (Borbély 1982). In this model, sleep and wake transition thresholds oscillate with a circadian rhythm such that the critical value of the homeostatic drive that will trigger the switch from wake to sleep (or vice-versa) varies across the 24-h day. Thus, there are times when it is very difficult to initiate sleep no matter how long you have been awake, and times when it is difficult to awaken even when you are well rested.

3.1 Homeostatic Regulation of Sleep: Adenosine

Accumulation of adenosine may drive the motivation to sleep (Porkka-Heiskanen et al. 2002; but see Blanco-Centurion et al. 2006). Adenosine inhibits wake-active neurons and allows sleep-active neurons to become active (Brown et al. 2012). Blocking adenosine receptor with drugs such as caffeine can transiently decrease the sleep drive and thus facilitate wakefulness. Histamine appears to play the opposite role, histamine agonists promote wakefulness (Brown et al. 2001, 2012), and histamine antagonists, such as those found in allergy medication and antiemetics, enhance the sleep drive (Krystal et al. 2013).

3.2 Circadian Modulation of Sleep

The motivation and drive to engage in sleep involves a large number of brain regions (Saper et al. 2005b; Antle 2015), many of which are regulated directly or indirectly by the circadian clock. Possibly, the circadian clock in the SCN regulates sleep and wake by regulating the activity of the subparaventricular zone (SPZ). The SPZ in turn innervates the dorsomedial hypothalamus (DMH) which provides major input to both hypocretin neurons and the ventrolateral preoptic area (VLPO, Saper et al. 2005b). A number of areas are active during wake, such as the noradrenergic locus coeruleus, serotonergic raphe, and histaminergic tuberomammillary nucleus. Other areas are active during sleep, notably the VLPO, which inhibits the activity of the wake-active areas. Lateral hypothalamic neurons containing hypocretin (also known as orexin) appear to drive activity in the wake-active neurons. Their loss in people with narcolepsy leads to sleep attacks (intrusions of sleepiness into periods of normal wake).

While the circadian clock in the SCN may provide the master organizing circadian signal within an organism, brain areas involved in regulation of sleep and wake may be intrinsically rhythmic as well. The histaminergic cells in the tuberomammillary nucleus express BMAL1, a key component of the circadian transcription and translation feedback loops. BMAL1 appears to regulate both synthesis of histamine and the circadian activity of these neurons (Yu et al. 2014). Hypocretin neurons in the lateral hypothalamus and noradrenergic neurons in the locus coeruleus exhibit circadian oscillations of *Per1* expression (Mahoney et al. 2013).

The problem of separating goal-directed and amplitude effects is well exemplified in studies of age-related sleep disruption. There is evidence that older people are more susceptible to the negative effects of circadian phase misalignment (such as occur with jet lag or shift work) than young people (Harma et al. 1994; Juda et al. 2013). This impacts the timing and display of motivated behaviors. Changes in sleep timing and duration (i.e., performance of sleep vs. other motivated behaviors), from adolescence to old age, and between the sexes, have been amply described (Roenneberg et al. 2007). Hypothetically, these might be a consequence of a reduction in the amplitude of the circadian aspect of the sleep–wake rhythm, or a reduction in the need/motivation to sleep. That is, changes in sleep duration with age may reflect reduced sleep "need," or may result from a reduced ability to sleep, due to unrelated causes. Separating the motivation/need to sleep from ability is important as our advice to an older person with short sleep will depend on whether we believe that sleep need declines with age.

3.3 Feedback of Sleep to Circadian System

Alterations in sleep may feedback and affect other motivational systems. In humans, sleep loss can lead to alterations in the activity of feeding-related areas of the brain (Greer et al. 2013). Additionally, the types of foods craved following sleep loss are different than when well rested. Specifically, after sleep loss, people eat more fats and carbohydrates (Brondel et al. 2010). In people, sleep restriction leads to an increase in the hunger hormone ghrelin and a decrease in the satiety hormone leptin (Spiegel et al. 2004). Furthermore, sleep deprivation leads to increased caloric intake in people (Brondel et al. 2010). Sleep loss can also affect the circadian system, with sleep deprivation phase shifting the hamster circadian clock (Antle and Mistlberger 2000) and impairing light-induced phase shifts of the circadian system (Mistlberger et al. 1997).

4 Exercise/Activity

4.1 Locomotor Activity is Rewarding

Many circadian studies employ the motivated behavior of wheel running to assess circadian phase in rodents. That said, there has been a concern that wheel running may be a laboratory artifact and represent a stereotypy rather than a natural behavior (Richter et al. 2014). In fact, wheel running is a strongly motivated behavior. Rodents will work to unlock or gain access to a wheel (reviewed in Sherwin 1998).

Furthermore, when wheels are available in a natural setting, wild animals will not only use them upon discovery, but will return repeatedly to run in them (Meijer and Robbers 2014), arguing that wheel running is not an artifact of laboratory housing. There even appears to be some degree of homeostasis for exercise, as when rats are deprived of a wheel for 1, 3, or 10 h, there is rebound activity afterward, proportional to the lost activity (Mueller et al. 1999). There is circadian regulation of this motivated behavior.

While it is not surprising that nocturnal animals will engage in running at night when they have 24-h access to a wheel, animals will nonetheless run even if the wheels are available only during the day. However, rats with wheel access only during the night will increase their level of running to up to 4 times the baseline level progressively over time, while those with wheel access during the daytime (i.e., light phase of the day–night cycle) will remain at a baseline level of activity (Eikelboom and Lattanzio 2003).

4.2 Feedback of Exercise to Circadian System

There are also feedback effects on the circadian clock from such exercise seen by motivating an animal to engage in activity in a phase-dependent manner. Scheduled daily wheel access or exercise can entrain the circadian clock (Edgar and Dement 1991; Marchant and Mistlberger 1996). In animals housed in light:dark cycles, scheduled confinement to wheels for 3 h, which typically induces running, can alter the alignment between the circadian clock and the light:dark cycle (Sinclair and Mistlberger 1997). Confinement to a wheel for 1–3 h often elicits running behavior which leads to phase advances when confinement occurs during the day and phase delays when confinement occurs during the late night (Bobrzynska and Mrosovsky 1998). The exercise per se may not be the critical feature, as enforced arousal through gentle handling can also elicit these same changes in phase (Antle and Mistlberger 2000), and it is possible that wheel running is simply a strongly motivating behavior that produces sufficient wakefulness.

A question raised by the foregoing studies is whether the motivation/arousal feeds back to alter clock function. Not all hamsters, particularly older hamsters, are motivated to run when simply presented with a novel wheel. However, these animals can be motivated to run by augmenting the stimulus. For instance, when given a novel wheel in the presence of a sexually receptive but inaccessible female, older male hamsters will run vigorously and will exhibit phase shifts of their circadian clocks (Janik and Mrosovsky 1993; Mrosovsky and Biello 1994). Cold exposure can also be used to motivate exercise in both younger and older hamsters. In this case, one study reported that the animals will shift (Mistlberger et al. 1996), while another study failed to observe large phase shifts despite high activity (Janik and Mrosovsky 1993). This suggested that the motivational context for the arousal/exercise (accessing a female or keeping warm) might be critical to its ability to feedback to the circadian clock.

4.3 Activity Influences Other Motivated Behaviors

Scheduled wheel access and exercise can also alter other motivated behaviors. For example, rats will decrease food intake for about a week after gaining access to a running wheel (Looy and Eikelboom 1989; Lattanzio and Eikelboom 2003). Eating a high fat diet can alter circadian rhythms of eating and activity, but these effects are mitigated by being able to exercise on a wheel (Pendergast et al. 2014).

5 Eating

Eating is a motivated behavior that is highly rewarding and has clear adaptive value. While some species, such as grazers, eat around the clock, others feed at particular times of their 24-h day. Notably, Dr. Fred Stephan, who studied circadian control of feeding, was fond of pointing out that "when food competes with light, food usually wins" (p. 290, Stephan 2002). That is, even though rats are preferentially nocturnal, when food is only available during the day, rats will adjust their circadian rhythms to exploit this resource. To a hungry rat, the *possibility* of death due to predation is outweighed by the *certainty* of death due to starvation. In Americans, the timing of meals is a partially learned and culturally determined phenomenon, but is often broken into a number of meals with periodic snacking in between (U.S. Department of Agriculture, Agricultural Research Service 2014). For other species, feeding occurs at species-typical times of the day (Siegel 1961) but may be adjusted by external factors (Kersten et al. 1980).

Timing not only influences when animals are motivated to eat, but also influences what they are motivated to eat. Meal size often differs across the day. For instance, in North American cultures, breakfast is frequently the smallest meal of the day (Kramer et al. 1992). In First World societies where food is abundant and available in wide varieties, foods typically craved and consumed for breakfast often differ from those craved and consumed for the late-day meal (Birch et al. 1984). While these cravings differ among cultures, it has been suggested that cultural preferences might have developed on top of daily oscillations of circulating insulin, glucagon, and other signals that influence appetite and craving for particular nutrients (Birch et al. 1984). Finally, time of day may also influence feeding behavior. For instance, rats will eat next to their food source during the night, but will take food back to their nest box to consume it during the day (Strubbe et al. 1986). While this might be interpreted as a response to the light, which nocturnal rodents avoid, this is not the case. When the full photoperiod (12 h of light followed by 12 h of dark) is switched to a skeleton photoperiod (1 h of light at dawn and again at dusk), animals will still exhibit subjective day (when rodents behave as if it were day and sleep) and subjective night (when rodents behave as if it were night, and are awake). Under such conditions, the behavior of returning to the nest box to consume the food during the daytime hours persists, suggesting circadian control of feeding behavior (Strubbe et al. 1986).

5.1 Feeding Duration

The times when rats consumed food most quickly are at the start and end of the dark phase (Whishaw et al. 1992). Additionally, rats will take longer to eat a specific amount of food in the light than in the dark. While hunger increases the rate of eating, the circadian effect persists with mealtime during the light lasting longer (Whishaw et al. 1992). This is not simply inhibition of behavior by light, as turning off the lights slows eating even further, while turning the lights on during the dark phase does not change eating speed (Whishaw et al. 1992).

5.2 Peripheral Factors Determining Eating

Aside from eating behavior, circadian factors can influence digestion. Many functions of the gastrointestinal tract (GIT) exhibit circadian rhythmicity, such as gut motility, gastric acid secretion, turnover of the mucosal barriers along the GIT, production of digestive enzymes, cell proliferation, and nutrient transport in the small intestines (Konturek et al. 2011). Under ad libitum feeding conditions, the circadian rhythm in eating behavior dovetails nicely with the circadian rhythm in digestive processes so that the animal can easily digest what it has eaten. Given these factors, there are times of the day when, despite motivation, the gut may simply not be prepared to receive a meal. This might be the case when arriving in a new time zone and eating with the local population even though it is not your own mealtime.

5.3 Anticipatory Behavior Entails Circadian Timing

When food is not available ad libitum, but rather is regularly available at a specific time of day, animals will reorganize their activity (and physiology) so as to anticipate these scheduled meals (Stephan et al. 1979; Antle and Silver 2009). An example of regularly timed food availability in nature is seen in the rabbit doe and her pups. In nature, the mother rabbit nurses her pups with a circadian rhythm, returning to her nest for only about 3 min once each day (reviewed in González-Mariscal et al. 2015). During this brief window, the pups must consume all the milk that they will need to sustain them for the following 24 h. Rabbit pups anticipate the opportunity to nurse, as can be measured by increased movements just prior to feeding time (Jilge 1993).

As discussed above, the SCN is the master circadian oscillator, and it is synchronized to our environmental cycles primarily by light. However, anticipation of scheduled feeding is not regulated by the SCN, as this motivated behavior persists even in rats in which the SCN has been lesioned (Stephan et al. 1979; Mistlberger

1994). If an animal with an SCN lesion is deprived of food after exposure to a restricted feeding schedule, these animals will initially exhibit a bout of activity in anticipation of the normal mealtime. This activity disappears when it is clear that the meal has been missed. On the subsequent day, the anticipatory activity re-emerges at the correct time, clearly indicating a circadian clock phenomenon, despite the loss of the master circadian clock in the SCN. Such evidence points to the existence of a food-entrainable circadian oscillator(s) in other tissues.

5.4 Extra-SCN Circadian Oscillators Support Anticipatory Behaviors

Feeding can uncouple light- and food-entrained circadian timing. The phase of the SCN does not appear strongly influenced by scheduled feeding (Damiola et al. 2000; Stokkan et al. 2001; Challet et al. 2003). In contrast, the phases of other oscillators in the body, such as the liver (Stokkan et al. 2001) and stomach (LeSauter et al. 2009), are strongly affected by scheduled feeding.

5.5 Oscillators Mediating Anticipation of Feeding

A number of organs and brain areas exhibit circadian rhythms in SCN-lesioned animals placed on a restricted feeding schedule. These include the DMH (Gooley et al. 2006; Mieda et al. 2006; Verwey et al. 2007, 2008), dorsal striatum and nucleus accumbens (Angeles-Castellanos et al. 2007; Verwey and Amir 2011), the cerebral cortex and hippocampus (Wakamatsu et al. 2001), the stomach (LeSauter et al. 2009), and the liver (Stokkan et al. 2001).

5.6 Brain Oscillators

In the brain, the DMH was an attractive area to serve as a node for regulating anticipation of daily meals. It receives input indirectly from the SCN and is responsive to a number of endocrine signals related to energy state (Chou et al. 2003). It also relays such signals to important sleep/wake areas such as the VLPO (Saper et al. 2005a). The circadian rhythm of expression of a number of genes is shifted in the DMH when animals are placed on a restricted feeding schedule or have ad libitum access to regular chow but, in addition, are given a "treat" or reward at the same time every day. Lesioning the DMH was initially reported to abolish anticipation of scheduled meals (Gooley et al. 2006); however, subsequent reports revealed that clear anticipation persisted in animals with unambiguous DMH

5.7 Peripheral Oscillators

In addition to oscillators in the brain, peripheral organs have circadian rhythms that can provide timing cues related to feeding. Daily signals participating in anticipation of scheduled daily meals may originate in the stomach's parietal cells (i.e., oxyntic or delomorphous cells) that release ghrelin. Ghrelin is a potent orexigenic, stimulating feeding behavior when administered to rats, mice, and humans (Nakazato et al. 2001; Wren et al. 2001a, b; LeSauter et al. 2009). Ghrelin rises in people in anticipation of their regular mealtimes (Cummings et al. 2001). Ghrelin-containing cells in the stomach exhibit circadian rhythms in the levels of PER1 and PER2 proteins, integral components of the intracellular clock (LeSauter et al. 2009). This rhythmic expression can be synchronized to scheduled mealtimes (LeSauter et al. 2009). Anticipation of a daily meal is reduced, but not eliminated, in ghrelin receptor knockout animals (Blum et al. 2009; LeSauter et al. 2009).

5.8 Disrupted Circadian Control: Night Eating Syndrome

While eating time varies among cultures, in all cases eating is regulated in a homeostatic and a circadian fashion. People generally consume three meals each day, and these usually occur at regular times of the day. Typically, feeding behavior ends in the evening. However, in patients with night eating syndrome, at least 25 % of their daily calories are consumed after the major evening meal, or they may interrupt their sleep multiple times a week to consume food (Depner et al. 2014). These patients also typically reduce their calorie intake in the mornings. These symptoms are consistent with a phase-delayed circadian clock. As their sleep cycles are not shifted relative to the typical population (O'Reardon et al. 2004), this may represent a uncoupling of various clock systems. The phasing and amplitude of a number of physiological and hormonal rhythms are altered in night eating syndrome (Goel et al. 2009). Morning anorexia, a frequent symptom of night eating syndrome, is associated with poor glycemic control and higher body mass index in diabetic patients (Reutrakul et al. 2014). Eating during the normal sleep phase has been associated with greater weight gain in mice (Arble et al. 2009). This is not surprising as the circadian system and metabolism must be aligned for optimal energy balance (Waterhouse et al. 2005). Consuming calories at night when the body is storing energy may lead to greater weight gain (Arble et al. 2009).

Interventions that realign the circadian system could treat night eating syndrome and improve health in these patients (Goel et al. 2009).

In summary, feeding appears to be the most important motivated behavior from a circadian perspective. Animals will leave their temporal niche and reorganize their daily behavioral rhythms when food is only available during the day (Stephan 2002). Feeding influences many of the oscillators of the body. While the master circadian clock in the SCN appears to remain relatively synchronized to the light:dark cycle, the phases of other oscillators appear to be more heavily influenced by scheduled feeding. Thus, the coupling between the SCN and extra-SCN oscillators is modified by scheduled feeding. While the location of the master circadian clock is known, evidence clearly demonstrates that the SCN is not the food clock. The location of the food-entrainable oscillator(s) is still unknown (Mistlberger and Antle 2011). Given that food is such a strong circadian signal, manipulations of mealtiming may be useful to expedite re-entrainment to new rotating shift work schedules, or following transmeridian travel. Shifting mealtimes to match up with a shift in the light:dark cycle can shorten the time needed to synchronize to a new time zone and overcome jet lag (Angeles-Castellanos et al. 2011). To do this, one can eat at the same time as the locals when arriving in a new time zone. If this is difficult as you may not feel hungry, it is recommended that you skip a meal to facilitate re-entrainment. However, given that your digestive system may not be prepared to digest a meal given the misalignment between your circadian clock and the local time, such meals should be smaller and should consist of foods that are easy to digest.

6 Mating/Sex

6.1 Sexual Behavior is Highly Motivated and is Under Temporal Control

Sexual behaviors in male and female animals are highly motivated, and various factors that influence sexual motivation are described in the chapter by Margarinos and Pfaff in this volume. The circadian system is a major regulator of sexual behavior. Given the differences in male and female sexual behavior, it should not be surprising that mating and reproduction are regulated by the circadian system in a sexually dimorphic manner.

Male sexual behavior exhibits a circadian rhythm. Under ad libitum access to sexually receptive females, male rats exhibit 80 % of their mounting, intromissions, and ejaculations during the night (Logan and Leavitt 1992). The timing of sexual behavior during the night may be species specific, with the peak of mating behavior occurring in the early (Logan and Leavitt 1992) or middle (Lisk 1969) portions of the night in rats, but late in the night in deer mice (Dewsbury 1981).

Motivated sexual behavior in female rodents is gated by two biological rhythms. It is most closely tied to the infradian estrus rhythm. Female rats are only sexually receptive once every 4–5 days. This sexual receptivity is gated by a luteinizing hormone (LH) surge that triggers ovulation and the transition from proestrus to estrus. However, timing of the LH surge is under control of the circadian clock (Williams and Kriegsfeld 2012). In free-running conditions where there are no external time cues, the LH surge tracks and precedes activity onset (Fitzgerald and Zucker 1976). As such, sexual receptivity in female rodents is greatest during their active phase (i.e., night in nocturnal rodents). If the temporal window during which the LH surge can occur is closed by treating the female with a barbiturate just prior to when the LH surge should occur, the surge and subsequent ovulation are both delayed until the gate opens at the same time the following day (Everett and Sawyer 1950; Alleva et al. 1971; Siegel et al. 1976; Stetson and Watson-Whitmyre 1977). However, when estradiol is held at a constantly high level, there does not appear to be a circadian rhythm in sexual receptivity, suggesting that the circadian clock controls the timing of the LH surge, while sexual receptivity is regulated by estradiol levels. While these are normally linked sequentially, the circadian clock does not appear to directly regulate sexual motivation in female rodents.

We have suggested that the circadian clock may independently regulate both specific goals and the associated arousal with respect to motivated behaviors. There is a clear interaction between the circadian cycle and the estrus cycle in terms of both timing and amplitude of motivated behaviors. Wheel running and intracranial self-stimulation both show highest levels during the night (Steiner et al. 1981). However, the levels of each of these behaviors are significantly higher on the night when female rats transition from diestrus to estrus (Steiner et al. 1981). This is not simply an increase in general arousal or activity, as other behaviors such as general locomotion, rearing, and grooming do not exhibit increases in their amplitude on the same night.

6.2 Interaction Between Sex and Other Motivated Behaviors

When the opportunity to engage in a variety of motivated behaviors is provided, some will take priority over others. When presented simultaneously with food and a receptive mate following long-term deprivation of both (6 days), male rats will choose to mate before eating (Sachs and Marsan 1972). Under situations where there is free access to a running wheel and sexually receptive females, males will reduce the amount and delay the onset of their wheel running and instead engage in sexual behavior (Logan and Leavitt 1992). This suggests that motivated behaviors are partitioned during the waking period and that pursuit of one goal (i.e., mating) occurs at the expense of another (i.e., wheel running and eating).

The circadian clock may *not* regulate sexual performance and sexual motivation to the same extent. Just as other example behaviors, anticipatory motivation and consummatory activity appear to involve distinct mechanisms. van Furth and van

Ree (1994) suggest that sexual performance in male rats is tightly controlled by the circadian clock, with poor performance and longer refractory periods during their daytime rest phase. Motivation for sexual behavior can be measured using the bi-level mating chamber (Mendelson and Pfaus 1989). This chamber has two levels, one above the other, connected at each end by stairs, thus ensuring that the female can always choose to either approach or avoid the male. Once experienced with mating in this chamber, male rats alone in the chamber show behaviors that suggest searching the chamber for the female. This anticipatory behavior manifests as constant changing between the levels. The changing between levels is not observed in males that have not associated this chamber with mating. Using this task, there does not appear to be a circadian rhythm of sexual motivation in male rats, in that they show just as many anticipatory level changes during the day as during the night (van Furth and van Ree 1994). In females hamsters, proceptive behaviors (i.e., exploration/approach, as scored by time spent with intact vs. castrated males, and used as a metric of sexual motivation) and consummatory behaviors (i.e., lordosis) are regulated by distinct brain regions; exploratory/approach sexual behavior, but not lordosis, is reduced by infusion of gonadotropin-inhibitory hormone (Piekarski et al. 2013). Given that gonadotropin-inhibitory hormone is involved in circadian regulation of the LH surge (Gibson et al. 2008; Williams et al. 2011), this mechanism enables synchronizing of ovulation and regulation of exploratory/approach sexual behavior.

6.3 Anticipatory Activity

Male rats will anticipate a scheduled daily opportunity to mate (Landry et al. 2012). When the daily opportunity to mate occurs at night, robust anticipation is observed. Rats also exhibited a strong conditioned place preference to the mating chamber when scheduled mating happens at night (Landry et al. 2012). Anticipation is also observed when the scheduled mating occurs during the light phase, a time of day when rats are much less likely to engage in mating (Logan and Leavitt 1992). In this case, it is not clear if it is the mating itself that is affecting the clock. In the daytime scheduled mating paradigm, the male rats will engage in post-coital feeding and running. As discussed above, the circadian system can anticipate daily opportunities to feed, and the circadian clock is influenced by exercise. When opportunities to engage in these post-coital motivated behaviors (feeding and exercise) were prevented, male rats rarely anticipated the scheduled daytime mating opportunity (Landry et al. 2012), suggesting that the observed anticipation of daytime mating may have been related to the exercise and feeding after sex, rather than the mating itself. Similarly, the male rats show no conditioned place preference to the mating chamber when mating is scheduled during the day. These data suggest that while the animal can anticipate many motivated behaviors, when these motivated behaviors fall outside their normal ecological time, only some of these behavior (i.e., feeding), and not other (i.e., sexual behavior) have a strong effect on the circadian system.

6.4 Feedback of Sexual Behavior to the Circadian System

Sociosexual cues can also influence the circadian clock. Exposure to a receptive female during the sleep phase can accelerate a male hamsters' re-entrainment to an advanced light:dark cycle (Honrado and Mrosovsky 1989) and can induce phase shifts (Mrosovsky 1988), but only if mating is prevented. It is likely that it is the associated enduring arousal (Antle and Mistlberger 2000) rather than the sociosexual cue itself that elicits the phase shift.

7 Maternal Behavior

Maternal care of offspring is a strongly motivated behavior in which the emergence of the motivation is closely tied to hormonal changes that accompany giving birth (Numan 2007). Maternal care in rodents includes nursing and licking/grooming of her pups. With respect to circadian rhythms, most attention has focused on nursing behavior. Dams engage in more nursing behavior during the day when they are inactive (Grota and Ader 1969; Pachón et al. 1995). Mice also crouch over their pups more during the day, but also show minor peaks in crouching behavior starting an hour after dark onset, and another peak beginning about midway through the dark phase. Prominent troughs in crouching behavior coincide with dawn and dusk (Hoshino et al. 2006).

7.1 Feedback of Maternal Behavior on the Circadian Clock

Nursing behavior can entrain the circadian clocks of neonatal mice, hamsters, and rabbits (Viswanathan and Chandrashekaran 1985; Viswanathan and Davis 1992; Jilge 1993; Viswanathan 1999; Caba and Gonzalez-Mariscal 2009). This phenomenon is most stunning in rabbits, where the doe only nurses her pups a single time each day for about 3–5 min. This nursing bout occurs at the same time each day, and the pups anticipate her arrival by increasing movements in the nest. They also show anticipatory increases in temperature, corticosterone secretion, and circulating ghrelin levels (Caba and Gonzalez-Mariscal 2009). During the brief feeding bout, they consume about 35 % of their body weight, leading to prominent distension of their stomach (Caba and Gonzalez-Mariscal 2009). All of this occurs in darkness in their burrow before pups are ever exposed to light.

7.2 Neural Basis of Anticipation of Maternal Care

This daily nursing can affect the brain of both the pups and the mothers. In female rabbits, there is a circadian rhythm of PER1 expression in dopaminergic neurons that regulate prolactin secretion. This rhythm peaks in the early night in non-nursing females, but shifts to align with the time of nursing in females that are nursing (Meza et al. 2011). While the circadian rhythm of PER1 protein in the SCN is not shifted in nursing does, the amplitude of the PER1 expression rhythm is lower relative to non-nursing does. In pups fed at the same time each day, clock gene expression (*Per1*, *Per2* and *Bmal1*) is entrained. When feeding time is shifted, SCN clock gene expression rhythms exhibit a corresponding shift to follow feeding time (Caldelas et al. 2009). Entrainment appears to use both food and social cues. The circadian clock can be entrained by artificial feeding through a tube surgically implanted into the stomach (Morgado et al. 2011). However, olfactory cues from the dam are also critical, as when fed milk formula by gavage at the same time every day, the pups only anticipate this feeding when it is accompanied by natural or artificial maternal pheromones (Montúfar-Chaveznava et al. 2013).

7.3 Cellular Clock and Maternal Care

While a circadian clock organizes maternal nursing, it is also important for proper maternal care. Mice that carry the delta19 *clock* gene mutation exhibit alterations in maternal care. Both homozygous and heterozygous *clock* mutant moms have been studied. The occurrence of even a single mutant *clock* allele can compromise maternal behavior; heterozygous *clock* mothers are more active during the light phase, resulting in less nursing behavior during the light phase (Koizumi et al. 2013). Wild-type pups raised by homozygous *clock* mice grow less quickly (Hoshino et al. 2006). This is likely due to the fact that homozygous *clock* mothers have no rhythms in prolactin and produce less milk (Hoshino et al. 2006). Furthermore, this altered care appears to lead to alterations in adult behavior of pups raised by such mothers, even if they themselves have a normal *clock* gene. Specifically, wild-type mice reared by *clock* mutant moms appear more anxious as adults as assessed using the open field and elevated plus-maze (Koizumi et al. 2013).

8 Circadian Basis of Motivation Disorders

Circadian disruption due to jet lag, shift work, or disease can have numerous adverse health consequences (Foster and Kreitzman 2014). Circadian disruption may lead to dysregulation of the amplitude and direction/goal of motivated behaviors and physiological systems underlying motivated behaviors. The circadian disruption may result from extrinsic social or environmental factors, such as is the case for shift work, jet lag, or work/school start times out of phase with our internal clock. In other cases, the circadian disruption may be due to an internal disease state, such as depression (Karatsoreos 2014). When examining motivation and circadian disruption, there are three considerations. First, given that circadian rhythms dictate the phase of particular motivated behaviors (Fig. 1), when our circadian clock is out of phase with our environment, the wrong motivated behavior may be cued at a particular time, such as insomnia associated with jet lag. Second, imposed sleep and circadian disruption may alter motivational attributes for a particular behavior, such as the increased hunger and caloric intake associated with sleep deprivation (Brondel et al. 2010), even in cases where feeding time is properly phased. Finally, some diseases involve clear disruption of both circadian rhythms and overall motivation levels. It is possible that circadian disruption may worsen adherent motivation in various diseases. We explore three disease states where circadian disruption may be a key component of the altered motivational states that underlie the disease.

8.1 Depression and Schizophrenia

Depression and schizophrenia are both associated with diminished motivational levels. A key symptom of depression is anhedonia, or diminished experience of pleasure or reward. Patients with schizophrenia appear to suffer less from anhedonia than patients with depression (see Reddy et al. in this volume). However, both conditions are associated with a reduction in goal seeking and anticipatory motivation (Gard et al. 2014; Barch et al. in this volume). In both depression and schizophrenia, there are clear disruptions of the circadian system (Bunney and Potkin 2008; Wulff et al. 2012), and in many cases, disruption of sleep and circadian rhythm may exacerbate the symptoms of some psychiatric diseases (Gruber et al. 2011; Jagannath et al. 2013).

8.2 Attention Deficit Hyperactivity Disorder

Individuals with attention deficit hyperactivity disorder (ADHD) also exhibit motivational problems. Lack of engagement of the motivational and reward systems has been suggested to underlie ADHD (Wasserman and Wasserman 2015). Patients with ADHD constantly shift their focus; thus, behaviors are often not goal directed, or fail to maintain a goal. Similarly, hyperactivity might be considered pure motivational amplitude or drive without a goal. Up to 80 % of patients with ADHD may experience sleep and circadian disruptions (Van Veen et al. 2010; Kooij and Bijlenga 2013). Furthermore, these circadian changes can be exacerbated by some

ADHD medications (Antle et al. 2012), and sleep problems can contribute to the major symptoms of ADHD, namely lack of attention and hyperactivity (Corkum et al. 2008).

8.3 Circadian Considerations for Treatment of Psychiatric Conditions

Many psychiatric disorders exhibit sleep and circadian problems, and treating the sleep problems does improve overall symptomatology (Jagannath et al. 2013; Karatsoreos 2014). Often, behavioral therapies for depression and insomnia include sleep hygiene components that include advice on when and when not to eat and exercise. Specifically, vigorous exercise, large meals, and alcohol consumption immediately before bedtime are contraindicated. That said engaging in motivated behaviors (i.e., eating and exercising) at the proper time of day may help diminish the symptoms of the disorder (Schroeder and Colwell 2013). For instance, remission from depression is much higher in patients that also engage in exercise (Belvederi Murri et al. 2015). Long-term exercise also helps improve insomnia (Passos et al. 2011). In this latter case, there was no difference between those who exercised in the morning versus the afternoon, suggesting that there may not be a best time of day to exercise, so long as you do not exercise right before sleep.

9 Conclusion

We have presented evidence that the circadian timing system is a critical feature of evolutionarily fundamental behaviors. Given that homeostatic behaviors that are critical to the survival of the organism and the species (e.g., feeding, drinking, sleeping, mating, and parenting) are highly pleasurable, it is interesting to consider the interface of the circadian system and limbic/prefrontal/striatal reward circuitry that regulates behavioral activation and effort-related functions (Robbins and Everitt 1996; Berridge and Kringelbach 2015).

The reward system of the brain participates in anticipation of regularly scheduled daily events, especially well documented for feeding behavior. For example, ghrelin has been implicated in full expression of anticipation of scheduled daily meals and appears to play a role in activating the reward system. Specifically, animals lacking ghrelin have reduced activation of cells (measured by FOS expression) in the ventral tegmental area and nucleus accumbens shell compared to wild-type animals when on feeding schedules (Lamont et al. 2012). Areas of the reward system (dorsal striatum and nucleus accumbens) rhythmically express clock genes (Wakamatsu et al. 2001), and regularly scheduled daytime meals can shift these rhythms (Angeles-Castellanos et al. 2007; Verwey and Amir 2011).

Perturbation of dopaminergic transmission alters the timing or amplitude of anticipatory activity (Liu et al. 2012; Smit et al. 2013). Loss of the D1 receptor attenuates anticipation of daily scheduled meals, even when the meal is palatable and high in fat (Gallardo et al. 2014). Viral replacement of D1 receptors to just the dorsal striatum restores anticipation of daily meals. Direct activation of D1 receptors in the dorsal striatum at the same time each day leads to anticipation of the injection time, as measured by high activity from a computer-scored video system (Gallardo et al. 2014). It appears that activation of D1 receptors in the dorsal striatum is both necessary and sufficient for anticipation of scheduled daily feeding, and possibly other rewarding situations. Consistent with these observations, daily rhythms in markers of dopaminergic activity have been observed in various mesolimbic and nigrostriatal target structures. "In rodents, rhythmic circadian clock gene expression has been observed in many reward-related brain regions, with the phase of peak expression depending both upon the gene and the brain structure examined" (see Table 3 in Webb et al. 2015). These studies suggest that dopamine signaling to D1R-expressing neurons in the dorsal striatum can synchronize circadian oscillators in the reward circuits, thereby modulating motivational processes and behavioral output.

Taken together, the evidence points to robust links between circadian clock genes, dopaminergic neurotransmission, and highly motivated responses. Optimally utilizing knowledge of these temporal parameters is likely to be useful in optimizing performance in a very broad array of behaviors.

References

Acosta-Galvan G, Yi CX, van der Vliet J, Jhamandas JH, Panula P, Angeles-Castellanos M, Del Carmen Basualdo M, Escobar C, Buijs RM (2011) Interaction between hypothalamic dorsomedial nucleus and the suprachiasmatic nucleus determines intensity of food anticipatory behavior. Proc Natl Acad Sci U S A 108:5813–5818

Alibhai FJ, Tsimakouridze EV, Reitz CJ, Pyle WG, Martino TA (2015) Consequences of circadian and sleep disturbances for the cardiovascular system. Can J Cardiol 31:860–872

Alleva JJ, Waleski MV, Alleva FR (1971) A biological clock controlling the estrous cycle of the hamster. Endocrinology 88:1368–1379

Angeles-Castellanos M, Mendoza J, Escobar C (2007) Restricted feeding schedules phase shift daily rhythms of c-Fos and protein Per1 immunoreactivity in corticolimbic regions in rats. Neuroscience 144:344–355

Angeles-Castellanos M, Amaya JM, Salgado-Delgado R, Buijs RM, Escobar C (2011) Scheduled food hastens re-entrainment more than melatonin does after a 6-h phase advance of the light-dark cycle in rats. J Biol Rhythms 26:324–334

Antle MC (2015) Sleep: neural systems. In: Wright JD (ed) International encyclopedia of the social & behavioral sciences, 2nd edn. Elsevier, Oxford, pp 87–93

Antle MC, Mistlberger RE (2000) Circadian clock resetting by sleep deprivation without exercise in the Syrian hamster. J Neurosci 20:9326–9332

Antle MC, Silver R (2005) Orchestrating time: arrangements of the brain circadian clock. Trends Neurosci 28:145–151

Antle MC, Silver R (2009) Neural basis of timing and anticipatory behaviors. Eur J Neurosci 30:1643–1649

Antle MC, van Diepen HC, Deboer T, Pedram P, Pereira RR, Meijer JH (2012) Methylphenidate modifies the motion of the circadian clock. Neuropsychopharmacology 37:2446–2455

Arble DM, Bass J, Laposky AD, Vitaterna MH, Turek FW (2009) Circadian timing of food intake contributes to weight gain. Obesity 17:2100–2102

Bailey MR, Jensen G, Taylor K, Mezias C, Williamson C, Silver R, Simpson EH, Balsam PD (2015) A novel strategy for dissecting goal-directed action and arousal components of motivated behavior with a progressive hold-down task. Behav Neurosci 129:269–280

Baron KG, Reid KJ (2014) Circadian misalignment and health. Int Rev Psychiatry 26:139–154

Belvederi Murri M, Amore M, Menchetti M, Toni G, Neviani F, Cerri M, Rocchi MB, Zocchi D, Bagnoli L, Tam E, Buffa A, Ferrara S, Neri M, Alexopoulos GS, Zanetidou S (2015) Physical exercise for late-life major depression. Br J Psychiatry. doi:10.1192/bjp.bp.114.150516

Berridge KC, Kringelbach ML (2015) Pleasure systems in the brain. Neuron 86:646–664

Birch LL, Billman J, Richards SS (1984) Time of day influences food acceptability. Appetite 5:109–116

Blanco-Centurion C, Xu M, Murillo-Rodriguez E, Gerashchenko D, Shiromani AM, Salin-Pascual RJ, Hof PR, Shiromani PJ (2006) Adenosine and sleep homeostasis in the basal forebrain. J Neurosci 26:8092–8100

Blum ID, Patterson Z, Khazall R, Lamont EW, Sleeman MW, Horvath TL, Abizaid A (2009) Reduced anticipatory locomotor responses to scheduled meals in ghrelin receptor deficient mice. Neuroscience 164:351–359

Bobrzynska KJ, Mrosovsky N (1998) Phase shifting by novelty-induced running: activity dose-response curves at different circadian times. J Comp Physiol A 182:251–258

Borbély AA (1982) A two process model of sleep regulation. Hum Neurobiol 1:195–204

Brondel L, Romer MA, Nougues PM, Touyarou P, Davenne D (2010) Acute partial sleep deprivation increases food intake in healthy men. Am J Clin Nutr 91:1550–1559

Brown RE, Stevens DR, Haas HL (2001) The physiology of brain histamine. Prog Neurobiol 63:637–672

Brown RE, Basheer R, McKenna JT, Strecker RE, McCarley RW (2012) Control of sleep and wakefulness. Physiol Rev 92:1087–1187

Bunney JN, Potkin SG (2008) Circadian abnormalities, molecular clock genes and chronobiological treatments in depression. Br Med Bull 86:23–32

Caba M, Gonzalez-Mariscal G (2009) The rabbit pup, a natural model of nursing-anticipatory activity. Eur J Neurosci 30:1697–1706

Cai A, Lehman MN, Lloyd JM, Wise PM (1997) Transplantation of fetal suprachiasmatic nuclei into middle-aged rats restores diurnal Fos expression in host. Am J Physiol 272:R422–R428

Caldelas I, Gonzalez B, Montufar-Chaveznava R, Hudson R (2009) Endogenous clock gene expression in the suprachiasmatic nuclei of previsual newborn rabbits is entrained by nursing. Dev Neurobiol 69:47–59

Challet E, Caldelas I, Graff C, Pevet P (2003) Synchronization of the molecular clockwork by light- and food-related cues in mammals. Biol Chem 384:711–719

Chou TC, Scammell TE, Gooley JJ, Gaus SE, Saper CB, Lu J (2003) Critical role of dorsomedial hypothalamic nucleus in a wide range of behavioral circadian rhythms. J Neurosci 23:10691–10702

Corkum P, Panton R, Ironside S, Macpherson M, Williams T (2008) Acute impact of immediate release methylphenidate administered three times a day on sleep in children with attention-deficit/hyperactivity disorder. J Pediatr Psychol 33:368–379

Cummings DE, Purnell JQ, Frayo RS, Schmidova K, Wisse BE, Weigle DS (2001) A preprandial rise in plasma ghrelin levels suggests a role in meal initiation in humans. Diabetes 50:1714–1719

Czeisler CA, Weitzman E, Moore-Ede MC, Zimmerman JC, Knauer RS (1980) Human sleep: its duration and organization depend on its circadian phase. Science 210:1264–1267

Damiola F, Le Minh N, Preitner N, Kornmann B, Fleury-Olela F, Schibler U (2000) Restricted feeding uncouples circadian oscillators in peripheral tissues from the central pacemaker in the suprachiasmatic nucleus. Genes Dev 14:2950–2961

Depner CM, Stothard ER, Wright KP Jr (2014) Metabolic consequences of sleep and circadian disorders. Curr Diab Rep 14:507

Dewsbury DA (1981) Social dominance, copulatory behavior, and differential reproduction in deer mice (Peromyscus maniculatus). J Comp Physiol Psych 95:880–895

Duffy E (1957) The psychological significance of the concept of "arousal" or "activation". Psychol Rev 64:265–275

Edgar DM, Dement WC (1991) Regularly scheduled voluntary exercise synchronizes the mouse circadian clock. Am J Physiol 261:R928–R933

Eikelboom R, Lattanzio SB (2003) Wheel access duration in rats: II. Day-night and within-session changes. Behav Neurosci 117:825–832

Everett JW, Sawyer CH (1950) A 24-hour periodicity in the "LH-release apparatus" of female rats, disclosed by barbiturate sedation. Endocrinology 47:198–218

Fitzgerald K, Zucker I (1976) Circadian organization of the estrous cycle of the golden hamster. Proc Natl Acad Sci U S A 73:2923–2927

Foster RG, Kreitzman L (2014) The rhythms of life: what your body clock means to you! Exp Physiol 99:599–606

Friedman L, Bergmann BM, Rechtschaffen A (1979) Effects of sleep deprivation on sleepiness, sleep intensity, and subsequent sleep in the rat. Sleep 1:369–391

Gallardo CM, Darvas M, Oviatt M, Chang CH, Michalik M, Huddy TF, Meyer EE, Shuster SA, Aguayo A, Hill EM, Kiani K, Ikpeazu J, Martinez JS, Purpura M, Smit AN, Patton DF, Mistlberger RE, Palmiter RD, Steele AD (2014) Dopamine receptor 1 neurons in the dorsal striatum regulate food anticipatory circadian activity rhythms in mice. Elife 3:e03781

Gard DE, Sanchez AH, Cooper K, Fisher M, Garrett C, Vinogradov S (2014) Do people with schizophrenia have difficulty anticipating pleasure, engaging in effortful behavior, or both? J Abnorm Psychol 123:771–782

Gibson EM, Humber SA, Jain S, Williams WP 3rd, Zhao S, Bentley GE, Tsutsui K, Kriegsfeld LJ (2008) Alterations in RFamide-related peptide expression are coordinated with the preovulatory luteinizing hormone surge. Endocrinology 149:4958–4969

Goel N, Stunkard AJ, Rogers NL, Van Dongen HP, Allison KC, O'Reardon JP, Ahima RS, Cummings DE, Heo M, Dinges DF (2009) Circadian rhythm profiles in women with night eating syndrome. J Biol Rhythms 24:85–94

González-Mariscal G, Caba M, Martinez-Gómez M, Bautista A, Hudson R (2015) Mothers and offspring: the rabbit as a model system in the study of mammalian maternal behavior and sibling interactions. Horm Behav, doi:10.1016/j.yhbeh.2015.05.011

Gooley JJ, Schomer A, Saper CB (2006) The dorsomedial hypothalamic nucleus is critical for the expression of food-entrainable circadian rhythms. Nat Neurosci 9:398–407

Gore BB, Zweifel LS (2013) Genetic reconstruction of dopamine D1 receptor signaling in the nucleus accumbens facilitates natural and drug reward responses. J Neurosci 33:8640–8649

Greer SM, Goldstein AN, Walker MP (2013) The impact of sleep deprivation on food desire in the human brain. Nat Commun 4:2259

Grota LJ, Ader R (1969) Continuous recording of maternal behaviour in *Rattus norvegicus*. Anim Behav 17:722–729

Gruber J, Miklowitz DJ, Harvey AG, Frank E, Kupfer D, Thase ME, Sachs GS, Ketter TA (2011) Sleep matters: sleep functioning and course of illness in bipolar disorder. J Affect Disord 134:416–420

Harma MI, Hakola T, Akerstedt T, Laitinen JT (1994) Age and adjustment to night work. Occup Environ Med 51:568–573

Hebb DO (1955) Drives and the C. N. S. (conceptual nervous system). Psychol Rev 62:243–254

Honrado G, Mrosovsky N (1989) Arousal by sexual stimuli accelerates the re-entrainment of hamsters to phase advanced light-dark cycles. Behav Ecol Sociobiol 25:57–63

Hoshino K, Wakatsuki Y, Iigo M, Shibata S (2006) Circadian clock mutation in dams disrupts nursing behavior and growth of pups. Endocrinology 147:1916–1923

Hurd MW, Zimmer KA, Lehman MN, Ralph MR (1995) Circadian locomotor rhythms in aged hamsters following suprachiasmatic transplant. Am J Physiol 269:R958–R968

Jagannath A, Peirson SN, Foster RG (2013) Sleep and circadian rhythm disruption in neuropsychiatric illness. Curr Opin Neurobiol 23:888–894

Janik D, Mrosovsky N (1993) Nonphotically induced phase shifts of circadian rhythms in the golden hamster: activity-response curves at different ambient temperatures. Physiol Behav 53:431–436

Jilge B (1993) The ontogeny of circadian rhythms in the rabbit. J Biol Rhythms 8:247–260

Juda M, Vetter C, Roenneberg T (2013) Chronotype modulates sleep duration, sleep quality, and social jet lag in shift-workers. J Biol Rhythms 28:141–151

Karatsoreos IN (2014) Links between circadian rhythms and psychiatric disease. Front Behav Neurosci 8:162

Kersten A, Strubbe JH, Spiteri NJ (1980) Meal patterning of rats with changes in day length and food availability. Physiol Behav 25:953–958

Klerman EB, Boulos Z, Edgar DM, Mistlberger RE, Moore-Ede MC (1999) Circadian and homeostatic influences on sleep in the squirrel monkey: sleep after sleep deprivation. Sleep 22:45–59

Koizumi H, Kurabayashi N, Watanabe Y, Sanada K (2013) Increased anxiety in offspring reared by circadian clock mutant mice. PLoS ONE 8:e66021

Konturek PC, Brzozowski T, Konturek SJ (2011) Gut clock: implication of circadian rhythms in the gastrointestinal tract. J Physiol Pharmacol 62:139–150

Kooij JJ, Bijlenga D (2013) The circadian rhythm in adult attention-deficit/hyperactivity disorder: current state of affairs. Expert Rev Neurother 13:1107–1116

Kramer FM, Rock K, Engell D (1992) Effects of time of day and appropriateness on food intake and hedonic ratings at morning and midday. Appetite 18:1–13

Kriegsfeld LJ, Leak RK, Yackulic CB, LeSauter J, Silver R (2004) Organization of suprachiasmatic nucleus projections in Syrian hamsters (*Mesocricetus auratus*): an anterograde and retrograde analysis. J Comp Neurol 468:361–379

Krout KE, Kawano J, Mettenleiter TC, Loewy AD (2002) CNS inputs to the suprachiasmatic nucleus of the rat. Neuroscience 110:73–92

Krystal AD, Richelson E, Roth T (2013) Review of the histamine system and the clinical effects of H1 antagonists: basis for a new model for understanding the effects of insomnia medications. Sleep Med Rev 17:263–272

Lamont EW, Patterson Z, Rodrigues T, Vallejos O, Blum ID, Abizaid A (2012) Ghrelin-deficient mice have fewer orexin cells and reduced cFOS expression in the mesolimbic dopamine pathway under a restricted feeding paradigm. Neuroscience 218:12–19

Landry GJ, Simon MM, Webb IC, Mistlberger RE (2006) Persistence of a behavioral food-anticipatory circadian rhythm following dorsomedial hypothalamic ablation in rats. Am J Physiol Regul Integr Comp Physiol 290:R1527–R1534

Landry GJ, Yamakawa GR, Webb IC, Mear RJ, Mistlberger RE (2007) The dorsomedial hypothalamic nucleus is not necessary for the expression of circadian food-anticipatory activity in rats. J Biol Rhythms 22:467–478

Landry GJ, Kent BA, Patton DF, Jaholkowski M, Marchant EG, Mistlberger RE (2011) Evidence for time-of-day dependent effect of neurotoxic dorsomedial hypothalamic lesions on food anticipatory circadian rhythms in rats. PLoS ONE 6:e24187

Landry GJ, Opiol H, Marchant EG, Pavlovski I, Mear RJ, Hamson DK, Mistlberger RE (2012) Scheduled daily mating induces circadian anticipatory activity rhythms in the male rat. PLoS ONE 7:e40895

Lattanzio SB, Eikelboom R (2003) Wheel access duration in rats: I. Effects on feeding and running. Behav Neurosci 117:496–504

Lehman MN, Silver R, Gladstone WR, Kahn RM, Gibson M, Bittman EL (1987) Circadian rhythmicity restored by neural transplant. Immunocytochemical characterization of the graft and its integration with the host brain. J Neurosci 7:1626–1638

LeSauter J, Hoque N, Weintraub M, Pfaff DW, Silver R (2009) Stomach ghrelin-secreting cells as food-entrainable circadian clocks. Proc Natl Acad Sci U S A 106:13582–13587

Lim MM, Gerstner JR, Holtzman DM (2014) The sleep-wake cycle and Alzheimer's disease: what do we know? Neurodegener Dis Manag 4:351–362

Lisk RD (1969) Cyclic fluctuations in sexual responsiveness in the male rat. J Exp Zool 171:313–319

Liu YY, Liu TY, Qu WM, Hong ZY, Urade Y, Huang ZL (2012) Dopamine is involved in food-anticipatory activity in mice. J Biol Rhythms 27:398–409

Logan FA, Leavitt F (1992) Sexual free behavior in male rats (*Rattus norvegicus*). J Comp Psychol 106:37–42

Looy H, Eikelboom R (1989) Wheel running, food intake, and body weight in male rats. Physiol Behav 45:403–405

Mahoney CE, Brewer JM, Bittman EL (2013) Central control of circadian phase in arousal-promoting neurons. PLoS ONE 8:e67173

Marchant EG, Mistlberger RE (1996) Entrainment and phase shifting of circadian rhythms in mice by forced treadmill running. Physiol Behav 60:657–663

Meijer JH, Robbers Y (2014) Wheel running in the wild. Proc Biol Sci 281

Mendelson SD, Pfaus JG (1989) Level searching: a new assay of sexual motivation in the male rat. Physiol Behav 45:337–341

Meza E, Waliszewski SM, Caba M (2011) Circadian nursing induces PER1 protein in neuroendocrine tyrosine hydroxylase neurones in the rabbit doe. J Neuroendocrinol 23:472–480

Mieda M, Williams SC, Richardson JA, Tanaka K, Yanagisawa M (2006) The dorsomedial hypothalamic nucleus as a putative food-entrainable circadian pacemaker. Proc Natl Acad Sci U S A 103:12150–12155

Mistlberger RE (1994) Circadian food-anticipatory activity: formal models and physiological mechanisms. Neurosci Biobehav Rev 18:171–195

Mistlberger RE (2005) Circadian regulation of sleep in mammals: role of the suprachiasmatic nucleus. Brain Res Brain Res Rev 49:429–454

Mistlberger RE, Antle MC (2011) Entrainment of circadian clocks in mammals by arousal and food. Essays Biochem 49:119–136

Mistlberger RE, Marchant EG, Sinclair SV (1996) Nonphotic phase-shifting and the motivation to run: cold exposure reexamined. J Biol Rhythms 11:208–215

Mistlberger RE, Landry GJ, Marchant EG (1997) Sleep deprivation can attenuate light-induced phase shifts of circadian rhythms in hamsters. Neurosci Lett 238:5–8

Model Z, Butler MP, LeSauter J, Silver R (2015) Suprachiasmatic nucleus as the site of androgen action on circadian rhythms. Horm Behav 73:1–7

Mollicone DJ, Van Dongen HP, Rogers NL, Banks S, Dinges DF (2010) Time of day effects on neurobehavioral performance during chronic sleep restriction. Aviat Space Environ Med 81:735–744

Montúfar-Chaveznava R, Trejo-Muñoz L, Hernández-Campos O, Navarrete E, Caldelas I (2013) Maternal olfactory cues synchronize the circadian system of artificially raised newborn rabbits. PLoS ONE 8:e74048

Moore RY, Eichler VB (1972) Loss of a circadian adrenal corticosterone rhythm following suprachiasmatic lesions in the rat. Brain Res 42:201–206

Morgado E, Juarez C, Melo AI, Dominguez B, Lehman MN, Escobar C, Meza E, Caba M (2011) Artificial feeding synchronizes behavioral, hormonal, metabolic and neural parameters in mother-deprived neonatal rabbit pups. Eur J Neurosci 34:1807–1816

Mrosovsky N (1988) Phase response curves for social entrainment. J Comp Physiol A 162:35–46

Mrosovsky N, Biello SM (1994) Nonphotic phase shifting in the old and the cold. Chronobiol Int 11:232–252

Mueller DT, Herman G, Eikelboom R (1999) Effects of short- and long-term wheel deprivation on running. Physiol Behav 66:101–107

Nakazato M, Murakami N, Date Y, Kojima M, Matsuo H, Kangawa K, Matsukura S (2001) A role for ghrelin in the central regulation of feeding. Nature 409:194–198

Numan M (2007) Motivational systems and the neural circuitry of maternal behavior in the rat. Dev Psychobiol 49:12–21

Ondo WG (2014) Sleep/wake problems in Parkinson's disease: pathophysiology and clinico-pathologic correlations. J Neural Transm 121(Suppl 1):S3–13

O'Reardon JP, Ringel BL, Dinges DF, Allison KC, Rogers NL, Martino NS, Stunkard AJ (2004) Circadian eating and sleeping patterns in the night eating syndrome. Obes Res 12:1789–1796

Pachón H, McGuire MK, Rasmussen KM (1995) Nutritional status and behavior during lactation. Physiol Behav 58:393–400

Passos GS, Poyares D, Santana MG, D'Aurea CV, Youngstedt SD, Tufik S, de Mello MT (2011) Effects of moderate aerobic exercise training on chronic primary insomnia. Sleep Med 12:1018–1027

Pauls S, Foley NC, Foley DK, LeSauter J, Hastings MH, Maywood ES, Silver R (2014) Differential contributions of intra-cellular and inter-cellular mechanisms to the spatial and temporal architecture of the suprachiasmatic nucleus circadian circuitry in wild-type, cryptochrome-null and vasoactive intestinal peptide receptor 2-null mutant mice. Eur J Neurosci 40:2528–2540

Pendergast JS, Branecky KL, Huang R, Niswender KD, Yamazaki S (2014) Wheel-running activity modulates circadian organization and the daily rhythm of eating behavior. Front Psychol 5:177

Pfaff DW, Martin EM, Faber D (2012) Origins of arousal: roles for medullary reticular neurons. Trends Neurosci 35:468–476

Piekarski DJ, Zhao S, Jennings KJ, Iwasa T, Legan SJ, Mikkelsen JD, Tsutsui K, Kriegsfeld LJ (2013) Gonadotropin-inhibitory hormone reduces sexual motivation but not lordosis behavior in female Syrian hamsters (*Mesocricetus auratus*). Horm Behav 64:501–510

Porkka-Heiskanen T, Alanko L, Kalinchuk A, Stenberg D (2002) Adenosine and sleep. Sleep Med Rev 6:321–332

Ralph MR, Foster RG, Davis FC, Menaker M (1990) Transplanted suprachiasmatic nucleus determines circadian period. Science 247:975–978

Ramkisoensing A, Meijer JH (2015) Synchronization of biological clock neurons by light and peripheral feedback systems promotes circadian rhythms and health. Front Neurol 6:128

Reinberg A, Motohashi Y, Bourdeleau P, Andlauer P, Levi F, Bicakova-Rocher A (1988) Alteration of period and amplitude of circadian rhythms in shift workers. With special reference to temperature, right and left hand grip strength. Eur J Appl Physiol Occup Physiol 57:15–25

Reutrakul S, Hood MM, Crowley SJ, Morgan MK, Teodori M, Knutson KL (2014) The relationship between breakfast skipping, chronotype, and glycemic control in type 2 diabetes. Chronobiol Int 31:64–71

Richard JM, Castro DC, Difeliceantonio AG, Robinson MJ, Berridge KC (2013) Mapping brain circuits of reward and motivation: in the footsteps of Ann Kelley. Neurosci Biobehav Rev 37:1919–1931

Richter SH, Gass P, Fuss J (2014) Resting is rusting: a critical view on rodent wheel-running behavior. Neuroscientist 20:313–325

Robbins TW, Everitt BJ (1996) Neurobehavioural mechanisms of reward and motivation. Curr Opin Neurobiol 6:228–236

Robinson I, Reddy AB (2014) Molecular mechanisms of the circadian clockwork in mammals. FEBS Lett 588:2477–2483

Roenneberg T, Kuehnle T, Juda M, Kantermann T, Allebrandt K, Gordijn M, Merrow M (2007) Epidemiology of the human circadian clock. Sleep Med Rev 11:429–438

Sachs B, Marsan E (1972) Male rats prefer sex to food after 6 days of food deprivation. Psychon Sci 28:47–49

Saini C, Brown SA, Dibner C (2015) Human peripheral clocks: applications for studying circadian phenotypes in physiology and pathophysiology. Front Neurol 6:95

Salamone J (1988) Dopaminergic involvement in activational aspects of motivation: Effects of haloperidol on schedule-induced activity, feeding, and foraging in rats. Psychobiology 16:196–206

Saper CB, Lu J, Chou TC, Gooley J (2005a) The hypothalamic integrator for circadian rhythms. Trends Neurosci 28:152–157

Saper CB, Scammell TE, Lu J (2005b) Hypothalamic regulation of sleep and circadian rhythms. Nature 437:1257–1263

Schmidt TM, Chen SK, Hattar S (2011) Intrinsically photosensitive retinal ganglion cells: many subtypes, diverse functions. Trends Neurosci 34:572–580

Schroeder AM, Colwell CS (2013) How to fix a broken clock. Trends Pharmacol Sci 34:605–619

Sherwin CM (1998) Voluntary wheel running: a review and novel interpretation. Anim Behav 56:11–27

Siegel PS (1961) Food intake in the rat in relation to the dark-light cycle. J Comp Physiol Psych 54:294–301

Siegel HI, Bast JD, Greenwald GS (1976) The effects of phenobarbital and gonadal steroids on periovulatory serum levels of luteinizing hormone and follicle-stimulating hormone in the hamster. Endocrinology 98:48–55

Sinclair SV, Mistlberger RE (1997) Scheduled activity reorganizes circadian phase of Syrian hamsters under full and skeleton photoperiods. Behav Brain Res 87:127–137

Smit AN, Patton DF, Michalik M, Opiol H, Mistlberger RE (2013) Dopaminergic regulation of circadian food anticipatory activity rhythms in the rat. PLoS ONE 8:e82381

Spiegel K, Tasali E, Penev P, Van Cauter E (2004) Brief communication: Sleep curtailment in healthy young men is associated with decreased leptin levels, elevated ghrelin levels, and increased hunger and appetite. Ann Intern Med 141:846–850

Steiner M, Katz RJ, Baldrighi G, Carroll BJ (1981) Motivated behavior and the estrous cycle in rats. Psychoneuroendocrinology 6:81–90

Stephan FK (2002) The "other" circadian system: food as a Zeitgeber. J Biol Rhythms 17:284–292

Stephan FK, Zucker I (1972) Circadian rhythms in drinking behavior and locomotor activity of rats are eliminated by hypothalamic lesions. Proc Natl Acad Sci U S A 69:1583–1586

Stephan FK, Swann JM, Sisk CL (1979) Anticipation of 24-hr feeding schedules in rats with lesions of the suprachiasmatic nucleus. Behav Neural Biol 25:346–363

Sterniczuk R, Rusak B, Rockwood K (2014) Sleep disturbance in older ICU patients. Clin Interv Aging 9:969–977

Stetson MH, Watson-Whitmyre M (1977) The neural clock regulating estrous cyclicity in hamsters: gonadotropin release following barbiturate blockade. Biol Reprod 16:536–542

Stokkan KA, Yamazaki S, Tei H, Sakaki Y, Menaker M (2001) Entrainment of the circadian clock in the liver by feeding. Science 291:490–493

Strubbe JH, Spiteri NJ, Alingh Prins AJ (1986) Effect of skeleton photoperiod and food availability on the circadian pattern of feeding and drinking in rats. Physiol Behav 36:647–651

Trifilieff P, Feng B, Urizar E, Winiger V, Ward RD, Taylor KM, Martinez D, Moore H, Balsam PD, Simpson EH, Javitch JA (2013) Increasing dopamine D2 receptor expression in the adult nucleus accumbens enhances motivation. Mol Psychiatry 18:1025–1033

U.S. Department of Agriculture, Agricultural Research Service (2014) Meals and snacks: distribution of meal patterns and snack occasions, by gender and age. In: What We Eat in America, NHANES 2011–2012

Uth K, Sleigh R (2014) Deregulation of the circadian clock constitutes a significant factor in tumorigenesis: a clockwork cancer. Part II. studies. Biotechnol Biotechnol Equip 28:379–386

van Furth WR, van Ree JM (1994) Endogenous opioids and sexual motivation and performance during the light phase of the diurnal cycle. Brain Res 636:175–179

Van Reeth O, Zhang Y, Zee PC, Turek FW (1994) Grafting fetal suprachiasmatic nuclei in the hypothalamus of old hamsters restores responsiveness of the circadian clock to a phase shifting stimulus. Brain Res 643:338–342

Van Veen MM, Kooij JJ, Boonstra AM, Gordijn MC, Van Someren EJ (2010) Delayed circadian rhythm in adults with attention-deficit/hyperactivity disorder and chronic sleep-onset insomnia. Biol Psychiatry 67:1091–1096

Verwey M, Amir S (2011) Nucleus-specific effects of meal duration on daily profiles of Period1 and Period2 protein expression in rats housed under restricted feeding. Neuroscience 192:304–311

Verwey M, Khoja Z, Stewart J, Amir S (2007) Differential regulation of the expression of Period2 protein in the limbic forebrain and dorsomedial hypothalamus by daily limited access to highly palatable food in food-deprived and free-fed rats. Neuroscience 147:277–285

Verwey M, Khoja Z, Stewart J, Amir S (2008) Region-specific modulation of PER2 expression in the limbic forebrain and hypothalamus by nighttime restricted feeding in rats. Neurosci Lett 440:54–58

Vida B, Deli L, Hrabovszky E, Kalamatianos T, Caraty A, Coen CW, Liposits Z, Kalló I (2010) Evidence for suprachiasmatic vasopressin neurones innervating kisspeptin neurones in the rostral periventricular area of the mouse brain: regulation by oestrogen. J Neuroendocrinol 22:1032–1039

Viswanathan N (1999) Maternal entrainment in the circadian activity rhythm of laboratory mouse (C57BL/6J). Physiol Behav 68:157–162

Viswanathan N, Chandrashekaran MK (1985) Cycles of presence and absence of mother mouse entrain the circadian clock of pups. Nature 317:530–531

Viswanathan N, Davis FC (1992) Maternal entrainment of tau mutant hamsters. J Biol Rhythms 7:65–74

Vujovic N, Gooley JJ, Jhou TC, Saper CB (in press) Projections from the subparaventricular zone define four channels of output from the circadian timing system. J Comp Neurol. doi:10.1002/cne.23812

Wakamatsu H, Yoshinobu Y, Aida R, Moriya T, Akiyama M, Shibata S (2001) Restricted-feeding-induced anticipatory activity rhythm is associated with a phase-shift of the expression of mPer1 and mPer2 mRNA in the cerebral cortex and hippocampus but not in the suprachiasmatic nucleus of mice. Eur J Neurosci 13:1190–1196

Wasserman T, Wasserman LD (2015) The misnomer of attention-deficit hyperactivity disorder. Appl Neuropsychol Child 4:116–122

Waterhouse J, Kao S, Edwards B, Weinert D, Atkinson G, Reilly T (2005) Transient changes in the pattern of food intake following a simulated time-zone transition to the east across eight time zones. Chronobiol Int 22:299–319

Webb IC, Lehman MN, Coolen LM (2015) Diurnal and circadian regulation of reward-related neurophysiology and behavior. Physiol Behav 143:58–69

Welsh DK, Takahashi JS, Kay SA (2010) Suprachiasmatic nucleus: cell autonomy and network properties. Annu Rev Physiol 72:551–577

Whishaw IQ, Dringenberg HC, Comery TA (1992) Rats (*Rattus norvegicus*) modulate eating speed and vigilance to optimize food consumption: effects of cover, circadian rhythm, food deprivation, and individual differences. J Comp Psychol 106:411–419

Williams WP 3rd, Kriegsfeld LJ (2012) Circadian control of neuroendocrine circuits regulating female reproductive function. Front Endocrinol (Lausanne) 3:60

Williams WP 3rd, Jarjisian SG, Mikkelsen JD, Kriegsfeld LJ (2011) Circadian control of kisspeptin and a gated GnRH response mediate the preovulatory luteinizing hormone surge. Endocrinology 152:595–606

Wren AM, Seal LJ, Cohen MA, Brynes AE, Frost GS, Murphy KG, Dhillo WS, Ghatei MA, Bloom SR (2001a) Ghrelin enhances appetite and increases food intake in humans. J Clin Endocrinol Metab 86:5992

Wren AM, Small CJ, Abbott CR, Dhillo WS, Seal LJ, Cohen MA, Batterham RL, Taheri S, Stanley SA, Ghatei MA, Bloom SR (2001b) Ghrelin causes hyperphagia and obesity in rats. Diabetes 50:2540–2547

Wulff K, Dijk DJ, Middleton B, Foster RG, Joyce EM (2012) Sleep and circadian rhythm disruption in schizophrenia. Br J Psychiatry 200:308–316

Yu X, Zecharia A, Zhang Z, Yang Q, Yustos R, Jager P, Vyssotski AL, Maywood ES, Chesham JE, Ma Y, Brickley SG, Hastings MH, Franks NP, Wisden W (2014) Circadian factor BMAL1 in histaminergic neurons regulates sleep architecture. Curr Biol 24:2838–2844

The Neural Foundations of Reaction and Action in Aversive Motivation

Vincent D. Campese, Robert M. Sears, Justin M. Moscarello,
Lorenzo Diaz-Mataix, Christopher K. Cain and Joseph E. LeDoux

Abstract Much of the early research in aversive learning concerned motivation and reinforcement in avoidance conditioning and related paradigms. When the field transitioned toward the focus on Pavlovian threat conditioning in isolation, this paved the way for the clear understanding of the psychological principles and neural and molecular mechanisms responsible for this type of learning and memory that has unfolded over recent decades. Currently, avoidance conditioning is being revisited, and with what has been learned about associative aversive learning, rapid progress is being made. We review, below, the literature on the neural substrates critical for learning in instrumental active avoidance tasks and conditioned aversive motivation.

Keywords Avoidance · Instrumental · Rat · Freezing

Contents

1	Introduction...	172
2	Taxonomy of Behavior...	172
3	Historical Context of Studies on Aversive Learning...	175

V.D. Campese (✉) · J.M. Moscarello · L. Diaz-Mataix · J.E. LeDoux
Center for Neural Science, NYU, New York, USA
e-mail: vc42@nyu.edu

J.M. Moscarello
e-mail: justin.moscarello@nyu.edu

L. Diaz-Mataix
e-mail: ldm5@nyu.edu

J.E. LeDoux
e-mail: ledoux@cns.nyu.edu

R.M. Sears · C.K. Cain · J.E. LeDoux
Emotional Brain Institute at NYU and Nathan Kline Institute, New York, USA
e-mail: robert.sears@nyu.edu

C.K. Cain
e-mail: ccain@NKI.RFMH.org

4	Neural Basis of Reactions	176
5	Diverse Functions of a Threat Stimulus: The Nature of Action and Reaction Learning in Aversive Motivation	180
6	The Neural Basis of Aversively Motivated Actions	181
	6.1 Neural Control of Signaled and Unsignaled Avoidance Behavior	181
	6.2 Neural Control of Escape from Threat	185
	6.3 Incentive Motivation: Neural Circuits of Aversive Pavlovian-to-Instrumental Transfer	186
7	Summary	189
References		189

1 Introduction

Close to the finish line of the 2013 Boston marathon, there were two explosions that killed three people and injured over two hundred others. These explosions caused a chaotic burst of activity in the group of people gathered in the area for the event. Immediately upon detonation of the bombs, people in the crowd oriented toward the explosions and froze, and then, they began to flee the site. This sequence of *re-action* and *action* is repeatedly seen in emotionally evocative experiences in human and non-human animals (LeDoux 1996a; LeDoux and Gorman 2001). While reactions are inflexible responses automatically elicited by a stimulus, actions are responses or behaviors that are produced by an organism. Reactive behaviors are genetically embedded into an organism's response repertoire as prepared consequences to significant environmental pressures (e.g., predators; Hirsch and Bolles 1980; Coss and Biardi 1997). On the other hand, actions are behaviors learned by their reinforcing consequences (for example, a child learning to cry in order to receive attention) performed in order to obtain some goal or reward.

Using animal models, research on the neural basis of emotion has uncovered a great deal on how threatening conditioned stimuli come to control defensive *reactions* (LeDoux 2014, 2015). Progress on the neural foundations of aversively motivated actions has been slower, though recent times have seen a resurgence of interest in deciphering the circuits and mechanisms underlying these behaviors. In this chapter, we will describe the relationship between reactions and actions insofar as they pertain to defensive motivation, and the neural substrates of these phenomena.

2 Taxonomy of Behavior

Among the forms of behaviors exhibited by humans, many can be seen as fitting into four categories: reflexes, reactions, actions and habits (see Balleine and Dickinson 1998; Cardinal et al. 2002; Lang and Davis 2006; Yin and Knowlton 2006).

Reflexes are genetically prepared, stimulus-evoked behaviors that involve simple mechanisms and limited muscle groups. For example, myotatic reflexes (e.g., monosynaptic stretch reflex) do not require control outside of a single spinal level. Examples of defensive reflexes include withdrawal reflexes seen when painful stimuli are applied to the skin, or eyelid closure when an air puff is delivered to the region (i.e., NMR preparation).

Reaction and reflex behaviors share the attribute that the response itself is unlearned. Both of these types of responses are innate and are typically elicited by stimuli that have, through evolutionary processes, come to be embedded in the genetic wiring of the nervous system. However, reactions are distinct from reflexes in that they involve the whole organism and require control from components of the central nervous system (CNS) outside of the processing level of the sensory stimulus. For example, freezing behavior, in which the organism crouches in a certain posture and non-respiratory-related movement ceases, occurs as an initial reaction to a threatening event in many organisms. This reaction requires sustained contraction of the skeletal musculature throughout the body to maintain the characteristic posture. Metabolic support for this energy-demanding activity is provided by changes in visceral organ physiology controlled by the autonomic nervous system under the direction of the CNS. Freezing reactions can be thought of as an example of a fixed action pattern (Lorenz and Tinbergen 1938; Tinbergen 1951), or more specifically, a species-typical defensive behavior (Blanchard and Blanchard 1972; Bolles 1970; Bolles and Fanselow 1980).

The stimuli that elicit reaction and reflex behaviors, called releasers by ethologists, are specific to the response they command: freezing, for example, is an innate reaction to predators (Lorenz and Tinbergen 1938; Tinbergen 1951; Thorpe 1963; Hinde 1966; Bolles 1970; Blanchard and Blanchard 1969b, 1972; Rosen 2004). For example, Hirsch and Bolles (1980) found that two strains of deer mice that were removed from their natural environment for two breeding generations still demonstrated responses specific to the particular predator present in that environment when encountered. These prepared and unlearned responses proved useful—subjects that were presented with their ancestral predator tended to fare better than those exposed to the historically unfamiliar one. Even more dramatic, Coss and Biardi (1997) reported that a group of squirrels living in the higher elevations of a mountain and separated from snakes at lower elevations by an earthquake millions of years ago still exhibited species-typical defense responses to snakes. While such responses require no prior personal experience with an innate releaser, they can come under the control of previously neutral stimuli paired with these events through Pavlovian associative learning processes (Blanchard and Blanchard 1969a, b, 1972; Bolles 1972; Fanselow 1980). When these responses are produced by acquired stimuli, they are referred to as conditioned or *learned defensive responses.* This is somewhat of a misnomer since the response itself is not learned or conditioned. Instead, the response comes to be conditional on the presence of the stimulus. Some therefore prefer then terminology conditional response over conditioned response (Pavlov 1927).

Actions are also complex behaviors, but they are not simply elicited by a stimulus. Rather, they are emitted in the combined presence of certain stimuli and internal factors such as motivation and arousal and performed in order to obtain a goal or reward (Skinner 1938; Estes and Skinner 1941; Estes 1948; Rescorla and LoLordo 1965; Rescorla 1968; Lovibond 1983; Balleine and Dickinson 1998; Holland and Gallagher 2003; Niv et al. 2006). These responses are understood associatively as response-outcome (R-O) learning and include hierarchical control by the surrounding stimuli. A further distinction of actions from reactions is that actions are flexible behaviors rather than fixed responses. Depending on the circumstances, the appropriate action one must take in order to reach a goal can be quite different (Fanselow and Lester 1988). Therefore, a certain degree of preparedness may allow for some responses to be acquired or performed more readily than others (see Cain and LeDoux 2007). Nevertheless, the response can take multiple physical forms. For example, in the description of the aftermath of the Boston bombings above, the general goal would be to escape the danger; thus, the appropriate response would be to run away from the area. However, if the explosion had taken place on a boat and survivors were in the water, the appropriate action would be to swim away. Furthermore, if one had friends or loved ones with them, the goal would be to find them before fleeing, regardless of the specific behavioral demands. Much of the research on action learning, especially in terms of brain mechanisms, has focused on appetitive motivation (i.e., food, sex, or drugs); the neural substrates for instrumental motivation based on aversive outcomes (e.g., shock omission, punishment) has received less attention. A major aim of this chapter is to summarize recent research on aversively motivated actions.

When an action has been performed so frequently that it has become automatic or inflexible, it is referred to as a habit (Thorndike 1898; Skinner 1938; Hull 1943; Killcross and Coutureau 2003). Initially, actions are guided by the outcome they obtain, which can be demonstrated using outcome revaluation manipulations (including devaluation and inflation). If a food reward is made less valuable through pairings with emetics or simply by free access, relevant instrumental actions are attenuated. However, if these manipulations are conducted after extended training, no changes in behavior are seen and subjects still respond despite the outcome value having been reduced. Thus, following extended training, the action becomes an inflexible habit and is not sensitive to outcome devaluation. At this point, the behavior is guided by the stimuli present around the response instead of the outcome and is referred to as a stimulus-response (S-R) habit. This makes habits similar to reactions in that they are elicited by trigger stimuli, but they are distinct in that the response is innate in the case of reactions and learned to varying degrees (e.g., shuttle responses vs. lever press) in the case of habits. The underlying associative nature of the learned action changes over time from the hierarchical stimulus[response-outcome] (S[R-O]) representation to an S-R habit. Behavioral automation can be beneficial as it frees cognitive resources to attend to other demands. For instance, perfecting a skill such as bike-riding can become second nature requiring less attention. Furthermore, turning off the light when one leaves a room can save money. However, habits have a well-known downside too, with consequences much worse than simply leaving a roommate in a dark room. Habits achieve pathological status when they become maladaptive (e.g., compulsive hand-washing or compulsive taking of addictive drugs), and are very resistant to extinction when contingencies change.

3 Historical Context of Studies on Aversive Learning

The background above defines reactions and actions among various classes of behavior. Below, we focus on reactions and actions. However, prior to discussing the current state of research on these behaviors, we will briefly review the history of this field.

In aversive conditioning a neutral conditioned stimulus (CS), such as a tone, is repeatedly paired with an aversive unconditioned stimulus (US), usually an electric shock. Following this episode, the CS alone elicits conditioned freezing responses (CRs; Blanchard and Blanchard 1969a; Bolles and Fanselow 1980). This learning process was originally studied by Pavlov (1927) who referred to it as '*defensive conditioning.*' And while there were some notorious studies of the basic phenomenon in humans such as John Watson's famous studies involving little Albert (Watson 1929), early psychologists were more interested in Thorndike's (1898) 'law of effect' and instrumental conditioning. As a result, in the first half of the twentieth century, psychologists interested in aversive motivation were more likely to use aversive instrumental tasks, such as *avoidance conditioning* (Miller 1948, 1951; Mowrer 1947; Mowrer and Lamoreaux 1946) instead of Pavlovian conditioning.

In avoidance conditioning, animals learn to perform (active avoidance) or withhold (passive avoidance) an action in order to avoid negative or harmful results (typically electric shock). Below, we focus on active avoidance, which we will refer to as avoidance conditioning throughout. Responses that are measured in avoidance studies include shuttling (moving between rooms), wheel turning, or pressing a lever. Avoidance conditioning has traditionally been understood as a two-stage process. First, stimuli become associated with the shock via Pavlovian conditioning, and then, instrumental responses are performed that terminate or escape (or avoid) those threatening stimuli, thus providing negative reinforcement of the response (Brown and Jacobs 1949; Kalish 1954; Levis 1989; McAllister and McAllister 1971; Miller 1948, 1951; Mowrer 1947; Mowrer and Lamoreaux 1946; Overmier and Lawry 1979; Solomon and Wynne 1954). Because it helps cope with a threat, the avoidance response can, therefore, be used to study aversive motivation. Typically, Pavlovian conditioning processes were not directly measured or studied outside of avoidance procedures in the heyday of avoidance.

Starting in the 1950s, avoidance tasks were used to explore the brain mechanisms of aversive behavior, typically involving imprecise tools to understand neural function. Because avoidance conditioning intermixes Pavlovian (reaction) and instrumental (action) learning, studies using these tasks (Weiskrantz 1956; Goddard

1964; Gabriel et al. 1983; Isaacson 1982) were not able to provide a clear picture of the neural mechanisms underlying either the Pavlovian or instrumental learning processes (for review, see LeDoux et al. 2009; Sarter and Markowitsch 1985).

While psychologists struggled to understand aversive motivation with avoidance studies, research on the neural basis of learning and memory turned to Pavlovian conditioning, especially in invertebrates (Alkon 1983; Carew et al. 1983; Dudai et al. 1976; Kandel and Spencer 1968; Walters et al. 1979). Success in identifying neural mechanisms with simple tasks in invertebrates leads vertebrate researchers to take a similar approach (Cohen 1974; Thompson 1976). In the 1980s and 1990s, considerable progress was made in identifying the neural systems underlying Pavlovian aversive conditioning (Davis 1986; Kapp et al. 1979; LeDoux et al. 1983, 1984, 1988; for review, see: Davis 1992; Kapp et al. 1992; LeDoux 1992). Because simpler behaviors depend on simpler mechanisms, progress was rapidly made in understanding the brain mechanisms of defensive learning, which we will review below.

4 Neural Basis of Reactions

Pavlov's defense conditioning procedure came to be called fear conditioning. However, because this term implies that a subjective mental state of fear is being acquired, we have recently argued that this procedure should be called Pavlovian threat conditioning (PTC; LeDoux 2014, 2015). This term is preferable to defense conditioning since the latter term implies that the defensive responses are being learned, whereas PTC implies that the stimulus is acquiring a new meaning.

PTC is a widely used paradigm for studying the brain because of its simplicity, robustness, and repeatability (Fig. 1). In most studies, the CS of choice is an auditory tone and the US is a mild footshock. Learning can occur following the first CS–US pairing as is evidenced by defensive responding (freezing) during the second presentation of the CS (before the shock US is presented; Blanchard and Blanchard 1969a, 1972; Bolles and Fanselow 1980. Moreover, these associations are long-lasting, as they can be observed hours, days, weeks, and even years following conditioning (Gale et al. 2004).

In addition to behavioral responses to the CS, physiological changes such as autonomic and endocrine responses also occur (Kapp et al. 1979; LeDoux et al. 1983; Schneiderman et al. 1974), indicating that the CS, like the US, can elicit an integrated organismic response to perceived danger.

Research over the past several decades has clearly identified the amygdala as a key site where CS–US associations are formed as well as a necessary site for the later expression of defensive reactions elicited by the CS (Davis et al. 1997; Fanselow and LeDoux 1999; LeDoux 1996b, 2000; Maren 2001; Maren and Fanselow 1996; Fig. 2). In particular, the neural circuitry of auditory cued PTC has been well established.

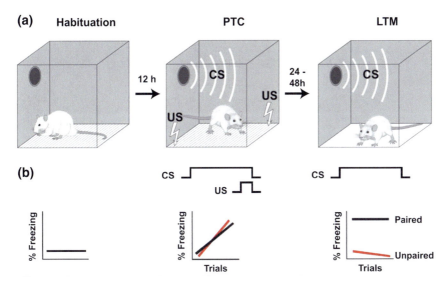

Fig. 1 Auditory Pavlovian threat conditioning: depiction of a standard Pavlovian threat conditioning (*PTC*) paradigm using an auditory cue as the CS (**a**) and typical results obtained (**b**). Prior to training, rats are habituated to the conditioning chamber in the absence of stimuli. The next day, animals are returned to the conditioning chamber and an unconditioned stimulus (*US*; a mild shock) is paired with the conditioned stimulus (*CS*; a tone), usually one to five times. After three hours for short-term memory (*STM*) or 24–48 h for long-term memory (*LTM*, depicted), animals are placed in a new context and exposed to multiple presentations of the CS alone. Innate defensive responses, i.e., 'freezing' behavior, are then scored as total time spent in this state over the course of the CS presentation. To test for non-associative effects, an 'unpaired' control group is often used in which CS and US do not co-occur. This unpaired control group shows very little CS-elicited freezing, indicating that associative learning processes are responsible for this defensive behavior

The amygdala may be separated into several regions, nuclei and subnuclei based on their emergence in evolution and neuroanatomical or cytoarchitectural features (Fig. 2). According to one scheme, the amygdala can be separated into an evolutionarily primitive division extending from subcortical areas (cortical, medial, and central nuclei) and a newer division associated with the neocortex (lateral, basal, and accessory basal nuclei) (Johnston 1923). The basal nucleus can be separated into basolateral and basomedial nuclei and the central nucleus into lateral, capsular, and medial subdivisions (Pitkänen 2000; Pitkänen et al. 2000). Considering the heterogeneity of these subnuclei, in terms of neuroanatomy, functional circuitry, and molecular expression, it is important to study these various subnuclei in isolation. Indeed, classic brain manipulations, including lesion studies and intracranial drug infusions, did not always account for the functional differences in these subareas (as highlighted in Romanski et al. 1993; Amano et al. 2011). Recent advances in molecular genetics, such as optogenetics, are allowing researchers to identify, with spatial and temporal precision, the distinct functions for subnuclear partitions within the amygdala (Ehrlich et al. 2009; Johansen et al. 2012).

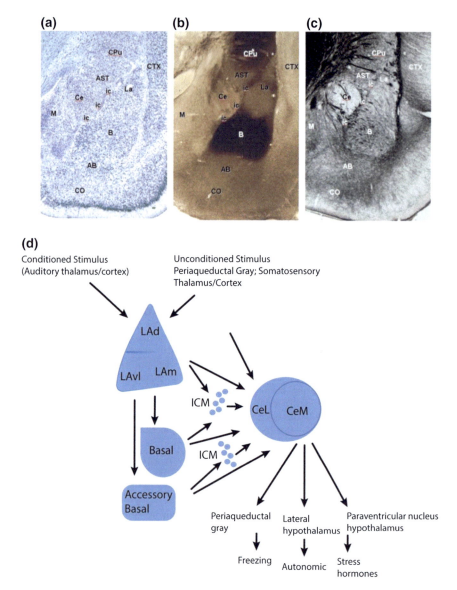

Fig. 2 Neural circuits underlying Pavlovian threat conditioning: a Anatomy of the amygdala: the rat amygdala may be separated into at least 12 distinct nuclei as identified using (a) Nissl cell body stain, b Acetylcholinesterase stain, and a c silver fiber stain. Abbreviations: Amygdala areas: *AB* accessory basal; *B* basal nucleus; *Ce* central nucleus; *CO* cortical nucleus; *ic* intercalated cells; *La* lateral nucleus; *M* medial nucleus. Non-amygdala areas: *AST* amygdalo-striatal transition area: *CPu* caudate putamen; *CTX* cortex. d Amygdala circuitry involved in the acquisition, consolidation, and expression of threat memories. The auditory CS (tone) and US (mild shock) converge in the lateral amygdala (*LA*) where synaptic plasticity and memory formation are thought to occur. The LA communicates with CeA directly and indirectly by way of connections in the basal, accessory basal, and intercalated cell masses (*ICM*). The CeA (CeM) sends projections to hypothalamic and brainstem areas to elicit the conditioned defensive response (freezing) as well as autonomic and hormonal responses. *CeL* lateral nucleus of Ce

Furthermore, advances in transgenic technology are allowing for manipulations of distinct cell subtypes (as defined by their molecular expression profile) within these subnuclei not only in mice, but also in rats and non-human primates (Gafford and Ressler 2015).

Decades of research indicate that these amygdala structures work together to coordinate defensive responses to threatening stimuli. Importantly, learning and expression of threat memories require the lateral nucleus of the amygdala (LA), the major input nucleus of the amygdala. Without the LA, and its efferent communication with the basal nucleus (B) and central nucleus (CeA), the CS cannot gain control of innate defensive responses.

Auditory PTC is thought to involve a Hebbian mechanism within the LA, whereby relatively weak auditory inputs on LA neurons are potentiated by concurrent strong depolarization produced by somatosensory inputs processing the US. The potentiation of auditory inputs increases the probability that LA neurons will show conditioned increases in cell firing when an auditory CS is presented. Indeed, this model is supported by significant work both in vitro (McKernan and Shinnick-Gallagher 1997; Rumpel et al. 2005; Clem and Huganir 2010; Tsvetkov et al. 2002; Schroeder and Shinnick-Gallagher 2004, 2005) and in vivo (Rogan and LeDoux 1995; Quirk et al. 1995; Grace and Rosenkranz 2002). With the development of new molecular tools for manipulating neural circuits, (i.e., optogenetics and chemogenetics), recent studies have provided more causal evidence for the importance of LA plasticity in PTC. For example, expanding on earlier studies shows that LA neurons can be selectively recruited to form memory by synthetically increasing their plasticity (i.e., infecting these neurons with the plasticity-related transcription factor CREB (cAMP Response Element Binding protein; Kida et al. 2002; Han et al. 2007); Josselyn and colleagues showed that activation of these specific neurons with DREADD (designer Receptors Exclusively Activated by Designer Drugs) increases PTC learning (Yiu et al. 2014). Moreover, it was recently shown that replacing footshock with light activation of LA principal cells expressing the excitatory opsin, channelrhodopsin-2 (ChR2), is sufficient for threat conditioning to an auditory tone (Johansen et al. 2010, 2014), and another recent study used optogenetics to 'engineer' a synthetic memory by inducing LTP at auditory synapses in LA paired with footshocks (Nabavi et al. 2014). Together, these data highlight the LA as a critical mediator of associative, Hebbian plasticity and memory formation.

The LA, in addition to receiving and integrating CS and US information for the formation of associative memories, also controls defense reactions via direct and indirect connections to CeA, a major output nucleus of the amygdala (Fig. 2). The indirect pathway is more prominent and involves synaptic connections to the basal nucleus (B) and the accessory basal nucleus (ABA), which make synaptic contacts on neurons in the central nucleus of the amygdala CeA. The CeA then coordinates the behavioral and autonomic responses to threat via long-range projections to the brainstem periaqueductal gray (PAG) and hypothalamus, respectively. Plasticity has also been shown to occur in CeA (Pascoe and Kapp 1985; Wilensky et al. 2006), and recent efforts have begun to dissect the microcircuitry within the CeA and the efferents that control defensive behaviors (see Ehrlich et al. 2009; Johansen et al. 2012 and Lüthi and Lüscher 2014 for a review of recent studies).

With the advent of modern molecular genetic techniques, recent work has corroborated and extended findings previously shown with more traditional techniques. The newfound ability to manipulate specific cell subtypes and circuits has allowed for detailed probing of circuitry to, within, and from the amygdala.

5 Diverse Functions of a Threat Stimulus: The Nature of Action and Reaction Learning in Aversive Motivation

The transition of the field toward the studies of purely Pavlovian conditioning led to considerable progress in understanding this fundamental phenomenon. However, studies of avoidance behavior are also valuable because they can provide additional insight into more complex forms of aversively motivated behavior. Knowledge obtained about the neural mechanisms involved in these processes could be very useful in understanding human disorders characterized by pathological avoidance. This is because in addition to eliciting anticipatory prepared responses, Pavlovian stimuli can do much more. As we have briefly pointed out earlier, Pavlovian stimuli play important roles in reinforcing instrumental responses during avoidance learning in tasks. Avoidance itself can be an adaptive way to cope with threats and exert control in dangerous situations, above and beyond the anticipatory responses produced by Pavlovian cues. Pavlovian stimuli can also function as incentives that motivate previously learned instrumental responses. Thus, Pavlovian processes can control the expression of maladaptive avoidance behaviors, and understanding this process on a neural level can help treat such behavioral problems. Below, we will review studies that have explored the neural circuits involved in supporting these different roles of aversive Pavlovian conditioned stimuli in avoidance and related behaviors. But first, we will briefly consider some conceptual issues regarding the psychological underpinnings of avoidance behavior.

A general feature of avoidance learning is negative reinforcement (Rescorla 1968) where some expected aversive event is removed contingent upon the performance of some target action (e.g., lever press or shuttling). Behaviors performed that remove these aversive events are subsequently more likely to be performed. While this would appear to reflect instrumental learning processes similar to those that guide actions when the reinforcer is a primary and appetitive one, there is some debate as to the exact nature of the underlying psychological substrates of this form of learning. An argument can be made that Pavlovian learning processes can explain avoidance behaviors without the involvement of goal-directed processes. This view is supported by observations that behaviors that closely approximate prepared defensive reactions (e.g., fleeing, rearing) are more readily acquired as avoidance responses than artificial responses (e.g., lever press; Bolles 1970). Additionally, some of these prepared behaviors (e.g., freezing) cannot be eliminated by punishment with shock, and this has been taken as evidence that they are not sensitive to control by instrumental contingencies and are, therefore, not truly instrumental behaviors (Fanselow and Lester 1988; Fanselow 1997). However, this is a complicated discussion that we will not focus on here. There has been evidence on both sides of this issue that suggest there is some basis for control of these behaviors by instrumental contingencies (Matthews et al. 1974; Killcross et al. 1997). For example, in addition to any modulation by Pavlovian associations, CS-escape and US-avoidance contingencies contribute independently to avoidance learning. Studies by Kamin (1956) as well as Herrnstein and Hineline (1966) show that learning is impaired if either contingency is discontinued. Furthermore, studies in appetitive learning suggest that similarly, some behaviors are more easily used as a means to food reinforcement (Shettleworth 1978). More germane to our current discussion, the neural control of these behaviors differs from basic Pavlovian learning processes, and differences also exist in neural control between different action-based tasks (i.e., different forms of avoidance or escape behavior). These differences will be considered below. Regardless of the exact nature of the representation developed through the learning of aversive actions, and specifically whether or not it is a true instance of instrumental learning, we believe unique processes underlie these actions and that further pursuit of their neural control is important. Through avoidance and related behaviors, organisms are able to reduce danger under a variety of circumstances, which attenuates defensive reactions and, therefore, has broad implications for understanding aversive motivation.

6 The Neural Basis of Aversively Motivated Actions

While a variety of preparations can be used to study avoidance and aversively motivated action learning, we focus on four main procedures here: signaled avoidance, unsignaled avoidance, escape from threat, and Pavlovian-to-instrumental transfer (PIT). We will discuss the underlying circuitry related to these tasks in the sections below.

6.1 Neural Control of Signaled and Unsignaled Avoidance Behavior

In a signaled avoidance procedure (SigA), a CS (e.g., tone) is first established as a threat via standard delay pairings of the CS with an aversive US (e.g., footshock). During later trials, presentation of the CS or the US can be terminated by performance of a target response. In many of our studies, subjects are trained to perform a shuttle response in a two-way chamber in order to control stimulus deliveries (e.g., see Fig. 3); we chose this seminatural response because of its similarity to prepared

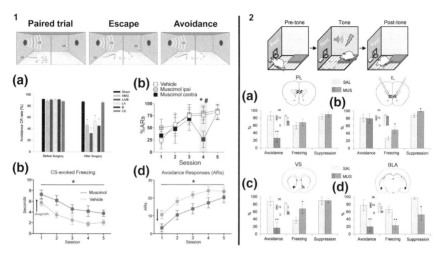

Fig. 3 Neural circuits underlying signaled active avoidance (SigA): *Left panel* shows data collected using a two-way shuttle response. Following an inescapable CS–US pairing, shuttle responses can terminate presentations of the CS and prevent the US. On trials where no response occurs during the CS, shuttling can escape the subsequent shock. **a** Lateral, basal, and central amygdala lesions impair SigA, from Choi et al. (2010). **b** Disconnection of basal amygdala and nucleus accumbens shell with muscimol impairs SigA, from Ramirez et al. (2015). Muscimol-induced inactivation of the infralimbic region of the prefrontal cortex increases CS-evoked freezing (**c**) and reduces avoidance responses (**d**) from Moscarello and LeDoux (2013). Right panel shows results using the step platform procedure (Bravo-Rivera et al. 2014, 2015). US delivery can be avoided by stepping onto the platform into the safety zone. Prelimbic (**a**) and not infralimbic (**b**) inactivation impairs SigA, as does inactivation of the ventral striatum (**c**) and basolateral complex of the amygdala (**d**)

fleeing behaviors among those found to have been used in the available literature. If the subject performs the shuttle response during the CS, the CS is removed, and additionally, the expected US is omitted. While both events (CS-offset and US-omission) possess the power to reinforce shuttling (see Kamin 1956), the contribution of each is difficult to disentangle in a SigA procedure because trials without responses result in further US deliveries.

Despite the limitations, early studies (see Gabriel et al. 2003) found that amygdala was important for avoidance, but what exact function it served could not be determined from these efforts. Using small lesions of amygdala subnuclei placed following baseline SigA training in rats, Choi et al. (2010) found that damage restricted to lateral (LA) or basal (BA) amygdala impaired performance, but that central (CeA) amygdala damage had no impact on SigA learning. Based on decades of studies in pure Pavlovian conditioning, the role of LA in SigA can be inferred; the Pavlovian learning presumed to be required for reinforcing and motivating the action via the CS is stored in LA. Without this associative knowledge, SigA learning cannot proceed. Given this role for LA, these findings suggest that BA is important for using the LA-based CS–US association to reinforce responding downstream. This pathway out of the amygdala as it relates to SigA will be further considered below.

It should be noted that a similar circuit is implicated when avoidance responses are acquired in the absence of any auditory CS. In unsignaled Sidman active avoidance (USA; see Sidman 1953), subjects are similarly placed in the two-way chamber, but are never exposed to a tone CS. Instead, in this procedure, footshock USs are presented at a fixed interval and any shuttle response that occurs postpones the next US, thus negatively reinforcing the response. Lazaro-Munoz et al. (2010) found that LA and BA but not CeA lesions impaired acquisition of USA behavior compared to sham controls. While the tone CS is absent from USA training procedures, the role of the context in promoting the response becomes more important, as do feedback cues presented when the response is performed. Because the context cannot be removed following a response in USA as a CS can be in SigA, the presence of discreet feedback cues has been found to help reinforce and shape USA. The feedback cues are negatively correlated with shock and may become safety signals that reinforce avoidance responding by counteracting the threat (Dinsmoor 2001; Rescorla 1968). Whether signaled by a discreet CS (as in SigA) or a context (as in USA), the associative component of avoidance learning appears to depend on LA, which after decades of Pavlovian conditioning studies may not appear surprising, but given that starting point, further analysis of avoidance circuitry can now proceed more quickly than before that was common knowledge. In both SigA and USA, BA appears critical for producing actions via LA-based associative knowledge regarding threats.

Before actions can be effectively promoted, incompatible reactions normally evoked by the CS must be sufficiently opposed to allow for the expression of action. There is evidence that this regulation is accomplished in a parallel manner via infra limbic ventromedial prefrontal cortex (PFC_{IL}) inhibition of CeA during signaled avoidance (Moscarello and LeDoux 2013; but see Bravo-Rivera et al. 2015). Preventing PFC_{IL} from normally functioning (permanently or temporarily) impairs avoidance behav2013ior. Subjects with lesions or muscimol inactivation of IL show reduced avoidance responding and increased freezing relative to control subjects (Moscarello and LeDoux 2013). Further evidence of this antagonistic process comes from subjects that fail to acquire SigA and USA, so-called *poor-performers*. These subjects exhibit more freezing and less avoidance relative to animals that acquire normally. If CeA is lesioned in these poorly performing subjects, normal responding emerges compared to sham-lesioned poor-performers (Choi et al. 2010; Lazaro-Munoz et al. 2010). This suggests that excessive CeA activity interferes with acquisition and expression of avoidance responses to cause poor performance in both tasks. Thus, in normal subjects, PFCiL comes to oppose CeA-mediated defensive reactions, which allows for expression of SigA. Whether this same regulatory process applies to USA has not been systematically investigated, but the findings regarding basic circuitry and response competition via CeA-induced freezing suggest convergence. Furthermore, Martinez et al. (2013) found that good USA performers showed more c-Fos in PFCiL compared to poor performers (however, see Bravo-Rivera et al. 2014, 2015 for results implicating PFC_{PL}).

While there has been a considerable amount of attention paid to the striatum in the context of appetitive instrumental conditioning and motivation (Cardinal et al. 2002; Setlow et al. 2002; Balleine and Killcross 2006; Shiflett and Balleine 2010), not much is known about how these structures contribute to aversive motivational processes. Cheer and colleagues (Oleson et al. 2012) have recently shown using fast-scan cyclic voltammetry that the amount of dopamine released during sub-second release events is increased in nucleus accumbens (NAcc) core following an avoidance response in a footshock SigA procedure. This result suggests that dopamine release in NAcc may play a critical role in reinforcement during SigA.

In agreement with these findings, Ramirez et al. (2015) found that expression of c-Fos in NAcc (but in the shell rather than the core) was significantly higher in subjects that were trained on SigA compared to yoked and box control subjects (see also Bravo-Rivera et al. 2014, 2015). Furthermore, this study also showed that muscimol inactivation of NAcc shell but not core impaired SigA performance. More powerful still temporary disconnection of BA and NAcc shell with muscimol impaired SigA. By inactivating BA and NAcc shell unilaterally, but on contralateral (or opposite) sides of the brain, communication between these structures, but not overall function is compromised. Because a single nucleus remains active in each hemisphere, this can mostly compensate for the inactivated counterpart. This manipulation impaired SigA compared to control subjects that received inactivation of these structures on the same side of the brain, thus leaving communication intact (Ramirez et al. 2015). Together, the Oleson et al. (2012) and Ramirez et al. (2015) findings clearly suggest a role for BA and NAcc in mediating acquisition and expression of SigA. There are many procedural and design difference between these studies; therefore, it is difficult to make easy comparisons and address discrepant findings regarding subregions of the NAcc (i.e., core vs. shell). For example, lever-press avoidance requires much more training than shuttle-based avoidance, and therefore, this could be a potential reason for the functional neuroanatomical difference. The NAcc shell may be required early on in training and the core later, when the behavior has transitioned to habit. Nevertheless, these findings provide strong evidence that a LA-BA-NAcc circuit is important for SigA behavior and that prefrontal regulation of CeA facilitates this process by reducing reactive freezing CRs. Whether this is also true of USA is currently not known.

A note on habitual avoidance behavior As briefly discussed above, actions that are repeatedly performed or overtrained are referred to as habits. Studies in appetitive motivation using reinforcer revaluation have found that early in training, instrumental actions are guided by the associated outcome and changes in the value of this outcome directly influence responding (Adams 1982; Balleine and Dickinson 1991, 1998). After extended training, performance of the response is no longer sensitive to the outcome value and persists despite manipulations that render the outcome devalued (e.g., taste aversion or selective satiety). The transition from an S[R-O] '*goal-directed*' action to an S-R '*habit*' has been found to depend on how the action is coordinated by the ventromedial prefrontal cortex. Killcross and Coutureau (2003) found that while sham control subjects showed reductions in lever-press behavior following pre-feeding only when given limited training, subjects with prelimbic lesions continued to demonstrate this devaluation effect regardless of training amount. This suggests that the transition to outcome independence and habitual action involved prelimbic function coming online to guide the response in a way that is unbound from the outcome representation. While the development of habits has not been thoroughly studied in the context of aversive motivation there is some evidence that a similar transition might apply. Poremba and Gabriel (1999) have shown that the passage of time renders avoidance behavior amygdala-independent. Using a discriminative footshock avoidance procedure in rabbits, muscimol inactivation of the amygdala impaired wheel-turning responses very early in training but not after extended training or following a seven-day rest period. In agreement with this result, using shuttling in rats, Lazaro-Munoz et al. (2010) found that amygdala lesions placed after extended USA training had no effect, while these lesions in another group impaired initial acquisition. This is in contrast to overtrained Pavlovian learning, which has been shown to not undergo this transition and remain amygdala dependent (Zimmerman et al. 2007). Avoidance behavior can be an effective means of coping with threats; however, habitual avoidance is maladaptive and disruptive. Whether the transition to amygdala independent avoidance discussed above reflects outcome independence as well is not known.

6.2 Neural Control of Escape from Threat

Acquisition of SigA responses requires BA, though it is not clear whether BA is sensitive to CS-offset, US-omission, or both. From a behavioral perspective, SigA does not permit isolation of these processes. As noted above, there have been studies that demonstrate the importance of the CS-escape and US-avoidance contingencies to learning (Kamin 1956) and also those that emphasized the importance of feedback in avoidance learning (Sidman 1953; Rescorla 1968). However, because Pavlovian and instrumental learning occur together in SigA, it is unclear which neural processes might pertain to which psychological phenomenon. In order to study this issue on a psychological level in the past, tasks were designed in which Pavlovian and instrumental conditioning occur in separate phases as opposed to the intermixed nature of SigA training described above. These tasks were referred to as '*escape from fear*' tasks. We will refer to them as '*escape from threat*' or EFT (see LeDoux 2014). EFT demonstrates that once a CS has been established as a threat, the removal of this threat alone can support acquisition of an instrumental response. This is because no shocks are presented during instrumental training (see Fig. 4). A trial never ends with a shock US during EFT training, even when no response occurs. Thus, the Pavlovian contingency is under extinction during the instrumental session. That responding is acquired under these conditions at all suggests that threat removal is a powerful reinforcement process on its own.

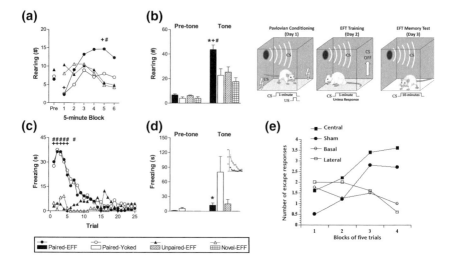

Fig. 4 Neural circuits underlying escape from threat (*EFT*): Experimental procedure developed by Cain and LeDoux (2007) (*Upper right corner*) Following standard Pavlovian conditioning, rearing can terminate CS presentations and retention of this response can be tested the following day. Over the course of EFT training, rearing frequency increases in subjects that received the R-O (i.e., CS-offset) contingency only, and yoked and unpaired control groups did not show this pattern. During EFT training, freezing decreases for both groups that received CS–US pairings during Pavlovian conditioning (**c**). More rearing is seen during the test session in EFT-trained subjects and (**b**) freezing spontaneously recovers in the yoked control group only (**d**). Using a one-way shuttle response, Amorapanth et al. (2000) found that EFT was impaired by lateral and basal amygdala lesions, but not central amygdala lesions

EFT has not been as thoroughly studied on a neural level as have SigA or USA due to complications with the task that have made it historically controversial. For example, the reliability and parametric uniformity of EFT learning have been sources of criticism for EFT as has a lack of testing procedures to demonstrate response retention (see Cain and LeDoux 2007). What research has been done suggests that the same amygdala circuitry is involved as in SigA. For example, using a one-way shuttle task, Amorapanth et al. (2000) found that LA and BA lesions impaired EFT, while CeA lesions did not. These findings suggest that a similar circuit may control both SigA and EFT forms of avoidance learning.

6.3 Incentive Motivation: Neural Circuits of Aversive Pavlovian-to-Instrumental Transfer

In addition to conditioned reinforcement, another function served by the CS in avoidance and related phenomena is conditioned motivation, which can be studied in isolation with Pavlovian-to-instrumental transfer (PIT). In PIT, an instrumental response is performed at some baseline rate, and presentation of the CS increases the rate with which this response is performed (Bolles and Popp 1964; Rescorla and LoLordo 1965; Rescorla 1968; Weisman and Litner 1969; Overmier and Payne 1971; Overmier and Brackbill 1977; Patterson and Overmier 1981; Campese et al. 2013; Nadler et al. 2011). The neural basis of PIT has been extensively studied in cases where the US is appetitive (e.g., food, or drugs; for a review see Holmes et al. 2010; Laurent et al. 2015). While some early studies used similar procedures in aversive motivation, only recently has there been progress on identifying the neural foundations of aversive PIT. Using early studies as a guide, we have developed a simple aversive PIT task for rats using USA shuttling in a two-way chamber as the instrumental response. As described earlier, subjects learn to avoid footshocks in the USA procedure independently from any discreet Pavlovian stimuli with an excitatory relationship to shock (i.e., in the absence of the tone CS). When a previously trained CS is presented during USA, subjects show little freezing to the CS and, instead, display enhanced USA behavior. An example of this PIT effect is shown below in Fig. 5 along with figures depicting the training procedure (see Campese et al. 2013).

We have begun to use this procedure to examine the neural circuits important for aversive PIT and have found evidence for a different pathway that processes incentive motivation than that seen for SigA, USA, and EFT acquisition. Using lesions, Campese et al. (2014) found that LA and CeA but not BA are required for

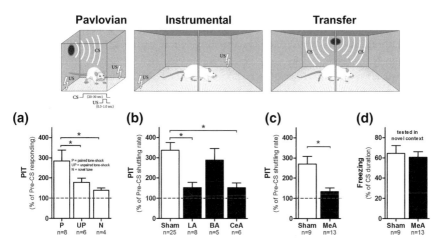

Fig. 5 Neural circuits underlying aversive Pavlovian-to-instrumental transfer (*PIT*): procedure used to study aversive PIT from Campese et al. (2013) (*top*). Following Pavlovian conditioning subjects undergo unsignaled Sidman active avoidance (USA) in the two-way shuttle chambers. During transfer testing, the CS is presented in this context and the effect on USA rate is measured. Subjects that received CS–US pairings during Pavlovian conditioning show an elevation of shuttle rate compared to the pre-CS period, while unpaired and naïve controls do not (**a** from Campese et al. 2013). Lateral, central but not basal amygdala lesions impair PIT (**b** from Campese et al. 2014). Medial amygdala lesions also impair PIT (**c**) but have no effect on CS-elicited freezing (**d** from McCue et al. 2014)

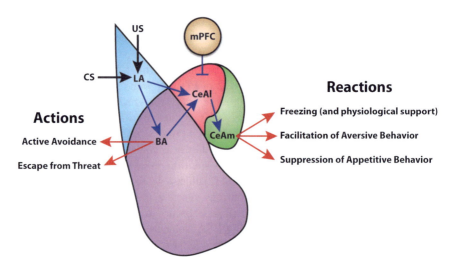

Fig. 6 Schematic representation of the circuits underlying aversively motivated instrumental reactions and actions: CS–US processing occurs in LA and 'informational' actions (SigA, EFT, USA) require the LA–BA pathway, while 'emotional' reactions (CS-elicited freezing, facilitation or suppression) engage the LA–CeA circuitry

aversive PIT. LA and CeA damage eliminated the facilitative effect of the CS, but BA lesions had no effect. Further studies (McCue et al. 2014) found that medial amygdala (MeA) is also required for PIT. One possibility suggested by these findings is that LA-CeA-MeA circuitry (see Fig. 6) is important for aversive PIT, departing from the LA-BA-NAcc circuit implicated in CS-based avoidance tasks discussed above. An important procedural difference between PIT and CS-based avoidance/escape (i.e., SigA) tasks is that avoidance/escape responses are learned in the repeated presence of the CS. This likely results in the CS becoming part of the associative structure during learning, where it comes to act as a discriminative or occasion-setting stimulus that signals when the instrumental (i.e., avoidance) contingency is in effect. This process engages extinction-like regulation of CeA defensive behavior (e.g., freezing) via the prefrontal cortex (Moscarello and LeDoux 2013; Bravo-Rivera et al. 2014, 2015), which allows for expression of active responding along the '*informational*' LA-BA-NAcc pathway. This is in contrast to PIT where the CS is not present during USA training and cannot come to influence the instrumental contingency. The CS is previously trained and is not presented again until transfer tests. Thus, the increased USA behavior observed in PIT tests can be considered a reaction because the CS has no control over the instrumental contingency. Therefore, during PIT, the CS modulates behavior through '*emotional/affective*' routes (mediated by CeA/MeA) rather than '*informational*' routes (mediated by BA). While studies in appetitive motivation have implicated NAcc (Shiflett and Balleine 2010; Laurent et al. 2015) in PIT, the contribution of this region to aversive PIT has not been explored as of yet (Figs. 5 and 6).

7 Summary

The systems described above depict a flexible and adaptive network capable of producing appropriate responses under a variety of circumstances motivated by threat or general aversive processes. These responses can be innate and reflexive, produced by evolutionarily selected stimuli or those given the power to convey threat through associative learning. The responses could alternatively be learned, artificial to some extent, but still prepared in other ways. It is clear that these dimensions of an aversively motivated response depend on unique circuitry more complex than that which accomplishes simple Pavlovian learning processes. There is evidence that microcircuits within the amygdala may contribute to conditioned motivation and conditioned reinforcement in different ways. These forms of aversively motivated learning have great relevance to clinical issues concerning coping strategies in treating anxiety, phobic disorders, and post-traumatic stress. Understanding the behavioral, neurobiological, and pharmacological basis of these phenomena can lead to profound progress in applications to these problems. While the fragility of extinction treatments are well known (e.g., spontaneous recovery, and renewal), avoidance learning produces reductions in defensive responding (i.e., active coping responses) that survive the passage of time and shifts in physical context (Bouton 2004; Cain and LeDoux 2007; LeDoux 2015). Additionally because maladaptive and inappropriate habitual avoidance behavior underlies emotional disorders in humans (e.g., agoraphobia, obsessive compulsive disorder), the importance of developing a rich understanding of these processes is twofold. Because of the great deal of knowledge obtained on PTC processes over decades of research, we are poised to learn more about these complex and elaborate circuits than was permitted in the past.

References

Adams C (1982) Variations in the sensitivity of instrumental responding to reinforcer devaluation. Q J Exp Psychol 34B:77–98

Alkon DL (1983) Learning in a marine snail. Sci Am 249:70–85

Amano T, Duvarci S, Popa D, Pare D (2011) The fear circuit revisited: contributions of the basal amygdala nuclei to conditioned fear. J Neurosci 31:15481–15489

Amorapanth P, LeDoux JE, Nader K (2000) Different lateral amygdala outputs mediate reactions and actions elicited by a fear-arousing stimulus. Nat Neurosci 3:74–79

Balleine BW, Dickinson A (1991) Instrumental performance following reinforcer devaluation depends upon incentive learning. Q J Exp Psychol 43B:279–296

Balleine BW, Dickinson A (1998) Goal-directed instrumental action: contingency and incentive learning and their cortical substrates. Neuropharmacology 37:407–419

Balleine BW, Killcross S (2006) Parallel incentive processing: an integrated view of amygdala function. Trends Neurosci 29:272–279

Blanchard RJ, Blanchard DC (1969a) Crouching as an index of fear. J Comp Physiol Psych 67:370–375

Blanchard RJ, Blanchard DC (1969b) Passive and active reactions to fear-eliciting stimuli. J Comp Physiol Psychol 68:129–135

Blanchard RJ, Blanchard DC (1972) Effects of hippocampal lesions on the Rat's reaction to a Cat. J Comp Physiol Psychol 78:77–82

Bolles RC (1970) Species-specific defense reactions and avoidance learning. Psychol Rev 77:32–48

Bolles RC (1972) Reinforcement, expectancy, and learning. Psychol Rev 79(5):394–409

Bolles RC, Popp RJ (1964) Parameters affecting the acquisition of sidman avoidance. J Exp Anal Behav 7(4):315

Bolles RC, Fanselow MS (1980) A perceptual-defensive-recuperative model of fear and pain. Behav Brain Sci 3:291–323

Bouton ME (2004) Context and behavioral processes in extinction. Learn Mem 11:485–494

Bravo-Rivera C, Roman-Ortiz C, Brignoni-Perez E, Sotres-Bayon F, Quirk GJ (2014) Neural structures mediating expression and extinction of platform-mediated avoidance. J Neurosci 34:9736–9742. PMC4099548

Bravo-Rivera C, Roman-Ortiz C, Montesinos-Cartagena M, Quirk GJ (2015) Persistent active avoidance correlates with activity in prelimbic cortex and ventral striatum. Front Behav Neurosci 9:184. http://doi.org/10.3389/fnbeh.2015.00184

Brown JS, Jacobs A (1949) The role of fear in the motivation and acquisition of responses. J Exp Psychol 39:747–759

Cain CK, LeDoux JE (2007) Escape from fear: a detailed behavioral analysis of two atypical responses reinforced by CS termination. J Exp Psychol Anim Behav Process 33:451–463

Campese V, McCue M, Lazaro-Munoz G, LeDoux JE, Cain CK (2013) Development of an aversive Pavlovian-to-instrumental transfer task in rat. Front Behav Neurosci 7:176. PMC3840425

Campese VD, Kim J, Lazaro-Munoz G, Pena L, LeDoux JE, Cain CK (2014) Lesions of lateral or central amygdala abolish aversive Pavlovian-to-instrumental transfer in rats. Front Behav Neurosci 8:161. PMC4019882

Cardinal RN, Parkinson JA, Hall J, Everitt BJ (2002) Emotion and motivation: the role of the amygdala, ventral striatum, and prefrontal cortex. Neurosci Biobehav Rev 26:321–352

Carew TJ, Hawkins RD, Kandel ER (1983) Differential classical conditioning of a defensive withdrawal reflex in Aplysia californica. Science 219:397–400

Choi JS, Cain CK, LeDoux JE (2010) The role of amygdala nuclei in the expression of auditory signaled two-way active avoidance in rats. Learn Mem 17:139–147

Clem RL, Huganir RL (2010) Calcium-permeable AMPA receptor dynamics mediate fear memory erasure. Science 330:1108–1112

Cohen DH (1974) The neural pathways and informational flow mediating a conditioned autonomic response. In: Di Cara, LV (ed) Limbic and autonomic nervous system research. Plenum Press, New York, pp 223–275

Coss RG, Biardi JE (1997) Individual variation in the antisnake behavior of California ground squirrels (Spermophilus beecheyi). J Mammal 78(2):294–310

Davis M (1986) Pharmacological and anatomical analysis of fear conditioning using the fear-potentiated startle paradigm. Behav Neurosci 100:814–824

Davis M (1992) The role of the amygdala in conditioned fear. In: Aggleton JP (ed) The amygdala: neurobiological aspects of emotion, memory, and mental dysfunction. Wiley-Liss, Inc, NY, pp 255–306

Davis M, Walker DL, Lee Y (1997) Roles of the amygdala and bed nucleus of the stria terminalis in fear and anxiety measured with the acoustic startle reflex. Possible relevance to PTSD. Ann N Y Acad Sci 21(821):305–331

Dinsmoor JA (2001) Stimuli inevitably generated by behavior that avoids electric shock are inherently reinforcing. J Exp Anal Behav 75(3):311–333

Dudai Y, Jan YN, Byers D, Quinn WG, Benzer S (1976) Dunce, a mutant of Drosophila deficient in learning. Proc Natl Acad Sci USA 73:1684–1688. PMC430364

Ehrlich I, Humeau Y, Grenier F, Ciocchi S, Herry C, Luthi A (2009) Amygdala inhibitory circuits and the control of fear memory. Neuron 62:757–771

Estes WK, Skinner BF (1941) Some quantitative properties of anxiety. J Exp Psychol 29:390–400

Estes WK (1948) Discriminative conditioning; effects of a Pavlovian conditioned stimulus upon a subsequently established operant response. J Exp Psychol 38:173–177

Fanselow MS, LeDoux JE (1999) Why we think plasticity underlying Pavlovian fear conditioning occurs in the basolateral amygdala. Neuron 23:229–232

Fanselow MS, Lester LS (1988) A functional behavioristic approach to aversively motivated behavior: predatory imminence as a determinant of the topography of defensive behavior. In: Bolles RC, Beecher MD (eds) Evolution and learning. Erlbaum, Hillsdale, N.J., pp 185–211

Fanselow MS (1980) Conditioned and unconditional components of post-shock freezing. Pavlovian J Biol Sci 15:177–182

Fanselow MS (1997) Species-specific defense reactions: retrospect and prospect. In: Bouton ME, Fanselow MS (eds) Learning, motivation, and cognition. American Psychological Association, Washington, D. C., pp 321–341

Gabriel M, Burhans L, Kashef A (2003) Consideration of a unified model of amygdalar associative functions. Ann N Y Acad Sci 985:206–217

Gabriel M, Lambert RW, Foster K, Orona E, Sparenborg S, Maiorca RR (1983) Anterior thalamic lesions and neuronal activity in the cingulate and retrosplenial cortices during discriminative avoidance behavior in rabbits. Behav Neurosci 97:675–696

Gale GD, Anagnostaras SG, Godsil BP, Mitchell S, Nozawa T, Sage JR, Wiltgen B, Fanselow MS (2004) Role of the basolateral amygdala in the storage of fear memories across the adult lifetime of rats. J Neurosci 24:3810–3815

Goddard G (1964) Functions of the amygdala. Psychol Rev 62:89–109

Gafford GM, Ressler KJ (2015). Mouse models of fear-related disorders: cell-type-specific manipulations in amygdala. Neuroscience doi: 10.1016/j.neuroscience.2015.06.019

Grace AA, Rosenkranz JA (2002) Regulation of conditioned responses of basolateral amygdala neurons. Physiol Behav 77:489–493

Han JH, Kushner SA, Yiu AP, Cole CJ, Matynia A, Brown RA, Neve RL, Guzowski JF, Silva AJ, Josselyn SA (2007) Neuronal competition and selection during memory formation. Science 316:457–460

Hinde RA (1966) Animal behaviour. McGraw-Hill, New York

Herrnstein RJ, Hineline PN (1966) Negative reinforcement as shock-frequency reduction. J Exp Anal Behav 9(4):421–30

Hirsch SM, Bolles RC (1980) Zeitschrift für Tierpsychologie 54(1):71–84

Holland PC, Gallagher M (2003) Double dissociation of the effects of lesions of basolateral and central amygdala on conditioned stimulus-potentiated feeding and Pavlovian-instrumental transfer. Eur J Neurosci 17:1680–1694

Holmes NM, Marchand AR, Coutureau E (2010) Pavlovian to instrumental transfer: a neurobehavioural perspective. Neurosci Biobehav Rev 34:1277–1295

Hull CL (1943) Principles of behavior. Appleton-Century-Crofts, New York

Isaacson RL (1982) The limbic system. Plenum Press, New York

Johansen JP, Diaz-Mataix L, Hamanaka H, Ozawa T, Ycu E, Koivumaa J, Kumar A, Hou M, Deisseroth K, Boyden ES, LeDoux JE (2014) Hebbian and neuromodulatory mechanisms interact to trigger associative memory formation. Proc Natl Acad Sci USA 111:E5584–E5592. PMC4280619

Johansen JP, Hamanaka H, Monfils MH, Behnia R, Deisseroth K, Blair HT, LeDoux JE (2010) Optical activation of lateral amygdala pyramidal cells instructs associative fear learning. Proc Natl Acad Sci USA 107(28):12692–12697. http://doi.org/10.1073/pnas.1002418107

Johansen JP, Wolff SB, Luthi A, LeDoux JE (2012) Controlling the elements: an optogenetic approach to understanding the neural circuits of fear. Biol Psychiatry 71:1053–1060

Johnston JB (1923) Further contribution to the study of the evolution of the forebrain. J Comp Neurol 35:337–481
Kalish HI (1954) Strength of fear as a function of the number of acquisition and extinction trials. J Exp Psychol 47:1–9
Kamin LJ (1956) The effects of termination of the CS and avoidance of the US on avoidance learning. J Comp Physiol Psychol 49:420–424
Kandel ER, Spencer WA (1968) Cellular neurophysiological approaches to the study of learning. Physiol Rev 48:65–134
Kapp BS, Frysinger RC, Gallagher M, Haselton JR (1979) Amygdala central nucleus lesions: effect on heart rate conditioning in the rabbit. Physiol Behav 23:1109–1117
Kapp BS, Whalen PJ, Supple WF, Pascoe JP (1992) Amygdaloid contributions to conditioned arousal and sensory information processing. In: Aggleton JP (ed) The amygdala: neurobiological aspects of emotion, memory, and mental dysfunction. Wiley-Liss, New York, pp 229–254
Kida S, Josselyn SA, de Ortiz SP, Kogan JH, Chevere I, Masushige S, Silva AJ (2002) CREB required for the stability of new and reactivated fear memories. Nat Neurosci 5:348–355
Killcross S, Coutureau E (2003) Coordination of actions and habits in the medial prefrontal cortex of rats. Cereb Cortex 13(4):400–408
Killcross S, Robbins TW, Everitt BJ (1997) Different types of fear-conditioned behaviour mediated by separate nuclei within amygdala. Nature 388:377–380
Lang PJ, Davis M (2006) Emotion, motivation, and the brain: reflex foundations in animal and human research. Prog Brain Res 156:3–29
Laurent V, Morse AK, Balleine BW (2015) The role of opioid processes in reward and decision-making. Br J Pharmacol 172(2):449–459. doi:10.1111/bph.12818
Lazaro-Munoz G, LeDoux JE, Cain CK (2010) Sidman instrumental avoidance initially depends on lateral and Basal amygdala and is constrained by central amygdala-mediated Pavlovian processes. Biol Psychiatry 67:1120–1127
LeDoux JE (1992) Emotion and the amygdala. In: Aggleton JP (ed) The amygdala: neurobiological aspects of emotion, memory, and mental dysfunction. Wiley-Liss, Inc, New York, pp 339–351
LeDoux JE (1996a) The emotional brain. Simon and Schuster, New York
LeDoux J (1996b) Related articles emotional networks and motor control: a fearful view. Prog Brain Res 107:437–446. Review
LeDoux JE (2000) Emotion circuits in the brain. Annu Rev Neurosci 23:155–184
LeDoux JE (2014) Coming to terms with fear. Proc Natl Acad Sci USA 111:2871–2878
LeDoux JE (2015) Anxious. Viking, New York
LeDoux JE, Gorman JM (2001) A call to action: overcoming anxiety through active coping. Am J Psychiatry 158:1953–1955
LeDoux JE, Iwata J, Cicchetti P, Reis DJ (1988) Different projections of the central amygdaloid nucleus mediate autonomic and behavioral correlates of conditioned fear. J Neurosci, 8 (7):2517–2529
LeDoux JE, Sakaguchi A, Reis DJ (1983a) Strain difference in fear between spontaneously hypertensive and normotensive rats. Brain Res 227:137–143
LeDoux JE, Thompson ME, Iadecola C, Tucker LW, Reis DJ (1983b) Local cerebral blood flow increases during auditory and emotional processing in the conscious rat. Science 221:576–578
LeDoux JE, Sakaguchi A, Reis DJ (1984) Subcortical efferent projections of the medial geniculate nucleus mediate emotional responses conditioned to acoustic stimuli. J Neurosci 4:683–698
LeDoux JE, Schiller D, Cain C (2009) Emotional reaction and action: from threat processing to goal-directed behavior. In: Gazzaniga MS (ed) The cognitive neurosciences. MIT Press, Cambridge, pp 905–924
Levis DJ (1989) The case for a return to a two-factor theory of avoidance: the failure of non-fear interpretations. In: Klein SB, Mowrer RR (eds) Contemporary learning theories: Pavlovian conditioning and the status of traditional learning theory. Lawrence Erlbaum Assn, Hillsdale, pp 227–277

Lorenz KZ, Tinbergen N (1938) Taxis und instinktbegriffe in der Eirollbewegung der Graugans. Z Tierpsych 2:1–29

Lovibond PF (1983) Facilitation of instrumental behavior by a Pavlovian appetitive conditioned stimulus. J Exp Psychol Anim Behav Process 9:225–247

Lüthi A, Lüscher C (2014) Pathological circuit function underlying addiction and anxiety disorders. Nat Neurosci 17(12):1635

Maren S, Fanselow MS (1996) The amygdala and fear conditioning: has the nut been cracked? Neuron 16:237–240

Maren S (2001) Neurobiology of Pavlovian fear conditioning. Annu Rev Neurosci 24:897–931

Martinez RC, Gupta N, Lazaro-Munoz G, Sears RM, Kim S, Moscarello JM, LeDoux JE, Cain CK (2013) Active vs. reactive threat responding is associated with differential c-Fos expression in specific regions of amygdala and prefrontal cortex. Learn Mem 20:446–452. PMCPMC3718200

Matthews TJ, McHugh TG, Carr LD (1974) Pavlovian and instrumental determinants of response suppression in the pigeon. J Comp Physiol Psychol 87(3):500–506

McAllister WR, McAllister DE (1971) Behavioral measurement of conditioned fear. In: Brush FR (ed) Aversive conditioning and learning. Academic Press, New York, pp 105–179

McCue MG, LeDoux JE, Cain CK (2014) Medial amygdala lesions selectively block aversive pavlovian-instrumental transfer in rats. Front Behav Neurosci 8:329. PMC4166994

McKernan MG, Shinnick-Gallagher P (1997) Fear conditioning induces a lasting potentiation of synaptic currents in vitro. Nature 390:607–611

Miller NE (1948) Studies of fear as an acquirable drive: I. Fear as motivation and fear reduction as reinforcement in the learning of new responses. J Exp Psychol 38:89–101

Miller NE (1951) Learnable drives and rewards. In: Stevens SS (ed) Handbook of experimental psychology. Wiley, New York, pp 435–472

Moscarello JM, LeDoux JE (2013) Active avoidance learning requires prefrontal suppression of amygdala-mediated defensive reactions. J Neurosci 33:3815–3823

Mowrer OH, Lamoreaux RR (1946) Fear as an intervening variable in avoidance conditioning. J Comp Psychol 39:29–50

Mowrer OH (1947) On the dual nature of learning: a reinterpretation of "conditioning" and "problem solving". Harvard Educ Rev 17:102–148

Nabavi S, Fox R, Proulx CD, Lin JY, Tsien RY, Malinow R (2014) Engineering a memory with LTD and LTP. Nature 511(7509):348–352. http://doi.org/10.1038/nature13294

Nadler N, Delgado MR, Delamater AR (2011) Pavlovian to instrumental transfer of control in a human learning task. Emotion 11:1112–1123. PMC3183152

Niv Y, Joel D, Dayan P (2006) A normative perspective on motivation. Trends Cogn Sci 10 (8):375–381. (Epub 2006 Jul 13)

Oleson EB, Gentry RN, Chioma VC, Cheer JF (2012) Subsecond dopamine release in the nucleus accumbens predicts conditioned punishment and its successful avoidance. J Neurosci 17;32 (42):14804–14808. doi:10.1523/JNEUROSCI.3087-12.2012

Overmier JB, Brackbill RM (1977) On the independence of stimulus evocation of fear and fear evocation of responses. Behav Res Ther 15:51–56

Overmier JB, Lawry JA (1979) Pavlovian conditioning and the mediation of avoidance behavior. In: Bower G (ed) The psychology of learning and motivation, vol 13. Academic Press, New York, pp 1–55

Patterson J, Overmier JB (1981) A transfer of control test for contextual associations. Anim Learn Behav 9:316–321

Overmier JB, Payne RJ (1971) Facilitation of instrumental avoidance learning by prior appetitive Pavlovian conditioning to the cue. Acta Neurobiol Exp (Wars) 31:341–349

Pascoe JP, Kapp BS (1985) Electrophysiological characteristics of amygdaloid central nucleus neurons in the awake rabbit. Brain Res Bull 14(4):331–338

Pavlov IP (1927) Conditioned reflexes. Dover, New York

Pitkänen A (2000) Connectivity of the rat amygdaloid complex. In: Aggleton JP (ed) The amygdala: a functional analysis. Oxford University Press, Oxford, pp 31–115

Pitkänen A, Pikkarainen M, Nurminen N, Ylinen A (2000) Reciprocal connections between the amygdala and the hippocampal formation, perirhinal cortex, and postrhinal cortex in rat. A review. Ann N Y Acad Sci 911:369–391

Poremba A., Gabriel M (1999) Amygdala neurons mediate acquisition but not maintenance of instrumental avoidance behavior in rabbits. J Neurosci 19:9635–9641

Quirk GJ, Repa C, LeDoux JE (1995) Fear conditioning enhances short-latency auditory responses of lateral amygdala neurons: parallel recordings in the freely behaving rat. Neuron 15:1029–1039

Ramirez F, Moscarello JM, LeDoux JE, Sears RM (2015) Active avoidance requires a serial Basal amygdala to nucleus accumbens shell circuit. J Neurosci 35:3470–3477

Rescorla RA, Lolordo VM (1965) Inhibition of avoidance behavior. J Comp Physiol Psychol 59:406–412

Rescorla RA (1968) Pavlovian conditioned fear in Sidman avoidance learning. J Comp Physiol Psychol 65(1):55–60

Rogan MT, LeDoux JE (1995) LTP is accompanied by commensurate enhancement of auditory-evoked responses in a fear conditioning circuit. Neuron 15:127–136

Romanski LM, Clugnet MC, Bordi F, LeDoux JE (1993) Somatosensory and auditory convergence in the lateral nucleus of the amygdala. Behav Neurosci 107(3):444–450

Rosen JB (2004) The neurobiology of conditioned and unconditioned fear: a neurobehavioral system analysis of the amygdala. Behav Cogn Neurosci Rev 3:23–41

Rumpel S, LeDoux J, Zador A, Malinow R (2005) Postsynaptic receptor trafficking underlying a form of associative learning. Science 308:83–88

Sarter MF, Markowitsch HJ (1985) Involvement of the amygdala in learning and memory: a critical review, with emphasis on anatomical relations. Behav Neurosci 99:342–380

Schneiderman N, Francis J, Sampson LD, Schwaber JS (1974) CNS integration of learned cardiovascular behavior. In: DiCara LV (ed) Limbic and autonomic nervous system research. Plenum, New York, pp 277–309

Schroeder BW, Shinnick-Gallagher P (2004) Fear memories induce a switch in stimulus response and signaling mechanisms for long-term potentiation in the lateral amygdala. Eur J Neurosci 20:549–556

Schroeder BW, Shinnick-Gallagher P (2005) Fear learning induces persistent facilitation of amygdala synaptic transmission. Eur J Neurosci 22(7):1775–1783

Setlow B, Holland PC, Gallagher M (2002) Disconnection of the basolateral amygdala complex and nucleus accumbens impairs appetitive pavlovian second-order conditioned responses. Behav Neurosci 116:267–275

Sidman M (1953) Avoidance conditioning with brief shock and no extero- ceptive warning signal. Science 118:157–158

Shettleworth SJ (1978) Reinforcement and the organization of behavior in golden hamsters: Pavlovian conditioning with food and shock USs. J Exp Psychol Anim Behav Process 4:152–169

Shiflett MW, Balleine BW (2010) At the limbic-motor interface: disconnection of basolateral amygdala from nucleus accumbens core and shell reveals dissociable components of incentive motivation. Eur J Neurosci 32(10):1735–1743. doi: 10.1111/j.1460-9568.2010.07439.x. (Epub Oct 7)

Skinner BF (1938) The behavior of organisms: an experimental analysis. Appleton-Century-Crofts, New York

Solomon RL, Wynne LC (1954) Traumatic avoidance learning: the principles of anxiety conservation and partial irreversibility. Psychol Rev 61:353

Thompson RF (1976) The search for the engram. Am Psychol 31:209–227

Thorndike EL (1898) Animal intelligence: an experimental study of the associative processes in animals. Psychol Monogr 2:109

Thorpe WH (1963) Learning and instinct in animals. Methuen, London

Tinbergen N (1951) The study of instinct. Oxford University Press, New York

Tsvetkov E, Carlezon WA, Benes FM, Kandel ER, Bolshakov VY (2002) Fear conditioning occludes LTP-induced presynaptic enhancement of synaptic transmission in the cortical pathway to the lateral amygdala. Neuron 34:289–300

Walters ET, Carew TJ, Kandel ER (1979) Classical conditioning in Aplysia californica. Proc Natl Acad Sci USA 76:6675–6679

Watson JB (1929) Behaviorism. W. W. Norton, New York

Weiskrantz L (1956) Behavioral changes associated with ablation of the amygdaloid complex in monkeys. J Comp Physiol Psychol 49:381–391

Weisman RG, Litner JS (1969) The course of Pavlovian excitation and inhibition of fear in rats. J Comp Physiol Psychol 69:667–672

Wilensky AE, Schafe GE, Kristensen MP, LeDoux JE (2006) Rethinking the fear circuit: the central nucleus of the amygdala is required for the acquisition, consolidation, and expression of pavlovian fear conditioning. J Neurosci 26:12387–12396

Yin HH, Knowlton BJ (2006) The role of the basal ganglia in habit formation. Nat Rev Neurosci 7(6):464–476

Yiu AP, Mercaldo V, Yan C, Richards B, Rashid AJ, Hsiang HL, Pressey J, Mahadevan V, Tran MM, Kushner SA, Woodin MA, Frankland P, Josselyn SA (2014) Neurons are recruited to a memory trace based on relative neuronal excitability immediately before training. Neuron 6;83(3):722–735. doi: 10.1016/j.neuron.2014.07.017

Zimmerman JM, Rabinak CA, McLachlan IG, Maren S (2007) The central nucleus of the amygdala is essential for acquiring and expressing conditional fear after overtraining. Learn Mem 14:634–644

Part II
Neural Measures and Correlates of Motivation Signals and Computations

Neurophysiology of Reward-Guided Behavior: Correlates Related to Predictions, Value, Motivation, Errors, Attention, and Action

Gregory B. Bissonette and Matthew R. Roesch

Abstract Many brain areas are activated by the possibility and receipt of reward. Are all of these brain areas reporting the same information about reward? Or are these signals related to other functions that accompany reward-guided learning and decision-making? Through carefully controlled behavioral studies, it has been shown that reward-related activity can represent reward expectations related to future outcomes, errors in those expectations, motivation, and signals related to goal- and habit-driven behaviors. These dissociations have been accomplished by manipulating the predictability of positively and negatively valued events. Here, we review single neuron recordings in behaving animals that have addressed this issue. We describe data showing that several brain areas, including orbitofrontal cortex, anterior cingulate, and basolateral amygdala signal reward prediction. In addition, anterior cingulate, basolateral amygdala, and dopamine neurons also signal errors in reward prediction, but in different ways. For these areas, we will describe how unexpected manipulations of positive and negative value can dissociate signed from unsigned reward prediction errors. All of these signals feed into striatum to modify signals that motivate behavior in ventral striatum and guide responding via associative encoding in dorsolateral striatum.

Keywords Reward · Value · Motivation · Attention · Prediction error · Decision-making

---

G.B. Bissonette (✉) · M.R. Roesch
Department of Psychology, University of Maryland,
College Park, MD, USA
e-mail: gbissone@umd.edu

M.R. Roesch
e-mail: mroesch@umd.edu

G.B. Bissonette · M.R. Roesch
Program in Neuroscience and Cognitive Science,
University of Maryland, College Park, MD, USA

© Springer International Publishing Switzerland 2015
Curr Topics Behav Neurosci (2016) 27: 199–230
DOI 10.1007/7854_2015_382
Published Online: 15 August 2015

Contents

1	Introduction	200
2	Value Versus Motivation and Salience	202
	2.1 Orbitofrontal Cortex (OFC)	203
	2.2 Nucleus Accumbens	206
	2.3 Parietal Cortex	207
3	Signed Prediction Error Versus Attention/Salience	208
	3.1 Midbrain Dopamine (DA) Neurons	209
	3.2 Basolateral Amygdala (ABL)	210
	3.3 Anterior Cingulate (ACC)	213
4	Correlates of Motivation and Associative Encoding in Striatum	214
5	Integration of Positive and Negative Information into a Common Output Signal	218
6	Conclusion	220
References		221

1 Introduction

Imagine that you ritualistically purchase your morning coffee from the same place every day, but one day, you have a bad experience and the coffee is subpar. What will you do? Will you continue to habitually follow your routine or will you take your business elsewhere? It seems like a relatively simple computation, but it is remarkable to think about how many different brain signals come into play in situations like these. At the very least, you need to break the habit and be motivated to pursue new goals. This involves detection of errors and allocation of attention so that new associations can be formed. Subsequently, you must consider potential options and their economic value, as well as the probability of achieving those options. In this chapter, we examine neural correlates of these functions in animals performing tasks during which past and potential experiences—both positive and negative—modify behavior.

Over the last two decades, the number of brain areas in which activity has been shown to be modulated by rewards and cues that predict reward has escalated dramatically. In fact, all the brain areas illustrated in Fig. 1a contain neurons that increase firing to cues that predict reward and to reward themselves (and even this is not an all-inclusive list). This raises an important question: Are all these brain areas encoding the exact same information or are they subserving different functional aspects of reward processing? Recent work has tried to tease apart these functions to determine what exactly is being encoded by nodes in this circuit. Below, we describe single neuron recordings in behaving animals that have addressed this issue.

This chapter is broken down into four sections. Each section describes experiments that have tried to parse "reward-related" neural activity into the functions illustrated in Fig. 1b. This figure tries to encapsulate all possible functions that go

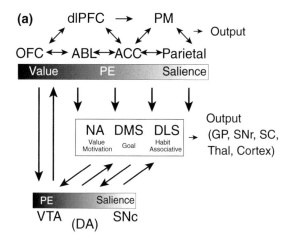

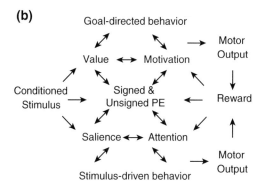

Fig. 1 **a** Circuit diagram demonstrating connectivity between brain regions involved in reward-guided decision-making. *Arrows* represent direction of information flow where *single-headed arrows* are unidirectional and *double-headed arrows* are reciprocal. Words in *shaded box* specify functions and *shading of box* provides a general idea of the role that nearby anatomical labels play in the strength of these functions. **b** Interplay of functions related to reward-guided decision-making. Orbitofrontal cortex *OFC*, dorsal-lateral prefrontal cortex *PFC*, basolateral amygdala *ABL*, anterior cingulate cortex *ACC*, parietal cortex *Parietal*, premotor cortex *PM*, nucleus accumbens *NA*, dorsal-medial striatum *DMS*, dorsolateral striatum *DLS*, ventral tegmental area *VTA*, dopamine *DA*, substantia nigra compacta *SNc*, globus pallidus *GP*, thalamus *Thal*, substantia nigra reticulata *SNr*, prediction error *PE*. Adapted from Bissonette et al. (2014) and Burton et al. (2014)

into making simple decisions based on anticipated reward and punishment. It is clear that this is an extremely complex computation! Several of the arrows linking proposed functions are bidirectional, illustrating that most mechanisms involved are highly interrelated and influence each other. When examining neural correlates of these functions, great care needs to be taken when linking firing patterns of single units to functions related to decision-making.

In the first section, we examine studies that distinguish *value* from other signals that covary with value, such as *motivation* and *salience*. Learned value can be defined as the relative anticipated worth that some cue predicts, either positive or negative. Motivation is the enhancement or decrement of motor output based on an increased or decreased level of arousal. More specifically, a stimulus is "salient" if it leads to a general increase in arousal or is *attention* grabbing, whereas something is "motivational" if it enhances motor behaviors. In the first section, we will focus on dissociating "value" signals from signals related to motivation and salience. Notably, the dissociation between the latter two is much more difficult to study and requires further investigation; however, signals directly related to motivated motor output should be observed in the period leading up the behavioral response, whereas signals related to attention or salience might solely be during the presentation of cues (i.e., salient events), but not necessarily the actions associated with them.

In the second section, we examine neurons that increase firing to the delivery of unexpected outcomes, which are critical for reporting surprising events so that learning can occur. Several brain areas respond to reward delivery but few specifically report errors in reward prediction. To uncover these correlates, experimental paradigms must violate expectations in both positive (outcome better than expected) and negative (outcome worse than expected) directions. This allows researchers to dissociate *signed* from *unsigned prediction error signals*. Signed prediction error signals lead to changes in the associate strength of conditioned stimuli that predict anticipated value, whereas unsigned prediction error signals lead to increased attention so that learning can occur.

All of these signals (value, motivation, prediction errors, attention, etc.) must modulate systems that guide behavior. The striatum is thought to be one major interface that integrates this information with motor output. In the third section, we will describe studies that demonstrate a basic trend along the diagonal of striatum (ventral-medial to dorsal-lateral) by which correlates are more reward-related in ventral-medial regions, whereas correlates are more associative and motor-related in more dorsal-lateral sections. Finally, in the fourth section, we discuss where all of these "reward-related" signals might be integrated into a common signal.

2 Value Versus Motivation and Salience

Neurons that increase firing prior to a desirable reward might be encoding value or they might reflect the increased motivation or salience associated with receiving this reward. It is not trivial to dissociate value from motivation or salience because they covary in most situations; that is, the more you value something, the more motivated you are to work to obtain it (Solomon and Corbit 1974; Daw et al. 2002; Lang and Davis 2006; Phelps et al. 2006; Anderson 2005; Anderson et al. 2011). The literature has extensively examined neural systems involved in making decisions based on potential outcomes, both good and bad, but whether these signals reflect value or motivation/salience is still not entirely clear. As described above, we define motivation as a process that invigorates motor responding and salience as cues that are attention grabbing or arousing. The two are intertwined; salient cues might lead to increased motivation, both of which might be triggered by cues that predict valued reward and induce faster responding. Importantly, neural correlates related with value can be dissociated from motivation and salience by examining firing to aversive stimuli, which have low value but are highly salient and motivating (Fig. 2a). A classic example is that of the carrot and the stick. A donkey finds both salient and motivating, but the carrot has high value, whereas the stick has low value.

By manipulating both anticipated appetitive and aversive events, experimental procedures can dissociate value from motivation and salience. That is, cues that predict appetitive and aversive outcomes have opposite values, but both are highly salient and motivational. Several studies have done just that, dissociating these signals by motivating behavior with both the promise of reward and the threat of punishment. In these experiments, animals learn that conditioned stimuli (CS) predict potential rewards or the possibility of punishment. Typically, there are three trial types where the CS predicts: (1) a large reward (e.g., sucrose, juice); (2) a neutral condition or a small (or no) reward; and (3) a small (or no) reward with the threat of an aversive outcome, such as delivery of a bitter quinine solution, electric shock, or air puff to the eye (Rolls et al. 1989; Roesch and Olson 2004; Roesch et al. 2010a; Bissonette et al. 2013; Matsumoto and Hikosaka 2009; Brischoux et al. 2009; Anstrom et al. 2009; Lammel et al. 2011). If neurons encode *value*, neural activity should show a monotonic relationship during appetitive, neutral, and aversive trials (Fig. 2a; theoretical neural signals). If activity is modulated by factors like motivation or salience that vary with the strength of appetitive and aversive stimuli, neurons should respond with the same "sign" for appetitive and aversive trials compared to neutral trials. This approach has been applied to several brain areas thought to represent value in one form or another.

2.1 Orbitofrontal Cortex (OFC)

Most evidence prior to the described work clearly suggested that activity in OFC reflected value. OFC is strongly connected with the limbic system and has proven to be a key associative structure (Pickens et al. 2003; Burke et al. 2008; Ostlund and Balleine 2007; Izquierdo et al. 2004). Neurons in OFC respond to cues that predict differently valued rewards, as well as to the anticipation and delivery of rewards themselves. Activity in OFC is modulated by a number of economic factors (e.g., probability, effort, delay, and size) for both appetitive and aversive stimuli. Its firing is also influenced by the availability of alternative rewards (i.e., relative reward value) and how satiated the animal is. Still, all these neural representations might reflect how salient or motivating the predicted reward is, not its value.

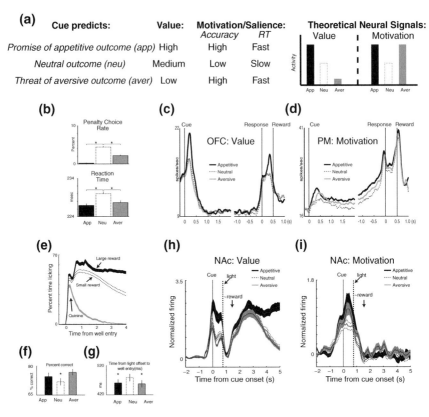

Fig. 2 a Tasks that dissociate value from motivation/salience correlate by varying appetitive and aversive outcomes. In these tasks, one trial type promises a large reward with little punishment (app = appetitive); another promises a small reward with little punishment (neu = neutral); and a third promises the small reward, but threatens the animal with a large punishment (aver = aversive). Both primates and rats performing tasks like these prefer large reward and dislike large punishment (aver) relative to neutral, but are highly motivated by both as indicated by faster reaction times and better performance. Thus, theoretically, if neurons in the brain participate in value computations, then they should fire highest for appetitive trial types and lowest for aversive trial types ("value"). If neurons participate in signals that reflect motivation/salience, their activity should be high for both appetitive and aversive trial types. **b** Monkey study that dissociated value from motivation. Accuracy and RT data from primates, illustrating that monkeys were faster and more accurate for large reward and punishment trials, versus neutral trials. **c** Neural recordings in primate OFC demonstrate higher activity for large over small reward cues, whereas activity in premotor cortex (PM) **d** reflected the level of motivation associated with those outcomes. **e–g**, behavior data from rats performing a similar task. Rats licked more for large, appetitive outcomes, and less for punishing trials and were more accurate **f** and faster **g** for large rewards and potential punishing trials compared to neutral trials. **h** Neural recordings in nucleus accumbens show higher activity for high-valued cues than for lower-valued cues in one neural population, while other NAc neurons **i** showed salience signals for both high-valued reward and possible punishment trial types. Firing rates were normalized by subtracting the baseline and dividing by the standard deviation. *Ribbons* represent standard error of the mean (*SEM*). *Gray-dashed* aversive (aver); *Black* appetitive (app); *Gray solid* neutral (Neu). Adapted from Bissonette et al. (2013)

To disentangle the two, Roesch and Olson recorded neural activity in OFC during performance of a task in which monkeys responded to cues that indicated the size of the reward the monkey would receive if correct (one or three drops of juice) and the size of a penalty that would be incurred upon failure (a 1 s or 8 s time-out). Figure 2a illustrates the 3 trial types. The first promises 3 drops of liquid reward with no threat of punishment. This trial type has high value and high motivation as demonstrated by high accuracy and fast reaction times. The second trial type (neutral) has low value and motivation because it only promises a small reward with no risk of punishment. The critical trial type is the third one. It also promises only a small reward but the threat of punishment is high; thus, it has low value but is highly motivational as evidenced by high accuracy and faster reaction times similar to those observed on high reward trials (Fig. 2b). Thus, choice rate and response latencies showed that monkeys were more motivated by large rewards and penalties as compared to smaller/neutral trials, allowing for the dissociation of value and motivation via simultaneous manipulation of appetitive and aversive outcomes (Roesch and Olson 2004).

In this study, OFC neural activity was found to encode the value associated with cues, as opposed to their motivational properties. That is, OFC neurons fired most strongly for cues that predicted large reward and least strongly for cues that predicted large penalty relative to neutral conditions (Fig. 2c). This was in stark contrast to neurons in an area of cortex more strongly associated with motor output, the premotor cortex, which fired at a higher rate during the preparation to move to achieve a large reward and to avoid the large penalty—i.e., they appeared to encode a factor related to the level of motivation to respond (Fig. 2d). Other studies have replicated these results in OFC and have further shown populations of neurons that encode potential rewards offers, the identity of specific rewards expected or obtained, and the option that is eventually chosen during performance of choice tasks (Hosokawa et al. 2007; Morrison et al. 2011; Padoa-Schioppa and Assad 2006, 2008).

From these and other studies in primates, it has been suggested that OFC encodes an abstract "common currency" (Padoa-Schioppa and Assad 2006; Padoa-Schioppa and Cai 2011; Rudebeck and Murray 2014); however, rodent studies have emphasized that OFC can also encode specific outcomes, representing the sensory qualities of available outcomes and potential maps of task/environmental space (Wilson et al. 2014; Schoenbaum et al. 2011a; Schoenbaum and Eichenbaum 1995). Some primate studies also support this notion, showing that OFC can convey sensory and informational aspects of reward in addition to their hedonic properties (Wallis and Miller 2003; Wallis et al. 2001; Tsujimoto et al. 2011, 2012). Nevertheless, all of these studies agree that OFC is critical for signaling predictions about future outcomes in the service of reward-guided decision-making.

2.2 Nucleus Accumbens

Next, we turn our discussion to the nucleus accumbens core (NAc), which receives strong glutamatergic projections from OFC pyramidal neurons. NAc has been described as the "critic" in actor-critic models of reinforcement learning, which model NAc as a generator of value predictions that are subsequently used by dopamine (DA) neurons to compute prediction errors necessary for updating actions polices in the "actor" (e.g., dorsal striatum) (Redish 2004; Joel et al. 2002; van der Meer and Redish 2011; Padoa-Schioppa 2011; Barto 1995; Niv and Schoenbaum 2008; Sutton and Barto 1998; Takahashi et al. 2008; Houk et al. 1995; Haber et al. 2000; Ikemoto 2007). In addition to this proposed role, NAc has been traditionally described as the "limbic-motor interface," motivating behaviors in response to both appetitive and aversive stimuli, and not for representing value per se. Consistent with both of these theories, pharmacological manipulations of NAc impact motivated behaviors dependent on value expectations during a variety of tasks (Cardinal et al. 2002a, b; Berridge and Robinson 1998; Di Chiara 2002; Ikemoto and Panksepp 1999; Salamone and Correa 2002; Di Ciano et al. 2001; Wadenberg et al. 1990; Wakabayashi et al. 2004; Yun et al. 2004; Gruber et al. 2009; Stopper and Floresco 2011; Ghods-Sharifi and Floresco 2010; Floresco et al. 2008; Blokland 1998; Giertler et al. 2003), including reward seeking (Ikemoto and Panksepp 1999), cost-benefit analysis (Stopper and Floresco 2011; Floresco et al. 2008), and delay/effort discounting (Ghods-Sharifi and Floresco 2010; Cardinal et al. 2001). Furthermore, single-unit recordings have clearly demonstrated that neural activity in NAc is modulated by the value associated with cues that predict reward in rats (Setlow et al. 2003; Janak et al. 2004; Carelli and Deadwyler 1994; Day et al. 2011; Ito and Doya 2009; Goldstein et al. 2012; Nicola et al. 2004; van der Meer et al. 2010; van der Meer and Redish 2009; Lansink et al. 2010; Kalenscher et al. 2010) and monkeys (Cromwell et al. 2005; Shidara and Richmond 2004; Schultz et al. 1992; Kim et al. 2009; Nakamura et al. 2012) performing a variety of instrumental tasks, including go/no-go (Setlow et al. 2003; Schultz et al. 1992), lever pressing (Janak et al. 2004; Carelli and Deadwyler 1994; Day et al. 2011; Cromwell et al. 2005; Shidara and Richmond 2004), discrimination (Goldstein et al. 2012; Nicola et al. 2004; van der Meer et al. 2010), maze running (van der Meer and Redish 2009; Lansink et al. 2010; Kalenscher et al. 2010), and eye movement paradigms (Kim et al. 2009; Nakamura et al. 2012).

Thus, the basic finding across many of these studies is that single-unit activity in NAc is modulated by cues that predict reward after an instrumental response is performed. As above, this activity might reflect value or motivation. Therefore, application of a paradigm similar to the one described above allowed for an exploration of these different potential roles. In a study by Bissonette et al., rats performed a task where they were motivated by both reward (sucrose solution) and threat of punishment (quinine delivery). Rats found sucrose and quinine appetitive and aversive, respectively, as illustrated by increased and decreased licking, (Fig. 2e), but were most strongly motivated by cues that predicted either a large possible sucrose reward or a possible quinine punishment, as illustrated by better accuracy and faster reaction times for these cues, compared to a neutral cue (Neu) which only predicted a small sucrose reward, similar to the neutral cue in the aforementioned primate study (Fig. 2f, g). In addition, cues that predicted potential reward and punishment (odor cues) were presented before stimuli (lights) that instructed the instrumental response to dissociate value signals from specific motor planning.

Interestingly, activity of separate populations of single neurons in NAc encoded either value or motivation prior to the response instruction (i.e., light). This result suggests that NAc might represent the value of the goal the rat is working for as well as the motivational level associated with differently signed outcomes. Activity of some neurons was stronger for conditioned stimuli that predicted large reward and weaker for punishment trials relative to neutral trials (i.e., value representation; Fig. 2h). Other NAc neurons fired strongly for cues that predicted reward and punishment, respectively (i.e., motivation; Fig. 2i). These results suggest that NAc fulfills both motivational and evaluative functions via separate neuronal populations and might be critical for integrating both types of information as the "limbic-motor interface," as well as the "critic" (Bissonette et al. 2013). Thus, NAc appears to be a common junction point for concurrently signaling value and motivation, leading to the invigoration of particular behavioral actions over others. This idea is consistent with pharmacological studies which suggest that DA in the NAc disrupts the ability to modify behavior based on the current value of predicted outcomes (Burton et al. 2013; Singh et al. 2010), as well as to interfere with behavioral measures of motivation or salience (Nunes et al. 2013; Salamone 1994; Salamone and Correa 2012; Salamone et al. 1991, 2012; Koch et al. 2000; Berridge 2007; Lex and Hauber 2010; Salamone 1986; McCullough and Salamone 1992).

2.3 Parietal Cortex

Finally, the most recent brain area to be scrutinized by separating value from salience is the parietal cortex. Parietal neural activity is dependent on the value of expected actions (action-value) (Platt and Glimcher 1999; Sugrue et al. 2004; Louie and Glimcher 2010) and is thought to be critical for making economic decisions (Louie and Glimcher 2010; Gold and Shadlen 2007; Sugrue et al. 2005; Rangel and Hare 2010). Others suggest that this signal reflects increased salience induced by the promise of a better reward, not the value associated with it. For example, Leathers and Olson reported that primate lateral intraparietal (LIP) neurons fire most strongly when a behavior is associated with a large versus small reward and more importantly, to cues that predicted a large versus small penalty (Leathers and Olson 2012). They argued that this signal reflected increased salience because activity was high for both large reward and penalty. They also suggest that this reflected salience (as opposed to motivation) because it did not span the delay between the cue and the behavioral response. This study has sparked considerable debate between influential leaders in the field (Leathers and Olson 2012, 2013; Newsome et al. 2013). Further work is necessary to determine whether salience versus value encoding in parietal cortex is task or procedurally specific. One possible middle ground in this debate might be that neurons in LIP, like NAc, encode both properties, and that one function is emphasized over the other depending on context.

3 Signed Prediction Error Versus Attention/Salience

The above discussion focused on neural firing leading to a decision and whether or not it reflects the value of potential goals or the motivation associated with obtaining those goals. Here, we consider what happens after the predicted outcome is or is not delivered, and whether it was better or worse than expected. These signals may be used to update previously described "value" and "motivation" signals when contingencies change.

Many brain areas increase activity in response to unexpected delivery of reward. Most commonly these signals are interpreted as "reward prediction signals." However, this signal might also reflect other functions, such as changes in attention that result due to the delivery of unexpected reward. Note that this activity cannot represent motivation because motivation usually increases with training as one learns to expect the more valued reward. Furthermore, signals related to motivation are maintained as long as the animal is not satiated and/or is actively pursuing reward. The signals that we will describe in this section *attenuate* with learning, as rewards become anticipated (i.e., no longer unexpected).

We will discuss two types of prediction errors, signed and unsigned. Signed prediction error (PE) signals strengthen or weaken associations in downstream brain areas during learning and adaptive decision-making. In the coffee example, if the coffee is bad, positive associations with the coffee shop must be attenuated (negative PE) so that it is no longer sought after, but if it is really good, these associations must be strengthened (positive PE) to promote coffee-seeking at that specific coffee shop. In contrast, unsigned prediction errors modulate attention. In the example, attention increases both when the coffee is really great and when it is very bad. Increased attention is necessary to determine what in the environment caused the deviation from reward expectations (e.g., new coffee brand or barista). Attention and prediction error signals likely work together in an intricate manner that has not been fully realized in the literature. Prediction errors are necessary to increase attention, and attention is needed to detect and learn from prediction errors. How these two interact in the brain is an interesting question, but as a first step, we must dissociate these two highly interrelated functions. Below we describe studies that accomplish this by manipulation of both positive and negative events, in an unpredictable manner.

3.1 Midbrain Dopamine (DA) Neurons

Many midbrain dopamine (DA) neurons signal signed PEs that are essential for reinforcement learning (Steinberg et al. 2013). Phasic bursting of DA neurons intensifies when an outcome occurs that is "better" than predicted (positive PE), and firing of DA neurons decreases or is inhibited when an outcome is "worse" than anticipated (negative PE). When rewards are accurately predicted, outcomes elicit "no" change in DA neural activity (Schultz 1997, 1999). Over the course of learning, phasic DA bursting "shifts" from occurring at the time of reward delivery to the time of reward-predicting cues that precede outcomes. Dopaminergic reward prediction error signals are commonly referred to as a "teaching signals," updating decision circuits about changes in contingencies so that behaviors can be modified when expectations are violated (Montague et al. 1996; Schultz 1998; Bromberg-Martin et al. 2010a).

Although PE signaling is frequently studied in the context of animals learning to approach cues that elicit an appetitive outcome, positive prediction errors appear to be elicited in any situation that is construed as being better than expected [e.g., short delay to reward, low effort to achieve reward, high probability (Roesch et al. 2010b)]. Positive PE signals are even induced when predicted negative events do not occur. This has been observed in primates, where DA neurons that increase firing to unexpected rewards also increase firing when an expected air puff is not delivered (Bromberg-Martin et al. 2010a), and when rats do not receive an anticipated foot shock during an avoidance procedure (Oleson et al. 2012).

The literature has recently turned its focus to how these DAergic prediction error signals may be derived. In its simplest form, reward prediction error computations require two pieces of information: the reward prediction and the reward that was actually received (actual minus predicted). As mentioned above, computational models refer to structures that signal predictions as the "critic." They calculate the expected value based on the summed values of environmental cues. Structures proposed to fill this role include the nucleus accumbens, prefrontal cortex, and/or amygdala (O'Doherty et al. 2004; Belova et al. 2008; Balleine et al. 2008; Daw et al. 2005). Indeed, several studies have shown that information from the orbitofrontal and prefrontal cortex is necessary for expectancy-related changes in phasic firing of midbrain dopamine neurons (Takahashi et al. 2009, 2011; Jo et al. 2013).

Others have demonstrated that an error signal also occurs upstream of DA neurons. The lateral habenula (LHb) (Bromberg-Martin et al. 2010a; Hikosaka 2010; Matsumoto and Hikosaka 2007), which is thought to receive error information from globus pallidus border (GPb) (Hong and Hikosaka 2008), also signals signed reward prediction errors but in the opposite sign. Activity of these neurons is excited or inhibited by negative or positive prediction errors, respectively. Furthermore, prediction error signaling in LHb occurs earlier than those in DA neurons and stimulation of LHb inhibits DA firing (Matsumoto and Hikosaka 2007; Bromberg-Martin et al. 2010b). DA neurons are likely to receive this information via the adjacent rostromedial tegmental nucleus (RMTg) and indirect connections through midbrain GABA neurons in the ventral tegmental area (VTA) of the midbrain (Hong et al. 2011; Ji and Shepard 2007; Omelchenko et al. 2009; Jhou et al. 2009; Kaufling et al. 2009; Brinschwitz et al. 2010). Together these results suggest that DA neurons receive widespread information regarding reward predictions and prediction errors themselves.

Signed prediction errors are not the only information conveyed by midbrain DA neurons. Some midbrain DA neurons also signal motivational salience and are excited by both rewarding and aversive events. This signal is thought to support systems for orienting, cognitive processing, and motivational drive (Matsumoto and Hikosaka 2009; Bromberg-Martin et al. 2010a). These neurons are identified on a gradient, being more prominently found in ventromedial substantia nigra pars compacta (SNc) (Matsumoto and Hikosaka 2009) and tapering toward the VTA. These salience signaling DA neurons show increased activity for cues which signal either positive or negative potential events, such as foot shock and air puff (Brischoux et al. 2009; Anstrom et al. 2009; Lammel et al. 2011). PE- or value-related signals are more commonly found in the ventromedial SNc area of the midbrain and the lateral VTA. In contrast, motivational salience signals are more predominantly found in the dorsolateral SNc (Matsumoto and Hikosaka 2009). Interestingly, DA neurons in these different areas of the midbrain project to different downstream regions. DA neurons in the salience predominant regions project preferentially to areas of the prefrontal cortex such as dorsolateral prefrontal cortex (DLPFC) and to NAc, in particular the core of the nucleus accumbens, while DA neurons in the value- and PE-predominant region project mainly to ventromedial PFC and to the shell of the nucleus accumbens in the NAc (Bromberg-Martin et al. 2010a). In this way, DA signaling is able to both modify representations of expected outcomes, and the need for motivation and attention to salient features in the environment when contingencies are violated (Bromberg-Martin et al. 2010a).

3.2 Basolateral Amygdala (ABL)

Like midbrain DA neurons, several reports have now suggested that ABL also signals reward prediction errors. Traditionally, ABL was thought to serve many of the same functions as OFC consistent with their reciprocal connections (Bechara et al. 1999; Malkova et al. 1997; Kesner and Williams 1995; Hatfield et al. 1996; Cousens and Otto 2003; Parkinson et al. 2001; Jones and Mishkin 1972; Winstanley et al. 2004; Churchwell et al. 2009; Ghods-Sharifi et al. 2009; Cardinal 2006). Studies examining the encoding of both appetitive and aversive signals in OFC and ABL have shown that the two are dependent on each other for normal encoding when rats discriminate stimuli that predict appetitive (sucrose) and aversive (quinine) outcomes during reversal learning (Roesch et al. 2007a, 2010a, b; Haney et al. 2010; Saddoris et al. 2005; Schoenbaum 2004; Schoenbaum et al. 1998, 1999, 2000, 2003a, 2006, 2007, 2009, 2011a; Schoenbaum and Esber 2010; Schoenbaum and

Roesch 2005; Stalnaker et al. 2007; Rudebeck and Murray 2008, 2011, 2014; Rudebeck et al. 2013a, b).

Although amygdala is important for signaling expected outcomes by acquiring and storing associative information related to both appetitive and aversive outcomes (LeDoux 2000; Murray 2007; Gallagher 2000; Schoenbaum et al. 2003b; Ambroggi et al. 2008), it also supports other functions related to associative learning, such as signaling attention to salient cues, uncertainty about outcome probability, and intensity of stimuli (Morrison et al. 2011; Saddoris et al. 2005; Tye and Janak 2007; Tye et al. 2010; Belova et al. 2007). These studies have shown that activity in ABL is modulated by the predictability of both appetitive and aversive events, specifically when predictions are violated (Roesch et al. 2010a, b; Tye et al. 2010; Belova et al. 2007). During events that are highly salient and attention grabbing, such as when an outcome is unexpectedly delivered or when an expected outcome is omitted, ABL neurons increase firing (Roesch et al. 2010a). In rats, it has been reported that activity in ABL increases when rats were expecting reward, but it was not delivered during extinction (Tye et al. 2010). Modulation of neural activity in lateral and basal nuclei of amygdala by expectation has also been described during fear conditioning in rats (Johansen et al. 2010). In primates, unexpected delivery of appetitive or aversive (air puff) outcomes caused amygdala neurons to fire more strongly than when they were completely predictable (Belova et al. 2007). Additionally, the same populations of ABL neurons which represent appetitive stimuli were also activated by aversive stimuli, regardless of the particular sensory modality from which the experience comes (Shabel and Janak 2009). These studies suggest a critical role for ABL in multiplexing signals related to attention, which may be modulated by salience or intensity, as well as signaling the associated value of cues and outcomes.

This hypothesis is consistent with data showing that outcome encoding ABL neurons exhibit increased firing to rewards that are unexpectedly delivered and omitted in a task in which reward expectations were violated by varying its size and the time to its delivery (Roesch et al. 2010a). Recall, these correlates are different than what we have described for DA neurons; DA neurons increase and decrease firing during unexpected delivery of reward, respectively. In this experiment, unexpected *up-shifts* in value occurred whenever a reward was larger than expected or arrived earlier than anticipated (Fig. 3a). On the other hand, *down-shifts* in value occurred whenever reward was unexpectedly smaller or was delayed (Fig. 3b). In this study, neurons in ABL tended to exhibit higher firing for both up- and down-shifts, whereas DA neurons exhibited increases and decreases in firing, respectively.

Two common models for interpreting neural signals relating to expected and actual outcomes are the Rescorla–Wagner (Rescorla and Wagner 1972) (R-W) model (Fig. 3c) and the Pierce–Hall (PH) (Pearce and Hall 1980) model (Fig. 3d). The R-W model uses errors to drive the change in associative strength, where larger errors cause larger changes in associative strength, smaller errors drive smaller changes, and if no error occurs, there is no change in associative strength. Since these error values are computed from the expected versus actual outcomes,

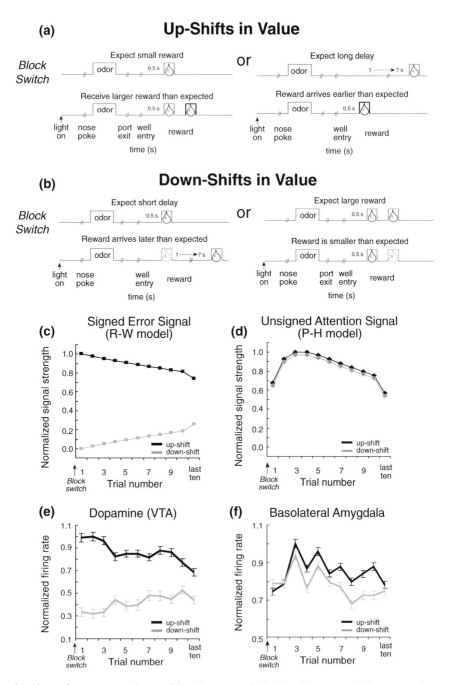

the sign of errors may be positive (outcome is better than expected) or negative (outcome is worse than expected) as observed in the firing of DA neurons (Fig. 3e). An alternative model presented by PH uses the absolute (unsigned) value of a

Fig. 3 Neural activity in VTA and ABL responses to unexpected reward and omission consistent with Rescorla–Wagner or Pearce–Hall attention models, respectively. **a** Example trial types representing up-shifts in value, with an unexpected increase in reward quantity (*left*) or unexpected decrease in wait time for reward (*right*). Deflections reflect time of events. *Heavy black lines* and *fluid drops* reflect unexpected reward delivery (i.e., up-shift). In this task, odors predicted short (0.5 s) or long (1–7 s) delays to reward during "delay" blocks. In "size" blocks, odors predicted large (2 boli) or small (1 bolus) reward. **b** Example trial types representing a down-shifting in value, with an unexpected decrease in reward quantity or increase in wait time for reward. Deflections reflect time of events. *Dashed gray lines* and *fluid drops* reflect unexpected reward omission (i.e., down-shift). **c, d** Signals predicted by the Rescorla–Wagner (**c**) and Pearce–Hall (**d**) models after unexpected delivery (*black*) and omission (*gray*) of reward. **e, f** Average firing during the 500 ms after reward delivery in dopamine neurons in VTA (**e**) and for ABL (**f**) during the first ten trials when value of delivered reward was unexpectedly higher (up-shifts = *black*) and in blocks when the value of the reward was unexpectedly lower (down-shifts = *gray*) normalized to the maximum firing rate. *Error bars* indicate SEMs. Adapted from Roesch et al. (2010a)

prediction error to determine the amount of attention needed on following trials. A large prediction error will warrant a large increase in attention as observed in the firing of ABL neurons (Fig. 3f), whereas a small prediction error will elicit a corresponding smaller change in attention. Notably, this signal would take several trials to develop because attention on previous trials must be taken into account. Importantly, both models involve changes that are proportional to the size of the prediction error, but because the R-W model uses signed PEs, the changes in strength of PEs over learning is opposite for better than expected versus worse than expected trials (Fig. 3c). In contrast, in the PH model the changes in strength over training are equivalent for both of those types of trials (Fig. 3d). Note, although there is strong evidence that ABL signals are unsigned, consistent with PH signals, there is also some evidence for a population of ABL neurons which instead encode signed prediction errors (Belova et al. 2007; Klavir et al. 2013).

3.3 Anterior Cingulate (ACC)

One brain area that likely receives and transmits prediction error-related information to ABL is the ACC (Klavir et al. 2013). ACC has strong reciprocal connections with ABL (Dziewiatkowski et al. 1998; Cassell and Wright 1986; Sripanidkulchai et al. 1984) and has been shown to be involved in a number of functions related to error processing, conflict monitoring, behavioral feedback, and attention (Totah et al. 2009; Walton et al. 2004; Rushworth et al. 2004, 2007; Wallis and Kennerley 2010; Rudebeck et al. 2008; Rushworth and Behrens 2008; Hillman and Bilkey 2010; Matsumoto et al. 2007; Amiez et al. 2005, 2006; Sallet et al. 2007; Hayden et al. 2011; Quilodran et al. 2008; Kennerley et al. 2006, 2009; Kennerley and Wallis 2009; Ito et al. 2003; Carter et al. 1998; Holroyd and Coles 2002; Oliveira et al. 2007; Paus 2001; Scheffers and Coles 2000; Rothe et al. 2011; Magno et al. 2006). During tasks that manipulate the predictability of reward, ACC neurons signal unsigned prediction errors (Hayden et al. 2011; Bryden et al. 2011a) consistent with theories put forth by Pearce and Hall as described above. Further, it has been shown that activity in ACC was elevated at the beginning of behavioral trials *after* reward contingencies changed unexpectedly and that these changes in firing were correlated with behavioral measures of attention (Bryden et al. 2011a). This is different than what is typically described in ABL, where modulation after violations in reward prediction induces attention-related changes at the time of reward delivery, not during subsequent presentations of cues. While these studies are consistent with the role that ACC might play in allocating attention when prediction errors occur, other primate studies have suggested that ACC can also signal prediction errors of the signed variety (Klavir et al. 2013; Matsumoto et al. 2007).

4 Correlates of Motivation and Associative Encoding in Striatum

In the above sections, we dissociate value from motivation and salience and define signals specifically related to signed prediction errors, which increase and decrease associative strength, and unsigned prediction errors, which increase attention so that learning can occur. How do these signals impact behavior? One conduit is striatum; however, even in striatum signals are not simply integrated and transformed into motor output. Striatum is a structure with many overlapping neural correlates related to reward-guided and stimulus-driven behaviors. Neurons there have been shown to be responsive to cues that predict reward, during initiation of actions, the anticipation of reward, and the delivery of reward (Yin and Knowlton 2006). Although there are many overlapping correlates, there appears to be a basic trend along the diagonal of striatum (ventral-medial to dorsal-lateral) by which correlates are more reward-related in ventral-medial regions, whereas correlates are more associative and motor-related in more dorsal-lateral sections.

This has been demonstrated by neural recordings in primates that have divided the caudate (one of three main regions of the primate striatum, of which the dorsal-medial striatum in rats is the closest anatomical homologue) into three subdivisions (dorsal, central, and ventral) (Nakamura et al. 2012) based on anatomical connections with cortical and subcortical structures as reported by Haber and Knutson (2010), and in accordance with tripartite subdivisions observed in humans (Fig. 4b) (Karachi et al. 2002). In this study, monkeys performed a task in which they fixated at central spot and then responded to a target to the left or the right of fixation. During different blocks of trials, one of those produced a large reward, the other a small reward (Fig. 4a).

As expected from the anatomy, the functional segregation as the recording electrode passed through ventral-medial to dorsolateral striatum reflected the progression of limbic to associative to sensorimotor afferents (Haber and Knutson 2010; Lau and Glimcher 2007, 2008; Samejima et al. 2005; Stalnaker et al. 2010).

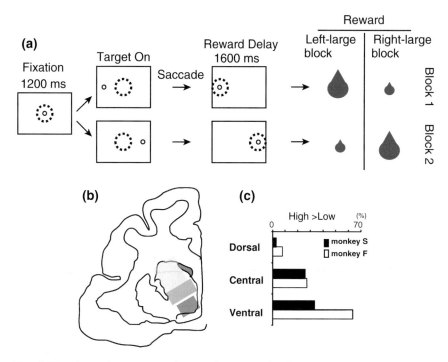

Fig. 4 Neural correlates across primate striatum. **a** Visually guided saccade task with an asymmetric reward schedule. After the monkey fixated on the *FP* (fixation point) for 1200 ms, the FP disappeared and a target cue appeared immediately on either the *left* or *right*, to which the monkey made a saccade to receive a liquid reward. The *dotted circles* indicate the direction of gaze. In a block of 20–28 trials (e.g., *left-big block*), one target position (e.g., *left*) was associated with a big reward and the other position (e.g., *right*) was associated with a small reward. The position–reward contingency was then reversed (e.g., *right-big block*). **b** Subdivisions of the primate striatum. **c** Percentage of neurons that showed large-reward preference for each subdivision of striatum. Modified from Nakamura et al. (2012), Roesch et al. (2009), Burton et al. (2014), Roesch and Bryden (2011)

Recordings obtained more ventral medially demonstrated that activity was driven by the expected size of the reward, with fewer representations related to the action necessary to achieve it. Ventral-medial portions of the caudate, including the central and ventral part, tended to fire more strongly for large versus small rewards than dorsal caudate (Fig. 4c) and fewer neurons were modulated by the direction of the behavioral response compared to dorsal caudate. This is in contrast to dorsal-lateral portions of caudate which exhibited high directionality and reward selectivity but showed no preference for large over small reward. This study suggests a continuum of correlates with ventral-medial aspects of striatum reflecting value, and more dorsal-lateral regions better reflecting associative and motor aspects of behavioral output.

Similar results have been obtained in studies which examine the difference across the ventral-medial/dorsal-lateral divide in monkeys (Hollerman et al. 1998;

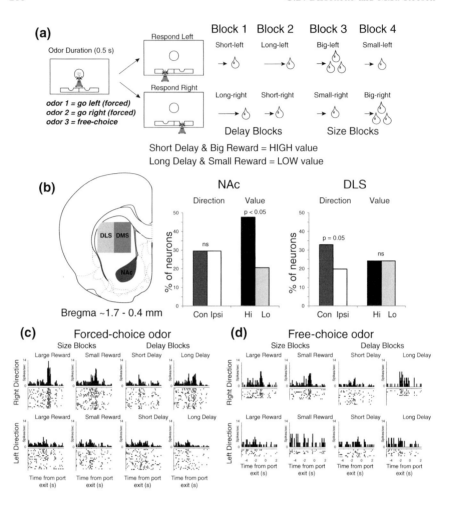

Apicella et al. 1991; Cai et al. 2011) and in rats (van der Meer et al. 2010; Takahashi et al. 2007; Wang et al. 2013). For example, in a task where odor cues predicted differently valued rewards and the direction necessary to achieve those rewards (Fig. 5a), the majority of NAc neurons fired significantly more strongly for cues that predicted high-value outcomes for actions made in one particular direction (into the cell's response field) (Roesch et al. 2009; Burton et al. 2015). Further, faster response times (a behavioral measure of more motivated behavior) were correlated with higher firing rates. These data suggested that activity in NAc represents the motivational value associated with chosen actions necessary for translating cue-evoked value signals into motivated behavior as discussed above (Bissonette et al. 2013; Roesch et al. 2009; McGinty et al. 2013; Catanese and van der Meer 2013). These findings are consistent with those described in Sect. 1, showing that separate NAc neurons encode both value and motivation.

◄ **Fig. 5** Neural correlates across rat NAc and DLS. Odor-guided choice task during which the delay to and size of reward were independently varied in ~60 trial blocks (i.e., "blocks 1–4"). Upon illumination of house lights, rats started the trial by poking into the central port. After 500 ms, an odor signaled the trial type. For odors 1 and 2, rats had to go to a left or right fluid well to receive reward (forced-choice trials). A third odor signaled that the rat was free to choose either well to receive the reward that was associated with that response direction during the given block of trials. In blocks 1–4, the length of the delay (blocks 1 and 2) to reward and the size of reward (blocks 3 and 4) were manipulated: short delay = 0.5 s wait before delivery of 1 bolus reward; Long delay = 1–7 s wait before 1 bolus reward; Big reward = 0.5 s wait for 2–3 boli reward; Small reward = 0.5 s wait before 1 bolus reward. Throughout each recording session, each of these trial types were associated with both directions and all three odors, allowing us to examine different associative correlates. **b** Locations of recording sites in rat NAc, DMS, and DLS and percentage of significantly modulated neurons. More NAc neurons encoded high-valued options (NAc, *black bar*). Representations of outcome were evenly distributed in DLS, while number of response direction encoding neurons was significantly elevated (contralateral *gray bar*). Chi-square was used to compare counts of neurons. *Hi* High value; *Lo* Low value; *Con* contralateral to recording site; *Ipsi* Ipsilateral to the recording site. **c** and **d** Example of a single neuron recorded in DLS for each of the trial types for forced (**c**) and free (**d**) choice odors. Modified from Nakamura et al. (2012), Roesch et al. (2009), Burton et al. (2014), Roesch and Bryden (2011)

To characterize neurons, we performed an ANOVA with value (high or low) and direction (contralateral or ipsilateral to the recording site) as factors on activity collected during the decision period (cue onset to response). In NAc, roughly equal proportions of neurons were selective for contralateral and ipsilateral response directions; however, of those selective for value, the large majority fired more strongly for high-value reward (Fig. 5b; NAc).

In contrast to NAc, encoding in dorsolateral striatum (DLS) was remarkably associative, representing all aspects of the task; outcome type, response direction, and the specific identity of conditioned stimuli (Burton et al. 2014, 2015). For example, the neuron in Fig. 5c responded the most strongly for odor cues that predicted a short delay after moving in the contralateral direction. Many DLS neurons were selective for very specific combinations of task events related to future outcomes, responses, and the stimuli that preceded them; however, there was not a preponderance of neurons that showed increased firing for any one combination over another, as was observed in NAc. If we characterize single neurons, the same as we did for NAc, we see that there is a preponderance of neurons showing a preference for contralateral movements, but of those that are value selective, equal proportions preferred high- and low-value reward (Fig. 5b; DLS). These results suggest that neural correlates in NAc are more closely tied to the motivational level associated with choosing high-value goals, whereas correlates in DLS are more associative, representing expected outcomes and response directions across a range of stimuli.

Although NAc and DLS are not directly connected, more ventral-medial regions like NAc likely impact more dorsal-lateral areas in striatum via the "spiraling" connectivity with dopamine (DA) neurons (Joel et al. 2002; van der Meer and Redish 2011; Niv and Schoenbaum 2008; Takahashi et al. 2008; Houk et al. 1995; Haber et al. 2000; Ikemoto 2007). The circuitry allows propagation of information from limbic networks to associative and sensorimotor networks so that behaviors can be more stimulus driven and therefore more efficient (Haber et al. 2000; Ikemoto 2007; Haber and Knutson 2010; Haber 2003; Balleine and O'Doherty 2010). This connectivity is consistent with ideas that behavior is first goal-directed and then become habitual with extended training (for more information on habitual action mechanisms, please see the chapter by O'Doherty in this volume). However, several studies have now shown that the NAc and DLS can function independent of each of each other, especially after learning. That is, these functions do not seem to always work in series, but can operate in parallel. For example, we have found that NAc lesions do not impair task-related correlates in DLS, but instead enhances neural correlates related to stimulus and response processing (Burton et al. 2013). These results suggest that rats normally use goal-directed mechanisms to perform the task, but rely more heavily on stimulus–response (S-R) processes after the loss of NAc (Lichtenberg et al. 2013). Consistent with this interpretation, rats recovered function after initial disruption observed after NAc lesions. Similar results have been described for NAc after loss of DLS function; rats recover function after an initial impairment was observed due to DLS interference (Nishizawa et al. 2012). This recovery was abolished with subsequent NAc lesions suggesting that it was NAc that was guiding behavior in DLS's absence (Nishizawa et al. 2012). Together, these results suggest that NAc and DLS both guide behavior normally during performance of reward-related tasks and that these two regions can compensate for each other if the need arises. Note that this is not to suggest that they do not normally interact, particular during learning or when reward contingencies change.

5 Integration of Positive and Negative Information into a Common Output Signal

How independent representations of positive and negative valences are converted to a unified representation of expected outcome value, which ultimately leads to motivated behavior, is still unclear. A number of brain areas are thought to represent abstract value in the service of making comparisons, allowing the brain to compare apples to oranges (or coffee to tea). Under choice paradigms, Lee and colleagues have found integrative encoding of value in several brain regions (dorsal striatum, ventral striatum, and DLPFC) using an intertemporal choice task (Cai et al. 2011; Kim et al. 2008). Other studies have clearly shown neural activity reflecting value in several frontal areas in primate cortex (Kim et al. 2008; Roesch and Olson 2005a, b). In primates, several prefrontal regions contain neurons that integrate several economic factors, including cost, size, delay, and probability (Padoa-Schioppa and Assad 2006; Padoa-Schioppa and Cai 2011; Rich and Wallis 2014; Wallis and Kennerley 2011; Kennerley et al. 2011; Wallis 2007; Padoa-Schioppa 2007). However, others have reported that distinct prefrontal areas encode rewards and punishments, with ventral and dorsal aspects being more active for appetitive and aversive trial types, respectively (Monosov and Hikosaka 2012). It is difficult to find one clear path by which predicted outcomes lead to motivated behavior. This is likely because there are multiple paths that work in parallel, which interact at different phases of behavior and are highly dependent on the context. For an in depth discussion on valuation systems in the brain in the context of decision-making processes, see Redish et al., in this volume. It has also been difficult to clearly separate value signals from signals that might represent attention, salience, and/or motivation. With that being said, there is a trend by which brain areas more closely tied to attention and motor networks appear to better reflect the integration of economic functions.

For example, both ACC and parietal cortex are thought to be critical for functions related to attention and motor planning, but they have also been described as encoding value in the service of making economic decisions. The fact that value and salience signals have been observed in these regions might reflect the need for increased attentional control when potential rewards are available. Increases in neural firing depending on the value of expected actions would help ensure that neural processes are prioritized depending on expected events (whether they are positive or negative). Interestingly, integration of value predictions and spatial attention has been shown to be integrated in clusters of neurons in primate prefrontal and parietal cortex (Kaping et al. 2011). Further research will need to be done to fully appreciate how specific nodes contribute to the process of transforming value signals into executive control signals.

Out of all the rat brain areas that we have recorded from in our laboratory, only the firing of DA neurons exhibited a strong correlation between delay and size of reward, reflecting a common representation of value. DA neurons fired more strongly to cues that predicted high reward, whether it be a short delay or a large-sized reward, and were inhibited by cues that predicted low reward, whether it be a long delay or a small sized reward. Importantly, these were the same neurons that encoded reward prediction errors during unexpected reward delivery and omission. This is interesting, considering that dopaminergic input is thought to build associations regarding value in these brains areas. In our rat studies, very few brain areas appear to be computing this function. That is, we found correlates related to delay length and reward magnitude in multiple brain areas including OFC, ABL, ACC, DMS, and DLS, all of which maintained dissociable activity patterns, which were signaled by different neurons. Even in NAc, where the population of neurons fired more strongly for cues that predicted shorter delays to reward and larger magnitude rewards, and whose activity was correlated with motivated output (i.e., reaction time), there was only a weak trend for single neurons to represent both size and delay components. Although some neurons did encode both factors at the single cell level, many more neurons represented one economic variable but not the other. This trend toward common encoding likely reflects the conversion of expected outcome information into appropriate motor signals at the level of NAc (i.e., limbic-motor interface). Consistent with this hypothesis, when looking even further downstream, activity in the substantia Nigra pars reticulata (SNr) does appear to reflect a common evaluation of goals, which likely better reflects its role as a motor output structure rather than a reporter of economic value (Bryden et al. 2011b). Even within SNr, significant correlations between delay and size were relatively weak, suggesting that we might have to move even deeper into the motor system to find a common output signal for similarly valued outcomes.

6 Conclusion

Conditioned stimuli that predict reward simultaneously set into motion several functions as illustrated in Fig. 1a. These functions are highly interrelated but can be distinguished through manipulations of reward certainty, positive and negative outcomes, and division into goal- and stimulus-driven behaviors. To a degree, these functions map on to specific brain areas in the reward/decision circuit depicted in Fig. 1b, but they clearly depend on each other and there is redundancy within the circuit. OFC appears to represent value expectancies necessary for guiding decision-making and learning. These signals are informed by ABL, which represents associative information as well as the intensity or salience of behavioral outcomes. Simultaneously, OFC and ABL broadcast their information to NAc and DA neurons. Prediction error and salience signals generated by DA neurons provide feed-forward information to more dorsal-medial and dorsal-lateral regions in striatum, which are critical for goal-directed and habitual behaviors, respectively. These regions receive specific inputs from cortex consistent with these roles. The medial regions of striatum receive dense projections from the medial prefrontal cortex (mPFC) and the anterior cingulate cortex (ACC). The DLS is densely innervated by sensorimotor cortex (SMC), but also receives projections from mPFC and ACC. Importantly, basal ganglia signals that loop back onto cortex are likely to be critical for modulating representations related to reward-guided decision-making in cortical areas and can drive behavior in their own right. Cortical input from parietal and ACC likely increase attentional control in conjunction with value encoding to ensure that neural processes are prioritized in downstream areas depending on expected actions and errors in reward prediction.

Well-designed behavioral tasks that continue to dissociate these highly related brain functions, combined with techniques that can probe for causality are necessary to continue to disentangle the roles that neural correlates play in converting positive and negative events into abstract representations of value and motivated behavior, and to elucidate critical relationships between nodes within the circuit. Understanding how these circuits work to produce behavior allows us to look for alterations in neural signals in animal models of human disorders, such as models of psychological disorders, drug addiction, and aging (Hernandez et al. 2015; Roesch et al. 2007b, 2012a, b; Gruber et al. 2010; Stalnaker et al. 2009). Future therapeutic methods (behavioral, psychological, pharmacological, etc.) should focus on restoring the lost or changed signals as observed in these animal models to restore lost functions observed in psychiatric disorders.

Acknowledgements This work was supported by grants from the NIDA (R01DA031695, MR).

References

Ambroggi F, Ishikawa A, Fields HL, Nicola SM (2008) Basolateral amygdala neurons facilitate reward-seeking behavior by exciting nucleus accumbens neurons. Neuron 59:648–661
Amiez C, Joseph JP, Procyk E (2005) Anterior cingulate error-related activity is modulated by predicted reward. Eur J Neurosci 21:3447–3452
Amiez C, Joseph JP, Procyk E (2006) Reward encoding in the monkey anterior cingulate cortex. Cereb Cortex 16:1040–1055
Anderson AK (2005) Affective influences on the attentional dynamics supporting awareness. J Exp Psychol Gen 134:258–281
Anderson BA, Laurent PA, Yantis S (2011) Value-driven attentional capture. Proc Natl Acad Sci USA 108:10367–10371
Anstrom KK, Miczek KA, Budygin EA (2009) Increased phasic dopamine signaling in the mesolimbic pathway during social defeat in rats. Neuroscience 161:3–12
Apicella P, Ljungberg T, Scarnati E, Schultz W (1991) Responses to reward in monkey dorsal and ventral striatum. Exp Brain Res. Experimentelle Hirnforschung. Experimentation cerebrale 85:491–500
Balleine BW, O'Doherty JP (2010) Human and rodent homologies in action control: corticostriatal determinants of goal-directed and habitual action. Neuropsychopharmacol: Official Publ Am Coll Neuropsychopharmacol 35:48–69
Balleine BW, Daw ND, O'Doherty JP (2008) Multiple forms of value learning and the function of dopamine. In: Glimcher PW, Camerer CF, Fehr E, Poldrack RA (eds) Neuroeconomics: decision making and the brain. Elsevier, Amsterdam
Barto A (ed) (1995) Adaptive critics and the basal ganglia. MIT Press, Cambridge
Bechara A, Damasio H, Damasio AR, Lee GP (1999) Different contributions of the human amygdala and ventromedial prefrontal cortex to decision-making. J Neurosci 19:5473–5481
Belova MA, Paton JJ, Morrison SE, Salzman CD (2007) Expectation modulates neural responses to pleasant and aversive stimuli in primate amygdala. Neuron 55:970–984
Belova MA, Patton JJ, Salzman CD (2008) Moment-to-moment tracking of state value in the amygdala. J Neurosci 28:10023–10030
Berridge KC (2007) The debate over dopamine's role in reward: the case for incentive salience. Psychopharmacology 191:391–431
Berridge KC, Robinson TE (1998) What is the role of dopamine in reward: hedonic impact, reward learning, or incentive salience? Brain Res Brain Res Rev 28:309–369
Bissonette GB et al (2013) Separate populations of neurons in ventral striatum encode value and motivation. PLoS ONE 8:e64673
Bissonette GB, Gentry RN, Padmala S, Pessoa L, Roesch MR (2014) Impact of appetitive and aversive outcomes on brain responses: linking the animal and human literatures. Frontiers Syst Neurosci 8:24
Blokland A (1998) Reaction time responding in rats. Neurosci Biobehav Rev 22:847–864
Brinschwitz K et al (2010) Glutamatergic axons from the lateral habenula mainly terminate on GABAergic neurons of the ventral midbrain. Neuroscience 168:463–476
Brischoux F, Chakraborty S, Brierley DI, Ungless MA (2009) Phasic excitation of dopamine neurons in ventral VTA by noxious stimuli. Proc Natl Acad Sci USA 106:4894–4899

Bromberg-Martin ES, Matsumoto M, Hikosaka O (2010a) Dopamine in motivational control: rewarding, aversive, and alerting. Neuron 68:815–834

Bromberg-Martin ES, Matsumoto M, Hikosaka O (2010b) Distinct tonic and phasic anticipatory activity in lateral habenula and dopamine neurons. Neuron 67:144–155

Bryden DW, Johnson EE, Tobia SC, Kashtelyan V, Roesch MR (2011a) Attention for learning signals in anterior cingulate cortex. J Neurosci: Official J Soc Neurosci 31:18266–18274

Bryden DW, Johnson EE, Diao X, Roesch MR (2011b) Impact of expected value on neural activity in rat substantia nigra pars reticulata. Eur J Neurosci 33:2308–2317

Burke KA, Franz TM, Miller DN, Schoenbaum G (2008) The role of the orbitofrontal cortex in the pursuit of happiness and more specific rewards. Nature 454:340–344

Burton AC, Bissonette GB, Lichtenberg NT, Kashtelyan V, Roesch MR (2013) Ventral striatum lesions enhance stimulus and response encoding in dorsal striatum. Biol Psychiatry

Burton AC, Bissonette GB, Lichtenberg NT, Kashtelyan V, Roesch MR (2014) Ventral striatum lesions enhance stimulus and response encoding in dorsal striatum. Biol Psychiatry 75:132–139

Burton AC, Nakamura K, Roesch MR (2015) From ventral-medial to dorsal-lateral striatum: neural correlates of reward-guided decision-making. Neurobiol Learn Mem 117:51–59

Cai X, Kim S, Lee D (2011) Heterogeneous coding of temporally discounted values in the dorsal and ventral striatum during intertemporal choice. Neuron 69:170–182

Cardinal RN (2006) Neural systems implicated in delayed and probabilistic reinforcement. Neural Netw 19:1277–1301

Cardinal RN, Pennicott DR, Sugathapala CL, Robbins TW, Everitt BJ (2001) Impulsive choice induced in rats by lesions of the nucleus accumbens core. Science 292:2499–2501

Cardinal RN, Parkinson JA, Hall J, Everitt BJ (2002a) Emotion and motivation: the role of the amygdala, ventral striatum, and prefrontal cortex. Neurosci Biobehav Rev 26:321–352

Cardinal RN et al (2002b) Effects of selective excitotoxic lesions of the nucleus accumbens core, anterior cingulate cortex, and central nucleus of the amygdala on autoshaping performance in rats. Behav Neurosci 116:553–567

Carelli RM, Deadwyler SA (1994) A comparison of nucleus accumbens neuronal firing patterns during cocaine self-administration and water reinforcement in rats. J Neurosci 14:7735–7746

Carter CS et al (1998) Anterior cingulate cortex, error detection, and the online monitoring of performance. Science 280:747–749

Cassell MD, Wright DJ (1986) Topography of projections from the medial prefrontal cortex to the amygdala in the rat. Brain Res Bull 17:321–333

Catanese J, van der Meer M (2013) A network state linking motivation and action in the nucleus accumbens. Neuron 78:753–754

Churchwell JC, Morris AM, Heurtelou NM, Kesner RP (2009) Interactions between the prefrontal cortex and amygdala during delay discounting and reversal. Behav Neurosci 123:1185–1196

Cousens GA, Otto T (2003) Neural substrates of olfactory discrimination learning with auditory secondary reinforcement. I. Contributions of the basolateral amygdaloid complex and orbitofrontal cortex. Integr Physiol Behav Sci 38:272–294

Cromwell HC, Hassani OK, Schultz W (2005) Relative reward processing in primate striatum. Exp Brain Res. Experimentelle Hirnforschung. Experimentation cerebrale 162:520–525

Daw ND, Kakade S, Dayan P (2002) Opponent interactions between serotonin and dopamine. Neural Networks: Official J Int Neural Network Soc 15:603–616

Daw ND, Niv Y, Dayan P (2005) Uncertainty-based competition between prefrontal and dorsolateral striatal systems for behavioral control. Nat Neurosci 8:1704–1711

Day JJ, Jones JL, Carelli RM (2011) Nucleus accumbens neurons encode predicted and ongoing reward costs in rats. Eur J Neurosci 33:308–321

Di Chiara G (2002) Nucleus accumbens shell and core dopamine: differential role in behavior and addiction. Behav Brain Res 137:75–114

Di Ciano P, Cardinal RN, Cowell RA, Little SJ, Everitt BJ (2001) Differential involvement of NMDA, AMPA/kainate, and dopamine receptors in the nucleus accumbens core in the acquisition and performance of pavlovian approach behavior. J Neurosci 21:9471–9477

Dziewiatkowski J et al (1998) The projection of the amygdaloid nuclei to various areas of the limbic cortex in the rat. Folia Morphol (Warsz) 57:301–308

Floresco SB, St Onge JR, Ghods-Sharifi S, Winstanley CA (2008) Cortico-limbic-striatal circuits subserving different forms of cost-benefit decision making. Cogn Affect Behav Neurosci 8:375–389

Gallagher M (2000) The amygdala and associative learning. In: Aggleton JP (ed) The amygdala: a functional analysis. Oxford University Press, Oxford, pp 311–330

Ghods-Sharifi S, Floresco SB (2010) Differential effects on effort discounting induced by inactivations of the nucleus accumbens core or shell. Behav Neurosci 124:179–191

Ghods-Sharifi S., St Onge, J.R. & Floresco, S.B. Fundamental contribution by the basolateral amygdala to different forms of decision making. J Neurosci 29, 5251-5259 (2009)

Giertler C, Bohn I, Hauber W (2003) The rat nucleus accumbens is involved in guiding of instrumental responses by stimuli predicting reward magnitude. Eur J Neurosci 18:1993–1996

Gold JI, Shadlen MN (2007) The neural basis of decision making. Annu Rev Neurosci 30:535–574

Goldstein BL et al (2012) Ventral striatum encodes past and predicted value independent of motor contingencies. J Neurosci: Official J Soc Neurosci 32:2027–2036

Gruber AJ, Hussain RJ, O'Donnell P (2009) The nucleus accumbens: a switchboard for goal-directed behaviors. PLoS ONE 4:e5062

Gruber AJ et al (2010) More is less: a disinhibited prefrontal cortex impairs cognitive flexibility. J Neurosci 30:17102–17110

Haber SN (2003) The primate basal ganglia: parallel and integrative networks. J Chem Neuroanat 26:317–330

Haber SN, Knutson B (2010) The reward circuit: linking primate anatomy and human imaging. Neuropsychopharmacol: Official Publ Am Coll Neuropsychopharmacol 35:4–26

Haber SN, Fudge JL, McFarland NR (2000) Striatonigrostriatal pathways in primates form an ascending spiral from the shell to the dorsolateral striatum. J Neurosci 20:2369–2382

Haney RZ, Calu DJ, Takahashi YK, Hughes BW, Schoenbaum G (2010) Inactivation of the central but not the basolateral nucleus of the amygdala disrupts learning in response to overexpectation of reward. J Neurosci 30:2911–2917

Hatfield T, Han JS, Conley M, Gallagher M, Holland P (1996) Neurotoxic lesions of basolateral, but not central, amygdala interfere with Pavlovian second-order conditioning and reinforcer devaluation effects. J Neurosci 16:5256–5265

Hayden BY, Heilbronner SR, Pearson JM, Platt ML (2011) Surprise signals in anterior cingulate cortex: neuronal encoding of unsigned reward prediction errors driving adjustment in behavior. J Neurosci 31:4178–4187

Hernandez A, Burton AC, O'Donnell P, Schoenbaum G, Roesch MR (2015) Altered basolateral amygdala encoding in an animal model of schizophrenia. J Neurosci 35:6394–6400

Hikosaka O (2010) The habenula: from stress evasion to value-based decision-making. Nat Rev Neurosci 11:503–513

Hillman KL, Bilkey DK (2010) Neurons in the rat anterior cingulate cortex dynamically encode cost-benefit in a spatial decision-making task. J Neurosci 30:7705–7713

Hollerman JR, Tremblay L, Schultz W (1998) Influence of reward expectation on behavior-related neuronal activity in primate striatum. J Neurophysiol 80:947–963

Holroyd CB, Coles MG (2002) The neural basis of human error processing: reinforcement learning, dopamine, and the error-related negativity. Psychol Rev 109:679–709

Hong S, Hikosaka O (2008) The globus pallidus sends reward-related signals to the lateral habenula. Neuron 60:720–729

Hong S, Jhou TC, Smith M, Saleem KS, Hikosaka O (2011) Negative reward signals from the lateral habenula to dopamine neurons are mediated by rostromedial tegmental nucleus in primates. J Neurosci 31:11457–11471

Hosokawa T, Kato K, Inoue M, Mikami A (2007) Neurons in the macaque orbitofrontal cortex code relative preference of both rewarding and aversive outcomes. Neurosci Res 57:434–445

Houk J, Adams JL, Barto AG (ed) (1995) A model of how the basal ganglia generate and use neural signals that predict reinforcement

Ikemoto S (2007) Dopamine reward circuitry: two projection systems from the ventral midbrain to the nucleus accumbens-olfactory tubercle complex. Brain Res Rev 56:27–78

Ikemoto S, Panksepp J (1999) The role of nucleus accumbens dopamine in motivated behavior: a unifying interpretation with special reference to reward-seeking. Brain Res Brain Res Rev 31:6–41

Ito M, Doya K (2009) Validation of decision-making models and analysis of decision variables in the rat basal ganglia. J Neurosci 29:9861–9874

Ito S, Stuphorn V, Brown JW, Schall JD (2003) Performance monitoring by the anterior cingulate cortex during saccade countermanding. Science 302:120–122

Izquierdo AD, Suda RK, Murray EA (2004) Bilateral orbital prefrontal cortex lesions in rhesus monkeys disrupt choices guided by both reward value and reward contingency. J Neurosci 24:7540–7548

Janak PH, Chen MT, Caulder T (2004) Dynamics of neural coding in the accumbens during extinction and reinstatement of rewarded behavior. Behav Brain Res 154:125–135

Jhou TC, Fields HL, Baxter MG, Saper CB, Holland PC (2009) The rostromedial tegmental nucleus (RMTg), a GABAergic afferent to midbrain dopamine neurons, encodes aversive stimuli and inhibits motor responses. Neuron 61:786–800

Ji H, Shepard PD (2007) Lateral habenula stimulation inhibits rat midbrain dopamine neurons through a GABA(A) receptor-mediated mechanism. J Neurosci 27:6923–6930

Jo YS, Lee J, Mizumori SJ (2013) Effects of prefrontal cortical inactivation on neural activity in the ventral tegmental area. J Neurosci 33:8159–8171

Joel D, Niv Y, Ruppin E (2002) Actor-critic models of the basal ganglia: new anatomical and computational perspectives. Neural Networks: Official J Int Neural Network Soc 15:535–547

Johansen JP, Tarpley JW, LeDoux JE, Blair HT (2010) Neural substrates for expectation-modulated fear learning in the amygdala and periaqueductal gray. Nat Neurosci 13:979–986

Jones B, Mishkin M (1972) Limbic lesions and the problem of stimulus-reinforcement associations. Exp Neurol 36:362–377

Kalenscher T, Lansink CS, Lankelma JV, Pennartz CM (2010) Reward-associated gamma oscillations in ventral striatum are regionally differentiated and modulate local firing activity. J Neurophysiol 103:1658–1672

Kaping D, Vinck M, Hutchison RM, Everling S, Womelsdorf T (2011) Specific contributions of ventromedial, anterior cingulate, and lateral prefrontal cortex for attentional selection and stimulus valuation. PLoS Biol 9:e1001224

Karachi C et al (2002) Three-dimensional cartography of functional territories in the human striatopallidal complex by using calbindin immunoreactivity. J Comp Neurol 450:122–134

Kaufling J, Veinante P, Pawlowski SA, Freund-Mercier MJ, Barrot M (2009) Afferents to the GABAergic tail of the ventral tegmental area in the rat. J Comp Neurol 513:597–621

Kennerley SW, Wallis JD (2009) Evaluating choices by single neurons in the frontal lobe: outcome value encoded across multiple decision variables. Eur J Neurosci 29:2061–2073

Kennerley SW, Walton ME, Behrens TE, Buckley MJ, Rushworth MF (2006) Optimal decision making and the anterior cingulate cortex. Nat Neurosci 9:940–947

Kennerley SW, Dahmubed AF, Lara AH, Wallis JD (2009) Neurons in the frontal lobe encode the value of multiple decision variables. J Cogn Neurosci 21:1162–1178

Kennerley SW, Behrens TE, Wallis JD (2011) Double dissociation of value computations in orbitofrontal and anterior cingulate neurons. Nat Neurosci 14:1581–1589

Kesner RP, Williams JM (1995) Memory for magnitude of reinforcement: dissociation between amygdala and hippocampus. Neurobiol Learn Mem 64:237–244

Kim S, Hwang J, Lee D (2008) Prefrontal coding of temporally discounted values during intertemporal choice. Neuron 59:161–172

Kim H, Sul JH, Huh N, Lee D, Jung MW (2009) Role of striatum in updating values of chosen actions. J Neurosci: Official J Soc Neurosci 29:14701–14712

Klavir O, Genud-Gabai R, Paz R (2013) Functional connectivity between amygdala and cingulate cortex for adaptive aversive learning. Neuron 80:1290–1300

Koch M, Schmid A, Schnitzler HU (2000) Role of muscles accumbens dopamine D1 and D2 receptors in instrumental and Pavlovian paradigms of conditioned reward. Psychopharmacology 152:67–73

Lammel S, Ion DI, Roeper J, Malenka RC (2011) Projection-specific modulation of dopamine neuron synapses by aversive and rewarding stimuli. Neuron 70:855–862

Lang PJ, Davis M (2006) Emotion, motivation, and the brain: reflex foundations in animal and human research. Prog Brain Res 156:3–29

Lansink CS, Goltstein PM, Lankelma JV, Pennartz CM (2010) Fast-spiking interneurons of the rat ventral striatum: temporal coordination of activity with principal cells and responsiveness to reward. Eur J Neurosci 32:494–508

Lau B, Glimcher PW (2007) Action and outcome encoding in the primate caudate nucleus. J Neurosci 27:14502–14514

Lau B, Glimcher PW (2008) Value representations in the primate striatum during matching behavior. Neuron 58:451–463

Leathers ML, Olson CR (2012) In monkeys making value-based decisions, LIP neurons encode cue salience and not action value. Science 338:132–135

Leathers ML, Olson CR (2013) Response to comment on "in monkeys making value-based decisions, LIP neurons encode cue salience and not action value". Science 340:430

LeDoux JE (2000) The amygdala and emotion: a view through fear. In: Aggleton JP (ed) The amygdala: a functional analysis. Oxford University Press, New York, pp 289–310

Lex B, Hauber W (2010) The role of nucleus accumbens dopamine in outcome encoding in instrumental and Pavlovian conditioning. Neurobiol Learn Mem 93:283–290

Lichtenberg NT, Kashtelyan V, Burton AC, Bissonette GB, Roesch MR (2013) Nucleus accumbens core lesions enhance two-way active avoidance. Neurosci

Louie K, Glimcher PW (2010) Separating value from choice: delay discounting activity in the lateral intraparietal area. J Neurosci 30:5498–5507

Magno E, Foxe JJ, Molholm S, Robertson IH, Garavan H (2006) The anterior cingulate and error avoidance. J Neurosci 26:4769–4773

Malkova L, Gaffan D, Murray EA (1997) Excitotoxic lesions of the amygdala fail to produce impairment in visual learning for auditory secondary reinforcement but interfere with reinforcer devaluation effects in rhesus monkeys. J Neurosci 17:6011–6020

Matsumoto M, Hikosaka O (2007) Lateral habenula as a source of negative reward signals in dopamine neurons. Nature 447:1111–1115

Matsumoto M, Hikosaka O (2009) Two types of dopamine neuron distinctly convey positive and negative motivational signals. Nature 459:837–841

Matsumoto M, Matsumoto K, Abe H, Tanaka K (2007) Medial prefrontal cell activity signaling prediction errors of action values. Nat Neurosci 10:647–656

McCullough LD, Salamone JD (1992) Involvement of nucleus accumbens dopamine in the motor activity induced by periodic food presentation: a microdialysis and behavioral study. Brain Res 592:29–36

McGinty VB, Lardeux S, Taha SA, Kim JJ, Nicola SM (2013) Invigoration of reward seeking by cue and proximity encoding in the nucleus accumbens. Neuron 78:910–922

Monosov IE, Hikosaka O (2012) Regionally distinct processing of rewards and punishments by the primate ventromedial prefrontal cortex. J Neurosci 32:10318–10330

Montague PR, Dayan P, Sejnowski TJ (1996) A framework for mesencephalic dopamine systems based on predictive Hebbian learning. J Neurosci: Official J Soc Neurosci 16:1936–1947

Morrison SE, Saez A, Lau B, Salzman CD (2011) Different time courses for learning-related changes in amygdala and orbitofrontal cortex. Neuron 71:1127–1140

Murray EA (2007) The amygdala, reward and emotion. Trends Cogn Sci 11:489–497

Nakamura K, Santos GS, Matsuzaki R, Nakahara H (2012) Differential reward coding in the subdivisions of the primate caudate during an oculomotor task. J Neurosci 32:15963–15982

Newsome WT, Glimcher PW, Gottlieb J, Lee D, Platt ML (2013) Comment on "in monkeys making value-based decisions, LIP neurons encode cue salience and not action value". Science 340:430

Nicola SM, Yun IA, Wakabayashi KT, Fields HL (2004) Cue-evoked firing of nucleus accumbens neurons encodes motivational significance during a discriminative stimulus task. J Neurophysiol 91:1840–1865

Nishizawa K et al (2012) Striatal indirect pathway contributes to selection accuracy of learned motor actions. J Neurosci: Official J Soc Neurosci 32:13421–13432

Niv Y, Schoenbaum G (2008) Dialogues on prediction errors. Trends in cognitive sciences 12:265–272

Nunes EJ, Randall PA, Podurgiel S, Correa M, Salamone JD (2013) Nucleus accumbens neurotransmission and effort-related choice behavior in food motivation: effects of drugs acting on dopamine, adenosine, and muscarinic acetylcholine receptors. Neurosci Biobehav Rev 37:2015–2025

O'Doherty J et al (2004) Dissociable roles of ventral and dorsal striatum in instrumental conditioning. Science 304:452–454

Oleson EB, Gentry RN, Chioma VC, Cheer JF (2012) Subsecond dopamine release in the nucleus accumbens predicts conditioned punishment and its successful avoidance. J Neurosci: Official J Soc Neurosci 32:14804–14808

Oliveira FT, McDonald JJ, Goodman D (2007) Performance monitoring in the anterior cingulate is not all error related: expectancy deviation and the representation of action-outcome associations. J Cogn Neurosci 19:1994–2004

Omelchenko N, Bell R, Sesack SR (2009) Lateral habenula projections to dopamine and GABA neurons in the rat ventral tegmental area. Eur J Neurosci 30:1239–1250

Ostlund SB, Balleine BW (2007) Orbitofrontal cortex mediates outcome encoding in Pavlovian but not instrumental learning. J Neurosci 27:4819–4825

Padoa-Schioppa C (2007) Orbitofrontal cortex and the computation of economic value. Ann N Y Acad Sci 1121:232–253

Padoa-Schioppa C (2011) Neurobiology of economic choice: a good-based model. Annu Rev Neurosci 34:333–359

Padoa-Schioppa C, Assad JA (2006) Neurons in the orbitofrontal cortex encode economic value. Nature 441:223–226

Padoa-Schioppa C, Assad JA (2008) The representation of economic value in the orbitofrontal cortex is invariant for changes of menu. Nat Neurosci 11:95–102

Padoa-Schioppa C, Cai X (2011) The orbitofrontal cortex and the computation of subjective value: consolidated concepts and new perspectives. Ann N Y Acad Sci 1239:130–137

Parkinson JA et al (2001) The role of the primate amygdala in conditioned reinforcement. J Neurosci 21:7770–7780

Paus T (2001) Primate anterior cingulate cortex: where motor control, drive and cognition interface. Nat Rev Neurosci 2:417–424

Pearce JM, Hall G (1980) A model for Pavlovian learning: variations in the effectiveness of conditioned but not of unconditioned stimuli. Psychol Rev 87:532–552

Phelps EA, Ling S, Carrasco M (2006) Emotion facilitates perception and potentiates the perceptual benefits of attention. Psychol Sci 17:292–299

Pickens CL et al (2003) Different roles for orbitofrontal cortex and basolateral amygdala in a reinforcer devaluation task. J Neurosci 23:11078–11084

Platt ML, Glimcher PW (1999) Neural correlates of decision variables in parietal cortex. Nature 400:233–238

Quilodran R, Rothe M, Procyk E (2008) Behavioral shifts and action valuation in the anterior cingulate cortex. Neuron 57:314–325

Rangel A, Hare T (2010) Neural computations associated with goal-directed choice. Curr Opin Neurobiol 20:262–270

Redish AD (2004) Addiction as a computational process gone awry. Science 306:1944–1947

Rescorla RA, Wagner AR (1972) A theory of Pavlovian conditioning: variations in the effectiveness of reinforcement and nonreinforcement. In: Black AH, Prokasy WF (eds) Classical conditioning II: current research and theory. Appleton-Century-Crofts, New York, pp 64–99

Rich EL, Wallis JD (2014) Medial-lateral organization of the orbitofrontal cortex. J Cogn Neurosci 26:1347–1362

Roesch MR, Bryden DW (2011) Impact of size and delay on neural activity in the rat limbic corticostriatal system. Frontiers Neurosci 5:130

Roesch MR, Olson CR (2004) Neuronal activity related to reward value and motivation in primate frontal cortex. Science 304:307–310

Roesch MR, Olson CR (2005a) Neuronal activity in primate orbitofrontal cortex reflects the value of time. J Neurophysiol 94:2457–2471

Roesch MR, Olson CR (2005b) Neuronal activity dependent on anticipated and elapsed delay in macaque prefrontal cortex, frontal and supplementary eye fields, and premotor cortex. J Neurophysiol 94:1469–1497

Roesch MR, Calu DJ, Burke KA, Schoenbaum G (2007a) Should I stay or should I go? Transformation of time-discounted rewards in orbitofrontal cortex and associated brain circuits. Ann N Y Acad Sci 1104:21–34

Roesch MR, Takahashi Y, Gugsa N, Bissonette GB, Schoenbaum G (2007b) Previous cocaine exposure makes rats hypersensitive to both delay and reward magnitude. J Neurosci 27:245–250

Roesch MR, Singh T, Brown PL, Mullins SE, Schoenbaum G (2009) Ventral striatal neurons encode the value of the chosen action in rats deciding between differently delayed or sized rewards. J Neurosci 29:13365–13376

Roesch MR, Calu DJ, Esber GR, Schoenbaum G (2010a) Neural correlates of variations in event processing during learning in basolateral amygdala. J Neurosci: Official J Soc Neurosci 30:2464–2471

Roesch MR, Calu DJ, Esber GR, Schoenbaum G (2010b) All that glitters ... dissociating attention and outcome expectancy from prediction errors signals. J Neurophysiol 104:587–595

Roesch MR et al (2012a) Normal aging alters learning and attention-related teaching signals in basolateral amygdala. J Neurosci 32:13137–13144

Roesch MR, Bryden DW, Cerri DH, Haney ZR, Schoenbaum G (2012b) Willingness to wait and altered encoding of time-discounted reward in the orbitofrontal cortex with normal aging. J Neurosci 32:5525–5533

Rolls ET, Sienkiewicz ZJ, Yaxley S (1989) Hunger modulates the responses to gustatory stimuli of single neurons in the caudolateral orbitofrontal cortex of the macaque monkey. EurJ Neurosci 1:53–60

Rothe M, Quilodran R, Sallet J, Procyk E (2011) Coordination of high gamma activity in anterior cingulate and lateral prefrontal cortical areas during adaptation. J Neurosci: Official J Soc Neurosci 31:11110–11117

Rudebeck PH, Murray EA (2008) Amygdala and orbitofrontal cortex lesions differentially influence choices during object reversal learning. J Neurosci: Official J Soc Neurosci 28:8338–8343

Rudebeck PH, Murray EA (2011) Balkanizing the primate orbitofrontal cortex: distinct subregions for comparing and contrasting values. Ann N Y Acad Sci 1239:1–13

Rudebeck PH, Murray EA (2014) The orbitofrontal oracle: cortical mechanisms for the prediction and evaluation of specific behavioral outcomes. Neuron 84:1143–1156

Rudebeck PH, Bannerman DM, Rushworth MF (2008) The contribution of distinct subregions of the ventromedial frontal cortex to emotion, social behavior, and decision making. Cogn Affect Behav Neurosci 8:485–497

Rudebeck PH, Saunders RC, Prescott AT, Chau LS, Murray EA (2013a) Prefrontal mechanisms of behavioral flexibility, emotion regulation and value updating. Nat Neurosci 16:1140–1145

Rudebeck PH, Mitz AR, Chacko RV, Murray EA (2013b) Effects of amygdala lesions on reward-value coding in orbital and medial prefrontal cortex. Neuron 80:1519–1531

Rushworth MF, Behrens TE (2008) Choice, uncertainty and value in prefrontal and cingulate cortex. Nat Neurosci 11:389–397

Rushworth MF, Walton ME, Kennerley SW, Bannerman DM (2004) Action sets and decisions in the medial frontal cortex. Trends Cogn Sci 8:410–417

Rushworth MF, Behrens TE, Rudebeck PH, Walton ME (2007) Contrasting roles for cingulate and orbitofrontal cortex in decisions and social behaviour. Trends Cogn Sci 11:168–176

Saddoris MP, Gallagher M, Schoenbaum G (2005) Rapid associative encoding in basolateral amygdala depends on connections with orbitofrontal cortex. Neuron 46:321–331

Salamone JD (1986) Different effects of haloperidol and extinction on instrumental behaviours. Psychopharmacology 88:18–23

Salamone JD (1994) The involvement of nucleus accumbens dopamine in appetitive and aversive motivation. Behav Brain Res 61:117–133

Salamone JD, Correa M (2002) Motivational views of reinforcement: implications for understanding the behavioral functions of nucleus accumbens dopamine. Behav Brain Res 137:3–25

Salamone JD, Correa M (2012) The mysterious motivational functions of mesolimbic dopamine. Neuron 76:470–485

Salamone JD et al (1991) Haloperidol and nucleus accumbens dopamine depletion suppress lever pressing for food but increase free food consumption in a novel food choice procedure. Psychopharmacology 104:515–521

Salamone JD, Correa M, Nunes EJ, Randall PA, Pardo M (2012) The behavioral pharmacology of effort-related choice behavior: dopamine, adenosine and beyond. J Exp Anal Behav 97:125–146

Sallet J et al (2007) Expectations, gains, and losses in the anterior cingulate cortex. Cogn Affect Behav Neurosci 7:327–336

Samejima K, Ueda Y, Doya K, Kimura M (2005) Representation of action-specific reward values in the striatum. Science 310:1337–1340

Scheffers MK, Coles MG (2000) Performance monitoring in a confusing world: error-related brain activity, judgments of response accuracy, and types of errors. J Exp Psychol Hum Percept Perform 26:141–151

Schoenbaum G (2004) Affect, action, and ambiguity and the amygdala-orbitofrontal circuit. Focus on "combined unilateral lesions of the amygdala and orbital prefrontal cortex impair affective processing in rhesus monkeys". J Neurophysiol 91:1938–1939

Schoenbaum G, Eichenbaum H (1995) Information coding in the rodent prefrontal cortex. I. Single-neuron activity in orbitofrontal cortex compared with that in pyriform cortex. J Neurophysiol 74:733–750

Schoenbaum G, Esber GR (2010) How do you (estimate you will) like them apples? Integration as a defining trait of orbitofrontal function. Curr Opin Neurobiol 20:205–211

Schoenbaum G, Roesch M (2005) Orbitofrontal cortex, associative learning, and expectancies. Neuron 47:633–636

Schoenbaum G, Chiba AA, Gallagher M (1998) Orbitofrontal cortex and basolateral amygdala encode expected outcomes during learning. Nat Neurosci 1:155–159

Schoenbaum G, Chiba AA, Gallagher M (1999) Neural encoding in orbitofrontal cortex and basolateral amygdala during olfactory discrimination learning. J Neurosci 19:1876–1884

Schoenbaum G, Chiba AA, Gallagher M (2000) Changes in functional connectivity in orbitofrontal cortex and basolateral amygdala during learning and reversal training. J Neurosci 20:5179–5189

Schoenbaum G, Setlow B, Nugent SL, Saddoris MP, Gallagher M (2003a) Lesions of orbitofrontal cortex and basolateral amygdala complex disrupt acquisition of odor-guided discriminations and reversals. Learn Mem 10:129–140

Schoenbaum G, Setlow B, Saddoris MP, Gallagher M (2003b) Encoding predicted outcome and acquired value in orbitofrontal cortex during cue sampling depends upon input from basolateral amygdala. Neuron 39:855–867

Schoenbaum G, Roesch MR, Stalnaker TA (2006) Orbitofrontal cortex, decision-making and drug addiction. Trends Neurosci 29:116–124

Schoenbaum G, Gottfried JA, Murray EA, Ramus SJ (2007) Linking affect to action: critical contributions of the orbitofrontal cortex. Preface. Ann N Y Acad Sci 1121:xi–xiii

Schoenbaum G, Roesch MR, Stalnaker TA, Takahashi YK (2009) A new perspective on the role of the orbitofrontal cortex in adaptive behaviour. Nat Rev Neurosci 10:885–892

Schoenbaum G, Takahashi Y, Liu TL, McDannald MA (2011a) Does the orbitofrontal cortex signal value? Ann N Y Acad Sci 1239:87–99

Schoenbaum G, Roesch MR, Stalnaker TA, Takahashi YK (2011b) Orbitofrontal cortex and outcome expectancies: optimizing behavior and sensory perception. In: Gottfried JA (ed) Neurobiology of sensation and reward. Boca Raton, Florida

Schultz W (1997) Dopamine neurons and their role in reward mechanisms. Curr Opin Neurobiol 7:191–197

Schultz W (1998) Predictive reward signal of dopamine neurons. J Neurophysiol 80:1–27

Schultz W (1999) The reward signal of midbrain dopamine neurons. News Physiol Sci: Int J Physiol Produced Jointly Int Union Physiol Sci Am Physiol Soc 14:249–255

Schultz W, Apicella P, Scarnati E, Ljungberg T (1992) Neuronal activity in monkey ventral striatum related to the expectation of reward. J Neurosci 12:4595–4610

Setlow B, Schoenbaum G, Gallagher M (2003) Neural encoding in ventral striatum during olfactory discrimination learning. Neuron 38:625–636

Shabel SJ, Janak PH (2009) Substantial similarity in amygdala neuronal activity during conditioned appetitive and aversive emotional arousal. Proc Natl Acad Sci USA 106:15031–15036

Shidara M, Richmond BJ (2004) Differential encoding of information about progress through multi-trial reward schedules by three groups of ventral striatal neurons. Neurosci Res 49:307–314

Singh T, McDannald MA, Haney RZ, Cerri DH, Schoenbaum G (2010) Nucleus accumbens core and shell are necessary for reinforcer devaluation effects on pavlovian conditioned responding. Front Integr Neurosci 4:126

Solomon RL, Corbit JD (1974) An opponent-process theory of motivation. I. Temporal dynamics of affect. Psychol Rev 81:119–145

Sripanidkulchai K, Sripanidkulchai B, Wyss JM (1984) The cortical projection of the basolateral amygdaloid nucleus in the rat: a retrograde fluorescent dye study. J Comp Neurol 229:419–431

Stalnaker TA, Franz TM, Singh T, Schoenbaum G (2007) Basolateral amygdala lesions abolish orbitofrontal-dependent reversal impairments. Neuron 54:51–58

Stalnaker TA, Takahashi Y, Roesch MR, Schoenbaum G (2009) Neural substrates of cognitive inflexibility after chronic cocaine exposure. Neuropharmacology 56(Suppl 1):63–72

Stalnaker TA, Calhoon GG, Ogawa M, Roesch MR, Schoenbaum G (2010) Neural correlates of stimulus-response and response-outcome associations in dorsolateral versus dorsomedial striatum. Frontiers Integr Neurosci 4:12

Steinberg EE et al (2013) A causal link between prediction errors, dopamine neurons and learning. Nat Neurosci 16:966–973

Stopper CM, Floresco SB (2011) Contributions of the nucleus accumbens and its subregions to different aspects of risk-based decision making. Cogn Affect Behav Neurosci 11:97–112

Sugrue LP, Corrado GS, Newsome WT (2004) Matching behavior and the representation of value in the parietal cortex. Science 304:1782–1787

Sugrue LP, Corrado GS, Newsome WT (2005) Choosing the greater of two goods: neural currencies for valuation and decision making. Nat Rev Neurosci 6:363–375

Sutton RA, Barto AG (ed) (1998) Reinforcement learning: an introduction

Takahashi Y, Roesch MR, Stalnaker TA, Schoenbaum G (2007) Cocaine exposure shifts the balance of associative encoding from ventral to dorsolateral striatum. Frontiers Integr Neurosci 1:11

Takahashi Y, Schoenbaum G, Niv Y (2008) Silencing the critics: understanding the effects of cocaine sensitization on dorsolateral and ventral striatum in the context of an actor/critic model. Front Neurosci 2:86–99

Takahashi YK et al (2009) The orbitofrontal cortex and ventral tegmental area are necessary for learning from unexpected outcomes. Neuron 62:269–280

Takahashi YK et al (2011) Expectancy-related changes in firing of dopamine neurons depend on orbitofrontal cortex. Nat Neurosci 14:1590–1597

Totah NK, Kim YB, Homayoun H, Moghaddam B (2009) Anterior cingulate neurons represent errors and preparatory attention within the same behavioral sequence. J Neurosci 29:6418–6426

Tsujimoto S, Genovesio A, Wise SP (2011) Comparison of strategy signals in the dorsolateral and orbital prefrontal cortex. J Neurosci 31:4583–4592

Tsujimoto S, Genovesio A, Wise SP (2012) Neuronal activity during a cued strategy task: comparison of dorsolateral, orbital, and polar prefrontal cortex. J Neurosci 32:11017–11031

Tye KM, Janak PH (2007) Amygdala neurons differentially encode motivation and reinforcement. J Neurosci: Official J Soc Neurosci 27:3937–3945

Tye KM, Cone JJ, Schairer WW, Janak PH (2010) Amygdala neural encoding of the absence of reward during extinction. J Neurosci: Official J Soc Neurosci 30:116–125

van der Meer MA, Redish AD (2009) Covert expectation-of-reward in rat ventral striatum at decision points. Front Integr Neurosci 3:1

van der Meer MA, Redish AD (2011) Ventral striatum: a critical look at models of learning and evaluation. Curr Opin Neurobiol 21:387–392

van der Meer MA, Johnson A, Schmitzer-Torbert NC, Redish AD (2010) Triple dissociation of information processing in dorsal striatum, ventral striatum, and hippocampus on a learned spatial decision task. Neuron 67:25–32

Wadenberg ML, Ericson E, Magnusson O, Ahlenius S (1990) Suppression of conditioned avoidance behavior by the local application of (-)sulpiride into the ventral, but not the dorsal, striatum of the rat. Biol Psychiatry 28:297–307

Wakabayashi KT, Fields HL, Nicola SM (2004) Dissociation of the role of nucleus accumbens dopamine in responding to reward-predictive cues and waiting for reward. Behav Brain Res 154:19–30

Wallis JD (2007) Orbitofrontal cortex and its contribution to decision-making. Annu Rev Neurosci 30:31–56

Wallis JD, Kennerley SW (2010) Heterogeneous reward signals in prefrontal cortex. Curr Opin Neurobiol 20:191–198

Wallis JD, Kennerley SW (2011) Contrasting reward signals in the orbitofrontal cortex and anterior cingulate cortex. Ann N Y Acad Sci 1239:33–42

Wallis JD, Miller EK (2003) Neuronal activity in primate dorsolateral and orbital prefrontal cortex during performance of a reward preference task. Eur J Neurosci 18:2069–2081

Wallis JD, Anderson KC, Miller EK (2001) Single neurons in prefrontal cortex encode abstract rules. Nature 411:953–956

Walton ME, Devlin JT, Rushworth MF (2004) Interactions between decision making and performance monitoring within prefrontal cortex. Nat Neurosci 7:1259–1265

Wang AY, Miura K, Uchida N (2013) The dorsomedial striatum encodes net expected return, critical for energizing performance vigor. Nat Neurosci 16:639–647

Wilson RC, Takahashi YK, Schoenbaum G, Niv Y (2014) Orbitofrontal cortex as a cognitive map of task space. Neuron 81:267–279

Winstanley CA, Theobald DEH, Cardinal RN, Robbins TW (2004) Contrasting roles of basolateral amygdala and orbitofrontal cortex in impulsive choice. J Neurosci 24:4718–4722

Yin HH, Knowlton BJ (2006) The role of the basal ganglia in habit formation. Nat Rev Neurosci 7:464–476

Yun IA, Wakabayashi KT, Fields HL, Nicola SM (2004) The ventral tegmental area is required for the behavioral and nucleus accumbens neuronal firing responses to incentive cues. J Neurosci 24:2923–2933

Mesolimbic Dopamine and the Regulation of Motivated Behavior

John D. Salamone, Marta Pardo, Samantha E. Yohn, Laura López-Cruz, Noemí SanMiguel and Mercè Correa

Abstract It has been known for some time that nucleus accumbens dopamine (DA) is involved in aspects of motivation, but theoretical approaches to understanding the functions of DA have continued to evolve based upon emerging data and novel concepts. Although it has become traditional to label DA neurons as "reward" neurons, the actual findings are more complicated than that, because they indicate that DA neurons can respond to a variety of motivationally significant stimuli. Moreover, it is important to distinguish between aspects of motivation that are differentially affected by dopaminergic manipulations. Studies that involve nucleus accumbens DA antagonism or depletion indicate that accumbens DA does not mediate primary food motivation or appetite. Nevertheless, DA is involved in appetitive and aversive motivational processes including behavioral activation, exertion of effort, sustained task engagement, and Pavlovian-to-instrumental transfer. Interference with accumbens DA transmission affects instrumental behavior in a manner that interacts with the response requirements of the task and also shifts effort-related choice behavior, biasing animals toward low-effort alternatives. Dysfunctions of mesolimbic DA may contribute to motivational symptoms seen in various psychopathologies, including depression, schizophrenia, parkinsonism, and other disorders.

Keywords Dopamine · Accumbens · Behavioral activation · Motivation · Reward · Depression · Fatigue · Anergia

J.D. Salamone (✉) · S.E. Yohn · M. Correa
Department of Psychology, University of Connecticut, Storrs, CT 06269-1020, USA
e-mail: john.salamone@uconn.edu

M. Pardo · L. López-Cruz · N. SanMiguel · M. Correa
Àrea de Psicobiologia, Universitat Jaume I, 12071 Castelló, Spain

Contents

1. Introduction 232
2. Dynamic Activity of DA Neurons: Multiple Modes of Responding 233
3. Behavioral Manifestations of Interference with Accumbens DA Transmission:
   Dissociation of Distinct Components of Motivation and Reinforcement 234
4. Manifestations of Interference with Accumbens DA Transmission: Behavioral
   Activation, Behavioral Economics, and Effort-Related Choice 239
5. Clinical Significance of Effort-Related Functions 242
6. Conclusions 246
References 247

1 Introduction

It has been known for some time that mesolimbic dopamine (DA) and its main projection target, the nucleus accumbens, are involved in aspects of motivation. Motivation has been defined in many ways, but for the purposes of this chapter, it is defined as the processes that enable organisms to regulate the proximity, probability, and availability of stimuli (Salamone 1992). Mogenson et al. (1980) considered nucleus accumbens to be a limbic–motor interface, at which brain areas processing information related to motivation and emotion can gain access to the basal ganglia motor system circuitry (Floresco 2015). In addition, mesolimbic DA appears to be a key part of the neural circuitry regulating phenomena such as intracranial self-stimulation and self-administration of drugs, especially stimulants (Wise 2008), though the specific processes being affected by dopaminergic manipulations are still a matter of some debate (e.g., Hernandez et al. 2010; Venugopalan et al. 2011).

Similarly, with natural reinforcers such as food, the question is not whether accumbens DA is involved in motivation, but how. In other words—which aspects of motivation require intact DA transmission? Mesolimbic DA neurons are activated during motivationally relevant situations, and interference with accumbens DA transmission can affect aspects of motivated behavior. Nevertheless, it should be recognized that motivation is a complex process involving multiple interacting functions mediated by various neural circuits. Classically, motivational theory has emphasized two distinct aspects of motivation: directional and activational aspects (Duffy 1963; Cofer and Apley 1964; Salamone 1988). Thus, behavior is said to be directed toward some stimuli (food, water, sex) and away from others (painful stimuli, predators). In addition to being directed toward or away from significant stimuli (see Chapter by Cornwell, Franks & Higgins), motivated behavior also is characterized by a high degree of behavioral activation, as demonstrated by the speed, vigor or persistence seen in both the instigation and maintenance of instrumental responding (Salamone 1988, 1992; Salamone and Correa 2002, 2012; Robbins and Everitt 2007; Croxson et al. 2009; Kurniawan et al. 2010; Nicola 2010; McGinty et al. 2013; Floresco 2015). As discussed below, considerable evidence indicates that interference with accumbens DA transmission can substantially affect activational aspects of motivation for food and other natural reinforcers, while leaving fundamental features of primary motivation intact (e.g., appetite, primary food motivation; see Salamone and Correa 2002, 2012). These findings have implications for understanding the neural circuitry underlying both normal and pathological aspects of motivation.

2 Dynamic Activity of DA Neurons: Multiple Modes of Responding

One way of studying the behavioral functions of DA systems is to investigate the responsiveness of these systems across an array of behaviorally relevant conditions. The dynamic activity of the mesolimbic DA system has been studied using a variety of methods, employing different neural markers and covering distinct time scales. Although mesolimbic DA neurons are sometimes referred to under the blanket term "reward neurons," and it is common to see the mesolimbic DA pathway referred to as "the reward system," it is clear from the literature that this is a gross oversimplification. For example, in experienced animals, the dopaminergic response in accumbens to primary food reinforcement or simple access to food is minimal or absent; this is seen as the lack of population response of ventral tegmental area (VTA) DA neurons in monkeys trained on a discrete trial discrimination task (Schultz et al. 1993), the lack of response to sucrose reinforcement in trained animals as measured by voltammetry (Roitman et al. 2004), the habituation of the extracellular DA response to palatable food (Bassareo and DiChiarra 1999a, b; Bassareo et al. 2002), and the lack of responsiveness to food in experienced animals as measured by microdialysis (Salamone et al. 1994b; Segovia et al. 2012). There appears to be general agreement that mesolimbic DA neurons respond to conditioned stimuli (Schultz et al. 1993; Schultz 2010). Moreover, microdialysis studies show that DA release and DA-related signal transduction is increased in nucleus accumbens during the performance of food-reinforced instrumental behavior (McCullough et al. 1993a; Salamone et al. 1994b; Sokolowski et al. 1998; Cousins et al. 1999; Roitman et al. 2004; Ostlund et al. 2011; Segovia et al. 2011, 2012). Finally, there is the frequently cited finding that the response of DA neurons provides a teaching signal that allows for the determination of reward prediction errors (Schultz et al. 1997; Bayer and Glimcher 2002; Niv 2009; Schultz 2010; Steinberg et al. 2013). The latter effect has led to a large number of physiological, computational, and theoretical papers.

Nevertheless, it has been evident for some time that DA neuron activity and accumbens DA release respond in a variety of different modes to a wide array of stimuli (Salamone and Correa 2012; Marinelli and McCutcheon 2014; Lammel et al. 2014). For example, electrophysiological and voltammetric studies have shown that putative VTA DA neurons can show increased responsiveness to stressful or aversive stimuli such as restraint stress (Anstrom and Woodward 2005), footshock (Brischoux et al. 2009), social defeat stress (Anstrom et al. 2009), and tail pinch (Zweifel et al. 2011; Budygin et al. 2012). In terms of responsiveness to aversive stimuli, there is heterogeneity based upon anatomical loci of the cell somata and the terminal projections (Brischoux et al. 2009; Roeper 2013; Lammel et al. 2014). Nevertheless, these electrophysiology and voltammetry studies are broadly consistent with the array of microdialysis reports showing increased extracellular DA in response to stressful or aversive conditions (McCullough and Salamone 1992; McCullough et al. 1993b; Salamone 1994, 1996; Tidey and Mizcek 1996; Young 2004). Thus, it is clear that one should not assume that all DA neurons are simply "reward neurons," nor should one link in a simple way enhanced DA release to subjective pleasure. Rather, it appears that activation of VTA DA neurons in specific subcircuits, and the release of DA in specific accumbens subregions, participates in a variety of motivational and learning processes in a rather complex manner that has yet to be fully characterized. For example, recent evidence indicates that phasic DA release in nucleus accumbens was related to preference for specific outcomes in choice procedures, but only when it was imbedded into the context of action selection (Saddoris et al. 2015). It seems likely that activation of mesolimbic DA transmission promotes the instigation of instrumental actions, modulates the vigor of these responses, increases energy expenditure, guides value-based action selection, and provides signals that are important for aspects of neuroplasticity involved in Pavlovian and instrumental learning.

3 Behavioral Manifestations of Interference with Accumbens DA Transmission: Dissociation of Distinct Components of Motivation and Reinforcement

Important though they are, studies of the dynamic activity of DA neurons only tell part of the story. Another critical part, which is the focus of the present chapter, is the behavioral manifestations of impaired DA transmission. Several decades ago, it was recognized that high doses of DA antagonists, or whole forebrain DA depletions, could reduce food intake (Ungerstedt 1971; Zigmond and Stricker 1972). However, evidence gradually accumulated demonstrating that this effect was due largely to the motoric effects of impaired DA transmission in the neostriatum (i.e., caudate/putamen, or dorsal striatum), and in particular, the lateral or ventrolateral neostriatum (Dunnett and Iversen 1982; Salamone et al. 1990, 1993; Bakshi and Kelley 1991). In terms of nucleus accumbens DA, several lines of evidence indicate that the impact of DA depletions or antagonism in that terminal region have minimal impact on food intake (Salamone and Correa 2002, 2009; Baldo and Kelley 2007). Ungerstedt (1971) and Koob et al. (1978) showed that selective depletions of nucleus accumbens DA did not reduce food intake. Salamone et al. (1993) showed that accumbens DA depletions induced by 6-hydroxydopamine did not reduce food intake, nor did they alter food handling or feeding rate. In DA-deficient mice, restoration of feeding behavior occurred after viral rescue of DA transmission in neostriatum, but not nucleus accumbens (Szczpka et al. 2001). Baldo et al. (2002) injected D1 and D2 family antagonists into core and shell subregions of the accumbens and found that doses of these drugs that suppressed locomotion failed to alter food intake.

Several lines of evidence also show that nucleus accumbens DA does not directly mediate hedonic reactivity to food stimuli. An enormous collection of studies from Berridge and colleagues has demonstrated that systemic administration of DA antagonists, as well DA depletions in whole forebrain or nucleus accumbens, does not blunt appetitive taste reactivity for food, which is a widely used measure of hedonic reactivity to sweet solutions (Berridge and Robinson 1998, 2003; Berridge 2007; Berridge and Kringelbach 2015; Robinson et al., this volume). DA D2 receptors in the shell subregion of the accumbens regulate aversive taste reactivity, and brainstem D2 receptor stimulation was shown to suppress sucrose consumption, but neither group of receptors mediated the hedonic display of taste (Sederholm et al. 2002). Furthermore, knockdown of the DA transporter (Peciña et al. 2003) and microinjections of amphetamine into nucleus accumbens (Smith et al. 2011), both of which elevate extracellular DA, failed to enhance appetitive taste reactivity for sucrose. In fact, some of the non-dopaminergic neurobiological processes that underlie hedonic reactions to appetitive rewards have been identified (for more information, see Robinson et al., this volume).

Another process that is sometimes assumed to be associated with nucleus accumbens DA transmission is reinforcement learning. But here again, the literature has yielded a complex set of findings. Reports indicating that intra-accumbens injections of stimulant drugs such as amphetamine can support self-administration, or that optogenetic stimulation of DA neurons can be reinforcing (Ilango et al. 2014; Steinberg et al. 2014), do not necessarily mean that accumbens DA release acts primarily as a "reinforcement system" that stamps in instrumental learning related to natural stimuli. As discussed previously, such a reinforcing effect could be an emergent property that results from the modulation of the various channels of information passing through the nucleus accumbens, including the enhancement of the impact of environmental cues (Salamone et al. 2005; Everitt and Robbins 2005; Steinberg et al. 2014). As noted by Ilango et al. (2014), a major outcome resulting from the phasic activation of DA neurons by optogenetic stimulation is the transient instigation of conditioned approach behavior, which is consistent with an earlier optogenetic study (Adamantidis et al. 2011). Selective genetic inactivation of NMDA receptors, which blunted burst firing in VTA DA neurons, impaired the acquisition of cue-dependent appetitive learning but did not disrupt the acquisition of responding on a progressive ratio schedule (Zweifel et al. 2009). Moreover, the research that has specifically focused on the potential role of striatal areas in mediating action–outcome associations suggests that neostriatum, rather than nucleus accumbens, is more critically involved (Yin et al. 2005; Corbit and Janak 2010; Corbit et al. 2013). In their comprehensive review of this literature, Yin et al. (2008) stated that "the accumbens is neither necessary nor sufficient for instrumental learning" (p. 1439).

Although there are many studies showing that cell body lesions, DA antagonists, or DA depletions can affect the learning-related outcomes in procedures such as place preference, acquisition of lever pressing, or other procedures, this does not directly demonstrate that accumbens neurons or mesolimbic DA transmission is essential for the action–outcome associations that are the basis of instrumental learning (Yin et al. 2008; Belin et al. 2009; Salamone and Correa 2012). Specific processes related to associative aspects of instrumental learning can be demonstrated by assessments of the effects of reinforcer devaluation or contingency degradation, which often are not conducted in pharmacology or lesion studies. These procedures assess the behavioral effects of reducing reinforcement value (e.g., pre-feeding or lacing food with quinine), or disrupting the response-reinforcer contingency [e.g., non-contingent presentation of the reinforcer; see Yin et al. (2008) for more details]. Thus, it is important to note that cell body lesions in either core or shell of the accumbens did not alter sensitivity to contingency degradation (Corbit et al. 2001). Furthermore, Lex and Hauber (2010) found that rats with nucleus accumbens DA depletions were still sensitive to reinforcer devaluation, indicating that accumbens core DA does not appear to be crucial for encoding action-outcome associations.

Considerable evidence indicates that accumbens DA is important for Pavlovian approach and Pavlovian-to-instrumental transfer (PIT; Parkinson et al. 2002; Wyvell and Berridge 2000; Dalley et al. 2005; Lex and Hauber 2008, 2010; Yin et al. 2008). PIT is a behavioral process that reflects the impact of Pavlovian-conditioned stimuli (CS) on instrumental responding. For example, presentation of a Pavlovian CS paired with food can increase output of food-reinforced instrumental behaviors, such as lever pressing. Outcome-specific PIT occurs when the Pavlovian unconditioned stimulus (US) and the instrumental reinforcer are the same stimulus, whereas general PIT is said to occur when the Pavlovian US and the reinforcer are different. Lex and Hauber (2008) showed that D1 or D2 family antagonists locally injected into either the core or shell subregions of nucleus accumbens reduced general PIT, which is consistent with the report of Corbit et al. (2007), who demonstrated that inactivation of the VTA suppressed both general and outcome-specific PIT. More recent evidence indicates that accumbens core and shell appear to mediate different aspects of PIT; shell lesions and inactivation reduced outcome-specific PIT, while core lesions and inactivation suppressed general PIT (Corbit and Balleine 2011). These core versus shell differences are likely due to the different anatomical inputs and pallidal outputs associated with these accumbens subregions (Root et al. 2015). These results led Corbit and Balleine (2011) to suggest that accumbens core mediates the general excitatory effects of reward-related cues. PIT provides a fundamental behavioral process by which conditioned stimuli can exert activating effects upon instrumental responding (Robbins and Everitt 2007; Salamone et al. 2007). The activating or arousing effects of conditioned stimuli can be a factor in amplifying an instrumental response that has already been acquired, but also could facilitate acquisition of instrumental learning by increasing response output and behavioral variability, thus setting the occasion for more opportunities to pair a response with reinforcement [Salamone and Correa 2012; see Rick et al. (2006) for a discussion of behavioral variability]. Further dissection and discussion of the behavioral and neural processes contributing to PIT are provided by Corbit and Balleine in this volume.

This discussion of the activating effects of conditioned stimuli, and the role of accumbens DA in these processes, leads one back to the distinction between directional and activational aspects of motivation that was discussed above. Indeed, an overwhelming body of evidence indicates that accumbens DA is involved in behavioral activation and energy expenditure (Salamone 1988, 1992; Salamone and Correa 2002, 2012; Robbins and Everitt 2005, 2007; Beeler et al. 2012). Accumbens DA depletions or antagonism suppress novelty-induced locomotion (Koob et al. 1978; Cousins et al. 1993; Baldo et al. 2002; Correa et al. 2002). Scheduled presentation of food pellets to food-restricted rats can induce various types of activity, including excessive drinking, wheel running, and locomotion; these schedule-induced activities have been shown to be accompanied by increases in accumbens DA release (McCullough and Salamone 1992) and suppressed by accumbens DA depletions (Robbins and Koob 1980; Wallace et al. 1983; McCullough and Salamone 1992). Knab et al. (2009) reported altered expression of dopaminergic genes in mice with high versus low levels of running wheel activity, which was independent of their previous wheel exposure. Recent studies have shown that inactivation of VTA DA neurons by DREADD (designer receptors exclusively activated by designer drugs) methods significantly suppressed locomotor activity (Marchant et al. 2015). Recent data from our laboratory have demonstrated that locomotor activity can be instigated by the presentation of cues associated with sucrose and that this effect was blocked by DA antagonism (Fig. 1).

Consistent with the known involvement of accumbens DA in behavioral activation and energy expenditure, several studies have demonstrated that the effects of nucleus accumbens DA depletions interact with the work requirements presented across various instrumental tasks. One way of varying the response requirements of instrumental behavior is to vary the ratio requirements of operant schedules. Studies have shown that accumbens DA depletions suppress operant lever pressing on ratio schedules in a manner that is directly related to the size of the ratio requirement. Fixed ratio (FR) 1 responding is only marginally and transiently affected by DA depletion, while rats responding on moderate size ratio schedules (FR 5, 16, 20) showed modest reductions in response rates, and animals tested on schedules with high ratios (e.g., FR16, 64, 300) were severely impaired (McCullough et al. 1993a; Aberman et al. 1998; Aberman and Salamone 1999; Salamone et al. 2001; Ishiwari et al. 2004). These ratio-related effects of accumbens DA depletion differed substantially from the pattern produced by pre-feeding to devalue the food reinforcement (Aberman and Salamone 1999). Moreover, these effects did not depend upon the time intervals or intermittency of reinforcement; rats responding on conventional variable interval schedules (e.g., VI 30, 60 or 120 s) were not affected by accumbens DA depletions that substantially suppressed responding when a ratio requirement (FR5 or 10) was attached to the interval requirement (Correa et al. 2002; Mingote et al. 2005). Thus, accumbens DA depletions appear to blunt the response-enhancing effects of moderate-sized ratios, and enhance the response suppressing effects of very large ratio requirements (Salamone and Correa 2002).

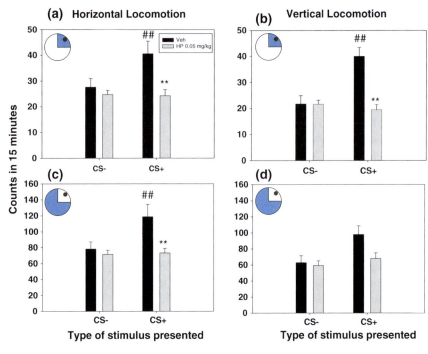

Fig. 1 Presentation of a CS + paired with sucrose instigates locomotor activity, and this effect is blocked by a very low dose of haloperidol (Correa et al. 2012). Mice ($N = 66$) were trained to associate an olfactory stimulus (CS+) with 10 % sucrose and another stimulus (CS−) with water during 30 min per day, 4-week training sessions. On the test day, after the conditioning period, one of the odors (CS+ or CS−) was present in the upper part of the locomotion chamber. The effect of haloperidol (0.0 or 0.05 mg/kg) on locomotion induced by CS+ or CS− presentation on horizontal locomotion (A and C) and vertical locomotion (B and D), in the quadrant where the CS (*black dot*) was located (*upper panels*, A and B; see *blue quadrant*) and in the other 3 quadrants (*lower panels*, C and D; *blue shading* indicates these quadrants). Mean (±SEM) number of counts in the open field during the 15 min test session. **$p < 0.01$ significantly different between doses in the same group; ##$p < 0.01$ significant difference between CS+ and CS− at the same dose. In the CS quadrant, the two-way factorial ANOVA for the horizontal locomotion showed a significant interaction ($F(1, 62) = 4.08, p < 0.05$), and for the vertical locomotion ($F(1, 62) = 13.05, p < 0.01$). In the quadrants with no CS, a two-way factorial ANOVA for horizontal locomotion showed a significant interaction ($F(1, 62) = 4.18, p < 0.05$), but not for the vertical locomotion ($F(1, 62) = 2.46, n.s.$). In a parallel study, vehicle-treated mice drank 0.94 ± 0.20 ml of 10 % sucrose in 60 min, and when they received 0.05 mg/kg haloperidol the amount of sucrose consumed was 1.07 ± 0.16 ml. The t-test for related samples yielded no significant differences ($t(20) = 0.97$; n.s.). Thus, the effect of haloperidol on the activational properties of the CS+ was not dependent upon an effect on sucrose consumption

4 Manifestations of Interference with Accumbens DA Transmission: Behavioral Activation, Behavioral Economics, and Effort-Related Choice

As evidence was emerging on the role of accumbens DA in behavioral activation, it was suggested several years ago that interference with DA transmission should affect cost/benefit decision making on tasks involving varying effort requirements (Salamone 1987, 1991, 1992). In complex environments, organisms continually make effort-related decisions based upon cost/benefit analyses, allocating behavioral resources into goal-directed actions based upon assessments of motivational value and response costs. It was hypothesized that behavioral processes related to activational aspects of motivation could be engaged to facilitate the ability of organisms to overcome work-related response costs that separate them from significant stimuli and that interference with DA transmission would alter the relative allocation of responses, biasing animals toward low-cost alternative (Salamone 1987, 1991, 1992). Such a function would be extremely important for animals foraging in the wild (e.g., for efficient investment of behavioral resources and energy expenditure) and could also be studied in laboratory experiments. Within the next few years, research was undertaken to study the role of DA in *effort-related choice behavior* (also called *effort-related* or *effort-based decision making*).

Effort-based decision making is generally studied using tasks that offer a choice between high-effort instrumental actions leading to more highly valued reinforcers versus low-effort options leading to less valued reinforcers. An operant concurrent choice task was developed, which offers rats a choice between fixed ratio 5 (FR5) lever pressing to obtain a relatively preferred food (high-carbohydrate pellets), versus approaching and consuming a less preferred food (lab chow) that is concurrently available (Salamone et al. 1991). Under control conditions, rats pressing on the FR5 schedule typically get most of their food by lever pressing and eat only small amounts of chow. Pre-feeding to devalue food reinforcement suppressed both lever pressing and chow intake (Salamone et al. 1991). In contrast, low doses of DA antagonists and depletions or antagonism of accumbens DA shift choice behavior, decreasing lever pressing but substantially increasing intake of the concurrently available chow (Salamone et al. 1991, 2002; Koch et al. 2000; Nowend et al. 2001; Sink et al. 2008; Farrar et al. 2010). Thus, despite the fact that lever pressing is decreased by accumbens DA antagonism or depletions, the rats show a compensatory reallocation of behavior and select a new path to an alternative food source. The use of this task as a measure of effort-related choice behavior has been validated in several ways. Drug treatment conditions that produced the shift in choice behavior did not alter food intake or preference in free-feeding choice tests (Salamone et al. 1991; Koch et al. 2000; Farrar et al. 2008; Nunes et al. 2013a, b; Pardo et al. 2015), indicating that DAergic manipulations were not simply changing food preference. Increasing the lever pressing work requirement (i.e., larger fixed ratios) resulted in a shift from lever pressing to chow intake (Salamone et al. 1997). Although DA antagonists reduce FR5 lever pressing and increase chow intake, appetite suppressants such as fenfluramine and cannabinoid CB1 antagonists do not increase chow intake at doses that suppress lever pressing (Salamone et al. 2002; Sink et al. 2008; Randall et al. 2012, 2015). Thus, interference with DA transmission does not simply reduce appetite for food (Salamone and Correa 2009, 2012).

A T-maze choice procedure also was developed to assess effort-related choice behavior (Salamone et al. 1994a). The two choice arms of the maze can have different reinforcement densities (e.g., 4 vs. 2 food pellets, or 4 vs. 0), and under some conditions, a vertical barrier is placed in the arm with the higher density of food to provide an effort-related challenge. DA antagonism and accumbens DA depletions bias animals toward the low-effort alternative, decreasing selection of the high-reward/high-cost arm with the barrier, but increasing selection of the low-reward/low-cost arm with no barrier (Salamone et al. 1994a; Cousins et al. 1996; Mott et al. 2009; Mai et al. 2012; Pardo et al. 2012). When no barrier is present in the arm with the high-reward density, or when both arms have a barrier, neither DA antagonism nor DA depletion alters response choice (Salamone et al. 1994a; Pardo et al. 2012). When the arm with the barrier contained 4 pellets, but the other arm contained no pellets, rats with impaired accumbens DA transmission still chose the high-density arm, climbed the barrier, and consumed the pellets (Cousins et al. 1996; Yohn et al. 2015a, b).

Effort-discounting tasks also have been developed, in which the relation between required effort and reinforcement value is systematically varied within a test session, and the animal is offered a variety of choices with different effort/reward trade-offs. Bardgett et al. (2009) developed an effort-discounting task based upon the T-maze barrier procedure described above and showed that D1 or D2 antagonism reduced selection of the high-effort arm with the barrier. Floresco and colleagues have established effort-discounting procedures based upon the choices between ratio schedules with different response requirements. Administration of the DA antagonist flupenthixol altered effort-related decision making and biased selection toward the lower ratio option, even if the time to completion of the ratio components was controlled for (Floresco et al. 2008). Moreover, local blockade of $GABA_{A/B}$ receptors in the core region of the accumbens, but not the shell, reduced selection of the higher effort alternative under both the standard and equivalent delay conditions (Ghods-Sharifi and Floresco 2010). Interestingly, the effects of dopaminergic receptor blockade on ratio discounting, which involves physical effort, are dissociable from the effects of DA antagonism on a cognitive effort-discounting task. Hosking et al. (2015) compared the effects of the DA D1 and DA D2 family antagonists on a ratio-discounting task that assesses physical effort versus a cognitive effort-discounting task in which animals can choose to allocate greater visuospatial attention to obtain a higher reward level. While DA antagonism altered decision making based upon physical effort, it had no effect on discounting based upon cognitive effort. These studies, together with the results of a recently developed progressive ratio (PROG)/chow feeding choice task (Randall et al. 2012, 2015), demonstrate that DA antagonism and accumbens DA depletions cause animals to reallocate their instrumental response selection based upon the response requirements of the task, and select lower cost alternatives (Salamone et al. 2007, 2012; Salamone and Correa 2012).

There also is evidence that DA systems exert a bidirectional influence over response output in tasks involving effort-related choice behavior. Cagniard et al. (2006) reported that DA transporter (DAT) knockdown mice showed increased lever pressing and decreased chow intake compared to wild-type mice. Consistent with this observation, Randall et al. (2015) recently reported that administration of the catecholamine uptake inhibitor bupropion, which elevates extracellular DA and increases expression of phosphorylated DARPP-32 in a manner consistent with increased D1 and D2 signaling, substantially increased selection of the high-effort (progressive ratio) option in rats tested on a concurrent choice procedure. Trifilieff et al. (2013) reported that selective overexpression of D2 receptors in the nucleus accumbens of adult mice also led to an increase in selection of high-effort alternatives in choice tasks. Interestingly, it also appears that individual differences in behavioral output on effort-related tasks can be correlated with neural markers of DA transmission. The progressive ratio/chow feeding choice task generates enormous individual variability in behavioral output, and Randall et al. (2012) found that rats with high lever pressing output showed greater expression of accumbens core DARPP-32 phosphorylated at the threonine 34 site compared to low performers.

Of course, nucleus accumbens DA is just one part of the broader neural circuitry involved in effort-related processes. DA interacts with other transmitters and neuromodulators to regulate effort-related functions. Systemic or local intra-accumbens core administration of adenosine A_{2A} antagonists can reverse the effort-related effects of DA antagonists (Farrar et al. 2007, 2010; Worden et al. 2009; Mott et al. 2009; Salamone et al. 2009; Nunes et al. 2010; Pardo et al. 2012; Randall et al. 2015; Yohn et al. 2015a). Intra-accumbens injections of adenosine A_{2A} agonists can induce effects on effort-related choice that resemble those induced by DA antagonism or depletion (Font et al. 2008; Mingote et al. 2008), and systemic administration of the adenosine A_{2A} antagonist to rats was shown to increase work output on the lever pressing component of the progressive ratio/chow feeding choice procedure (Randall et al. 2012). A number of papers, including some that have employed disconnection methods (i.e., combined contralateral manipulation of two different parts of the circuit), have shown that there is a distributed neural circuitry that regulates effort-based decision making, which involves basolateral amygdala, prefrontal/anterior cingulate cortex, nucleus accumbens, and ventral pallidal GABA (Salamone et al. 1994a, 1997, 2007; Walton et al. 2003; Floresco and Ghods-Sharifi 2007; Farrar et al. 2008; Mingote et al. 2008; Hauber and Sommer 2009; see Fig. 2). (Further discussion of these and related circuits is found in the O'Doherty chapter and the Bissonette and Roesch chapter in this volume.)

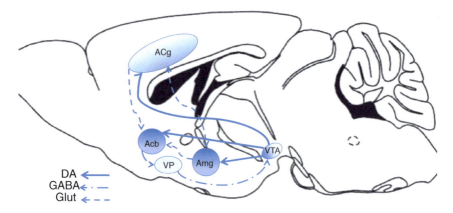

Fig. 2 Schematic showing anatomical connections between structures involved in effort-related choice behavior in the rodent brain. *Acb* nucleus accumbens; *ACg* anterior cingulate gyrus; *VP* ventral pallidum; *Amg* amygdala; *VTA* ventral tegmental area; *DA* dopamine; *GABA* gamma aminobutyric acid; *Glut* glutamate

5 Clinical Significance of Effort-Related Functions

There are widespread reports in the clinical literature of human pathologies involving activational or psychomotor impairments. Motivational dysfunctions are reported to be some of the most common psychiatric symptoms seen in general medicine (Demyttenaere et al. 2005). In addition, motivational/psychomotor symptoms variously labeled as anergia, fatigue, psychomotor retardation, or lassitude are frequently observed in patients with major depression and related disorders (Stahl 2002; Caligiuri et al. 2003; Demyttenaere et al. 2005; Salamone et al. 2006, 2014; Bella et al. 2010; Treadway and Zald 2011; Fava et al. 2014; Soskin et al. 2013). Effort-related symptom severity in depressed people is correlated with problems involving social function, employment, and response to treatment (Tylee et al. 1999; Stahl 2002). Gullion and Rush (1998) conducted a correlational and factor analytic study of data from depressed patients and identified a "lack of energy" factor that was related to problems such as low energy/increased fatigability, inability to work, and psychomotor retardation; this was also the factor that loaded most strongly onto a second-order general depression factor. Depressed patients can have core impairments in exertion of effort during reward seeking that do not simply depend upon any problems that they may have with experiencing pleasure in response to a primary motivational stimulus (Treadway and Zald 2011; Treadway et al. 2012; Argyropoulos and Nutt 2013). (More information on motivation-related phenotypes and biomarkers in depression can be found in Treadway et al. in this volume.) Because motivational symptoms in depression and other disorders can be highly resistant to treatment (Stahl 2002; Fava et al. 2014), this is an important unmet need in psychiatry.

Within the last few years, human tasks assessing effort-related decision making have been developed (Treadway et al. 2009; Gold et al. 2013), and there has been a wave of publications reporting on the presence of altered effort-based function across various clinical populations. People with major depression show reduced selection of high-effort alternatives (Treadway et al. 2012; Yang et al. 2014). Several reports involving various behavioral procedures have shown that schizophrenic patients also show "effort shyness," and tend to select low-effort alternatives when tested on choice tasks (Gold et al. 2013, 2015; Fervaha et al. 2013; Green and Horan 2015; Green et al. 2015; Hartmann et al. 2015). The clinical assessment of motivational deficits in schizophrenia is described in detail in this volume by Reddy et al. In addition, an elegant series of experiments that have identified some of the behavioral components of motivational deficits in schizophrenia are provided by Waltz and Gold in this volume. Patients with Parkinson's disease also showed reduced selection of high-effort alternatives compared to control subjects, and these deficits were significantly reduced when patients were on their dopaminergic medication (Chong et al. 2015). In contrast, autistic patients showed increased selection of high-effort choices (Damiano et al. 2012).

Because of the clinical significance of deficits in behavioral activation and exertion of effort, another line of work has emerged in which effort-related dysfunctions are being studied using explicit animal models that are related to psychopathology. Thus, the rodent tasks described above can be employed to study the effects of conditions associated with depression, and this type of task can be used to assess the effects of putative and well-established therapeutic agents. Recent studies have shown that conditions associated with depression in humans can shift effort-related choice behavior and reduce selection of high-effort choices in rats. These conditions include stress (Shafiei et al. 2012), the proinflammatory cytokine interleukin 1-β (IL1-β; Nunes et al. 2014), and administration of tetrabenazine (TBZ). TBZ inhibits the vesicular monoamine transporter type 2 (VMAT-2), which leads to a blockade of vesicular storage and a depletion of monoamines, with its greatest effects at low doses being on striatal DA (Pettibone et al. 1984; Tanra et al. 1995). TBZ is used to treat Huntington's disease and other movement disorders, but major side effects include depressive symptoms (Frank 2009, 2010; Guay 2010; Chen et al. 2012). Moreover, TBZ has been employed in studies involving traditional animal models of depression (Preskorn et al. 1984; Kent et al. 1986; Wang et al. 2010). Recent research has demonstrated that low doses of TBZ that decreased DA release and DA-related signal transduction in the accumbens could alter effort-related choice behavior as assessed by concurrent lever pressing/chow feeding choice procedures (Nunes et al. 2013b; Randall et al. 2015), as well as the T-maze barrier choice task (Yohn et al. 2015b). The low doses of TBZ that decreased selection of FR5 or progressive ratio lever pressing did not alter relative preference for high-carbohydrate pellets (the reinforcer for the high-effort option) versus chow intake (Nunes et al. 2013b) and did not produce effects similar to pre-feeding or appetite suppressant drugs (Randall et al. 2012, 2015). These effects of systemic TBZ were also shown after local injections of the drug into nucleus accumbens core, but not overlying medial dorsal striatum (Nunes et al. 2013b).

Recently, a version of the concurrent lever pressing/chow intake task was developed in which different sucrose concentrations were used (Pardo et al. 2015). TBZ reduced lever pressing for the strongly preferred higher concentration of sucrose, but actually increased selection of the low concentration of sucrose that was obtained with low effort. Nevertheless, the same doses of TBZ had no effect on sucrose preference or appetitive taste reactivity (Pardo et al. 2015). In another recent study (Yohn et al. 2015c), TBZ altered effort-related decision making in rats responding on the T-maze barrier task, as marked by a reduction in the selection of the barrier arm that contained the high density of food reinforcement (4 pellets), and increased selection of the arm with 2 pellets but no barrier. In the dose range tested (0.25–0.75 mg/kg), TBZ did not affect arm selection when there was no barrier in either arm, or when the arm with the barrier had 4 reinforcement pellets but the other arm had no pellets. This pattern of evidence indicates that TBZ was not reducing selection of the high-effort alternative (i.e., the barrier arm in the 4–2 barrier condition) because it was impairing sensitivity to reinforcement density, preference for 4 pellets versus 2, reference memory, left/right discrimination, or because of an absolute inability to climb the barrier or a ceiling level of barrier crossings (Yohn et al. 2015c).

Several drugs have been evaluated for their ability to reverse the deficits in effort-related choice described above. Adenosine A_{2A} antagonists have antiparkinsonian effects in animal models and human clinical studies and can produce behavioral effects in rodents that are consistent with antidepressant actions as assessed by classical behavioral models such as the forced swimming and tail suspension tests (Hodgson et al. 2009; Hanff et al. 2010; Yamada et al. 2013, 2014). The adenosine A_{2A} antagonist MSX-3 has been shown to reverse the effects of IL-1β on concurrent FR5/chow feeding choice performance (Nunes et al. 2014). This drug also reversed the effort-related effects of TBZ in rats responding across multiple tasks (Nunes et al. 2013b; Randall et al. 2015; Yohn et al. 2015a). These findings are consistent with previous research showing that adenosine A_{2A} antagonists can reverse the effort-related effects of DA D2 family antagonists (Farrar et al. 2007; Salamone et al. 2009; Worden et al. 2009; Mott et al. 2009; Nunes et al. 2010; Santerre et al. 2012; Pardo et al. 2012). Anatomical studies have shown that adenosine A_{2A} receptors are co-localized with DA D2 family receptors on enkephalin-positive medium spiny neurons in both neostriatum and accumbens (Rosin et al. 1998; Svenningson et al. 1999). Adenosine A_{2A} and DA D2 receptors can form heteromeric complexes, and they also converge onto the same c-AMP/protein kinase A signal transduction pathway (Ferré et al. 2008; Santerre et al. 2012). Recently, it was shown that 0.75 mg/kg TBZ reduced DA-related signal transduction mediated by D1 and D2 receptors (e.g., changes in cFos and DARPP-32 expression) and that 2.0 mg/kg MSX-3 could reverse the cellular effects of diminished D2 transmission (Nunes et al. 2013b).

Bupropion (Wellbutrin) is a widely prescribed antidepressant (Milea et al. 2010), which has been shown to produce antidepressant-like effects in traditional rodent models such as the forced swim and tail suspension tests (Cryan et al. 2004; Bourin et al. 2005; Kitamura et al. 2010). Bupropion inhibits catecholamine uptake and has been shown to occupy DA transporters in humans at doses that are clinically useful for treating depression (Learned-Coughlin et al. 2003). This drug also elevates extracellular DA and norepinephrine (NE) in rats as measured by microdialysis methods (Hudson et al. 2012; Randall et al. 2015). Bupropion fully reversed the effects of TBZ in rats tested on the T-maze barrier choice task, increasing selection of the barrier arm in TBZ-treated rats (Yohn et al. 2015a). These results are consistent with previous research showing that bupropion increased lever pressing performance in TBZ-treated rats tested on the fixed ratio 5/chow feeding choice (Nunes et al. 2013b) and progressive ratio/chow feeding choice tasks (Randall et al. 2015). Furthermore, doses of bupropion that increase extracellular DA and DARPP-32 expression in nucleus accumbens core also increased progressive ratio output in rats responding on the progressive ratio/chow feeding choice task (Randall et al. 2015). This is consistent with previous reports showing that the novel DA uptake inhibitor MRZ-9547 increased progressive ratio choice lever pressing output (Sommer et al. 2014) and that amphetamine increased selection of the high-effort alternative in humans responding on an effort-related decision-making task (Wardle et al. 2011). Considering the known antidepressant actions of catecholamine uptake inhibitors such as bupropion in humans, these results serve to validate the hypothesis that tests of effort-related choice behavior can be used to assess some of the motivational effects of well-known or putative therapeutic agents.

Clinical data indicate that catecholamine uptake inhibitors are moderately efficacious for treating psychomotor retardation and fatigue symptoms of depression (Fabre et al. 1983; Rampello et al. 1991; Pae et al. 2007; Cooper et al. 2014) and can be more effective than 5-HT uptake blockers for treating motivational dysfunction in depressed people (Papakostas et al. 2006; Cooper et al. 2014). Recent research has studied the effects of various monoamine uptake inhibitors for their ability to reverse the effects of TBZ (Yohn et al. 2015c). The effort-related effects of TBZ in rats tested on the concurrent FR5/chow feeding choice task were attenuated by bupropion, and this effect of bupropion was reversed by either D1 or D2 family antagonism. The effort-related effects of TBZ also were attenuated by the selective DA transport inhibitor GBR12909. However, the 5-HT uptake inhibitor fluoxetine and the norepinephrine uptake inhibitor desipramine failed to reverse the effects of TBZ, and higher doses of these drugs, when given alone or in combination with TBZ, led to further behavioral impairments (Yohn et al. 2015c). Thus, drugs acting on DA transmission appear to be relatively effective at reversing the effort-related effects of TBZ and for enhancing work-related behavioral output. These findings are consistent with the hypothesis that drugs that enhance DA transmission may be effective at treating effort-related psychiatric symptoms in humans.

Tests of effort-related decision making also have been used to model features of schizophrenia. Although local overexpression of DA D2 receptors in adult animals leads to increased behavioral activation and effort expenditure (Trifilieff et al. 2013), several studies have shown that overexpression of D2 receptors in striatal medium spiny neurons throughout development leads to the opposite effect, reducing behavioral activation and exertion of effort in motivated behavior (Ward et al. 2012).

These D2 overexpressing mice show reductions in progressive ratio break responding (Drew et al. 2007; Simpson et al. 2011) and reduced selection of the high-effort alternative in a test of effort-based choice (Ward et al. 2012). However, these animals do not show alterations in hedonic reactivity to food rewards, or changes in appetite or food preference. It has been hypothesized that these motivational impairments in D2 receptor overexpressing mice could be useful for modeling some of the negative symptoms of schizophrenia (i.e., avolition, amotivation, apathy; Simpson et al. 2012; see also the chapter by Ward in this volume).

6 Conclusions

As reviewed above, considerable evidence indicates that mesolimbic DA is a critical component of the neural circuitry that regulates behavioral activation, energy expenditure, and effort-related processes. This research is important for understanding the brain mechanisms involved in regulating important aspects of motivation, but it also has considerable clinical significance. Tests of effort-related decision making can be used for the preclinical assessment of novel drugs that are targeted for the treatment of effort-related motivational symptoms, and future research should assess the effects of a broad range of potential therapeutic agents. Although potentially useful for drug development, and for identifying specific neurochemical mechanisms regulating activational aspects of motivation, tests of effort-based decision making in rodents do not represent animal models of depression or any other specific disorder, and their potential utility is not limited to the assessment of antidepressant drugs. Impairments in behavioral activation and effort-related functions appear across multiple disorders and psychiatric conditions (Winograd-Gurvich et al. 2006; Treadway and Zald 2011; Treadway et al. 2012; Gold et al. 2013; Markou et al. 2013). There are some known similarities, and also some differences, in the mechanisms that underlie motivational deficits in depression and schizophrenia (see Barch et al., this volume). Therefore, rodent tests of effort-related dysfunction are likely to reflect a component of depression, rather than a global measure, and represent models of a set of motivational symptoms that cross multiple diagnostic categories, rather than being a model of a specific disorder. This view is consistent with the Research Domain Criterion (RDoC) approach, which suggests that researchers should focus on psychiatric symptoms and their associated neural circuits, in addition to the traditional emphasis on specific disorders (Cuthbert and Insel 2013).

Acknowledgements This work was supported by a grant to J.S. from the National Institute of Mental Health (MH094966), and to Mercè Correa from U.J.I. P1.1A2013-01.

Disclosure/Conflict of Interest J. Salamone has received grants from Merck-Serrono, Pfizer, Roche, Shire, and Prexa.

References

Aberman JE, Ward SJ, Salamone JD (1998) Effects of dopamine antagonists and accumbens dopamine depletions on time-constrained progressive ratio performance. Pharmacol Biochem Behav 61:341–348

Aberman JE, Salamone JD (1999) Nucleus accumbens dopamine depletions make animals more sensitive to high ratio requirements but do not impair primary food reinforcement. Neuroscience 92:545–552

Adamantidis AR, Tsai HC, Boutrel B, Zhang F, Stuber GD, Budygin EA, Touriño C, Bonci A, Deisseroth K, de Lecea L (2011) Optogenetic interrogation of dopaminergic modulation of the multiple phases of reward-seeking behavior. J Neurosci 31(30):0829–10835

Anstrom KK, Woodward DJ (2005) Restraint increases dopaminergic burst firing in awake rats. Neuropsychopharmacology 30:1832–1840

Anstrom KK, Miczek KA, Budygin EA (2009) Increased phasic dopamine signaling in the mesolimbic pathway during social defeat in rats. Neuroscience 161:3–12

Argyropoulos SV, Nutt DJ (2013) Anhedonia revisited: is there a role for dopamine-targeting drugs for depression? J Psychopharmacol 27(10):869–877

Bardgett ME, Depenbrock M, Downs N, Points M, Green L (2009) Dopamine modulates effort-based decision making in rats. Behav Neurosci 123(2):242–251

Bassareo V, Di Chiara G (1999a) Modulation of feeding-induced activation of mesolimbic dopamine transmission by appetitive stimuli and its relation to motivational state. Eur J Neurosci 11:4389–4397

Bassareo V, Di Chiara G (1999b) Differential responsiveness of dopamine transmission to food-stimuli in nucleus accumbens shell/core compartments. Neuroscience 89:637–641

Bassareo V, De Luca MA, Di Chiara G (2002) Differential expression of motivational stimulus properties by dopamine in nucleus shell versus core and prefrontal cortex. J Neurosci 22:4709–4719

Bakshi VP, Kelley AE (1991) Dopaminergic regulation of feeding behavior: I. Differential effects of haloperidol microinjection in three striatal subregions. Psychobiology 19:223–232

Baldo BA, Kelley AE (2007) Discrete neurochemical coding of distinguishable motivational processes: insights from nucleus accumbens control of feeding. Psychopharmacology 191:439–459

Baldo BA, Sadeghian K, Basso AM, Kelley AE (2002) Effects of selective dopamine D1 or D2 receptor blockade within nucleus accumbens subregions on ingestive behavior and associated motor activity. Behav Brain Res 137:165–177

Bayer HM, Glimcher PW (2002) Midbrain dopamine neurons encode a quantitative reward prediction error signal. Neuron 47:129–141

Beeler JA, Frazier CR, Zhuang X (2012) Putting desire on a budget: dopamine and energy expenditure, reconciling reward and resources. Front Integr Neurosci 6:49

Belin D, Jonkman S, Dickinson A, Robbins TW, Everitt BJ (2009) Parallel and interactive learning processes within the basal ganglia: relevance for the understanding of addiction. Behav Brain Res 199:89–102

Bella R, Pennisi G, Cantone M, Palermo F, Pennisi M, Lanza G, Zappia M, Paolucci S (2010) Clinical presentation and outcome of geriatric depression in subcortical ischemic vascular disease. Gerontology 56(3):298–302

Berridge KC, Robinson TE (1998) What is the role of dopamine in reward: hedonic impact, reward learning, or incentive salience? Brain Res Brain Res Rev 28(3):309–369

Berridge KC, Robinson TE (2003) Parsing reward. Trends Neurosci 26(9):507–513

Berridge KC (2007) The debate over dopamine's role in reward: the case for incentive salience. Psychopharmacology 191(3):391–431

Berridge KC, Kringelbach ML (2015) Pleasure systems in the brain. Neuron 86(3):646–664

Bourin M, Chenu F, Ripoll N, David DJ (2005) A proposal of decision tree to screen putative antidepressants using forced swim and tail suspension tests. Behav Brain Res 164(2):266–269

Brischoux F, Chakraborty S, Brierley DI, Ungless MA (2009) Phasic excitation of dopamine neurons in ventral VTA by noxious stimuli. Proc Natl Acad Sci USA 106:4894–4899

Budygin EA, Park J, Bass CE, Grinevich VP, Bonin KD, Wightman RM (2012) Aversive stimulus differentially triggers subsecond dopamine release in reward regions. Neuroscience 201:331–337

Cagniard B, Balsam PD, Brunner D, Zhuang X (2006) Mice with chronically elevated dopamine exhibit enhanced motivation, but not learning, for a food reward. Neuropsychopharmacology 31:1362–1370

Caligiuri MP, Gentili V, Eberson S, Kelsoe J, Rapaport M, Gillin JC (2003) A quantitative neuromotor predictor of antidepressant non-response in patients with major depression. J Affect Disord 77:135–141

Chen JJ, Ondo WG, Dashtipour K, Swope DM (2012) Tetrabenazine for the treatment of hyperkinetic movement disorders: a review of the literature. Clin Ther 34(7):1487–1504

Chong TT, Bonnelle V, Manohar S, Veromann KR, Muhammed K, Tofaris GK, Hu M, Husain M (2015) Dopamine enhances willingness to exert effort for reward in Parkinson's disease. Cortex 69:40–46

Cofer CN, Appley MH (1964) Motivation: theory and research. Wiley, New York

Cooper JA, Tucker VL, Papakostas GI (2014) Resolution of sleepiness and fatigue: a comparison of bupropion and selective serotonin reuptake inhibitors in subjects with major depressive disorder achieving remission at doses approved in the European Union. J Psychopharmacol 28:118–124

Corbit LH, Muir JL, Balleine BW (2001) The role of the nucleus accumbens in instrumental conditioning: evidence of a functional dissociation between accumbens core and shell. J Neurosci 21(9):3251–3260

Corbit LH, Janak PH (2010) Posterior dorsomedial striatum is critical for both selective instrumental and Pavlovian reward learning. Eur J Neurosci 31(7):1312–1321

Corbit LH, Janak PH, Balleine BW (2007) General and outcome-specific forms of Pavlovian-instrumental transfer: the effect of shifts in motivational state and inactivation of the ventral tegmental area. Eur J Neurosci 26(11):3141–3149

Corbit LH, Balleine BW (2011) The general and outcome-specific forms of Pavlovian-instrumental transfer are differentially mediated by the nucleus accumbens core and shell. J Neurosci 31(33):11786–11794

Corbit LH, Leung BK, Balleine BW (2013) The role of the amygdala-striatal pathway in the acquisition and performance of goal-directed instrumental actions. J Neurosci 3(45):17682–17690

Correa M, Pardo M, Lopez-Cruz L, Doñate T, Carbó-Gas M, Monferrer L, Salamone JD (2012) Impact of dopamine D2 receptor antagonism on the activational effects produced by conditioned stimuli and on the preference for primary reinforcers based on their effort requirements. Society for Neuroscience 923.06/EEE82. https://www.researchgate.net/publication/280304157_Correa_et_al._2012_SFN_abstract

Correa M, Carlson BB, Wisniecki A, Salamone JD (2002) Nucleus accumbens dopamine and work requirements on interval schedules. Behav Brain Res 137:179–187

Cousins MS, Sokolowski JD, Salamone JD (1993) Different effects of nucleus accumbens and ventrolateral striatal (DA) depletions on instrumental response selection in the rat. Pharmacol Biochem Behav 46:943–951

Cousins MS, Atherton A, Turner L, Salamone JD (1996) Nucleus accumbens (DA) depletions alter relative response allocation in a T-maze cost/benefit task. Behav Brain Res 74:189–197

Cousins MS, Trevitt J, Atherton A, Salamone JD (1999) Different behavioral functions of dopamine in the nucleus accumbens and ventrolateral striatum: a microdialysis and behavioral investigation. Neuroscience 91:925–934

Croxson PL, Walton ME, O'Reilly JX, Behrens TE, Rushworth MF (2009) Effort-based cost-benefit valuation and the human brain. J Neurosci 29(14):4531–4541

Cryan JF, O'Leary OF, Jin SH, Friedland JC, Ouyang M, Hirsch BR, Page ME, Dalvi A, Thomas SA, Lucki I (2004) Norepinephrine-deficient mice lack responses to antidepressant drugs, including selective serotonin reuptake inhibitors. Proc Natl Acad Sci USA 101 (21):8186–8191

Cuthbert BN, Insel TR (2013) Toward the future of psychiatric diagnosis: the seven pillars of RDoC. BMC Med 11(126)

Dalley JW, Laane K, Theobald DE, Armstrong HC, Corlett PR, Chudasama Y et al (2005) Time-limited modulation of appetitive Pavlovian memory by D1 and NMDA receptors in the nucleus accumbens. Proc Natl Acad Sci 102:6189–6194

Damiano CR, Aloi J, Treadway M, Bodfish JW, Dichter GS (2012) Adults with autism spectrum disorders exhibit decreased sensitivity to reward parameters when making effort-based decisions. J Neurodev Disord 4(1):13

Demyttenaere K, De Fruyt J, Stahl SM (2005) The many faces of fatigue in major depressive disorder. Int J Neuropsychopharmacol 8:93–105

Drew MR, Simpson EH, Kellendonk C, Herzberg WG, Lipatova O, Fairhurst S, Kandel ER, Malapani C, Balsam PD (2007) Transient overexpression of striatal D2 receptors impairs operant motivation and interval timing. J Neurosci 27(29):7731–7739

Duffy E (1963) Activation and behavior. Wiley, New York

Dunnett SB, Iversen SD (1982) Regulatory impairments following selective 6-OHDA lesions of the neostriatum. Behav Brain Res 4:195–202

Everitt BJ, Robbins TW (2005a) Neural systems of reinforcement for drug addiction: from actions to habits to compulsion. Nat Neurosci 8:1481–1489

Everitt BJ, Robbins TW (2005b) Neural systems of reinforcement for drug addiction: from actions to habits to compulsion. Nat Neurosci 8(11):1481–1489

Fabre LF, Brodie HK, Garver D, Zung WW (1983) A multicenter evaluation of bupropion versus placebo in hospitalized depressed patients. J Clin Psychiatry 44:88–94

Farrar AM, Pereira M, Velasco F, Hockemeyer J, Muller CE, Salamone JD (2007) Adenosine A (2A) receptor antagonism reverses the effects of (DA) receptor antagonism on instrumental output and effort-related choice in the rat: implications for studies of psychomotor slowing. Psychopharmacology 191:579–586

Farrar AM, Font L, Pereira M, Mingote SM, Bunce JG, Chrobak JJ, Salamone JD (2008) Forebrain circuitry involved in effort-related choice: injections of the GABAA agonist muscimol into ventral pallidum alters response allocation in food-seeking behavior. Neuroscience 152:321–330

Farrar AM, Segovia KN, Randall PA, Nunes EJ, Collins LE, Stopper CM, Port RG, Hockemeyer J, Müller CE, Correa M, Salamone JD (2010) Nucleus accumbens and effort-related functions: behavioral and neural markers of the interactions between adenosine A_{2A} and (DA) D_2 receptors. Neuroscience 166:1056–1067

Fava M, Ball S, Nelson JC, Sparks J, Konechnik T, Classi P, Dube S, Thase ME (2014) Clinical relevance of fatigue as a residual symptom in major depressive disorder. Depress Anxiety 31 (3):250–257

Ferré S, Quiroz C, Woods AS, Cunha R, Popoli P, Ciruela F, Lluis C, Franco R, Azdad K, Schiffmann SN (2008) An update on adenosine A2A-dopamine D2 receptor interactions: implications for the function of G protein-coupled receptors. Curr Pharm Des 14(15):1468–1474

Fervaha G, Foussias G, Agid O, Remington G (2013) Neural substrates underlying effort computation in schizophrenia. Neurosci Biobehav Rev 37:2649–2665

Floresco SB (2015) The nucleus accumbens: an interface between cognition, emotion, and action. Annu Rev Psychol 66:25–252

Floresco SB, Ghods-Sharifi S (2007) Amygdala-prefrontal cortical circuitry regulates effort-based decision making. Cereb Cortex 17(2):251–260

Floresco SB, Tse MT, Ghods-Sharifi S (2008) Dopaminergic and glutamatergic regulation of effort- and delay-based decision making. Neuropsychopharmacology 33(8):1966–1979

Frank S (2009) Tetrabenazine as anti-chorea therapy in Huntington disease: an open-label continuation study. Huntington Study Group/TETRA-HD Investigators. BMC Neurol 9:62

Frank S (2010) Tetrabenazine: the first approved drug for the treatment of chorea in US patients with Huntington's disease. Neuropsychiatr Dis Treat 5(6):657–665

Font L, Mingote S, Farrar AM, Pereira M, Worden L, Stopper C, Port RG, Salamone JD (2008) Intra-accumbens injections of the adenosine A2A agonist CGS 21680 affect effort-related choice behavior in rats. Psychopharmacology 199:515–526

Ghods-Sharifi S, Floresco SB (2010) Differential effects on effort discounting induced by inactivations of the nucleus accumbens core or shell. Behav Neurosci 124(2):179–191

Gold JM, Strauss GP, Waltz JA, Robinson BM, Brown JK, Frank MJ (2013) Negative symptoms of schizophrenia are associated with abnormal effort-cost computations. Biol Psychiatry 74 (2):130–136

Gold JM, Waltz JA, Frank MJ (2015) Effort cost computation in schizophrenia: a commentary on the recent literature. Biol Psychiatry [Epub ahead of print]. doi:10.1016/j.biopsych.2015.05.005

Green MF, Horan WP (2015) Effort-based decision making in schizophrenia: evaluation of paradigms to measure motivational deficits. Schizophr Bull [Epub ahead of print]. pii: sbv084

Green MF, Horan WP, Barch DM, Gold JM (2015) Effort-based decision making: a novel approach for assessing motivation in schizophrenia. Schizophr Bull [Epub ahead of print]. pii: sbv071

Guay DR (2010) Tetrabenazine, a monoamine-depleting drug used in the treatment of hyperkinectic movement disorders. Am J Geriatr Pharmacother 8(4):331–373

Gullion CM, Rush AJ (1998) Toward a generalizable model of symptoms in major depressive disorder. Biol Psychiatry 44(10):959–972

Hanff TC, Furst SJ, Minor TR (2010) Biochemical and anatomical substrates of depression and sickness behavior. Isr J Psychiatry Relat Sci 47(1):64–71

Hartmann MN, Hager OM, Reimann AV, Chumbley JR, Kirschner M, Seifritz E, Tobler PN, Kaiser S (2015) Apathy but not diminished expression in schizophrenia is associated with discounting of monetary rewards by physical effort. Schizophr Bull 41(2):503–512

Hauber W, Sommer S (2009) Prefrontostriatal circuitry regulates effort-related decision making. Cereb Cortex 19(10):2240–2247

Hernandez G, Breton YA, Conover K, Shizgal P (2010) At what stage of neural processing does cocaine act to boost pursuit of rewards? PLoS ONE 5:e15081

Hodgson RA, Bertorelli R, Varty GB, Lachowicz JE, Forlani A, Fredduzzi S (2009) Characterization of the potent and highly selective A2A receptor antagonists preladenant and SCH 412348 [7-[2-[4-2,4-difluorophenyl]-1-piperazinyl]ethyl]-2-(2-furanyl)-7H-pyrazolo [4,3-e][1,2,4]triazolo[1,5-c]pyrimidin-5-amine] in rodent models of movement disorders and depression. J Pharmacol Exp Ther 330(1):294–303

Hosking JG, Floresco SB, Winstanley CA (2015) Dopamine antagonism decreases willingness to expend physical, but not cognitive, effort: a comparison of two rodent cost/benefit decision-making tasks. Neuropsychopharmacology 40(4):1005–1015

Hudson AL, Lalies MD, Silverstone P (2012) Venlafaxine enhances the effect of bupropion on extracellular dopamine in rat frontal cortex. Can J Physiol Pharmacol 90(6):803–809

Ilango A, Kesner AJ, Broker CJ, Wang DV, Ikemoto S (2014) Phasic excitation of ventral tegmental dopamine neurons potentiates the initiation of conditioned approach behavior: parametric and reinforcement-schedule analyses. Front Behav Neurosci 8:155

Ishiwari K, Weber SM, Mingote S, Correa M, Salamone JD (2004) Accumbens dopamine and the regulation of effort in food-seeking behavior: modulation of work output by different ratio or force requirements. Behav Brain Res 151:83–91

Kent TA, Preskorn SH, Glotzbach RK, Irwin GH (1986) Amitriptyline normalizes tetrabenazine-induced changes in cerebral microcirculation. Biol Psychiatry 21:483–491

Kitamura Y, Yagi T, Kitagawa K, Shinomiya K, Kawasaki H, Asanuma M, Gomita Y (2010) Effects of bupropion of the forced swim test and release of dopamine in the nucleus accumbens in ACTH-treated rats. Naunyn Schmiedebergs Arch Pharmacol 382(2):151–158

Knab AM, Bowen RS, Hamilton AT, Gulledge AA, Lightfoot JT (2009) Altered dopaminergic profiles: implications for the regulation of voluntary physical activity. Behav Brain Res 204 (1):147–152

Koch M, Schmid A, Schnitzler HU (2000) Role of nucleus accumbens (DA) D1 and D2 receptors in instrumental and Pavlovian paradigms of conditioned reward. Psychopharmacology 152:67–73

Koob GF, Riley SJ, Smith SC, Robbins TW (1978) Effects of 6-hydroxydopamine lesions of the nucleus accumbens septi and olfactory tubercle on feeding, locomotor activity, and amphetamine anorexia in the rat. J Comp Physiol Psychol 92:917–927

Kurniawan IT, Seymour B, Talmi D, Yoshida W, Chater N, Dolan RJ (2010) Choosing to make an effort: the role of striatum in signaling physical effort of a chosen action. J Neurophysiol 104 (1):313–321

Lammel S, Lim BK, Malenka RC (2014) Reward and aversion in a heterogeneous midbrain dopamine system. Neuropharmacology 76:351–359

Learned-Coughlin SM, Bergström M, Savitcheva I, Ascher J, Schmith VD, Långstrom B (2003) In vivo activity of bupropion at the human dopamine transporter as measured by positron emission tomography. Biol Psychiatry 54:800–805

Lex A, Hauber W (2008) Dopamine D1 and D2 receptors in the nucleus accumbens core and shell mediate Pavlovian-instrumental transfer. Learn Mem 15:483–491

Lex B, Hauber W (2010) The role of nucleus accumbens dopamine in outcome encoding in instrumental and Pavlovian conditioning. Neurobiol Learn Mem 93:283–290

Mai B, Sommer S, Hauber W (2012) Motivational states influence effort-based decision making in rats: the role of dopamine in the nucleus accumbens. Cogn Affect Behav Neurosci 12:74–84

Marchant NJ, Whitaker LR, Bossert JM, Harvey BK, Hope BT, Kaganovsky K, Adhikary S, Prisinzano TE, Vardy E, Roth BL, Shaham Y (2015) Behavioral and physiological effects of a novel kappa-opioid receptor-based DREADD in rats. Neuropsychopharmacology [Epub ahead of print]. doi:10.1038/npp.2015.149

Marinelli M, McCutcheon JE (2014) Heterogeneity of dopamine neuron activity across traits and states. Neuroscience 282:176–197

Markou A, Salamone JD, Bussey TJ, Mar AC, Brunner D, Gilmour G, Balsam P (2013) Measuring reinforcement learning and motivation constructs in experimental animals: relevance to the negative symptoms of schizophrenia. Neurosci Biobehav Rev 37(9):2149–2165

McCullough LD, Salamone JD (1992) Involvement of nucleus accumbens dopamine in the motor activity induced by periodic food presentation: a microdialysis and behavioral study. Brain Res 592:29–36

McCullough LD, Cousins MS, Salamone JD (1993a) The role of nucleus accumbens dopamine in responding on a continuous reinforcement operant schedule: a neurochemical and behavioral study. Pharmacol Biochem Behav 46:581–586

McCullough LD, Sokolowski JD, Salamone JD (1993b) A neurochemical and behavioral investigation of the involvement of nucleus accumbens dopamine in instrumental avoidance. Neuroscience 52:919–925

McGinty VB, Lardeux S, Taha SA, Kim JJ, Nicola SM (2013) Invigoration of reward seeking by cue and proximity encoding in the nucleus accumbens. Neuron 78:910–922

Milea D, Guelfucci F, Bent-Ennakhil N, Toumi M, Auray JP (2010) Antidepressant monotherapy: a claims database analysis of treatment changes and treatment duration. Clin Ther 32 (12):2057–2072

Mingote S, Font L, Farrar AM, Vontell R, Worden LT, Stopper CM, Port RG, Sink KS, Bunce JG, Chrobak JJ, Salamone JD (2008) Nucleus accumbens adenosine A2A receptors regulate exertion of effort by acting on the ventral striatopallidal pathway. J Neurosci 28:9037–9046

Mingote S, Weber SM, Ishiwari K, Correa M, Salamone JD (2005) Ratio and time requirements on operant schedules: effort-related effects of nucleus accumbens dopamine depletions. Eur J Neurosci 21(6):1749–1757

Mogenson G, Jones D, Yim CY (1980) From motivation to action: functional interface between the limbic system and the motor system. Prog Neurobiol 14:69–97

Mott AM, Nunes EJ, Collins LE, Port RG, Sink KS, Hockemeyer J, Muller CE, Salamone JD (2009) The adenosine A2A antagonist MSX-3 reverses the effects of the (DA) antagonist haloperidol on effort-related decision making in a T-maze cost/benefit procedure. Psychopharmacology 204:103–112

Nicola SM (2010) The flexible approach hypothesis: unification of effort and cue-responding hypotheses for the role of nucleus accumbens dopamine in the activation of reward-seeking behavior. J Neurosci 30(49):16585–16600

Niv Y (2009) Reinforcement learning in the brain. J Math Psychol 53:139–154

Nowend KL, Arizzi M, Carlson BB, Salamone JD (2001) D_1 or D_2 antagonism in nucleus accumbens core or dorsomedial shell suppresses lever pressing for food but leads to compensatory increases in chow consumption. Pharmacol Biochem Behav 69:373–382

Nunes EJ, Randall PA, Santerre JL, Given AB, Sager TN, Correa M, Salamone JD (2010) Differential effects of selective adenosine antagonists on the effort-related impairments induced by (DA) D1 and D2 antagonism. Neuroscience 170:268–280

Nunes EJ, Randall PA, Podurgiel S, Correa M, Salamone JD (2013a) Nucleus accumbens neurotransmission and effort-related choice behavior in food motivation: Effects of drugs acting on dopamine, adenosine, and muscarinic acetylcholine receptors. Neurosci Biobehav Rev 37:2015–2025

Nunes EJ, Randall PA, Hart EE, Freeland C, Yohn SE, Baqi Y, Müller CE, López-Cruz L, Correa M, Salamone JD (2013b) Effort-related motivational effects of the VMAT-2 inhibitor tetrabenazine: implications for animal models of the motivational symptoms of depression. J Neurosci 33(49):19120–19130

Nunes EJ, Randall PA, Estrada A, Epling B, Hart E, Lee CE, Baqi Y, Müller CE, Correa M, Salamone JD (2014) Effort-related motivational effects of the pro-inflammatory cytokine interleukin 1-beta: studies with the concurrent fixed ratio 5/ chow feeding choice task. Psychopharmacology 231:727–736

Ostlund SB, Wassum KM, Murphy NP, Balleine BW, Maidment NT (2011) Extracellular dopamine levels in striatal subregions track shifts in motivation and response cost during instrumental conditioning. J Neurosci 31:200–207

Pae CU, Lim HK, Han C, Patkar AA, Steffens DC, Masand PS, Lee C (2007) Fatigue as a core symptom in major depressive disorder: overview and the role of bupropion. Expert Rev Neurother 7(10):1251–1263

Papakostas GI, Nutt DJ, Hallett LA, Tucker VL, Krishen A, Fava M (2006) Resolution of sleepiness and fatigue in major depressive disorder: a comparison of bupropion and the selective serotonin reuptake inhibitors. Biol Psychiatry 60(12):1350–1355

Pardo M, Lopez-Cruz L, Valverde O, Ledent C, Baqi Y, Muller CE, Salamone JD, Correa M (2012) Adenosine A2A receptor antagonism and genetic deletion attenuate the effects of dopamine D2 antagonism on effort-related decision making in mice. Neuropharmacology 62 (5–6):2068–2077

Pardo M, López-Cruz L, Miguel NS, Salamone JD, Correa M (2015) Selection of sucrose concentration depends on the effort required to obtain it: studies using tetrabenazine, D1, D2, and D3 receptor antagonists. Psychopharmacology 232(13):2377–239

Parkinson JA, Dalley JW, Cardinal RN, Bamford A, Fehnert B, Lachenal G, Rudarakanchana N, Halkerston KM, Robbins TW, Everitt BJ (2002) Nucleus accumbens dopamine depletion impairs both acquisition and performance of appetitive Pavlovian approach behaviour: implications for mesoaccumbens dopamine function. Behav Brain Res 137:149–163

Peciña S, Cagniard B, Berridge KC, Aldridge JW, Zhuang X (2003) Hyperdopaminergic mutant mice have higher "wanting" but not "liking" for sweet rewards. J Neurosci 23:9395–9402

Pettibone DJ, Totaro JA, Pflueger AB (1984) Tetrabenazine-induced depletion of brain monoamines: characterization and interaction with selected antidepressants. Eur J Pharmacol 102:425–430

Preskorn SH, Kent TA, Glotzbach RK, Irwin GH, Solnick JV (1984) Cerebromicrocirculatory defects in animal model of depression. Psychopharmacology 84:196–199

Rampello L, Nicoletti G, Raffaele R (1991) Dopaminergic hypothesis for retarded depression: a symptom profile for predicting therapeutical responses. Acta Psychiatry Scand 84(6):552–554

Randall PA, Pardo M, Nunes EJ, López Cruz L, Vemuri VK, Makriyannis A, Baqi Y, Müller CE, Correa M, Salamone JD (2012) Dopaminergic modulation of effort-related choice behavior as assessed by a progressive ratio chow task: pharmacological studies and role of individual differences. PLoS ONE 7(10):e47934

Randall PA, Lee CA, Podurgiel SJ, Hart E, Yohn SE, Jones M, Rowland M, López-Cruz L, Correa M, Salamone JD (2015) Bupropion increases selection of high effort activity in rats tested on a progressive ratio/chow feeding choice procedure: implications for treatment of effort-related motivational symptoms. Int J Neuropsychopharmacol 18(2). doi:10.1093/ijnp/pyu017

Rick JH, Horvitz JC, Balsam PD (2006) Dopamine receptor blockade and extinction differentially affect behavioral variability. Behav Neurosci 120:488–492

Roeper J (2013) Dissecting the diversity of midbrain dopamine neurons. Trends Neurosci 36(6):336–342

Robbins TW, Koob GF (1980) Selective disruption of displacement behaviour by lesions of the mesolimbic dopamine system. Nature 285:409–412

Robbins TW, Everitt BJ (2007) A role for mesencephalic dopamine in activation: commentary on Berridge (2006). Psychopharmacology 191:433–437

Roitman MF, Stuber GD, Phillips PEM, Wightman RM, Carelli RM (2004) Dopamine operates as a subsecond modulator of food seeking. J Neurosci 24:1265–1271

Root DH, Melendez RI, Zaborszky L, Napier TC (2015) The ventral pallidum: subregion-specific functional anatomy and roles in motivated behaviors. Prog Neurobiol 130:29–70

Rosin DL, Robeva A, Woodard RL, Guyenet PG, Linden J (1998) Immunohistochemical localization of adenosine A2A receptors in the rat central nervous system. J Comp Neurol 401:163–186

Saddoris MP, Sugam JA, Stuber GD, Witten IB, Deisseroth K, Carelli RM (2015) Mesolimbic dopamine dynamically tracks, and is causally linked to, discrete aspects of value-based decision making. Biol Psychiatry 77(10):903–911

Salamone JD (1987) The actions of neuroleptic drugs on appetitive instrumental behaviors. In: Iversen LL, Iversen SD, Snyder SH (eds) Handbook of psychopharmacology. Plenum Press, New York, pp 575–608

Salamone JD (1988) Dopaminergic involvement in activational aspects of motivation: effects of haloperidol on schedule induced act-ivity, feeding and foraging in rats. Psychobiology 16:96–206

Salamone JD (1991) Behavioral pharmacology of dopamine systems: A new synthesis. In: Willner P, Scheel Kruger J (eds) The mesolimbic dopamine system: from motivation to action. Cambridge University Press: Cambridge, England, pp 598–613

Salamone JD (1992) Complex motor and sensorimotor functions of accumbens and striatal dopamine: involvement in instrumental behavior processes. Psychopharmacology 107:160–174

Salamone JD (1994) Involvement of nucleus accumbens dopamine in appetitive and aversive motivation. Behav Brain Res 61:117–133

Salamone JD (1996) The behavioral neurochemistry of motivation: methodological and conceptual issues in studies of the dynamic activity of nucleus accumbens dopamine. J Neurosci Meth 64:137–149

Salamone JD, Correa M (2002) Motivational views of reinforcement: implications for understanding the behavioral functions of nucleus accumbens dopamine. Behav Brain Res 137(1–2):3–25

Salamone JD, Correa M (2009) Dopamine/adenosine interactions involved in effort-related aspects of food motivation. Appetite 53(3):422–425

Salamone JD, Correa M (2012) The mysterious motivational functions of mesolimbic dopamine. Neuron 76(3):470–485

Salamone JD, Steinpreis RE, McCullough LD, Smith P, Grebel D, Mahan K (1991) Haloperidol and nucleus accumbens (DA) depletion suppress lever pressing for food but increase free food consumption in a novel food choice procedure. Psychopharmacology 104:515–521

Salamone JD, Cousins MS, Bucher S (1994a) Anhedonia or anergia? Effects of haloperidol and nucleus accumbens (DA) depletion on instrumental response selection in a T-maze cost/benefit procedure. Behav Brain Res 65:221–229

Salamone JD, Cousins MS, McCullough LD, Carriero DL, Berkowitz RJ (1994b) Nucleus accumbens dopamine release increases during instrumental lever pressing for food but not free food consumption. Pharmacol Biochem Behav 49:25–31

Salamone JD, Cousins MS, Snyder BJ (1997) Behavioral functions of nucleus accumbens DA: empirical and conceptual problems with the anhedonia hypothesis. Neurosci Biobehav Rev 21:341–359

Salamone JD, Wisniecki A, Carlson BB, Correa M (2001) Nucleus accumbens dopamine depletions make animals highly sensitive to high fixed ratio requirements but do not impair primary food reinforcement. Neuroscience 105:863–870

Salamone JD, Arizzi MN, Sandoval MD, Cervone KM, Aberman JE (2002) (DA) antagonists alter response allocation but do not suppress appetite for food in rats: contrast between the effects of SKF 83566, raclopride, and fenfluramine on a concurrent choice task. Psychopharmacology 160:371–380

Salamone JD, Correa M, Mingote SM, Weber SM, Farrar AM (2006) Nucleus Accumbens (DA) and the forebrain circuitry involved in behavioral activation and effort-related decision making: implications for understanding anergia and psychomotor slowing in depression. Curr Psychiatry Rev 2:267–280

Salamone JD, Correa M, Farrar A, Mingote SM (2007) Effort-related functions of nucleus accumbens (DA) and associated forebrain circuits. Psychopharmacology 191:461–482

Salamone JD, Correa M, Nunes EJ, Randall PA, Pardo M (2012) The behavioral pharmacology of effort-related choice behavior: dopamine, adenosine and beyond. J Exp Anal Behav 97:125–146

Salamone JD, Farrar AM, Font L, Patel V, Schlar DE, Nunes EJ, Collins LE, Sager TN (2009) Differential actions of adenosine A1 and A2A antagonists on the effort-related effects of (DA) D2 antagonism. Behav Brain Res 201:216–222

Salamone JD, Koychev I, Correa M, McGuire P (2014) Neurobiological basis of motivational deficits in psychopathology. Eur Neuropsychopharmacol [Epub ahead of print]

Salamone JD, Zigmond MJ, Stricker EM (1990) Characterization of the impaired feeding behavior in rats given haloperidol or dopamine-depleting brain lesions. Neuroscience 39:17–24

Salamone JD, Mahan K, Rogers S (1993) Ventrolateral striatal dopamine depletions impair feeding and food handling in rats. Pharmacol Biochem Behav 44:605–610

Salamone JD, Correa M, Mingote SM, Weber SM (2005) Beyond the reward hypothesis: alternative functions of nucleus accumbens dopamine. Curr Opin Pharmacol 5:34–41

Salamone JD (1991) Behavioral pharmacology of dopamine systems: a new synthesis. In: Willner P, Scheel-Kruger J (eds) The mesolimbic dopamine system: from motivation to action. Cambridge University Press; Cambridge, England, vol 1, pp 599–613

Santerre JL, Nunes EJ, Randall PA, Baqi Y, Müller CE, Salamone JD (2012) Behavioral studies with the novel adenosine A2A antagonist MSX-4: reversal of the effects of (DA) D2 antagonism. Pharamcol Biochem Behav 102(4):477–487

Schultz W (2010) Dopamine signals for reward value and risk: basic and recent data. Behav Brain Funct 6:24

Schultz W, Apicella P, Ljungberg T (1993) Responses of monkey dopamine neurons to reward and conditioned stimuli during successive steps of learning a delayed response task. J Neurosci 13:900–913

Schultz W, Dayan P, Montague RR (1997) A neural substrate of prediction and reward. Science 275:1593–1599

Sederholm F, Johnson AE, Brodin U, Södersten P (2002) Dopamine D(2) receptors and ingestive behavior: brainstem mediates inhibition of intraoral intake and accumbens mediates aversive taste behavior in male rats. Psychopharmacology 160:161–169

Segovia KN, Correa M, Salamone JD (2011) Slow phasic changes in nucleus accumbens dopamine release during fixed ratio acquisition: a microdialysis study. Neuroscience 196:178–188

Segovia KN, Correa M, Lennington JB, Conover JC, Salamone JD (2012) Changes in nucleus accumbens and neostriatal c-Fos and DARPP-32 immunoreactivity during different stages of food-reinforced instrumental training. Eur J Neurosci 35:1354–1367

Shafiei N, Gray M, Viau V, Floresco SB (2012) Acute stress induces selective alterations in cost/benefit decision-making. Neuropsychopharmacology 37(10):2194–2209

Smith KS, Berridge KC, Aldridge JW (2011) Disentangling pleasure from incentive salience and learning signals in brain reward circuitry. Proc Natl Acad Sci USA 108:E255–E264

Simpson EH, Kellendonk C, Ward RD, Richards V, Lipatova O, Fairhurst S, Kandel ER, Balsam PD (2011) Pharmacologic rescue of motivational deficit in an animal model of the negative symptoms of schizophrenia. Biol Psychiatry 69(10):928–935

Simpson EH, Waltz JA, Kellendonk C, Balsam PD (2012) Schizophrenia in translation: dissecting motivation in schizophrenia and rodents. Schizophr Bull 38(6):1111–1117

Sink KS, Vemuri VK, Olszewska T, Makriyannis A, Salamone JD (2008) Cannabinoid CB1 antagonists and dopamine antagonists produce different effects on a task involving response allocation and effort-related choice in food-seeking behavior. Psychopharmacology 196:565–574

Sokolowski JD, Conlan AN, Salamone JD (1998) A microdialysis study of nucleus accumbens core and shell dopamine during operant responding in the rat. Neuroscience 86:1001–1009

Sommer S, Danysz W, Russ H, Valastro B, Flik G, Hauber W (2014) The dopamine reuptake inhibitor MRZ-9547 increases progressive ratio responding in rats. Int J Neuropsychopharmacol 17(12):2045–2056

Soskin DP, Holt DJ, Sacco GR, Fava M (2013) Incentive salience: novel treatment strategies for major depression. CNS Spectr 1:1–8

Stahl SM (2002) The psychopharmacology of energy and fatigue. J Clin Psychiatry 63(1):7–8

Steinberg EE, Keiflin R, Boivin J, Witten IB, Deisseroth K, Janak PH (2013) A causal link between prediction errors, dopamine neurons and learning. Nat Neurosci 16(7):966–973

Steinberg EE, Boivin JR, Saunders BT, Witten IB, Deisseroth K, Janak PH (2014) Positive reinforcement mediated by midbrain dopamine neurons requires D1 and D2 receptor activation in the nucleus accumbens. PLoS ONE 9(4):e94771

Svenningsson P, Le Moine C, Fisone G, Fredholm BB (1999) Distribution, biochemistry and function of striatal adenosine A_{2A} receptors. Prog Neurobiol 59:355–396

Szczypka MS, Kwok K, Brot MD, Marck BT, Matsumoto AM, Donahue BA, Palmiter RD (2001) Dopamine production in the caudate putamen restores feeding in dopamine-deficient mice. Neuron 30:819–828

Tanra AJ, Kagaya A, Okamoto Y, Muraoka M, Motohashi N, Yamawaki S (1995) TJS-010, a new prescription of oriental medicine, antagonizes tetrabenazine-induced suppression of spontaneous locomotor activity in rats. Prog Neuropsychopharmacol Biol Psychiatry 19(5):963–971

Tidey JW, Miczek KA (1996) Social defeat stress selectively alters mesocorticolimbic dopamine release: an in vivo microdialysis study. Brain Res 721:140–149

Treadway MT, Buckholtz JW, Schwartzman AN, Lambert WE, Zald DH (2009) Worth the 'EEfRT'? The effort expenditure for rewards task as an objective measure of motivation and anhedonia. PLoS ONE 4(8):e6598

Treadway MT, Zald DH (2011) Reconsidering anhedonia in depression: lessons from translational neuroscience. Neurosci Biobehav Rev 35:537–555

Treadway MT, Bossaller NA, Shelton RC, Zald DH (2012) Effort-based decision making in major depressive disorder: a translational model of motivational anhedonia. J Abnorm Psychol 121(3):553–558

Trifilieff P, Feng B, Urizar E, Winiger V, Ward RD, Taylor KM, Martinez D, Moore H, Balsam PD, Simpson EH, Javitch JA (2013) Increasing dopamine D2 receptor expression in the adult nucleus accumbens enhances motivation. Mol Psychiatry 18:1025–1033

Tylee A, Gastpar M, Lepine JP, Mendlewicz J (1999) DEPRES II (Depression Research in European Society II): a patient survey of the symptoms, disability and current management of depression in the community. Int Clin Psychopharmacol 14:139–151

Ungerstedt U (1971) Aphagia and adipsia after 6 hydroxydopamine induced degeneration of the nigro striatal dopamine system. Acta Physiol Scand 82:95–122

Venugopalan VV, Casey KF, O'Hara C, O'Loughlin J, Benkelfat C, Fellows LK, Leyton M (2011) Acute phenylalanine/tyrosine depletion reduces motivation to smoke cigarettes across stages of addiction. Neuropsychopharmacology 36:2469–2476

Wallace M, Singer G, Finlay J, Gibson S (1983) The effect of 6-OHDA lesions of the nucleus accumbens septum on schedule-induced drinking, wheelrunning and corticosterone levels in the rat. Pharmacol Biochem Behav 18(1):129–136

Walton ME, Bannerman DM, Alterescu K, Rushworth MFS (2003) Functional specialization within medial frontal cortex of the anterior cingulate for evaluating effort-related decisions. J Neurosci 23:6475–6479

Wang H, Chen X, Li Y, Tang TS, Bezprozvanny I (2010) Tetrabenazine is neuroprotective in Huntington's disease mice. Mol Neurodegener 5:18

Ward RD, Simpson EH, Richards VL, Deo G, Taylor K, Glendinning JI, Kandel ER, Balsam PD (2012) Dissociation of hedonic reaction to reward and incentive motivation in an animal model of the negative symptoms of schizophrenia. Neuropsychopharmacology 37(7):1699–1707

Wardle MC, Treadway MT, Mayo LM, Zald DH, de Wit H (2011) Amping up effort: effects of d-amphetamine on human effort-based decision-making. J Neurosci 31(46):16597–16602

Winograd-Gurvich C, Fitzgerald PB, Georgiou-Karistianis N, Bradshaw JL, White OB (2006) Negative symptoms: A review of schizophrenia, melancholic depression and Parkinson's disease. Brain Res Bull 70:312–321

Wise RA (2008) Dopamine and reward: the anhedonia hypothesis—30 years on. Neurotox Res 14:169–183

Worden LT, Shahriari M, Farrar AM, Sink KS, Hockemeyer J, Muller CE, Salamone JD (2009) The adenosine A2A antagonist MSX-3 reverses the effort-related effects of (DA) blockade: differential interaction with D1 and D2 family antagonists. Psychopharmacology 203:489–499

Wyvell CL, Berridge KC (2000) Intra-accumbens amphetamine increases the conditioned incentive salience of sucrose reward: enhancement of reward "wanting" without enhanced "liking" or response reinforcement. J Neurosci 20(21):8122–8130

Yamada K, Kobayashi M, Mori A, Jenner P, Kanda T (2013) Antidepressant-like activity of the adenosine A(2A) receptor antagonist, istradefylline (KW-6002), in the forced swim test and the tail suspension test in rodents. Pharmacol Biochem Behav 114–115:23–30

Yamada K, Kobayashi M, Shiozaki S, Ohta T, Mori A, Jenner P, Kanda T (2014) Antidepressant activity of the adenosine A(2A) receptor antagonist, istradefylline (KW-6002) on learned helplessness in rats. Psychopharmacology 231(14):2839–2849

Yang XH, Huang J, Zhu CY, Wang YF, Cheung EF, Chan RC, Xie GR (2014) Motivational deficits in effort-based decision making in individuals with subsyndromal depression, first-episode and remitted depression patients. Psychiatry Res 220(3):874–882

Yin HH, Knowlton BJ, Balleine BW (2005) Blockade of NMDA receptors in the dorsomedial striatum prevents action-outcome learning in instrumental conditioning. Eur J Neurosci 22:505–512

Yin HH, Ostlund SB, Balleine BW (2008) Reward-guided learning beyond dopamine in the nucleus accumbens: the integrative functions of cortico-basal ganglia networks. Eur J Neurosci 28:1437–1448

Yohn SE, Thompson C, Randall PA, Lee CA, Müller CE, Baqi Y, Correa M, Salamone JD (2015a) The VMAT-2 inhibitor tetrabenazine alters effort-related decision making as measured by the T-maze barrier choice task: reversal with the adenosine A2A antagonist MSX-3 and the catecholamine uptake blocker bupropion. Psychopharmacology 232(7):1313–1323

Yohn SE, Santerre JL, Nunes EJ, Kozak R, Podurgiel SJ, Correa M, Salamone JD (2015b) The role of dopamine D1 receptor transmission in effort-related choice behavior: Effects of D1 agonists. Pharmacol Biochem Behav 135:217–226

Yohn SE, Collins SL, Contreras-Mora HM, Errante EL, Rowland MA, Correa M, Salamone JD (2015c) Not all antidepressants are created equal: differential effects of monoamine uptake inhibitors on effort-related choice behavior. Neuropsychopharmacology [Epub ahead of print]. doi:10.1038/npp.2015.188

Young AM (2004) Increased extracellular dopamine in nucleus accumbens in response to unconditioned and conditioned aversive stimuli: studies using 1 min microdialysis in rats. J Neurosci Methods 138:57–63

Zigmond MJ, Stricker EM (1972) Deficits in feeding behavior after intraventricular injection of 6-hydroxydopamine in rats. Science 177:1211–1214

Zweifel LS, Parker JG, Lobb CJ, Rainwater A, Wall VZ, Fadok JP, Darvas M, Kim MJ, Mizumori SJY, Paladini CA, Phillips PEM, Palmiter RD (2009) Disruption of NMDAR-dependent burst firing by dopamine neurons provides selective assessment of phasic dopamine-dependent behavior. Proc Natl Acad Sci USA 106:7281–7288

Zweifel LS, Fadok JP, Argilli E, Garelick MG, Dickerson TMK, Allen J, Mizumori SJY, Bonci A, Palmiter RD (2011) Dopamine neuronal activity imbalance evoked generalized anxiety following aversive experience. Nat Neurosci 14:620–626

Learning and Motivational Processes Contributing to Pavlovian–Instrumental Transfer and Their Neural Bases: Dopamine and Beyond

Laura H. Corbit and Bernard W. Balleine

Abstract Pavlovian stimuli exert a range of effects on behavior from simple conditioned reflexes, such as salivation, to altering the vigor and direction of instrumental actions. It is currently accepted that these distinct behavioral effects stem from two sources (i) the various associative connections between predictive stimuli and the component features of the events that these stimuli predict and (ii) the distinct motivational and cognitive functions served by cues, particularly their arousing and informational effects on the selection and performance of specific actions. Here, we describe studies that have assessed these latter phenomena using a paradigm that has come to be called Pavlovian–instrumental transfer. We focus first on behavioral experiments that have described distinct sources of stimulus control derived from the general affective and outcome-specific predictions of conditioned stimuli, referred to as general transfer and specific transfer, respectively. Subsequently, we describe research efforts attempting to establish the neural bases of these transfer effects, largely in the afferent and efferent connections of the nucleus accumbens (NAc) core and shell. Finally, we examine the role of predictive cues in examples of aberrant stimulus control associated with psychiatric disorders and addiction.

Keywords Instrumental learning · Pavlovian learning · Nucleus accumbens · Amygdala · Reward · Incentive

Contents

1	Introduction	260
2	The Functional Importance of Pavlovian Conditioning	260
3	The Pavlovian Control of Instrumental Action	262

L.H. Corbit
Department of Psychology, University of Sydney, Sydney, NSW 2006, Australia

B.W. Balleine (✉)
Brain and Mind Research Institute, University of Sydney, 94 Mallett Street, Camperdown, NSW 2050, Australia
e-mail: bernard.balleine@sydney.edu.au

	3.1 Instrumental and Pavlovian Incentive Processes	262
	3.2 General Versus Specific Transfer	264
	3.3 Specific Transfer and Choice	269
4	Neural Bases of Transfer	270
	4.1 The Limbic-Striatal Circuit	270
	4.2 Dopamine and the Modulation of the Nucleus Accumbens	272
	4.3 Beyond Dopamine	274
	4.4 Beyond the Accumbens: The Extended Circuit	276
5	Pavlovian Incentives and Pathologies of Decision Making	281
	5.1 Exposure to Drugs of Abuse	281
	5.2 Obesogenic Food Exposure	282
	5.3 Stress	283
6	Summary and Conclusions	284
References		284

1 Introduction

We rely heavily on environmental stimuli to guide our day-to-day actions. As an obvious example, different colored traffic lights tell us when to stop and when to go in traffic, but the effects can also be more subtle. For example, as we are rushing to an appointment, the smell wafting from a nearby bakery may remind us that we are hungry or even guide us through the bakery's door even when we were not planning to eat. Such stimuli acquire predictive and motivational power through our previous experiences and through associative learning. Such learning can be highly adaptive; the predictive properties of stimuli can aid rapid decision making, automatize routine behaviors, and help us navigate a complex world. However, stimulus influences can be problematic when conditions change and the behaviors provoked by stimuli are no longer appropriate or when the influence of such stimuli is excessive and at the expense of flexible, goal-directed control. For example, to someone trying to stick to a strict diet, environmental stimuli that remind us of cakes and cookies and the cravings that they elicit could be quite unwanted and difficult to control. Indeed, the intrusive thoughts and actions elicited by environmental stimuli are often argued to contribute to psychiatric disorders including substance use disorders and obsessive–compulsive disorder, examples of where stimuli contribute to behaviors that are either unwanted or excessive. The following chapter will review research into the ways that environmental events, particularly those predictive of reward, can influence our actions and what is known about the associative and neural control of these effects.

2 The Functional Importance of Pavlovian Conditioning

Although, descriptively speaking, Pavlovian conditioning can be characterized as the pairing of two events (Dickinson 1980), in practice it is usual for at least one of those events, the unconditioned stimulus (US), to have biological consequences;

which means to say that because the US is tied to, and its significance defined by, a motivational system, the responses that it evokes will also be sensitive to the subject's motivational state. This motivational sensitivity extends to conditioned responses (e.g., Holland and Rescorla 1975; Winterbauer and Balleine 2005), and as such, there have been a number of theories over the years proposing that, as a consequence of their pairing with the US, conditioned stimuli (CSs) can themselves acquire motivational properties (see, e.g., Konorski 1967; Bindra 1974, 1978; Mowrer 1960; Rescorla and Solomon 1967; Toates 1986).

The nature of those properties depends, however, both on the type of event predicted and that specific component of an event with which the predictive stimulus is particularly strongly associated. So, for example, as Pavlov (1927) found, pairing a bell with meat powder will make hungry dogs salivate to the bell, but they will salivate more when it is paired with food than non-food outcomes and more too when they are hungry than when they are not. In fact, the component events acquiring these motivational properties can be exquisitely specific; Zener (1937) found, again using dogs, that when he associated a bell with Sargent's biscuits and a tone with Spratt's Ovals, satiating the dog on Sargent's biscuits reduced salivation to the bell but not to the tone.

This is not always the case; however, in other situations predictive stimuli, while motivationally relevant, can elicit more general predictions; for example, Ganesan and Pearce (1988) found that, in hungry and thirsty rats, a stimulus paired with food can block learning about another stimulus when the compound is subsequently paired with water, something that could only occur if the rats were learning about some general 'good' rather than a motivationally specific one. Furthermore, this learning, manifest as approach to a food magazine during the conditioned stimuli, was evoked equally strongly when the stimuli were paired with food or with water; i.e., this approach response constitutes a motivationally general 'preparatory' reflex, as Konorski (1967) termed it, and the Ganesan and Pearce (1988) transreinforcer blocking effect provides some of the best evidence for preparatory conditioning. Generally speaking, therefore, Pavlovian conditioning can be divided into consummatory, motivational, and preparatory forms and each of these appears to reflect associations between the conditioned stimuli and some specific elements of the US (see Fig. 1).

It is clear, therefore, that Pavlovian conditioning provides animals with both general and highly specific information about forthcoming events. Nevertheless, although such information could allow planning and other cognitive processes to intercede to modify behavior to maximize beneficial and avoid harmful consequences, such information is not in itself much use in modifying conditioned reflexes. It can alter their frequency, but their form, being a product of evolutionary pressure, is not subjected to individual learning. In unstable environments, evolutionary solutions have fewer advantages; rapid changes in environmental contingencies require animals to be able to modify basic reflexes, i.e., to exert behavioral control. Contemporary research on decision making has focused attention on two such control processes, both a product of instrumental conditioning. One process is based on the encoding and deployment of action–outcome associations in the form of goal-directed actions (cf. Dickinson and Balleine 1994 for review). The second is

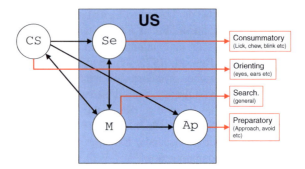

Fig. 1 Motivational control of Pavlovian conditioning. Conditioned stimuli (CS) can become associated (*black arrows*) with one or more components of the unconditioned stimulus (US) and those associations determine specific conditioned responses (*red arrows* and *boxes*). Specific motivational states (M; such as hunger and thirst) can modulate the activation of stimulus representations, e.g., CS and the sensory properties of the US (Se), as well as of a general appetitive state (Ap)

engaged during the acquisition of simple stimulus–response habits established through a process of selective reinforcement (Dickinson 1985).

If predictive information is of value in determining action selection, then instrumental conditioning should work closely with Pavlovian predictive learning to increase the adaptive significance of an animal's actions. Understanding the motivational determinants of adaptive behavior requires, therefore, an understanding of the way Pavlovian and instrumental learning systems are integrated, both behaviorally and at a neural level, and the role this integration plays in the biological interests of species. In this chapter, we will review one of the best-studied forms of integration of this kind; the phenomenon of Pavlovian–instrumental transfer.

3 The Pavlovian Control of Instrumental Action

3.1 *Instrumental and Pavlovian Incentive Processes*

Goal-directed actions allow us and other animals to act flexibly to obtain desired outcomes and avoid events that are aversive. The acquisition of such actions relies on encoding the contingent relationships between actions and their consequences, whereas their performance requires the integration of that learning with the current value of those consequences. Thus, goal-directed actions typically track changes in the value of outcomes, responding increases or decreases as the instrumental outcome becomes more or less valuable (Dickinson and Balleine 1994). In addition to this instrumental incentive process, Pavlovian incentive learning, which mediates the excitatory and inhibitory effects of stimuli based on learned associations, can also contribute importantly to various aspects of instrumental performance (Estes 1943; Lovibond 1983). The impact of this latter incentive process can be readily examined using a behavioral paradigm typically referred to as Pavlovian–instrumental transfer (cf. Holmes et al. 2010 for a review).

Pavlovian–instrumental transfer experiments typically involve three stages. In the first, Pavlovian conditioning is conducted pairing a stimulus or a number of stimuli with an outcome or outcomes (such as different foods). In the second stage, instrumental conditioning is conducted in which one or more instrumental actions are trained, typically using the same outcomes presented in stage one but now in the absence of any Pavlovian stimuli. In a final stage, the Pavlovian–instrumental transfer test is conducted in which any instrumental actions are made available, and for the first time, the Pavlovian stimuli are presented periodically allowing their influence on instrumental action to be assessed. The influence or transfer of control by the Pavlovian stimuli onto performance of any available action constitutes the Pavlovian–instrumental transfer effect (Fig. 2).

While there are often parametric differences between studies, including differences in the order of Pavlovian and instrumental training or the use of discriminative stimuli versus Pavlovian CSs, these appear to have little affect on the overall pattern of responding. Nevertheless, there are a number of important features of the design that have implications for understanding this effect. Firstly, the stimulus and target response are encountered together for the first time at test; i.e., there has been no prior opportunity for the stimulus and response to become directly associated. If the stimulus affects responding, therefore, it cannot be through a direct S-R association, which has not had an opportunity to form. Furthermore, the stimulus is unrelated to performance of the action preventing opportunities for conditioned reinforcement. In this way, the task provides a relatively pure test of the incentive value of Pavlovian stimuli and their ability to motivate instrumental performance. Finally, testing usually occurs under extinction conditions, i.e., no outcomes are delivered during test. This ensures that performance is not influenced by the outcome either by producing new learning during the test or by generating response competition.

Experiments using this task have shown that appetitive Pavlovian stimuli greatly enhance instrumental performance for appetitive outcomes, and this elevation defines the Pavlovian–instrumental transfer effect in this context. This effect is not restricted to appetitive paradigms; however, presentation of a stimulus that predicts an aversive outcome, such as shock, can enhance a previously established avoidance response (e.g., Solomon and Turner 1962; Rescorla and LoLordo 1965) and inhibit instrumental responding for appetitive outcomes such as food (often shown in the well-established conditioned emotional response (CER) paradigm, e.g., Ray and Stein 1959). Generally, these kinds of effect have been important in establishing the breadth of the role of a Pavlovian incentive process in transfer, particularly the role of general appetitive and aversive affective processes in such effects (e.g., Konorski 1967; Rescorla and Solomon 1967; Dickinson and Dearing 1979). For the purposes of this chapter, however, we will focus on the effects of appetitive stimuli on reward-related actions, which are the more prevalent in the recent literature (see Campese et al. 2013 for some recent investigations of aversive transfer or Rescorla and Solomon 1967 for a review of early findings which largely utilized aversive paradigms).

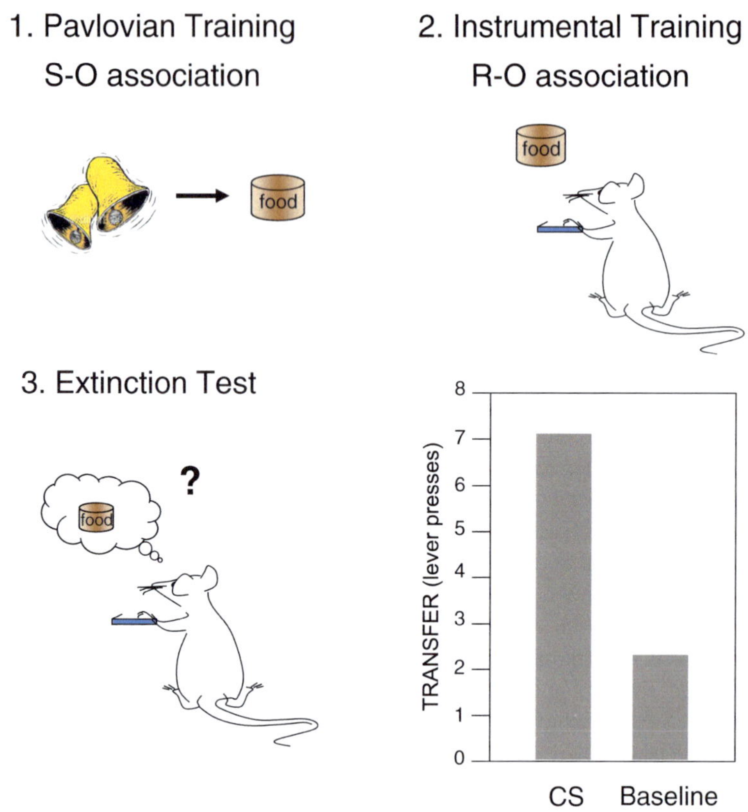

Fig. 2 Pavlovian–instrumental transfer. *1* Animals initially undergo Pavlovian training wherein a discrete stimulus is paired with delivery of food. *2* Instrumental training is then conducted wherein the animal must make a response, such a pressing a lever, to earn food reward. *3* The Pavlovian–instrumental transfer test is then conducted, typically in extinction meaning that no reward is actually delivered. The lever is available for the animal to press, and the previously trained stimulus is presented periodically. Elevation of responding in the presence of the stimulus, relative to the absence of the stimulus (baseline period), describes the Pavlovian–instrumental transfer effect

3.2 General Versus Specific Transfer

Whereas early demonstrations of Pavlovian–instrumental transfer were largely interpreted in terms of the ability of stimuli to activate quite diffuse affective processes (e.g., Rescorla and Solomon 1967), the influence of Pavlovian stimuli can often be quite specific within a particular primary motivational state, such as hunger or thirst (Dickinson and Dawson 1987; Dickinson and Balleine 2002). For example, stimuli associated with appetitive outcomes such as water often have little effect unless animals are water deprived; likewise, the excitatory effect of stimuli associated with nutritive outcomes often depends on animals being hungry. Such effects suggest that specific motivational states gate the arousing effects of Pavlovian incentives processes on instrumental performance (cf. Balleine 1994).

Other transfer effects have resisted explanation in affective or motivational terms. For example, some early studies found that a stimulus previously paired with, say, sucrose increased the performance of instrumental actions that earned sucrose more than actions earning a different outcome, even if both outcomes were appetitive and relevant to the same motivational state (e.g., Baxter and Zamble 1982; Kruse et al. 1983; Colwill and Motzkin 1994). These specific transfer effects lead to suggestions that predictive stimuli can activate an expectancy of a specific outcome that, through a form of S-R process, could elevated the performance of an associated action; that is, presentation of the stimulus recalls the associated outcome which, in turn, recalls responses with which it has also been associated; thus, although somewhat indirectly, the stimulus provokes the response (S(O)-R, e.g., Trapold and Overmier 1972; Corbit and Balleine 2005). This possibility has been explored more fully using designs in which two stimuli are paired with distinct outcomes (e.g., grain pellets and sucrose solution), and then, the effect of the stimuli assessed on the performance of two distinct actions trained with the rewards used in Pavlovian training. In such experiments, a response is typically enhanced by the presentation of a stimulus that predicts the same outcome relative to stimuli predicting other outcomes. In fact, it is important to note that no excitatory effect is typically observed when the stimulus and response are associated with different outcomes; e.g., presentation of the stimulus that predicts pellets does not enhance, and in some examples inhibits, responding for sucrose (Colwill and Motzkin 1994; Corbit et al. 2001). These specific transfer effects are not readily explicable in terms of general affective or motivational states; to a hungry rat, stimuli predicting both sucrose and pellets are relevant to the state of hunger and should be expected to have similar effects on responding for a food reward but, as just described, this is not the result observed with these training conditions.

As a consequence, two main accounts have been advanced of the effects of reward-related stimuli on instrumental performance (cf. Fig. 3): First, a stimulus may produce a general enhancement (or suppression) of responding as a result of the general arousal that the stimulus elicits through its association with reinforcement generally (hereafter referred to as *general transfer*). Alternatively, a stimulus may generate an outcome-specific enhancement of a particular action as a result of the relationship between the stimulus and the unique sensory properties of the particular outcome earned by the action (hereafter referred to *as specific transfer*)— see Fig. 3 for a summary of the associative basis for these transfer effects. Data and theory can be found to support both of these effects (Rescorla and Solomon 1967; Trapold and Overmier 1972; Overmier and Lawry 1979; Colwill and Motzkin 1994; Corbit and Balleine 2005), and indeed, behavioral evidence suggests that they are mediated by distinct processes. For example, Corbit and Balleine (2005) were able to establish a procedure that generated both general and specific transfer in the same subject (see Fig. 4). In that paradigm, rats underwent Pavlovian training in which three distinct stimuli were paired with the delivery of three unique food

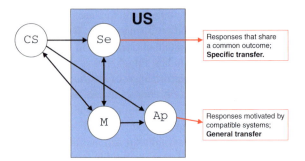

Fig. 3 Pavlovian influences on instrumental responding. Conditioned stimuli (CS) can become associated (*black arrows*) with one or more components of the unconditioned stimulus (US), and those associations determine the nature of the influence on instrumental behaviors (*red arrows*). Stimuli associated with the specific sensory attributes of a US or outcome (Se, e.g., taste, smell, texture) can activate representations of that unique outcome and, in turn, responses with which that outcome is associated, thus generating specific transfer. Stimuli associated with the properties of USs relevant to particular motivational states (M; such as calories relevant under hunger) can influence instrumental performance generally, promoting responding driven by compatible motivational states (e.g., hunger) and suppressing behaviors driven by incompatible motivational states (e.g., fear). Training parameters, such as the use of multiple food outcomes that differ primarily in their sensory properties, may encourage rats to focus on these distinct sensory properties, at least more than in single outcome paradigms in which these sensory properties have little relevance. See text for further discussion

outcomes. Next, the rats were trained in separate sessions to perform two different instrumental actions that earned only two of the three outcomes used in the Pavlovian training phase. Finally, the influence of the three stimuli on performance of each of the responses was assessed in two extinction tests (one for each lever). As a result of this training, for a particular response one stimulus predicted the same outcome as the response (referred to as the 'same' stimulus), another stimulus predicted the outcome earned by a different response ('different' stimulus), and the third stimulus predicted a third outcome. Using this, Corbit and Balleine (2005) found evidence of specific transfer, by assessing the effect of same versus different stimuli against baseline (stimulus 'same' increased performance relative to 'different' and baseline (no stimulus) periods) and, of general transfer, by assessing performance induced by the third stimulus against baseline (both actions were found to be elevated above baseline to a similar degree). Importantly, using this same procedure, Corbit, Janak and Balleine (2007) assessed the effect of a shift in deprivation state on these transfer effects by allowing free access to laboratory chow in the home cages for 24 h prior to testing. They found that, although the difference between same and different stimuli was the same whether rats were hungry or not on test, the third stimulus no longer generated general transfer; i.e., general satiety did not influence specific transfer but abolished general transfer.

This effect of a motivational shift is consistent with the view that, whereas general transfer relies on the ability of the stimuli to activate a motivational state relevant to the available response, specific transfer does not and instead relies on an

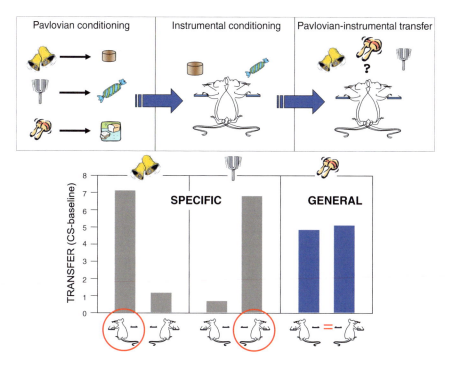

Fig. 4 Pavlovian–instrumental transfer: the three stimulus design. This paradigm provides a procedure for revealing both specific and general transfer in the same subject. Rats are given Pavlovian conditioning with three auditory cues each paired with a different food outcome. In a second phase, the same rats are trained to perform two different lever-press actions each earning one of the outcomes used on Pavlovian training. As a consequence, one Pavlovian stimulus predicts an outcome not earned during instrumental conditioning. On test, each of the three stimuli is presented in a situation in which the rats are allowed to perform each instrumental action. The *bottom* panel shows the results for an experiment published using this procedure presented for each action (Corbit and Balleine 2005). Stimuli that predicted outcomes that were earned by the instrumental actions promoted performance of a response earning the same but not a different outcome, and thus generated specific transfer (*left* two panels at *bottom*). In contrast, the third stimulus, that predicted an outcome that was not earned by the actions, elevated both actions so producing general transfer

association between the CS and the sensory rather than motivational aspects of the outcome (cf. Fig. 3). This failure to influence specific transfer is also consistent with previous failures to find an effect of outcome devaluation on specific transfer (Colwill and Rescorla 1990; Rescorla 1994). For example, Colwill and Rescorla (1990) trained rats to anticipate two outcomes based on distinct auditory cues and then to perform two actions for those outcomes before one of the outcomes was devalued. The results showed that, although devaluation reduced baseline responding for the response associated with the devalued outcome, the ability of a stimulus to augment the performance of the action predicting the same outcome as the stimulus was maintained whether or not that shared outcome was devalued.

Nevertheless, given that transfer reflects responding relative to baseline, the devaluation-induced reduction in baseline responding makes interpretation of the transfer effect somewhat problematic. To address this potential issue, in addition to training two responses with unique outcomes that would both ultimately be devalued, Rescorla (1994) added a phase to training in which both responses were reinforced with a common reward that would not be devalued but serve, at test, to preserve the instrumental baseline. At test, rats showed similar performance of the responses paired with the devalued and non-devalued outcome in the baseline period because of this additional training. Importantly, specific transfer was still observed and the magnitude of the effect was not affected by devaluation of the outcomes. Thus, this finding provides evidence not only that the cueing properties of a stimulus by virtue of a shared outcome are preserved despite devaluation of that outcome, but also that the magnitude of this effect appears to be unaffected by devaluation. Confirming this result, Holland (2004) has also reported that specific transfer was unaffected by outcome devaluation.

Such results speak to the nature of the learning that underlies specific transfer. Although it might be assumed that the transfer effect reflects stimulus-induced augmentation of the effects of an existing action–outcome association, such a mechanism should be expected to be sensitive to outcome devaluation. Specific transfer requires that the presentation of a stimulus evokes some representation of its associated outcome; however, demonstrations that such transfer effects are impervious to devaluation encourage the view that it is not the retrieval of the outcome as a goal of a specific action that is the critical determinant of transfer. Rather, the stimulus appears to function to retrieve actions with which the outcome is associated, and perhaps, as suggested by Rescorla (1994), the outcome is simply a mediating step in a chain of events that activates the response (S-O-R). This could be through retrieval of the R-O association demonstrated to be critical for goal-directed responding (Dickinson and Balleine 1994), or specific transfer may rely on an O-R association, which could also be formed during free-operant training. Thus, whereas specific transfer does not rely on the current value of the outcome, the ability to activate a representation of the response through either of these possible associations appears to be sufficient to promote performance of that particular response. Further, while learning about the consequences of an action and so formulating specific R-O associations is a critical aspect of goal-directed learning, it is reasonable to expect that under some circumstances, when animals activate representations of outcomes, they can use previously learned associations to recall the action that produce those outcomes. A very good example of exactly this kind of recall is outcome-specific reinstatement of instrumental performance, i.e., the demonstration that an extinguished action is reinstated by the outcome it earned during training but not by a different outcome (Ostlund and Balleine 2007). The ability of the outcome representation to initiate its associated action is consistent with it acting more as a stimulus retrieving a response than as a goal to which that action is directed and does not require that the current value of the outcome play any particular role.

3.3 Specific Transfer and Choice

A final behavioral issue worth addressing is perhaps best appreciated by reconsidering the results of the three stimulus design introduced by Corbit and Balleine (2005) described above and shown in Fig. 4. Given that the 'general' stimulus was paired with an outcome that differed from those earned by the two actions, one might imagine it would produce an effect similar to the 'different' stimulus and fail to influence performance. The fact that it generated an excitatory effect on performance suggests, however, that specific transfer is as much a product of the 'different' stimulus failing to elevate performance as it is the 'same' stimulus succeeding in doing so. Why, then, does the 'different' stimulus fail to affect performance despite having been paired previously with reward?

The finding that the 'general' stimulus excites performance suggests that the lack of effect of the 'different' stimulus has to do with the fact that it predicts an outcome that is earned by an alternative action, something most likely to occur in choice situations in which multiple actions are paired with multiple outcomes. The use of different food outcomes in instrumental conditioning may encourage the rats to focus on their distinct sensory properties, at least more than in single outcome paradigms in which these sensory properties have little relevance. This idea was explored to some extent by Holland (2004) who found that, when rats were trained with two stimuli each predicting a unique outcome (pellets and sucrose solution) and underwent instrumental training where two responses earned only one of those outcomes, both stimuli enhanced responding at test. If, however, the rats underwent the same Pavlovian conditioning but, during instrumental training, the two actions earned two different outcomes, then, at test, transfer was specific; i.e., stimulus presentation enhanced performance of the action that earned the outcome predicted by the stimulus relative to the other action.

The introduction of a second R-O association during instrumental training seems, therefore, to be particularly important for generating specific transfer and one reason for this is the opportunity for rats to learn not only what outcomes their actions produce (R1 $\rightarrow$ O1) but also what outcomes they do not produce (R1 $\rightarrow$ no O2). It is these inhibitory R-O associations that could be critical in determining performance in the presence of the 'different' stimuli, during which a prediction such as S2 $\rightarrow$ O2 could suppress the performance of actions (like R1 in this example) associated with 'no O2'. Direct evidence for this account of specific transfer comes from the demonstration that conditioned inhibitors (stimuli predicting the unavailability of a particular outcome, e.g., S2 $\rightarrow$ no O2) act contrary to excitors and elevate the performance of actions not associated with that outcome (Laurent et al. 2015). Thus, when animals are trained with multiple actions, outcomes, and stimuli, in addition to learning the associations between these events, they learn about what outcomes their action do not produce and they use this information to inform choice.

In summary, although the exact associative underpinnings of transfer effects deserve further study, it cannot be that stimulus presentations serve merely to gate what is otherwise general motivational arousal or make representations of the outcome more accessible as has sometimes been claimed (Dickinson and Dearing 1979; Rescorla 1994). Rather, it appears the influence of outcome-related stimuli on choice depends on the information that those stimuli provide about outcome delivery; that stimulus presentation promotes retrieval of the sensory features of an outcome, which in turn influences instrumental performance by retrieving the action associated with that outcome. When multiple stimuli, outcomes and in particular, actions are trained, the influence of stimuli can be quite specific. Finally, while the term 'general' PIT has widely been adopted to describe experiments where the impact of a single CS+ (and often neutral stimulus or CS− as controls) on performance of a single instrumental response is assessed, there has been relatively little systematic research into the nature of the learning that underlies this effect. Whether it is equivalent to the general transfer generated by a third stimulus in the three stimulus design will be an important area for future study.

4 Neural Bases of Transfer

4.1 The Limbic-Striatal Circuit

Based on its hypothesized role as a limbic-motor interface, (Mogenson et al. 1980) initial studies into the neural control of transfer effects focused on the role of the nucleus accumbens (NAc) as a structure likely to be important for the ability of Pavlovian cues carrying information about motivationally relevant events to modulate performance instrumental actions. Whereas there is now strong support for such a role, early studies found conflicting results. The heterogeneous structure of the NAc as well as the training procedures used was eventually demonstrated to explain this conflict and highlights how the details of the training procedures and an ever-evolving understanding of the neural control of motivational systems can be important contributors to the understanding of motivational control of behavior.

For example, Corbit et al. (2001) created cell-body lesions of either the NAc core or shell and, after surgical recovery, trained rats to perform two instrumental responses (left and right lever-press) each earning a unique food reward (sucrose and pellets). Rats then underwent Pavlovian conditioning wherein two auditory stimuli were each paired with one of the same two foods. In the test sessions, the effects of the two stimuli on performance of each of the actions were assessed in extinction. Lesions of the NAc shell eliminated specific transfer in this paradigm, whereas lesions of the core were without effect. In the same year, using a general transfer design, Hall et al. (2001) reported that lesions of the core and not shell eliminated transfer, complicating understanding of the role of the NAc in transfer effects. Importantly, none of these lesions impaired the acquisition of Pavlovian magazine entry behavior or the performance of instrumental actions such as lever-pressing, suggesting that these effects were specific to the ability of Pavlovian cues to modulate instrumental performance. However, substantial differences in the training procedures and thus potentially in the forms of transfer generated in these experiments likely account for these differences (see also de Borchgrave et al. 2002). Hall et al. (2001) assessed the influence of a single excitatory cue on performance on a single lever, a procedure likely to maximize the influence of the general motivational or activating effects of the cue on performance. In contrast, Corbit et al. (2001) assessed the effects of two excitatory CSs on the performance of two different actions each rewarded with unique outcomes, a procedure likely to maximize the influence of the outcome-specific predictions on performance as well as invoke inhibitory processes to suppress responding during the different stimulus.

To further examine the role of the core and shell in transfer, Corbit and Balleine (2011) utilized the 3-stimulus paradigm, described in Sect. 2.2. They found that lesions of the core reduced the impact of the general stimulus but left specific transfer intact. Conversely, lesions of the shell abolished specific transfer while leaving the excitatory impact of the third, general stimulus intact. Furthermore, whereas the majority of studies examining the NAc in transfer have used pre-training lesions, leaving open the possibility that the lesioned rats learn something different during training that could ultimately bias the expression of transfer at test, Corbit and Balleine (2011) also examined the effects of temporary inactivation of the core or shell immediately prior to the test, which replicated the results of the pre-training lesion experiment. Thus, the evidence supports the suggestion that the NAc is essential for the ability of Pavlovian stimuli to modulate ongoing instrumental performance. Importantly, however, the core and shell regions serve distinct functions in this regard.

As with the NAc core and shell, the basolateral (BLA) and central (CeN) subnuclei of the amygdala have been found to play important but distinct roles in transfer and, at least in the case of the BLA, have been found to do so via connections with the accumbens. Hall et al. (2001) examined the role of the BLA and CeN in a general transfer design and found evidence of control by the CeN but not BLA. Similar results were also reported by Holland and Gallagher (2003) using a similar design. Again, these results were in contrast to others, notably those of Blundell et al. (2001) who found that lesions of the BLA impaired transfer. Importantly, Blundell et al. (2001) used an outcome-specific transfer design and found that the typical outcome specificity of transfer was abolished in animals with lesions of the BLA although the generally arousing effects of the stimuli appeared intact as animals responded more in the presence of the stimuli regardless of whether they predicted the same or different outcome as the response. Corbit and Balleine (2005) tested the effects of BLA and CeN lesions on PIT using the 3-stimulus design that can assess the contribution of any generally arousing impact of the Pavlovian cues separately from the specific cuing function of the stimuli when they predict delivery of an outcome also earned by instrumental performance. As might be expected from the previous studies, pre-training lesions of the BLA were found to eliminate specific transfer while leaving the impact of the third stimulus, and thus general transfer, intact. The reverse was true following lesions of the CeN; general transfer was eliminated, but specific transfer was intact.

The similarity of the effects of BLA lesions to those of the NAc shell on specific transfer suggests that the two structures may form a part of a circuit involved in the specific transfer effect. To address this question, Shiflett and Balleine (2010) used asymmetrical lesions to disconnect the BLA from either the accumbens shell or the accumbens core; i.e., rats were given a lesion of the BLA in one hemisphere and of either the shell or core in the contralateral hemisphere prior to Pavlovian training, instrumental training, and then a transfer test. Two other control groups received lesions of both the BLA and the shell or core but in the same hemisphere, leaving one amygalostriatal connection intact. Groups given either ipsilateral lesions or BLA-core disconnection showed similar levels of specific transfer to unlesioned controls. In contrast, rats given the disconnection lesion with the BLA on one side and the NAc shell on the other showed no specific transfer. The clear prediction from this finding is that disconnection of the CeN from the accumbens core would selectively abolish general but not specific transfer; however, that experiment has yet to be conducted.

4.2 Dopamine and the Modulation of the Nucleus Accumbens

General transfer. Many theories link the function of the mesolimbic dopamine system to the behavioral processes related to reward. Dopaminergic manipulations have been shown to affect a variety of motivated behaviors, yet the specific role of dopamine in reward-related learning remains a matter of debate and this volume contains several chapters on this topic. It has been hypothesized that, in instrumental conditioning, dopamine controls the exertion of effort (Salamone et al. 2007) and see Salamone et al., in this volume for a detailed discussion. It has also been hypothesized that dopamine signals a reward prediction error (Schultz 2006), a hypothesis that is examined in the chapter by Bissonette and Roesch, focused on evidence from rodents and non-human primates, and the chapter by O'Doherty, which is focused on evidence from the human literature. It is further hypothesized that dopamine mediates the attribution of incentive salience to reward-related stimuli (Berridge 2007), and this function of dopamine is discussed in detail by Robinson et al., in this volume. It is possible that all of these potential functions of dopamine (and others) contribute to the generation of the various transfer effects.

Initial pharmacological investigations of transfer examined the effects of drugs that either block or augment dopamine function. Dickinson, Smith, and Mirenowicz (2000) found that systemic administration of the dopamine receptor antagonist pimozide or alpha flupentixol eliminated general transfer without affecting instrumental incentive learning. Wyvell and Berridge (2001) found complementary effects; the magnitude of general transfer was enhanced in animals previously receiving sensitizing injections of the indirect dopamine agonist amphetamine or if amphetamine was infused directly into the NAc prior to testing (Wyvell and

Berridge 2000, 2001; Peciña and Berridge 2013). Importantly, baseline responding or responding in the presence of a neutral stimulus was unaffected in these same animals.

More recently, studies using fast-scan cyclic voltammetry have observed phasic dopamine release in the NAc in response to presentations of a reward-paired cue, and the amplitude of the measured dopamine transients was found to correlate with the magnitude of the general transfer effect (Wassum et al. 2013), providing direct evidence that dopamine is released in the NAc during stimulus presentations that generate transfer. Together, these results have been interpreted as support for the idea that dopamine plays an essential role in incentive salience allowing reward-paired stimuli to elicit behaviors such as approach and to invigorate instrumental behaviors that yield reward. More selectively, Lex and Hauber (2008) investigated the effects of D1 versus D2 antagonists infused into the NAc core or shell on general PIT. The D1 antagonist SCH23390 infused into the core or shell eliminated general transfer and also decreased baseline responding. Similarly, the D2 antagonist raclopride attenuated but did not completely eliminate transfer at doses that significantly decreased baseline responding. Nevertheless, the effects on dopamine blockade within the NAc core are consistent with lesion and inactivation studies implicating the core in control of general transfer (Hall et al. 2001; Corbit and Balleine 2011) and suggest that the direct D1 containing pathway plays a particularly important role in generating these effects. The effects of dopamine antagonists in the shell are somewhat surprising given lesion and inactivation data that suggest that the shell is involved in generating specific rather than general transfer effects (Corbit et al. 2001; Corbit and Balleine 2011; Hall et al. 2001). Whether these common effects of infusions into core and shell on general transfer reflect their individual effects or diffusion of the drug remains an open question.

Dopamine and specific transfer. Less ambiguity has ultimately emerged from studies of the role of accumbens dopamine in specific transfer. In what amounts to the first investigation of dopamine involvement in specific transfer, Ostlund and Maidment (2012) trained rats on two Pavlovian stimulus–outcome and two instrumental response–outcome relationships and found that systemic administration of the nonspecific dopamine antagonist flupentixol prior to tests of outcome-specific transfer reduced the specific transfer effect, although the ability of a particular cue to bias responding toward a response that earned the same outcome was preserved to some degree.

More focal investigations have found, however, that it is the D1 receptor-expressing neurons in the accumbens shell that play a particularly central role in specific transfer. For example, based on the finding that roughly half of the medium spiny neurons in the accumbens express dopamine D1 or D2 receptors (Gerfen and Surmeier 2011), Laurent et al. (2014) trained two groups of mice, expressing enhanced green fluorescent protein under the DRD2 promoter (D2-eGFP mice) allowing the D2 containing population of neurons to be visualized, on two actions and two stimuli for different outcomes. One group was given a transfer test after which they were killed and slices of accumbens tissue processed for a marker revealing recent cellular activity, i.e., phosphorylation of extracellular signal regulated kinase (pERK), in both GFP positive (i.e., D2 neurons) and GFP negative (i.e., D1 neurons) neurons in the accumbens shell and core. Relative to animals not given the transfer test but only exposed to the context, big increases in pERK activity were observed in the shell but not the core of the accumbens. Second, that increase in activity was found to be exclusively in D1 neurons.

Laurent et al. (2014) followed up this finding in a subsequent study looking at the effect of D1 (SCH23390) and D2 (raclopride) antagonists infused into the accumbens shell or core prior to the specific transfer test relative to vehicle control infusions. This study provided clear evidence for D1 receptor involvement in the accumbens shell in specific transfer; the D1 antagonist abolished specific transfer when infused into the shell but had no effect when infused in the core, and the D2 antagonist did not have any effect on specific transfer when infused into either structure relative to controls.

4.3 Beyond Dopamine

4.3.1 Opioid–Dopamine Interactions and the Role of Acetylcholine

One particular challenge for understanding transfer at the neural level is to establish how two distinct forms of learning-related neural plasticity (resulting from Pavlovian and instrumental conditioning) occurring at different times are integrated to promote choice at a much later time. Interestingly, recent research suggests that the dynamics of δ-opioid receptor (DOR) trafficking in the accumbens shell could be pivotal, at least as far as specific transfer is concerned (Laurent et al. 2012, 2014; Bertran-Gonzalez et al. 2013). First, unlike their wild-type littermate controls, DOR knockout mice show a deficit in specific transfer but no impairment in either Pavlovian or instrumental training (Laurent et al. 2012—Experiment 1). Second, local infusion of the DOR antagonist naltrindole into the shell prior to the transfer test blocks specific transfer, whereas it has no effect when infused into the core, and the infusion of the μ-opioid receptor antagonist CTAP has no affect on specific transfer in either region (Laurent et al. 2012—Experiment 2).

Taken together with the effects of the dopamine manipulations described above, it is clear that DOR and D1 receptor antagonists have similar effects on specific transfer, and as such, it is logical to ask whether DOR and D1 receptor-related processes interact in the accumbens shell to generate this transfer effect. To address this question, Laurent et al. (2014) gave unilateral infusions of naltrindole and SCH23390 into the shell but in opposite hemispheres, antagonising D1-related processes in one hemisphere and DOR-related processes in the other. These infusions into opposing hemispheres abolished specific transfer, whereas controls given unilateral infusion of one or other drug did not.

How do dopaminergic and opioidergic processes interact in the shell? A clue was found to lie in the distribution of DOR receptors. To assess this distribution, Bertran-Gonzalez et al. (2013) used DOR-eGFP knock-in mice. These mice express a fluorescent but functional form of the DOR (Scherrer et al. 2006). Initial experiments revealed substantial levels of DOR in the shell (Bertran-Gonzalez et al. 2013), higher than the core, and mostly if not exclusively expressed postsynaptically. However, Bertran-Gonzalez et al. (2013) could not find any clear relationship between the primary striatal output neurons, i.e., the medium spiny neurons, and DOR expression. Rather, they observed that DOR were largely localized in the shell on the perisomatic membrane of cholinergic interneurons (CINs). This observation was particularly interesting given that CINs have been shown to be involved in predictive learning and theories have been developed concerning their role in promoting the influence of contextual cues on decision-making processes (Apicella 2007; Brown et al. 2012; Stocco 2012).

Bertran-Gonzalez et al. (2013) assessed the role of DOR expression on CINs by exposing food-deprived DOR-eGFP mice to specific transfer; i.e., two stimuli (S1 and S2) predicted two distinct food outcomes (O1 and O2, respectively) that were then used to reward two responses (R1 and R2). Mice given these two learning stages showed specific transfer during a subsequent choice extinction test (i.e., they showed S1: R1 > R2 and S2: R1 < R2). Importantly, these mice also displayed higher levels of DOR expression on the membrane of CINs in the NAc shell than control mice that had only been exposed to the conditioning chambers. Furthermore, this increase in DOR expression was not restricted to mice that received the transfer test; DOR accumulation on the membrane of CINs was present in mice given Pavlovian conditioning but in which either instrumental training or the test had been omitted (Bertran-Gonzalez et al. 2013). These findings suggest that Pavlovian training produced the change in DOR expression observed on shell CINs. And, indeed, consistent with this suggestion, Bertran-Gonzalez et al. (2013) found a strong positive correlation between the levels of DOR expression and conditioned responding across Pavlovian training; i.e., the higher the level of conditioned responding, the more DOR accumulated on CINs in the shell (Bertran-Gonzalez et al. 2013). Importantly, a similar effect was not observed when the mice were given non-contingent training; i.e., presentations of S1 and S2 and deliveries of O1 and O2 were uncorrelated preventing the mice from learning the predictive S-O relationship. As expected, non-contingently trained mice displayed lower levels of conditioned responding than those contingently trained. More importantly, they also exhibited lower levels of DOR expression on shell CINs compared to contingently trained mice. In fact, their levels were similar to those displayed by control mice that had simply been exposed to the conditioning chambers. These effects were specific to CINs located in the accumbens shell. No such differences were found in the dorsal striatum or in the core (Bertran-Gonzalez et al. 2013).

Generally, these data suggest that the influence of dopaminergic processes on specific transfer is modulated both by opioid and cholinergic processes. First, the acquisition of specific stimulus-outcome contingencies across Pavlovian training triggers DOR accumulation on the membrane of CINs in the shell. Second, this accumulation is long lasting and survives at least for the time between Pavlovian training and the test (i.e., 11 days), substantially alters the firing pattern of CINs (Bertran-Gonzalez et al. 2013) and is not necessary for Pavlovian conditioning per se (Laurent et al. 2012) but is required for the influence of Pavlovian stimuli on choice between actions; i.e., for Pavlovian–instrumental transfer. By regulating acetylcholine release, DOR is also in position to regulate D1 neuron activity; this is because acetylcholine inhibits D1 neuron activity in the striatum (Goldberg et al. 2012). Indeed, blockade of a specific muscarinic M4 receptor specifically expressed on D1 neurons in the shell blocks the effect of the DOR antagonist on transfer (Laurent et al. 2014), confirming its modulatory role on striatal projection neurons. Given its temporal and functional characteristics, therefore, DOR-related plasticity constitutes a convincing neural substrate in the accumbens shell through which two forms of learning, Pavlovian and instrumental conditioning, established at different times, are subsequently integrated in a manner that promotes the expression of predictive learning on choice.

In the course of these experiments, Bertran-Gonzalez et al. (2013) were also able to obtain evidence of functional and physiological changes associated with DOR accumulation, which appeared to powerfully alter the firing pattern of CINs by increasing the variance of action potential activity, an effect that was potentiated by a DOR agonist. This increase and its potentiation by the agonist differed in mice showing high levels of conditioned responding compared to those showing much lower levels and in contingently trained versus non-contingently trained animals. The change in the physiological properties of the CINs is of interest as it is widely agreed that CIN activity is a critical modulator of MSN function (Ding et al. 2010; Goldberg et al. 2012). This suggests that DOR-mediated changes in CIN firing could quickly alter shell activity to express the effect of predictive learning on choice. Consistent with this suggestion, DOR-eGFP mice trained on two action–outcome associations showed similar behavioral effects to rats when a specific conditioned inhibitor was presented during the transfer test; i.e., they showed the reversed transfer effect elevating the different relative to the same action consistent with the inhibitor activating the inhibitory, or counterfactual, action–outcome association (Laurent et al. 2015). Importantly, as with the effects of specific excitors, this effect of the inhibitor was both associated with an increase in DOR translocation to the membrane of CINs in the accumbens shell and could be blocked by the peripheral injection of naltrindole, suggesting that this same mechanism mediates the effect of excitatory and inhibitory action–outcome mappings on choice.

4.4 Beyond the Accumbens: The Extended Circuit

4.4.1 Dorsal Striatum

Pavlovian–instrumental transfer is an integrative process; instrumental learning allows animals to learn about their actions and the consequences they produce, and by integrating this learning with Pavlovian predictive information, animals can evaluate the predicted value of those consequences to choose between competing courses of action. As multiple associations must, therefore, be formed and utilized to direct motor output, it is not surprising that transfer is mediated by a complex network of structures which are summarized in Fig. 5. At a neural level many of the structures that have been implicated in transfer are also involved in the acquisition and/or expression of instrumental response–outcome or Pavlovian stimulus–outcome associations (Corbit and Balleine 2005; Corbit and Janak 2007a; Ostlund and Balleine 2007, 2008). For example, several studies have implicated the dorsal striatum in transfer effects. Using a specific transfer design, an initial study found that inactivation of the DMS eliminated the outcome specificity of the effect, but left the excitatory effects of the stimuli intact; that is, transfer was not eliminated but the different as well as same stimulus enhanced responding to a similar extent (Corbit and Janak 2007a). In the context of previous work that has established that the DMS is critical for learning and expression of specific R-O associations (Yin et al. 2005, Corbit and Janak 2010), this result was interpreted as secondary to an instrumental learning deficit; i.e., the inability to utilize specific R-O associations that would be expected following DMS inactivation would prevent the expression

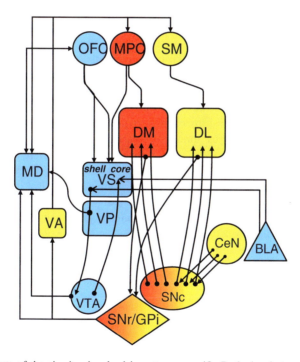

Fig. 5 Elements of the circuitry involved in outcome-specific Pavlovian–instrumental transfer. The critical circuit is in *blue* centered on connections between the BLA-VS-VP. The circuit governing goal-directed action is in *red* and that governing habits in *yellow*. See text for details. *BLA* basolateral amygdala; *CeN* central nucleus of amygdala; *DL* dorsolateral striatum; *DMS* dorsolateral striatum; *OFC* orbitofrontal cortex, *MPC* medial prefrontal cortex (includes prelimbic cortex); *GPi* internal globus pallidus; *SNc* substantia nigra pars compacta; *VA/MD* ventral anterior/medial dorsal nuclei of the thalamus; *VS* ventral striatum; *VP* ventral pallidum; and *VTA* ventral tegmental area

of specific transfer, which is the result observed. Furthermore, it does not appear that dopamine in the DMS is important for specific transfer; dopamine depletion was found not to impact either general or specific PIT (Pielock et al. 2011). In contrast, inactivation of the DLS was found to attenuate specific transfer; presentation of the same stimulus still elevated responding relative to baseline, while the different stimulus did not; however, this effect was greatly reduced compared to controls (Corbit and Janak 2007a). This finding is consistent with other demonstrations that DLS manipulations disrupt the ability of stimuli to trigger responding under conditions where the stimulus has been directly paired with performance of the response, such as in cue-induced reinstatement and conditioned reinforcement paradigms, and is consistent with the suggestion that the DLS is important for S-R learning (Fuchs et al. 2006; Vanderschuren et al. 2005). Within the striatum, only the nucleus accumbens shell (NAc-S) has been found to play a unique and critical role in the integration of the associative processes involved in specific transfer (Corbit et al. 2001; Corbit and Balleine 2011; Laurent et al. 2012).

4.4.2 Ventral Pallidum

Given the established role of the NAc shell in specific transfer, it is of considerable interest to understand how the output of this structure interacts with basal ganglia circuitry as a whole to direct motor performance. The NAc shell projects prominently to the ventromedial ventral pallidum (VPm), and recent evidence suggests that the NAc shell to VPm projection is required for the expression of specific transfer in rats. Using c-Fos immunohistochemistry combined with retrograde tracing, Leung and Balleine (2013) found that VPm neurons were activated by the specific transfer test. Further, both bilateral inactivation of the VPm and functional disconnection of the VPm from the NAc shell using asymmetrical inactivation both eliminated specific transfer.

The VP functions as both an intrinsic component and an output relay of the basal ganglia, and the VPm maintains both a direct and an indirect output pathways via its projections to the mediodorsal thalamus (MD) and ventral tegmental area (VTA) (Zahm and Heimer 1990; Heimer et al. 1991; Smith et al. 2013), allowing projection neurons from this structure to exert both an excitatory and an inhibitory effect on performance. Connections between the VPm and the VTA and MD have previously been implicated in reward-related actions (McAlonan et al. 1993; Kalivas et al. 1999; Corbit et al. 2003; Geisler et al. 2008; Mahler and Aston-Jones 2012; Mahler et al. 2014; Parnaudeau et al. 2015), and more importantly, both MD and VTA have been implicated in different aspects of the specific transfer effect; whereas damage to the MD abolishes the effects of predictive learning on choice evidenced by disruption of specific transfer (Corbit et al. 2003; Ostlund and Balleine 2008), inactivation of the VTA appears to attenuate the impact of Pavlovian stimuli on response vigor quite generally while preserving their effect on choice (Corbit et al. 2007).

More recently, Leung and Balleine (2015) assessed activity in the direct thalamic and midbrain projections of the VPm by infusing retrograde tracers in the MD and VTA and comparing specific transfer-related c-fos activity in the VPm. Retrograde labeling in the VPm produced by the two tracers is consistent with previous findings showing comparable numbers of neurons in the VP projection to the MD and VTA and, in addition, a small population of neurons that send collaterals to both structures (Zahm 2000; Tripathi et al. 2013). Examination of c-Fos found a selective increase in c-Fos during specific transfer in MD-projecting neurons as well as in neurons innervating both MD and VTA compared to controls. In contrast, c-Fos activity in VTA-projecting neurons correlated with the size of the specific transfer effect.

The disconnection of the VPm from the MD or VTA during specific transfer also produced differential behavioral effects. Disconnection of the MD from the VPm produced an increase in responding on both levers during stimulus presentations such that the rats did not show any response bias during the stimuli. This pattern of results is similar to that seen after MD lesions (Ostlund and Balleine 2008) and suggests that the removal of the inhibitory input from the VPm to the MD eliminates the ability of rats to use specific stimulus–outcome associations to control choice. In stark contrast, VPm–VTA disconnection preserved specific transfer, however, this disconnection resulted in general disinhibition of performance; i.e., rats responded significantly more on the levers in both the presence and absence of the stimuli. This pattern of responding was sustained throughout the entire specific transfer test and suggests that animals without the VPm–VTA connection may lack the ability to inhibit responding when a stimulus is no longer present.

Together with previous observations (Leung and Balleine 2013), these results suggest that neurons in the VPm that receive input from the NAc shell are normally activated during specific transfer and, in turn, act to inhibit both the MD and VTA. The suggestion that the former pathway mediates the response bias in choice performance during specific PIT whereas the latter pathway mediates the relative size of that bias agrees generally with current views of distinct functions of direct and indirect pathways through the basal ganglia. Although the traditional view of basal ganglia circuitry suggests that activation of the direct pathway tends to promote action and activation of the indirect pathway suppresses action, there have also been reports of cooperation between the two pathways during coordinated action, where activation of the direct pathway promotes a specific action and activation of the indirect pathway promotes the suppression of competing actions (Mink 1996; Brown 2007; Kravitz et al. 2012; Cui et al. 2013). Nevertheless, specific hypotheses as to whether the deficits observed following manipulation of VP output reflect one or another process are complicated by the fact that the projection from the NAc-S to the VPm contains neurons belonging to both a direct and indirect pathway (Lu et al. 1998; Zhou et al. 2003). Given that MSNs expressing D1 receptors appear to govern specific transfer (Laurent et al. 2015), it is likely that the shell to VPm pathway is driven by the D1-MSN projection to the MD-projecting VPm neurons and so satisfies definitions of direct pathway neurons (Gerfen and Surmeier 2011). Similarly, MSNs in the NAc-S innervating

VTA-projecting VP-m neurons could reflect an indirect pathway—using a multi-synaptic circuit to reach the output nuclei of the basal ganglia—responsible for the general inhibition of competing or inappropriate actions (Gerfen and Surmeier 2011).

4.4.3 Ventral Tegmental Area

Although the effects of direct manipulation of dopamine and its receptors are described above, the VTA, which is the predominant source of dopaminergic inputs to limbic structures, is comprised of heterogeneous cell types and cells other than those releasing dopamine may contribute to transfer effects. Murschall and Hauber (2006) conducted an initial investigation into the role of the VTA in transfer. In their study, rats received training in which a single excitatory stimulus was paired with delivery of a food pellet. Rats were then trained to perform a single instrumental response earning food pellets. They observed that VTA inactivation, achieved by microinfusion of the GABA(A) and GABA(B) receptor agonist muscimol and baclofen prior to testing, dose-dependently eliminated general transfer in this design. Corbit et al. (2007) extended this initial finding, using the more complex, three stimulus design, and found that both general and outcome-specific forms of PIT were greatly reduced by VTA inactivation, although some residual responding to the same and general stimuli was still observed. Importantly, baseline responding was also reduced by VTA inactivation. Whereas Murschall and Hauber (2006) reported no effect of VTA inactivation on baseline responding in their paradigm, there was a numeric decrease at their highest dose consistent with the pattern observed by Corbit et al. (2007). Furthermore, baseline responding was already very low (1–2 responses per minute) which may have made further decreases difficult to detect statistically.

The residual transfer observed after VTA inactivation reported by Corbit et al. (2007) can be explained if the infusion failed to cover the entire area, e.g., if the single injection site used or the volume of the infusion was insufficient, particularly across the rostro-caudal extent of the VTA. As such, it remains possible that a more complete inactivation of the VTA would have entirely eliminated transfer. However, given the effects on baseline performance, it is likely that more extensive inactivation would also have yielded greater performance deficits unrelated specifically to stimulus-directed performance. Instead, the reduction in baseline responding suggests that inactivation of the VTA using muscimol/baclofen may affect motivated responding in general in addition to any effects on stimulus-driven behavior. As such, although it appears that the VTA mediates the facilitation of responding by reward-related stimuli, it likely also plays a more general role in response initiation that does not depend explicitly on the presence of predictive stimuli. In this regard, the effects of VTA inactivation on stimulus control are similar to the ability of systemic dopamine antagonists to reduce general transfer (Dickinson et al. 2000; Ostlund and Maidment 2012). However, at the right dose, pharmacological manipulations of dopamine can be specific to the effects of stimulus presentation eliminating (or in the case of agonists enhancing) transfer without altering baseline responding, suggesting that the baseline effects following VTA inactivation may reflect a more general involvement of the VTA in performance and response initiation; that inactivation of the VTA should not be considered equivalent to dopaminergic blockade.

It is not currently known how the shell-VPm-MD and the shell-VPm-VTA circuits that control the response bias and the level of response-specific motivation, respectively, are integrated with instrumental performance. One suggestion is that these processes are integrated in the MD-prefrontal cortex feedback circuit (Balleine et al. 2015); however, at present, there are very little data to support this hypothesis.

5 Pavlovian Incentives and Pathologies of Decision Making

While predictive stimuli play an important role in guiding our actions, there is increasing interest in the role of stimuli in guiding or invigorating actions that are considered impulsive or compulsive and where, in general, there is a perceived dysregulation of goal-directed control. For example, many individuals with substance use disorders relapse after prolonged abstinence despite a stated desire to quit, and similarly, many dieters fail over the long term even when strongly motivated to lose weight. The ability of Pavlovian stimuli to invigorate and/or bias responding may contribute importantly to these effects. For example, as noted above, specific transfer effects are impervious to outcome devaluation, suggesting that the cuing effects of stimuli do not require that the consequences of the invigorated response be a current goal. Furthermore, transfer effects appear to grow under conditions that favor habitual responding, such as following extended training (Holland 2004) or drug exposure (Saddoris et al. 2011; LeBlanc et al. 2013), again providing evidence of a dissociation between the ability of stimuli to invigorate responding and an evaluative process concerned with the consequences of that responding.

5.1 *Exposure to Drugs of Abuse*

Direct evidence that stimulus influences exemplified by transfer can contribute to maintaining unwanted actions, such as drug use, comes from recent reports that the strength of transfer effects may be an indicator of relapse risk (Garbusow et al. 2014, 2015). In these studies, recently detoxified alcoholics showed a stronger transfer effect than healthy controls. Furthermore, the activation of the NAc during transfer was greater in patients that went on to relapse, than in those that successfully abstained or in healthy controls. Although the transfer design in these studies used monetary rather than alcohol reward and did not directly assess effects of stimuli on alcohol seeking, other work has implicated NAc activation in cue reactivity and relapse in alcoholics (Heinz et al. 2004; Beck et al. 2012; Wiers et al. 2014) and these results suggest that transfer, as an index of susceptibility to Pavlovian influences, may prospectively help identify patients at greater risk of relapse.

Other studies have shown that alcohol stimuli can directly affect performance of an alcohol-seeking response in both rats and humans (Corbit and Janak 2007b; Martinovic et al. 2014) and that the magnitude and specificity of transfer effects may be altered following drug exposure. For example, Corbit and Janak (2007b) trained rats in a specific PIT design with alcohol and sucrose reward and found that, whereas the sucrose stimulus generated the expected specific transfer effect, enhancing responding for sucrose but not alcohol, the same was not true for the alcohol stimulus which was found to enhance both responses in a manner reminiscent of general transfer (see also Shiflett 2012). The specificity of PIT in alcoholics, as well as the direct effects of alcohol stimuli on alcohol seeking, would be of significant interest for future study. Additional information about subregions of the NAc activated by transfer in this population as well as the role of other neural structures would also be of interest and may help dissociate the type of transfer generated.

5.2 Obesogenic Food Exposure

Another potential application of Pavlovian–instrumental transfer to impaired decision making is the demonstration that stimuli previously paired with food reward can increase responding for food independently of satiation. In a study by Watson et al. (2014), participants learned to press keys for two food rewards (chocolate and popcorn) and to associate specific stimuli with the delivery of those rewards. Prior to testing, some subjects were satiated on one of the food rewards, whereas others remained hungry. At test, subjects showed outcome-specific transfer, and although specific satiety reduced baseline response rates and preference ratings for the pre-fed food, the ability of a cue to increase responding for the signaled food remained intact. Such results may not be surprising in the context of related animal studies (Rescorla 1994); however, their extension to humans increases potential clinical utility. Indeed, such findings suggest that food-associated cues, which are prevalent in many peoples' day-to-day environments, may override satiety to promote overeating and so contribute to the obesity epidemic and failures to change eating behaviors. For more information on the many factors that regulate the motivation to eat, please see the chapter by Woods and Begg in this volume.

5.3 Stress

Finally, stress has been shown to increase the magnitude of transfer effects, particularly general transfer, suggesting a shift in cognitive control under stress conditions. For example, Pool et al. (2015) reported that general transfer was enhanced by acute stress in humans. In this study, subjects first underwent instrumental conditioning and then Pavlovian conditioning each reinforced with delivery of a chocolate odor. Prior to testing, the subjects were divided into groups and underwent either an acute stressor (cold stressor) or control conditions. Subjects exposed to the stressor showed a greater enhancement in responding during the stimuli paired with chocolate odor although stress had no effect on the rated pleasantness of the chocolate odor. The authors interpreted these results within an incentive salience framework, and although higher baseline responding in controls may have affected the ability to detect stimulus effects in the control group, they again suggest that the outcome, or at least its motivationally relevant components, plays little role in transfer effects. These findings also accord with previous reports in animals, suggesting that stress-related hormones can modulate transfer effects. For example, corticotropin-releasing factor (CRF) administered directly into the NAc shell enhances transfer in a dose-dependent fashion without affecting baseline lever-press performance (Pecina et al. 2006), suggesting that CRF amplifies the motivational impact of cued rewards in much the same manner as dopamine.

An important factor in determining the effects of stress on transfer is the nature of the stressor. In general, acute treatments seem to enhance transfer effects, whereas chronic treatments decrease transfer (Pool et al. 2015; Pecina et al. 2006; Morgado et al. 2012; Soares-Cunha et al. 2014). For example, prenatal exposure to glucocorticoids, which is known to produce morphological changes in the NAc and amygdala as well as to reduce dopaminergic innervation, eliminates rather than enhances transfer (Soares-Cunha et al. 2014). This treatment affected both specific and general transfer and is related to changes in the NAc and the dopaminergic innervation of this structure; normalization of dopamine levels with levodopa/carbidopa treatment was able to rescue transfer in animals previously exposed to glucocorticoids. Further, the same rescue effect emerged using the D2 receptor agonist quinpirole (Soares-Cunha et al. 2014).

In summary, a variety of factors, such as exposure to drugs of abuse or stress, can enhance the expression of transfer. As with animal studies, transfer appears to be disconnected from evaluative processes. Indeed, the factors that enhance transfer tend to be those that have also been shown to diminish the role of outcome value in the control of instrumental action in general. Extended training (Holland 2004), alcohol or stimulant exposure (Nelson and Killcross 2006; Corbit et al. 2012, 2014; LeBlanc et al. 2013), chronic access to obesogenic foods (Furlong et al. 2014), and stress (Dias-Ferreira et al. 2009; Schwabe and Wolf 2010) have each been demonstrated to promote habitual control of behavior. Together, such findings point to a fundamental shift in behavioral control away from factors associated with the evaluation our actions in terms of the outcomes that they produce toward control based on previously established, and relatively inflexible, associations between actions and antecedent stimuli.

6 Summary and Conclusions

Predictive information carried by environmental stimuli as a result of Pavlovian learning can be utilized to inform when, how much, and what to do to maximize beneficial and avoid harmful outcomes. A substantial literature examining the influence of Pavlovian stimuli on instrumental responding has developed in recent years. Behavioral experiments have demonstrated that Pavlovian stimuli play an important modulatory role in the expression of instrumental action. This influence can either be general, increasing the vigor of instrumental responding, or specific, cuing particular actions that produce an outcome shared with the stimulus. Behavioral findings are supported by evidence that distinct neural circuits centered on the NAc core and shell mediate the general and specific forms of transfer, respectively, and ongoing work is beginning to explain how Pavlovian and instrumental learning processes that occur independently and at separate times are integrated within neural circuits that govern behavioral control. While under stable environmental conditions, the influence of predictive stimuli is adaptive as it invigorates and directs instrumental actions to optimize reward, important evidence demonstrates that these stimulus influences are disconnected from evaluative processes that otherwise guide flexible goal-directed action. Thus, while general transfer has been demonstrated to be sensitive to changes in motivational state, the ability of stimuli to bias instrumental choice is not affected by a motivational shift or even devaluation of the very outcome predicted by the stimulus. As such, the influence of Pavlovian learning may help explain why certain behaviors persist even when individuals want to invoke change. Recent translational studies indicate that transfer paradigms may have utility for understanding and identifying the role of predictive stimuli in driving unwanted behaviors which will be important for ultimately producing behavior change.

Acknowledgments This preparation of this chapter was supported by a Project Grant (APP1050137) to LHC and by a Senior Principal Research Fellowship to BWB, each from the National Health and Medical Research Council of Australia.

References

Apicella P (2007) Leading tonically active neurons of the striatum from reward detection to context recognition. Trends Neurosci 30:299–306

Balleine BW (1994) Asymmetrical interactions between thirst and hunger in Pavlovian-instrumental transfer. Q J Exp Psychol B 47:211–231

Balleine BW, Morris RW, Leung BK (2015) Thalamocortical integration of instrumental learning and performance and their disintegration in addiction. Brain Res (Epub ahead of print)

Baxter DJ, Zamble E (1982) Reinforcer and response specificity in appetitive transfer of control. Anim Learn Behav 10:201–210

Beck A, Wustenberg T, Genauck A, Wrase J, Schlagenhauf F, Smolka et al (2012) Effect of brain structure, brain function, and brain connectivity on relapse in alcohol-dependent patients. Arch Gen Psychiatry 69:842–852

Berridge KC (2007) The debate over dopamine's role in reward: the case for incentive salience. Psychopharmacology 191:391–431

Bertran-Gonzalez J, Laurent V, Chieng BC, Christie MJ, Balleine BW (2013) Learning-related translocation of δ-opioid receptors on ventral striatal cholinergic interneurons mediates choice between goal-directed actions. J Neurosci 33:16060–16071

Bindra D (1974) A motivational view of learning, performance, and behaviour modification. Psychol Rev 81:199–213

Bindra D (1978) How adaptive behavior is produced: a perceptual motivational alternative to response-reinforcement. Behav Brain Sci 1:41–91

Blundell P, Hall G, Killcross S (2001) Lesions of the basolateral amygdala disrupt selective aspects of reinforcer representation in rats. J Neurosci 21(22):9018–9026

Brown P (2007) Abnormal oscillatory synchronisation in the motor system leads to impaired movement. Curr Opin Neurobiol 17:656–664

Brown MT, Tan KR, O'Connor EC, Nikonenko I, Muller D, Lüscher C (2012) Ventral tegmental area GABA projections pause accumbal cholinergic interneurons to enhance associative learning. Nature 492:452–456

Campese V, McCue M, Lázaro-Muñoz G, Ledoux JE, Cain CK (2013) Development of an aversive Pavlovian-to-instrumental transfer task in rat. Front Behav Neurosci 7:176

Colwill RM, Rescorla RA (1990) Effect of reinforcer devaluation on discriminative control of instrumental behavior. J Exp Psychol Anim Behav Process 16(1):40

Colwill RM, Motzkin DK (1994) Encoding of the unconditioned stimulus in Pavlovian conditioning. Anim Learn Behav 22:384–394

Corbit LH, Balleine BW (2005) Double dissociation of basolateral and central amygdala lesions on the general and outcome-specific forms of Pavlovian-instrumental transfer. J Neurosci 25:962–970

Corbit LH, Balleine BW (2011) The general and outcome-specific forms of Pavlovian-instrumental transfer are differentially mediated by the nucleus accumbens core and shell. J Neurosci 31:11786–11794

Corbit LH, Chieng BC, Balleine BW (2014) Effects of repeated cocaine exposure on habit learning and reversal by N-acetylcysteine. Neuropsychopharmacology 39:1893–1901

Corbit LH, Janak PH (2007a) Inactivation of the lateral but not medial dorsal striatum eliminates the excitatory impact of Pavlovian stimuli on instrumental responding. J Neurosci 27:13977–13981

Corbit LH, Janak PH (2007b) Ethanol-associated cues produce general Pavlovian-instrumental transfer. Alcohol Clin Exp Res 31:766–774

Corbit LH, Janak PH (2010) Posterior dorsomedial striatum is critical for both selective instrumental and Pavlovian reward learning. Eur J Neurosci 31:1312–1321

Corbit LH, Janak PH, Balleine BW (2007) General and outcome-specific forms of Pavlovian-instrumental transfer: the effect of shifts in motivational state and inactivation of the ventral tegmental area. Eur J Neurosci 26:3141–3149

Corbit LH, Muir JL, Balleine BW (2001) The role of the nucleus accumbens in instrumental conditioning: evidence for a functional dissociation between core and shell. J Neurosci 21:3251–3260

Corbit LH, Muir JL, Balleine BW (2003) Lesions of mediodorsal thalamus and anterior thalamic nuclei produce dissociable effects on instrumental conditioning in rats. Eur J Neurosci 18:1286–1294

Corbit LH, Nie H, Janak PH (2012) Habitual alcohol seeking: time course and the contribution of subregions of the dorsal striatum. Biol Psychiatry 72:389–395

Cui G, Jun SB, Jin X, Pham MD, Vogel SS, Lovinger DM, Costa RM (2013) Concurrent activation of striatal direct and indirect pathways during action initiation. Nature 494:238–242 de Borchgrave R, Rawlins JN, Dickinson A, Balleine BW (2002) Effects of cytotoxic nucleus accumbens lesions on instrumental conditioning in rats. Exp Brain Res 144:50–68

Dias-Ferreira E, Sousa JC, Melo I, Morgado P, Mesquita AR, Cerqueira JJ, Costa RM, Sousa N (2009) Chronic stress causes frontostriatal reorganization and affects decision-making. Science 325:621–625

Dickinson A (1980) Contemporary animal learning theory. Cambridge University Press, Cambridge

Dickinson A (1985) Actions and habits—the development of behavioral autonomy. Philos Trans R Soc Lond Ser B Biol Sci 308:67–78

Dickinson A, Balleine BW (1994) Motivational control of goal-directed action. Anim learn Behav 22:1–18

Dickinson A, Balleine BW (2002) Steven's handbook of experimental psychology: learning, motivation and emotion. In: Gallistel C (ed) The role of learning in the operation of motivational systems, vol 3. Wiley, New York, pp 497–534

Dickinson A, Dawson GR (1987) Pavlovian processes in the motivational control of instrumental performance. Q J Exp Psychol Sect B Comp Physiol Psychol 39:201–213

Dickinson A, Dearing MF (1979) Appetitive-aversive interactions and inhibitory processes. In: Dickinson A, Boakes RA (eds) Mechanism of learning and motivation. Lawrence Erlbaum Associates, Hillsdale, NJ, pp 203–231

Dickinson A, Smith J, Mirenowicz J (2000) Dissociation of Pavlovian and instrumental incentive learning under dopamine antagonists. Behav Neurosci 114:468–83

Ding JB, Guzman JN, Peterson JD, Goldberg JA, Surmeier DJ (2010) Thalamic gating of corticostriatal signaling by cholinergic interneurons. Neuron 67:294–307

Estes WK (1943) Discriminative conditioning I: a discriminative property of conditioned anticipation. J Exp Psychol 32:150–155

Fuchs RA, Branham RK, See RE (2006) Different neural substrates mediate cocaine seeking after abstinence versus extinction training: a critical role for the dorsolateral caudate-putamen. J Neurosci 26:3584–3588

Furlong TM, Jayaweera HK, Balleine BW, Corbit LH (2014) Binge-like consumption of a palatable food accelerates habitual control of behavior and is dependent on activation of the dorsolateral striatum. J Neurosci 34:5012–5022

Ganesan R, Pearce JM (1988) Effect of changing the unconditioned stimulus on appetitive blocking. J Exp Psychol Anim Behav Proc 14(3):280–291

Garbusow M, Schad DJ, Sebold M, Friedel E, Bernhardt N, Koch SP et al (2015) Pavlovian-to-instrumental transfer effects in the nucleus accumbens relate to relapse in alcohol dependence. Addict Biol (Epub ahead of print)

Garbusow M, Schad DJ, Sommer C, Jünger E, Sebold M, Friedel E, Wendt J, Kathmann N, Schlagenhauf F, Zimmermann US, Heinz A, Huys QJ, Rapp MA (2014) Pavlovian-to-instrumental transfer in alcohol dependence: a pilot study. Neuropsychobiology 70:111–121

Geisler S, Marinelli M, Degarmo B, Becker ML, Freiman AJ, Beales M, Meredith GE, Zahm DS (2008) Prominent activation of brainstem and pallidal afferents of the ventral tegmental area by cocaine. Neuropsychopharmacology 33:2688–2700

Gerfen CR, Surmeier DJ (2011) Modulation of striatal projection systems by dopamine. Annu Rev Neurosci 34:441–466

Goldberg JA, Ding JB, Surmeier DJ (2012) Muscarinic modulation of striatal function and circuitry. Handb Exp Pharmacol 208:223–241

Hall J, Parkinson JA, Connor TM, Dickinson A, Everitt BJ (2001) Involvement of the central nucleus of the amygdala and nucleus accumbens core in mediating Pavlovian influences on instrumental behavior. Eur J Neurosci 13:1984–1992

Heimer L, Zahm DS, Churchill L, Kalivas PW, Wohltmann C (1991) Specificity in the projection patterns of acumbal core and shell in the rat. Neuroscience 41:89–125

Heinz A, Siessmeier T, Wrase J, Hermann D, Klein S, Grusser S, Flor H, Braus D, Buchholz HG, Grunder G, Schreckenberger M, Smolka M, Rosch F, Mann K, Bartenstein P (2004)

Correlation between dopamine D2 receptors in the ventral striatum and central processing of alcohol cues and craving. Am J Psychiatry 161:1783–1789

Holland PC (2004) Relations between Pavlovian-instrumental transfer and reinforcer devaluation. J Exp Psychol Anim Behav Process 30:104–117

Holland PC, Gallagher M (2003) Double dissociation of the effects of lesion of basolateral and central amygdala on conditioned stimulus-potentiated feeding and Pavlovian-instrumental transfer. Eur J Neurosci 17:1680–1694

Holland PC, Rescorla RA (1975) The effect of two ways of devaluing the unconditioned stimulus after first- and second-order appetitive conditioning. J Exp Psychol Anim Behav Process 1:355–363

Holmes NM, Marchand AR, Coutureau E (2010) Pavlovian to instrumental transfer: a neurobehavioural perspective. Neurosci Biobehav Rev 34:1277–1295

Kalivas PW, Churchill L, Romanides A (1999) Involvement of the pallidal-thalamocortical circuit in adaptive behavior. Ann N Y Acad Sci 877:64–70

Konorski J (1967) Integrative activity of the brain. University of Chicago Press, Chicago

Kravitz AV, Tye LD, Kreitzer AC (2012) Distinct roles for direct and indirect pathway striatal neurons in reinforcement. Nat Neurosci 15:816–818

Kruse JM, Overmier JB, Konz WA, Rokke E (1983) Pavlovian conditioned stimulus effects upon instrumental choice behavior are reinforcer specific. Learn Motiv 14(2):165–181

Laurent V, Leung B, Maidment N, Balleine BW (2012) Mu- and delta-opioid-related processes in the accumbens core and shell differentially mediate the influence of reward-guided and stimulus-guided decisions on choice. J Neurosci 32:1875–1883

Laurent V, Bertran-Gonzalez J, Chieng BC, Balleine BW (2014) Delta-opioid and dopaminergic processes in accumbens shell modulate the cholinergic control of predictive learning and choice. J Neurosci 34:1358–1369

Laurent V, Wong FL, Balleine BW (2015) δ-opioid receptors in the accumbens shell mediate the influence of both excitatory and inhibitory predictions on choice. Br J Pharmacol 172:562–570

LeBlanc KH, Maidment NT, Ostlund SB (2013) Repeated cocaine exposure facilitates the expression of incentive motivation and induces habitual control in rats. PLoS One 8(4):e61355

Leung BK, Balleine BW (2013) The ventral striato-pallidal pathway mediates the effect of predictive learning on choice between goal-directed actions. J Neurosci 33:13848–13860

Leung BK, Balleine BW (2015) Ventral pallidal projections to mediodorsal thalamus and ventral tegmental area play distinct roles in outcome-specific Pavlovian-instrumental transfer. J Neurosci 35:4953–4964

Lex A, Hauber W (2008) Dopamine D1 and D2 receptors in the nucleus accumbens core and shell mediate Pavlovian-instrumental transfer. Learn Mem. 15:483–491

Lovibond PF (1983) Facilitation of instrumental behavior by a Pavlovian appetitive conditioned stimulus. J Exp Psychol Anim Behav Process 9:225–247

Lu XY, Ghasemzadeh MB, Kalivas PW (1998) Expression of D1 receptor, D2 receptor, substance P and enkephalin messenger RNAs in the neurons projecting from the nucleus accumbens. Neuroscience 82:767–780

Mahler SV, Aston-Jones GS (2012) Fos activation of selective afferents to ventral tegmental area during cue-induced reinstatement of cocaine seeking in rats. J Neurosci 32:13026–13309

Mahler SV, Vazey EM, Beckley JT, Keistler CR, McGlinchey EM, Kaufling J, Wilson SP, Deisseroth K, Woodward JJ, Aston-Jones G (2014) Designer receptors show role for ventral pallidum input to ventral tegmental area in cocaine seeking. Nat Neurosci 17:577–585

Martinovic J, Jones A, Christiansen P, Rose AK, Hogarth L, Field M (2014) Electrophysiological responses to alcohol cues are not associated with Pavlovian-to-instrumental transfer in social drinkers. PLoS One 9(4):e94605

McAlonan GM, Robbins TW, Everitt BJ (1993) Effects of medial dorsal thalamic and ventral pallidal lesions on the acquisition of a conditioned place preference: further evidence for the involvement of the ventral striatopallidal system in reward-related processes. Neuroscience 52:605–620

Mink JW (1996) The basal ganglia: focused selection and inhibition of competing motor programs. Prog Neurobiol 50:381–425
Mogenson GJ, Jones DL, Yim CY (1980) From motivation to action: functional interface between the limbic system and the motor system. Prog Neurobiol 14:69–97
Morgado P, Silva M, Sousa N, Cerqueira JJ (2012) Stress transiently affects Pavlovian-to-instrumental transfer. Front Neurosci 6:93
Mowrer OH (1960) Learning theory and behavior, Wiley, New York
Nelson A, Killcross S (2006) Amphetamine exposure enhances habit formation. J Neurosci 26:3805–3812
Murschall A, Hauber W (2006) Inactivation of the ventral tegmental area abolished the general excitatory influence of Pavlovian cues on instrumental performance. Learn Mem 13:123–126
Ostlund SB, Balleine BW (2007) Instrumental reinstatement depends on sensory- and motivationally-specific features of the instrumental outcome. Learn Behav 35(1):43–52
Ostlund SB, Balleine BW (2008) Differential involvement of the basolateral amygdala and mediodorsal thalamus in instrumental action selection. J Neurosci 28:4398–4405
Ostlund SB, Maidment NT (2012) Dopamine receptor blockade attenuates the general incentive motivational effects of noncontingently delivered rewards and reward-paired cues without affecting their ability to bias action selection. Neuropsychopharmacology 37:508–519
Overmier JB, Lawry JA (1979) Pavlovian conditioning and the mediation of behavior. Psychology of Learning and Motivation 13:1–55
Pavlov IP (1927) Conditioned reflexes. Oxford University Press, Oxford, UK
Parnaudeau S, Taylor K, Bolkan SS, Ward RD, Balsam PD, Kellendonk C (2015) Mediodorsal thalamus hypofunction impairs flexible goal-directed behavior. Biol Psychiatry 77:445–453
Peciña S, Schulkin J, Berridge KC (2006) Nucleus accumbens corticotropin-releasing factor increases cue-triggered motivation for sucrose reward: paradoxical positive incentive effects in stress? BMC Biol 4:8
Peciña S, Berridge KC (2013) Dopamine or opioid stimulation of nucleus accumbens similarly amplify cue-triggered 'wanting' for reward: entire core and medial shell mapped as substrates for PIT enhancement. Eur J Neurosci 37(9):1529–1540
Pielock SM, Lex B, Hauber W (2011) The role of dopamine in the dorsomedial striatum in general and outcome-selective Pavlovian-instrumental transfer. Eur J Neurosci 33:717–725
Pool E, Brosch T, Delplanque S, Sander D (2015) Stress increases cue-triggered "wanting" for sweet reward in humans. J Exp Psychol Anim Learn Cogn 41:128–136
Ray OS, Stein L (1959) Generalization of conditioned suppression. J Exp Anal Behav 2:357–361
Rescorla RA (1994) Transfer of instrumental control mediated by a devalued outcome. Anim Learn Behav 22:27–33
Rescorla RA, LoLordo VM (1965) Inhibition of avoidance behavior. J Comp Physiol Psychol 59:406–412
Rescorla RA, Solomon RL (1967) Two-process learning theory: relationship between Pavlovian conditioning and instrumental learning. Psychol Rev 74:151–182
Saddoris MP, Stamatakis A, Carelli RM (2011) Neural correlates of Pavlovian-to-instrumental transfer in the nucleus accumbens shell are selectively potentiated following cocaine self-administration. Eur J Neurosci 33:2274–2287
Salamone JD, Correa M, Farrar A, Mingote SM (2007) Effort-related functions of nucleus accumbens dopamine and associated forebrain circuits. Psychopharmacology 191:461–482
Scherrer G, Tryoen-Tóth P, Filliol D, Matifas A, Laustriat D, Cao YQ, Basbaum AI, Dierich A, Vonesh JL, Gavériaux-Ruff C, Kieffer BL (2006) Knockin mice expressing fluorescent delta-opioid receptors uncover G protein-coupled receptor dynamics in vivo. Proc Natl Acad Sci 103:9691–9696
Schwabe L, Wolf OT (2010) Socially evaluated cold pressor stress after instrumental learning favors habits over goal-directed action. Psychoneuroendocrinology. 35:977–986
Schultz W (2006) Behavioral theories and the neurophysiology of reward. Annu Rev Psychol 2006(57):87–115

Shiflett MW (2012) The effects of amphetamine exposure on outcome-selective Pavlovian-instrumental transfer in rats. Psychopharmacology 223(3):361–370

Shiflett MW, Balleine BW (2010) At the limbic-motor interface: disconnection of basolateral amygdala from nucleus accumbens core and shell reveals dissociable components of incentive motivation. Eur J Neurosci 32:1735–1743

Smith RJ, Lobo MK, Spencer S, Kalivas PW (2013) Cocaine-induced adaptations in D1 and D2 accumbens projection neurons (a dichotomy not necessarily synonymous with direct and indirect pathways). Curr Opin Neurobiol 23:546–552

Soares-Cunha C, Coimbra B, Borges S, Carvalho MM, Rodrigues AJ, Sousa N (2014) The motivational drive to natural rewards is modulated by prenatal glucocorticoid exposure. Transl Psychiatry. 2014(4):e397

Solomon RL, Turner LH (1962) Discriminative classical conditioning in dogs paralyzed by curare can later control discriminative discriminative avoidance responses in the normal state. Psychol Rev 69:202–219

Stocco A (2012) Acetylcholine-based entropy in response selection: a model of how striatal interneurons modulate exploration, exploitation, and response variability in decision-making. Front Neurosci 6:18

Toates FM (1986) Motivational systems. Cambridge University Press, Cambridge

Trapold MA, Overmier JB (1972) The second learning process in instrumental learning. In: Black AA, Prokasy WF (eds) Classical conditioning II: current research and theory. Appleton-Century-Crofts, New York, pp 427–452

Tripathi A, Prensa L, Mengual E (2013) Axonal branching patterns of ventral pallidal neurons in the rat. Brain Struct Funct 218:1133–1157

Vanderschuren LJ, Di Ciano P, Everitt BJ (2005) Involvement of the dorsal striatum in cue-controlled cocaine seeking. J Neurosci 25:8665–8670

Wassum KM, Ostlund SB, Loewinger GC, Maidment NT (2013) Phasic mesolimbic dopamine release tracks reward seeking during expression of Pavlovian-to-instrumental transfer. Biol Psychiatry 73(8):747–755

Watson P, Wiers RW, Hommel B, de Wit S (2014) Working for food you don't desire cues interfere with goal-directed food-seeking. Appetite 79:139–148

Wiers CE, Stelzel C, Park SQ, Gawron CK, Ludwig VU, Gutwinski S, Heinz A, Lindenmeyer J, Wiers RW, Walter H, Bermpohl F (2014) Neural correlates of alcohol-approach bias in alcohol addiction: the spirit is willing but the flesh is weak for spirits. Neuropsychopharmacology 39:688–697

Winterbauer NE, Balleine BW (2005) Motivational control of second-order conditioning. J Exp Psychol Anim Behav Process 31(3):334–340

Wyvell CL, Berridge KC (2000) Intra-accumbens amphetamine increases the conditioned incentive salience of sucrose reward: enhancement of reward "wanting" without enhanced "liking" or response reinforcement. J Neurosci 20:8122–8130

Wyvell CL, Berridge KC (2001) Incentive sensitization by previous amphetamine exposure: increased cue-triggered "wanting" for sucrose reward. J Neurosci 21:7831–7840

Yin HH, Ostlund SB, Knowlton BJ, Balleine BW (2005) The role of the dorsomedial striatum in instrumental conditioning. Eur J Neurosci 22:513–523

Zahm DS (2000) An integrative neuroanatomical perspective on some subcortical substrates of adaptive responding with emphasis on the nucleus accumbens. Neurosci Biobehav Rev 24:85–105

Zahm DS, Heimer L (1990) Two transpallidal pathways originating in the rat nucleus accumbens. J Comp Neurol 302:437–446

Zener K (1937) The significance of behavior accompanying conditioned salivary secretion for theories of the conditioned response. Am J Psychol 50:384–403

Zhou L, Furuta T, Kaneko T (2003) Chemical organization of projection neurons in the rat accumbens nucleus and olfactory tubercle. Neuroscience 120:783–798

Multiple Systems for the Motivational Control of Behavior and Associated Neural Substrates in Humans

John P. O'Doherty

Abstract In this chapter, we will review evidence about the role of multiple distinct systems in driving the motivation to perform actions in humans. Specifically, we will consider the contribution of goal-directed action selection mechanisms, habitual action selection mechanisms and the influence of Pavlovian predictors on instrumental action selection. We will further evaluate evidence for the contribution of multiple brain areas including ventral frontal and dorsal cortical areas and several distinct parts of the striatum in these processes. Furthermore, we will consider circumstances in which adverse interactions between these systems can result in the decoupling of motivation from incentive valuation and performance.

Keywords Goal-directed · Habitual · Pavlovian · fMRI · Decision-making · Valuation

Contents

1	Introduction	292
2	The Neural Representation of Goal Values	292
3	Outcome Values	293
4	Prechoice Goal-Value Signals	293
5	Common and Distinct Goal Values for Different Goods	294
6	From Goal Values to Actions	296
7	A Role for Dorsal Cortical Areas in Encoding Action–Outcome Probabilities and Effort	297
8	Action Values	297
9	An Alternative Route to Action: Habitual Mechanisms	298
10	Pavlovian Effects on Motivation	300
11	Pavlovian and Habitual Interactions and Paradoxical Motivational Effects for Non-valued Outcomes	302

J.P. O'Doherty (✉)
California Institute of Technology, Pasadena, USA
e-mail: jdoherty@hss.caltech.edu

12	Other Approaches to Demonstrating the Role of the Ventral Striatum in Motivation Beyond PIT	302
13	"Over-Arousal" and Choking Effects on Instrumental Responding	303
14	How Is the Control of These Systems Over Behavior Regulated: The Role of Arbitration	306
15	Translational Implications	307
16	Conclusions	308
References		308

1 Introduction

In this chapter, I will review our current state of understanding of how it is the human brain makes it possible for an individual to be motivated to perform actions. I will define motivation as the vigor with which a particular action is implemented. The motivation to perform a particular action will often scale in proportion to the expected value or utility of the outcome engendered by that action, in that an individual will be more motivated to perform a particular action if it leads to a more valued outcome. In the first part of this chapter, I will detail what is known about how the human brain represents the value of potential goals at the time of decision-making, which could then be used to motivate actions to attain those goals. I will then review evidence about how actions and goal values might get bound together so that when a particular action is being considered, the value of the corresponding goal associated with successful performance of that action can be evaluated. However, while the current value or utility of a goal state is often going to be an important source of motivation, we will also consider that under a number of circumstances the motivation to perform an action might get decoupled from the expected value of its outcome. These situations involve the influence of stimulus-response habits and the effects of Pavlovian cues on action selection. Overall, we will show that there are multiple ways in which the vigor of an action can be modulated in the brain. Some of these are dependent on the current value of the goal, while others are less so. Understanding how these different potential mechanisms interact has the potential to yield novel insights into why on occasion people might be prompted to take actions that may not always be in their best interests.

2 The Neural Representation of Goal Values

There is evidence to suggest from human neuroimaging studies to indicate that ventromedial prefrontal cortex (vmPFC) which includes the ventral aspects of the medial prefrontal cortex and the adjacent orbitofrontal cortex (OFC) is involved in representing the value of goal states at several stages in the action selection process (O'Doherty 2007, 2011). First of all, this region is involved in encoding the value of the goal outcome as it is experienced, once the action has been performed.

Secondly, this region is involved in encoding the value of prospective goals at the point of decision-making. Thirdly, the region is involved in encoding the value of the goal that is ultimately chosen.

3 Outcome Values

There is now an extensive literature implicating the vmPFC in responding to goal outcomes as they are attained following the performance of an action, particularly involving the receipt of monetary rewards (see, e.g., O'Doherty et al. 2001; see also Knutson et al. 2001; Smith et al. 2010). This region is also known to be involved in responding to other types of reward outcomes, irrespective of whether an action was performed to attain the outcome or not. For example, regions of medial and central OFC are known to be correlated with the subjective value of food and other odor outcomes as they are presented, while participants remain passive in the scanner (de Araujo et al. 2003b; Rolls et al. 2003). Other kinds of rewarding stimuli such as attractive faces also recruit the medial OFC and adjacent medial prefrontal cortex (O'Doherty et al. 2003). OFC representations to outcomes are also strongly influenced by changes in underlying motivational states. Activity decreases in this region to food or odor or even water outcomes as motivational states change from hungry or thirsty to satiated, in a manner that parallels changes in the subjective pleasantness experienced to the stimulus (O'Doherty et al. 2000; Small et al. 2001; de Araujo et al. 2003a; Kringelbach et al. 2003). Not only can such representations be modulated as a function of changes in internal motivational state, but value-related activity in this region can also be influenced by cognitive factors such as the provision of price information or merely the use of semantic information or even semantic labels (de Araujo et al. 2005; Plassmann et al. 2008). Thus, the online computation of outcome value in the OFC is highly flexible and can be directly influenced by a variety of internal and external factors. Overall, these findings which are highly reproducible and consistent across studies (Bartra et al. 2013) implicate the ventromedial prefrontal cortex as a whole in representing the value of experienced outcomes.

4 Prechoice Goal-Value Signals

In order to select an action in a goal-directed manner, a goal-directed agent needs to be able to retrieve a representation of the goal associated with each possible available action at the time of decision-making. A number of experiments have examined the representation of goal values at the time of decision-making. Plassmann et al. (2007) used a procedure from behavioral economics to assay goal-value representations. In this paradigm, hungry human participants are scanned while being presented with pictures denoting a variety of foods while indicating their "willingness-to-pay" (WTP) for each of the food items, out of an initial endowment of four dollars available for each item. After the experiment was over, one of the trials is selected at random, and if the reported WTP exceeds a random draw from a lottery, then subjects are provided with the good and invited to consume it (and their endowment is drawn on); otherwise, they keep the endowment and do not receive the good. This procedure is designed to ensure that the participants give their true underlying valuation for each of the items. Activity in a region of vmPFC was found to be correlated with trial-by-trial variations in WTP, suggesting a role for this region in encoding the goal value of the potential outcome. A follow-up experiment compared and contrasted goal-value representations for appetitive food goods that participants would pay to obtain as well as aversive food goods that participants would pay to avoid (Plassmann et al. 2010). The same region of ventromedial prefrontal cortex was found to correlate with goal-value representations for both appetitive and aversive food items, with activity increasing in this region in proportion to the value of goods with positive goal values, and with activity decreasing in proportion to the value of goods with negative goal values. These findings suggest that goal-value codes are represented in vmPFC on a single scale ranging from negative to positive value, indicating that positive and negative goals are encoded using a common coding mechanism.

5 Common and Distinct Goal Values for Different Goods

The finding of a role for a region of ventromedial prefrontal cortex in encoding goal values for food items then invites the question of whether the same region is also involved in the value of other categories of goods. Chib et al. (2009) used the same WTP paradigm to identify areas of brain activation correlating with the goal value of three distinct categories of good: food items, non-food consumer items (such as DVDs and clothing) and monetary gambles. An overlapping region of vmPFC (just above the orbital surface) was found to be correlated with the value of all three classes of items, suggesting that goal-value codes for many different categories of goods may all converge within the same region of vmPFC. Such a region would, therefore, be an excellent candidate for mediating coding of the utilities assigned to diverse types of goal stimuli (Fig. 1a). The finding of an overlapping representation for the goal value of different classes of goods raises the question of whether the value of these goods are being coded in a common currency in which the value of diverse goods are represented using a common scale. The existence of a common currency would facilitate decisions to be made between very different classes of goods that otherwise would not be comparable. For example, a common currency would enable decisions to be made between the prospects of going to the movies versus opting to go for a steak dinner. Levy and Glimcher (2011) found evidence consistent with the existence of a common currency by giving people explicit choices between different types of goods, specifically money versus food. Activity in vmPFC was found to scale similarly for the value of food items and monetary

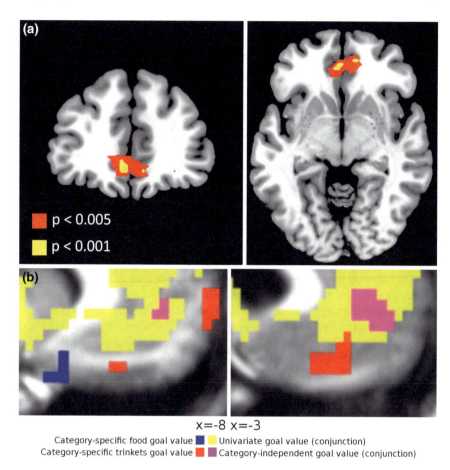

Fig. 1 Role of the ventromedial prefrontal cortex in encoding the value of a goal at the time of decision-making. **a** Region of ventromedial prefrontal cortex correlating with the value of 3 different categories of goal stimuli: monetary gambles, food items and non-consumable consumer items. From Chib et al. (2009). **b** Region of medial prefrontal cortex exhibiting category independent coding of goal values is shown in *purple*, while a food-category-specific goal-value signal is shown in *blue* and a consumer-item category specific non-food value is shown in *red*. From McNamee et al. (2013)

items across participants, suggesting that these items were being represented on the same scale whereby the subjective value of a food item was associated with a similar level of activation for a comparable subjectively valued monetary item.

Another issue about the encoding of a common currency is whether or not the same distributed representations within the area of vmPFC found to show overlapping activations are elicited for the value of different goods. It could be the case that while the same brain region is activated for the value of different goods, these overlapping activations at the group level belie entirely separate (or largely non-overlapping) representations of value at the voxel level, or alternatively the same voxels might provide further evidence that the value of these goods is being encoded in a common currency. To address this question, McNamee et al. (2013) used a very similar paradigm to that deployed by Chib et al. in which participants made decisions about how much to pay to obtain food, monetary and non-food consumer items. These authors trained multivariate pattern classifiers (which are statistical tools that can detect patterns in the fMRI data and relate those patterns to specific perceptual, cognitive or behavioral states) to detect distributed patterns of voxels corresponding to the encoding of high versus low goal values for each category of good separately. To find out whether the distributed value codes generalized across category, the authors tested whether the classifier trained on one class of good (say food items) could decode the value of another class of good (say money or consumer items). In a very circumscribed region of vmPFC above the orbital surface, evidence for a very general value code was found in which a classifier trained on one category could successfully decode the value of a good from another category (Fig. 1b).

Taken together, these findings provide strong support for the existence of a common currency in this region in which the value code for a given good scales in a manner proportional to its subjective value irrespective of the category from which it is drawn. Intriguingly, in addition to finding evidence for a common value representation in vmPFC, McNamee et al. also found evidence for the existence of value codes that are more selective to particular categories of good along the medial orbital surface. In posterior medial OFC, a region was found to have a distributed encoding of the value of a food items, but not of other categories of items, while a more anterior region of mOFC was found to encode the value of non-food consumer items but not food or monetary goods (Fig. 1b). The apparent category specificity of coding along the medial orbital surface could suggest that these regions represent a precursor stage to the encoding of a common currency in which the value of particular item categories is encoded in a manner specific to that category of good before being combined together to make a common currency in more dorsal parts of vmPFC. Another possibility (which is not incompatible with the above suggestion) is that these signals could correspond to the point at which the value of individual items is computed from its underlying sensory features in the first place. Interestingly, no region was found to uniquely encode the distributed value of monetary items, but instead the value of monetary goods was only represented within the vmPFC area found to encode general value signals. This might be because money is a generalized reinforcer that can be exchanged for many different types of goods.

6 From Goal Values to Actions

Ultimately in order to attain a goal within the goal-directed system, it is necessary to establish which actions can be taken in order to attain a particular goal. In order to establish which action to take, it is necessary to compute an **action value** for each available action in a given state, which encodes the overall expected utility that would follow from taking each action. At the point of decision-making, a goal-directed agent can then compare and contrast available action values in order to choose to pursue the action with the highest expected utility. How can an action value be computed? Clearly, it is necessary to integrate the current incentive value of the goal state as discussed above, with the probability of attaining the goal if performing a given action, while discounting the action value by the expected effort cost that will ensue from performing the same action.

7 A Role for Dorsal Cortical Areas in Encoding Action–Outcome Probabilities and Effort

There is now emerging evidence to suggest that information about action–outcome probability and action effort is encoded not in the ventromedial prefrontal cortex and adjacent orbitofrontal cortex that we previously described as being important in the computation of goal values and outcome values, but instead these action-related variables appear to be represented in dorsal areas of the cortex, ranging from the posterior parietal cortex all the way to dorsolateral and dorsomedial prefrontal cortex. With regard to action–outcome probabilities, Liljeholm et al. (2011) reported that a region of inferior parietal lobule is involved in computing the action contingency, which is the probability of obtaining an outcome if an action is performed less than the probability of obtaining an outcome if the action is not performed. Liljeholm et al. (2013) extended this work by showing that inferior parietal lobule appears more generally to be involved in encoding the divergence in the probability distributions of outcomes over available actions, which is suggested to be valence independent (i.e. not modulated as a function of differences in the expected value across outcomes) representation of the extent to which different actions lead to different distributions of outcomes.

There is now emerging evidence to suggest that the effort associated with performing an action is represented in parts of the dorsomedial prefrontal cortex alongside other areas such as insular cortex (Prévost et al. 2010). Taken together, these findings indicate that much of the key information required to compute an overall action value is represented in dorsal parts of cortex.

8 Action Values

A number of studies have also examined the representation of prechoice action values that could be used at the point of decision-making. Studies in rodents and monkeys examining single-neuron responses have found candidate action-value signals to be encoded in the dorsal parts of the striatum, as well as in dorsal cortical areas of the brain including parietal and supplementary motor cortices (Platt and

Glimcher 1999; Samejima et al. 2005; Lau and Glimcher 2007; Sohn and Lee 2007). Consistent with the animal studies, human fMRI studies have also found evidence that putative action-value signals are present in dorsal cortical areas, including the supplementary motor cortex, as well as lateral parietal and dorsolateral cortex (Wunderlich et al. 2009; Hare et al. 2011; Morris et al. 2014). However, few studies to date have examined the integration of all of the variables needed to compute an overall action value, incorporating action effort costs, goal values and contingencies. In monkeys, Hosokawa et al. (2013) found that some neurons in the anterior cingulate cortex are involved in encoding an integrated value signal that summed over expected costs and benefits for an action. Hunt et al. (2014) also reported a region of dorsomedial prefrontal cortex to be involved in encoding integrated action values. A pretty consistent finding in both humans and animals is that the ventromedial prefrontal and adjacent orbitofrontal cortices appear to not be involved in computing integrated action values or in encoding action effort costs and that instead as indicated earlier, this region is more concerned with representing the value of potential outcomes or goals and not the value of actions required to attain those goals. Thus, the tentative evidence to date points to the possibility that integration of goals with action probabilities and costs occurs in the dorsal cortex and parts of the dorsal striatum and not in anterior ventral parts of the cortex. This supports the possibility that goal-directed valuation involves an interaction between multiple brain systems, and that goal-value representations in the vmPFC are ultimately integrated with action information in dorsal cortical regions in order to compute an overall action value (Fig. 2).

9 An Alternative Route to Action: Habitual Mechanisms

So far we have considered the role that regions of the cortex play in computing the value of goal-directed actions. When an individual is behaving in a goal-directed manner, the vigor with which a particular action is performed will be strongly influenced by the overall value of that action. However, it has long been known that there exists another mechanism for controlling actions in the mammalian brain: habits. In contrast to goal-directed actions that are sensitive to the current value of an associated goal, habitual actions are not sensitive to current goal values, but instead are selected on the basis of the past history of reinforcement of that action. Evidence for the existence of habits as being distinct from goal-directed actions was first uncovered in rodents using a devaluation manipulation in which the (food) goal associated with a particular action is rendered no longer desirable to the animal by feeding it to satiety on that food or pairing the food with illness (Dickinson et al. 1983; Balleine and Dickinson 1998). After being exposed to modest amounts of experience on a particular action–outcome relationship, animals exhibit evidence of being goal-directed in that they flexibly respond with lower response rates on the action associated with the now devalued outcome. However, after extensive experience with a particular action, animals exhibit evidence of being insensitive to

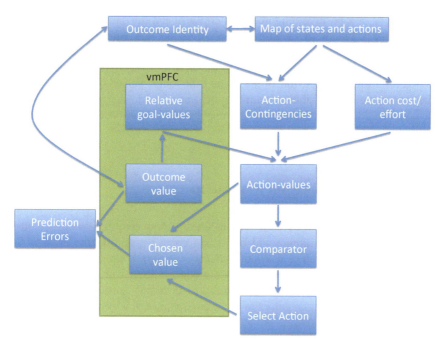

Fig. 2 Illustration of some of the key component processes involved in goal-directed decision-making. An associative map of stimuli and actions bound together by transition probabilities, together with a representation of the outcome stimuli, and forms a representation of a world "model." Within this model representation through learned associations, stimuli or actions can elicit a representation of the associated outcome that in turn retrieves a representation of the incentive value for that outcome (goal). This is done over all-known available outcomes, enabling a relative goal-value signal to be computed, taking into account the range of possible outcomes. Action contingencies are then calculated for a given outcome using the forward model machinery, which, when combined with the relative goal-value signal and effort cost, constitute an action value. This signal is in turn fed into a decision comparator that enables selection of a particular action, which is then fed to the motor output system as well as enabling the chosen value to be elicited. The vmPFC is suggested to be involved in encoding a number of these signals (illustrated by the *green area*), particularly those pertaining to encoding the value of goal outcomes. Yet, the process of computing action values and ultimately generating a choice is suggested to occur outside of this structure, in dorsal cortical areas as discussed in the text

the value of the associated goal, indicating that behavior eventually transitions to habitual control. Similar experience-dependent effects have been found to occur in humans (Tricomi et al. 2009). There is now converging evidence to implicate a circuit involving the posterior putamen and premotor cortex in habitual actions in humans (Tricomi et al. 2009; Wunderlich et al. 2012; Lee et al. 2014; de Wit et al. 2012; McNamee et al. 2013), which suggests a considerable degree of overlap with the circuits identified as being involved in habitual control in the rodent brain (Balleine and O'Doherty, 2010). Associative learning theories propose that habits depend on the formation of stimulus-response associations (Dickinson 1985). In these theories, the stimulus-response association gets stamped in or strengthened by the delivery of a positive reinforcer, but critically and in contrast to goal-directed actions, the value of the reinforcer itself does not get embedded in with the stimulus-response association. Thus, in contrast to goal-directed actions, when habits control behavior, it is possible for an individual to vigorously perform an action even when the outcome of the action is no longer valued. As a consequence, the invocation of the habitual control system can lead to apparently paradoxical behavior in which animals appear to compulsively pursue outcomes that are not currently valued.

10 Pavlovian Effects on Motivation

In addition to the goal-directed and habitual action control systems, another important associative learning and behavioral control system in the brain is the Pavlovian system (Dayan et al. 2006; Balleine et al. 2008). Unlike its instrumental counterparts, the Pavlovian system does not concern itself with learning to select actions in order to increase the probability of obtaining particular rewards, but instead exploits statistical regularities in the environment in which particular stimuli provide information about the probability of particular appetitive or aversive outcomes occurring. The Pavlovian system then initiates reflexive actions such as approach or withdrawal as well as preparatory and consummatory responses both skeletomotor and physiological that have come to be selected over an evolutionary timescale as adaptive responses to particular classes of outcome. In associative learning terms, the Pavlovian system can be considered to be concerned with learning associations between stimuli and outcomes based on the contingent relationship between those stimuli and the subsequent delivery of a particular outcome. There is now a large body of evidence in animals to implicate specific neural circuits in Pavlovian learning and behavioral expression, including the amygdala, ventral striatum and parts of the orbitofrontal cortex (Parkinson et al. 1999; LeDoux 2003; Ostlund and Balleine 2007). Furthermore in humans, similar brain systems have also been consistently implicated as playing a role in Pavlovian learning and control in both reward-related and aversive contexts (LaBar et al. 1998; Gottfried et al. 2002, 2003).

Of particular interest to the present discussion is the fact that Pavlovian associations and their effects on behavior do not occur in isolation, but instead appear to interact with instrumental actions in interesting and important ways. Pavlovian to instrumental transfer (or PIT) is the term typically given to these types of interactions in associative learning theory (Lovibond 1983). PIT effects are typically manifested as a modulation in instrumental response rates in the presence of a stimulus that has by virtue of a learned Pavlovian association acquired a predictive relationship to the subsequent delivery of a particular outcome. In the presence of a stimulus that has predicted the subsequent delivery of a reward, the vigor of responding on an instrumental action for a reward can be increased. Thus, in the appetitive domain, Pavlovian stimuli can exert energizing effects, appearing to result in an increased motivation to respond on an instrumental action. A distinction has been made between general Pavlovian to instrumental transfer in which a Pavlovian cue associated with even an unrelated reward can elicit increased responding on an instrumental action, and specific transfer effects in which a Pavlovian cue is associated with the same specific outcome as that engendered by the instrumental action on which the agent is responding. In general PIT, an appetitive Pavlovian stimulus can result in increased responding on an instrumental action irrespective of the identity of the outcome signaled by the Pavlovian stimulus, whereas in specific PIT, instrumental responding is specifically enhanced on an instrumental action associated with the same outcome predicted by the Pavlovian stimulus.

Lesion studies in animals have implicated specific regions of the amygdala and ventral striatum in general and specific transfer effects. The basolateral amygdala and the shell of the nucleus accumbens have been implicated in specific PIT, while the centromedial complex of the amygdala and the core of the accumbens have been implicated in general PIT (Hall et al. 2001; Corbit and Balleine 2005, 2011). Studies in humans have similarly found partly distinct circuits to play a role in these different types of PIT (Fig. 3). General PIT effects may depend on the nucleus accumbens

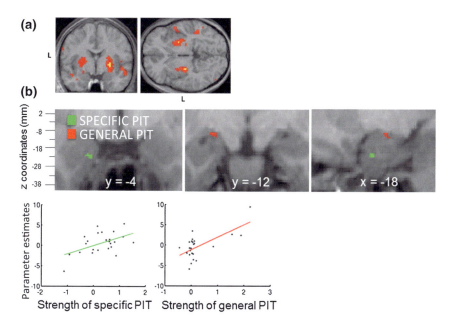

Fig. 3 Regions of the human striatum and amygdala contributing to Pavlovian to instrumental interactions. **a** Region of the ventrolateral putamen involved in specific PIT effects in humans (data from Bray et al. 2008). **b** Distinct regions of human amygdala contributing to specific and general PIT effects in humans. Basolateral amygdala is involved in specific PIT while centromedial parts of the amygdala contribute to general PIT effects. The graphs at the bottom show the relationship between the strength of the specific and general PIT effects in the behavior of each individual participant in the experiment and the degree of activation (parameter estimates) in the basolateral and centromedial amygdala, respectively, in each participant. Data from Prevost et al. (2012)

proper in humans as well as the centromedial amygdala, whereas specific PIT effects appear to involve the basolateral amygdala and ventrolateral parts of the putamen in humans (Bray et al. 2008; Talmi et al. 2008; Prevost et al. 2012).

11 Pavlovian and Habitual Interactions and Paradoxical Motivational Effects for Non-valued Outcomes

While in many cases, the presence of a Pavlovian cue will energize responding and may constructively facilitate the performance of instrumental actions to obtain rewards, Pavlovian effects on instrumental responding can also promote maladaptive behavior whereby responding persists under situations in which the goal of an instrumental action is no longer valued by an organism. This is because specific PIT effects appear to be immune to reinforcer devaluation effects (Holland 2004). That is, Pavlovian cues continue to exert an energizing effect on instrumental actions even under situations where the outcome predicted by the Pavlovian cue has been devalued. This effect initially shown in rodents has also recently been demonstrated in humans (Watson et al. 2014; though see Allman et al. 2010), further demonstrating that not only do specific PIT effects persist after outcome devaluation, but so do general PIT effects. One interpretation of these results is that PIT effects may serve to selectively engage the habitual system, increasing the engagement of a devaluation insensitive habitual action at the expense of its goal-directed counterpart. Thus, Pavlovian cues can through a putative interaction with the habitual system promote the motivation to respond to obtain an outcome that is not considered valuable by the organism.

12 Other Approaches to Demonstrating the Role of the Ventral Striatum in Motivation Beyond PIT

A number of fMRI studies have implicated the ventral striatum in mediating the effects of incentives on the motivation to perform an action beyond the PIT paradigm described above. For instance, Pessiglione et al. (2007) used a paradigm in which the amount of physical effort exerted for performing an action correlated with the amount of incentive provided. While activity in the motor cortex scaled with the magnitude of force exerted, activity in the ventral striatum was found to be correlated with the magnitude of the potential incentive available and the strength of the correlation in the activity patterns between the ventral striatum and motor cortex was found to be related to the translation from incentive to motor action, in that the greater the coupling of activity between these two areas the greater the effects of incentive amounts on the effort exerted. In follow-up work, Schmidt et al. (2012) also found activity in ventral striatum scaled with incentive motivation not only during performance of a motor task but also during performance of a cognitively demanding task, and that increases in connectivity between the ventral striatum and brain areas involved in motor control versus cognition were differentially engaged as a function of which task the participant was performing on a given trial, suggesting that ventral striatum may be a common node for facilitating the transfer of incentives to motivations for task performance irrespective of the nature of the task being performed.

Miller et al. (2014) used a task in which trials varied both in the difficulty of responding successfully to obtain an outcome and the magnitude of the outcome was varied for performing a speeded response. An interesting property of this task is that the speed of responding (a proxy for the motivation to respond) was actually greater on more difficult trials even though this type of trial was associated with a lower probability of reward. While activity in the nucleus accumbens scaled with the expected value, activity in putamen and caudate within the striatum scaled instead with difficulty, perhaps suggesting a role for those regions in representing the level of motivation to respond (which increased with greater difficulty). Thus, the above-mentioned studies do largely converge with the PIT paradigms mentioned earlier in implicating parts of the striatum (particularly its ventral aspects) in motivational processes. These studies could be taken as evidence for a more general role for this structure in motivation beyond Pavlovian to instrumental interactions. Or, alternatively (as favored by the present author), the results of these studies could also be considered to reflect a manifestation of Pavlovian to instrumental interactions whereby the monetary incentive cues are acting as Pavlovian cues which in turn influence instrumental motor performance.

13 "Over-Arousal" and Choking Effects on Instrumental Responding

The classical relationship between incentives, motivation and instrumental responding is that the provision of higher incentives results in increased motivation to respond which results in the more vigorous and ultimately more successful performance of an action. However, under certain circumstances, increased incentives can result in the counter-intuitive effect of resulting in less efficacious instrumental performance, an effect known in the cognitive psychology and behavioral economics literature as "choking under pressure." In a stark demonstration of this effect, Ariely et al. (2009) provided participants in rural India with the prospect of winning very large monetary amounts relative to their average monthly salaries, contingent on successful performance on a range of motor and cognitive tasks. Compared to a group offered smaller incentive amounts, the performance of the high incentive group was markedly reduced, suggesting the paradoxical effect of reduced performance under a situation where the motivation to respond is likely to be very high.

Mobbs et al. (2009) examined the neural correlates of this effect in humans, in which participants could sometimes obtain a relatively large incentive ($\sim$$10) for successfully completing a reward-related pursuit task and on other occasions obtain a smaller incentive ($\sim$$1) for completing the same task. Performance on the task decreased in the high incentive condition indicative of choking, and this was also associated with an increase of activity in the dopaminergic midbrain. Relatedly, Aarts et al. (2014) measured individual differences in dopamine synthesis in the striatum using FMT PET and demonstrated that performance decrements in response to monetary incentives on a simple cognitive task was associated with the degree of dopamine synthesis capacity in the striatum. Another study by Chib et al. (2012) also explored the paradoxical relationship between incentives and performance and the role of dopaminoceptive striatal circuits. In the study by Chib et al., participants were offered the prospect of different incentives ranging from $5 to $100 in order to complete a complex motor skill (moving a ball attached to a virtual spring to a target location). Whereas small to moderate incentives resulted in improved performance, once incentives became too large, performance started to decrease, consistent with a choking effect. Activity in the ventral striatum was directly associated with the propensity for choking to manifest (Fig. 4). Whereas at the time in the trial when the available incentive was signaled to the participant but before the motor response itself was triggered, ventral striatum activity correlated positively with the magnitude of the incentive presented, at the time that the motor response itself was being implemented by the participant, activity in the ventral striatum changed markedly and instead of correlating positively with the incentive amount, began to correlate negatively with the amount of incentive available, i.e. activity decreased in the ventral striatum in proportion to the incentive amount. Notably, the slope of the decrease in the ventral striatum during motor task performance was correlated with individual differences in the decrease of susceptibility to the behavioral choking effect, while no such correlation was found between ventral striatal activity and choking effects at the time of initial cue presentation. Given the role of ventral striatum in Pavlovian learning and in the influence of Pavlovian cues on instrumental responding discussed earlier, one interpretation of these findings is that Pavlovian skeletomotor reflexes mediated by the striatum may interfere with the production of a skilled motor response. Further supporting this interpretation was the finding that individual differences in loss aversion (the extent to which an individual is prepared to avoid a loss relative to obtaining a gain of a similar amount) were also correlated both with ventral striatal activity and behavioral manifestations of choking. This led Chib et al. to suggest that the engagement of specifically aversive Pavlovian avoidance-related responses could be responsible for mediating behavioral choking. According to this perspective, adverse interactions between the Pavlovian system and instrumental actions are responsible for producing behavioral choking effects. In a subsequent follow-up study, Chib et al. (2014) demonstrated that the behavioral choking effect was strongly dependent on the framing of the incentive structure in terms of losses and gains and that the correlation with behavioral loss aversion was more nuanced than previously suspected, as it interacted in a complex way with the extrinsic framing manipulation:

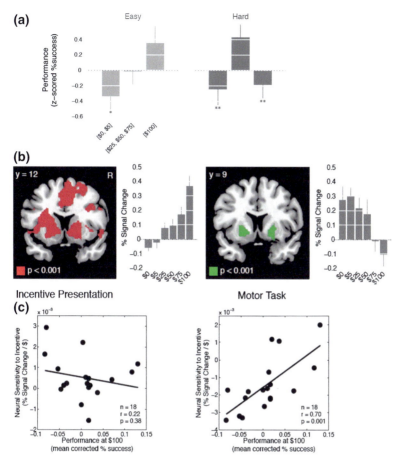

Fig. 4 Relationship between ventral striatum activity responses to incentives and susceptibility to behavioral choking effects for large incentives. **a** When performing a skilled motor task in a hard (60 % performance level) compared to an easy (80 % performance level) condition, large incentives ($100) resulted in an increased susceptibility to performance decrements. **b** Activity in the ventral striatum correlated positively with the incentive amount during the initial trial onset when the incentive available for successfully performing the task was indicated, but during performance of the motor task itself activity instead flipped and correlated negatively with incentive amount. **c** The degree of the activity decrease as a function of incentive during the motor task was correlated with the decrement in behavioral performance across participants: those participants who showed greater performance decrements for large incentives showed a greater deactivation effect for large incentives in the ventral striatum during motor task performance but not during the time of the initial incentive presentation. Data from Chib et al. (2012)

whereas individual with high loss aversion showed increased choking in response to high incentives when those incentives were initially framed as a gain as found in the initial Chib et al. study, individuals with low loss aversion paradoxically showed increased susceptibility to choking when the initial outcomes were framed as a loss. These puzzling behavioral results can potentially be explained by an account in which individuals shift their internal frame of reference when moving from the initial incentive phase to the phase when they are performing the motor task: if initially the incentive is framed as a gain, then participants start focusing on the prospect of losing that potential gain when performing the task, whereas if the initial incentive is framed as a potential loss, then the individual focuses on the prospective of avoiding that loss (and hence incurring a relative gain) during the motor task. However, in the ventral striatum, a relatively more straightforward relationship was observed between incentives and performance which is that irrespective of the extrinsic framing manipulation, activity in the ventral striatum decreased as a function of increased incentives during the motor task, and this activity correlated directly with performance decrements for high incentives whether potential gains or to be avoided losses. Importantly, connectivity between the ventral striatum and premotor cortex was significantly decreased on trials in which choking effects were manifested. These results, when taken together, suggest that large incentives whether prospective gains or losses can result in increased choking effects. Thus, in the context of a Pavlovian to instrumental interaction account for this phenomenon, it appears unlikely that this effect is mediated exclusively by aversive Pavlovian to instrumental interactions, but instead may result from the effects of increased arousal generated by either aversive or appetitive Pavlovian predictions, resonating with an "over-arousal" account advocated by Mobbs et al. and others (Broadhurst 1959; Mobbs et al. 2009).

14 How Is the Control of These Systems Over Behavior Regulated: The Role of Arbitration

Given these different systems, all appear to be exerting effects on instrumental motivation, and a natural question that arises is what factors influence which of these systems is going to influence behavior at any one moment in time. With respect to goal-directed and habitual control, an influential hypothesis is that there exists an arbitrator that deputizes the influence of these systems over behavior based on a number of criteria. The first of these is the relative precision or accuracy of the estimates about which action should be selected within the two systems. All else being equal, behavior should be controlled by the system with the most accurate prediction (Daw et al. 2005). Using a computational framework and a computational account for goal-directed and habitual control described in detail elsewhere (Daw et al. 2005), Lee et al. (2014) found evidence for the existence of an arbitrator which allocates the relative amount of control over goal-directed and habitual systems as a function of which system is predicted to have the most reliable (a proxy for precision) estimates. Specifically, Lee et al. (2014) implicated the ventrolateral prefrontal cortex bilaterally as well as a region of right frontopolar cortex in this function. In addition to precision, other variables are also likely to be important for the arbitration between these strategies, including the amount of cognitive effort that needs to be exerted. While goal-directed actions may require considerable cognitive effort to implement, habitual actions require much less effort, and therefore, there is likely a trade-off between the amount of cognitive effort needed by the two systems and the relative precision of their estimates (FitzGerald et al. 2014). Subsequent work will need to follow on precisely which variables are involved in the arbitration process between goal-directed and habitual action selection mechanisms.

In addition, much less is known about how arbitration occurs when determining the degree of influence that a Pavlovian predictor exerts on behavior. It is known that changes in cognitive strategies or appraisal implemented via prefrontal cortex can influence the degree of Pavlovian expression of both aversive and appetitive Pavlovian conditioned responses, which are suggested to be manifested via down-regulatory effects on the amygdala and ventral striatum (Delgado et al. 2008a, b; Staudinger et al. 2009). For example, Delgado et al. (2008a) had participants either attend to cues that were associated with a subsequent reward by focusing on the reward expected, or to "regulate" their processing of such cues (in essence by distracting themselves and thinking of something other than the imminent reward), when regulating, activity was increased in lateral prefrontal cortex, while reward-related activity in the ventral striatum was modulated as a function of the regulatory strategy. Thus, some type of "top-down" control mechanism clearly is being implemented via the prefrontal cortex on the expression of Pavlovian behaviors. However, the nature of the computations mediating this putative arbitration process is not well understood. Also, unknown is how the Pavlovian system interacts with the habitual and goal-directed system to influence instrumental motivation. As suggested earlier, the devaluation insensitive nature of PIT effects support a specific role for interactions with the habitual system in governing the effects of Pavlovian stimuli on motivation. However, direct neural evidence for this suggestion is currently lacking, and a mechanistic account for how this interaction occurs is not yet in place.

15 Translational Implications

The body of research reviewed in this chapter has the potential not only to shed light on the nature of human motivation, but also has clear and obvious applications toward improving understanding of the nature of the dysfunctions in motivation and behavioral control that can emerge in psychiatric, neurological and other disorders. For example, some theories of addiction have emphasized the possibility that drugs of abuse could hijack the habitual system in particular, resulting in the stamping in of very strong drug-taking habits (Robbins and Everitt 1999), although it is likely that a complete understanding of the brain mechanisms of addiction will need to take into account effects of drugs of abuse on goal-directed and Pavlovian systems too (see chapter in this volume by Robinson et al.). Relatedly, psychiatric disorders involving compulsive behavioral symptoms such as obsessive compulsive disorder may also be associated with dysregulation in habitual control, or in the arbitration of goals and habits (Gillan and Robbins 2014; Voon et al. 2014; Gruner et al. 2015). These examples are only the tip of the iceberg in terms of the applicability of the framework reviewed in this chapter to clinical questions. The basic research summarized in this chapter has influenced the emergence of a new movement in translational research that has come to be known as "Computational Psychiatry," in which formal computational models of brain processes are used to aid classification and fundamental understanding of psychiatric disorders (see Montague et al. 2012; Maia and Frank 2011 for reviews).

16 Conclusions

Here, we have considered the role of multiple systems implemented in at least partly distinct neural circuits in governing the motivation to respond on an instrumental action: the goal-directed action selection system, the habitual action selection system and the Pavlovian system. Under many and perhaps most circumstances, the vigor and effectiveness of an instrumental action will scale appropriately in proportion to the underlying incentive value of the goal of the action. However, we have also seen that a number of "bugs" or quirks can also emerge as a result of the interactions between these systems such that the overall motivation to respond on an instrumental action or the effectiveness of the action energized by such motivational processes can sometimes end up being inappropriately mis-matched to the incentive value of the associated outcome.

Ultimately in order to fully understand when and how human actions are selected effectively or ineffectively as a function of incentives it will be necessary to gain a more refined understanding of how these distinct action control systems interact at the neural level, and in particular to gain better insights into the nature of the neural computations underlying such interactions.

References

Aarts E, Wallace DL, Dang LC, Jagust WJ, Cools R, D'Esposito M (2014) Dopamine and the cognitive downside of a promised bonus. Psychol Sci. doi:10.1177/0956797613517240

Allman MJ, DeLeon IG, Cataldo MF, Holland PC, Johnson AW (2010) Learning processes affecting human decision making: an assessment of reinforcer-selective pavlovian-to-instrumental transfer following reinforcer devaluation. J Exp Psychol Anim Behav Process 36:402

Ariely D, Gneezy U, Loewenstein G, Mazar N (2009) Large stakes and big mistakes. Rev Eco Stud 76:451–469

Balleine BW, Dickinson A (1998) Goal-directed instrumental action: contingency and incentive learning and their cortical substrates. Neuropharmacology 37:407–419

Balleine BW, O'Doherty JP (2010) Human and rodent homologies in action control: corticostriatal determinants of goal-directed and habitual action. Neuropsychopharmacology 35:48–69

Balleine BW, Daw ND, O'Doherty JP (2008) Multiple forms of value learning and the function of dopamine. In: Glimcher PW, Camerer C, Fehr E, Poldrack RA (eds) Neuroeconomics: decision making and the brain. Elsevier, New York, pp 367–385

Bartra O, McGuire JT, Kable JW (2013) The valuation system: a coordinate-based meta-analysis of BOLD fMRI experiments examining neural correlates of subjective value. Neuroimage 76:412–427

Bray S, Rangel A, Shimojo S, Balleine B, O'Doherty JP (2008) The neural mechanisms underlying the influence of pavlovian cues on human decision making. J Neurosci 28:5861–5866

Broadhurst P (1959) The interaction of task difficulty and motivation: the Yerkes-Dodson Law revived. Acta Psychol 16:321–338

Chib VS, Rangel A, Shimojo S, O'oherty JP (2009) Evidence for a common representation of decision values for dissimilar goods in human ventromedial prefrontal cortex. J Neurosci 29:12315–12320

Chib VS, De Martino B, Shimojo S, O'Doherty JP (2012) Neural mechanisms underlying paradoxical performance for monetary incentives are driven by loss aversion. Neuron 74:582–594

Chib VS, Shimojo S, O'Doherty JP (2014) The effects of incentive framing on performance decrements for large monetary outcomes: behavioral and neural mechanisms. J Neurosci 34:14833–14844

Corbit LH, Balleine BW (2005) Double dissociation of basolateral and central amygdala lesions on the general and outcome-specific forms of pavlovian-instrumental transfer. J Neurosci 25:962–970

Corbit LH, Balleine BW (2011) The general and outcome-specific forms of Pavlovian-instrumental transfer are differentially mediated by the nucleus accumbens core and shell. J Neurosci 31:11786–11794

Daw ND, Niv Y, Dayan P (2005) Uncertainty-based competition between prefrontal and dorsolateral striatal systems for behavioral control. Nat Neurosci 8:1704–1711

Dayan P, Niv Y, Seymour B, Daw ND (2006) The misbehavior of value and the discipline of the will. Neural Netw 19:1153–1160

de Araujo IE, Kringelbach ML, Rolls ET, McGlone F (2003a) Human cortical responses to water in the mouth, and the effects of thirst. J Neurophysiol 90:1865–1876

de Araujo IE, Rolls ET, Kringelbach ML, McGlone F, Phillips N (2003b) Taste-olfactory convergence, and the representation of the pleasantness of flavour, in the human brain. Eur J Neurosci 18:2059–2068

de Araujo IE, Rolls ET, Velazco MI, Margot C, Cayeux I (2005) Cognitive modulation of olfactory processing. Neuron 46:671–679

de Wit S, Watson P, Harsay HA, Cohen MX, van de Vijver I, Ridderinkhof KR (2012) Corticostriatal connectivity underlies individual differences in the balance between habitual and goal-directed action control. J Neurosci 32:12066–12075

Delgado MR, Gillis MM, Phelps EA (2008a) Regulating the expectation of reward via cognitive strategies. Nat Neurosci 11:880–881

Delgado MR, Nearing KI, LeDoux JE, Phelps EA (2008b) Neural circuitry underlying the regulation of conditioned fear and its relation to extinction. Neuron 59:829–838

Dickinson A (1985) Actions and habits: the development of a behavioural autonomy. Philos Trans R Soc Lond Ser B Biol Sci 308:67–78

Dickinson A, Nicholas D, Adams CD (1983) The effect of the instrumental training contingency on susceptibility to reinforcer devaluation. Q J Exp Psychol 35:35–51

FitzGerald TH, Dolan RJ, Friston KJ (2014) Model averaging, optimal inference, and habit formation. Front Hum Neurosci 8:457

Gillan CM, Robbins TW (2014) Goal-directed learning and obsessive–compulsive disorder. Philos Trans R Soc B Biol Sci 369:20130475

Gottfried JA, O'Doherty J, Dolan RJ (2002) Appetitive and aversive olfactory learning in humans studied using event-related functional magnetic resonance imaging. J Neurosci 22: 10829–10837

Gottfried JA, O'Doherty J, Dolan RJ (2003) Encoding predictive reward value in human amygdala and orbitofrontal cortex. Science 301:1104–1107

Gruner P, Anticevic A, Lee D, Pittenger C (2015) Arbitration between action strategies in obsessive-compulsive disorder. Neuroscientist, 1073858414568317

Hall J, Parkinson JA, Connor TM, Dickinson A, Everitt BJ (2001) Involvement of the central nucleus of the amygdala and nucleus accumbens core in mediating Pavlovian influences on instrumental behaviour. Eur J Neurosci 13:1984–1992

Hare TA, Schultz W, Camerer CF, O'Doherty JP, Rangel A (2011) Transformation of stimulus value signals into motor commands during simple choice. Proc Natl Acad Sci 108: 18120–18125

Holland PC (2004) Relations between Pavlovian-instrumental transfer and reinforcer devaluation. J Exp Psychol Anim Behav Process 30:104

Hosokawa T, Kennerley SW, Sloan J, Wallis JD (2013) Single-neuron mechanisms underlying cost-benefit analysis in frontal cortex. J Neurosci 33:17385–17397

Hunt LT, Dolan RJ, Behrens TE (2014) Hierarchical competitions subserving multi-attribute choice. Nat Neurosci 17:1613–1622

Knutson B, Fong GW, Adams CM, Varner JL, Hommer D (2001) Dissociation of reward anticipation and outcome with event-related fMRI. NeuroReport 12:3683–3687

Kringelbach ML, O'Doherty J, Rolls ET, Andrews C (2003) Activation of the human orbitofrontal cortex to a liquid food stimulus is correlated with its subjective pleasantness. Cereb Cortex 13:1064–1071

LaBar KS, Gatenby JC, Gore JC, LeDoux JE, Phelps EA (1998) Human amygdala activation during conditioned fear acquisition and extinction: a mixed-trial fMRI study. Neuron 20: 937–945

Lau B, Glimcher PW (2007) Action and outcome encoding in the primate caudate nucleus. J Neurosci 27:14502–14514

LeDoux J (2003) The emotional brain, fear, and the amygdala. Cell Mol Neurobiol 23:727–738

Lee SW, Shimojo S, O'Doherty JP (2014) Neural computations underlying arbitration between model-based and model-free learning. Neuron 81:687–699

Levy DJ, Glimcher PW (2011) Comparing apples and oranges: using reward-specific and reward-general subjective value representation in the brain. J Neurosci 31:14693–14707

Liljeholm M, Tricomi E, O'Doherty JP, Balleine BW (2011) Neural correlates of instrumental contingency learning: differential effects of action-reward conjunction and disjunction. J Neurosci 31:2474–2480

Liljeholm M, Wang S, Zhang J, O'Doherty JP (2013) Neural correlates of the divergence of instrumental probability distributions. J Neurosci 33:12519–12527

Lovibond PF (1983) Facilitation of instrumental behavior by a Pavlovian appetitive conditioned stimulus. J Exp Psychol Anim Behav Process 9:225

Maia TV, Frank MJ (2011) From reinforcement learning models to psychiatric and neurological disorders. Nat Neurosci 14:154–162

McNamee D, Rangel A, O'Doherty JP (2013) Category-dependent and category-independent goal-value codes in human ventromedial prefrontal cortex. Nat Neurosci 16:479–485

Miller EM, Shankar MU, Knutson B, McClure SM (2014) Dissociating motivation from reward in human striatal activity. J Cogn Neurosci 26:1075–1084

Mobbs D, Hassabis D, Seymour B, Marchant JL, Weiskopf N, Dolan RJ, Frith CD (2009) Choking on the money reward-based performance decrements are associated with midbrain activity. Psychol Sci 20:955–962

Montague PR, Dolan RJ, Friston KJ, Dayan P (2012) Computational psychiatry. Trends Cogn Sci 16:72–80

Morris RW, Dezfouli A, Griffiths KR, Balleine BW (2014) Action-value comparisons in the dorsolateral prefrontal cortex control choice between goal-directed actions. Nat Commun 5

O'Doherty JP (2007) Lights, camembert, action! The role of human orbitofrontal cortex in encoding stimuli, rewards, and choices. Ann N Y Acad Sci 1121:254–272

O'Doherty JP (2011) Contributions of the ventromedial prefrontal cortex to goal-directed action selection. Ann N Y Acad Sci 1239:118–129

O'Doherty J, Kringelbach ML, Rolls ET, Hornak J, Andrews C (2001) Abstract reward and punishment representations in the human orbitofrontal cortex. Nat Neurosci 4(1):95–102

O'Doherty J, Rolls ET, Francis S, Bowtell R, McGlone F, Kobal G, Renner B, Ahne G (2000) Sensory-specific satiety-related olfactory activation of the human orbitofrontal cortex. NeuroReport 11:893–897

O'Doherty J, Winston J, Critchley H, Perrett D, Burt DM, Dolan RJ (2003) Beauty in a smile: the role of medial orbitofrontal cortex in facial attractiveness. Neuropsychologia 41:147–155

Ostlund SB, Balleine BW (2007) Orbitofrontal cortex mediates outcome encoding in Pavlovian but not instrumental conditioning. J Neurosci 27:4819–4825

Parkinson JA, Olmstead MC, Burns LH, Robbins TW, Everitt BJ (1999) Dissociation in effects of lesions of the nucleus accumbens core and shell on appetitive pavlovian approach behavior and the potentiation of conditioned reinforcement and locomotor activity byd-amphetamine. J Neurosci 19:2401–2411

Pessiglione M, Schmidt L, Draganski B, Kalisch R, Lau H, Dolan RJ, Frith CD (2007) How the brain translates money into force: a neuroimaging study of subliminal motivation. Science 316:904–906

Plassmann H, O'Doherty J, Rangel A (2007) Orbitofrontal cortex encodes willingness to pay in everyday economic transactions. J Neurosci 27:9984–9988

Plassmann H, O'Doherty J, Shiv B, Rangel A (2008) Marketing actions can modulate neural representations of experienced pleasantness. Proc Natl Acad Sci U S A 105:1050–1054

Plassmann H, O'Doherty JP, Rangel A (2010) Appetitive and aversive goal values are encoded in the medial orbitofrontal cortex at the time of decision making. J Neurosci 30:10799–10808

Platt ML, Glimcher PW (1999) Neural correlates of decision variables in parietal cortex. Nature 400:233–238

Prevost C, Liljeholm M, Tyszka JM, O'Doherty JP (2012) Neural correlates of specific and general Pavlovian-to-Instrumental Transfer within human amygdalar subregions: a high-resolution fMRI study. J Neurosci 32:8383–8390

Prévost C, Pessiglione M, Météreau E, Cléry-Melin M-L, Dreher J-C (2010) Separate valuation subsystems for delay and effort decision costs. J Neurosci 30:14080–14090

Robbins TW, Everitt BJ (1999) Drug addiction: bad habits add up. Nature 398:567–570

Rolls ET, Kringelbach ML, De Araujo IE (2003) Different representations of pleasant and unpleasant odours in the human brain. Eur J Neurosci 18:695–703

Samejima K, Ueda Y, Doya K, Kimura M (2005) Representation of action-specific reward values in the striatum. Science 310:1337–1340

Schmidt L, Lebreton M, Cléry-Melin M-L, Daunizeau J, Pessiglione M (2012) Neural mechanisms underlying motivation of mental versus physical effort. PLoS Biol 10:e1001266

Small DM, Zatorre RJ, Dagher A, Evans AC, Jones-Gotman M (2001) Changes in brain activity related to eating chocolate: from pleasure to aversion. Brain 124:1720–1733

Smith DV, Hayden BY, Truong T-K, Song AW, Platt ML, Huettel SA (2010) Distinct value signals in anterior and posterior ventromedial prefrontal cortex. J Neurosci 30:2490–2495

Sohn JW, Lee D (2007) Order-dependent modulation of directional signals in the supplementary and presupplementary motor areas. J Neurosci 27:13655–13666

Staudinger MR, Erk S, Abler B, Walter H (2009) Cognitive reappraisal modulates expected value and prediction error encoding in the ventral striatum. Neuroimage 47:713–721

Talmi D, Seymour B, Dayan P, Dolan RJ (2008) Human Pavlovian-instrumental transfer. J Neurosci 28:360–368

Tricomi E, Balleine BW, O'Doherty JP (2009) A specific role for posterior dorsolateral striatum in human habit learning. Eur J Neurosci 29:2225–2232

Voon V, Derbyshire K, Rück C, Irvine M, Worbe Y, Enander J, Schreiber L, Gillan C, Fineberg N, Sahakian B (2014) Disorders of compulsivity: a common bias towards learning habits. Mol Psychiatry 20:345–352

Watson P, Wiers R, Hommel B, de Wit S (2014) Working for food you don't desire. Cues interfere with goal-directed food-seeking. Appetite 79:139–148

Wunderlich K, Rangel A, O'Doherty JP (2009) Neural computations underlying action-based decision making in the human brain. Proc Natl Acad Sci U S A 106:17199–17204

Wunderlich K, Dayan P, Dolan RJ (2012) Mapping value based planning and extensively trained choice in the human brain. Nat Neurosci 15:786–791

The Computational Complexity of Valuation and Motivational Forces in Decision-Making Processes

A. David Redish, Nathan W. Schultheiss and Evan C. Carter

Abstract The concept of value is fundamental to most theories of motivation and decision making. However, value has to be measured experimentally. Different methods of measuring value produce incompatible valuation hierarchies. Taking the agent's perspective (rather than the experimenter's), we interpret the different valuation measurement methods as accessing different decision-making systems and show how these different systems depend on different information processing algorithms. This identifies the translation from these multiple decision-making systems into a single action taken by a given agent as one of the most important open questions in decision making today. We conclude by looking at how these different valuation measures accessing different decision-making systems can be used to understand and treat decision dysfunction such as in addiction.

Keywords Neuroeconomonics · Valuation · Multiple Decision Theory · Decision-Making

Contents

1	What is Value?	314
	1.1 Measuring Value	315
2	Taking the Subject's Point of View	316
	2.1 Information Processing in Decision-Making Systems	316
3	Testing the Theory	322

A.D. Redish (✉) · N.W. Schultheiss
Department of Neuroscience, University of Minnesota, Minneapolis, USA
e-mail: redish@umn.edu

N.W. Schultheiss
e-mail: nschulth@umn.edu

E.C. Carter
Department of Ecology, Evolution, and Behavior, University of Minnesota,
St. Paul, USA
e-mail: evan.c.carter@gmail.com

© Springer International Publishing Switzerland 2015
Curr Topics Behav Neurosci (2016) 27: 313–334
DOI 10.1007/7854_2015_375
Published Online: 17 May 2015

4 Summary and Implications	326
5 Is Value Still a Valuable Hypothetical Construct?	328
References	328

Value-based decision-making processes are integral to adaptive behavior, and accordingly, the concept of value is ubiquitous across decision-making studies from diverse perspectives, including studies of neuroeconomic choice (Glimcher et al. 2008; Kable and Glimcher 2009), optimal foraging theory (Stephens and Krebs 1987), reinforcement learning (Sutton and Barto 1998), and deliberative decision making (Rangel et al. 2008; Rangel and Hare 2010). Operational definitions of value within each of these fields often point to "value" as a common currency that can be applied to objects, actions, and experiences (Kable and Glimcher 2009; Levy and Glimcher 2012). Theoretically, using value as a common currency may ease an agent's difficulty making choices between very different options. However, value and motivation are hypothetical constructs (MacCorquodale and Meehl 1954), and, as such, cannot be directly measured. Instead, they must be inferred from interpretations of observations. This review will start from the observation that measuring value in different ways can lead to incompatible orderings of valuation—how you measure value changes what things are valued more than others. Valuation is not trans-situational; instead, it is context-dependent.

1 What is Value?

The concept of value can be defined as a quantification of a resource in terms of its costs and benefits, as well as the subjective desire or preference for some quantity of one resource over another. Typically, value is considered to be a derivative property of the relationship between an agent and a given object of desire (Glimcher et al. 2008). Because both human and non-human animals make decisions using similar systems dependent on similar neural structures (Redish 2013), we draw evidence from both human and non-human decision-making studies to reach our conclusions. Among all animals, the needs of the individual may modulate the value of candidate actions in the world, as hunger predisposes an agent for consumable rewards, but even though it may be context-dependent, the valuation step is typically taken as a metric of external features of the world in the present condition of the agent (Glimcher et al. 2008, but see Niv et al. 2006b).

Fundamentally, the value of an object to an agent is a multidimensional entity. We rarely choose where to eat dinner based solely on the nutritional content of a particular dish, but rather we integrate the convenience and expense of different restaurants or our own kitchen, the taste or style of foods at local restaurants or the ingredients in our cupboards, the speed of service at different restaurants or the preparation time of foods, our mood, restaurant atmospheres, and other social factors, as well as what we may have had for lunch today or dinner yesterday. Potential costs, as in these examples, are determined by temporal, energetic, and resource constraints; while benefits range from social and hedonic experiences to energetic and health-related issues. Despite the numerous factors that may contribute to the valuations of different restaurants and our kitchen at home, we ultimately choose where to have dinner based on some integration of these factors. Current theories suggest that this process entails value as a means of reaching a common currency to make the choice (Glimcher 2003; Levy and Glimcher 2012; Wunderlich et al. 2012). Taking that idea to its logical conclusion suggests that the choice itself should be understood as the measure of the valuation process (Samuelson 1937; Redish 2013).

1.1 Measuring Value

This observation suggests that we need to take value as revealed through the actions of an agent. Different computational processes may well underlie different decision processes and different effective motivational measures of valuation (Daw et al. 2005; Niv et al. 2006a, b; Rangel et al. 2008; Redish et al. 2008; Montague et al. 2012; van der Meer et al. 2012). There are at least three simple ways of measuring value—*revealed preference*, *willingness to pay*, and by *approach/avoidance*. Interestingly, these three measures can produce different orderings of the same choices.

Revealed preferences. The simplest means of measuring whether something is more valued than another is to provide the two (or more) options as a choice and see which one is selected. Preferentially selecting option A over option B implies that value(A) > value(B). Logically, this simple conceptualization implies that value should be transitive [value(A) > value(B) and value(B) > value(C) implies that value (A) > value(C)], at least within the bounds of noise. However, as we will see below, the algorithm used when calculating value in a revealed preferences situation is more complex and the transitive property does not necessarily apply.

Willingness to pay. One can directly measure the value of a single thing by asking how much effort or sacrifice one is willing to make to get that thing. Typically, this is measured by requiring a physical effort to obtain an object (such as lever presses to obtain food or drug) or by demanding a sacrifice (such as asking how much money one would be willing to pay for a given item, such as a car). Measuring how much one is willing to pay for each of several options should lead to an ordering of preferences—logically, if one is willing to pay more for option A than one is willing to pay for option B, then one would expect A to be preferred to B, but this is not always the case (Lichtenstein and Slovic 2006; Ahmed 2010; Perry et al. 2013). Interestingly, as we will see below, the algorithm used when selecting actions that lead to paying effort to achieve a goal does not apparently apply to selling that goal back (Kahneman et al. 1991).

Humans, with their explicitly accessible linguistic abilities, can report their willingness to pay directly. (You can ask them.) Linguistically reported willingness-to-pay can diverge from a behaviorally observed willingness to pay (Kahneman and Tversky 2000; Lichtenstein and Slovic 2006). For example, many drug users will deny their willingness to pay high costs for drugs, but will, when faced with the drug, pay that high cost (Goldstein 2000).

Approach/avoidance. Even simpler than measuring the willingness to pay or the revelation of preferences, one can measure whether an agent will approach or avoid the option in question. This can be taken as a binary value applied to an option—approach it (positive) or avoid it (negative), but one can also measure the speed and vigor with which one approaches the option, which can provide a quantitative measure of that positive or negative component. Interestingly, the behaviors that agents do when in simple approach or avoidance tasks tend to reflect species-specific behaviors (such as a rat gnawing on a handle that predicts food reward) (Breland and Breland 1961; Dayan et al. 2006; Rangel et al. 2008; Redish 2013).

Interactions and modulations. It is, of course, possible to construct experiments in which these measures interact with each other—for example, if one measures simple approach to a reward and then places a shock before it, one is measuring the willingness to pay that shocks for that reward. Higher shock levels will (of course) lead to less likelihood of taking an option. One can also measure motivation in the modulations of these measures of value, such as the fact that a reminder of a potential reward in one context changes the revealed preferences toward that outcome (a phenomenon known as "Pavlovian-to-instrumental transfer" or PIT, Kruse et al. 1983; Corbit and Janak 2007; Talmi et al. 2008). As another example, it is possible to change the dimensions on which options are compared in a revealed preference task by guiding attention (Plous 1993; Gilovich et al. 2002; Hill 2008), by changing one's emotional state (Dutton and Aron 1974; Andrade and Ariely 2009), or by making one of the options more concrete (Peters and Büchel 2010; Benoit et al. 2011).

2 Taking the Subject's Point of View

The problem with the logic laid out in the start of this chapter is that it is derived from the experimenter's point of view—it assumes that each of the three experimental paradigms measures something explicitly different. (This gets particularly complicated when we start to look at interactions and modulations.) Rather than working backwards from the behavioral experiments to hypothesized constructs, let us take the subject's point of view to first consider how each of these means of measuring value reflects aspects of the agent's decision-making process and then work forward from current taxonomies of the decision-making systems.

2.1 Information Processing in Decision-Making Systems

Decisions arise from information processing applied to a combination of (1) information about the current world (processed inferences from perception), (2) past experience (memory, history), and (3) goals and motivations (valuation). Although it is hypothetically possible that all decisions arise from a single fundamental algorithm applied to these three inputs (such as the maximization of subjective value, Samuelson 1937; Glimcher et al. 2008; Kable and Glimcher 2009), this is not necessarily true. Computational analyses of information processing that include calculation time, memory storage and access requirements, and the willingness to generalize suggest that different algorithms will be optimal in different situations (O'Keefe and Nadel 1978; Nadel 1994; Daw et al. 2005; Niv et al. 2006b; Rangel et al. 2008; Redish et al. 2008; van der Meer et al. 2012; Montague et al. 2012; Redish 2013).

Current analyses of the information processing that occurs within decision making suggest a taxonomy of three different processes[1]—*deliberation* between imagined options, *procedural action chains*, and *Pavlovian action-selection systems* (Rangel et al. 2008; Redish et al. 2008; Montague et al. 2012; van der Meer et al. 2012; Redish 2013). As a first-order description, the three ways of measuring value tap into each of these three systems. However, the information processing that goes into each of these three systems implies a complexity of valuation that will need to be addressed. Additionally, the three systems can interact to lead to interesting situational dependencies of valuation.

2.1.1 Algorithms of Revealed Preference (Deliberation)

Computationally, deliberation entails the imagination and evaluation (comparison) of future outcomes, a process related to episodic future thinking. Episodic future thinking is the ability to imagine oneself into a specific potential future (Atance and O'Neill 2001; Buckner and Carroll 2007). Humans with hippocampal lesions do not create fully integrated episodic futures (Hassabis et al. 2007), and creating an integrated imagined episodic future activates prefrontal cortex and hippocampus as revealed by fMRI signals (Hassabis and Maguire 2011; Schacter et al. 2008; Schacter and Addis 2011). In rats, the primary evidence for episodic future thinking lies in decoding sequences of firing in hippocampal place cells, which have been revealed to fire in a sequence representing a serial path to the next goal (Pfeiffer and Foster 2013; Wikenheiser and Redish 2015). At choice points, rats will sometimes pause (Tolman 1932) and the hippocampus will represent the sequences to potential goals serially (i.e., episodically) (Johnson and Redish 2007).

Once the representation of that future is created, it must be evaluated. In rats, this evaluation process depends on the ventral striatum or nucleus accumbens (Smith et al. 2009; van der Meer and Redish 2011; Jones et al. 2012). In humans, this

---

[1]Actually, we have argued elsewhere (Redish 2013) that decision making needs to be defined by action selection and that one should thus include reflexes as well in our taxonomy of decision-making systems. However, it is unlikely that the valuation and motivation mechanisms discussed here influence the information processing that goes on in reflexes, and we will leave the Reflex system out of our analyses here.

evaluation process is thought to include the orbitofrontal and ventromedial prefrontal cortex (O'Doherty et al. 2001; O'Doherty 2004; Coricelli et al. 2005; Hare et al. 2011; Winecoff et al. 2013); however, these interpretations have depended on fMRI signals, which do not have the resolution to determine whether these signals occur early enough to be actually involved in the decision process itself. In rats, fine time-scale analyses have found that orbitofrontal representations do not represent future goals until after a decision has been reached (Steiner and Redish 2012; Stott and Redish 2014). It remains unknown whether the human cortical components are also only active post-decision, or whether this difference is a species-specific difference. Species-specific differences can occur through how the computation is performed (humans tend to be more cortically dependent than rats, Streidter 2005), or they can occur because the tasks being used are different (spatial vs. non-spatial), or they can occur due to anatomical differences in the source of the signal (e.g., medial vs. lateral orbitofrontal cortex).

Nevertheless, the basic computation underlying deliberation is clear: To choose between options A and B, specific potential futures incorporating each option are imagined, evaluated, and compared. This computation requires sufficient understanding of the world to search forward to those imagined outcomes. (Thus, this decision process is often termed "model-based," as it uses the model of the world to create a forward/imagined outcome.) Deliberation allows for fast learning because it is inherently flexible (knowing that option A will lead to consequence α does not require one to take option A), but it is also slow and computationally expensive because one needs to infer consequence α from one's knowledge of the world.

At this time, the selection process that underlies deliberation remains unknown —Is it a test-and-evaluate system in which hypotheses are generated serially and the first one that is good enough is chosen or is there direct comparison between options?[2] Importantly, the inconsistency between revealed-preference and willingness-to-pay measures of value suggests that deliberation depends on a more direct comparison between options. A serial test-and-evaluate system should produce similar measures whether one or two options are offered, which would lead to similar valuations between the revealed-preference and willingness-to-pay measures, but, as noted at the top of this chapter, these valuation measures often reveal incompatible valuation orderings.

Humans, other primates, and rats have all been observed to change their choices dependent on the set of available choices—the set of choices available changes the options chosen, sometimes incompatibly (such as in the classic case of extremeness aversion, where humans tend to select the middle option, Simonson and Tversky 1992, see Gallistel 1990; Plous 1993; Tremblay and Schultz 1999; Padoa-Schioppa 2009; Rangel and Clithero 2012). Incompatible choices is a case of *intransitivity*, in which choice A is chosen over C, C over B, and B over A. Neurophysiological

---

[2]Work under stressful situations (such as fire ground commanders) suggest a serial test-and-evaluate system (Klein 1999), but it is unclear whether these experienced commanders are using deliberative or procedural mechanisms.

recordings have found that neural representations of value in monkeys are transitive within a block, but not between blocks, a process known as "renormalization" (Tremblay and Schultz 1999; Padoa-Schioppa 2009). One process that can create this effect is for the value to be *normalized* within block (i.e., value is divided by a function of the average or maximum value available within that block). Neural representations are known to be dependent on excitatory–inhibitory networks that show content-addressable properties where inhibitory networks enforce limitations in the total activation of the excitatory cells (Hertz et al. 1991). Renormalization would be an obvious consequence of these excitatory–inhibitory networks.

Algorithmically, the fact that attention to specific features changes the selected option in a multivalued comparison (Plous 1993; Hill 2008; Hare et al. 2011) further suggests that deliberation entails a direct comparison between options. This would, of course, make the deliberative process non-transitive. In deliberation, a specific future is imagined and options within that future are compared (Johnson et al. 2007; Schacter et al. 2008). Imagination, like great art, consists of painting a few specific strokes—thus, the imagined comparison of *A* and *B* might focus on one dimension that the two options share, while the comparison of *B* and *C* may focus on another dimension, and the comparison of *A* and *C* on yet a third.

Further support for the idea that the selection process in deliberation is an actual comparison comes from the fact that concrete futures that are easier to imagine (Trope and Liberman 2003; Schacter et al. 2008) are preferred (Peters and Büchel 2010; Benoit et al. 2011). Consistent with this comparison hypothesis, working memory is related to deliberative abilities—agents with better working memory abilities are more likely to deliberate when given the chance (Burks et al. 2009; Bickel et al. 2011), consider more options (Franco-Watkins et al. 2006), and look further into the future (Bickel et al. 2011). These effects are a direct prediction of the search–evaluate–compare model of deliberative decision making (Kurth-Nelson and Redish 2012).

2.1.2 Algorithms of Willingness to Pay (Procedural)

Deliberation is a slow and laborious process; if you have to act quickly, or if the situation is not changing, then it would be more efficacious and more efficient to cache the best action, so it can be directly recalled. Algorithmically, this process entails a combination of recognition (categorization) processes and associated action chains (Klein 1999; Redish et al. 2007; Dezfouli and Balleine 2012; Redish 2013).

The recognition process is a form of categorization and parameterization of the world. Anatomically, this categorization can be seen in cortical signals that integrate information to identify the most likely situation (Yang and Shadlen 2007; Redish et al. 2007). Once the parameterized situation has been identified, an action chain can be released through learned mechanisms (Jog et al. 1999; Dezfouli and Balleine 2012; Smith and Graybiel 2013).

In much of the literature, the procedural system is identified as "model-free" because it does not require a model of the transitions that can occur within the world

(Daw et al. 2005; Gläscher et al. 2010; Lee et al. 2014), but this is a misnomer. The recognition component of the procedural system requires development of a schema that defines the parameters of the task (Charness 1991; Klein 1999; Redish et al. 2007; Redish 2013; Schmidhuber 2014). An incorrect parameterization will prevent learning and severely reduce the efficacy of the actions selected. As noted by Klein (1999), these learned schemas are a form of expertise, which requires an implicit model of the structure of the world.

Economically, the question asked by the procedural system is fundamentally different from that asked by the deliberative system—the deliberative system asks *Which choice is better?*, while the procedural system asks *How sure am I that this is the right action at this time? How expensive is it?*.

As with the deliberative system, attention can modulate procedural actions, by identifying the specific cues and parameters that define a situation (Klein 1999; Redish et al. 2007). And motivational components can guide attention to specific aspects of a given situation.

2.1.3 Algorithms of Approach and Avoidance (Pavlovian)

Computationally, the third decision-making system entails stored situation-recognition and associative processes that release species-specific behaviors (see Rangel et al. 2008 and Redish 2013, for review).[3] For example, Pavlov's dogs learned to salivate on hearing the bell using this system, but could not have learned to apply an arbitrary action without using one of the other systems. Anatomically, this system depends on associations made within the amygdala (LeDoux 2000; Janak and Tye 2015) driving species-specific behaviors in the periaqueductal gray (Bandler and Shipley 1994; McNally et al. 2011), as well as simple approach/avoidance behaviors involving the shell of the accumbens (Flagel et al. 2009; Laurent et al. 2012; Robinson et al. 2014). The specific survival circuits that underlie individual species-specific behaviors are likely to access different neural substrates (LeDoux 2012), but the general Pavlovian learning system can be seen as a valuation system in its own right, depending on different learning and valuation algorithms than deliberative or procedural processes (Rangel et al. 2008; Redish 2013).

As with the deliberative and procedural processes, Pavlovian action-selection systems reveal an underlying valuation process, because response strength is modulated by the magnitude of the available reward and by the needs of the animal. Classic species-specific behaviors include approach and avoidance, freezing or

---

[3]We have chosen to call this system the "Pavlovian" system because it is what Pavlov's dogs were doing (an association between the bell and the food led to the prewired species-specific salivation behavior on hearing the bell, Pavlov 1927), but it should not be confused with classical definitions of Pavlovian learning based on the experimenter-defined task parameters (that an animal does not need to act in order to receive reward or punishment, Bouton 2007), nor should it be confused with the recent definitions of state [situation] versus state-action reinforcement-learning algorithms (Dayan et al. 2006; Cavanagh et al. 2013).

grooming, fighting or fleeing, eating or not, etc. When these behaviors are put in conflict, they can create a value hierarchy, approaching the palatable food, avoiding the unpalatable, fleeing from a larger opponent, fighting a smaller one. Such experimentally-imposed behavioral conflicts are representative of ubiquitous approach-avoidance conflicts in nature as an animal navigates in pursuit of good things and avoiding threats. Since an animal cannot simultaneously approach two good things or avoid two bad things in opposite directions, valuation (relative valuations) may be a necessary part of natural approach-avoidance decisions, and the needs or learning of the animal may modulate these valuations.

Importantly, this value hierarchy is only revealed in situations that activate the Pavlovian action-selection system, which is based on immediately available sensory cues (such as the sound of Pavlov's bell or the smell of baking bread). This increased valuation of immediate cues can modulate other behaviors (providing, for example, a preference for immediately available, concrete options[4]).

2.1.4 The Role of Motivation

An important (and open) question is whether there is a separate motivation system that modulates all three of these decision-making systems (deliberative, procedural, and Pavlovian) or whether each of the systems has their own specific modulation system.

Most economic theories suggest that there is a separate valuation system that is called upon by action-selection systems. This separate system is usually identified with the orbitofrontal and ventromedial prefrontal cortices and the nucleus accumbens, modulated by dopamine and other neuromodulator signals (Doya 2000; O'Doherty 2004; Dayan and Niv 2008; van der Meer et al. 2012; Winecoff et al. 2013). Obviously, some (presumably hypothalamic) systems need to identify the intrinsic needs of the animal (such as indicated by hunger or thirst), and there is evidence that this motivational system is sensitive to cues (pictures of pizza make us hungry) and learning (such as in conditioned taste-aversion, which is why hospitals provide strangely flavored foods to chemotherapy patients, Bernstein 1978, 1999). The complexity of how and when motivational factors are learned or can occur in response to visceral sensations of homeostatic changes is beyond the scope of this chapter, but can be found in other chapters in this volume, including Waltz and Gold, and Woods and Begg.

---

[4]Tempting as it is to try to use this concrete modulation as an explanation for discounting phenomena (in which more temporally proximal options are preferred to equivalently valuable temporally distant options), the data suggest that all three systems have discounting effects within them that interact to produce the delay discounting phenomenon.

2.1.5 The Macro-Agent

Although these decision-making systems entail different computational processes, in the end, there is a single agent that needs to take the action. An open question in the field of decision making is how conflicts between these different decision-making systems are resolved. It is not clear yet whether there is a separate executive that selects between systems or whether there is a mechanism by which the components directly compete for behavioral expression, for example, by intrinsic components within each system that make it more or less likely to be "listenable to" by downstream motor areas.

Most theories have suggested that the decision of which subsystem is allowed to drive behavior depends on an external valuation system that takes the most-valued option from each of the components (Levy and Glimcher 2012; Wunderlich et al. 2012, but see van der Meer et al. 2012), but the inconsistency of value under the different experimental paradigms laid out at the start of this chapter belies this hypothesis. More likely, some other parameter is being used to decide between systems. Suggestions have included expected calculation time (Simon 1955; Gigerenzer and Goldstein 1996; Keramati et al. 2011), or an internal representation of reliability or uncertainty in the valuation calculation (Daw et al. 2005). Interestingly, reliability in valuation could be represented by the intrinsic self-consistency of the value representation, which can vary due to the distributed nature of representation in neural systems. Because neural representations are distributed, the activity of individual neurons within a representation can agree with each other about what is specifically represented or they can disagree with each other. It is possible to quantitatively measure the self-consistency of those representations, and it is possible for downstream structures to use that self-consistency to control the influence of a representation on the activity of that downstream structure (Jackson and Redish 2003; Johnson et al. 2008).

However the process is resolved, it is clear that these different algorithmic processes are called upon (win-out) under different experimental conditions. We can use this to explain the inconsistencies in the valuation function—a simple willingness-to-pay experiment can be solved by any of the three systems, but providing a choice forces the agent to deliberate, and making a choice concrete tends to access Pavlovian action-selection systems. Similarly, the act of perception and conscious reporting in a linguistic version of the willingness-to-pay task may drive the decision process into more deliberative cognitive processes that can change the valuation.

3 Testing the Theory

This multiple decision-making system explanation of valuation and the observed effects of motivation can explain a number of different phenomena that have been identified over the years.

What is motivated behavior? The fact that these decision-making systems can create action plans that are in conflict and that the action plans take time to execute implies that there will be internal states of motivation in which an action is desired but not necessarily released yet. We can refer to this state as the "urge" to act.

Particularly intriguing in this light is dysfunction of motivated behaviors such as might occur in Tourette's syndrome (Kurlan 1993; Leckman and Riddle 2000) or Obsessive-Compulsive Disorder (Goodman et al. 2000). Contrary to popular belief, Tourette's and OCD do not manifest as dystonic actions that are released before the subject is aware of them, but rather they manifest as "urges" which can be suppressed (Kathmann et al. 2005; Maia and McClelland 2012). Eventually, the effort expended (presumably by one of the other systems) to suppress the urge becomes too much, the urge becomes overwhelming, and the macro-agent releases the motivated action.

Intriguingly, most of the dysfunctional behaviors seen in Tourette's syndrome are Pavlovian in the sense used here (cursing[5], facial expressions, etc.), while the control and suppression of them is effortful, conscious, cognitive, and depends on limited cognitive resources, which suggests a more deliberative process. However, some have found treatment in their Tourette's tics by channeling them into rhythmic actions and action chains (a presumably procedural process) (Sacks 1985). Whether the motor tics of Tourette's and other motoric dysfunctions with identifiable urges is a dysfunction in Pavlovian, procedural, or some motor component remains unknown.

Sign-tracking and goal-tracking. Another classic example of conflict between decision-making systems is that in a task in which a cue signals reward delivery, but the animal does not need to do anything at the cue to receive the reward,[6] some animals approach the cue (sign trackers), while others go directly to the reward (goal trackers) (Flagel et al. 2009). Sign-trackers are presumably using Pavlovian action-selection systems to approach things that the system has identified as valuable, while goal trackers are using the cue to identify the appropriate action (using either deliberative or procedural systems). A direct prediction of this would be that there should be specific neurophysiological differences between sign trackers and goal trackers, particularly in neural structures associated with valuation in the Pavlovian action-selection systems. Indeed, such differences exist—sign-trackers show dopamine shifts to the cue (reward-prediction-errors), but goal-trackers do not—in sign trackers but not goal trackers, transient dopamine bursts that initially occur at the time of reward transition to occurring at the time of the cue

---

[5]While a specific curse word is presumably not pre-wired within the human motor plan, all human languages have curse words that humans release at times of stress and pain and that are not supposed to be used in social company. This abstract behavior may well be a part of the species-specific human social construct.

[6]This makes such a task "Pavlovian" in the classic sense of the word—the reward is delivered whether the animal acts or not. The fact that the task can be solved by any of the three systems described above and that animals act differently under each decision-making system shows the importance of looking at behavior from the animal's perspective rather than the experimenters.

(Flagel et al. 2011; Lesaint et al. 2014). *Does the sign have value? How does the sign motivate actions?* It depends on which decision system is being used to take those actions in response to the sign.

Pavlovian and Deliberative Morality. Even social behavior is fundamentally about decision making, and even the most human behaviors (such as questions of morality) are driven by interactions between these multiple decision-making systems. One of the most interesting recent discoveries of the past several decades has been that human social interactions are fundamentally Pavlovian in the sense used in this chapter—they are species-specific behaviors that we learn the appropriate situations for (Singer et al. 2006, 2009; Hein et al. 2011; Greene 2013). Moral decisions (such as whether to allow one person to die to save five others (Greene et al. 2001), or whether to provide a shock to another person, Milgram 1974) depend greatly on the immediacy of the social interaction (Milgram 1974; Greene et al. 2004; Zak 2008; Haslam and Reicher 2012; Rand et al. 2012). Subjects are more likely to refuse to kill one person to save five others and more likely to refuse to shock another person if there is a social bond between them (Milgram 1974; Greene et al. 2001; Haslam and Reicher 2012). Manipulations that push subjects into more deliberative modes (such as forcing the subject to make the decision in a foreign language) drive subjects into being more willing to apply utilitarian (and nonsocial) calculations (Hoffman et al. 1994; Sanfey 2007; Smith 2009; Costa et al. 2014).

Concrete preferences. A classic motivation task is to put immediate and future rewards in conflict with each other, for example, in the marshmallow task, in which a subject (usually a young child) is offered the choice of one marshmallow immediately or two marshmallows if the first marshmallow remains uneaten for 15 min (Mischel et al. 1989; Mischel 2014). It is likely that the marshmallow task is an example of conflict between Pavlovian and deliberative decision-making systems. While it is tempting to suggest that the ability to wait for the future is fundamentally deliberative, computational models suggest that both procedural and deliberative systems need to have their own discounting functions within them—preferring temporally proximate to temporally distant options (Sutton and Barto 1998; Kurth-Nelson and Redish 2012).

Looking at the algorithm underlying deliberation suggests that the fundamental reason for the preference for temporally proximate options in deliberation may reflect the ability to imagine and positively evaluate that future outcome (Kurth-Nelson and Redish 2012). One direct prediction of that hypothesis is that making the future option more concrete will make subjects more likely to choose it, effectively reducing the discounting function (making subjects discount time more slowly). Experiments have conclusively shown this to be true (Peters and Bu¨chel 2010, 2011; Benoit et al. 2011).

Attention to cues. Perception and valuation are fundamentally intertwined throughout the decision-making processes. Valuation judgments accomplished by deliberative, procedural, and Pavlovian processes of decision-making systems allow flexible decision making under multiple different conditions. Each of these processes, however, contains some form of situation-recognition step either prior to or as an early stage within the decision-making process. During this situation-recognition step, information is derived from the environmental, interoceptive, and proprioceptive cues that define the decision-making context, the identification of available goals to be sought, the potential threats to be avoided, and the state or needs of the agent, as well as numerous other features of the environment registered through varying degrees of attention, including specific details about objects or events or more abstract impressions such as social atmosphere.

The aspect of situation recognition where available goals and threats are identified parallels the approach-avoidance decision-making process and is likely to be modulated by Pavlovian processes (Phelps et al. 2014). This can be seen by processes in which Pavlovian associations drive deliberative actions, such as in Pavlovian-to-instrumental transfer (Corbit and Balleine 2005; Talmi et al. 2008), or in the influence of concrete options on decisions (Trope and Liberman 2003; Peters and Büchel 2010; Kang et al. 2011).

An agent's perception of a given situation, in addition to the concept of value applied to items within the situation, is fundamentally multidimensional and depends both on the state of the agent as well as the features of the environment. An agent's emotional or physiological state colors their percepts of both ongoing and remembered experiences. Sexual arousal, for example, changes choice behavior when choosing potential mates and influences what risky behaviors are considered acceptable (Wilson and Daly 2004; Ariely and Loewenstein 2006). Likewise, the emotional state of an individual influences an agent's responses to a given situation (Dutton and Aron 1974; Andrade and Ariely 2009). Both physiological and emotional factors can be understood to reflect the needs of the agent and influence the decision-making process by impacting the perceived set of available or acceptable courses of action, the expected outcomes of those actions, and the costs that an agent is willing to endure for those outcomes.

In addition to information about one's own state, sensory cues about external factors (the environment or situation) also fundamentally influence choice behavior by impinging on valuation. The features of an environment that color an agent's experience of the situation (multidimensional situation recognition) must be integrated from a diversity of perceived situational constraints and available courses of action to be taken including pursuit or avoidance of goals and threats. The fact that environmental cues that are not relevant to the decision can nonetheless influence choice behavior suggests that situation recognition and the concomitant attention to cues necessary for categorizing situations are integral parts of the decision-making process itself.

The endowment effect. The price at which people are willing to sell an attained item is higher than the price at which people are willing to buy that same item (Kahneman et al. 1991). This effect may be another example of an interaction between decision systems affecting valuation. Specifically, it may be that the Pavlovian system provides a greater contribution to the valuation of an item when that item has already been obtained, a situation in which, most likely, the cues associated with it are more immediately apparent and concrete than they would be if it were not yet owned.

This logic might also explain the well-documented finding that patch-foraging (human and non-human) animals will deviate from the rate-maximizing behavior that is predicted by optimal foraging theory by "over-staying" in a patch of resources rather than leaving to search for a new patch (Nonacs 2001). An intriguing possibility is that the Pavlovian system induces longer patch residence by contributing to the valuation of staying (but not leaving), much as it might when considering selling (but not buying). Interestingly, human participants have been found to display this overstaying bias during a computerized patch-foraging task for which the leave option did not require additional costs (e.g., energy and time spent traveling), suggesting that overstaying may occur due to an overvaluation of the impending reward rather than an undervaluation of leaving (Carter et al. 2015). A similar process may explain the observation that rats' aversion to rejecting an offer is positively correlated with the overall quality of the offers in the environment, which is incompatible with optimal foraging theory and standard delay discounting models (Wikenheiser et al. 2013). One possibility is that there is an increased Pavlovian preference to staying at a patch of food (because impending food is cued). This preference may be increased in rich environments, in which the rewards are more easily available.

4 Summary and Implications

In this chapter, we have argued that because valuation is a hypothetical construct, it cannot be directly measured, and it must be inferred from observed behaviors. Following from current theories of motivated behavior based on multiple interacting decision-making systems, we have argued that valuation is multidimensional. This multidimensionality complicates decision making, but these interactions can also be useful when targeting treatment and other interventions.

Craving. For example, craving is often used as a particular example of motivation, but it is useful to ask what craving is within this computational conceptualization. Craving is the computational recognition of a potential outcome with very high value, which means that it must either come from the Pavlovian system (which recognizes a situation and outcome to release a motivated action related to that outcome—such as salivating to the expectation of food) or from the deliberative system (which imagines a future situation).

Fundamentally, craving is transitive, one must crave something. Craving is always goal-directed. Some researchers have suggested that craving is Pavlovian (in the sense used in this chapter) (Skinner and Aubin 2010) and others have suggested that it is deliberative (Tiffany 1999; Tiffany and Wray 2009; Redish and Johnson 2007; Redish et al. 2008). Craving may well be an interaction between the two systems—a deliberative (model-based) computation recognizes a path to a high-value goal, which leads to the motivation of Pavlovian approach.

Because relapse in addiction can occur from any of the three systems, but craving only from the Pavlovian or deliberative, craving should be dissociable from relapse (Sayette et al. 2000; Tiffany and Wray 2009; Redish et al. 2008; Redish 2009). It should be possible to relapse without craving and to crave without relapse. This follows directly from the observations that only drug-seeking arising from Pavlovian (and possibly deliberative) processes will co-occur with craving; procedural drug-seeking will not. This means that a true "habit," one that does not devalue (Balleine and Dickinson 1998), and one that is often done "non-cognitively" (Tiffany 1990; Redish 2013), will likely be resistant to treatments aimed at reducing craving.

Contingency Management. In contrast, it might be possible to use the different decision-making systems to provide treatment. Another implication of the multi-decision-making system theory is that if we can shift the decision-making question from one valuation measure to another, we might be able to change the decision. One place in which this may be occurring is in the success of *contingency management*, a treatment used for drug addiction and other behavioral modification processes. In contingency management, a drug user is rewarded for coming in clean to the clinic[7] (Petry 2011, and see Walter and Petry in this volume). Historically, the efficacy of contingency management has been explained as an alternate reward which increases the opportunity cost of drug use (Higgins et al. 2002; Stitzer and Petry 2006); however, this depends on how quickly drug-taking falls off as drugs increase in price, an economic concept known as *elasticity*. Drug-taking is generally far too inelastic to explain the success of contingency management (Bruner and Johnson 2013). We have recently suggested that contingency management provides an opportunity for the user to engage more deliberative decision processes in their decision making (Regier and Redish 2012).

This hypothesis suggests that alternate rewards that are easier to remember and to episodically imagine would provide stronger effects in contingency management. Making the reward more concrete should thus improve contingency management, as should making it more temporally proximal, or larger. Similarly, training working memory (and other methods that provide cognitive resources) that improve episodic future thinking should improve contingency management. A simple first step would be to provide explicit reminders.

To go back to the measures of value that we began the chapter with, we are suggesting that contingency management shifts the valuation of drugs from a willing-to-pay (or an approach/avoid) valuation process to a revealed-preference valuation process. Animal drug self-administration experiments have found that shifting from willing-to-pay or approach/avoid tasks to decision-between-options (revealed-preference) tasks reduces drug-taking and drug-seeking, even at very low costs (LeSage et al. 2004; Lenoir et al. 2007; Cantin et al. 2010; Ahmed 2010; Perry et al. 2013).

---

[7]Usually this needs to be verified by a drug-negative urine sample.

5 Is Value Still a Valuable Hypothetical Construct?

Because value is not directly measurable, it must be inferred from behavior. As noted above, valuation is inconsistent. In order to understand the underlying microeconomics of decision making, we need to take into account the information processing that underlies decision making in humans and other animals (Padoa-Schioppa 2008; Rangel et al. 2008; Redish 2013). We can explain the inconsistency of valuation through these separate information processing algorithms, each of which provide a different path to motivation and valuation. These theories suggest that the process of valuation may continue to be a useful construct but that the construct of value as a means of identifying a common currency may no longer be useful.

References

Ahmed SH (2010) Validation crisis in animal models of drug addiction: beyond non-disordered drug use toward drug addiction. Neurosci Biobehav Rev 35(2):172–184
Andrade EB, Ariely D (2009) The enduring impact of transient emotions on decision making. Organ Behav Hum Decis Process 109(1):1–8
Ariely D, Loewenstein G (2006) The heat of the moment: the effect of sexual arousal on sexual decision making. J Behav Decis Mak 19:87–98
Atance CM, O'Neill DK (2001) Episodic future thinking. Trends Cogn Sci 5(12):533–539
Balleine BW, Dickinson A (1998) Goal-directed instrumental action: contingency and incentive learning and their cortical substrates. Neuropharmacology 37(4–5):407–419
Bandler R, Shipley MT (1994) Columnar organization in the midbrain periaqueductal gray: modules for emotional expression? Trends Neurosci 17(9):379–389
Benoit RG, Gilbert SJ, Burgess PW (2011) A neural mechanism mediating the impact of episodic prospection on farsighted decisions. J Neurosci 31(18):6771–6779
Bernstein IL (1978) Learned taste aversions in children receiving chemotherapy. Science 200 (4347):1302–1303
Bernstein IL (1999) Taste aversion learning: a contemporary perspective. Nutrition 15(3):229–234
Bickel WK, Yi R, Landes RD, Hill PF, Baxter C (2011) Remember the future: working memory training decreases delay discounting among stimulant addicts. Biol Psychiatry 69(3):260–265
Bouton ME (2007) Learning and behavior: a contemporary synthesis. Sinauer Associates, Massachusetts
Breland K, Breland M (1961) The misbehavior of organisms. Am Psychol 16(11):682–684
Bruner N, Johnson M (2013) Demand curves for hypothetical cocaine in cocaine-dependent individuals. Psychopharmacology 1–9
Buckner RL, Carroll DC (2007) Self-projection and the brain. Trends Cogn Sci 11(2):49–57
Burks SV, Carpenter JP, Goette L, Rustichini A (2009) Cognitive skills affect economic preferences, strategic behavior, and job attachment. Proc Nat Acad Sci 106(19):7745–7750
Cantin L, Lenoir M, Augier E, Vanhille N, Dubreucq S, Serre F, Vouillac C, Ahmed SH (2010) Cocaine is low on the value ladder of rats: possible evidence for resilience to addiction. PLoS ONE 5(7):e11592
Carter EC, Pedersen EJ, McCullough ME (2015) Reassessing intertemporal choice: human decision-making is more optimal in a foraging task than in a self-control task. Frontiers Psychol Decis Neurosci 6:95

Cavanagh JF, Eisenberg I, Guitart-Masip M, Huys Q, Frank MJ (2013) Frontal theta overrides Pavlovian learning biases. J Neurosci 33(19):8541–8548

Charness N (1991) Expertise in chess: the balance between knowledge and search. In: Ericsson KA, Smith J (eds) Toward a general theory of expertise: prospects and limits (Chap. 2). Cambridge University Press, Cambridge, pp 39–63

Corbit LH, Balleine BW (2005) Double dissociation of basolateral and central amygdala lesions on the general and outcome-specific forms of Pavlovian-instrumental transfer. J Neurosci 25(4):962–970

Corbit LH, Janak PH (2007) Inactivation of the lateral but not medial dorsal striatum eliminates the excitatory impact of Pavlovian stimuli on instrumental responding. J Neurosci 27(51):13977–13981

Coricelli G, Critchley HD, Joffily M, O'Doherty JP, Sirigu A, Dolan RJ (2005) Regret and its avoidance: a neuroimaging study of choice behavior. Nat Neurosci 8:1255–1262

Costa A, Foucart A, Hayakawa S, Aparici M, Apesteguia J, Heafner J, Keysar B (2014) Your morals depend on language. PLoS ONE 9(4):e94842. doi:10.1371/journal.pone.0094842

Daw ND, Niv Y, Dayan P (2005) Uncertainty-based competition between prefrontal and dorsolateral striatal systems for behavioral control. Nat Neurosci 8:1704–1711

Dayan P, Niv Y (2008) Reinforcement learning: the good, the bad and the ugly. Curr Opin Neurobiol 18(2):185–196

Dayan P, Niv Y, Seymour B, Daw ND (2006) The misbehavior of value and the discipline of the will. Neural Networks 19:1153–1160

Dezfouli A, Balleine B (2012) Habits, action sequences and reinforcement learning. Eur J Neurosci 35(7):1036–1051

Doya K (2000) Metalearning, neuromodulation, and emotion. In: Hatano G, Okada N, Tanabe H (eds) Affective Minds. Elsevier, Amsterdam

Dutton DG, Aron AP (1974) Some evidence for heightened sexual attraction under conditions of high anxiety. J Pers Soc Psychol 30(4):510–517

Flagel SB, Akil H, Robinson TE (2009) Individual differences in the attribution of incentive salience to reward-related cues: implications for addiction. Neuropharmacology 56(Suppl. 1):139–148

Flagel SB, Clark JJ, Robinson TE, Mayo L, Czuj A, Willuhn I, Akers CA, Clinton SM, Phillips PEM, Akil H (2011) A selective role for dopamine in stimulus-reward learning. Nature 469(7328):53–57

Franco-Watkins AM, Pashler H, Rickard TC (2006) Does working memory load lead to greater impulsivity? Commentary on Hinson, Jameson and Whitney (2003). J Exp Psychol Learn Mem Cogn 32(2):443–447

Gallistel CR (1990) The organization of learning. MIT Press, Cambridge

Gigerenzer G, Goldstein DG (1996) Reasoning the fast and frugal way: models of bounded rationality. Psychol Rev 103:650–669

Gilovich T, Griffin D, Kahneman D (eds) (2002) Heuristics and biases: the psychology of intuitive judgement. Cambridge University Press, Cambridge

Gläscher J, Daw N, Dayan P, O'Doherty JP (2010) States versus rewards: dissociable neural prediction error signals underlying model-based and model-free reinforcement learning. Neuron 66(4):585–595

Glimcher PW (2003) Decisions, uncertainty, and the brain: the science of neuroeconomics. MIT Press, Cambridge

Glimcher PW, Camerer C, Poldrack RA (eds) (2008) Neuroeconomics: decision making and the brain. Academic Press, Massachusetts

Goldstein A (2000) Addiction: from biology to drug policy. Oxford, New York

Goodman WK, Rudorfer MV, Maser JD (eds) (2000) Obsessive-compulsive disorder: contemporary issues in treatment. Lawrence Earlbaum, Hillsdale

Greene J (2013) Moral tribes: emotion, reason, and the gap between us and them. Penguin

Greene JD, Sommerville RB, Nystrom LE, Darley JM, Cohen JD (2001) An fMRI investigation of emotional engagement and moral judgement. Science 293(5537):2105–2108

Greene JD, Nystrom LE, Engell AD, Darley JM, Cohen JD (2004) The neural basis of cognitive conflict and control in moral judgement. Neuron 44(2):389–400

Hare TA, Malmaud J, Rangel A (2011) Focusing attention on the health aspects of foods changes value signals in vmPFC and improves dietary choice. J Neurosci 31(30):11077–11087

Haslam SA, Reicher SD (2012) Contesting the "nature" of conformity: what Milgram and Zimbardo's studies really show. PLoS Biol 10(11):e1001426

Hassabis D, Maguire EA (2011) The construction system in the brain. In: Bar M (ed) Predictions in the brain: using our past to generate a future. Oxford University Press, Oxford, pp 70–82

Hassabis D, Kumaran D, Vann SD, Maguire EA (2007) Patients with hippocampal amnesia cannot imagine new experiences. Proc Natl Acad Sci 104:1726–1731

Hein G, Lamm C, Brodbeck C, Singer T (2011) Skin conductance response to the pain of others predicts later costly helping. PLoS ONE 6(8):e22759

Hertz J, Krogh A, Palmer RG (1991) Introduction to the theory of neural computation. Addison-Wesley, Reading

Higgins ST, Alessi SM, Dantona RL (2002) Voucher-based incentives: a substance abuse treatment innovation. Addict Behav 27:887–910

Hill C (2008) The rationality of preference construction (and the irrationality of rational choice). Minn J Law Sci Technol 9(2):689–742

Hoffman E, McCabe K, Shachat K, Smith V (1994) Preferences, property rights, and anonymity in bargaining games. Game Econ Behav 7:346–380

Jackson JC, Redish AD (2003) Detecting dynamical changes within a simulated neural ensemble using a measure of representational quality. Network Comput Neural Syst 14:629–645

Janak P, Tye K (2015) From circuits to behaviour in the amygdala. Nature 517:284–292

Jog MS, Kubota Y, Connolly CI, Hillegaart V, Graybiel AM (1999) Building neural representations of habits. Science 286:1746–1749

Johnson A, Redish AD (2007) Neural ensembles in CA3 transiently encode paths forward of the animal at a decision point. J Neurosci 27(45):12176–12189

Johnson A, van der Meer MAA, Redish AD (2007) Integrating hippocampus and striatum in decision-making. Curr Opin Neurobiol 17(6):692–697

Johnson A, Jackson J, Redish AD (2008) Measuring distributed properties of neural representations beyond the decoding of local variables—implications for cognition. In: Hölscher C, Munk MHJ (eds) Mechanisms of information processing in the brain: encoding of information in neural populations and networks. Cambridge University Press, Cambridge, pp 95–119

Jones JL, Esber GR, McDannald MA, Gruber AJ, Hernandez A, Mirenzi A, Schoenbaum G (2012) Orbitofrontal cortex supports behavior and learning using inferred but not cached values. Science 338(6109):953–956

Kable JW, Glimcher PW (2009) The neurobiology of decision: consensus and controversy. Neuron 63(6):733–745

Kahneman D, Tversky A (eds) (2000) Choices, values, and frames. Cambridge University Press, Cambridge

Kahneman D, Knetsch JL, Thaler RH (1991) The endowment effect, loss aversion, and status quo bias. J Econ Perspect 5(1):193–206

Kang MJ, Rangel A, Camus M, Camerer CF (2011) Hypothetical and real choice differentially activate common valuation areas. J Neurosci 31(2):461–468

Kathmann N, Rupertseder C, Hauke W, Zaudig M (2005) Implicit sequence learning in obsessive-compulsive disorder: further support for the fronto-striatal dysfunction model. Biol Psychiatry 58(3):239–244

Keramati M, Dezfouli A, Piray P (2011) Speed/accuracy trade-off between the habitual and the goal-directed processes. PLoS Comput Biol 7(5):e1002055

Klein G (1999) Sources of power: how people make decisions. MIT Press, Cambridge

Kruse JM, Overmier JB, Konz WA, Rokke E (1983) Pavlovian conditioned stimulus effects upon instrumental choice behavior are reinforcer specific. Learn Motiv 14(2):165–181

Kurlan R (ed) (1993) Handbook of tourette's syndrome and related tic and behavioral disorders. Marcel Dekker, New York

Kurth-Nelson Z, Redish AD (2012) A theoretical account of cognitive effects in delay discounting. Eur J Neurosci 35:1052–1064
Laurent V, Leung B, Maidment N, Balleine BW (2012) μ- and δ-opioid-related processes in the accumbens core and shell differentially mediate the influence of reward- guided and stimulus-guided decisions on choice. J Neurosci 32(5):1875–1883
Leckman JF, Riddle MA (2000) Tourette's syndrome: when habit-forming systems form habits of their own? Neuron 28:349–354
LeDoux JE (2000) Emotion circuits in the brain. Annu Rev Neurosci 23:155–184
LeDoux JE (2012) Rethinking the emotional brain. Neuron 73:653–676
Lee SW, Shimoko S, O'Doherty JP (2014) Neural computations underlying arbitration between model-based and model-free learning. Neuron 81(3):687–699
Lenoir M, Serre F, Cantin L, Ahmed SH (2007) Intense sweetness surpasses cocaine reward. PLoS ONE 2(8):e698
LeSage MG, Burroughs D, Dufek M, Keyler DE, Pentel PR (2004) Reinstatement of nicotine self-administration in rats by presentation of nicotine-paired stimuli, but not nicotine priming. Pharmacolol Biochem Behav 79(3):507–513
Lesaint F, Sigaud O, Flagel SB, Robinson TE, Khamassi M (2014) Modelling individual differences in the form of Pavlovian conditioned approach responses: a dual learning systems approach with factored representations. PLoS Comput Biol 10(2):e1003466. doi:10.1371/journal.pcbi.1003466
Levy DJ, Glimcher PW (2012) The root of all value: a neural common currency for choice. Curr Opin Neurobiol 22:1–12
Lichtenstein S, Slovic P (eds) (2006) The construction of preference. Cambridge University Press, Cambridge
MacCorquodale K, Meehl PE (1954) Edward C. Tolman. In: Estes W (ed) Modern learning theory. Appleton-Century-Crofts, New York, pp 177–266
Maia TV, McClelland JL (2012) A neurocomputational approach to obsessive-compulsive disorder. Trends Cogn Sci 16(1):14–15
McNally GP, Johansen JP, Blair HT (2011) Placing prediction into the fear circuit. Trends Neurosci 34(6):283–292
Milgram S (1974/2009) Obedience to authority: an experimental view. Harper Collins, New York
Mischel W (2014) The marshmallow test: mastering self-control. Little, Brown, and Co, New York
Mischel W, Shoda Y, Rodriguez ML (1989) Delay of gratification in children. Science 244 (4907):933–938
Montague PR, Dolan RJ, Friston KJ, Dayan P (2012) Comput Psychiatry. Trends Cogn Sci 16 (1):72–80
Nadel L (1994) Multiple memory systems: what and why, an update. In: Schacter DL, Tulving E (eds) Memory systems 1994. MIT Press, Cambridge, pp 39–64
Niv Y, Daw ND, Dayan P (2006a) Choice values. Nat Neurosci 9:987–988
Niv Y, Joel D, Dayan P (2006b) A normative perspective on motivation. Trends Cogn Sci 10 (8):375–381
Nonacs P (2001) State dependent patch use and the marginal value theorem. Behav Ecol 12:71–83
O'Doherty JP (2004) Reward representations and reward-related learning in the human brain: insights from neuroimaging. Curr Opin Neurobiol 14:769–776
O'Doherty JP, Kringelbach ML, Rolls ET, Hornak J, Andrews C (2001) Abstract reward and punishment representations in the human orbitofrontal cortex. Nat Neurosci 4:95–102
O'Keefe J, Nadel L (1978) The hippocampus as a cognitive map. Clarendon Press, Oxford
Padoa-Schioppa C (2008) The syllogism of neuro-economics. Econ Philos 24:449–457
Padoa-Schioppa C (2009) Range-adapting representation of economic value in the orbitofrontal cortex. J Neurosci 29(44):14004–14014
Pavlov I (1927) Conditioned reflexes. Oxford University Press, Oxford
Perry AN, Westenbroek C, Becker JB (2013) The development of a preference for cocaine over food identifies individual rats with addiction-like behaviors. PLoS ONE 8(11):e79465

Peters J, Büchel C (2010) Episodic future thinking reduces reward delay discounting through an enhancement of prefrontal-mediotemporal interactions. Neuron 66(1):138–148

Peters J, Büchel C (2011) The neural mechanisms of inter-temporal decision-making: understanding variability. Trends Cogn Sci 15(5):227–239

Petry NM (2011) Contingency management for substance abuse treatment: a guide to implementing this evidence-based practice. Routledge, London

Pfeiffer BE, Foster DJ (2013) Hippocampal place-cell sequences depict future paths to remembered goals. Nature 497:74–79

Phelps E, Lempert KM, Sokol-Hessner P (2014) Emotion and decision making: multiple modulatory circuits. Annu Rev Neurosci 37:263–287

Plous S (1993) The psychology of judgement and decision-making. McGraw-Hill, New York

Rand DG, Greene JD, Nowak MA (2012) Spontaneous giving and calculated greed. Nature 489:427–430

Rangel A, Clithero JA (2012) Value normalization in decision making: theory and evidence. Curr Opin Neurobiol 22(6):970–981

Rangel A, Hare T (2010) Neural computations associated with goal-directed choice. Curr Opin Neurobiol 20(2):262–270

Rangel A, Camerer C, Montague PR (2008) A framework for studying the neurobiology of value-based decision making. Nat Rev Neurosci 9:545–556

Redish AD (2009) Implications of the multiple-vulnerabilities theory of addiction for craving and relapse. Addiction 104(11):1940–1941

Redish AD (2013) The mind within the brain: how we make decisions and how those decisions go wrong. Oxford University Press, Oxford

Redish AD, Johnson A (2007) A computational model of craving and obsession. Ann New York Acad Sci 1104(1):324–339

Redish AD, Jensen S, Johnson A, Kurth-Nelson Z (2007) Reconciling reinforcement learning models with behavioral extinction and renewal: implications for addiction, relapse, and problem gambling. Psychol Rev 114(3):784–805

Redish AD, Jensen S, Johnson A (2008) A unified framework for addiction: vulnerabilities in the decision process. Behav Brain Sci 31:415–487

Regier PS, Redish AD (2012) What is the role of decision-making systems in contingency management? Society for Neuroscience Abstracts

Robinson MJ, Warlow SM, Berridge KC (2014) Optogenetic excitation of central amygdala amplifies and narrows incentive motivation to pursue one reward above another. J Neurosci 34(50):16567–16580

Sacks O (1985). Witty ticcy ray. In The Man Who Mistook His Wife for a Hat. Simon and Schuster

Samuelson PA (1937) A note on measurement of utility. Rev Econ Stud 4(2):155–161

Sanfey AG (2007) Social decision-making: insights from game theory and neuroscience. Science 318(5850):598–602

Sayette MA, Shiffman S, Tiffany ST, Niaura RS, Martin CS, Shadel WG (2000) The measurement of drug craving. Addiction 95(Suppl. 2):S189–S210

Schacter DL, Addis DR (2011) On the nature of medial temporal lobe contributions to the constructive simulation of future events. In: Bar M (ed) Predictions in the brain: using our past to generate a future. Oxford University Press, Oxford, pp 58–69

Schacter DL, Addis DR, Buckner RL (2008) Episodic simulation of future events: concepts, data, and applications. Ann New York Acad Sci 1124:39–60

Schmidhuber J (2014) Deep learning in neural networks. Neural Networks 61:85–117

Simon H (1955) A behavioral model of rational choice. Q J Econ 69:99–118

Simonson I, Tversky A (1992) Choice in context: tradeoff contrast and extremeness aversion. J Mark Res 29(3):281–295

Singer T, Seymour B, O'Doherty JP, Stephan KE, Dolan RJ, Frith CD (2006) Empathic neural responses are modulated by the perceived fairness of others. Nature 439(7075):466–469

Singer T, Critchley HD, Preuschoff K (2009) A common role of insula in feelings, empathy and uncertainty. Trends Cogn Sci 13(8):334–340
Skinner MD, Aubin H-J (2010) Craving's place in addiction theory: contributions of the major models. Neurosci Biobehav Rev 34(4):606–623
Smith V (2009) Rationality in economics: constructivist and ecological forms. Cambridge University Press, Cambridge
Smith KS, Graybiel AM (2013) A dual operator view of habitual behavior reflecting cortical and striatal dynamics. Neuron 79(2):361–374
Smith KS, Tindell AJ, Aldridge JW, Berridge KC (2009) Ventral pallidum roles in reward and motivation. Behav Brain Res 196(2):155–167
Steiner A, Redish AD (2012) Orbitofrontal cortical ensembles during deliberation and learning on a spatial decision-making task. Front Decis Neurosci 6:131
Stephens DW, Krebs JR (1987) Foraging theory. Princeton University Press, Princeton
Stitzer M, Petry N (2006) Contingency management for treatment of substance abuse. Annu Rev Clin Psychol 2:411–434
Stott JJ, Redish AD (2014) A functional difference in information processing between orbitofrontal cortex and ventral striatum during decision-making behavior. Philos Trans R Soc B 369(1655). 10.1098/rstb.2013.0472
Streidter G (2005) Principles of brain evolution. Sinauer Associates, Sunderland
Sutton RS, Barto AG (1998) Reinforcement learning: an introduction. MIT Press, Cambridge
Talmi D, Seymour B, Dayan P, Dolan RJ (2008) Human Pavlovian-instrumental transfer. J Neurosci 28(2):360–368
Tiffany ST (1990) A cognitive model of drug urges and drug-use behavior: role of automatic and nonautomatic processes. Psychol Rev 97(2):147–168
Tiffany ST (1999) Cognitive concepts of craving. Alcohol Res Health 23(3):215–224
Tiffany ST, Wray J (2009) The continuing conundrum of craving. Addiction 104:1618–1619
Tolman EC (1932) Purposive behavior in animals and men. Appleton-Century-Crofts, New York
Tremblay L, Schultz W (1999) Relative reward preference in primate orbitofrontal cortex. Nature 398(6729):704–708
Trope Y, Liberman N (2003) Temporal construal. Psychol Rev 110(3):403–421
van der Meer MAA, Redish AD (2011) Ventral striatum: a critical look at models of learning and evaluation. Curr Opin Neurobiol 21(3):387–392
van der Meer MAA, Kurth-Nelson Z, Redish AD (2012) Information processing in decision-making systems. Neuroscientist 18(4):342–359
Wikenheiser AM, Redish AD (2015) Hippocampal theta sequences reflect current goals. Nat Neurosci 18:289–294
Wikenheiser AM, Stephens DW, Redish AD (2013) Subjective costs drive overly-patient foraging strategies in rats on an intertemporal foraging task. Proc Natl Acad Sci USA 110(20):8308–8313
Wilson M, Daly M (2004) Do pretty women inspire men to discount the future? Proc R Soc Lond B 271:S177–S179
Winecoff A, Clithero JA, Carter RM, Bergman SR, Wang L, Huettel SA (2013) Ventromedial prefrontal cortex encodes economic value. J Neurosci 33(27):11032–11039
Wunderlich K, Dayan P, Dolan RJ (2012) Mapping value based planning and extensively trained choice in the human brain. Nat Neurosci 15(5):786–791
Yang T, Shadlen MN (2007) Probabilistic reasoning by neurons. Nature 447:1075–1080
Zak PJ (ed) (2008) Moral markets: the critical role of values in the economy. Princeton University Press, Princeton

Part III
Apathy and Pathological Deficits in Motivation

The Neurobiology of Motivational Deficits in Depression—An Update on Candidate Pathomechanisms

Michael T. Treadway

Abstract Anhedonia has long been recognized as a central feature of major depression, yet its neurobiological underpinnings remain poorly understood. While clinical definitions of anhedonia have historically emphasized reductions in pleasure and positive emotionality, there has been growing evidence that motivation may be substantially impaired as well. Here, we review recent evidence suggesting that motivational deficits may reflect an important dimension of symptomatology that is discrete from traditional definitions of anhedonia in terms of both behavior and pathophysiology. In summarizing this work, we highlight two candidate neurobiological mechanisms—elevated inflammation and reduced synaptic plasticity—that may underlie observed reductions in motivation and reinforcement learning in depression.

Keywords Depression · Motivation · Effort-based decision-making · Reinforcement learning · Dopamine · Inflammation · Neuroplasticity

Contents

1 Introduction	338
2 Motivation, Reinforcement, and Dopamine	340
3 Motivation and Reinforcement in Depression—Implications for DA Dysfunction	343
4 Candidate Pathophysiological Mechanisms of Motivational and Reinforcement Deficits in Depression	345
4.1 A Role for Inflammation in Motivational Deficits in Depression	345
5 Synaptic Plasticity Alterations May Impact Motivation in Patients with Depression	347
6 Conclusion	348
References	349

M.T. Treadway (✉)
Department of Psychology, Emory University, Atlanta, USA
e-mail: mtreadway@emory.edu
URL: http://tinyurl.com/TReADLab

© Springer International Publishing Switzerland 2015

1 Introduction

The term anhedonia—introduced to the clinical literature over 100 years ago—describes a debilitating psychological state that reflects an almost complete lack of enjoyment and positive emotionality. Activities such as meals, work, and social interaction are left devoid of the normal pleasures associated with appetite, motivation, or connection. Though anhedonia was originally intended to describe a state of severe despair and lack of pleasure (Ribot 1896), it has been used by clinicians and clinical researchers as a "catch-all" term related to patient impairments across a range of components that underlie approach behavior, including motivation, enjoyment, optimism, and positive mood states. As discussed in more detail below, this "big tent" definition for anhedonia can raise challenges when attempting to study its pathophysiology. Consequently, the current chapter uses the term anhedonia to denote a symptom domain that has a variety of possible subcomponents rather than a single construct and focuses primarily on what is known about pathomechanisms for deficits related to motivation and impaired reinforcement learning as compared to a pure loss of pleasure.

This stance is consistent with a number of recent reviews that have called for a critical reexamination of the anhedonia construct [(Foussias and Remington 2008; Barch and Dowd 2010; Treadway and Zald 2011; Strauss and Gold 2012) also see chapters in this volume by Barch et al., Reddy et al., Waltz and Gold]. A central question raised by this work has been whether anhedonia describes a primary deficit in the capacity to experience positive emotions, or does it include deficiencies in a number of reward-related domains? While there is no debate that depressive states are associated with reduced motivation for and failure to anticipate rewarding experiences, the former hypothesis assumes that these are more or less normative responses to reduced hedonic capacity. In contrast, the alternative hypothesis is that, at least for some depressed individuals, the pathophysiology of the disorder may diminish motivation directly. The answer to this question has substantial implications for the theoretical conceptualization of anhedonia and related constructs, for the assessment of psychopathology involving these symptoms, for understanding the neural substrates of psychiatric symptoms, and for the treatment or reward processing abnormalities.

As currently defined by the DSM-5, anhedonia is one of two symptoms required for diagnosis of a depressive episode. A patient is considered to meet the anhedonia criterion by reporting either the loss of pleasure in previously enjoyable activities or a loss of interest/motivation in pursuing them. If pleasure and interest are reflections of a singular process, then this reduction to a single symptom is not a problem. However, if reductions in pleasure and interest reflect different pathophysiologies, then underlying neurobiological mechanisms may differ across anhedonic patients, thereby eluding their detection in research studies that treat both manifestations of anhedonia as being equivalent (Barch and Dowd 2010; Treadway and Zald 2011). Indeed, most traditional measures of psychopathology and dimensional assessments of anhedonia fail to discriminate between these various domains of reward processing. While such measures have had a useful place in the context of clinical assessment and care, they may mask important behavioral and biological distinctions that are critical toward understanding pathophysiology. It has long been recognized that reinforcement involves multiple subprocesses, such as anticipation, motivation, prediction, subjective pleasure, and satiety. It has only been more recently, however, that investigators have been able to clearly show that these subcomponents are neurobiologically dissociable (see Robinson et al., in this volume). That is, manipulations of distinct circuits and neurochemicals can produce isolated effects on a single dimension of reward-related behavior, such as an abolition of motivation without any change in hedonic responsiveness. This finding suggests that a reduction in reward-seeking behavior may result from impairments in one or many subcomponent processes, which in turn implies that they may have shared or unshared neurobiological origins across different individuals. Despite this new understanding of the biological divisions involved in reward and reinforcement in the preclinical literature, current clinical methods have largely continued to conceptualize anhedonic symptoms along a unitary dimension of pleasure and positive emotions (Gold et al. 2008; Treadway and Zald 2011).

It is worth highlighting that this debate has occurred against a backdrop of growing interest in the clinical significance of anhedonia. Following the publication of DSM-3, which prominently featured anhedonia in conditions of major depressive disorder (MDD) and schizophrenia (Klein 1974; Meehl 1975), empirical research devoted to the understanding and treatment of this symptom domain has grown rapidly. Further augmenting the focus on anhedonic symptoms has been the observation of comparatively poorer treatment outcomes for patients with an anhedonic presentation (Shelton and Tomarken 2001), as well as a steep rise in preclinical discoveries regarding the molecular and system-level mechanisms underlying reward processing generally [for reviews, see Salamone et al. 2007; Berridge and Kringelbach 2008; Rushworth et al. 2011 as well as chapters in this volume by Corbit and Balleine, Robinson et al., Roesch et al., and Salamone et al.]. This latter trend is of particular importance as the field of psychiatry has increasingly turned toward translational neuroscience as a means of understanding the etiopathophysiology of mental disorders (Insel et al. 2010; Insel and Cuthbert 2015).

Despite this heightened focus, many fundamental questions remain regarding the nature of anhedonic symptoms, their etiology, phenomenology, biological underpinnings, and specificity to psychiatric illness. In this chapter, we briefly review the known neural circuitry evidence supporting motivated behavior, highlight recent behavioral evidence suggesting that impairments in these processes are associated with depressive symptoms, and discuss candidate pathomechanisms that may underlie the expression and etiology of these deficits. Importantly, we believe that this work can help begin to isolate subtypes of depressive disorders that may be defined by distinct pathophysiologies rather than symptoms, which has long been a goal of psychiatric medicine.

2 Motivation, Reinforcement, and Dopamine

Over the past two decades, animal models have found robust evidence linking mesolimbic dopamine circuitry to motivated behavior. The mesolimbic DA system encompasses a specific subpopulation of DA neurons that innervate the ventral striatum (VS), an integrative hub involved in translating value-related information into motivated action (Haber and Knutson 2010; Floresco 2015). Evidence for the role of mesolimbic DA in motivation was first provided through the use of effort-based decision-making tasks in rodents (see Salamone et al., this volume). In these paradigms, animals must choose whether to exert physical effort in exchange for greater or more palatable food rewards (High Effort) or to consume freely available, but less desirable food rewards (Low Effort). Across different paradigms, healthy rats on a food-restricted diet typically show a strong preference for the High-Effort option, while attenuation or blockade of DA—especially in the VS— results in a behavioral shift toward Low-Effort options (Cousins and Salamone 1994; Salamone et al. 2007). Importantly, DA blockade does not reduce overall consumption, suggesting that these manipulations do not impair primary motivation for food, but rather selectively ablate willingness to work for larger or more preferred rewards. Additionally, potentiation of DA through drugs such as *d*-amphetamine produces the opposite effects, resulting in an increased willingness to work for preferred rewards (Bardgett et al. 2009). In contrast to this strong evidence for DA in motivation, attenuation or even complete absence of DA appears to have little effect on measures of hedonic response, including sucrose preference and hedonic facial reactions (for a review, see Berridge and Kringelbach 2008).

In humans, similar results have been observed using effort-based decision-making tasks in which participants choose how much (physical or mental) effort to invest in order to obtain a reward (typically money). Previously, our laboratory developed the effort expenditure for reward task (EEfRT, pronounced "effort"), which has been used to examine neural substrates of effort mobilization in humans. During this task, participants perform a series of trials in which they are asked to choose between completing a "High Effort" task and completing a "Low Effort" task in exchange for monetary compensation, where the required effort is in the form of speeded button presses (see Fig. 1). Other groups have developed similar tasks for physical effort using handgrip paradigms (Pessiglione et al. 2007; Hartmann et al. 2013) or cognitive effort in the form of task switching (McGuire and Botvinick 2010), attentional control (Croxson et al. 2009) or working memory (Westbrook et al. 2013).

Using these tasks, human studies have begun to map out the role of mesolimbic DA circuitry in normal and abnormal reward motivation. Mirroring the effects of DA potentiation in rats, one study found that administration of the DA agonist d-amphetamine produced a dose-dependent increase in the willingness to work for rewards as assessed by the EEfRT (Wardle et al. 2011) (see Fig. 2a). Similar effects of DA enhancement using the DA precursor L-Dopa have been observed on

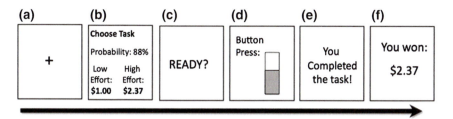

Fig. 1 Schematic diagram of a single trial of the effort expenditure for reward task (EEfRT) (Treadway et al. 2009). **a** Trial begins with a 1-s fixation cue, followed by **b** a 5-s choice period in which subjects are presented with information regarding the reward magnitude of the hard task for that trial, and the probability of receiving a reward. After making a choice, **c** a 1-s "ready" screen is displayed, after which **d** subjects make rapid button presses to complete the chosen task for 7 s (easy task) or 21 s (hard task). **e** Subjects receive feedback on whether they have completed the task. **f** Subjects receive reward feedback as to whether they received any money for that trial

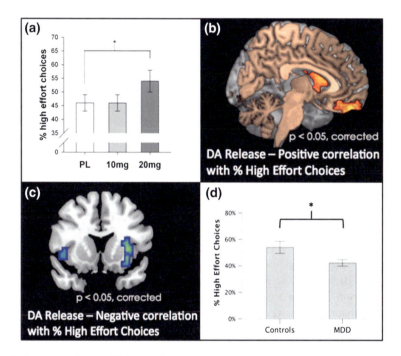

Fig. 2 Summary of recent EEfRT studies in humans (adapted from Treadway and Zald (2013)). **a** Administration of amphetamine produces a dose-dependent increase in willingness to expend greater effort for larger rewards (Wardle et al. 2011). **b** Proportion of High-Effort choices during low-probability trials shows positive associations with amphetamine-induced change in D2/D3 binding potential in striatum and vmPFC (Treadway et al. 2012). **c** Proportion of High-Effort choices is *inversely* associated with amphetamine-induced change in D2/D3 binding potential in bilateral insula. **d** Depressed patients choose fewer High-Effort options than matched controls (Treadway et al. 2012)

measures of vigorous effortful responding (Beierholm et al. 2013) as well reward anticipation (Sharot et al. 2009).

While these studies have suggested that direct manipulation of DAergic systems may alter motivated behavior in humans, they do not address the questions of whether endogenous variability in DA function may serve as a substrate for individual differences. This latter issue is particularly important if DA dysfunction is to serve as a pathomechanism for depressive symptoms. To investigate this question, a follow-up study from our group used positron emission tomography (PET) imaging to test associations between amphetamine-induced DA release (a probe of DA system reactivity) and willingness to work for rewards on during the EEfRT (Treadway 2012). Here, we found that the magnitude of DA release in dorsomedial and ventral aspects of the striatum positively predicted the proportion of High-Effort choices subjects made during low-probability trials (see Fig. 2b). Localization to this region is consistent with preclinical findings (Cousins and Salamone 1994; Salamone et al. 2007) as well as human functional neuroimaging studies (Croxson et al. 2009; Kurniawan et al. 2010; Schmidt et al. 2012, also see O'Doherty, this volume). Intriguingly, our study also found a negative relationship between percentage of High-Effort choices and DA release in the insula (see Fig. 2c). While insula DA function has not traditionally been a focus for rodent models of effort-based decision-making, recent work suggests that insula DA receptor mRNA expression is predictive of effort-related behaviors (Simon et al. 2013). Moreover, other human imaging studies have observed insula activation when participants chose not to expend effort (Prevost et al. 2010). Although further investigation is necessary, these data suggest that the insula and striatum may play somewhat antagonistic roles in determining whether an individual is willing to overcome effort costs.

In addition to the role of DA in motivation, DA signaling in the striatum–especially phasic signaling–has been heavily linked to reinforcement learning. Consistent evidence from DA cell recordings (Schultz 2007) and fast-scan cyclic voltammetry of DA projection targets in the striatum (Hart et al. 2014) support the hypothesis that phasic DA signaling in the striatum reflects the difference between expected rewards and received rewards, often referred to as a "prediction error" (Sutton and Barto 1998). In humans, corroborative results have been obtained using high-resolution fMRI of the midbrain (D'Ardenne et al. 2008), as well as pharmacologic manipulations (Pessiglione et al. 2006).

Taken together, a solid body of evidence has implicated DA signaling in the striatum in both reward motivation (effort expenditure) and reinforcement learning. Using these data as a foundation, a number of studies in depression over the last ten years have investigated whether this disorder is associated with alterations in these behaviors, possibly implicating a corticostriatal DAergic mechanism.

3 Motivation and Reinforcement in Depression—Implications for DA Dysfunction

In a relatively recent literature, studies of motivation and reinforcement in depression have been largely consistent in detecting differences as compared to healthy controls (Whitton et al. 2015). In several studies using the effort expenditure for reward task (EEfRT), patients with MDD expended less effort for rewards when compared with controls (Treadway et al. 2012; Yang et al. 2014) (see Fig. 2d). Further evidence suggests that the longer the depressive episode is, the more impaired this decision-making is (Treadway et al. 2012a), and when in remission, this deficit normalizes (Yang et al. 2014). However, a similarly designed study did not detect a main effect of depression on willingness to expend effort for rewards (Sherdell et al. 2011). It is worth noting, however, that this study utilized an EEfRT requiring significantly fewer button presses for its High-Effort condition (an average of 42 mouse clicks as compared to the EEfRT, which requires 100 button presses in 21 s with the non-dominant pinky finger); these varying levels of required effort may have influenced the sensitivity to these two tasks for detecting group differences. For reinforcement learning, Pizzagalli and colleagues have used a signal detection task that has consistently demonstrated reduced implicit reinforcement learning in depression (Pizzagalli et al. 2008; Vrieze et al. 2013). Like the EEfRT, performance on this task has similarly been linked to DA function in humans (Vrieze et al. 2011). Supporting these behavioral findings, functional neuroimaging studies have found that depression is associated with reduced prediction error signaling in the striatum during reinforcement learning (Kumar et al. 2008) as well as striatal responses to reward feedback (Pizzagalli et al. 2009).

In sum, clinical, behavioral, and a handful of imaging studies support the hypothesis that motivation and reinforcement learning are impaired in MDD and that DA function may serve as a primary substrate. Several factors complicate this hypothesis, however, as most studies focused on direct imaging of DA have produced ambiguous results (Treadway and Pizzagalli 2014). First, PET studies of DA receptor distribution—a common tool for measuring pathological alteration of DA systems—have found some mixed evidence for DA involvement in depression. An early study using single-photon emission tomography (SPECT) found that MDD patients exhibited reduced DA synthesis capacity as measured by L-Dopa uptake (Agren and Reibring 1994). In subsequent PET and SPECT studies of the DA transporter (DAT), MDD has been associated with both lower (Meyer et al. 2001) and higher (Laasonen-Balk et al. 1999; Amsterdam and Newberg 2007; Yang et al. 2008) binding potential in the striatum. These data should be taken with the caveat that all studies that observed an increase in striatal DAT binding in MDD used SPECT, which has much lower sensitivity than PET (Rahmim and Zaidi 2008), and postmortem studies also suggest a decrease, rather than increase, in DAT availability in depression (Klimek et al. 2002).

PET studies of DA receptor availability have yielded similarly mixed results. PET measures of striatal D2/D3 receptor binding potential have observed increases in several depressed samples (D'Haenen and Bossuyt 1994; Shah et al. 1997), a finding that conflicts with predictions based on some preclinical animal data (Gershon et al. 2007). It should be noted, however, that striatal D2/D3 PET ligands are unable to differentiate between pre- versus postsynaptic receptors. Given that pre- and postsynaptic D2/D3 receptors are known to exert distinct—even oppositional—effects on postsynaptic DA signaling, the inability to resolve this difference in clinical studies may limit interpretability. Other studies using medication-naïve or medication-free patients have failed to find group differences in striatal receptor binding (Parsey et al. 2001; Hirvonen et al. 2008). Interestingly, one additional small study showed variable changes in D2-like binding following treatment with SSRIs such that patients who showed increased binding exhibiting greater clinical improvement than those who did not (Klimke et al. 1999). With respect to the D1 receptor, fewer studies have examined this system given the lack of available ligands that reliably distinguish between D1 and serotonin 5-HT$_{2A}$ receptors, especially in extrastriatal areas where the receptor density of D1 and 5-HT$_{2A}$ is roughly equivalent. One study reported reduced D1 availability in left middle caudate (Cannon et al. 2009), but this finding has not yet been replicated. Taken together, while behavioral and clinical evidence suggests that depression affects motivational and reinforcement behaviors that are known to depend heavily on DAergic function, the evidence for a primary DAergic deficit is undeniably equivocal, particularly when compared to other DA-linked neuropsychiatric disorders, such as schizophrenia, Parkinson's disease, or substance use.

One explanation for these discrepancies is the possibility of distinct subtypes of depression, only some of which may involve alterations to DA signaling. Supporting this is the observation of slightly more consistent effects when MDD samples are selected on the basis of a particular symptom profile. For example, one study that restricted its MDD patient sample to individuals with symptoms of affective flattening as assessed by the Snaith–Hamilton Pleasure Scale reported decreased DAT binding (Sarchiapone et al. 2006). In addition, decreases in [^{18}F] Dopa binding—a marker of DA synthesis capacity—have been observed in the striatum of depressed individuals with flat affect or psychomotor slowing as measured by the Depression Retardation Scale (Martinot et al. 2001; Bragulat et al. 2007). As with behavioral studies, these data suggest that some—but not all—patients with depression may exhibit abnormalities in DA signaling, which may manifest as anhedonic symptoms. If true, however, this hypothesis begs the question as to what possible mechanisms may account for this selective effect.

4 Candidate Pathophysiological Mechanisms of Motivational and Reinforcement Deficits in Depression

In this next section, we present two candidate mechanisms that may be partially responsible for motivational and reinforcement learning impairments observed in depression.

4.1 A Role for Inflammation in Motivational Deficits in Depression

One candidate mechanism for motivation-related impairments in MDD is inflammation. An extensive literature has now shown that compared to controls, a subset of depressed patients exhibit elevated inflammatory proteins and gene expression in both peripheral tissue and cerebrospinal fluid (CSF), as well as increased peripheral blood acute-phase proteins, chemokines, and adhesion molecules (Miller et al. 2009a, b). Meta-analyses of this literature have identified that the most reliable inflammatory biomarkers in depression are increases in peripheral blood inflammatory cytokine tumor necrosis factor (TNF) and interleukin (IL)-6 as well as increases in the acute-phase protein C-reactive protein (CRP) (Howren et al. 2009; Miller et al. 2009a, b; Dowlati et al. 2010). Finally, non-depressed individuals who develop a primary immune disorder show substantially higher rates of anhedonic symptoms on commonly used symptom inventories than the general population (Pincus et al. 1996; Dickens and Creed 2001; Blume et al. 2011).

While inflammation may affect a variety of brain areas, significant data highlight the striatum as a primary site of inflammation-induced CNS dysfunction (Capuron et al. 2007; Miller et al. 2009a, b). Inflammatory cytokines are known to disrupt DA neurotransmission including DA synthesis and release in rodents and non-human primates (Felger et al. 2007; Qin et al. 2007; Miller et al. 2009a, b; Dantzer et al. 2012; Felger et al. 2013), leading to impairments in effort expenditure and anticipation.

Of note, these reductions in motivation are similar to those observed by direct DA antagonism in the striatum as described above and suggest that cytokine interference with DA synthesis capacity may be partially responsible for these effects. Moreover, the effects of inflammation on DA signaling may worsen over time; DA plays an important anti-inflammatory role in the brain (Sarkar et al. 2010; Yan et al. 2015), and decreased DA availability may further exacerbate inflammatory effects via a positive feedback loop, resulting in a chronically inflamed, hypo-dopaminergic state.

As a consequence of inflammation-mediated DA interference, corticostriatal networks may become dysfunctional. Supporting this notion, human fMRI studies have identified associations between peripheral cytokine levels and fMRI measures of neural processing in the DA-rich striatum and regions of medial prefrontal cortex (mPFC) in healthy volunteers exposed to typhoid vaccination (Harrison et al. 2009). Additionally, administration of inflammatory cytokines [interferon (IFN) alpha] or cytokine inducers (endotoxin) is associated with blunted ventral striatal responses to reward anticipation (Eisenberger et al. 2010; Capuron et al. 2012), as well as decreased DA release within the striatum as measured by [18F] Dopa binding in humans and in vivo microdialysis in non-human primates (Capuron et al. 2012; Felger et al. 2013). These data clearly demonstrate that increasing inflammation can reduce DA availability and impair corticostriatal circuit function.

Further supporting an inflammation–dopamine subtype of depression, growing evidence suggests that inflammation may specifically induce symptoms related to motivation. In animal models of effort-based decision-making, administration of cytokines or cytokine inducers reduces willingness to work for rewards (Nunes et al. 2013; Vichaya et al. 2014), an effect which is reversible through pharmacologic stimulation of striatal pathways (Nunes et al. 2013). Motivational deficits have also been observed in rodents and non-human primates following IFN-α manipulations (Couch et al. 2013; Felger et al. 2007; Thorne et al. 2008; Felger and Miller 2014). Finally, in humans, IFN-α and cytokine inducers rapidly and robustly produce motivational complaints (apathy, lassitude) in the majority of recipients, while cognitive and affective symptoms develop later, and are often more pronounced in individuals with a predisposing diathesis, such as elevated trait neuroticism (Capuron et al. 2004; Capuron and Miller 2011).

Blockade of inflammation selectively improves motivation symptoms *only* in patients with high baseline inflammation. Recently, a randomized, placebo-controlled clinical study was conducted to determine the effects of inflammation blockade on depressive symptoms (Raison et al. 2013). Blockade was achieved using infliximab, a monoclonal antibody to TNF with minimal "off-target" effects compared to other anti-inflammatory compounds. Following a single infusion of this highly selective TNF antagonist, a robust decrease in the plasma inflammatory biomarker, high-sensitivity C-reactive protein (hs-CRP), was observed and a subsequent improvement in reported symptoms of motivation and engagement in activities. Notably, the magnitude of change in this symptom domain following infliximab was double that of any other symptoms, and this effect was only present in patients with high inflammation at baseline (CRP > 5 mg/L).

In sum, there is strong evidence that inflammation is elevated in a subset of depressed patients and that increased inflammatory signaling may deplete DA availability, reduce DA-moderated regulation of inflammation, and produce motivational impairments in animals that mirror depressed phenotypes. Moreover, stimulation of cytokines results in motivational symptoms, and these symptoms are selectively ameliorated by anti-inflammatory treatments in patients with elevated inflammatory profiles. These data suggest that inflammation-mediated decreases in DA synthesis capacity may underlie reduced motivation in a subset of depressed patients.

5 Synaptic Plasticity Alterations May Impact Motivation in Patients with Depression

Clearly, however, many patients meeting criteria for depression and apathy do not express gross alterations in immune signaling (Raison and Miller 2011). Consequently, a second candidate mechanism underlying motivational deficits in depression is impaired synaptic plasticity. Growing evidence from preclinical and computational modeling work suggests that dopamine may contribute to reinforcement in large part through altering synaptic plasticity within corticostriatal circuits; that is, DA signaling—particularly phasic DA bursts or dips—helps strengthen corticostriatal synaptic connections that link reward-related cues to rewarding outcomes (Reynolds et al. 2001; Frank et al. 2004; Wieland et al. 2015). Alterations of postsynaptic plasticity mechanisms may therefore manifest as blunting of DA-related reinforcement signals, thereby contributing to dysfunction in DAergic circuitry without reflecting a primary deficit in DA-releasing neurons per se.

A variety of data support the hypothesis that neuroplasticity is affected in MDD. Early evidence comes from structural neuroimaging studies, demonstrating diminished gray matter volume in the hippocampus (Sheline et al. 1999; MacQueen et al. 2003)—a key region involved in neurogenesis—and these findings have been confirmed and replicated in meta-analyses and subsequent large-sample studies (Kempton et al. 2011; Schmaal et al. 2015). Further evidence comes from postmortem studies, where decreases in cellular density (Cotter et al. 2001; Chana et al. 2003; Monkul et al. 2006) and reduced expression of proteins involved in neurogenesis and synaptic plasticity (Kempermann and Kronenberg 2003; Dwivedi et al. 2005; Pittenger and Duman 2007) have been observed. More recently, serum measures of brain-derived neurotrophic factor (BDNF) have been found to be significantly decreased in patients during a depressive episode (Molendijk et al. 2014; Bus et al. 2015). This latter finding is of particular interest, as BDNF is a well-characterized neurotrophin that is known to stimulate growth of new synapses and dendrites throughout the life span via stimulation of tropomyosin kinase B (Trk-B) receptors (Yoshii and Constantine-Paton 2010; Autry and Monteggia 2012).

Finally, the discovery of ketamine as an efficacious antidepressant with remarkably rapid onset (Zarate et al. 2006; Diazgranados et al. 2010; Ibrahim et al. 2011; Zarate et al. 2012; Murrough et al. 2013a, b) has suggested possible plasticity-dependent mechanisms. Administration of ketamine at therapeutic doses is believed to stimulate synaptic plasticity (Duman and Aghajanian 2012, Monteggia and Zarate 2015). Evidence for this hypothesis comes from a recent study showing that ketamine may stimulate BDNF expression via inhibition of a eukaryotic elongation factor 2 signaling pathway that ultimately results in reduced suppression of BDNF translation (Autry et al. 2011). Moreover, this study found that administration of ketamine failed to produce antidepressant effects in BDNF knockout animals, suggesting that enhanced BDNF translation may be a necessary component for ketamine's antidepressant effects.

In sum, two plausible biological pathways are proposed that may result in disruption of DAergic corticostriatal circuitry and subsequent reductions in motivation and reinforcement learning. Both of these pathways have found significant support in preclinical and clinical studies of depression, though we note that only one involves a direct effect on DA itself, which may partially account for some of the heterogeneity observed in DA imaging studies in depression. It is worth noting that some studies have suggested that these pathways may interact, as inflammation may directly impair peripheral BDNF levels in humans (Lotrich et al. 2013), as well as disrupt neurogenesis in rodents (Monje et al. 2003). Conversely, DA is believed to play a role in synaptic plasticity via promoting long-term potentiation or depression within striatal circuits, as described above (Reynolds et al. 2001; Wieland et al. 2015). Therefore, it may be the case that inflammation and impaired plasticity represent distinct points of vulnerability within a common circuit and may act independently or in concert to produce deficits in motivation and reinforcement learning associated with depression.

6 Conclusion

Anhedonia is a complex symptom domain that may include multiple facets associated with approach behavior. Over the years, there have been various efforts to introduce new terminology to distinguish the narrow definition of anhedonia as "loss of pleasure" from the much broader connotation used in the clinical literature (Klein 1987; Salamone et al. 1994; Treadway and Zald 2011), but a new consensus has yet to emerge. This can hamper translational efforts, as studies performed at different levels may use the same terms to describe distinct processes. As described in this chapter, the various components of anhedonia are likely instantiated by distinct neural circuits, and isolating these factors biologically may require parsing this symptom domain more finely than has been achieved by many clinical measures.

In this review, we have focused primarily on possible mechanisms related to deficits in motivation and reinforcement learning, as these are two possible components of anhedonia that have received empirical support in recent years. This should not be taken to imply, however, that other aspects of anhedonic symptoms, such as loss of pleasure, are unimportant or less prevalent. Indeed, these questions have yet to be fully addressed in part because the relational structure of different components within the anhedonic symptom domain remains unknown. Rather, a possible advantage of the focus on motivation and reinforcement learning is that there is a rich preclinical literature upon which to draw. As summarized above, this approach has already begun to bear fruit, as studies linking elevated neuroinflammation, decreased dopamine signaling, and reduced motivation have begun to point to the targeted administration of anti-inflammatory treatments for individuals with depression and high inflammation. In the not-too-distant future, one can even imagine that screening for high inflammation may become a routine part of selecting an antidepressant treatment. While more needs to be done, progress has been made in our understanding of the pathophysiology of motivational symptoms in depression.

Acknowledgements and Disclosures The author wishes to thank Amanda Arulpragasam for helpful comments. The author declares no conflict of interest, financial or otherwise. The author has served as a paid consultant to Avanir Pharmaceuticals and the Boston Consulting Group. No funding or sponsorship was provided by these companies for the current work, and all views expressed herein are solely those of author.

References

Agren H, Reibring L (1994) PET studies of presynaptic monoamine metabolism in depressed patients and healthy volunteers. Pharmacopsychiatry 27(1):2–6

Amsterdam JD, Newberg AB (2007) A preliminary study of dopamine transporter binding in bipolar and unipolar depressed patients and healthy controls. Neuropsychobiology 55(3–4): 167–170

Autry AE, Adachi M, Nosyreva E, Na ES, Los MF, Cheng P-F, Kavalali ET, Monteggia LM (2011) NMDA receptor blockade at rest triggers rapid behavioural antidepressant responses. Nature 475(7354):91–95

Autry AE, Monteggia LM (2012) Brain-derived neurotrophic factor and neuropsychiatric disorders. Pharmacol Rev 64(2):238–258

Barch DM, Dowd EC (2010) Goal representations and motivational drive in schizophrenia: the role of prefrontal-striatal interactions. Schizophr Bull 36(5):919–934

Bardgett ME, Depenbrock M, Downs N, Points M, Green L (2009) Dopamine modulates effort-based decision making in rats. Behav Neurosci 123(2):242–251

Beierholm U, Guitart-Masip M, Economides M, Chowdhury R, Düzel E, Dolan R, Dayan P (2013) Dopamine modulates reward-related vigor. Neuropsychopharmacology 38:1495–1503

Berridge KC, Kringelbach ML (2008) Affective neuroscience of pleasure: reward in humans and animals. Psychopharmacology 199(3):457–480

Blume J, Douglas SD, Evans DL (2011) Immune suppression and immune activation in depression. Brain Behav Immun 25(2):221–229

Bragulat V, Paillere-Martinot ML, Artiges E, Frouin V, Poline JB, Martinot JL (2007) Dopaminergic function in depressed patients with affective flattening or with impulsivity: [18F] fluoro-L-dopa positron emission tomography study with voxel-based analysis. Psychiatry Res 154(2):115–124

Bus B, Molendijk M, Tendolkar I, Penninx B, Prickaerts J, Elzinga B, Voshaar R (2015) Chronic depression is associated with a pronounced decrease in serum brain-derived neurotrophic factor over time. Mol Psychiatry 20(5):602–608

Cannon DM, Klaver JM, Peck SA, Rallis-Voak D, Erickson K, Drevets WC (2009) Dopamine type-1 receptor binding in major depressive disorder assessed using positron emission tomography and [11C]NNC-112. Neuropsychopharmacology 34(5):1277–1287

Capuron L, Miller AH (2011) Immune system to brain signaling: neuropsychopharmacological implications. Pharmacol Ther 130(2):226–238

Capuron L, Pagnoni G, Demetrashvili MF, Lawson DH, Fornwalt FB, Woolwine B, Berns GS, Nemeroff CB, Miller AH (2007) Basal ganglia hypermetabolism and symptoms of fatigue during interferon-alpha therapy. Neuropsychopharmacology 32(11):2384–2392

Capuron L, Pagnoni G, Drake DF, Woolwine BJ, Spivey JR, Crowe RJ, Votaw JR, Goodman MM, Miller AH (2012) Dopaminergic mechanisms of reduced Basal Ganglia responses to hedonic reward during interferon alfa administration. Arch Gen Psychiatry 69(10):1044–1053

Capuron L, Ravaud A, Miller AH, Dantzer R (2004) Baseline mood and psychosocial characteristics of patients developing depressive symptoms during interleukin-2 and/or interferon-alpha cancer therapy. Brain Behav Immun 18(3):205–213

Chana G, Landau S, Beasley C, Everall IP, Cotter D (2003) Two-dimensional assessment of cytoarchitecture in the anterior cingulate cortex in major depressive disorder, bipolar disorder, and schizophrenia: evidence for decreased neuronal somal size and increased neuronal density. Biol Psychiatry 53(12):1086–1098

Cotter D, Mackay D, Landau S, Kerwin R, Everall I (2001) Reduced glial cell density and neuronal size in the anterior cingulate cortex in major depressive disorder. Arch Gen Psychiatry 58(6):545–553

Couch Y, Anthony DC, Dolgov O, Revischin A, Festoff B, Santos AI, Steinbusch HW, Strekalova T (2013) Microglial activation, increased TNF and SERT expression in the prefrontal cortex define stress-altered behaviour in mice susceptible to anhedonia. Brain Behav Immun 29:136–146

Cousins MS, Salamone JD (1994) Nucleus accumbens dopamine depletions in rats affect relative response allocation in a novel cost/benefit procedure. Pharmacol Biochem Behav 49(1):85–91

Croxson PL, Walton ME, O'Reilly JX, Behrens TE, Rushworth MF (2009) Effort-based cost-benefit valuation and the human brain. J Neurosci 29(14):4531–4541

D'Ardenne K, Nystrom LE, Cohen JD (2008) BOLD responses reflecting dopaminergic signals in the human ventral tegmental area. Science 5867(319):1264–1267

D'Haenen HA, Bossuyt A (1994) Dopamine D2 receptors in depression measured with single photon emission computed tomography. Biol Psychiatry 35(2):128–132

Dantzer R, Meagher MW, Cleeland CS (2012) Translational approaches to treatment-induced symptoms in cancer patients. Nat Rev Clin Oncol 9(7):414–426

Diazgranados N, Ibrahim L, Brutsche NE, Newberg A, Kronstein P, Khalife S, Kammerer WA, Quezado Z, Luckenbaugh DA, Salvadore G, Machado-Vieira R, Manji HK, Zarate CA Jr (2010) A randomized add-on trial of an N-methyl-D-aspartate antagonist in treatment-resistant bipolar depression. Arch Gen Psychiatry 67(8):793–802

Dickens C, Creed F (2001) The burden of depression in patients with rheumatoid arthritis. Rheumatology 40(12):1327–1330

Dowlati Y, Herrmann N, Swardfager W, Liu H, Sham L, Reim EK, Lanctot KL (2010) A meta-analysis of cytokines in major depression. Biol Psychiatry 67(5):446–457

Duman RS, Aghajanian GK (2012) Synaptic dysfunction in depression: potential therapeutic targets. Science 338(6103):68–72

Dwivedi Y, Rizavi HS, Conley R, Pandey G (2005) ERK MAP kinase signaling in post-mortem brain of suicide subjects: differential regulation of upstream Raf kinases Raf-1 and B-Raf. Mol Psychiatry 11(1):86–98

Eisenberger NI, Berkman ET, Inagaki TK, Rameson LT, Mashal NM, Irwin MR (2010) Inflammation-induced anhedonia: endotoxin reduces ventral striatum responses to reward. Biol Psychiatry 68(8):748–754

Felger JC, Alagbe O, Hu F, Mook D, Freeman AA, Sanchez MM, Kalin NH, Ratti E, Nemeroff CB, Miller AH (2007) Effects of interferon-alpha on rhesus monkeys: a nonhuman primate model of cytokine-induced depression. Biol Psychiatry 62(11):1324–1333

Felger JC, Miller AH (2014) Cytokine effects on the basal ganglia and dopamine function: the subcortical source of inflammatory malaise. Front Neuroendocrinol

Felger JC, Mun J, Kimmel HL, Nye JA, Drake DF, Hernandez CR, Freeman AA, Rye DB, Goodman MM, Howell LL (2013) Chronic interferon-α decreases dopamine 2 receptor binding and striatal dopamine release in association with anhedonia-like behavior in nonhuman primates. Neuropsychopharmacology 38(11):2179–2187

Floresco SB (2015) The nucleus accumbens: an interface between cognition, emotion, and action. Annu Rev Psychol 66:25–52

Foussias G, Remington G (2008) Negative symptoms in schizophrenia: avolition and Occam's razor. Schizophr Bull 36(2):359–369

Frank MJ, O'Reilly RC (2006) A mechanistic account of striatal dopamine function in human cognition: psychopharmacological studies with cabergoline and haloperidol. Behav Neurosci 120(3):497

Frank MJ, Seeberger LC, O'Reilly RC (2004) By carrot or by stick: cognitive reinforcement learning in parkinsonism. Science 306(5703):1940–1943

Gershon AA, Vishne T, Grunhaus L (2007) Dopamine D2-like receptors and the antidepressant response. Biol Psychiatry 61(2):145–153

Gold JM, Waltz JA, Prentice KJ, Morris SE, Heerey EA (2008) Reward processing in schizophrenia: a deficit in the representation of value. Schizophr Bull 34(5):835–847

Haber SN, Knutson B (2010) The reward circuit: linking primate anatomy and human imaging. Neuropsychopharmacology 35(1):4–26

Harrison NA, Brydon L, Walker C, Gray MA, Steptoe A, Critchley HD (2009) Inflammation causes mood changes through alterations in subgenual cingulate activity and mesolimbic connectivity. Biol Psychiatry 66(5):407–414

Hart AS, Rutledge RB, Glimcher PW, Phillips PEM (2014) Phasic dopamine release in the rat nucleus accumbens symmetrically encodes a reward prediction error term. J Neurosci 34 (3):698–704

Hartmann MN, Hager OM, Tobler PN, Kaiser S (2013) Parabolic discounting of monetary rewards by physical effort. Behav Process 100:192–196

Hirvonen J, Karlsson H, Kajander J, Markkula J, Rasi-Hakala H, Nagren K, Salminen JK, Hietala J (2008) Striatal dopamine D2 receptors in medication-naive patients with major depressive disorder as assessed with [11C]raclopride PET. Psychopharmacology 197(4):581–590

Howren MB, Lamkin DM, Suls J (2009) Associations of depression with C-reactive protein, IL-1, and IL-6: a meta-analysis. Psychosom Med 71(2):171–186

Ibrahim L, Diazgranados N, Luckenbaugh DA, Machado-Vieira R, Baumann J, Mallinger AG, Zarate CA Jr (2011) Rapid decrease in depressive symptoms with an N-methyl-d-aspartate antagonist in ECT-resistant major depression. Prog Neuropsychopharmacol Biol Psychiatry 35(4):1155–1159

Insel T, Cuthbert B, Garvey M, Heinssen R, Pine DS, Quinn K, Sanislow C, Wang P (2010) Research domain criteria (RDoC): toward a new classification framework for research on mental disorders. Am J Psychiatry 167(7):748–751

Insel TR, Cuthbert BN (2015) Brain disorders? Precisely. Science 348(6234):499–500

Kempermann G, Kronenberg G (2003) Depressed new neurons?, Adult hippocampal neurogenesis and a cellular plasticity hypothesis of major depression. Biol Psychiatry 54(5):499–503

Kempton MJ, Salvador Z, Munafo MR, Geddes JR, Simmons A, Frangou S, Williams SC (2011) Structural neuroimaging studies in major depressive disorder. Meta-analysis and comparison with bipolar disorder. Arch Gen Psychiatry 68(7):675–690

Klein DF (1974) Endogenomorphic depression. A conceptual and terminological revision. Arch Gen Psychiatry 31(4):447–454

Klein DN (1987) Depression and Anhedonia. Anhedonia Affect Deficit States 31:1–14

Klimek V, Schenck JE, Han H, Stockmeier CA, Ordway GA (2002) Dopaminergic abnormalities in amygdaloid nuclei in major depression: a postmortem study. Biol Psychiatry 52(7):740–748

Klimke A, Larisch R, Janz A, Vosberg H, Muller-Gartner HW, Gaebel W (1999) Dopamine D2 receptor binding before and after treatment of major depression measured by [123I]IBZM SPECT. Psychiatry Res 90(2):91–101

Kravitz AV, Tye LD, Kreitzer AC (2012) Distinct roles for direct and indirect pathway striatal neurons in reinforcement. Nat Neurosci 15(6):816–818

Kumar P, Waiter G, Ahearn T, Milders M, Reid I, Steele JD (2008) Abnormal temporal difference reward-learning signals in major depression. Brain 131(Pt 8):2084–2093

Kurniawan IT, Seymour B, Talmi D, Yoshida W, Chater N, Dolan RJ (2010) Choosing to make an effort: the role of striatum in signaling physical effort of a chosen action. J Neurophysiol 104 (1):313–321

Laasonen-Balk T, Kuikka J, Viinamaki H, Husso-Saastamoinen M, Lehtonen J, Tiihonen J (1999) Striatal dopamine transporter density in major depression. Psychopharmacology 144 (3):282–285

Lotrich FE, Albusaysi S, Ferrell RE (2013) Brain-derived neurotrophic factor serum levels and genotype: association with depression during interferon-α treatment. Neuropsychopharmacology 38(6):985–995

MacQueen GM, Campbell S, McEwen BS, Macdonald K, Amano S, Joffe RT, Nahmias C, Young LT (2003) Course of illness, hippocampal function, and hippocampal volume in major depression. Proc Natl Acad Sci 100(3):1387–1392

Martinot M, Bragulat V, Artiges E, Dolle F, Hinnen F, Jouvent R, Martinot J (2001) Decreased presynaptic dopamine function in the left caudate of depressed patients with affective flattening and psychomotor retardation. Am J Psychiatry 158(2):314–316

McGuire JT, Botvinick MM (2010) Prefrontal cortex, cognitive control, and the registration of decision costs. Proc Natl Acad Sci USA 107(17):7922–7926

Meehl PE (1975) Hedonic capacity: some conjectures. Bull Menninger Clin 39(4):295–307

Meyer JH, Kruger S, Wilson AA, Christensen BK, Goulding VS, Schaffer A, Minifie C, Houle S, Hussey D, Kennedy SH (2001) Lower dopamine transporter binding potential in striatum during depression. NeuroReport 12(18):4121–4125

Miller AH, Maletic V, Raison CL (2009a) Inflammation and its discontents: the role of cytokines in the pathophysiology of major depression. Biol Psychiatry 65(9):732–741

Miller AH, Maletic V, Raison CL (2009b) Inflammation and its discontents: the role of cytokines in the pathophysiology of major depression. Biol Psychiatry 65(9):732–741

Molendijk M, Spinhoven P, Polak M, Bus B, Penninx B, Elzinga B (2014) Serum BDNF concentrations as peripheral manifestations of depression: evidence from a systematic review and meta-analyses on 179 associations (N= 9484). Mol Psychiatry 19(7):791–800

Monje ML, Toda H, Palmer TD (2003) Inflammatory blockade restores adult hippocampal neurogenesis. Science 302(5651):1760–1765

Monkul E, Hatch JP, Nicoletti MA, Spence S, Brambilla P, Lacerda AL, Sassi RB, Mallinger A, Keshavan M, Soares JC (2006) Fronto-limbic brain structures in suicidal and non-suicidal female patients with major depressive disorder. Mol Psychiatry 12(4):360–366

Monteggia LM, Zarate C (2015) Antidepressant actions of ketamine: from molecular mechanisms to clinical practice. Curr Opin Neurobiol 30:139–143

Murrough JW, Iosifescu DV, Chang LC, Al Jurdi RK, Green CE, Perez AM, Iqbal S, Pillemer S, Foulkes A, Shah A, Charney DS, Mathew SJ (2013a) Antidepressant efficacy of ketamine in treatment-resistant major depression: a two-site randomized controlled trial. Am J Psychiatry 170(10):1134–1142

Murrough JW, Perez AM, Pillemer S, Stern J, Parides MK, aan het Rot M, Collins KA, Mathew SJ, Charney DS, Iosifescu DV (2013b) Rapid and longer-term antidepressant effects of repeated ketamine infusions in treatment-resistant major depression. Biol Psychiatry 74 (4):250–256

Nunes EJ, Randall PA, Estrada A, Epling B, Hart EE, Lee CA, Baqi Y, Mueller CE, Correa M, Salamone JD (2013) Effort-related motivational effects of the pro-inflammatory cytokine interleukin 1-beta: studies with the concurrent fixed ratio 5/chow feeding choice task. Psychopharmacology 1–10

Parsey RV, Oquendo MA, Zea-Ponce Y, Rodenhiser J, Kegeles LS, Pratap M, Cooper TB, Van Heertum R, Mann JJ, Laruelle M (2001) Dopamine D(2) receptor availability and amphetamine-induced dopamine release in unipolar depression. Biol Psychiatry 50(5):313–322

Pessiglione M, Schmidt L, Draganski B, Kalisch R, Lau H, Dolan RJ, Frith CD (2007) How the brain translates money into force: a neuroimaging study of subliminal motivation. Science 316 (5826):904–906

Pessiglione M, Seymour B, Flandin G, Dolan RJ, Frith CD (2006) Dopamine-dependent prediction errors underpin reward-seeking behaviour in humans. Nature 442(7106):1042–1045

Pincus T, Griffith J, Pearce S, Isenberg D (1996) Prevalence of self-reported depression in patients with rheumatoid arthritis. Br J Rheumatol 35(9):879–883

Pittenger C, Duman RS (2007) Stress, depression, and neuroplasticity: a convergence of mechanisms. Neuropsychopharmacology 33(1):88–109

Pizzagalli DA, Holmes AJ, Dillon DG, Goetz EL, Birk JL, Bogdan R, Dougherty DD, Iosifescu DV, Rauch SL, Fava M (2009) Reduced caudate and nucleus accumbens response to rewards in unmedicated individuals with major depressive disorder. Am J Psychiatry 166 (6):702–710

Pizzagalli DA, Iosifescu D, Hallett LA, Ratner KG, Fava M (2008) Reduced hedonic capacity in major depressive disorder: evidence from a probabilistic reward task. J Psychiatr Res 43 (1):76–87

Prevost C, Pessiglione M, Metereau E, Clery-Melin ML, Dreher JC (2010) Separate valuation subsystems for delay and effort decision costs. J Neurosci 30(42):14080–14090

Qin L, Wu X, Block ML, Liu Y, Breese GR, Hong JS, Knapp DJ, Crews FT (2007) Systemic LPS causes chronic neuroinflammation and progressive neurodegeneration. Glia 55(5):453–462

Rahmim A, Zaidi H (2008) PET versus SPECT: strengths, limitations and challenges. Nucl Med Commun 29(3):193–207

Raison CL, Miller AH (2011) Is depression an inflammatory disorder? Current Psychiatry Reports 13(6):467–475

Raison CL, Rutherford RE, Woolwine BJ, Shuo C, Schettler P, Drake DF, Haroon E, Miller AH (2013) A randomized controlled trial of the tumor necrosis factor antagonist infliximab for treatment-resistant depression. JAMA Psychiatry 70(1):31–41

Reynolds JN, Hyland BI, Wickens JR (2001) A cellular mechanism of reward-related learning. Nature 413(6851):67–70

Ribot T (1896) La psychologie des sentiment (The psychology of feelings). Felix Alcan, Paris

Rushworth MF, Noonan MP, Boorman ED, Walton ME, Behrens TE (2011) Frontal cortex and reward-guided learning and decision-making. Neuron 70(6):1054–1069

Salamone JD, Correa M, Farrar A, Mingote SM (2007) Effort-related functions of nucleus accumbens dopamine and associated forebrain circuits. Psychopharmacology 191(3):461–482

Salamone JD, Cousins MS, Bucher S (1994) Anhedonia or anergia? Effects of haloperidol and nucleus accumbens dopamine depletion on instrumental response selection in a T-maze cost/benefit procedure. Behav Brain Res 65(2):221–229

Sarchiapone M, Carli V, Camardese G, Cuomo C, Di Giuda D, Calcagni ML, Focacci C, De Risio S (2006) Dopamine transporter binding in depressed patients with anhedonia. Psychiatry Res 147(2–3):243–248

Sarkar C, Basu B, Chakroborty D, Dasgupta PS, Basu S (2010) The immunoregulatory role of dopamine: an update. Brain Behav Immun 24(4):525–528

Schmaal L, Veltman DJ, van Erp TG, Sämann P, Frodl T, Jahanshad N, Loehrer E, Tiemeier H, Hofman A, Niessen W (2015) Subcortical brain alterations in major depressive disorder: findings from the ENIGMA Major Depressive Disorder working group. Molecular Psychiatry

Schmidt L, Lebreton M, Clery-Melin ML, Daunizeau J, Pessiglione M (2012) Neural mechanisms underlying motivation of mental versus physical effort. PLoS Biol 10(2):e1001266

Schultz W (2007) Multiple dopamine functions at different time courses. Annu Rev Neurosci 30:259–288

Shah PJ, Ogilvie AD, Goodwin GM, Ebmeier KP (1997) Clinical and psychometric correlates of dopamine D2 binding in depression. Psychol Med 27(6):1247–1256

Sharot T, Shiner T, Brown AC, Fan J, Dolan RJ (2009) Dopamine enhances expectation of pleasure in humans. Curr Biol 19(24):2077–2080

Sheline YI, Sanghavi M, Mintun MA, Gado MH (1999) Depression duration but not age predicts hippocampal volume loss in medically healthy women with recurrent major depression. J Neurosci 19(12):5034–5043

Shelton RC, Tomarken AJ (2001) Can recovery from depression be achieved? Psychiatr Serv 52 (11):1469–1478
Sherdell L, Waugh CE, Gotlib IH (2011) Anticipatory pleasure predicts motivation for reward in major depression. J Abnorm Psychol
Simon NW, Beas BS, Montgomery KS, Haberman RP, Bizon JL, Setlow B (2013) Prefrontal cortical–striatal dopamine receptor mRNA expression predicts distinct forms of impulsivity. Eur J Neurosci 37:1779–1788
Strauss GP, Gold JM (2012) A new perspective on anhedonia in Schizophrenia. Am J Psychiatry
Sutton RS, Barto AG (1998) Reinforcement learning: an introduction. MIT Press, Cambridge
Thorne R, Hanson L, Ross T, Tung D, Frey Ii W (2008) Delivery of interferon-β to the monkey nervous system following intranasal administration. Neuroscience 152(3):785–797
Treadway MT, Bossaller NA, Shelton RC, Zald DH (2012a) Effort-based decision-making in major depressive disorder: a translational model of motivational anhedonia. J Abnorm Psychol 121(3):553
Treadway MT, Buckholtz JW, Cowan RL, Woodward ND, Li R, Ansari MS, Baldwin RM, Schwartzman AN, Kessler RM, Zald DH (2012b) Dopaminergic mechanisms of individual differences in human effort-based decision-making. J Neurosci 32(18):6170–6176
Treadway MT, Buckholtz JW, Schwartzman AN, Lambert WE, Zald DH (2009) Worth the 'EEfRT'? The effort expenditure for rewards task as an objective measure of motivation and anhedonia. PLoS ONE 4(8):e6598
Treadway MT, Pizzagalli DA (2014) Imaging the pathophysiology of major depressive disorder-from localist models to circuit-based analysis. Biol Mood Anxiety Disord 4(5)
Treadway MT, Zald DH (2011) Reconsidering anhedonia in depression: lessons from translational neuroscience. Neurosci Biobehav Rev 35(3):537–555
Treadway MT, Zald DH (2013) Parsing anhedonia translational models of reward-processing deficits in psychopathology. Curr Dir Psychol Sci 22(3):244–249
Tsai HC, Zhang F, Adamantidis A, Stuber GD, Bonci A, de Lecea L, Deisseroth K (2009) Phasic firing in dopaminergic neurons is sufficient for behavioral conditioning. Science 324 (5930):1080–1084
Vichaya EG, Hunt SC, Dantzer R (2014) Lipopolysaccharide Reduces incentive motivation while boosting preference for high reward in mice. Neuropsychopharmacology
Vrieze E, Ceccarini J, Pizzagalli DA, Bormans G, Vandenbulcke M, Demyttenaere K, Van Laere K, Claes S (2011) Measuring extrastriatal dopamine release during a reward learning task. Hum Brain Mapp
Vrieze E, Pizzagalli DA, Demyttenaere K, Hompes T, Sienaert P, de Boer P, Schmidt M, Claes S (2013) Reduced reward learning predicts outcome in major depressive disorder. Biol Psychiatry
Wardle MC, Treadway MT, Mayo LM, Zald DH, de Wit H (2011) Amping up effort: effects of d-amphetamine on human effort-based decision-making. J Neurosci 31(46):16597–16602
Westbrook A, Kester D, Braver TS (2013) What is the subjective cost of cognitive effort? Load, trait, and aging effects revealed by economic preference. PLoS ONE 8(7):e68210
Whitton AE, Treadway MT, Pizzagalli DA (2015) Reward processing dysfunction in major depression, bipolar disorder and schizophrenia. Curr Opin Psychiatry 28(1):7–12
Wieland S, Schindler S, Huber C, Köhr G, Oswald MJ, Kelsch W (2015) Phasic dopamine modifies sensory-driven output of striatal neurons through synaptic plasticity. J Neurosci 35 (27):9946–9956
Yan Y, Jiang W, Liu L, Wang X, Ding C, Tian Z, Zhou R (2015) Dopamine Controls Systemic Inflammation through Inhibition of NLRP3 Inflammasome. Cell 160(1):62–73
Yang X-H, Huang J, Zhu C-Y, Wang Y-F, Cheung EFC, Chan RCK, Xie G-R (2014) Motivational deficits in effort-based decision making in individuals with subsyndromal depression, first-episode and remitted depression patients. Psychiatry Res 220(3):874–882
Yang YK, Yeh TL, Yao WJ, Lee IH, Chen PS, Chiu NT, Lu RB (2008) Greater availability of dopamine transporters in patients with major depression–a dual-isotope SPECT study. Psychiatry Res 162(3):230–235

Yoshii A, Constantine-Paton M (2010) Postsynaptic BDNF-TrkB signaling in synapse maturation, plasticity, and disease. Dev Neurobiol 70(5):304–322

Zarate CA Jr, Brutsche NE, Ibrahim L, Franco-Chaves J, Diazgranados N, Cravchik A, Selter J, Marquardt CA, Liberty V, Luckenbaugh DA (2012) Replication of ketamine's antidepressant efficacy in bipolar depression: a randomized controlled add-on trial. Biol Psychiatry 71(11):939–946

Zarate CA Jr, Singh JB, Carlson PJ, Brutsche NE, Ameli R, Luckenbaugh DA, Charney DS, Manji HK (2006) A randomized trial of an N-methyl-D-aspartate antagonist in treatment-resistant major depression. Arch Gen Psychiatry 63(8):856–864

Motivational Deficits and Negative Symptoms in Schizophrenia: Concepts and Assessments

L. Felice Reddy, William P. Horan and Michael F. Green

Abstract Recent years have seen a resurgence of interest in motivational disturbances in schizophrenia. This is largely driven by the recognition that these disturbances are central to the "experiential" subdomain of negative symptoms and are particularly important determinants of functional disability. Research into the causes and treatment of experiential negative symptoms is therefore a high priority. This chapter reviews findings from experimental psychopathology and affective science relevant to understanding the neurobehavioral processes that underlie these negative symptoms. We focus on abnormalities in four processes that have received the most attention as likely contributors: anticipatory pleasure, reward learning, effort-based decision-making, and social motivation. We also review the research literature on pharmacological and psychosocial approaches to reduce functional deficits attributable to negative symptoms. Translational research is beginning to inform the development of new treatments specifically designed to target the experiential subdomain of negative symptoms.

Keywords Negative symptoms · Motivation · Schizophrenia · Neurobehavioral · Reward learning · Effort-based decision-making · Social motivation

Contents

1	Introduction	358
	1.1 Historical Context	358
	1.2 Types of Negative Symptoms	358
	1.3 The Anhedonia Paradox	359
2	Possible Factors that Contribute to Motivational Negative Symptoms	360
	2.1 Abnormalities in Anticipatory Pleasure	360
	2.2 Alterations in Reward Learning	362
	2.3 Effort-Based Decision-Making	363

L.F. Reddy (✉) · W.P. Horan · M.F. Green
VA Greater Los Angeles Healthcare System, University of California, MIRECC 210A, Bldg. 210, 11301 Wilshire Blvd., Los Angeles, CA 90073, USA
e-mail: lenafelice@ucla.edu

© Springer International Publishing Switzerland 2015

```
2.4   Social Motivation (Approach/Avoidance) ............................................... 364
3  Interventions for Motivational Negative Symptoms ........................................ 365
   3.1   Conventional Approaches ...................................................................... 366
   3.2   Recent Developments............................................................................. 367
4  Conclusions ..................................................................................................... 369
References............................................................................................................. 369
```

1   Introduction

1.1   Historical Context

Motivation disturbances, predominantly considered in the context of negative symptoms, have long been acknowledged as a core clinical feature of schizophrenia (Meehl 1962). In Kraepelin's original characterization of schizophrenia, he described avolition as a primary cause of the progressive deterioration he observed in this disorder (Kraepelin 1971). Similarly, Bleuler described schizophrenia as reflecting a breakdown in emotion and motivation, with affective indifference as the primary marker (Bleuler 1950). Recent years have seen a resurgence of interest in motivational disturbances in schizophrenia. This renewed interest is largely driven by the recognition that negative symptoms and associated motivational disturbances are key contributors to the profound functional disability that is a hallmark of schizophrenia. In addition, the emergence of concepts and methods from the burgeoning field of affective neuroscience has facilitated translational research into the neurobehavioral correlates of these disturbances. It is now understood that motivational disturbances are central to a particular subcomponent of negative symptoms.

1.2   Types of Negative Symptoms

Negative symptoms broadly refer to the absence or diminution of normal functions in the areas of emotion, sociality, productive goal-directed behavior, and communication. Although a number of clinical ratings scales have been used for several decades to assess negative symptoms, these scales were based on various definitions of negative symptoms and differed in the specific domains that were assessed. In 2006, an NIMH-sponsored conference was held with the goal of achieving consensus on the optimal assessment of negative symptoms and identifying the most impactful ways to advance novel treatment development for these disabling symptoms (Kirkpatrick, Fenton et al. 2006). Based on a comprehensive literature review and input from diverse stakeholders, five core consensus-based negative symptoms were defined. These include *avolition*, reflecting diminished engagement in intellectual, occupational, and social pursuits; *anhedonia*, a reduction in the range and intensity of pleasant emotions; *asociality,* social withdrawal or avoidance, and a lack of meaningful interpersonal connections; *restricted affect*, reduced emotional expression including diminished facial expressions, gestures, and vocal intonation; *alogia*, diminished verbal production and spontaneous speech.

The 2006 conference members evaluated the overall structure of negative symptoms after a comprehensive review of the literature on the factor structure of negative symptoms across a variety of clinical ratings scales. These studies provided consistent support for two separable factors: (1) an experiential dimension reflecting disturbances in motivation and pleasure that are central to avolition, anhedonia, and asociality; and (2) an expressivity dimension comprised of restricted affect and poverty of speech. It was also established that these factors were separable from other aspects of psychopathology, including cognitive impairment (which was included on some negative symptom scales) and positive symptoms (delusions, hallucinations, disorganization) (Blanchard and Cohen 2006). As discussed further below, the conference led to the development of two new state-of-the-art negative symptom assessment scales that are organized in terms of the experiential and expressive domains of negative symptoms (Kane 2013; Marder and Kirkpatrick 2014).

The recognition that negative symptoms consist of two components spurred interest in the question of whether they show similar or different relations to poor community outcome. Available research indicates that the experiential component shows stronger and more consistent relations to functioning, including the areas of work, independent living, and social networks. This pattern was demonstrated in studies using older negative scales (e.g., SANS) (e.g., Mueser 1994; Sayers et al. 1996; Milev et al. 2005; Siegel et al. 2006; Dowd and Barch 2010; Strauss et al. 2013). It has also been replicated in studies using the two more recently developed scales (Horan et al. 2011; Kirkpatrick et al. 2011; Strauss et al. 2012; Kring et al. 2013). Thus, the experiential negative symptoms, which are defined by motivational disturbances, comprise the subdomain that appears most strongly related to community functioning.

1.3 The Anhedonia Paradox

An important, and as yet unanswered, question is: Exactly what specific neurobehavioral process(es) contribute(s) to the motivational/emotion disturbances that manifest clinically as experiential negative symptoms? Historically, it was believed that diminished engagement in rewarding and productive activities reflected a fundamental inability to experience pleasure. Indeed, the available clinical interview and self-report measures available up to the 1990s consistently indicated that patients reported experiencing diminished levels of pleasure in their daily lives. However, a major challenge to this assumption emerged in the early 1990s when laboratory studies using affective science methods were first applied to individuals with schizophrenia and revealed an apparently paradoxical set of findings. The studies showed that when individuals with schizophrenia are actually presented with evocative stimuli (e.g., pictures, food, film clips), they experience in-the-moment affective responses comparable to those of controls (for reviews see, Horan et al. 2006a; Strauss et al. 2014). Further, this intact "consummatory pleasure" has been verified using other research techniques including neuroimaging (Dowd and Barch 2010), physiological measures (Curtis et al. 1999; Volz et al. 2003; Horan et al. 2010), and implicit preference measures (Herbener 2009; Waltz et al. 2009).

These findings converge to indicate that although individuals with schizophrenia show diminished engagement in rewarding and productive activities in their daily lives, this disengagement does not reflect a basic inability to experience pleasure. If individuals can and do experience normal levels of pleasure, why do not they seek out opportunities to engage in rewarding activities? Researchers have started to approach this question through a variety of conceptual frameworks and methods grounded in basic affective neuroscience. The following section reviews several of the main approaches that have been used.

2 Possible Factors that Contribute to Motivational Negative Symptoms

2.1 Abnormalities in Anticipatory Pleasure

An important insight from affective neuroscience is that "pleasure" is not a unitary construct. Instead, hedonic experience consists of multiple, distinct components. One key distinction that has been well established in basic animal research is between "liking" or consummatory, in-the-moment pleasure, and "wanting" or anticipatory motivation for future rewards. These involve dissociable neural substrates—liking involves serotoninergic/opioid systems in the nucleus accumbens shell, the ventral palladium, and the orbitofrontal cortex (Smith and Berridge 2007), whereas wanting involves the mesolimbic dopamine system and projections to ventral and dorsal striatal regions of the basal ganglia (Schultz et al. 1997; Barch and Dowd 2010). For a detailed review of the roles of wanting and liking in motivating behavior, see Robinson et al., in this volume. Although, as noted above, individuals with schizophrenia show relatively normal liking, that does not appear to be the case for anticipatory pleasure.

Experimental paradigms used in clinical research indicate neural and behavioral deficits in anticipatory pleasure in schizophrenia. One novel paradigm used an experimental approach to separately measure behavioral responses in a condition with an immediately available pleasant reward (repeated button presses could prolong exposure to rewarding pleasant images) versus a condition in which the pleasant reward is only imagined, or anticipated (Heerey and Gold 2007). There were no differences between patients and controls in the immediately available reward condition, but there were significant group differences in the anticipatory condition, such that patients responded at a lower rate (Heerey and Gold 2007).

This separation between wanting and liking has also been measured in schizophrenia patients using experience sampling (Gard et al. 2007), neuroimaging

(Dowd and Barch 2012), and event-related potentials (Wynn et al. 2010). For example, the ERP study by Wynn and colleagues used a cued, reaction-time contingent picture viewing task to assess two types of anticipatory ERP's, one involving motor response preparation [contingent negative variation (CNV)] and one during anticipation of emotionally significant stimuli [not involving motor preparation; stimulus-preceding negativity (SPN)]. In this task, participants were instructed to make a button press as quickly as possible when a cue appeared on the computer screen (+, 0, or −) that signified the valence (pleasant, neutral, unpleasant) of a forthcoming picture; faster reaction times resulted in longer subsequent picture presentations, whereas slow reaction times resulted in shorter picture presentations. Patients and healthy controls demonstrated similar patterns of reaction time, as well as self-reported emotion while viewing the different types of pictures (Fig. 1). However, patients demonstrated generally lower CNV (prior to button pressing) and SPN (prior to picture viewing) amplitudes than controls across the picture conditions. These results further support a separation between wanting and liking processes in individuals with schizophrenia. Taken together, the experience sampling, neuroimaging, and electrophysiological findings converge to indicate intact responsiveness to immediately present rewards versus impaired responsiveness to future rewards, implicating a diminution of anticipatory pleasure that impedes motivated behavior. Further supporting the notion that the anticipatory deficit may have real implications for functional outcome are studies that have reported significant correlations between diminished anticipatory pleasure and clinically rated negative symptoms (Gold et al. 2012; Kring and Barch 2014).

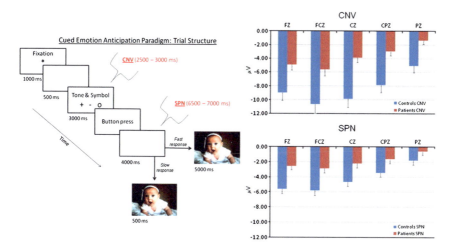

Fig. 1 Left panel: schematic of the trial structure for the cued emotion anticipation task. The CNV was measured as activity in the last 500 ms of the cue slide and the SPN as activity in the last 500 ms of the anticipatory period. Right panel: mean amplitudes (with standard error bars) for the CNV (top panel) and SPN (bottom panel) at electrodes Fz, FCz, Cz, CPz, and Pz for schizophrenia patients (*red columns*) and healthy controls (*blue columns*)

2.2 Alterations in Reward Learning

Another aspect of hedonic experience that has garnered significant attention in research aimed at understanding pathways to motivational deficits in schizophrenia is reward learning. Reward learning broadly involves the process of using feedback information to guide adaptive behavioral responding. A key aspect of reward learning involves the ability to adjust response tendencies in accordance with shifting reinforcement and punishment contingencies, which largely depends on functions associated with the ventral prefrontal cortex. A disturbance in the ability to correctly monitor and respond to positive or negative feedback signals would impede efforts to carry out goal-directed activities and could therefore contribute to motivational deficits in schizophrenia. Investigations into such deficits and their underlying neural mechanisms are grounded in well-established animal models, particularly those focusing on dopamine-mediated reward learning tasks. Reward learning deficits have been replicated across a range of paradigms in individuals with schizophrenia and have shown correlations with overall negative symptoms (Pantelis et al. 1999; Gold et al. 2008; Murray et al. 2008; Weiler et al. 2009; Waltz et al. 2013).

Deficits in reward learning may be particularly evident when individuals are required to continually and flexibly adapt behavior. Reversal learning is a prime example of a situational paradigm that requires continual response adaptations. Reversal learning refers to tasks in which stimulus–response contingencies are initially learned with responses to "correct" stimuli reinforced more often than those to incorrect stimuli (e.g., 80 vs. 20 % reinforcement, respectively). Mental flexibility is required because the reward contingencies are reversed throughout the task after a certain number of correct responses, such that a different stimulus becomes the "correct" choice and previously correct choices intermittently become "incorrect." Reversal learning tasks include an initial discrimination phase, in which subjects learn to detect correct versus incorrect stimuli, which is followed by a series of reversal phases, in which the response contingencies are inverted. Prediction errors are reflected in the dopaminergic neural signaling processes that underlie the initial learning and the relearning of reward contingencies following reversals. Prediction errors are termed "positive" when an unexpected reward triggers dopaminergic firing; prediction errors are "negative" when an expected reward is not delivered, and there is a decrease in or absence of dopaminergic firing. Functionally, prediction errors facilitate flexible reward learning of the alternating reward contingencies that occur throughout this type of task (Wise 2004). In both animal and human models, it is well established that reversal learning depends on dopamine-mediated striatal networks (e.g., Frank and Claus 2006; Pessiglione et al. 2006; Lee et al. 2007).

Reversal learning paradigms, as described above, are designed to sensitively test participants' ability to quickly and flexibly change response patterns in the context of reversed reward contingencies. Individuals with schizophrenia consistently show impairments on these tasks (Waltz and Gold 2007; Murray et al. 2008; Prentice et al. 2008; Leeson et al. 2009; McKirdy et al. 2009; Waltz et al. 2013; Schlagenhauf et al. 2014). The common finding among the studies published to date is a significant deficit in ability to reverse learned contingencies, above and beyond difficulties learning initial reward–response relationships. Several mechanisms of action have been considered to explain this deficit, including problems with value representation and reduced reward sensitivity, more errors related to frequent response switching, and aberrant salience. Although there is no consensus on causal mechanism, it does appear that difficulties with flexibly adapting behavioral responses in accordance with new feedback precede illness onset, persist in periods of remission, are apparent in medicated and non-medicated samples, and are not solely caused by generalized cognitive deficits. Reversal learning deficits are likely an important contributor to motivational deficits. For a more detailed discussion on reward learning deficits in patients with schizophrenia, and how such deficits contribute to motivational impairment, see Waltz et al. in this volume.

2.3 Effort-Based Decision-Making

When considering whether to engage in activities that may be rewarding, we evaluate not only the potential benefits, but also the associated costs. Across non-human and human animals, the costs associated with the effort required to obtain a reward are a major factor in the decisional balance. Effort-based decision-making refers to an individual's evaluation of the amount of effort required to obtain a given reward, and the subsequent decision of whether or not to engage in the effortful behavior. Animal studies show there is a general law of least effort in which organisms choose the least amount of effort expenditure necessary to obtain a given reward (Solomon 1948), but when a larger potential reward alternative is available, the decisional balance becomes more complex. Indeed, several factors contribute to the decision of how much effort to expend for a given reward. These factors include valuation of potential rewards, the perceived effort required to complete the associated task, and the likelihood that the reward will actually be received if the task is successfully completed. Thus, effort-based decision-making paradigms attempt to objectively assess the culmination of these processes—that is, motivated behavior defined as how much effort one is willing to exert for different levels of reward. Deficits in effort-based decision-making reflect impairments in the dopamine-rich regions of the prefrontal cortex, the anterior cingulate cortex, and the ventral striatum—this neural system mediates how the cost of effort is weighed against possible benefits (Kurniawan et al. 2010; Salamone and Correa 2012). Effort-based decision-making has been studied in different domains in animals and humans, including physical and cognitive effort, which appear to involve separable circuits (Hosking et al. 2014).

Abnormalities in effort-based decision-making have been examined in schizophrenia as potential contributors to negative symptoms. This has primarily been examined using tasks that measure physical effort. In the most commonly used type of task, participants must repeatedly press a button on a computer keyboard to earn various amounts of monetary reward. Participants make a series of choices between performing either low effort (few button presses)/low reward or higher effort (up to 100 button presses)/higher reward. A similar paradigm that involves choosing between low-effort/low-reward and higher effort/higher reward tasks using a handgrip device has also been used. In four out of six published studies of physical effort-based decision-making, individuals with schizophrenia were significantly less likely than healthy participants to select hard tasks as monetary reward, or probability of reward, increased, reflecting decreased willingness to exert physical effort for rewards (Fervaha, Graff-Guerrero et al. 2013; Gold et al. 2013; Barch et al. 2014; Hartmann et al. 2014; Docx et al. 2015; Treadway et al. 2015). All six studies examined associations between effort task performance and negative symptoms, and four found significant relations such that participants with more negative symptoms exerted less effort. An additional study used a progressive ratio break point task to examine cognitive effort-based decision-making (Wolf et al. 2014). In this task, participants chose when to cease exerting effort (in this case, making rapid numerical judgments) for additional increments of monetary reward—this cease juncture represents an effort "break point." Individuals with schizophrenia had significantly lower break points than healthy controls, and lower break points reflecting less willingness to exert cognitive effort correlated with higher levels of experiential negative symptoms. Thus, there is emerging evidence that people with schizophrenia tend to avoid activities with higher perceived effort demands, which could contribute to diminished motivation to seek out and engage in many types of potentially rewarding activities.

2.4 Social Motivation (Approach/Avoidance)

Another construct that has been used to investigate neurobehavioral processes that contribute to experiential negative symptoms is social approach/avoidance motivation. Across several models of motivation, a basic distinction is made between behavioral approach and behavioral avoidance (Gray 1987; Gable and Gosnell 2013; Spielberg et al. 2013). Behavioral approach relies on a reward system (i.e., behavioral activation system; BAS) sensitive to appetitive stimuli and the termination of punishment. Behavioral avoidance (i.e., behavioral inhibition system; BIS), in contrast, is sensitive to aversive stimuli and activated by anxiety, novelty, and innate fear stimuli and is responsible for ceasing or inhibiting behavior. Neurophysiological research in animals implicates corticolimbic circuitry including amygdala, insula, and prefrontal cortex in avoidance responses and corticostriatal circuitry (including ventral striatum and nucleus accumbens) in approach motivation (Aupperle and Paulus 2010). Similar to basic motivation, social motivation depends on both approach and avoidance mechanisms and the resolution of the two opposing systems. In healthy adults, social motivation refers to achieving a sense of belonging by pursuing and maintaining healthy social ties with others. Such motivation is associated with well-being and has both a need for affiliation component and a need to avoid rejection aspect (Gable 2006).

Clinical studies in schizophrenia indicate that there may be different types of motivational drives that lead to social functioning deficits. Some patients show a profound disinterest in social interactions in the apparent absence of loneliness, while others are interested in social connections but avoid them because of social anxiety or concerns about the harmful intentions of others (Horan and Blanchard 2003; Horan et al. 2006b). A recently published study further supports the notion that social motivation disturbances can be caused by lack of approach *or* motivated avoidance in schizophrenia (Reddy et al. 2014). This study involved 151 individuals with schizophrenia who completed measures of approach and avoidance motivation. A cluster analysis identified four distinguishable subgroups, two of which had particularly poor social functioning. One was characterized by diminished interest in people and diminished drive to develop close interpersonal attachments. The other subgroup, with elevated avoidance motivation, was characterized by avoidance of social interactions and social aloofness attributable to anxiety and/or fear of rejection. These findings suggest that distinct types of social motivational drives can lead to asociality in schizophrenia and that considering different subtypes may help address the vexing issue of heterogeneity and guide efforts to make treatments for social dysfunction more personalized.

In summary, negative symptoms associated with motivational disturbances are major determinants of impaired functioning for many individuals with schizophrenia. We know people with schizophrenia do not lack the basic capacity to experience pleasure. Although individuals with schizophrenia can and do experience pleasure from rewarding stimuli, they report and demonstrate markedly decreased tendencies to seek out and engage in productive, rewarding activities in their daily lives. Translational neuroscience research points to four primary processes that likely contribute to motivational deficits: anticipatory pleasure, reward learning, effort-based decision-making, and social motivation. Several lines of translational research are currently providing new insights into neurobehavioral processes that underlie these motivational disturbances, for details see the chapter by Ward in this volume. Advances in this area hold considerable promise for facilitating the development of more effective functional recovery-oriented treatments. In the final section, we review the current status of pharmacological and psychosocial interventions that have been evaluated as potential treatments for negative symptoms.

3 Interventions for Motivational Negative Symptoms

As noted above, there is widespread agreement that negative symptoms are an unmet therapeutic need in a large proportion of individuals with schizophrenia (Kirkpatrick et al. 2006). The efficacy of various conventional pharmacological and psychosocial approaches has been evaluated in terms of their effects on negative symptoms, and results have, unfortunately, been generally modest. However, there have been a number of recent developments in clinical trial design and assessment methodology, as well as novel pharmacological and psychosocial treatments that are encouraging.

3.1 Conventional Approaches

Over the past several decades, negative symptoms were often assessed within studies of treatments that were primarily developed to manage the positive symptoms of schizophrenia. In this context, the impact of several classes of medications on negative symptoms has been examined. First- and second-generation antipsychotics, as well as antidepressants, have been evaluated in numerous clinical trials, but the majority of studies failed to produce clinically significant improvements in primary negative symptoms (Rummel et al. 2005; Singh et al. 2010). In addition, several key methodological and assessment issues complicate the interpretability of these studies. Studies often enrolled patients in the midst of acute psychotic symptom exacerbations and used various assessment measures based on different definitions of negative symptoms. Study designs also rarely addressed possible secondary sources of negative symptoms, such as depression, extrapyramidal side effects, and paranoia (Davis et al. 2014). Furthermore, expressive and experiential negative symptoms were not differentiated as treatment targets or in outcome assessments.

Studies of conventional psychosocial approaches, while subject to the same study design and assessment limitations as pharmacological interventions, were somewhat more promising for negative symptoms. Social skills training (SST) was one of the first manualized interventions that targeted functional deficits (usually social communication). A large meta-analysis of SST revealed small-to-medium improvements in negative symptoms (effect size = 0.40) (Kurtz and Mueser 2008). Although gains in social communication skills are relevant to social motivation, SST does not directly address motivational impairments. Another intervention designed to improve functional outcomes, cognitive enhancement training (CET), combines cognitive remediation with SST and also showed significant improvements in negative symptoms at post-treatment and at follow-up (Eack et al. 2013). Finally, a number of studies of cognitive behavioral therapy (CBT) for psychosis have included assessments of negative symptoms as secondary outcomes. Some earlier studies reported significant benefits for negative symptoms at the end of treatment and/or at post-treatment follow-up (Wykes et al. 2008; Klingberg et al. 2011; Grant et al. 2012). However, a more recent meta-analytic review indicated that the overall effects were small and nonsignificant, suggesting that CBT studies focused on psychotic symptoms may not reduce negative symptom as well as previously thought (Velthorst et al. 2015). Importantly, neither SST, CET, nor CBT directly address any of the four processes hypothesized to contribute to motivational negative symptoms. The closest relation is SST and social motivation, but it is not a therapeutic target of the intervention. It has been recognized that motivation-enhancing techniques yield treatment-related improvements within cognitive remediation therapy. This work is reviewed by Saperstein & Medalia in this volume. In summary, conventional approaches inadequately address negative symptoms and novel treatment development that directly targets this symptom domain is sorely needed.

3.2 Recent Developments

Since the National Institute of Mental Health's Consensus Development Conference on Negative Symptoms in 2006, it has become widely recognized that improved methodologies are required to convincingly demonstrate the efficacy of new treatments. The methodological issues fall in two categories: assessment tools and clinical trial design features. Regarding assessment tools, two new clinical interview assessment measures, the Brief Negative Symptom Scale (BNSS; Strauss et al. 2012) and the Clinical Assessment Interview for Negative Symptoms (CAINS; Horan et al. 2011), have recently been developed. The BNSS is a 13 items' measure that yields six subscales: blunted affect, alogia, asociality, anhedonia, distress, and avolition. For the subscales assessing experiential symptoms, items are included to separately measure relevant subjective reports and objective assessments. The CAINS was created through a multisite, multiphase scale development process (Kring et al. 2013). The final version also includes 13 items, which are organized into separate motivation and pleasure [including avolition, anhedonia (current and expected pleasure), and asociality items] and expression (affective blunting, alogia) subscales. The motivation and pleasure items combine information about both subjective experience and actual level of engagement in relevant activities, rather than measuring these separately as in the BNSS. For both the BNSS and CAINS, evidence supporting their psychometric properties (inter-rater reliability, test-retest reliability) and two-factor structures has been documented. Furthermore, evidence supporting the external validity of both scales, particularly relations between experiential negative symptoms and community functioning, has been reported (Kring et al. 2013). For the CAINS, the scale, detailed manual, and training videos with gold-standard ratings are available on the Internet (http://www.med.upenn.edu/bbl/downloads/CAINSVideos.shtml) and the BNSS scale and manual are available from its developers. Both scales have been translated into several languages and distributed internationally.

Regarding clinical trial design issues, a recent international meeting established consensus on several design parameters for future pharmacological trials of novel compounds. These parameters include the following: (1) Study subjects should be under the age of 65; (2) subjects should be excluded for symptoms of depression that do not overlap with negative symptoms; (3) functional measures should not be required as a coprimary in negative symptom trials; (4) information from informants should be included for ratings when available; (5) Phase 2 negative symptom trials should be 12 weeks, and 26 weeks is preferred for Phase 3 trials; (6) prior to entry into a negative symptom study, subjects should demonstrate clinical stability for a period of 4–6 months by collection of retrospective information; and (7) prior to entry, the stability of negative and positive symptoms should be confirmed prospectively for 4 weeks or longer (Marder et al. 2013). With these consensus-based guidelines and the availability of new assessment measures, state-of-the-art methodological standards are now available to evaluate the efficiency of new intervention approaches.

Aside from these methodological advances, there has been considerable recent activity in the areas of psychopharmacological and psychosocial treatment development. For pharmacological treatments, a number of compounds have been considered with some showing greater promise than others. Agents showing the most promise, and in relatively late stages in pharmaceutical development, include those that target the glutamatergic system, the cholinergic system, and the hormone oxytocin (Davis et al. 2014). For example, regarding glutamate, the strongest evidence for decreasing negative symptoms comes from agents that increase NMDA glutamatergic receptor activity (e.g., glycine site agonists such as D-serine and the GlyT1 inhibitor sarcosine). For the cholinergic system, drugs that activate cholinergic receptors, including partial alpha 7-nicotinic agonists, currently show the most promise. Finally, exogenous oxytocin (delivered by intranasal spray) has been found to improve negative symptoms in two studies (Feifel 2012; Modabbernia et al. 2013). It should be noted, however, that the relevant studies typically used older trial design and assessment methods, and many of them targeted cognition or social cognition, rather than negative symptoms, as the primary treatment outcome.

There have also been a few promising psychosocial treatment developments. Progress in the area of cognitive therapy (CT) for negative symptoms has been particularly encouraging. This approach is grounded in Beck and colleagues' recently proposed CT-based conceptualization of negative symptoms as stemming from certain dysfunctional attitudes (Grant and Beck 2009). For example, within this framework, defeatist performance beliefs (e.g., "Why bother trying, I always fail," "It's not worth the effort"; "If you can't do something well, there's little point in doing it at all") about one's self and ability to perform productive activities are posited to contribute to avolition and asociality. A handful of recent studies has attempted to target these types of dysfunctional beliefs through standard CT methods such as Socratic questioning, reality testing, and cognitive restructuring. Notably, these therapeutic techniques were originally developed for the treatment of depression (see Barch et al., this volume, for a review of similarities and differences in motivational deficits between schizophrenia and depression). These initial studies provide support for the efficacy of CT for improving experiential negative symptoms and functional deficits, with effect sizes of approximately 0.50 (Klingberg et al. 2011; Grant et al. 2012; Staring et al. 2013), although they used older symptom assessment measures.

Preliminary investigations have also started to explore some alternative methods for treating experiential negative symptoms. For example, a pilot study of six sessions of Loving Kindness Meditation, focused on mindfully directing compassion toward the self and others, led to improvements in motivational negative symptoms as measured by CAINS, and these benefits remained at a three-month follow-up (Johnson et al. 2011). The feasibility of a cognitive/pleasure skills training approach to enhance anticipatory pleasure has also been considered (Favrod et al. 2010), and improvements were found on a self-report measure of anticipatory pleasure (Gard et al. 2007) and a measure of daily activities. Thus, from a psychosocial treatment perspective, recent studies indicate that motivational deficits may be responsive to interventions that address defeatist beliefs, social connections, general compassion for oneself and others, and prospection.

4 Conclusions

Motivational deficits have long been understood to be a core component of schizophrenia, and the recent resurgence of interest in negative symptoms is elucidating the nature of these deficits. The experiential negative symptoms, avolition, anhedonia, and asociality, are driven by low motivation to engage in goal-oriented behaviors. Because these negative symptoms are consistently associated with poor functioning, research into their causes and treatment is a high priority. Translational research, based largely on animal models of motivation and reward processing, has allowed our field to make considerable strides in conceptualizing the likely causes of the deficits. This chapter reviewed the four factors that have received the most attention in clinical neuroscience research as likely contributors to motivation deficits in schizophrenia, which include anticipatory pleasure, reward learning, effort-based decision-making, and social motivation. This research is beginning to inform the development of new pharmacological and psychosocial treatments specifically designed to target negative symptoms. In line with recent advances in clinical assessment technology and clinical trial design guidelines, our field is poised to make important breakthroughs in treating the disabling motivational impairments associated with schizophrenia.

References

Aupperle RL, Paulus MP (2010) Neural systems underlying approach and avoidance in anxiety disorders. Dialogues Clin Neurosci 12(4):517–531

Barch DM, Dowd EC (2010) Goal representations and motivational drive in schizophrenia: the role of prefrontal-striatal interactions. Schizophr Bull 36(5):919–934

Barch DM, Treadway MT et al (2014) Effort, anhedonia, and function in schizophrenia: reduced effort allocation predicts amotivation and functional impairment. J Abnorm Psychol 123 (2):387–397

Blanchard JJ, Cohen AS (2006) The structure of negative symptoms within schizophrenia: implications for assessment. Schizophr Bull 32:238–245

Bleuler E (1950) Dementia praecox or the group of schizophrenias. International Universities Press, New York

Curtis CE, Lebow B et al (1999) Acoustic startle reflex in schizophrenia patients and their first-degree relatives: evidence of normal emotional modulation. Psychophysiology 36(4):469–475

Davis MC, Horan WP et al (2014) Psychopharmacology of the negative symptoms: current status and prospects for progress. Eur Neuropsychopharmacol 24(5):788–799

Docx L, de la Asuncion J et al (2015) Effort discounting and its association with negative symptoms in schizophrenia. Cogn Neuropsychiatry 20(2):172–185

Dowd EC, Barch DM (2010) Anhedonia and emotional experience in schizophrenia: neural and behavioral indicators. Biol Psychiatry 67(10):902–911

Dowd EC, Barch DM (2012) Pavlovian reward prediction and receipt in schizophrenia: relationship to anhedonia. PLoS ONE 7(5):e35622

Eack SM, Mesholam-Gately RI et al (2013) Negative symptom improvement during cognitive rehabilitation: results from a 2-year trial of Cognitive Enhancement Therapy. Psychiatry Res 209(1):21–26

Favrod J, Giuliani F et al (2010) Anticipatory pleasure skills training: a new intervention to reduce anhedonia in schizophrenia. Perspect Psychiatr Care 46(3):171–181

Feifel D (2012) Oxytocin as a potential therapeutic target for schizophrenia and other neuropsychiatric conditions. Neuropsychopharmacology 37(1):304–305

Fervaha G, Graff-Guerrero A et al (2013) Incentive motivation deficits in schizophrenia reflect effort computation impairments during cost-benefit decision-making. J Psychiatr Res 47(11):1590–1596

Frank MJ, Claus ED (2006) Anatomy of a decision: striato-orbitofrontal interactions in reinforcement learning, decision making, and reversal. Psychol Rev 113(2):300–326

Gable SL (2006) Approach and avoidance social motives and goals. J Pers 74(1):175–222

Gable SL, Gosnell CL (2013) Approach and avoidance behavior in interpersonal relationships. Emot Rev 5(3):269–274

Gard DE, Kring AM et al (2007) Anhedonia in schizophrenia: distinctions between anticipatory and consummatory pleasure. Schizophr Res 93(1–3):253–260

Gold JM, Strauss GP et al (2013) Negative symptoms of schizophrenia are associated with abnormal effort-cost computations. Biol Psychiatry 74(2):130–136

Gold JM, Waltz JA et al (2012) Negative symptoms and the failure to represent the expected reward value of actions: behavioral and computational modeling evidence. Arch Gen Psychiatry 69(2):129–138

Gold JM, Waltz JA et al (2008) Reward processing in schizophrenia: a deficit in the representation of value. Schizophr Bull 34(5):835–847

Grant PM, Beck AT (2009) Defeatist beliefs as a mediator of cognitive impairment, negative symptoms, and functioning in schizophrenia. Schizophr Bull 35(4):798–806

Grant PM, Huh GA et al (2012) Randomized trial to evaluate the efficacy of cognitive therapy for low-functioning patients with schizophrenia. Arch Gen Psychiatry 69(2):121–127

Gray JA (1987) The psychology of fear and stress. Cambridge University Press, Cambridge

Hartmann MN, Hager OM et al (2014) Apathy but not diminished expression in schizophrenia is associated with discounting of monetary rewards by physical effort. Schizophr Bull 41:503–512

Heerey EA, Gold JM (2007) Patients with schizophrenia demonstrate dissociation between affective experience and motivated behavior. J Abnorm Psychol 116(2):268–278

Herbener ES (2009) Impairment in long-term retention of preference conditioning in schizophrenia. Biol Psychiatry 65:1086–1090

Horan WP, Blanchard JJ (2003) Emotional responses to psychosocial stress in schizophrenia: the role of individual differences in affective traits and coping. Schizophr Res 60(2–3):271–283

Horan WP, Green MF et al (2006a) Does anhedonia in schizophrenia reflect faulty memory for subjectively experienced emotions? J Abnorm Psychol 115(3):496–508

Horan WP, Kring AM et al (2006b) Anhedonia in schizophrenia: a review of assessment strategies. Schizophr Bull 32:259–273

Horan WP, Kring AM et al (2011) Development and psychometric validation of the clinical assessment interview for negative symptoms (CAINS). Schizophr Res 132(2–3):140–145

Horan WP, Wynn JK et al (2010) Electrophysiological correlates of emotional responding in schizophrenia. J Abnorm Psycholol 119(1):18–30

Hosking JG, Cocker PJ et al (2014) Dissociable contributions of anterior cingulate cortex and basolateral amygdala on a rodent cost/benefit decision-making task of cognitive effort. Neuropsychopharmacology 39(7):1558–1567

Johnson DP, Penn DL et al (2011) A pilot study of loving-kindness meditation for the negative symptoms of schizophrenia. Schizophr Res 129(2–3):137–140

Kane JM (2013) Tools to assess negative symptoms in schizophrenia. J Clin Psychiatry 74(6):e12

Kirkpatrick B, Fenton W et al (2006) The NIMH-MATRICS consensus statement on negative symptoms. Schizophr Bull 32:296–303

Kirkpatrick B, Strauss GP et al (2011) The brief negative symptom scale: psychometric properties. Schizophr Bull 37(2):300–305

Klingberg S, Wolwer W et al (2011) Negative symptoms of schizophrenia as primary target of cognitive behavioral therapy: results of the randomized clinical TONES study. Schizophr Bull 37(2):S98–S110

Kraepelin E (1971) Dementia praecox and paraphrenia. Krieger Publishing Co., Inc., Huntington

Kring AM, Barch DM (2014) The motivation and pleasure dimension of negative symptoms: neural substrates and behavioral outputs. Eur Neuropsychopharmacol 24(5):725–36

Kring AM, Gur RE et al (2013) The clinical assessment interview for negative symptoms (CAINS): final development and validation. Am J Psychiatry 170:165–172

Kurniawan IT, Seymour B et al (2010) Choosing to make an effort: the role of striatum in signaling physical effort of a chosen action. J Neurophysiol 104(1):313–321

Kurtz MM, Mueser KT (2008) A meta-analysis of controlled research on social skills training for schizophrenia. J Consult Clin Psychol 76(3):491–504

Lee B, Groman S et al (2007) Dopamine D2/D3 receptors play a specific role in the reversal of a learned visual discrimination in monkeys. Neuropsychopharmacology 32(10):2125–2134

Leeson VC, Robbins TW et al (2009) Discrimination learning, reversal, and set-shifting in first-episode schizophrenia: stability over six years and specific associations with medication type and disorganization syndrome. Biol Psychiatry 66(6):586–593

Marder SR, Alphs L et al (2013) Issues and perspectives in designing clinical trials for negative symptoms in schizophrenia. Schizophr Res 150(2–3):328–333

Marder SR, Kirkpatrick B (2014) Defining and measuring negative symptoms of schizophrenia in clinical trials. Eur Neuropsychopharmacol 24(5):737–743

McKirdy J, Sussmann JED et al (2009) Set shifting and reversal learning in patients with bipolar disorder or schizophrenia. Psychol Med 39(08):1289–1293

Meehl P (1962) Schizotaxia, schizotypy, schizophrenia. Am Psychol 17:827–838

Milev P, Ho BC et al (2005) Predictive values of neurocognition and negative symptoms on functional outcome in schizophrenia: a longitudinal first-episode study with 7-year follow-up. Am J Psychiatry 162(3):495–506

Modabbernia A, Rezaei F et al (2013) Intranasal oxytocin as an adjunct to risperidone in patients with schizophrenia. CNS Drugs 27(1):57–65

Mueser KT, Sayers SL et al (1994) A multisite investigation of the reliability of the scale for the assessment of negative symptoms. Am J Psychiatry 151(10):1453–1462

Murray GK, Cheng F et al (2008) Reinforcement and reversal learning in first-episode psychosis. Schizophr Bull 34(5):848–855

Pantelis C, Barber FZ et al (1999) Comparison of set-shifting ability in patients with chronic schizophrenia and frontal lobe damage. Schizophr Res 37(3):251–270

Pessiglione M, Seymour B et al (2006) Dopamine-dependent prediction errors underpin reward-seeking behaviour in humans. Nature 442(7106):1042–1045

Prentice KJ, Gold JM et al (2008) The Wisconsin Card Sorting impairment in schizophrenia is evident in the first four trials. Schizophr Res 106(1):81–87

Reddy LF, Green MF et al (2014) Behavioral inhibition and activation systems in schizophrenia: an evaluation of motivational profiles. Schizophr Res 159(1):164–170

Rummel C, Kissling W et al (2005) Antidepressants as add-on treatment to antipsychotics for people with schizophrenia and pronounced negative symptoms: a systematic review of randomized trials. Schizophr Res 80(1):85–97

Salamone JD, Correa M (2012) The mysterious motivational functions of mesolimbic dopamine. Neuron 76(3):470–485

Sayers SL, Curran PJ et al (1996) Factor structure and construct validity of the Scale for the Assessment of Negative Symptoms. Psychol Assess 8:269–280

Schlagenhauf F, Huys QJM et al (2014) Striatal dysfunction during reversal learning in unmedicated schizophrenia patients. NeuroImage 89:171–180

Schultz W, Dayan P et al (1997) A neural substrate of prediction and reward. Science 275 (5306):1593–1599

Siegel SJ, Irani F et al (2006) Prognostic variables at intake and long-term level of function in schizophrenia. Am J Psychiatry 163(3):433–441

Singh SP, Singh V et al (2010) Efficacy of antidepressants in treating the negative symptoms of chronic schizophrenia: meta-analysis. Br J Psychiatry 197(3):174–179

Smith KS, Berridge KC (2007) Opioid limbic circuit for reward: interaction between hedonic hotspots of nucleus accumbens and ventral pallidum. J Neurosci 27(7):1594–1605

Solomon RL (1948) The influence of work on behavior. Psychol Bull 45(1):1–40

Spielberg JM, Heller W et al (2013) Hierarchical brain networks active in approach and avoidance goal pursuit. Front Hum Neurosci 7(284)

Staring ABP, ter Huurne M-AB et al (2013) Cognitive behavioral therapy for negative symptoms (CBT-n) in psychotic disorders: a pilot study. J Behav Ther Exp Psychiatry 44(3):300–306

Strauss GP, Hong LE et al (2012) Factor structure of the brief negative symptom scale. Schizophr Res 142(1–3):96–98

Strauss GP, Horan WP et al (2013) Deconstructing negative symptoms of schizophrenia: avolition-apathy and diminished expression clusters predict clinical presentation and functional outcome. J Psychiatr Res 47(6):783–790

Strauss GP, Waltz JA et al (2014) A review of reward processing and motivational impairment in schizophrenia. Schizophr Bull 40(Suppl 2):S107–S116. doi:10.1093/schbul/sbt197 Epub 2013 Dec 27

Treadway MT, Peterman JS et al (2015) Impaired effort allocation in patients with schizophrenia. Schizophr Res 161(2–3):382–385

Velthorst E, Koeter M et al (2015) Adapted cognitive–behavioural therapy required for targeting negative symptoms in schizophrenia: meta-analysis and meta-regression. Psychol Med 45 (03):453–465. doi:10.1017/S0033291714001147

Volz M, Hamm AO et al (2003) Temporal course of emotional startle modulation in schizophrenia patients. Int J Psychophysiol 49(2):123–137

Waltz JA, Gold JM (2007) Probabilistic reversal learning impairments in schizophrenia: further evidence of orbitofrontal dysfunction. Schizophr Res 93(1–3):296–303

Waltz JA, Kasanova Z et al (2013) The roles of reward, default, and executive control networks in set-shifting impairments in schizophrenia. PLoS ONE 8(2):e57257

Waltz JA, Schweitzer JB et al (2009) Patients with schizophrenia have a reduced neural response to both unpredictable and predictable primary reinforcers. Neuropsychopharmacology 34 (6):1567–1577

Weiler JA, Bellebaum C et al (2009) Impairment of probabilistic reward-based learning in schizophrenia. Neuropsychology 23(5):571–580

Wise RA (2004) Dopamine, learning and motivation. Nat Rev Neurosci 5(6):483–494

Wolf DH, Satterthwaite TD et al (2014) Amotivation in schizophrenia: integrated assessment with behavioral, clinical, and imaging measures. Schizophr Bull 40(6):1328–1337

Wykes T, Steel C et al (2008) Cognitive behavior therapy for schizophrenia: effect sizes, clinical models, and methodological rigor. Schizophr Bull 34(3):523–537

Wynn JK, Horan WP et al (2010) Impaired anticipatory event-related potentials in schizophrenia. Int J Psychophysiol 77(2):141–149

Motivational Deficits in Schizophrenia and the Representation of Expected Value

James A. Waltz and James M. Gold

Abstract Motivational deficits (avolition and anhedonia) have historically been considered important negative symptoms of schizophrenia (SZ). Numerous studies have attempted to identify the neural substrates of avolition and anhedonia in schizophrenia, but these studies have not produced much agreement. Deficits in various aspects of reinforcement processing have been observed in individuals with schizophrenia, but it is not exactly clear which of these deficits actually engender motivational impairments in SZ. The purpose of this chapter is to examine how various reinforcement-related behavioral and neural signals could contribute to motivational impairments in both schizophrenia and psychiatric illness, in general. In particular, we describe different aspects of the concept of expected value (EV), such as the distinction between the EV of stimuli and the expected value of actions, the acquisition of value versus the estimation of value, and the discounting of value as a consequence of time or effort required. We conclude that avolition and anhedonia in SZ are most commonly tied to aberrant signals for expected value, in the context of learning. We discuss implications for further research on the neural substrates of motivational impairments in psychiatric illness.

Keywords Reinforcement · Basal ganglia · Orbitofrontal · Avolition · Anhedonia

Contents

1	Introduction	376
2	Identifying a Relationship Between EV and Avolition: Considerations	377
	2.1 How Do We Quantify the Severity of Motivational Deficits in Schizophrenia?	377
	2.2 What Constitutes a Behavioral EV Signal?	378
	2.3 What Constitutes a Neural EV Signal?	380
	2.4 Distinguishing EV Signals from General Salience Signals	382

J.A. Waltz (✉) · J.M. Gold
Maryland Psychiatric Research Center, Department of Psychiatry, University of Maryland School of Medicine, P.O. Box 21247, Baltimore, MD 21228, USA
e-mail: jwaltz@mprc.umaryland.edu

© Springer International Publishing Switzerland 2015
Curr Topics Behav Neurosci (2016) 27: 375–410
DOI 10.1007/7854_2015_385
Published Online: 15 September 2015

2.5 Acquisition of Incentive Salience Versus on-the-Fly Computation of EV 383
2.6 EV of Stimuli Versus EV of Actions 384
3 Integrating EV with Cost Considerations to Guide Decision Making 385
3.1 Distinguishing Wanting from Willingness to Work 385
3.2 Distinguishing Wanting from Willingness to Wait 386
4 Evidence for Faulty EV Signaling in SZ 386
4.1 Behavioral and Modeling Evidence for Faulty EV Signaling in SZ 386
4.2 Neural Evidence for Faulty EV Signaling in SZ 390
5 Avolition and Outcome Processing in Schizophrenia 395
5.1 Behavioral Studies of Outcome Processing in Schizophrenia 396
5.2 Neuroimaging Studies of Reward and Punishment Receipt in Schizophrenia 397
5.3 Neuroimaging Studies of RPE Signaling in SZ 398
5.4 Do Abnormalities in Consummatory Hedonics or RPE Signaling Account
   for Abnormalities in EV Signaling in SZ? 400
6 General Conclusions 401
References 403

1 Introduction

The reduced tendency to initiate goal-oriented behavior (avolition; Goldberg and Weinberger 1988; Saykin et al. 1994) is an aspect of negative symptomatology thought to typify many patients with schizophrenia (SZ). While there are many factors which could lead to avolition/motivational deficits in SZ, the preponderance of studies indicate that SZ patients do not differ from controls in their self-reported *experience* of pleasure ("consummatory hedonics"; Cohen and Minor 2008; Gard et al. 2007). Partially, based on this evidence, we (Gold et al. 2008) hypothesized that avolition results from a failure to look forward to pleasurable outcomes ("anticipatory hedonics"), by virtue of the assignment of incentive salience to cues. As defined by Berridge and Robinson (1998), a stimulus becomes imbued with incentive salience when it is transformed from a neutral object into an object of attraction that animals will work to acquire. This is the essential outcome of reinforcement learning (RL), and it is thought to be a primary functional role of dopamine in the nervous system (Berridge and Robinson 1998). The updating of the incentive value of a stimulus is thought to occur via the signaling of reward prediction errors (RPEs), which are mismatches between expected and obtained outcomes. Thus, a *failure* to update the incentive value of a stimulus could happen for at least three reasons: (1) the signal of the expected outcome is degraded or inaccurate; (2) the signal of the obtained outcome is degraded or inaccurate; or (3) the mechanism for computing the RPE is dysfunctional. Given the evidence that signals related to reward receipt in schizophrenia are intact (Cohen and Minor 2010), considerable attention has been focused on the other two possibilities: that the signal of the expected outcome is degraded or inaccurate, and that the mechanism for computing the RPE is dysfunctional.

In fact, there is considerable evidence that acutely ill patients (particularly those that are unmedicated) have genuinely disrupted RPE signaling (Murray et al. 2007;

Schlagenhauf et al. 2009, 2014), with important implications for RL and belief formation. Furthermore, there have been numerous findings of correlations between measures of both positive symptoms in schizophrenia and supposed RPE signals in the brain (Gradin et al. 2011). It is, however, much less certain that RPE signaling is abnormal in chronic, medicated patients (Walter et al. 2009, 2010), despite clear evidence of reinforcement learning deficits in these patients (Farkas et al. 2008; Waltz et al. 2007). Furthermore, measures of RL performance have been shown to correlate with the severity of motivational deficits in chronic SZ patients. Were RL deficits to persist in stably medicated SZ patients, despite evidence of intact RPE signaling, it would suggest that aberrant RPE-driven learning observed in medicated SZs may be more a problem of faulty *input* to the PE computation than dysfunction in the mechanism itself. In this chapter, our purpose is to evaluate the data arguing for and against the idea that the signaling of expected value (EV) in chronic SZ patients relates to motivational deficits, which are thought to persist throughout the illness and be largely unaffected by antipsychotic medications. This area has been the focus of numerous basic and clinical studies. Prior to discussing clinical findings, we will first review the basic concepts and methods that have served to guide the field.

2 Identifying a Relationship Between EV and Avolition: Considerations

2.1 How Do We Quantify the Severity of Motivational Deficits in Schizophrenia?

The first step in linking an aspect of behavioral performance or a purported neural signal to the severity of motivational deficits in a psychiatric population is to establish how one quantifies the severity of motivational deficits. In the field of schizophrenia research, motivational deficits are commonly thought of as a component of "negative symptoms," or areas of subnormal function (Peralta and Cuesta 1995; Sayers et al. 1996). Thus, clinical interviews for assessing the severity of negative symptoms in schizophrenia—such as the Scale for the Assessment of Negative Symptoms (SANS; Andreasen 1984)—involve questions about motivation. In previous studies (Gold et al. 2012; Strauss et al. 2011; Waltz et al. 2009), we have used both the individual avolition/role functioning and anhedonia/asociality subscores, as well as a combination of the two, to quantify motivational deficits in SZ. Another scale used to quantify the severity of motivational deficits in SZ is the Schedule for the Deficit Syndrome (SDS; Kirkpatrick et al. 1989). The deficit syndrome has been described as a separate disease within the syndrome of schizophrenia (Kirkpatrick et al. 2001), whereby patients are characterized *primarily* by negative symptoms (such as avolition and anhedonia) not attributable to psychotic symptoms or antipsychotic medications. The SDS is an instrument designed to identify this subset of SZ patients, the prevalence of which ranges from 25 to 30 % of chronic SZ patients (Kirkpatrick et al. 2001).

Two additional scales have recently been developed, in an attempt to separately ascertain capacities for consummatory and anticipatory hedonics: the Brief Negative Symptom Scale (Kirkpatrick et al. 2011) and the Clinical Assessment Interview for Negative Symptoms (Horan et al. 2011). Validation studies (Kring et al. 2013; Strauss et al. 2012) have demonstrated that these measures show good convergent validity in their relationships with other symptom rating scales, but are not redundant with them (for more detail on these newer assessment tools, please read the chapter by Reddy, Horan, and Green in this volume). Aside from clinical rating scales, there are self-report measures, meant to quantify the severity of anhedonia, specifically, such as the Chapman Scales for Physical and Social Anhedonia (Chapman et al. 1976). The Chapman Scales for Physical and Social Anhedonia can be, and have been, administered to both psychiatric patients and controls and are thought to serve as "trait" measures of anhedonia, rather than in-the-moment "state" measures.

One of our major goals has been to provide a mechanistic account of motivational deficits in schizophrenia. The first step in that enterprise has been to identify component processes in reward processing and reinforcement learning and to describe their relationships with clinical assessments of motivational deficits. It is possible, however, that experimental measures might do a better job of capturing the phenomenon/disability than a clinical rating scale. Thus, an additional goal of studying motivational deficits in SZ from an experimental standpoint is to develop better assessment tools.

2.2 What Constitutes a Behavioral EV Signal?

In order to argue the claim that degraded value representations contribute to avolition in schizophrenia, we need to be able to isolate the use of value representations experimentally. From a behavioral standpoint, there are several ways to specifically assess the intactness or aberrance of value representations in human subjects. One way is to probe for preferences between conditioned stimuli, following a feedback-driven acquisition period. This approach is used in many operant learning paradigms, such as probabilistic response (or Go/NoGo) learning (Frank and O'Reilly 2006; Holroyd et al. 2004; Pessiglione et al. 2006), probabilistic stimulus selection (PSS; Frank et al. 2004; Shanks et al. 2002), and probabilistic reversal learning (Cools et al. 2002). Learning, in the context of such paradigms, is driven by mismatches between expected and obtained outcomes, called reward prediction errors (RPEs; Glimcher 2011; Montague et al. 2004). Choices that lead to better-than-expected outcomes, or positive RPEs, facilitate those choices in those contexts—a process called "Go-learning" (Frank et al. 2004). Choices that lead to *worse*-than-expected outcomes, or negative RPEs, decrease the likelihood of making those choices in those contexts—a process called "NoGo-learning"

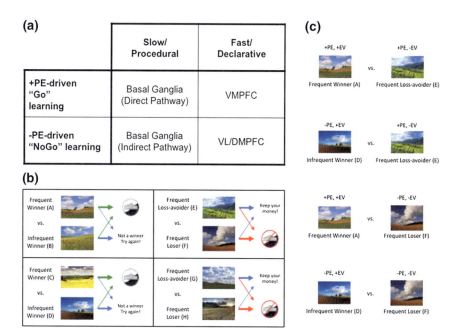

Fig. 1 a Taxonomy of reinforcement learning mechanisms and their hypothesized neural substrates (see Frank and Claus 2006, for details). Abbreviations: *PE* prediction error, *PFC* prefrontal cortex, *VM* ventromedial, *VL* ventrolateral, *DM* dorsomedial. **b** Sample acquisition pairs from the gain versus loss-avoidance probabilistic stimulus selection task. Individuals learn to choose the best stimulus from each of four pairs, based on the reinforcement probabilities of each stimulus. In two pairs, choices resulted in either a gain or a neutral outcome. In two other pairs, choices resulted in either a loss or a neutral outcome. *Arrow thickness* reflects probability of specific outcome. *Green arrows* denote gain, *red arrows* denote loss, and *blue arrows* denote neutral outcome. **c** Sample transfer pairs from the gain versus loss-avoidance probabilistic stimulus selection task. By prompting subjects to choose between stimuli associated with the same frequency of positive prediction errors, but difference expected values (e.g., between "frequent winner" stimuli vs. "frequent loss-avoider" stimuli), one can isolate the influence of expected value on choice. Abbreviations: *PE* prediction error, *EV* expected value

(Frank et al. 2004). Prediction-error-driven learning happens on multiple timescales, with rapid, often single-trial learning known to rely on intact functioning of the prefrontal cortex, or PFC, and gradual, more procedural processes, reliant of signaling in the basal ganglia, or BG (Collins and Frank 2012; Frank and Claus 2006; see Fig. 1a). Additionally, one can isolate the contribution of expected value to performance deficits by manipulating the expected value of a choice, while holding RPE valence and magnitude constant.

Gold et al. (2012) have provided one example of how this might be done, using a paradigm where individuals learn to choose the best stimulus from each of four pairs, based on the reinforcement probabilities of each stimulus. This task involved the presentation, in an acquisition phase (Fig. 1b), of two kinds of probabilistic discriminations in a pseudorandom order: (1) "Gain" pairs, involving learning to choose a frequently-rewarded stimulus over one frequently leading to a neutral outcome, and (2) "loss-avoidance" pairs, involving learning to choose a stimulus frequently leading to a neutral outcome rather than a loss. By prompting subjects to choose, in a transfer phase (Fig. 1c), between stimuli associated with the same frequency of positive prediction errors, but difference expected values (e.g., between "frequent winner" stimuli versus "frequent loss-avoider" stimuli), one can isolate the influence of *expected* value on choice.

Alternatively, one can isolate the contribution of expected value to performance deficits by using a computational model of reinforcement learning to model RPE valence and magnitude, as well as expected value, on a trial-by-trial basis. These algorithms are typically derived from the work of Rescorla and Wagner (1972) and Sutton and Barto (1998). In computational models of reward learning and decision making (DM), the contribution of the basal ganglia system is often formalized using an "actor-critic" framework or a "Q-learning" framework (Sutton and Barto 1998). In an actor-critic model, a "critic" evaluates the reward values of particular states, and the "actor" selects responses as a function of learned stimulus-response weights. In a Q-learning framework, by contrast, the model learns expected quality ("Q value") of each action separately, as opposed to representing the value of the "state." Actions are then selected by comparing the various Q values of each candidate action and probabilistically choosing the largest one. Both algorithms depend on the signaling of prediction errors (mismatches between expected and obtained outcomes) to update value representations. However, whereas the actor in the actor-critic scheme does not consider the outcome values of competing actions, the Q-learning scheme makes these fundamental, updating value representations for alternative choices in context (Gold et al. 2012). There is compelling evidence that the orbitofrontal cortex (OFC) has a critical role in subserving these kinds of value representations (Plassmann et al. 2010; Roesch and Olson 2007) see also the chapter by Bissonette and Roesch in this volume. Certain algorithms allow researchers to estimate prediction error valence and magnitude, as well as expected value, on a trial-by-trial basis. As a consequence, it is possible to distinguish the contribution of an aberrant RPE-driven learning mechanism from the contribution of faulty input to that mechanism, in observed nonnormative behavior.

2.3 What Constitutes a Neural EV Signal?

The representation of value in the brain—like most representations and processes in the brain—is likely to be distributed across multiple nodes (Knutson et al. 2005). A growing scientific literature has, in fact, supported the idea that value representations reside in fronto-striatal circuits, centered on ventral striatum (VS) and the ventral and medial aspects of prefrontal cortex (PFC), in particular (Kahnt et al. 2010, 2014; Smith et al. 2014; Takahashi et al. 2009). Whereas studies in nonhuman subjects have primarily used electrophysiology to identify value representations in the brain, studies with human subjects have largely used functional

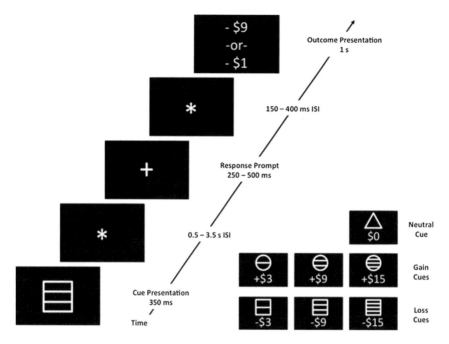

Fig. 2 Schematic of a single trial from a Monetary Incentive Delay (MID) task, with example time intervals. The subject is first presented with a cue, depicting the valence and magnitude of the potential outcome (see inset). Following a variable inter-stimulus interval (ISI; filled in this case by a *black screen* with a *fixation star*), a target stimulus (e.g., a *white cross* on a *black background*) appears, prompting the participant to respond within a target time window. Following another ISI, the actual outcome is displayed. If a participant responds within the acceptable response window on a gain trial, the total amount of money increments by the associated amount; if not, the total increments by a nominal (small) amount, or not at all. If a participant responds within the acceptable response window on a loss trial, the total amount of money decrements by a nominal (small) amount, or not at all; if the subject is too slow, the subject loses the full predicted amount. Because the interval between the cue and the outcome is variable and long, the cue and outcome regressors are not collinear, and thus, brain responses to the cue (thought to be associated with outcome anticipation) and outcome receipt can be modeled separately in functional imaging experiments

Magnetic Resonance Imaging (fMRI), with Monetary Incentive Delay (MID) paradigms being the most commonly used (Knutson et al. 2001, 2003). MID paradigms are thought to isolate representations of expected value in the brain, because they are designed with temporal intervals between cues predictive of outcomes and deliveries of outcomes that are long enough and variable enough to model each separately in a functional MRI study (Fig. 2). The success of MID paradigms in linking EV representations to model blood-oxygen-level-dependent (BOLD) signals in the VS (Knutson et al. 2001, 2003) led to the application of similar paradigms to the study of reward anticipation in schizophrenia (Juckel et al. 2006b; Walter et al. 2009; Waltz et al. 2010). It should be noted, however, that both studies in nonhuman animals (Roesch and Olson 2004; Schoenbaum and Roesch

2005; Schoenbaum et al. 2009) and more recent studies with human subjects (Kahnt et al. 2010, 2014) have emphasized the role of OFC in the representation of value. While OFC has historically been a difficult brain region to capture with fMRI, newer behavioral paradigms and imaging techniques have made this task easier. As a consequence, current and future studies of the neural representations of value in schizophrenia should account for within- and between-group variability in task-related OFC signals.

A major consideration in isolating EV signals in the human brain is how the task-related MRI data are modeled. Typically, EV signals in fMRI tasks have been identified through experimental manipulations and in contrasts between beta values for conditions in either whole-brain or regions-of-interest (ROI) analyses. Such studies, for example, might employ a paradigm with conditions where the participant might anticipate a monetary gain or loss, or a monetary gain or neutral outcome, or a large gain and a small gain. An effect of such a manipulation on associated beta values is typically taken as evidence for a region's role in coding for value, and a group difference in the effect of such a manipulation on associated beta values is typically taken as evidence for an abnormality in neural representations of value in a given group.

More recently, however, neuroimaging researchers have begun to take advantage of the kinds of computational modeling techniques mentioned above, in order to estimate EV, as well as RPE valence and magnitude, on a trial-by-trial basis. This allows one to model BOLD signals with parametric regressors (O'Doherty et al. 2007). This has also been done in the case of studies of reinforcement learning in schizophrenia (Gradin et al. 2011), perhaps allowing researchers to pinpoint neural signals associated with EV with more precision. This methodology also enables one to distinguish cue-evoked, or outcome-anticipation, signals, from feedback-evoked, or outcome-integration, signals. While outcome processing is associated with neural activity in multiple systems, RPE signaling, in particular (at least in the case of the delivery and omission of appetitive stimuli), has consistently been linked to neural activity in the dopaminergic midbrain and its targets in the striatum (McClure et al. 2003; Schultz 1998; Takahashi et al. 2009). Importantly, while the axiomatic approach described by Caplin and Dean (2008) clearly implicates the striatum in the signaling of appetitive RPEs (Rutledge et al. 2010), it casts doubt on a role for the striatum in the signaling of aversive RPEs, a function more likely to be subserved by the periaqueductal gray (Roy et al. 2014). That is to say, one cannot say with 100 % certainty whether punishment-evoked activity in the VS can be interpreted as an RPE signal.

2.4 Distinguishing EV Signals from General Salience Signals

It is one matter to identify neural responses to reward-predicting cues in the brain; it is another matter to link these neural responses specifically to reward prediction and to distinguish them from other sorts of neural signals, such as salience signals, which might be generated simultaneously (for more information on this topic see Salamone et al. in this volume). Because of the difficulty in isolating specific cognitive representations and processes, and because of the overlap in functions attributed to specific brain regions, there is obvious ambiguity as to what can be called an "expected value" signal, and what not. As Kahnt et al. (2014) have noted, for example, in tasks examining only appetitive stimuli, or only aversive stimuli, value signals are perfectly correlated with salience signals.

Multiple research groups have sought to disambiguate *value* from salience signals in the human brain. In fact, the findings of several groups (Jensen et al. 2007; Zink et al. 2003, 2006) point to a role for the ventral striatum in general salience signaling, rather than exclusively representing the valence of prediction errors. This finding does not preclude a role for VS in signaling RPEs, as prediction errors could be understood as a particular *kind* of salient event. Ventral striatum may also participate in expected value representation by virtue of inputs from OFC; however, OFC appears to play a more specific role in value representation and does not appear to play a large role in signaling mismatches between expectations and outcomes (Kahnt et al. 2014; Takahashi et al. 2009).

Additional work suggests a role for inferior parietal cortex in the signaling of stimulus-driven salience, as well (Geng and Mangun 2009; Kahnt et al. 2014). Finally, recent work indicates that anterior insula serves as a hub in a global salience network (with the amygdalae and other structures as nodes; Harsay et al. 2012; Seeley et al. 2007) and is not thought to signal either expected value or signed prediction errors. Thus, while neural responses in anterior insula, the amygdalae, and parietal cortices are often evoked by salient stimuli with incentive value, these responses are more likely to reflect their salience than their incentive value.

2.5 Acquisition of Incentive Salience Versus on-the-Fly Computation of EV

It is possible to talk about value-based DM in multiple contexts. One sense, which was described above, involved DM in the context of reinforcement learning, where stimuli acquire incentive value by virtue of their frequent temporal contiguity with valenced outcomes. Another kind of value-based DM, with which many people are familiar, is the sort of hypothetical DM based on the instantaneous integration of reward probabilities and magnitudes, first formalized by von Neumann and Morgenstern (1947). This sort of decision making is the subject of Kahneman and Tversky's (1979) "Prospect Theory," the main point of which is to illustrate ways in which human choice is guided by irrational biases. It has also been the subject of several recent neuroimaging studies (De Martino et al. 2006; Tom et al. 2007), which have pinpointed the neural processes associated with online EV computation and the irrational biases that lead to deviations from purely EV-based DM. In particular, these studies have shown that individual differences in susceptibility to irrational biases (such as disproportionate loss aversion) correspond to individual differences in decision-related activity in areas such as medial PFC and amygdala. Although we know of no imaging studies of SZ patients, involving hypothetical decision making situations, such as those described above, several recent *behavioral* studies (Brown et al. 2013; Heerey et al. 2008; Tremeau et al. 2008) have used such paradigms to investigate the ability to estimate expected value on the fly, as well as the influence of irrational biases on this ability, in schizophrenia. The fact that it is possible to make a distinction between the acquisition of incentive salience and the on-the-fly computation of EV begs the question of whether variables related to one type of decision are likely to be more closely tied to clinically-ratable motivational deficits than to the other. We discuss relationships between clinical measures of motivational deficits and variables related to both types of decisions below.

2.6 EV of Stimuli Versus EV of Actions

Another important distinction to consider in identifying relationships between expected value signaling and clinical ratings of motivational deficits in SZ is whether one means the expected value of a stimulus that has acquired incentive value, or the expected value of a choice. Several recent reviews (Noonan et al. 2012; Rushworth 2008) have addressed the issue, providing evidence that the expected value of stimuli and choices is subserved by somewhat different neural networks. One version of this dissociation is that ventrolateral areas of PFC are responsible for representing the value of stimuli, whereas medial areas of PFC are responsible for representing the value of actions (Noonan et al. 2010; but see Glascher et al. 2009). A slightly different hypothesis regarding the lateral/medial dissociation in OFC is that the lateral OFC is concerned with the linking of specific stimulus representations to representations of specific types of reward outcome, whereas VMPFC/medial OFC is concerned with evaluation, value-guided decision making, and maintenance of a choice over successive decisions (Noonan et al. 2012). A role for VLPFC in "linking of specific stimulus representations to representations of specific types of reward outcome" may account for the established role of VLPFC in the successful performance of reversal learning tasks (Cools et al. 2002). Noonan et al. (2012) demonstrated that lesions in neither orbitofrontal subdivision caused perseveration; rather, lesions in the lOFC made animals switch more. By contrast, lesions in the mOFC caused animals to lose their normal predisposition to repeat previously successful choices, suggesting that the mOFC does not just mediate value comparison in choice but also facilitates maintenance of the same choice if it has been successful. Furthermore, recent findings from Guitart-Masip et al. (2011) argue against the claim that cue-evoked activity in the VS reflects cue-evoked reward anticipation (i.e., the value assigned to a stimulus);

rather, work from this group indicates that cue-evoked striatal activation reflects *action* anticipation.

From an experimental standpoint, the type of EV signal being probed may be a function of whether reward delivery in a task is action dependent, or not. As an example, in the context of a Pavlovian paradigm, a stimulus can acquire value, without the need to respond during training. In the context of an operant learning paradigm, a response is required, and the response, in context, is assigned value. For that reason, it is likely that, when we are talking about EV, with regard to probabilistic response learning, or probabilistic reversal learning, paradigms, we are talking about the EV of actions in context. Furthermore, if VS and medial PFC are more critically involved in action valuation than stimulus valuation, then aberrant cue-evoked responses in VS and medial PFC are more likely to point to an abnormal ability to assign value to actions than assign "incentive salience" to stimuli. Furthermore, this suggests that an abnormal ability to assign value to actions would likely be a better model for clinical avolition then an abnormal ability to assign "incentive salience" to stimuli.

3 Integrating EV with Cost Considerations to Guide Decision Making

3.1 *Distinguishing Wanting from Willingness to Work*

Decision making requires one to estimate not only the benefits of actions, but also the costs of actions. Thus, even stimuli that have acquired incentive salience—that have become the objects of wanting—require a willingness to overcome the estimated effort cost of an action. The willingness to expend effort has been the subject of recent investigation, with regard to motivational deficits, in general (Treadway et al. 2009), as well as avolition/anhedonia within psychiatric disorders, such as major depressive disorder (Treadway et al. 2012, chapter by Treadway in this Volume) and schizophrenia (Gold et al. 2013). From a brain mapping standpoint, however, it is very difficult to distinguish the neural substrates of "reward anticipation," or "wanting," from the neural substrates of "willingness to work." Dorsal anterior cingulate cortex (dACC) has been implicated in the representation of both the costs and benefits of actions (Kennerley et al. 2006; Walton et al. 2006), as well as conflict monitoring and resolution, with regard to action selection (Ridderinkhof et al. 2004; Rushworth et al. 2004; van Veen and Carter 2002). The implication of these studies is that the task of isolating a "selective deficit in representing the expected value of actions," or the specific contribution of action value representations to avolition, will likely be very difficult. By contrast, isolating the neural substrates of "a deficit in the ability to integrate the prospective costs and benefits of actions" may prove more tractable.

3.2 Distinguishing Wanting from Willingness to Wait

Related to the concept of the "effort cost" of an action, and the "effort discounting" of the value of an action, is the concept of the "delay discounting (DD)" of the value of an appetitive stimulus. This refers to the rate at which the value of a stimulus is discounted as a function of time, and is usually estimated by way of queries such as, "Would you rather have $10 now, or $100 in two weeks?" Like effort cost paradigms and devaluation paradigms, these measures assess not only in-the-moment valuation of stimuli, but also valuation as a dynamic, time-varying process. Steep devaluation of stimuli with time is often taken as a measure of impulsivity (Holt et al. 2003; Kirby and Santiesteban 2003), associated with OFC lesions, in particular (Rudebeck et al. 2006). Because schizophrenia has also been hypothesized to involve maladaptive DM, possibly brought on by the abnormal valuation of stimuli and actions, DD paradigms have also been administered to SZ patients (Avsar et al. 2013; Heerey et al. 2007). Such studies are described in the following section.

4 Evidence for Faulty EV Signaling in SZ

Thus, as Fig. 1 illustrates, one can be speaking of a number of different cognitive and physiological processes, when one refers to expected value signaling:

- the ability to estimate EV on the fly (by integrating the probabilities and magnitudes of various outcomes);
- the signaling of the acquired incentive value of a stimulus in the environment;
- the signaling of the learned value of an action/choice; and
- the signaling of the learned value of an action/choice, relative to its estimated effort cost (Fig. 3).

In the following sections, we will discuss evidence for and against the disruption of each of these aspects of valuation in schizophrenia patients, in general, as well as in subgroups of SZ patients with prevalent primary negative symptoms.

4.1 Behavioral and Modeling Evidence for Faulty EV Signaling in SZ

4.1.1 Probabilistic Discrimination Learning in Schizophrenia

A number of behavioral studies (Waltz et al. 2007, 2011) have used probabilistic RL paradigms like the probabilistic stimulus selection task (described in Sect. 2.2) to investigate possible relationships between aspects of RL performance and

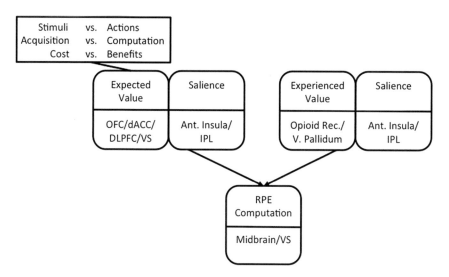

Fig. 3 Aspects of reinforcement learning and expected value signaling. Signals for expected value, experienced value RPE computation, and stimulus salience (both cue and outcome) are associated with distinct neural substrates. The signaling of expected value can be further parsed along a number of dimensions. Abbreviations: *OFC* orbitofrontal cortex, *dACC* dorsal anterior cingulate cortex, *DLPFC* dorsolateral prefrontal cortex, *VS* ventral striatum, *Ant.* anterior, *IPL* inferior parietal lobule, *Rec.* receptors, and *V.* ventral

clinically rated avolition/anhedonia in schizophrenia. These studies have generally found that SZ patients, as a group, have shown impaired (positive-RPE-driven) Go-learning, of the sort thought to rely on the basal ganglia. However, these impairments in Go-learning have not always been found to be most severe in SZ patients with the highest clinical ratings for motivational deficits—as would be theoretically attractive (Waltz et al. 2007, 2011). Systematic relationships with clinical ratings for motivational deficits *have*, however, been observed with measures of (supposedly PFC-driven) trial-to-trial adjustments, such as win-stay and lose-shift rates (Waltz et al. 2007, 2011). An association between an aspect of RL performance and negative symptoms (which include motivational deficits) was also observed by Farkas et al. (2008), who administered the Rutgers Acquired Equivalence Test (AET) to SZ patients and controls. Specifically, these authors found that feedback-driven acquisition performance correlated with negative symptom scores and was impaired only in deficit syndrome SZ patients. Additional evidence for a specific impairment in feedback-driven acquisition performance in deficit SZ patients comes from a related study by the same group (Polgar et al. 2008), using the "chaining" association task originally developed by Shohamy et al. (2005) and Nagy et al. (2007).

The Iowa gambling task (IGT; Bechara et al. 1994, 1997) is another canonical probabilistic RL task, where subjects receive variable rewards and punishments based on their choices. In this task, subjects can choose gambles from four different decks. Two of the decks offer $100 rewards, and two offer $50 rewards. However, the decks offering the higher rewards also involve large potential losses, and choosing these higher paying decks is ultimately disadvantageous. Thus, the choice of the decks with smaller rewards turns out to be more advantageous. Learning these contingencies typically requires an extended period of sampling across decks, as the frequency and magnitude of the punishments vary across decks. Rather remarkably, the initial sample of patients with extensive frontal lesions showed a robust preference for the higher paying but ultimately disadvantageous decks and appeared to be almost totally indifferent to punishment (Bechara et al. 1994). Thus, it appeared that their behavior was driven by reward seeking alone, as if the punishments simply failed to occur.

The IGT has been administered in numerous studies of DM and reinforcement learning in schizophrenia (Kester et al. 2006; Kim et al. 2009, 2012; Lee et al. 2007; Ritter et al. 2004; Sevy et al. 2007; Shurman et al. 2005; Wilder et al. 1998), with mixed results. While SZ patients do not show the dramatic insensitivity to punishments observed in OFC lesion patients in the early studies of Bechara et al. (1994), SZ patients *often* show a reduced ability to learn to choose good decks more frequently over time (Kester et al. 2006; Shurman et al. 2005). However, genuine insensitivity to punishments—which may be characteristic of some patient groups, such as those with bipolar affective disorder (Brambilla et al. 2013; Burdick et al. 2014) or orbitofrontal lesions (Bechara et al. 1994, 1999)—does not appear to be the best explanation of poor IGT performance in SZ patients. Rather, recent evidence (Brambilla et al. 2013) indicates that poor performance on the IGT most likely stems from impairment in the ability to use feedback to adaptively update estimations of choice value. Specifically, SZ patients appear to show intact sensitivity to the frequency of losses, but are less able than control subjects to integrate information about the frequency and magnitude of gains and losses in order to form adaptive representations of expected value (Brown et al. 2015).

Finally, the Balloon Analog Risk Task (BART; Lejuez et al. 2003) is another experimental paradigm designed to examine DM under risk. In this task, subjects "inflate" a balloon using the space bar on the computer. As the balloon gets bigger, the potential reward gets bigger. The balloon pops randomly somewhere between the first and the 128th potential press, such that the optimal strategy to maximize gains would be to press 64 times each time, thereby ensuring the fewest pops coupled with the maximal retained gains. Subjects are presented with three blocks of 30 trials, during which they can learn to optimize their behavior. Thus, unlike the IGT, eventual loss is certain in the BART, and the question is how much risk subjects are willing to take to increase the magnitude of their reward. Prior studies with the BART have found that several clinical populations with impulse control deficits (such as individuals with substance dependence and individuals with attention deficit disorder; Aklin et al. 2012; Mantyla et al. 2012) show abnormal risk seeking, whereby reward seeking drives behavior in the face of ultimately certain punishment. In three studies using the BART in SZ, patients showed reduced tolerance for risk (fewer pumps), relative to controls (Brown et al. 2015; Cheng et al. 2012; Reddy et al. 2014). That is, they appeared to be abnormally sensitive to the prospect of a punishment and settled for reduced gains.

In sum, findings from the above-described studies point to deficits in multiple processes subserving the successful performance of probabilistic RL in SZ patients, but leave open the questions of how closely these RL deficits relate to motivational impairments in SZ. Recent findings, however, appear to indicate that motivational impairments in SZ relate more closely to PFC-driven rapid/explicit RL processes than BG-driven gradual/procedural RL processes. These observations fit with observations that: (1) rapid/explicit RL processes relate closely to PFC-dependent working memory processes (Collins and Frank 2012); and (2) strong correlations between performance on RL tasks and standard measures of working memory/executive function (Brown et al. 2015).

4.1.2 Studies of in-the-Moment Value Estimation in Schizophrenia

A limited set of findings (Brown et al. 2013; Heerey et al. 2008; Tremeau et al. 2008) points to an impaired ability in schizophrenia patients to estimate expected value on the fly, in hypothetical decision making situations. However, while data support a connection between the instantaneous computation of expected value and intellectual function in schizophrenia (Brown et al. 2013; Heerey et al. 2008), we are not aware of data supporting a link between the instantaneous computation of expected value and *motivational deficits* in schizophrenia. When the time of reward delivery is a factor in DM, however, data do support a relationship between decision making performance and *motivational deficits* in schizophrenia (Heerey et al. 2007). That is, when the decision making situation calls for the adjustment of value representation, based on the fact that potential rewards would be delivered at some point in the future, rather than immediately, one sees systematic relationships between decision making performance and ratings of motivational deficits in SZ patients (Heerey et al. 2007). This effect was revealed in a study of DD, in which Heerey et al. (2007) found that patients discounted more steeply than did comparison participants and that discounting among patients related to both memory capacity and negative symptoms. Taken as a whole, studies of in-the-moment value estimation in SZ suggest that the online estimation of EV depends heavily on intellectual resources (such as working memory), whereas the discounting of the value of future rewards is a process that relates closely to avolition and anhedonia.

4.1.3 Devaluation and Extinction Experiments

Another approach to investigating the integrity of value representations in schizophrenia has been to use devaluation experiments, whereby a rewarding stimulus, or a stimulus or action *predictive* of a reward, becomes less valued, as a consequence of satiety or an aversive event. Such a paradigm may involve conditioning, whereby a stimulus, or a stimulus-response pairing, becomes associated with an aversive outcome across encounters with the stimulus. Alternatively, a reinforcer (usually a primary reinforcer) may lose value as a consequence of satiety—not because the stimulus has become associated with an aversive event.

Holt et al. (2009) used a contextual fear conditioning paradigm to examine extinction learning and recall in SZ patients. In this paradigm, individuals were conditioned to expect a finger shock following a CS (a lit lampshade) in one context (e.g., an office), but not in another (e.g., a conference room). Among individuals showing autonomic responsivity, Holt et al. (2009) found no group differences in conditioning. That is, patients learned to associate a CS in context with a US (the finger shock) just as well as controls. However, responders in the SZ group showed impaired extinction recall, relative to controls, in that they exhibited a stronger association between the CS+ and the US in the extinction learning context than controls did. This abnormality in extinction recall in SZs was found to correlate with the severity of psychotic symptoms, but the authors reported no significant relationships with negative symptoms. In a second study, Waltz et al. (2015) used a sensory-specific satiety (SSS) paradigm to examine associations between measures of value updating in schizophrenia and clinical ratings of negative symptoms. The paradigm involved feeding subjects small squirts of liquid foods (V8 juice and chocolate hazelnut drink), as well as a control solution, in a pseudorandom order, using syringes. In each of 2 sessions, subjects received 16 squirts of each rewarding food and 32 squirts of the control solution. In between the 2 sessions, each subject was instructed to drink one of the foods (determined through counterbalancing) until he/she felt "full, but not uncomfortable." At ten regular intervals, interspersed throughout the 2 sessions, subjects rated each liquid from 0 to 100, using a Likert-type scale. We observed group differences in SSS effects, such that controls showed an effect of satiety that was sensory specific, but patients showed an effect of satiety that was not. Furthermore, in SZ patients, we observed correlations between the magnitude of SSS effects and measures of anhedonia and avolition. Both of these results indicate that patients with SZ have an impairment in the ability to flexibly and rapidly update representations of the value of stimuli and actions. Our study of SSS in SZ patients, however, specifically indicates that the ability to flexibly and rapidly update representations of the value of stimuli figures critically in the adaptive motivation of goal-directed behavior.

4.2 Neural Evidence for Faulty EV Signaling in SZ

Neural correlates of reward (or punishment) anticipation are often operationalized as brain responses to reward- (or punishment-) predicting cues. Many of these reports are on studies using MID paradigms, but other studies assessed the expression of preferences among stimuli following conditioning. Some of these conditioning studies involved the application of computational model to behavioral data, in order to estimate expected value on a trial-by-trial basis. In this section, we review these different kinds of studies.

4.2.1 Investigating Hypothetical Decision Making in SZ with MRI

We know of a single study of hypothetical DM in SZ with MRI. In this study, involving the use of a DD paradigm, Avsar et al. (2013) found that, when compared with controls matched for consistency of preferences, SZ patients showed reduced activation during performance of the DD task in executive function and reward areas, such as the inferior frontal gyrus, dACC, posterior parietal cortex, and VS, thalamus, and midbrain. Furthermore, SZ patients had abnormal activation of lateral and medial frontal regions in relation to trial difficulty.

4.2.2 Making Sense of MID Results

As noted above, MID paradigms have been used numerous times in MRI studies involving schizophrenia patients, with findings seeming to depend, to some extent, on the medication status of patients (Juckel et al. 2006a, b; Walter et al. 2009). Studies involving chronic, medicated SZ patients have not always found evidence of blunted reward anticipation responses (most commonly sought in VS) in entire samples of SZ patients, although Walter et al. (2009) found that patients and controls differed in their anticipatory responses in anterior cingulate cortex, consistent with a deficit in action valuation. Several studies (Simon et al. 2010; Waltz et al. 2010) found that the blunting of reward anticipation responses correlated with clinical ratings of avolition and anhedonia in SZ patients. More recently, Mucci et al. (2015) found the reward anticipation signals in the dorsal caudate correlated with avolition in schizophrenia.

In general, one could interpret the results of studies using MID paradigms as pointing to aberrant EV-related signals in patients with schizophrenia. More specifically, however, reports on studies using MID paradigms have revealed correlations between measures of avolition and anhedonia and EV-related signals in both striatum and PFC.

4.2.3 Conditioning Experiments

Another set of studies attempted to examine expected value signaling in schizophrenia by using neuroimaging in conjunction with conditioning paradigms, such that stimuli or actions *acquired* value over the course of the experiment. In classical/Pavlovian conditioning paradigms, stimuli acquired value, because they predict/co-occur with rewards or punishments throughout a learning phase of an experiment. In such experiments, there is nothing that the subject needs to do in order to receive a reward or punishment; they only observe the temporal associations between stimuli and outcomes.

In an fMRI study of classical conditioning in schizophrenia, Jensen et al. (2008) trained participants to expect an aversive outcome after the presentation of a circle of a certain color. Whereas one colored circle (the CS+) was followed by a loud noise on 50 % of trials, another colored circle (the CS−) was followed by the visual presentation of a star on 50 % of trials. These authors observed that SZ patients showed inappropriately strong activations in the VS in response to the neutral stimulus (CS−) as compared to the healthy controls. That is, patients with SZ activated VS in such a way as to indicate that they had assigned motivational salience to the neutral cue, and not the cue paired with the salient/aversive event. Despite the fact that all patients in the study were stably medicated, and no associations were observed between VS responses and clinical symptoms, this finding was primarily interpreted by the authors to indicate that the aberrant processing of salient events predisposes patients to psychosis. No group differences were observed in adaptive salience signaling in the VS.

Romaniuk et al. (2010) designed an experiment in which stimuli (a yellow or blue background) were associated with neutral or aversive images from the International Affective Pictures Set (IAPS; Lang et al. 2005). One conditioned stimulus CS (CSav) was paired on 50 % of trials with the presentation of an aversive IAPS picture. The second CS (CSneu) was paired on 50 % of trials with an emotionally neutral IAPS picture. The authors included parametric regressor for both cues and outcomes to assess brain signals for both expected value and reward prediction errors. Romaniuk et al. (2010) found that patients with schizophrenia showed abnormal activation of the amygdala, midbrain, and VS during conditioning, such that these brain regions failed to distinguish between cues to aversive outcomes and cues to neutral outcomes (as they did in healthy volunteers). Furthermore, activation of the midbrain in response to neutral rather than aversive cues during conditioning was correlated with the severity of delusional symptoms in the patient group.

In a study using two separate classical conditioning paradigms (one involving a primary reinforcer and one involving a symbolic/monetary reinforcer), Dowd and Barch (2012) found that patients with schizophrenia exhibited striatal responses to appetitive cues that were largely intact. However, these authors (Dowd and Barch 2012) observed that striatal responses to appetitive cues correlated with self-reports of trait anhedonia in both patients *and* controls. Furthermore, Individual difference analyses in patients revealed an association between physical anhedonia and activity in ventromedial prefrontal cortex during anticipation of reward, such that greater anhedonia severity was associated with reduced activation to money versus no-money cues, in both controls and patients. These findings suggest that attenuated EV signaling may be a marker for anhedonia, regardless of diagnosis, as opposed to motivational deficits in schizophrenia, in particular.

Several groups have used operant conditioning paradigms to assess the intactness or disruption of value signals in individuals with schizophrenia. Koch et al. (2010), for example, used a reinforcement learning paradigm in which participants were presented with three sets of probabilistic contingencies: one highly uncertain condition that did not allow any outcome prediction (i.e., 50 % stimulus-outcome contingency), one condition permitting full prediction (i.e., 100 % stimulus-outcome contingency), and one condition where prediction was partly possible (i.e., 81 % stimulus-outcome contingency). Each probabilistic contingency involved the presentation of a card with a geometrical figure on it (i.e., circle, cross, half-moon, triangle, square, or pentagon). Participants were told that each figure was associated with an unknown value ranging from 1 to 9 and that each figure predicted the respective value (higher or lower than five) with a certain probability. The participant was asked to guess whether the figure on the card predicted a value higher or lower than the number five and told that each correct guess was followed by a monetary reward (+0.50 €), whereas each wrong guess was followed by a punishment (−0.50 €). The whole paradigm consisted of two stimuli in each stimulus category, and 16 trials with each stimulus, for a total of 96 trials. Consistent with the results of many of the studies listed above, Koch et al. (2010) found that patients' ability to learn contingencies on the basis of feedback and reward was significantly impaired. Furthermore, these researchers observed that the effects of reward probability on neural responses in ACC and DLPFC were modulated by group, such that controls showed steeper (inverse) relationships between BOLD signal activation and reward likelihood. That is, controls showed significantly stronger activation, compared to patients, in association with decreasing predictability, whereas patients showed weaker effects of reward likelihood on the BOLD signal in ACC and DLPFC.

Gradin et al. (2011) used a probabilistic stimulus selection paradigm with a juice reward, in conjunction with computational modeling, to estimate brain responses to both EV and RPE, finding that patients with SZ showed attenuated MRI activity associated with EV in amygdala-hippocampus, as well as posterior hippocampal gyrus, bilaterally. Of note, SZ patients and controls did not differ in EV-associated activity in VS. In fact, EV-associated activity in the left VS was significantly *greater* in SZ patients than in controls. Furthermore, EV-associated activity in amygdala-hippocampus was found to correlate with psychotic symptoms in SZ patients.

There is also one published report in the literature describing the results of a neuroimaging study using an extinction paradigm in schizophrenia. In this study, Holt et al. (2012) used a paradigm similar to the one described above in their behavior study (Holt et al. 2009). In their neuroimaging study, Holt et al. (2012) found that, during contextual fear conditioning, SZ patients showed abnormal BOLD responses, relative to control participants, within the posterior cingulate gyrus (PCG), hippocampus, and thalamus, with PCG abnormalities linked to negative symptoms. Although SZ patients and controls showed comparable neural responses during extinction learning, patients and controls differed in their brain responses 24 h after the learning phases. Whereas controls showed increased vmPFC responses in the extinction (safe) context (indicating successful retention of the extinction memory), SZ patients showed blunted vmPFC responses in the safe context. This attenuation was especially pronounced in delusional patients.

Finally, there is one published report in the literature describing the results of a neuroimaging study using a devaluation paradigm in schizophrenia. In this study, Morris et al. (2015) used a Pavlovian–instrumental transfer (PIT) paradigm, in order to evoke responding to cues predictive of a food reward. In a later phase of the experiment, the food rewards (crackers) were devalued, by the presentation of images in which cockroaches were crawling on the food rewards. These authors found that, whereas controls developed a strong preference for foods that had not been devalued, SZ patients chose the devalued food just as much as the alternative. This result indicated that SZ patients had a starkly reduced ability to update the values of actions when actions were guided by experienced outcomes. Accordingly, patients in the study showed reduced activity in the caudate, relative to controls, during choices of appetitive stimulus, with the greatest attenuations in caudate activity observed in patients with the most severe negative symptoms.

As shown in Table 1, some neuroimaging studies of conditioning (Dowd and Barch 2012; Morris et al. 2015) suggest that disruptions in fronto-striatal circuits related to EV signaling do, in fact, correlate with negative symptoms in SZ patients. In other studies, correspondences between aberrant EV-related fronto-striatal signals and positive symptoms were reported (Holt et al. 2012; Romaniuk et al. 2010). It is not clear whether correspondences between aberrant EV-related fronto-striatal signals and negative symptoms were investigated in these studies. In any case,

Table 1 Neuroimaging studies of reward anticipation/expected value in schizophrenia: effects of group and symptom dimension

Paradigm	Group-diff. in PFC/HC	Group-diff. in VS/MB	Neg. Sx. effect in PFC/HC	Neg. Sx. effect in VS/MB	Pos. Sx. effect in PFC/HC	Pos. Sx. effect in VS/MB
MID	Walter et al. (2009)	Juckel et al. (2006b)[b] Juckel et al. (2006a)[a]		Mucci et al. (2015) Waltz et al. (2010) Simon et al. (2009)	Walter et al. (2009)	
Pavlovian conditioning	Holt et al. (2012)	Romaniuk et al. (2010) Jensen et al. (2008)	Dowd & Barch (2012)	Dowd & Barch (2012)	Holt et al. (2012)	Romaniuk et al. (2010)
Instrumental conditioning	Gradin et al. (2011) Koch et al. (2010)				Gradin et al. (2011)	
Pavlovian-instrumental transfer		Morris et al. (2015)		Morris et al. (2015)		
Delay discounting	Avsar et al. (2013)	Avsar et al. (2013)				

[a]Study in unmedicated psychosis patients
[b]Comparison of patients medicated with first- and second-generation antipsychotics
Abbreviations: *diff.* difference, *PFC* prefrontal cortex, *HC* hippocampus, *VS* ventral striatum, *MB* midbrain, *Neg. Sx.* negative symptoms, *Pos. Sx.* positive symptoms, *MID* Monetary Incentive Delay

neuroimaging studies of conditioning provide some of the best evidence that fronto-striatal circuit activity associated with EV signaling plays a role in schizophrenia psychopathology.

4.2.4 Conclusions Regarding Evidence for Faulty EV Signaling in SZ

When we revisit the various ways in which value representations can influence behavior, we conclude that the ability to estimate EV on the fly (by integrating the probabilities and magnitudes of various outcomes) is disrupted in schizophrenia, but does not systematically relate to motivational deficits. By contrasts, there is considerable behavioral and neuroimaging evidence in the literature that abnormalities in the signaling of the acquired incentive value of stimuli in the environment relate systematically to clinical and self-report ratings for motivational deficits, in medicated patients with schizophrenia. It should be noted, however, that abnormalities in the signaling of the acquired incentive value of stimuli in the environment have not been found to relate solely to the measures of motivational deficits. A finding common to multiple studies (e.g., Jensen et al. 2008) was that SZ patients activated dopaminergic brain regions to neutral stimuli in a way indicating that inappropriate motivational salience had been assigned to those stimuli. Although not every neuroimaging study of value representation in SZ study reported results bearing on the question of whether the ability to adaptively assign value to stimuli related to motivational deficits (Holt et al. 2012; Romaniuk et al. 2010), some systematic relationships between behavioral/brain signals and clinical measures of motivational deficits in SZ have been reported—particularly in the context of paradigms specifically probing the signaling of the learned value of an action or choice (Morris et al. 2015). Finally, there is behavioral evidence that motivational deficits in SZ are associated with deficits in the ability to estimate the costs of actions (Gold et al. 2013).

5 Avolition and Outcome Processing in Schizophrenia

While a deficit in hedonic experience was long considered a core feature of schizophrenia (Bleuler 1950/1911; Kraepelin 1919), this view has been challenged in recent years by researchers who argue that the hedonic deficit in schizophrenia is primary one of pleasure anticipation (Kring and Neale 1996). As noted in the introduction, the preponderance of behavioral self-report studies support this view, indicating that SZ patients do not differ from controls in their self-reported experience of pleasure (Cohen and Minor 2008). Nonetheless, more recent studies, especially those using neuroimaging techniques, have painted a more complicated picture. While a deficit in hedonic experience is still not thought to be a core feature of schizophrenia, there is some evidence that measures of consummatory anhedonia and neural correlates thereof relate to *avolition* in schizophrenia. Whether a deficit in hedonic experience might be the primary driver of avolition in SZ (rather than a deficit in expected value—the ability to anticipate rewards) is the subject of the following sections.

5.1 Behavioral Studies of Outcome Processing in Schizophrenia

The interpretation of the synthesis of self-report studies, by Cohen and Minor (2008), is based on studies using both experience sampling (Gard et al. 2007) and experimental behavioral paradigms. A study from our group (Heerey and Gold 2007), for example, involved the use of a paradigm where participants were presented with images from the IAPS (Lang et al. 2005), and prompted to respond in multiple ways: (1) by rating the degree to which each slide was experienced as pleasurable and arousing using 9-point Likert scales anchored by "extremely [unpleasant/calm]" and "extremely [pleasant/arousing]"; (2) by repeatedly pressing one set of keys if they wanted to see pleasant images again, or repeatedly pressing another set of keys if they did *not* want to see unpleasant images again; and (3) by repeatedly pressing one set of keys to extend the display time of pleasant images, or repeatedly pressing another set of keys to reduce the display time of unpleasant images. In this study, we found that, relative to controls, SZ patients showed a weaker correspondence between their hedonic ratings of images and their goal-directed responding, both when it was "representational" (for the purpose of making images return, or stay away) and "evoked" (for the purpose of extending or reducing the display time of images). Additionally, representational responding was predicted by self-reported social anhedonia, confirming our hypothesis that anhedonia might relate to faulty representations of reward value.

It should be noted that the results of experimental studies do not *exclusively* point to an intact experience of pleasure in schizophrenia. First, a fraction of the studies described in the Cohen and Minor (2010) meta-analysis actually are suggestive of a consummatory hedonic deficit in schizophrenia patients. Second, a number of studies reporting the lack of a group difference between the entire sample of SZ patients and the entire sample of controls in consummatory hedonic ratings include subsets of patients who do differ from controls in consummatory hedonic ratings. Third, the experience of reinforcers through the gustatory and olfactory modalities may be specifically disrupted in schizophrenia (Crespo-Facorro et al. 2001; Moberg et al. 1999). This may be partially due to degraded discriminatory abilities by these modalities (Turetsky et al. 2003), although, as described below, there is also evidence of attenuated hedonic responses in the presence of intact sensory discrimination in SZ (Plailly et al. 2006).

5.2 Neuroimaging Studies of Reward and Punishment Receipt in Schizophrenia

Results from neuroimaging studies are perhaps more mixed than those from behavioral studies. In fact, findings from a number of neuroimaging studies suggest that brain responses underlying pleasurable emotional experiences are aberrant in SZ (Paradiso et al. 2003; Plailly et al. 2006; Taylor et al. 2005). Plailly et al. (2006), for example, presented subjects with 48 different odorants during 8 positron emission tomography (PET) scans. Subjects had either to detect odor, or to judge odor familiarity or hedonicity. Plailly et al. (2006) found that SZ patients and controls did not differ in their ability to detect suprathreshold odorants, but patients found odors less familiar, and pleasant odors less pleasant than controls. These behavioral results were related to reduced regional cerebral blood flow (rCBF) in patients, in posterior piriform cortex and orbital regions for familiarity judgments, the insular gyrus for hedonicity judgments, and the left inferior frontal gyrus and anterior piriform cortex/putamen region for the three olfactory tasks. In another study, Dowd and Barch (2010) had SZ patients and controls undergo fMRI scanning while making valence and arousal ratings in response to emotional pictures, words, and faces. These authors found that functional activity was largely intact in the sample of patients, as a whole, except for regions in VS and left putamen, which showed reduced responses to positive stimuli.

Studies using MID paradigms have also been a source of findings regarding brain responses to outcomes in schizophrenia. Note that, in studies using the MID, it is difficult to specifically label outcome responses as RPE signals, because unexpected outcomes are also generally salient, and participants are trained on reward contingencies prior to MRI scanning, and thus, there is no learning to be modeled. Group differences in outcome-evoked brain responses have been observed in several of these studies, but these group differences have tended to be localized in cortical areas, such as the anterior insula/VLPFC, medial PFC, lateral temporal cortex, and amygdala (Walter et al. 2009; Waltz et al. 2010).

The main issue, however, for the purpose of this review, is whether reward-related signals in fronto-striatal circuits have been shown to correlate with the severity of symptoms in schizophrenia. Multiple studies have observed significant correlations between outcome responses in the brain and ratings of delusional symptoms (e.g., Schlagenhauf et al. 2009). Multiple studies have observed significant correlations between outcome responses in the brain and ratings of motivational deficits in SZ, as well. Work from our group (Waltz et al. 2009), for example, has tied the experience of a gustatory reinforcer (juice) to motivational deficits in SZ. We observed that avolition scores from the SANS correlated significantly with activity in the primary gustatory cortex and putamen at the time of unsurprising deliveries of the reinforcer (Waltz et al. 2009). In a study using an MID task, Simon et al. (2010) found that depressive symptoms (from the Calgary Depression Scale; Addington et al. 1992; Muller et al. 1999) were predictive of the magnitudes of the [REWARD RECEIVED—REWARD OMITTED] contrast in the VS in individual SZ patients,

VS responses in patients with higher ratings for depression distinguished less between reward deliveries and omissions than they did in less depressed patients. In our study, using an MID paradigm (Waltz et al. 2010), described above, we also assessed correlations between negative symptoms in SZ and neural responses to outcomes, finding that ratings of avolition/anhedonia (from the SANS) in patients correlated with sensitivity to obtained losses in medial PFC (Waltz et al. 2010). Thus, in at least two studies, neural responses to the experience of pleasant stimuli in the striatum were correlated with measures of motivational deficits in SZ, while in a third, neural responses to the experience of pleasant stimuli in VMPFC were correlated with measures of motivational deficits. In a more recent study, our group *also* observed a correlation between clinical ratings of avolition/anhedonia and the magnitude of [PUNISHMENT—REWARD] contrasts in the VS (Waltz et al. 2013). Specifically, in the context of the performance of a probabilistic reversal learning task, patients with higher ratings for avolition/anhedonia showed less VS deactivation for punishment patients with lower ratings. Finally, in their study using emotional pictures, words, and faces as stimuli, (Dowd and Barch 2010) found that higher anhedonia scores were associated with reduced activation to positive versus negative stimuli in bilateral amygdala and right VS in SZ patients.

Findings from at least one recent neuroimaging study provide strong evidence that, in the absence of behavioral response requirements, immediate brain responses to pleasurable visual stimuli are largely intact in medicated individuals with chronic schizophrenia. Ursu et al. (2011) examined the brain activity of SZ patients and controls during trials in which they viewed an affective picture and, after a delay, reported their emotional experience while viewing it. These authors (Ursu et al. 2011) found that, in the presence of emotional stimuli, SZ patients and controls exhibited brain activity that was similar. By contrast, patients showed decreased activation, relative to controls, in dorsolateral, medial, and ventrolateral prefrontal cortices, during the delay. Importantly, the delay-related response of the DLPFC to pleasant stimuli correlated negatively with SANS anhedonia scores.

5.3 Neuroimaging Studies of RPE Signaling in SZ

The observation of aberrant signals related to both anticipatory and consummatory hedonics in SZ does not preclude the possibility that prediction error computation is *also* aberrant in SZ (and vice versa). In theory, inaccurate, or maladaptive, EV representations could reflect a reduced ability to translate hedonic experience into the expectation of value. That is, a reduced ability to update representations of expected value could be the result of faulty learning mechanisms (presumably reinforcement learning algorithms driven by reward prediction errors, or RPEs). Establishing, however, that the actual mechanism of computing the mismatch between expected and experienced outcomes is abnormal, despite the frequent input of inaccurate or degraded representations, becomes a more difficult task. Isolating these three processes, again, is made possible through experimental design and computational modeling. While the literature, summarized below, does not point to general sparing of RPE signaling in SZ, modeling work indicates that RL deficits persist in the presence of intact RPE signaling.

Studies of acutely ill patients have found evidence of disrupted RPE signaling in the midbrain (Murray et al. 2007) and VS (Schlagenhauf et al. 2014), with potentially important implications for RL and belief formation (Deserno et al. 2013; Heinz and Schlagenhauf 2010). Neuroimaging studies with chronically ill schizophrenia patients clearly point to abnormal outcome-related signals in schizophrenia (Jensen et al. 2008; Walter et al. 2009, 2010; Waltz et al. 2009, 2010), but, as noted above, it is not entirely certain that those abnormal outcome-related signals definitively represent RPE signals. Walter et al. (2009, 2010), for example, have demonstrated that healthy volunteers use anterior insula to signal unsigned prediction errors, in the context of MID task performance. That is, in controls, anterior insula indicates when outcomes are *either* better or worse than expected. By contrast, patients with schizophrenia fail to show this U-shaped response pattern, as a function of RPE valence and magnitude—a result that can be interpreted as a blunted general salience signal in schizophrenia.

The results of several studies using RL paradigms point specifically to the disruption of RPE signals in the striatum in chronic, medicated SZ patients. Koch et al. (2010), for example, showed that brain responses scale with the magnitude of appetitive RPEs in healthy volunteers in the putamen, and to a much lesser degree in patients with SZ. In the MRI study of learning from Gradin et al. (2011), SZ patients were found to show abnormal outcome-related brain responses in the midbrain and caudate. In this study, in particular, the authors used model-derived trial-by-trial estimates of RPE valence and magnitude in order to construct amplitude-modulated regressor for fMRI data analysis.

Findings from multiple groups indicate that negative RPE signals in the striatum may be intact in chronic, medicated SZ patients—at least for worse-than-expected outcomes (Walter et al. 2009, 2010; Waltz et al. 2009, 2010). In their study of classical conditioning, mentioned above, Dowd and Barch (2012) also examined responses to outcomes, finding that, at the time of receipt, SZ patients showed largely intact responses to receipt of reward versus nonreward in striatum, midbrain, and frontal cortex. Right anterior insula demonstrated greater activation for nonreward than reward cues in controls and for reward than nonreward cues in patients. At the time of receipt, robust responses to receipt of reward versus nonreward were seen in striatum, midbrain, and frontal cortex in both groups. Furthermore, both groups demonstrated enhanced responses to unexpected versus expected outcomes in cortical areas including bilateral dorsolateral prefrontal cortex. Responses to reward receipt in dorsal PFC (BA 6; but not in VS) were correlated with physical anhedonia in both patients and controls.

In short, a large body of evidence indicates that the aberrance of RPE signals relates to the severity of psychotic symptoms in SZ, thereby supporting the idea that the inappropriate signaling of mismatches between expected and experienced outcomes may be a path to abnormal learning and, subsequently, the formation of delusions that comprise psychosis (Kapur 2003). There is some evidence that the aberrance of RPE signaling relates to the severity of negative symptoms in SZ. More work is needed to determine whether the aberrant outcome signals associated with negative symptoms in SZ can be labeled RPE signals by the strictest measures.

5.4 Do Abnormalities in Consummatory Hedonics or RPE Signaling Account for Abnormalities in EV Signaling in SZ?

The results cited above present a complicated picture: More often than not, patients appear to report stimuli as pleasurable or aversive in a similar fashion as healthy controls. Frequently, however, neural responses involved in the actual experience of pleasure are found to be altered in SZ patients. This raises the possibility that abnormal neural responses to pleasurable experiences, at the physiological level, may contribute to the motivational deficits observed in schizophrenia, even if patients report normal hedonic experience. Given evidence of both aberrant EV and aberrant outcome signals in medicated patients with schizophrenia, one would be right to wonder whether the two abnormalities ever occur independently, or if one is only epiphenomenal to the other. Presumably, a deficit in "wanting" would reflect a reduced ability to translate hedonic experience into the expectation of value. This reduced ability to update representations of expected value could, in turn, be the result of faulty learning mechanisms (presumably reinforcement learning algorithms driven by reward prediction errors, or RPEs), or the consequence of reinforcement learning algorithms receiving faulty inputs.

Experimental studies of RL in SZ, where expected value was either specifically manipulated as an independent variable, or modeled on a trial-by-trial basis, or both, have been rare. In particular, rare have been studies where expected value, as a condition, was crossed with outcome value, and orthogonal with outcome salience. We sought to dissociate the contributions of EV and RPE signaling to RL deficits in SZ through the use of computational modeling of behavioral data from an experimental reinforcement learning paradigm (Gold et al. 2012). In this study (Gold et al. 2012), SZ patients and controls performed a probabilistic reinforcement learning task, requiring subjects to learn to choose the best option from each of four pairs of stimuli. In two pairs, the best option resulted in a frequent monetary gain; in the other two pairs, the best option resulted in the frequent avoidance of a monetary loss. Once the pairs were acquired, stimuli were presented in novel combinations in a test phase to assess preferences among stimuli, based on their valence and reinforcement probability. In order to investigate relationships with negative symptoms, patients were split into two subgroups, based on the median rating for avolition/anhedonia in the sample (from the SANS). We (Gold et al. 2012) found that patients with mild negative symptoms exhibited normal preferences, during both acquisition and test, for gain stimuli, relative to neutral stimuli, and for neutral stimuli, relative to loss stimuli. This result pointed to an intact ability in patients with low negative symptoms to learn from reward prediction errors (RPEs). However, when patients with high negative symptoms were compared to either controls or patients with mild negative symptoms, they showed normal acquisition for loss-avoidance stimuli, but slower acquisition and lower accuracy for gain stimuli. High negative patients also showed a reduced preference for "frequent-gain" stimuli, relative to "frequent loss-avoidance" stimuli. Therefore, negative symptoms were not associated with a general deficit in learning from all reward prediction errors, but more specifically with a reduced ability to represent or utilize expected positive values to guide decisions—a capacity consistently associated with function in the ventral and medial aspects of PFC (Noonan et al. 2012; Rushworth 2008).

The behavioral findings were elaborated upon by computational modeling analyses of the data. These analyses made use of a hybrid model integrating an (hypothetically striatally driven) actor-critic mechanism with a Q-learning component (meant to capture the role of OFC in RL; see Sect. 2.2). Computational modeling of reinforcement learning performance (Gold et al. 2012) suggested that the mechanisms for integrating expected value estimates with information derived from RPE calculations were disabled in patients with severe negative symptoms. That is, the addition of a Q-learning component to the model did not help to account for the behavior of avolitional SZ patients, beyond that of a simple actor-critic model (Gold et al. 2012). It was as if avolitional SZ patients relied entirely on striatally driven RL mechanisms, with little contribution from OFC-dependent representations of choice value. This account is consistent with the idea that what is disrupted in avolitional SZ patients, with regard to reinforcement learning, is the ability to signal the expected values of choices, and not the ability to compute RPEs by comparing expected to obtained outcomes.

The results of this study are consistent with the recent findings of a separate group, using a similar task (Reinen et al. 2014). This group (Reinen et al. 2014) found that, when compared to controls, the acquisition performance of SZ patients was impaired, especially in a gain condition, with learning in this also inversely correlated with negative symptom severity.

6 General Conclusions

Our purpose in discussing the issues pertaining to experimental studies of reinforcement learning was to pinpoint the sort of neural signal that one could say, with confidence, constituted an "expected value," for purposes of relating that signal to clinical measures. The bulk of evidence from basic neuroscience ties the representation of expected value to neural activity in fronto-striatal networks centered on ventral and medial regions of prefrontal cortex. Surprisingly—while many of the results recounted above are consistent with a contribution of OFC dysfunction to motivational deficits in SZ, by way of degraded representations of value, direct evidence of such a mechanism is limited (Barch and Dowd 2010). Based on the present state of the literature, we have reached the following conclusions and recommendations, regarding the contributions of neural representations of value to motivational deficits in schizophrenia.

Based on behavioral and neuroimaging results from the literature, there is reason to believe that the severity of motivational deficits in schizophrenia relates to the ability to precisely represent value in the brain and flexibly update these representations. Many of the reports cited above describe relationships between clinical ratings of motivational deficits and aspects of reinforcement learning related to updating of the prospective value of stimuli and actions. In our view, observations of systematic relationships between clinical ratings of motivational deficits in SZ and abnormalities in other aspects of reinforcement learning, such as blunted consummatory hedonics and aberrant RPE signals, have been more rare. We do not mean to imply that other factors (such as blunted consummatory hedonics or aberrant RPE signals) *never* contribute to motivational deficits in individuals with schizophrenia. That is, there is considerable imaging evidence that blunted consummatory hedonics and aberrant RPE signals contribute to motivational deficits in at least some individuals with serious mental illness (including schizophrenia). In our own work, we have found evidence of abnormal RPE signaling in patients with severe negative symptoms (Waltz et al. 2013). Whether abnormalities in RPE signaling in subsets of patients with psychiatric illness mediate relationships between EV signals and avolition is an issue that requires further study to resolve.

1. There are multiple findings, from neuroimaging studies, of abnormal striatal reward anticipation signals in SZ patients (or, at least, *avolitional* SZ patients). Functional imaging studies from the last two decades have most commonly implicated two regions the representation of the expected value in the brain: ventral striatum and ventral PFC (Kahnt et al. 2010; Knutson et al. 2005). In SZ patients, abnormal representations of expected value have most commonly been tied to dysfunction in ventral striatum (Morris et al. 2015; Mucci et al. 2015; Waltz et al. 2010). This could, to some extent, reflect the difficulty in extracting a strong signal from ventral PFC (Deichmann et al. 2003), and it is possible that neuroimaging studies with SZ patients would be more able to evaluate expected value signals in ventral and medial areas of PFC with newer imaging and analytic techniques (Kahnt et al. 2010). There is at least one instance of an observed relationship between a cue-evoked PFC signal (in the context of a classical conditioning paradigm) and a measure of a motivational deficit in SZ (Dowd and Barch 2012). While theoretically attractive, we are not aware of any direct evidence tying a cue-evoked PFC signal in the context of an operant learning task (in the dACC, in particular), and a measure of avolition/anhedonia in SZ.
2. Neural signals related to reward anticipation have been found to correlate with measures of motivational deficits in samples other than schizophrenia patients, including patients with major depressive disorder (MDD) (Remijnse et al. 2009; Smoski et al. 2009) and healthy volunteers (Dowd and Barch 2012). Thus, it

remains an open question, as to whether there are unique mechanisms that contribute to avolition in schizophrenia. A reduced ability to precisely represent and flexibly update value representations is a probable source of motivational deficits in general, and not just in individuals with schizophrenia. For a comprehensive review of the similarities and differences in mechanisms of motivational deficits in schizophrenia and depression, see Barch et al. in this volume.

3. Motivation appears to be driven not just by representations of value, but also by representations of the costs of actions (Treadway et al. 2009). For that reason, motivational deficits in SZ are likely driven not just by a failure to accurately represent the value of actions, but also by a failure to accurately represent the costs of actions. Schizophrenia has been associated with deficits in representing both the costs (Gold et al. 2013) and expected value (Gold et al. 2008) of actions—as well as with attenuated neural signals in dACC (Kerns et al. 2005; Minzenberg et al. 2009; but see Wilmsmeier et al. 2010), the brain region most frequently associated with the representation of the costs and benefits of actions (Kennerley et al. 2006; Walton et al. 2006). There is additional evidence that *negative symptoms* in schizophrenia are associated with deficits in representing both the costs (Gold et al. 2013) and expected value (Gold et al. 2012) of actions, though not necessarily with neural signals in dACC related to conflict monitoring.

4. Treatments for motivational deficits in SZ should have, as a goal, the enhancement of neural signals related to valuation. Whether this goal can be attained by pharmacological treatment, or nonpharmacological therapy, is not yet clear. The findings recounted in this chapter further suggest that treatments should seek to enhance not only the valuation of rewards, but also the ability to adaptively modulate the value attached to stimuli and actions (based, in part, on the instability of the environment being experienced). Finally, one could also imagine that psychosocial interventions aimed at helping people better weigh the costs and benefits of choices could lead to improvements in clinical status and real-world functioning.

References

Addington D, Addington J, Maticka-Tyndale E, Joyce J (1992) Reliability and validity of a depression rating scale for schizophrenics. Schizophr Res 6:201–208

Aklin WM, Severtson SG, Umbricht A, Fingerhood M, Bigelow GE, Lejuez CW, Silverman K (2012) Risk-taking propensity as a predictor of induction onto naltrexone treatment for opioid dependence. J Clin Psychiatry 73:e1056–e1061

Andreasen NC (1984) The scale for the assessment of negative symptoms (SANS). University of Iowa, Iowa City

Avsar KB, Weller RE, Cox JE, Reid MA, White DM, Lahti AC (2013) An fMRI investigation of delay discounting in patients with schizophrenia. Brain Behav 3:384–401

Barch DM, Dowd EC (2010) Goal representations and motivational drive in schizophrenia: the role of prefrontal-striatal interactions. Schizophr Bull 36:919–934

Bechara A, Damasio AR, Damasio H, Anderson SW (1994) Insensitivity to future consequences following damage to human prefrontal cortex. Cognition 50:7–15

Bechara A, Damasio H, Damasio AR, Lee GP (1999) Different contributions of the human amygdala and ventromedial prefrontal cortex to decision-making. J Neurosci 19:5473–5481

Bechara A, Damasio H, Tranel D, Damasio AR (1997) Deciding advantageously before knowing the advantageous strategy. Science 275:1293–1295

Berridge KC, Robinson TE (1998) What is the role of dopamine in reward: hedonic impact, reward learning, or incentive salience? Brain Res Brain Res Rev 28:309–369

Bleuler E (1950/1911) Dementia praecox or the group of schizophrenias. International UP, New York

Brambilla P, Perlini C, Bellani M, Tomelleri L, Ferro A, Cerruti S, Marinelli V, Rambaldelli G, Christodoulou T, Jogia J, Dima D, Tansella M, Balestrieri M, Frangou S (2013) Increased salience of gains versus decreased associative learning differentiate bipolar disorder from schizophrenia during incentive decision making. Psychol Med 43:571–580

Brown EC, Hack SM, Gold JM, Carpenter WT Jr, Fischer BA, Prentice KP, Waltz JA (2015) Integrating frequency and magnitude information in decision-making in schizophrenia: an account of patient performance on the Iowa Gambling Task. J Psychiatr Res 66–67:16–23

Brown JK, Waltz JA, Strauss GP, McMahon RP, Frank MJ, Gold JM (2013) Hypothetical decision making in schizophrenia: the role of expected value computation and "irrational" biases. Psychiatry Res 209:142–149

Burdick KE, Braga RJ, Gopin CB, Malhotra AK (2014) Dopaminergic influences on emotional decision making in euthymic bipolar patients. Neuropsychopharmacology 39:274–282

Caplin A, Dean M (2008) Axiomatic methods, dopamine and reward prediction error. Curr Opin Neurobiol 18:197–202

Chapman LJ, Chapman JP, Raulin ML (1976) Scales for physical and social anhedonia. J Abnorm Psychol 85:374–382

Cheng GL, Tang JC, Li FW, Lau EY, Lee TM (2012) Schizophrenia and risk-taking: impaired reward but preserved punishment processing. Schizophr Res 136:122–127

Cohen AS, Minor KS (2008) Emotional experience in patients with schizophrenia revisited: meta-analysis of laboratory studies. Schizophr Bull

Cohen AS, Minor KS (2010) Emotional experience in patients with schizophrenia revisited: meta-analysis of laboratory studies. Schizophr Bull 36:143–150

Collins AGE, Frank MJ (2012) How much of reinforcement learning is working memory, not reinforcement learning? a behavioral, computational, and neurogenetic analysis. Eur J Neurosci 35:1024–1035

Cools R, Clark L, Owen AM, Robbins TW (2002) Defining the neural mechanisms of probabilistic reversal learning using event-related functional magnetic resonance imaging. J Neurosci 22:4563–4567

Crespo-Facorro B, Paradiso S, Andreasen NC, O'Leary DS, Watkins GL, Ponto LL, Hichwa RD (2001) Neural mechanisms of anhedonia in schizophrenia: a PET study of response to unpleasant and pleasant odors. JAMA 286:427–435

De Martino B, Kumaran D, Seymour B, Dolan RJ (2006) Frames, biases, and rational decision-making in the human brain. Science 313:684–687

Deichmann R, Gottfried JA, Hutton C, Turner R (2003) Optimized EPI for fMRI studies of the orbitofrontal cortex. Neuroimage 19:430–441

Deserno L, Boehme R, Heinz A, Schlagenhauf F (2013) Reinforcement learning and dopamine in schizophrenia: dimensions of symptoms or specific features of a disease group? Front Psychiatry 4:172

Dowd EC, Barch DM (2010) Anhedonia and emotional experience in schizophrenia: neural and behavioral indicators. Biol Psychiatry 67:902–911

Dowd EC, Barch DM (2012) Pavlovian reward prediction and receipt in schizophrenia: relationship to anhedonia. PLoS ONE 7:e35622

Farkas M, Polgar P, Kelemen O, Rethelyi J, Bitter I, Myers CE, Gluck MA, Keri S (2008) Associative learning in deficit and nondeficit schizophrenia. NeuroReport 19:55–58

Frank MJ, Claus ED (2006) Anatomy of a decision: striato-orbitofrontal interactions in reinforcement learning, decision making, and reversal. Psychol Rev 113:300–326

Frank MJ, O'Reilly RC (2006) A mechanistic account of striatal dopamine function in cognition: psychopharmacological studies with cabergoline and haloperidol. Behav Neurosci 120:497–517

Frank MJ, Seeberger LC, O'Reilly RC (2004) By carrot or by stick: cognitive reinforcement learning in parkinsonism. Science 306:1940–1943

Gard DE, Kring AM, Gard MG, Horan WP, Green MF (2007) Anhedonia in schizophrenia: distinctions between anticipatory and consummatory pleasure. Schizophr Res 93:253–260

Geng JJ, Mangun GR (2009) Anterior intraparietal sulcus is sensitive to bottom-up attention driven by stimulus salience. J Cogn Neurosci 21:1584–1601

Glascher J, Hampton AN, O'Doherty JP (2009) Determining a role for ventromedial prefrontal cortex in encoding action-based value signals during reward-related decision making. Cereb Cortex 19:483–495

Glimcher PW (2011) Understanding dopamine and reinforcement learning: the dopamine reward prediction error hypothesis. Proc Natl Acad Sci U S A 108(Suppl 3):15647–15654

Gold JM, Strauss GP, Waltz JA, Robinson BM, Brown JK, Frank MJ (2013) Negative symptoms of schizophrenia are associated with abnormal effort-cost computations. Biol Psychiatry 74:130–136

Gold JM, Waltz JA, Kasanova Z, Strauss GP, Matveeva TM, Herbener ES, Frank MJ (2012) Negative symptoms and the failure to represent the expected reward value of actions: behavioral and computational modeling evidence. Archives Gen Psychiatry 69:129–138

Gold JM, Waltz JA, Prentice KJ, Morris SE, Heerey EA (2008) Reward processing in schizophrenia: a deficit in the representation of value. Schizophr Bull 34:835–847

Goldberg TE, Weinberger DR (1988) Probing prefrontal function in schizophrenia with neuropsychological paradigms. Schizophr Bull 14:179–183

Gradin VB, Kumar P, Waiter G, Ahearn T, Stickle C, Milders M, Reid I, Hall J, Steele JD (2011) Expected value and prediction error abnormalities in depression and schizophrenia. Brain 134:1751–1764

Guitart-Masip M, Fuentemilla L, Bach DR, Huys QJ, Dayan P, Dolan RJ, Duzel E (2011) Action dominates valence in anticipatory representations in the human striatum and dopaminergic midbrain. J Neurosci 31:7867–7875

Harsay HA, Spaan M, Wijnen JG, Ridderinkhof KR (2012) Error awareness and salience processing in the oddball task: shared neural mechanisms. Front Hum Neurosci 6:246

Heerey EA, Bell-Warren KR, Gold JM (2008) Decision-making impairments in the context of intact reward sensitivity in schizophrenia. Biol Psychiatry 64:62–69

Heerey EA, Gold JM (2007) Patients with schizophrenia demonstrate dissociation between affective experience and motivated behavior. J Abnorm Psychol 116:268–278

Heerey EA, Robinson BM, McMahon RP, Gold JM (2007) Delay discounting in schizophrenia. Cognit Neuropsychiatry 12:213–221

Heinz A, Schlagenhauf F (2010) Dopaminergic dysfunction in schizophrenia: salience attribution revisited. Schizophr Bull 36:472–485

Holroyd CB, Nieuwenhuis S, Yeung N, Nystrom L, Mars RB, Coles MG, Cohen JD (2004) Dorsal anterior cingulate cortex shows fMRI response to internal and external error signals. Nat Neurosci 7:497–498

Holt DD, Green L, Myerson J (2003) Is discounting impulsive?. Evidence from temporal and probability discounting in gambling and non-gambling college students. Behav Process 64:355–367

Holt DJ, Coombs G, Zeidan MA, Goff DC, Milad MR (2012) Failure of neural responses to safety cues in schizophrenia. Arch Gen Psychiatry 69:893–903

Holt DJ, Lebron-Milad K, Milad MR, Rauch SL, Pitman RK, Orr SP, Cassidy BS, Walsh JP, Goff DC (2009) Extinction memory is impaired in schizophrenia. Biol Psychiatry 65:455–463

Horan WP, Kring AM, Gur RE, Reise SP, Blanchard JJ (2011) Development and psychometric validation of the clinical assessment interview for negative symptoms (CAINS). Schizophr Res 132:140–145

Jensen J, Smith AJ, Willeit M, Crawley AP, Mikulis DJ, Vitcu I, Kapur S (2007) Separate brain regions code for salience vs. valence during reward prediction in humans. Hum Brain Mapp 28:294–302

Jensen J, Willeit M, Zipursky RB, Savina I, Smith AJ, Menon M, Crawley AP, Kapur S (2008) The formation of abnormal associations in schizophrenia: neural and behavioral evidence. Neuropsychopharmacology 33:473–479

Juckel G, Schlagenhauf F, Koslowski M, Filonov D, Wustenberg T, Villringer A, Knutson B, Kienast T, Gallinat J, Wrase J, Heinz A (2006a) Dysfunction of ventral striatal reward prediction in schizophrenic patients treated with typical, not atypical, neuroleptics. Psychopharmacology 187:222–228

Juckel G, Schlagenhauf F, Koslowski M, Wustenberg T, Villringer A, Knutson B, Wrase J, Heinz A (2006b) Dysfunction of ventral striatal reward prediction in schizophrenia. Neuroimage 29:409–416

Kahneman D, Tversky A (1979) Prospect theory: an analysis of decision under risk. Econometrica 47:263–291

Kahnt T, Heinzle J, Park SQ, Haynes JD (2010) The neural code of reward anticipation in human orbitofrontal cortex. Proc Natl Acad Sci U S A 107:6010–6015

Kahnt T, Park SQ, Haynes JD, Tobler PN (2014) Disentangling neural representations of value and salience in the human brain. Proc Natl Acad Sci U S A 111:5000–5005

Kapur S (2003) Psychosis as a state of aberrant salience: a framework linking biology, phenomenology, and pharmacology in schizophrenia. Am J Psychiatry 160:13–23

Kennerley SW, Walton ME, Behrens TE, Buckley MJ, Rushworth MF (2006) Optimal decision making and the anterior cingulate cortex. Nat Neurosci 9:940–947

Kerns JG, Cohen JD, MacDonald AW 3rd, Johnson MK, Stenger VA, Aizenstein H, Carter CS (2005) Decreased conflict- and error-related activity in the anterior cingulate cortex in subjects with schizophrenia. Am J Psychiatry 162:1833–1839

Kester HM, Sevy S, Yechiam E, Burdick KE, Cervellione KL, Kumra S (2006) Decision-making impairments in adolescents with early-onset schizophrenia. Schizophr Res 85:113–123

Kim YT, Lee KU, Lee SJ (2009) Deficit in decision-making in chronic, stable schizophrenia: from a reward and punishment perspective. Psychiatry Investig 6:26–33

Kim YT, Sohn H, Kim S, Oh J, Peterson BS, Jeong J (2012) Disturbances of motivational balance in chronic schizophrenia during decision-making tasks. Psychiatry Clin Neurosci 66:573–581

Kirby KN, Santiesteban M (2003) Concave utility, transaction costs, and risk in measuring discounting of delayed rewards. Journal of experimental psychology. Learn Mem Cognit 29:66–79

Kirkpatrick B, Buchanan RW, McKenney PD, Alphs LD, Carpenter WT Jr (1989) The schedule for the deficit syndrome: an instrument for research in schizophrenia. Psychiatry Res 30:119–123

Kirkpatrick B, Buchanan RW, Ross DE, Carpenter WT Jr (2001) A separate disease within the syndrome of schizophrenia. Arch Gen Psychiatry 58:165–171

Kirkpatrick B, Strauss GP, Nguyen L, Fischer BA, Daniel DG, Cienfuegos A, Marder SR (2011) The brief negative symptom scale: psychometric properties. Schizophr Bull 37:300–305

Knutson B, Fong GW, Adams CM, Varner JL, Hommer D (2001) Dissociation of reward anticipation and outcome with event-related fMRI. NeuroReport 12:3683–3687

Knutson B, Fong GW, Bennett SM, Adams CM, Hommer D (2003) A region of mesial prefrontal cortex tracks monetarily rewarding outcomes: characterization with rapid event-related fMRI. Neuroimage 18:263–272

Knutson B, Taylor J, Kaufman M, Peterson R, Glover G (2005) Distributed neural representation of expected value. J Neurosci 25:4806–4812

Koch K, Schachtzabel C, Wagner G, Schikora J, Schultz C, Reichenbach JR, Sauer H, Schlosser RG (2010) Altered activation in association with reward-related trial-and-error learning in patients with schizophrenia. Neuroimage 50:223–232

Kraepelin E (1919) Dementia praecox and paraphrenia. Livingston, Amsterdam

Kring AM, Gur RE, Blanchard JJ, Horan WP, Reise SP (2013) The Clinical Assessment Interview for Negative Symptoms (CAINS): final development and validation. Am J Psychiatry 170:165–172

Kring AM, Neale JM (1996) Do schizophrenic patients show a disjunctive relationship among expressive, experiential, and psychophysiological components of emotion? J Abnorm Psychol 105:249–257

Lang PJ, Bradley MM, Cuthbert BN (2005) International affective picture system (IAPS): Digitized photographs, instruction manual and affective ratings. University of Florida, Gainesville

Lee Y, Kim YT, Seo E, Park O, Jeong SH, Kim SH, Lee SJ (2007) Dissociation of emotional decision-making from cognitive decision-making in chronic schizophrenia. Psychiatry Res 152:113–120

Lejuez CW, Aklin WM, Zvolensky MJ, Pedulla CM (2003) Evaluation of the balloon analogue risk task (BART) as a predictor of adolescent real-world risk-taking behaviours. J Adolesc 26:475–479

Mantyla T, Still J, Gullberg S, Del Missier F (2012) Decision making in adults with ADHD. J Atten Disord 16:164–173

McClure SM, Berns GS, Montague PR (2003) Temporal prediction errors in a passive learning task activate human striatum. Neuron 38:339–346

Minzenberg MJ, Laird AR, Thelen S, Carter CS, Glahn DC (2009) Meta-analysis of 41 functional neuroimaging studies of executive function in schizophrenia. Arch Gen Psychiatry 66:811–822

Moberg PJ, Agrin R, Gur RE, Gur RC, Turetsky BI, Doty RL (1999) Olfactory dysfunction in schizophrenia: a qualitative and quantitative review. Neuropsychopharmacology 21:325–340

Montague PR, Hyman SE, Cohen JD (2004) Computational roles for dopamine in behavioural control. Nature 431:760–767

Morris RW, Quail S, Griffiths KR, Green MJ, Balleine BW (2015) Corticostriatal control of goal-directed action is impaired in schizophrenia. Biol Psychiatry 77:187–195

Mucci A, Dima D, Soricelli A, Volpe U, Bucci P, Frangou S, Prinster A, Salvatore M, Galderisi S, Maj M (2015) Is avolition in schizophrenia associated with a deficit of dorsal caudate activity? A functional magnetic resonance imaging study during reward anticipation and feedback. Psychol Med 1–14

Muller MJ, Marx-Dannigkeit P, Schlosser R, Wetzel H, Addington D, Benkert O (1999) The calgary depression rating scale for schizophrenia: development and interrater reliability of a German version (CDSS-G). J Psychiatr Res 33:433–443

Murray GK, Corlett PR, Clark L, Pessiglione M, Blackwell AD, Honey G, Jones PB, Bullmore ET, Robbins TW, Fletcher PC (2007) Substantia nigra/ventral tegmental reward prediction error disruption in psychosis. Mol Psychiatry 13:267–276

Nagy H, Keri S, Myers CE, Benedek G, Shohamy D, Gluck MA (2007) Cognitive sequence learning in Parkinson's disease and amnestic mild cognitive impairment: Dissociation between sequential and non-sequential learning of associations. Neuropsychologia 45:1386–1392

Noonan MP, Kolling N, Walton ME, Rushworth MF (2012) Re-evaluating the role of the orbitofrontal cortex in reward and reinforcement. Eur J Neurosci 35:997–1010

Noonan MP, Walton ME, Behrens TE, Sallet J, Buckley MJ, Rushworth MF (2010) Separate value comparison and learning mechanisms in macaque medial and lateral orbitofrontal cortex. Proc Natl Acad Sci U S A 107:20547–20552

O'Doherty JP, Hampton A, Kim H (2007) Model-based fMRI and its application to reward learning and decision making. Ann N Y Acad Sci 1104:35–53

Paradiso S, Andreasen NC, Crespo-Facorro B, O'Leary DS, Watkins GL, Boles Ponto LL, Hichwa RD (2003) Emotions in unmedicated patients with schizophrenia during evaluation with positron emission tomography. Am J Psychiatry 160:1775–1783

Peralta V, Cuesta MJ (1995) Negative symptoms in schizophrenia: a confirmatory factor analysis of competing models. Am J Psychiatry 152:1450–1457

Pessiglione M, Seymour B, Flandin G, Dolan RJ, Frith CD (2006) Dopamine-dependent prediction errors underpin reward-seeking behaviour in humans. Nature 442:1042–1045

Plailly J, d'Amato T, Saoud M, Royet JP (2006) Left temporo-limbic and orbital dysfunction in schizophrenia during odor familiarity and hedonicity judgments. Neuroimage 29:302–313

Plassmann H, O'Doherty JP, Rangel A (2010) Appetitive and aversive goal values are encoded in the medial orbitofrontal cortex at the time of decision making. J Neurosci 30:10799–10808

Polgar P, Farkas M, Nagy O, Kelemen O, Rethelyi J, Bitter I, Myers CE, Gluck MA, Keri S (2008) How to find the way out from four rooms? The learning of "chaining" associations may shed light on the neuropsychology of the deficit syndrome of schizophrenia. Schizophr Res 99:200–207

Reddy LF, Lee J, Davis MC, Altshuler L, Glahn DC, Miklowitz DJ, Green MF (2014) Impulsivity and risk taking in bipolar disorder and schizophrenia. Neuropsychopharmacology 39:456–463

Reinen J, Smith EE, Insel C, Kribs R, Shohamy D, Wager TD, Jarskog LF (2014) Patients with schizophrenia are impaired when learning in the context of pursuing rewards. Schizophr Res 152:309–310

Remijnse PL, Nielen MM, van Balkom AJ, Hendriks GJ, Hoogendijk WJ, Uylings HB, Veltman DJ (2009) Differential frontal-striatal and paralimbic activity during reversal learning in major depressive disorder and obsessive-compulsive disorder. Psychol Med 39:1503–1518

Rescorla RA, Wagner AR (1972) A theory of Pavlovian conditioning: variations in the effectiveness of reinforcement and nonreinforcement. In: Black AH, Prokasy WF (eds) Classical conditioning II. Appleton-century-crofts, pp 64–99

Ridderinkhof KR, Ullsperger M, Crone EA, Nieuwenhuis S (2004) The role of the medial frontal cortex in cognitive control. Science 306:443–447

Ritter LM, Meador-Woodruff JH, Dalack GW (2004) Neurocognitive measures of prefrontal cortical dysfunction in schizophrenia. Schizophr Res 68:65–73

Roesch MR, Olson CR (2004) Neuronal activity related to reward value and motivation in primate frontal cortex. Science 304:307–310

Roesch MR, Olson CR (2007) Neuronal activity related to anticipated reward in frontal cortex: does it represent value or reflect motivation? Ann N Y Acad Sci 1121:431–446

Romaniuk L, Honey GD, King JR, Whalley HC, McIntosh AM, Levita L, Hughes M, Johnstone EC, Day M, Lawrie SM, Hall J (2010) Midbrain activation during Pavlovian conditioning and delusional symptoms in schizophrenia. Arch Gen Psychiatry 67:1246–1254

Roy M, Shohamy D, Daw N, Jepma M, Wimmer GE, Wager TD (2014) Representation of aversive prediction errors in the human periaqueductal gray. Nat Neurosci 17:1607–1612

Rudebeck PH, Walton ME, Smyth AN, Bannerman DM, Rushworth MF (2006) Separate neural pathways process different decision costs. Nat Neurosci 9:1161–1168

Rushworth MF, Walton ME, Kennerley SW, Bannerman DM (2004) Action sets and decisions in the medial frontal cortex. Trends Cogn Sci 8:410–417

Rushworth MFS (2008) Intention, Choice, and the Medial Frontal Cortex. Ann N Y Acad Sci 1124:181–207

Rutledge RB, Dean M, Caplin A, Glimcher PW (2010) Testing the reward prediction error hypothesis with an axiomatic model. J Neurosci 30:13525–13536

Sayers SL, Curran PJ, Mueser KT (1996) Factor structure and construct validitiy of the scale for the assessment of negative symptoms. Psychol Assess 8:269–280

Saykin AJ, Shtasel DL, Gur RE, Kester DB, Mozley LH, Stafiniak P, Gur RC (1994) Neuropsychological deficits in neuroleptic naive patients with first-episode schizophrenia. Arch Gen Psychiatry 51:124–131

Schlagenhauf F, Huys QJ, Deserno L, Rapp MA, Beck A, Heinze HJ, Dolan R, Heinz A (2014) Striatal dysfunction during reversal learning in unmedicated schizophrenia patients. Neuroimage 89:171–180

Schlagenhauf F, Sterzer P, Schmack K, Ballmaier M, Rapp M, Wrase J, Juckel G, Gallinat J, Heinz A (2009) Reward feedback alterations in unmedicated schizophrenia patients: relevance for delusions. Biol Psychiatry 65:1032–1039

Schoenbaum G, Roesch M (2005) Orbitofrontal cortex, associative learning, and expectancies. Neuron 47:633–636

Schoenbaum G, Roesch MR, Stalnaker TA, Takahashi YK (2009) A new perspective on the role of the orbitofrontal cortex in adaptive behaviour. Nat Rev Neurosci 10:885–892

Schultz W (1998) Predictive reward signal of dopamine neurons. J Neurophysiol 80:1–27

Seeley WW, Menon V, Schatzberg AF, Keller J, Glover GH, Kenna H, Reiss AL, Greicius MD (2007) Dissociable intrinsic connectivity networks for salience processing and executive control. J Neurosci 27:2349–2356

Sevy S, Burdick KE, Visweswaraiah H, Abdelmessih S, Lukin M, Yechiam E, Bechara A (2007) Iowa gambling task in schizophrenia: a review and new data in patients with schizophrenia and co-occurring cannabis use disorders. Schizophr Res 92:74–84

Shanks DR, Tunney RJ, McCarthy JD (2002) A re-examination of probability matching and rational choice. J Behav Decis Mak 15:233–250

Shohamy D, Myers CE, Grossman S, Sage J, Gluck MA (2005) The role of dopamine in cognitive sequence learning: evidence from Parkinson's disease. Behav Brain Res 156:191–199

Shurman B, Horan WP, Nuechterlein KH (2005) Schizophrenia patients demonstrate a distinctive pattern of decision-making impairment on the Iowa Gambling Task. Schizophr Res 72:215–224

Simon JJ, Biller A, Walther S, Roesch-Ely D, Stippich C, Weisbrod M, Kaiser S (2010) Neural correlates of reward processing in schizophrenia—relationship to apathy and depression. Schizophr Res 118:154–161

Smith DV, Hayden BY, Truong TK, Song AW, Platt ML, Huettel SA (2014) Distinct value signals in anterior and posterior ventromedial prefrontal cortex. J Neurosci 30:2490–2495

Smoski MJ, Felder J, Bizzell J, Green SR, Ernst M, Lynch TR, Dichter GS (2009) fMRI of alterations in reward selection, anticipation, and feedback in major depressive disorder. J Affect Disord 118:69–78

Strauss GP, Frank MJ, Waltz JA, Kasanova Z, Herbener ES, Gold JM (2011) Deficits in positive reinforcement learning and uncertainty-driven exploration are associated with distinct aspects of negative symptoms in schizophrenia. Biol Psychiatry 69:424–431

Strauss GP, Keller WR, Buchanan RW, Gold JM, Fischer BA, McMahon RP, Catalano LT, Culbreth AJ, Carpenter WT, Kirkpatrick B (2012) Next-generation negative symptom assessment for clinical trials: validation of the brief negative symptom scale. Schizophr Res 142:88–92

Sutton RS, Barto AG (1998) Reinforcement learning: an introduction. MIT Press, Cambridge

Takahashi YK, Roesch MR, Stalnaker TA, Haney RZ, Calu DJ, Taylor AR, Burke KA, Schoenbaum G (2009) The orbitofrontal cortex and ventral tegmental area are necessary for learning from unexpected outcomes. Neuron 62:269–280

Taylor SF, Phan KL, Britton JC, Liberzon I (2005) Neural response to emotional salience in schizophrenia. Neuropsychopharmacology 30:984–995

Tom SM, Fox CR, Trepel C, Poldrack RA (2007) The neural basis of loss aversion in decision-making under risk. Science 315:515–518

Treadway MT, Bossaller NA, Shelton RC, Zald DH (2012) Effort-based decision-making in major depressive disorder: a translational model of motivational anhedonia. J Abnorm Psychol 121:553–558

Treadway MT, Buckholtz JW, Schwartzman AN, Lambert WE, Zald DH (2009) Worth the 'EEfRT'? the effort expenditure for rewards task as an objective measure of motivation and anhedonia. PLoS ONE 4:e6598

Tremeau F, Brady M, Saccente E, Moreno A, Epstein H, Citrome L, Malaspina D, Javitt D (2008) Loss aversion in schizophrenia. Schizophr Res 103:121–128

Turetsky BI, Moberg PJ, Owzar K, Johnson SC, Doty RL, Gur RE (2003) Physiologic impairment of olfactory stimulus processing in schizophrenia. Biol Psychiatry 53:403–411

Ursu S, Kring AM, Gard MG, Minzenberg MJ, Yoon JH, Ragland JD, Solomon M, Carter CS (2011) Prefrontal cortical deficits and impaired cognition-emotion interactions in schizophrenia. Am J Psychiatry 168:276–285

van Veen V, Carter CS (2002) The anterior cingulate as a conflict monitor: fMRI and ERP studies. Physiol Behav 77:477–482

von Neumann J, Morgenstern O (1947) Theory of games and economic behavior, 2nd edn. Princeton University Press, Princeton

Walter H, Heckers S, Kassubek J, Erk S, Frasch K, Abler B (2010) Further evidence for aberrant prefrontal salience coding in schizophrenia. Front Behav Neurosci 3:62

Walter H, Kammerer H, Frasch K, Spitzer M, Abler B (2009) Altered reward functions in patients on atypical antipsychotic medication in line with the revised dopamine hypothesis of schizophrenia. Psychopharmacology 206:121–132

Walton ME, Kennerley SW, Bannerman DM, Phillips PE, Rushworth MF (2006) Weighing up the benefits of work: behavioral and neural analyses of effort-related decision making. Neural Netw 19:1302–1314

Waltz JA, Brown JK, Gold JM, Ross TJ, Salmeron BJ, Stein EA (2015) Probing the dynamic updating of value in schizophrenia using a sensory-specific satiety paradigm. Schizophr Bull

Waltz JA, Frank MJ, Robinson BM, Gold JM (2007) Selective reinforcement learning deficits in schizophrenia support predictions from computational models of striatal-cortical dysfunction. Biol Psychiatry 62:756–764

Waltz JA, Frank MJ, Wiecki TV, Gold JM (2011) Altered probabilistic learning and response biases in schizophrenia: behavioral evidence and neurocomputational modeling. Neuropsychology 25:86–97

Waltz JA, Kasanova Z, Ross TJ, Salmeron BJ, McMahon RP, Gold JM, Stein EA (2013) The roles of reward, default, and executive control networks in set-shifting impairments in schizophrenia. PLoS ONE 8:e57257

Waltz JA, Schweitzer JB, Gold JM, Kurup PK, Ross TJ, Salmeron BJ, Rose EJ, McClure SM, Stein EA (2009) Patients with schizophrenia have a reduced neural response to both unpredictable and predictable primary reinforcers. Neuropsychopharmacology 34:1567–1577

Waltz JA, Schweitzer JB, Ross TJ, Kurup PK, Salmeron BJ, Rose EJ, Gold JM, Stein EA (2010) Abnormal responses to monetary outcomes in cortex, but not in the basal ganglia, in schizophrenia. Neuropsychopharmacology 35:2427–2439

Wilder KE, Weinberger DR, Goldberg TE (1998) Operant conditioning and the orbitofrontal cortex in schizophrenic patients: unexpected evidence for intact functioning. Schizophr Res 30:169–174

Wilmsmeier A, Ohrmann P, Suslow T, Siegmund A, Koelkebeck K, Rothermundt M, Kugel H, Arolt V, Bauer J, Pedersen A (2010) Neural correlates of set-shifting: decomposing executive functions in schizophrenia. J Psychiatry Neurosci JPN 35:321–329

Zink CF, Pagnoni G, Chappelow J, Martin-Skurski M, Berns GS (2006) Human striatal activation reflects degree of stimulus saliency. Neuroimage 29:977–983

Zink CF, Pagnoni G, Martin ME, Dhamala M, Berns GS (2003) Human striatal response to salient nonrewarding stimuli. J Neurosci 23:8092–8097

Mechanisms Underlying Motivational Deficits in Psychopathology: Similarities and Differences in Depression and Schizophrenia

Deanna M. Barch, David Pagliaccio and Katherine Luking

Abstract Motivational and hedonic impairments are core aspects of a variety of types of psychopathology. These impairments cut across diagnostic categories and may be critical to understanding major aspects of the functional impairments accompanying psychopathology. Given the centrality of motivational and hedonic systems to psychopathology, the Research Domain Criteria (RDoC) initiative includes a "positive valence" systems domain that outlines a number of constructs that may be key to understanding the nature and mechanisms of motivational and hedonic impairments in psychopathology. These component constructs include initial responsiveness to reward, reward anticipation or expectancy, incentive or reinforcement learning, effort valuation, and action selection. Here, we review behavioral and neuroimaging studies providing evidence for impairments in these constructs in individuals with psychosis versus in individuals with depressive pathology. There are important differences in the nature of reward-related and hedonic deficits associated with psychosis versus depression that have major implications for our understanding of etiology and treatment development. In particular, the literature strongly suggests the presence of impairments in in-the-moment hedonics or "liking" in individuals with depressive pathology, particularly among those who experience anhedonia. Such deficits may propagate forward and contribute to impairments in other constructs that are dependent on hedonic responses, such as anticipation, learning, effort, and action selection. Such hedonic impairments could reflect alterations in dopamine and/or opioid signaling in the striatum related to depression or specifically to anhedonia in depressed populations. In contrast, the literature points to relatively intact in-the-moment hedonic processing in psychosis, but provides much evidence for impairments in other components involved in translating reward to action selection. Particularly, individuals

D.M. Barch (✉)
Departments of Psychology, Psychiatry and Radiology, Washington University, Box 1125, One Brookings Drive, St. Louis, MO 63130, USA
e-mail: dbarch@wustl.edu

D. Pagliaccio · K. Luking
Neurosciences Program, Washington University, Box 1125,
One Brookings Drive, St. Louis, MO 63130, USA

with schizophrenia exhibit altered reward prediction and associated striatal and prefrontal activation, impaired reward learning, and impaired reward-modulated action selection.

Keywords Depression · Motivation · Prefontal cortex · Reward · Schizophrenia · Striatum

Contents

1 Introduction ... 412
   1.1 Translating Hedonic Experience into Motivated Behavior ... 413
2 Hedonics and Liking ... 417
   2.1 Schizophrenia ... 417
   2.2 Depression ... 419
   2.3 Hedonics in Schizophrenia versus Depression ... 422
3 Reward Prediction, Anticipation, and Reinforcement Learning ... 422
   3.1 Schizophrenia ... 422
   3.2 Depression ... 425
   3.3 Summary of Reward Prediction, Anticipation, and Reinforcement Learning in Schizophrenia and Depression ... 427
4 Value Computations and OFC Function ... 428
   4.1 Schizophrenia ... 428
   4.2 Depression ... 429
5 Effort Computations ... 429
   5.1 Schizophrenia ... 430
   5.2 Depression ... 431
   5.3 Summary of Effort Allocation in Schizophrenia and Depression ... 431
6 Goal-Directed Action ... 432
   6.1 Schizophrenia ... 432
   6.2 Depression ... 433
7 Summary of Reward and Motivational Neuroscience in Schizophrenia and Depression ... 433
References ... 436

1 Introduction

Motivational and hedonic impairments are core aspects of psychopathology that cut across diagnostic categories. In particular, motivational and hedonic impairments are part of the diagnostic criteria for several disorders, like schizophrenia and depression, and deficits in these domains may be critical to understanding the functional impairments that often accompany these forms of psychopathology. The Research Domain Criteria (RDoC) initiative has recognized the importance of studying motivation and hedonic processing in psychopathology and includes a "positive valence" systems domain (Cuthbert and Insel 2010; Insel et al. 2010; Cuthbert and Kozak 2013). This domain includes numerous component constructs that may be central to understanding the nature and mechanisms of motivational impairments in psychopathology, including initial responsiveness to reward, reward anticipation or expectancy, incentive learning, effort valuation, and action selection. One goal of the RDoC initiative is to understand whether there are core brain–behavior systems with common deficits that cut across the current diagnostic categories or whether there are truly unique or differential deficits associated with different facets of psychopathology.

An important instantiation of this goal is to understand whether there are common psychological and neurobiological mechanisms contributing to motivational and hedonic processing impairments associated with both psychosis and mood pathology or whether there are unique, differentiable mechanisms contributing to impairments in each disorder. The existing clinical literature provides support for both alternatives. On the one hand, depression and psychosis frequently co-occur and there are a number of ways in which motivational and hedonic impairments operate similarly in psychosis and mood pathology. For example, motivational/hedonic impairments can be present in individuals at risk for developing psychosis (Delawalla et al. 2006; Glatt et al. 2006; Juckel et al. 2012; Grimm et al. 2014; Schlosser et al. 2014) or at risk for developing depression (Gotlib et al. 2010; Foti et al. 2011a, b, c; McCabe et al. 2012; Kujawa et al. 2014; Macoveanu et al. 2014; Olino et al. 2014; Sharp et al. 2014). Further, there is evidence that the presence or severity of motivational/hedonic impairments is associated with the development of manifest illness for both psychosis (Chapman et al. 1994; Kwapil et al. 1997; Gooding et al. 2005; Velthorst et al. 2009) and depression (Bress et al. 2013; Morgan et al. 2013). In addition, motivation/hedonic impairments are associated with functional impairment and/or treatment non-response in psychosis (Fenton and McGlashan 1991; Herbener et al. 2005; Bowie et al. 2006; Ventura et al. 2009; Kurtz 2012) and depression (Downar et al. 2014). On the other hand, there are important *differences* in the manifestation of motivational/reward impairments across psychosis and mood pathology. Most critically, there is much more evidence of an episodic pattern of hedonic impairments associated with mood pathology than psychosis. More specifically, elevated anhedonia typically resolves along with acute depression, whereas anhedonia is much more likely to still be present among individuals with psychosis even when their acute psychotic symptoms have resolved (Blanchard et al. 2001; Herbener et al. 2005).

1.1 Translating Hedonic Experience into Motivated Behavior

These types of clinical data highlight the importance of mechanistically understanding the similarities and differences in motivational/hedonic impairments across psychopathology. In prior work primarily focused on psychosis (Barch and Dowd 2010; Kring and Barch 2014), we have used a heuristic model of the psychological processes and neural systems thought to link experienced or anticipated rewards/incentives with the action plans that need to be generated and maintained in order to obtain these rewards. As this literature is quite large, we have simplified and

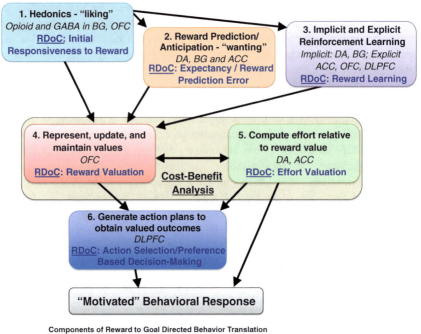

Fig. 1 Components of the translation between experience reward/pleasure and goal-directed behavior. *Notes ACC* Anterior cingulate cortex, *BG* Basal ganglia, *DA* Dopamine, *DLPFC* Dorsolateral prefrontal cortex, *OFC* Orbital frontal cortex, *RDoC* Research Domain Criteria

focused on six major components in the translation of appetitive or reward information into behavioral responses (Schultz et al. 1997; Berridge 2004; Schultz 2004, 2007; Wallis 2007) (see Fig. 1). The first component, **hedonics or liking** (component 1 in Fig. 1), reflects the ability to "enjoy" a stimulus or event that may provide pleasure or reward. In the RDoC Positive Valence system, this aligns with *initial responsiveness to reward*. Traditionally researchers have focused on the neurotransmitter dopamine (DA) as a primary substrate of liking (Berridge 2004), but more recent research instead suggests that hedonic responses (at least to primary sensory stimuli) seem to be mediated by activation of the opioid and GABA-ergic systems in the nucleus accumbens shell and its projections to the ventral pallidum, as well as in the orbital frontal cortex (OFC) (Richardson et al. 2005; Burgdorf and Panksepp 2006; Pecina et al. 2006; Smith and Berridge 2007; Berridge et al. 2009). For more information on the neurobiology of liking and the role of liking in motivated behaviors, see Robinson et al in this volume.

A second component, **reward prediction and wanting** (component 2 in Fig. 1), is mediated at least in part by the midbrain DA system, particularly projections to ventral and dorsal striatum (Berridge 2004; Schultz 2007). In the RDoC Positive Valence system, this aligns with *expectancy/reward predictor error*. Many DA neurons in the substantia nigra and ventral tegmental area (VTA) respond to stimuli that *predict* reward, as well as to rewards themselves. Importantly, the degree of response depends on predictability—if the reward was not predicted, then the DA neurons fire strongly (positive prediction error), whereas there is a transient depression in DA neuron firing (negative prediction error) if a predicted reward does not occur (Schultz 1992, 2004, 2007; Schultz et al. 1993, 1997). Over time, DA neurons begin to fire to the predictive cues rather than to rewards themselves (Schultz 2007). Similar effects have been found in humans using functional magnetic resonance imaging (fMRI) of the ventral and dorsal striatum (Knutson et al. 2000, 2001; McClure et al. 2003; Abler et al. 2006). These types of DA/striatal responses have been captured by temporal difference models that simulate learning about stimuli that predict rewards (Montague and Sejnowski 1994; Montague et al. 1996). A prominent, though slightly different theory emphasizes the role of the DA-learning process in transferring incentive salience from the reward itself to reward-predicting cues, thus imbuing these cues with motivational properties themselves [e.g., a "wanting" response (Berridge 2004)].

A third component is **reward or reinforcement learning** (component 3 in Fig. 1), both implicit (i.e., outside of conscious awareness) and explicit (i.e., including the use of explicit representations about potential reward associations). In the RDoC Positive Valence system, this aligns with *reward learning*. The types of DA/striatal responses described above for reward prediction are thought to underlie basic aspects of reinforcement learning that may occur without conscious awareness (Dayan and Balleine 2002; Frank et al. 2004). However, there is also evidence that the development of explicit representations that are accessible to conscious awareness can also drive reinforcement learning, albeit with a potentially different timecourse (Frank et al. 2001; Frank and O'Reilly 2006; Hazy et al. 2007; Gold et al. 2012). These more explicit forms of reinforcement learning also including neural systems involved in cognitive control and value representations, such as dorsal frontal and parietal regions and the OFC (Frank et al. 2001; Frank and O'Reilly 2006; Hazy et al. 2007; Gold et al. 2012). By cognitive control, we mean the ability to maintain goal or task representations in order to focus attentional resources on task-relevant information while filtering out task-irrelevant information (Miller and Cohen 2001; Braver 2012).

Another critical aspect of translating rewards to actions is *cost–benefit analysis*, or balancing the value of an outcome with the effort it would take to achieve that outcome. Thus, a necessary fourth component is the ability to **represent, maintain, and update value information** (component 4 in Fig. 1), which is thought to be mediated, at least in part, by OFC (Padoa-Schioppa and Cai 2011; Rudebeck and Murray 2011). In the RDoC Positive Valence system, this aligns with *reward valuation*. This construct takes into account, not only the hedonic properties of a stimulus, but also the internal state of the organism (e.g., value of juice when thirsty versus not) (Rolls et al. 1989), the delay before the reward occurs (Roesch and Olson 2005; Rudebeck et al. 2006a, b), the available reward options (e.g., juice versus wine after a hard day) (Padoa-Schioppa and Assad 2006; Padoa-Schioppa 2007), and the changing consequences associated with a stimulus (e.g., a previously rewarded response is now punished) (Dias et al. 1996; Cools et al. 2002). Human functional neuroimaging studies also highlight activation of OFC under conditions requiring value representations (O'Doherty et al. 2003; O'Doherty 2007), particularly those in which response contingencies need to be updated, such as reversal learning (Cools et al. 2002, 2007; O'Doherty et al. 2003). In addition, humans with OFC lesions can show reversal learning impairments (Fellows and Farah 2003, 2005; Hornak et al. 2004).

A fifth component, that is also a part of *cost–benefit analysis*, is the ability to **compute effort relative to reward value** (component 5 in Fig. 1), i.e., determining the cost of engaging in actions necessary to obtain a desired outcome. In the RDoC Positive Valence system, this aligns with *effort valuation/willingness to work*. For example, you may really want to eat some chocolate candy and may perceive eating candy as rewarding, but you may not want to put forth the effort to go to the store to get the candy. Research suggests that the dorsal anterior cingulate cortex (ACC) may be important for evaluating the cognitive and physical effort associated with different action plans (Shenhav et al. 2013) with contributions of DA input from the nucleus accumbens and related forebrain circuitry (Salamone 2007; Salamone et al. 2007; Botvinick et al. 2009; Croxson et al. 2009). For example, ACC lesions, as well as depletions of accumbens DA, lead animals to choose low effort, but low reward options over higher reward, but higher effort options (Rudebeck et al. 2006a, b, 2007; Rushworth et al. 2007; Salamone 2007; Salamone et al. 2007; Walton et al. 2007; Hosking et al. 2014a, b, 2015).

Lastly, a sixth component is the ability to **generate and execute goal-directed action plans necessary to achieve the valued outcome** (component 6 in Fig. 1). In the RDoC Positive Valence system, this aligns with *action selection/preference-based decision making*. Many researchers have argued for the role of the lateral PFC in relation to this component in the context of reward and motivation (in particular dorsolateral PFC) (Braver and Cohen 1999; Miller and Cohen 2001; Wallis 2007). This is consistent with a number of other lines of research and theory, including the following: (1) the role of the DLPFC in top-down control of cognitive processing; (2) models suggesting that the DLPFC provides a bias signal that helps to facilitate goal-directed behavior (Miller and Cohen 2001); (3) evidence for impaired action planning following lateral prefrontal lesions (Zalla et al. 2001; Manes et al. 2002); and (4) evidence that increases in DLFPC activity mediate "motivated" cognitive control enhancements that occur with the provision of incentives in both animals (Watanabe 1996; Kobayashi et al. 2006; Krawczyk et al. 2007; Sakagami and Watanabe 2007) and humans (Tsujimoto and Sawaguchi 2005; Beck et al. 2010; Jimura et al. 2010; Savine and Braver 2010). In other words, intact DLPFC function may be necessary to translate information about value into goal representations and to maintain such information so that it can be implemented as action plans to achieve the desired outcome.

Here, we review evidence for psychosis- and depression-related impairments in these six components of mechanisms that translate hedonic experiences of rewards into goal-directed actions that then allow individuals to obtain such rewards. Further, where available, we review neuroimaging evidence as to the neurobiological correlates of each type of impairment. As a preview, this review will suggest many impairments common across individuals with schizophrenia and depression, including consistent evidence for impairments in reward prediction and prediction errors, as well as effort allocation for incentives. However, this review will also reveal important differences that point to potentially divergent etiological pathways to impairments in motivated behavior associated with psychotic versus depressive pathology.

2 Hedonics and Liking

2.1 Schizophrenia

Numerous individual studies (e.g., Berenbaum and Oltmanns 1992; Kring et al. 1993; Dowd and Barch 2010) and several recent reviews (Kring and Moran 2008; Cohen and Minor 2010) have found that individuals with schizophrenia show relatively intact self-reported emotional responses to affect-eliciting stimuli. Further, individuals with schizophrenia show intact responses in emotion-modulated startle paradigms during the presentation of pleasant stimuli, when given sufficient time to process the stimuli (Schlenker et al. 1995; Curtis et al. 1999; Volz et al. 2003; Kring et al. 2011) and intact memory enhancement for positive stimuli (Mathews and Barch 2004; Horan et al. 2006; Hall et al. 2007) though see (Herbener et al. 2007).

2.1.1 Monetary Rewards

Neuroimaging studies examining striatal responses to the receipt of monetary rewards in schizophrenia have also shown a consistent pattern of intact responses, with robust ventral striatal responses to the receipt of money in unmedicated patients (Nielsen et al. 2012a, b) and patients treated with either typical or atypical antipsychotics (Kirsch et al. 2007; Simon et al. 2009; Walter et al. 2009; Dowd and Barch 2012; Morris et al. 2012; Gilleen et al. 2014; Wolf et al. 2014; Mucci et al. 2015). Further, studies have also shown intact feedback negativity responses, an ERP component in response to explicit feedback, to the receipts of rewards and losses in schizophrenia (Horan et al. 2011; Morris et al. 2011). However, while striatal responses to reward receipt seem to be largely intact in schizophrenia, some of these studies did report abnormal cortical responses to reward receipt. Particularly, prior work has noted reduced reward-related responses in medial PFC (Schlagenhauf et al. 2009), abnormal responses in both medial and lateral PFC (Waltz et al. 2010), and reduced salience coding in ventrolateral PFC in schizophrenia patients, which was correlated with negative symptom severity (Walter et al. 2010).

2.1.2 Primary Rewards

A more mixed picture has arisen from functional neuroimaging studies examining brain responses to other types of pleasurable or rewarding stimuli in schizophrenia (Crespo-Facorro et al. 2001; Paradiso et al. 2003). Plailly et al. (2006) found reduced activation in schizophrenia within the insula and OFC during hedonicity judgments of positive and negative odors. Schneider et al. (2007) also found reduced activation of the insula during the experience of positive olfactory stimuli in schizophrenia. Taylor et al. (2005) showed reduced phasic ventral striatal responses comparing positive versus neutral picture viewing in both medicated and unmedicated individuals with schizophrenia. In a large sample of medicated patients, we found that individuals with schizophrenia showed the same pattern of brain activation as controls in response to both negative and positive stimuli in a range of brain regions associated with the *perception and experience of emotion,* including medial frontal cortex, insula, OFC, and the amygdala (Dowd and Barch 2010). However, we did find some evidence for reduced ventral and dorsal striatal responses to positive stimuli among individuals with schizophrenia, and the severity of these deficits correlated with the magnitude of self-reported anhedonia. Other research has found evidence for reduced striatal responses to the receipt of juice, with the magnitude of this reduction associated with the severity of anhedonia (Waltz et al. 2009), as well as reduced striatal responses to food cues (Grimm et al. 2012), though medications may have been a confound in both of these studies.

2.1.3 Summary of Hedonics and Liking in Schizophrenia:

In sum (see Table 1), the self-report literature in schizophrenia provides relatively consistent evidence for intact self-reports of "liking" in schizophrenia, though there is evidence that greater self-reports of anhedonia or negative symptom ratings are associated with reduced "liking" (Blanchard et al. 1994; Burbridge and Barch 2007; Herbener et al. 2007; Dowd and Barch 2010). In addition, the functional neuroimaging literature provides fairly consistent evidence for intact responses to the receipt of monetary rewards. However, the functional neuroimaging literature on responses to other types of rewarding stimuli provides a more muddled picture, with some evidence for reduced insular responses and mixed evidence for altered striatal responses. Further, studies examining individual differences in negative symptom severity do suggest an important relationship between the magnitude of striatal responses to rewarding or pleasurable stimuli and anhedonia among individuals with schizophrenia (Waltz et al. 2009; Dowd and Barch 2010).

Table 1 Summary of impairments across constructs in schizophrenia and depression

Construct	Depression	Schizophrenia	Comments
1. Hedonic response to positive stimuli			
Primary rewards	Mixed	Mixed	Need clearer evaluation of the potential role of smoking and medications
Secondary rewards	Impaired	Intact	–
2. Reward anticipation/prediction			
Reward anticipation	Impaired	Impaired	Need to determine whether these reflect impairments in DA-learning systems or problems with representation or maintenance of value representations (schizophrenia) or hedonics (depression)
Prediction errors	Impaired	Impaired	
3. Reinforcement learning			
Implicit	Impaired	Intact	Need to determine whether these reflect impairments in DA-learning systems or problems with representation or maintenance of value representations (schizophrenia) or hedonics (depression)
Explicit	Intact	Impaired	
4. Value representation			
	Impaired	Impaired	Not clear whether there is a distinct impairment in value representation in either schizophrenia or depression, or whether impairment is due to problems in other components of the system
5. Effort allocation			
Physical effort	Impaired	Impaired	–
Cognitive effort	Untested	Mixed	Additional research is needed that includes assessments of perceived cognitive effort
6. Action plans/goal-directed action			
	Unclear	Impaired	Additional research is needed on the role of goal-directed action selection in motivational impairments in depression

2.2 Depression

There is a growing and consistent literature demonstrating that adults and adolescents with or at risk for depression have impaired hedonic responses to both pleasurable stimuli (primary reward) and monetary (secondary) rewards. Such group differences have been reported using behavioral measures as well as event-related potential (ERP) and fMRI measures of brain function (Foti and Hajcak 2009; Foti et al. 2011a, b, c; Bress et al. 2012, 2013; Zhang et al. 2013) and have been related to elevated levels of anhedonia.

2.2.1 Monetary Rewards

Behaviorally, depressed individuals show reduced reward-related biases (Henriques et al. 1994; Pizzagalli et al. 2008; Pechtel et al. 2013) and reduced ability to learn from reward (Herzallah et al. 2010; Maddox et al. 2012). Reduced reward sensitivity has been specifically related to self-reported anhedonia (Pizzagalli et al. 2008; Vrieze et al. 2013) and can be predictive of treatment response (Vrieze et al. 2013). Using ERPs, a number of studies have examined feedback-related negativity (FN) in MDD. FN is an ERP component elicited by reward or loss feedback and is thought to reflect activity in the ventral striatum, caudate, and the dorsal ACC (Carlson et al. 2011; Foti et al. 2011a, b, c). Depressed adults show decreased FN to rewards (Foti et al. 2014). Increased depressive symptoms are also related to reduced FN in children (Bress et al. 2012) and prospectively predict future onset of depression in adolescents (Bress et al. 2013). In addition, a reduced FN in depressed adults is specifically related to the severity of clinically rated anhedonia (Liu et al. 2014).

fMRI studies of reward processing in adolescents and adults have also found that depression is associated with decreased activation following positive feedback (i.e., reward) in reward-related brain areas such as the caudate, the putamen, the ACC, and the insula (Knutson et al. 2008; Kumar et al. 2008; Pizzagalli et al. 2009; Remijnse et al. 2009; Smoski et al. 2009; Gradin et al. 2011; Robinson et al. 2012 Zhang et al. 2013). Such reductions have been specifically associated with anhedonia symptoms (Gradin et al. 2011; Stoy et al. 2012), have been detected in individuals at risk for depression (Gotlib et al. 2010; McCabe et al. 2012; Olino et al. 2013), and predict the development of depression in adolescents (Morgan et al. 2013). Further, research in adults has shown increased ventral striatal responses to reward following successful treatment (Stoy et al. 2012). These findings suggest that depression and risk for depression are robustly associated with reduced behavioral and neural responsivity to monetary (secondary) rewards.

2.2.2 Primary Rewards

A meta-analysis examining emotional responsivity in depression suggested that depressed patients tend to show blunted self-reported and physiological responses to positive emotional stimuli (Bylsma et al. 2008) and recent evidence suggests that this type of blunting relates specifically to elevated anhedonia, rather than depressed mood, in a non-clinical sample (Saxena et al. submitted). Further, evidence for physiological blunting (decreased attenuation of startle blink by positive stimulus viewing) in subclinical depression (e.g., individuals reporting high levels of depression on clinical scales but whom do not meet diagnostic criteria) seems to be specific to in-the-moment experience of emotional stimuli rather than anticipation (Moran et al. 2012). Further, blunted behavioral and physiological (heart rate) reactivity to amusing film clips predicts poor recovery from depression (Rottenberg et al. 2002). However, studies examining self-reported hedonic responses to tastes/

odors have been more mixed and generally do not find strong evidence for behavioral differences associated with depression. Particularly, hedonic response ratings of sucrose and odor stimuli are generally not different when comparing MDD patients and healthy controls (Berlin et al. 1998; Clepce et al. 2010; Dichter et al. 2010) and depressive symptom severity in a non-clinical sample did not correlate with pleasantness ratings of sweet, sour, salty, or bitter tastes (Scinska et al. 2004). Yet, while examining group differences has typically yielded negative results, there is some evidence that elevated levels of anhedonia negatively predict hedonic responses to sucrose across depressed, schizophrenic, and healthy individuals (Berlin et al. 1998) and that measures of anticipatory anhedonia negatively predict anticipated hedonic responses to chocolate but not actual or recalled responses (Chentsova-Dutton and Hanley 2010).

fMRI studies provide more consistent evidence for reduced reactivity to primary rewards/pleasant stimuli in MDD than the self-report studies discussed above. For example, in one study, individuals with remitted depression showed no difference in the rating of pleasant food images/tastes as compared to controls, but did show decreased ventral striatal response relative to never-depressed controls (McCabe et al. 2009). Further, adolescents/young adults at elevated risk for depression based on a parental history of MDD showed lower OFC and ACC responses to pleasant food images/tastes as compared to those with no parental history (McCabe et al. 2012). fMRI studies utilizing other types of pleasant stimuli, such as happy faces or pleasant scenes, have also found reduced striatal responses in MDD patients as compared to controls (Gotlib 2005; Smoski 2011). Importantly, reduced striatal responses to pleasant stimuli specifically related to elevated levels of anhedonia, rather than to general depressive symptom severity (Keedwell et al. 2005).

2.2.3 Summary of Hedonics and Liking in Depression:

In summary (see Table 1), across behavioral, ERP, and fMRI methodologies in-the-moment hedonic response to both primary and monetary rewards are reduced in MDD and relate to anhedonic symptoms. However, such effects are most reliably observed when using fMRI methods (versus self-report) for odor/taste stimuli and behavior (versus physiology) for positive images/film clips (Bylsma et al. 2008). This could be due to the fact that self-reports can be more influenced by expectancy effects, such that individuals realize what the "normative" response to a specific stimulus should be and report that response, rather than their actual experience. Further, while similar subcortical regions show differences between MDD and control groups in fMRI across studies examining monetary versus primary rewards (Zhang et al. 2013), there is some evidence that group differences may be larger when using primary than monetary rewards (Smoski 2011). Given that primary rewards are innately and immediately rewarding they are more closely tied to "hedonic" experience than monetary rewards, where stimulus value is abstractly linked to future reward attainment, these results offer even stronger evidence linking a core in-the-moment hedonic or "liking" deficit to anhedonia in MDD.

2.3 Hedonics in Schizophrenia versus Depression

These results in depression stand in fairly strong contrast to the data on individuals with schizophrenia, which much more consistently supports *intact* in the in-the-moment hedonic or "liking," with some exceptions, particularly in the literature on primary reward processing in psychosis. This distinction may be critical, as it suggests potentially very different fundamental pathways and mechanisms to impairments in motivation and goal-directed behavior in the context of psychosis versus depressive mood pathology. In particular, it sets the stage for the hypothesis that in depressive pathology, deficits in motivated behavior may be traced back to impairments in hedonic or liking responses to both primary and secondary rewards/positive stimuli. In contrast, the data suggest that the mechanisms giving rise to impaired motivation behavior in schizophrenia occur subsequent to immediate hedonic responses and instead may reflect alterations in the way information about hedonic experience is stored, represented, maintained, or used.

3 Reward Prediction, Anticipation, and Reinforcement Learning

3.1 Schizophrenia

3.1.1 Reward Anticipation

There is a mixed self-report literature on anticipated pleasure in schizophrenia, with some studies suggesting impairments (Gard et al. 2006, 2007; Wynn et al. 2010; Mote et al. 2014) and others not (Tremeau et al. 2010, 2014; Gard et al. 2014). However, outside of self-reports, there are relatively few behavioral studies in schizophrenia that directly measure reward anticipation/prediction, though one such study did find evidence for reduced anticipation (Heerey and Gold 2007). As such, much of the focus has been on neuroimaging studies of reward prediction ("wanting"), which have primarily examined neural responses to reward-predicting cues, sometimes following conditioning trials and sometimes through explicit instruction, such as in the monetary incentive delay (MID) task. A number of studies have reported reduced ventral striatum activity to cues predicting reward in schizophrenia. These results have been found in unmedicated individuals with schizophrenia (Juckel et al. 2006a, b; Schlagenhauf et al. 2009; Esslinger et al. 2012; Nielsen et al. 2012a, b) and medicated individuals (Juckel et al. 2006a, b; Schlagenhauf et al. 2008; Simon et al. 2009; Walter et al. 2009; Grimm et al. 2012). There is some suggestion that these deficits are not present in individuals taking atypical medication (Juckel et al. 2006a, b) nor in prodromal individuals (Juckel et al. 2012), though some of these results are in small samples and need replication. For example, (Kirsch et al. 2007) found reduced ventral striatal responses to reward cues in individuals with schizophrenia taking typical compared to atypical antpsychotic mediation. Other work has noted reduced ventral striatal responses to anticipation cues in antipsychotic-naïve schizophrenia patients, which improved following atypical antipsychotic treatment (Nielsen et al. 2012a, b). Importantly, a number of studies also showed a relationship between negative symptom severity and deficits in anticipatory ventral striatal activity. Juckel et al. (2006a, b) showed that the severity of negative symptoms predicted the reduction in ventral striatal responses in unmedicated and typically medicated patients, Simon et al. (2009) showed that the magnitude of this response was inversely correlated with apathy ratings, and Waltz et al. (2010) showed a relationship between negative symptom severity and ventral striatal activation during anticipated gains. Reward prediction has also been studied using a Pavlovian task and found a relationship between reduced striatal response to reward-predictive cues and greater anhedonia among individuals with schizophrenia (Dowd and Barch 2012).

3.1.2 Reward Prediction Error

A number of studies have also examined the role of the striatum in reward prediction by looking at prediction error responses—an increase in striatal (potentially dopaminergic) responses to unexpected rewards and a decrease in striatal responses when predicted rewards do not occur. Several studies have now shown altered prediction error responses in schizophrenia, both in terms of reductions in responses to unpredicted rewards and larger than expected responses to predicted rewards (Murray et al. 2008; Morris et al. 2012; Schlagenhauf et al. 2014). Gradin et al. (2011) found reduced prediction error responses in the caudate, but increased activation associated with expected reward value in the ventral striatum. Waltz et al. (2009) examined positive and negative prediction error responses in a passive paradigm that required participants to learn about the timing of a potential reward. These researchers found evidence for reduced positive prediction error responses in a range of regions that included the striatum (dorsal and ventral) as well as insula, but relatively intact negative prediction errors in these same regions. Interestingly, Waltz et al. (2009) did find that the magnitude of prediction errors in the basal ganglia among patients was negatively correlated with avolition scores, suggesting a link to clinically relevant symptoms. In other work, Walter et al. (2009) found intact prediction error responses in the striatum for both positive and negative prediction errors, though this was a population with relatively low-level negative symptom, in contrast to the high-level negative symptom population in the Waltz et al. 2009 study. There is again suggestion that medication may have an important influence; Insel and colleagues found that individuals with chronic schizophrenia taking higher doses of medication showed smaller prediction error responses (Insel et al. 2014). However, the fact that reduced prediction error responses have also been seen in unmedicated individuals (Schlagenhauf et al. 2014) argues against such abnormalities resulting only from medication effects in schizophrenia. For further discussion on neuroimaging studies of RPE-signaling in patients with schizophrenia, see Waltz and Gold in this volume.

3.1.3 Reinforcement Learning

There are several possible mechanisms that could be contributing to altered reward prediction error and anticipation responses in psychosis. The most common interpretation is that they reflect abnormalities in the learning mechanisms supported by DA in the ventral striatum, suggesting that individuals with schizophrenia cannot appropriately learn what cues predict reward and do not update stimulus–response associations via striatal learning mechanisms. Such an interpretation would predict that individuals with schizophrenia would show deficits on reinforcement learning tasks that also tap into these mechanisms. However, the evidence suggests surprisingly intact performance on a range of tasks in which learning is either relatively easy or relatively implicit (Elliott et al. 1995; Hutton et al. 1998; Joyce et al. 2002; Turner et al. 2004; Tyson et al. 2004; Jazbec et al. 2007; Waltz and Gold 2007; Ceaser et al. 2008; Heerey et al. 2008; Weiler et al. 2009; Somlai et al. 2011), though with some exceptions (Oades 1997; Pantelis et al. 1999). Further, individuals with schizophrenia show intact learning rates on the weather prediction task, a probabilistic category-learning task frequently used to measure reinforcement learning, though with overall impaired performance (Keri et al. 2000, 2005a, b; Weickert et al. 2002, 2009; Beninger et al. 2003). There is some evidence that reinforcement learning may be more intact for patients on atypical than typical antipsychotics, though it has been found in those on typicals as well (Beninger et al. 2003; Keri et al. 2005a, b). Such intact reinforcement learning on more implicit tasks argues against the explanation that anticipation and prediction error deficits in schizophrenia are due solely to DA deficits in the striatum, though of course DA is involved in more than just prediction error signaling.

An alternative explanation is that such anticipatory and prediction error deficits may reflect impairments in more explicit learning and representation processes that engage cognitive control regions such as the DLPFC, the dorsal parietal cortex, and the ACC and/or the OFC. Consistent with this hypothesis, when the reinforcement learning paradigms become more difficult and require the explicit use of representations about stimulus–reward contingencies, individuals with schizophrenia show more consistent evidence of impaired reinforcement learning (Waltz et al. 2007; Morris et al. 2008; Koch et al. 2009; Gold et al. 2012; Yilmaz et al. 2012; Cicero et al. 2014). Interestingly, these impairments may be greater when individuals with schizophrenia must learn from reward versus from punishment (Waltz et al. 2007; Cheng et al. 2012; Gold et al. 2012; Reinen et al. 2014), though some studies also find impaired learning from punishment (Fervaha et al. 2013a, b; Cicero et al. 2014). Further, there is recent work suggesting that working memory impairments may make a significant contribution to reinforcement learning deficits in schizophrenia (Collins et al. 2014). Further, there is a growing literature suggesting altered activity in cortical regions involved in cognitive control during anticipation/prediction error (Walter et al. 2009; Gilleen et al. 2014) and during reinforcement learning (Waltz et al. 2013; Culbreth et al. in submission). Such findings are consistent with the larger literature suggesting altered cognitive control function in schizophrenia and are also consistent with the growing basic science literature suggesting important interactions between what have been referred to as "model-free" learning systems (e.g., DA in the striatum) and "model-based" learning systems that engage prefrontal and parietal systems that support representations of action–outcome models (Glascher et al. 2010; Daw et al. 2011; Doll et al. 2012; Lee et al. 2014; Otto et al. 2015). Taken together, these data and literatures point to the need to examine interactions between these systems and dopamine-mediated reinforcement learning systems.

3.2 Depression

3.2.1 Reward Anticipation

A number of studies have examined reward prediction and reward anticipation in individuals with depression. Given the literature reviewed above on the abnormalities in self-report, behavioral, and neural responses to the processing of positive stimuli and rewards, one would anticipate that individuals with depression should also show altered responses to the anticipation of prediction of reward. The literature supports this hypothesis, at least in part. For example, work by McFarland and Klein demonstrated reduced self-reports of joy in response to anticipated reward among individuals with depression (McFarland and Klein 2009). Similarly, other work supports reduced self-reports of anticipated pleasure (Sherdell et al. 2012). In addition, individuals with depression or with a family history of depression show reduced frontal EEG asymmetries during reward anticipation (Shankman et al. 2007, 2013; Nelson et al. 2013, 2014). Further, a recent meta-analysis on neural processing of rewards in depression found evidence for <reduced>? reward anticipation responses in the left caudate, along with evidence for increased activity in the right ACC (Zhang et al. 2013). Yet, there is some variability in these responses across individual imaging studies. A number of studies have found reduced activation in various regions of the striatum during reward anticipation among individuals with current depression or individuals at risk for depression (Dichter et al. 2009; Forbes et al. 2009; Pizzagalli et al. 2009; Smoski et al. 2009; Gotlib et al. 2010; Olino et al. 2011, 2014; Stoy et al. 2012; Ubl et al. 2015), as well as increased activity in the ACC (Dichter et al. 2012; Gorka et al. 2014). However, some other studies have found no differences in striatal activation during reward anticipation between healthy individuals and those with current (Knutson et al. 2008; Gorka et al. 2014) or remitted depression (Dichter et al. 2012), although in at least one case it was not clear that any participant showed activity in the striatum during reward anticipation (Chase et al. 2013). Further, at least one study found reduced ACC responses during reward anticipation (Chase et al. 2013).

3.2.2 Reward Prediction Error

There is also a small but growing literature on striatal reward prediction errors associated with depression. The majority of studies have found evidence for reduced positive prediction errors in depression in the striatum, including the caudate and the nucleus accumbens (Kumar et al. 2008; Gradin et al. 2011; Robinson et al. 2012). Further, the magnitude of these reductions was associated with the severity of anhedonia. However, two other studies did not find reduced positive prediction errors in the striatum in depression (Chase et al. 2013; Ubl et al. 2015), and one of the studies that found reduced prediction errors in the striatum also found increased prediction error responses in the VTA (Kumar et al. 2008). It is not obvious why these two studies found differing results, as they did not differ systematically form the other studies in terms of type of population, symptom severity, or medication use.

Across the reward anticipation and prediction error literatures, the findings provide support for the hypotheses that a dysfunction in striatal responses to reward, potentially reflecting altered DA function, is an important component of altered hedonic processing in depression. The evidence for altered dorsal ACC responses is also intriguing. In recent work, Shankman and others have posited the hypothesis that this increased activation may actually reflect "conflict" that individuals with depression experience when asked to anticipate processing hedonically positive stimuli that conflict with their current negative internal state (Gorka et al. 2014). If so, this would suggest that altered ACC activation is an outcome of the phenomenology of depression rather than potentially playing a causal role in anticipatory pleasure impairments. However, there is also a large literature on altered error-related negativities in depression (Olvet and Hajcak 2008; Vaidyanathan et al. 2012), thought to reflect, at least in part, altered activity in the ACC. Given these accumulated findings, more work is needed to establish what role ACC alterations may play in experienced or anticipated hedonic processing deficits associated with depressive pathology.

3.2.3 Reinforcement Learning

In contrast to the literature on reinforcement learning in schizophrenia, and consistent with an important role for striatally mediated reward processing abnormalities, there is good evidence for impairments in implicit reinforcement learning in depression. A number of studies have shown that individuals with depression show reduced biases in response to reward on a probabilistic learning task (Henriques et al. 1994; Pizzagalli et al. 2008; Vrieze et al. 2013). This is seen in remitted depression (Pechtel et al. 2013), as well as current depression, and is worse in individuals with depression who have higher anhedonia (Vrieze et al. 2013; Liverant et al. 2014). Similar impairments in depression have been found on a reinforcement learning task similar to the weather prediction task (Herzallah et al. 2010, 2013a, b), where it is typically difficult to develop explicit representations of the reward contingencies. In contrast, the literature on explicit reinforcement learning in depression suggests surprisingly intact performance. For example, there are a number of studies showing that individuals with depression perform similarly to healthy controls on the same probabilistic selection task that shows clear impairments in schizophrenia (Chase et al. 2010; Anderson et al. 2011; Cavanagh et al. 2011; Whitmer et al. 2012). The literature on the effects of subclinical depression on reinforcement learning also provides evidence for intact learning from reward (Beevers et al. 2013) though other work observed impaired reward learning (Kunisato et al. 2012; Maddox et al. 2012). As noted above, there are at least two pathways to impaired reinforcement learning—altered striatal-mediated stimulus–response learning and the use of cognitive control processes to develop and maintain explicit representations of action–outcome contingencies that can guide behavior. One intriguing possibility is that the former is impaired in depression but that the relatively more intact function of cognitive control processes in depression (as compared to schizophrenia) may allow individuals with depression to compensate for such implicit learning impairments, but only when they can develop explicit representations to support learning.

3.3 Summary of Reward Prediction, Anticipation, and Reinforcement Learning in Schizophrenia and Depression

As with the literature on hedonics and liking, this literature provides intriguing hints about potentially different mechanisms associated with motivational impairments in psychosis and depressive pathology (see Table 1). The literature on reinforcement learning and reward prediction in schizophrenia suggests relatively intact learning on simple reinforcement learning paradigms that may be implicit in nature. On difficult tasks that may also engage explicit learning mechanisms, there is more consistent evidence for impaired performance. An open question is the degree to which these impairments reflect differences in striatally mediated implicit learning mechanisms versus more cortically mediated explicit learning mechanisms. A growing number of studies in the imaging literature suggest reduced ventral striatal reward prediction/"wanting" responses in unmedicated and typically medicated individuals with schizophrenia (though with mixed evidence in those taking atypical antipsychotics) and evidence for reduced positive prediction errors. However, not all studies have found impaired striatal responses to reward prediction cues or to prediction error, and there is also evidence that the magnitude of these striatal impairments may be related to the severity of negative symptoms, pointing to the importance of examining individual difference relationships among individuals with schizophrenia. Further, a number of studies have also found altered activation in frontal regions during reward prediction or reinforcement learning, suggesting a potentially important role for cortically mediated mechanisms in schizophrenia.

The literature on depression also provides evidence for impairments in both self-report and neural indicators of reward anticipation and for deficits in striatal prediction error responses. In contrast to the literature on schizophrenia, there is robust evidence for impairments in "implicit" reinforcement learning in depression on tasks that are thought to reflect slow striatally mediated reinforcement learning, consistent with the idea that impairments in hedonic responses to reward may propagate forward to impair other components of reward processing. Interestingly, there is much less evidence for impaired performance on more explicit reinforcement learning tasks in depression, raising the intriguing possibility that individuals with depression are able to recruit more intact cognitive control or other explicit learning mechanisms to compensate for impaired reward responsivity.

4 Value Computations and OFC Function

As described above, one hypothesis is that the OFC supports the computation of value, or the integration of the reinforcing properties of the stimulus with the internal state of the organism, which includes updating changes in the reinforcing properties of the stimulus. One prominent theory suggests that reward processing deficits in schizophrenia reflect impairments in the representation of value (Gold et al. 2008). Although many different paradigms can be interpreted in the context of value representations (Gold et al. 2012), there are two experimental paradigms in particular that have been frequently used as probes of lateral and medial OFC function: probabilistic reversal learning and the Iowa Gambling Task. Both tasks require individuals to integrate information about rewards and punishments across trials and to use such information to update value representations appropriately.

4.1 Schizophrenia

The literature on the Iowa Gambling Task in schizophrenia provides evidence for impairment (Shurman et al. 2005; Kester et al. 2006; Lee et al. 2007; Martino et al. 2007; Sevy et al. 2007; Premkumar et al. 2008; Kim et al. 2009; Yip et al. 2009), though with some exceptions (Wilder et al. 1998; Evans et al. 2005; Rodriguez-Sanchez et al. 2005; Turnbull et al. 2006). In addition, several studies suggest impaired reversal learning in schizophrenia (Elliott et al. 1995; Oades 1997; Pantelis et al. 1999; Tyson et al. 2004; Turnbull et al. 2006; Waltz and Gold 2007; Ceaser et al. 2008), though a few studies using the intra-dimensional/extra-dimensional task did not find simple reversal learning deficits in schizophrenia (Hutton et al. 1998; Joyce et al. 2002; Jazbec et al. 2007). These reversal learning impairments are present even when individuals with schizophrenia and controls are matched on initial acquisition performance (Weiler et al. 2009). However, the imaging studies on reversal learning in schizophrenia do not point to altered activation of the OFC in relationship to these deficits, instead pointing to either alterations in striatal prediction error responses (Schlagenhauf et al. 2014), deactivation of default-mode regions (Waltz et al. 2013), or impaired activation of cognitive control networks (Culbreth et al. in submission). Thus, while there may be impairments in value computations associated with schizophrenia, there is yet little direct evidence that such impairments reflect OFC dysfunction (see Table 1).

4.2 Depression

In depression, there is also evidence for impaired performance on the Iowa Gambling Task (Must et al. 2006, 2013; Cella et al. 2010; Han et al. 2012), though with some studies not finding impairment (Westheide et al. 2007; Smoski et al. 2008). There is also evidence of impairments in reversal learning in depression (Murphy et al. 2003; Robinson et al. 2012; Hall et al. 2014). However, like the literature on schizophrenia, there is no evidence directly linking such impairments to OFC function (see Table 1), and the small imaging literature on reversal learning in depression points to altered striatal responses associated with impaired reversal learning (Robinson et al. 2012; Hall et al. 2014).

5 Effort Computations

There is a growing literature on the neurobiological mechanisms that regulate effort allocation and expenditure in both humans and animals (Salamone et al. 2007, 2009; Walton et al. 2007; McGuire and Botvinick 2010; Salamone and Correa 2012; Shenhav et al. 2013; Botvinick and Braver 2015). This literature makes distinctions between effort that needs to be allocated for cognitive demands and effort that needs to be allocated for physical demands, with evidence for both common and distinct mechanisms. In the animal literature, there is robust evidence that DA plays a key role in regulating physical effort allocation, in that blockade of DA, especially in the accumbens, reduces physical effort allocation (Salamone et al. 2009, 2012; Farrar et al. 2010; Salamone and Correa 2012), and increased D2 receptor expression in the nucleus accumbens of adult mice increases physical effort expenditure (Trifilieff et al. 2013). There is also recent evidence for important interactions between DA and adenosine (Salamone et al. 2012) in regulating effort. Consistent with this work in animals, Treadway et al. (2012a, b) found that, in humans, increased DA release in response to d-amphetamine in the left striatum and the left ventromedial PFC was associated with increased willingness to expend physical effort. However, there is recent evidence from animal work that DA antagonism may not reduce willingness to expend cognitive effort (Hosking et al. 2014a, b), though human work has shown that activity in the ventral striatum (which may reflect DA activity) predicts effort allocation for both physical and cognitive domains (Schmidt et al. 2012). It is important to note, though, that the developmental timing of DA function may be critical for understanding the role of DA in effort allocation. Particularly, research has shown that mice with D2 receptor overexpression throughout development (developed as a murine model of the negative symptoms of schizophrenia) actually show a decrease in effort expenditure (Simpson et al. 2011; Ward et al. 2012). This overexpression across the course of development may lead to alterations in other parts of the system, such as the prefrontal cortex (Kellendonk et al. 2006; Li et al. 2011), that in turn lead to reduced effort allocation.

There is also a large literature pointing to an important role for the medial prefrontal cortex, particularly the dorsal ACC, in regulating effort allocation. Recent computational work has posited a role for dorsal ACC in computing the expected value of control (Shenhav et al. 2013), arguing that the dorsal ACC integrates information about the expected value of the outcome, the expected cognitive control needed to obtain that outcome, and the expected cost of that cognitive control, in order to make decisions about the utility of expending effort. This hypothesized function of the dorsal ACC is consistent with the rodent and primate literature showing that lesions/inactivation of the dorsal ACC reduced both physical and mental effort allocation (Walton et al. 2003; Rudebeck et al. 2006a, b; Croxson et al. 2014; Hosking et al. 2014a, b, 2015), with evidence that rodent ACC neurons encode cost–benefit computations (Hillman and Bilkey 2010, 2012), and with the human literature showing activation of the dorsal ACC during effort based decision making (Croxson et al. 2009; Prevost et al. 2010).

5.1 Schizophrenia

The vast majority of the literature on effort allocation in schizophrenia has focused on physical effort, using paradigms that either involve finger-tapping (Treadway task Treadway et al. 2009) or a balloon-popping task (Gold et al. 2013) or grip strength as metrics of physical effort allocation. The paradigms using finger taping have quite consistently found a specific pattern of reduced effort allocation on the part of individuals with schizophrenia—they do not differ from controls at low levels of reward or low levels of probability of receiving the outcome, but do not show the same increase in effort allocation as either reward or probability increase (Fervaha et al. 2013a, b; Gold et al. 2013; Barch et al. 2014; Treadway et al. 2015). Further, the majority of the studies found that the degree of reduction in effort allocation was associated with either negative symptoms (Fervaha et al. 2013a, b; Gold et al. 2013; Treadway et al. 2015) or functional status (Barch et al. 2014). The two studies using grip strength showed differing results—one found a significant reduction in effort allocation among individuals with schizophrenia rated clinically as having higher apathy (Hartmann et al. 2014), while the other study found no effects of either diagnosis or symptom severity (Docx et al. 2015). Two recent studies have also examined cognitive effort allocation. One study using a progressive ratio task found evidence for reduced effort allocation in schizophrenia, although the design of the task was such that cognitive effort was confounded with physical effort (Wolf et al. 2014). In contrast, Gold et al., found little evidence of reduced cognitive effort in schizophrenia across three studies, though these studies did suggest that individuals with schizophrenia had difficulty detecting variations in cognitive effect among conditions (Gold et al. 2014).

5.2 Depression

All of the studies to date on effort allocation in depression have focused on physical effort, either using the Treadway finger-tapping task, a grip strength task, or a finger-tapping task with humorous cartoons. These studies provide a relatively consistent picture. Individuals with current depression show reduced effort allocation as a function of increasing monetary incentives (Clery-Melin et al. 2011; Treadway et al. 2012a, b; Yang et al. 2014). In other words, individuals with current depression are less likely to increase their likelihood of choosing harder tasks as the reward associated with the harder task increases. Further, there is some evidence that individual differences in self-reported anticipatory and/or consumatory pleasure relate to individual differences in the severity of effort allocation impairments (Yang et al. 2014). Individuals with remitted depression do not show effects as a group, though they do still show these individual difference relationships (Yang et al. 2014). In a novel study using viewing of humor cartoons as the incentive, Sherdell et al. did not find group differences in effort allocation, though they did find that those individuals with major depression who self-reported increased anticipatory anhedonia did show reduced effort allocation (Sherdell et al. 2012).

5.3 Summary of Effort Allocation in Schizophrenia and Depression

Taken together (see Table 1), these studies point to consistent evidence of reduced physical effort allocation in both schizophrenia and depression, but do not yet suggest a clear consensus on cognitive effort allocation. As of yet, there is no neuroimaging literature examining potential neural alterations associated with these impairments. In particular, it will be important to examine whether such abnormalities in schizophrenia or depression are associated with abnormal activity in the ACC and/or the ventral striatum and to also evaluate the degree to which abnormalities in effort allocation are associated with other components of reward processing.

6 Goal-Directed Action

6.1 Schizophrenia

Numerous reviews have outlined the strong evidence for impairments in goal representation and cognitive control in schizophrenia from a variety of sources (Barch and Ceaser 2012; Lesh et al. 2013), as well as the evidence for altered activation, connectivity, and structure of brain regions, such as the DLPFC, in schizophrenia (Glahn et al. 2005, 2008; Minzenberg et al. 2009; Fornito et al. 2011). Thus, a key question is, whether some of the motivational impairments observed in schizophrenia at least in part reflect problems translating reward information into goal representations that can be used and maintained in DLPFC to guide goal-directed behavior? One means of examining this issue would be to determine how motivational incentives impact cognitive performance, potentially via modulation of DLPFC activity. Several studies suggest that individuals with schizophrenia are not able to improve their performance on cognitive tasks when offered monetary incentives (Green et al. 1992; Vollema et al. 1995; Hellman et al. 1998; Roiser et al. 2009). While there is also at least some evidence for performance improvements with reward (Kern et al. 1995; Penn and Combs 2000; Rassovsky et al. 2005), these studies have not examined executive control tasks.

There is also work on the use of token economies in schizophrenia suggesting that functioning can be improved through an explicit reward system, though token economies provide a number of "external" supports for maintaining reward-related information that could compensate for deficits in the ability to translate reward information into action plans. A recent study has examined whether or not individuals with schizophrenia could improve cognitive control on a response inhibition task. Patients were able to speed their responses when presented with specific cues about winning reward and to a certain extent could speed their responses on trials in the reward "context" even when they could not earn money, an effect thought to reflect the maintenance of reward information through proactive control mechanisms. However, the individuals with schizophrenia showed a significantly smaller incentive context effect than controls, suggesting a reduction in the use of proactive control and a greater reliance on the use of "just-in-time," reactive control strategies (Mann et al. 2013). To date, there are no published fMRI studies examining whether or not incentives modulate DLPFC activity during cognitive control or working memory tasks in schizophrenia. This is a line of work that would be critical in helping to understand the relative contributions that prefrontal cognitive control deficits versus striatal incentive processing deficits make to motivational impairments in schizophrenia (see Table 1).

6.2 Depression

There is evidence in the literature that individuals with depression may show deficits on a range of cognitive control tasks (Rock et al. 2013), though the magnitude of the deficits are typically not as large as one sees among individuals with schizophrenia. However, there is much more mixed evidence for altered activation of cognitive control regions during working memory (at least without affective challenges), cognitive control, or goal maintenance tasks in depression without psychosis, with some studies finding little or no alterations in cognitive control regions (Barch et al. 2003; Walter et al. 2007; Schoning et al. 2009) and other studies finding some evidence (Siegle et al. 2007; Halari et al. 2009). There is much more evidence for altered prefrontal activity in depression during emotion regulation paradigms, though the pattern of activation alterations varies across studies (Beauregard et al. 2006; Johnstone et al. 2007; Sheline et al. 2009; Erk et al. 2010; Kanske et al. 2012; Perlman et al. 2012). To our knowledge, there are only two studies that have looked at incentive-modulated cognitive control in depression, both of which examined adolescents. Both studies found intact effects of incentives (rewards and punishments) on reducing anti-saccade errors among depressed adolescents, but reduced effects of incentives on latencies (Jazbec et al. 2005; Hardin et al. 2007). Such reduced effects would be expected if reward were experienced as less hedonically pleasurable for depressed individuals (i.e., hedonic/ "liking" deficits feeding forward to produce other deficits), but further work is needed to understand the degree to which cognitive control impairments might also contribute to altered goal-directed action in depression (see Table 1).

7 Summary of Reward and Motivational Neuroscience in Schizophrenia and Depression

Above, we reviewed evidence for impairments in six components of the mechanisms that translate hedonic experiences of rewards into goal-directed actions that allow individuals to obtain such rewards. This review provided evidence for a number of common impairments across individuals with schizophrenia and depression, including consistent evidence for impairments in both self-report and neuroimaging indicators of reward prediction and prediction errors, as well as robust evidence for impairments in effort allocation. However, this review also revealed critical differences in the nature of incentive processing impairments that point to differential etiological pathways leading to impairments in motivated behavior associated with psychotic versus depressive pathology. Specifically, individuals with schizophrenia show relatively *intact* in-the-moment hedonic experiences as well as relatively *intact* explicit reinforcement learning, where dorsal frontal and parietal cognitive control systems may contribute to deficits in reward anticipation, reversal learning, and goal maintenance/action selection. In contrast,

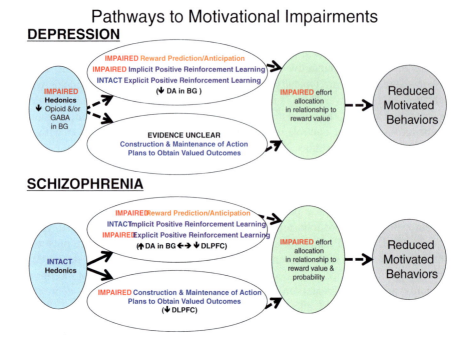

Fig. 2 Pathways to motivational impairments in schizophrenia and depressive pathology. *Dashed lines* indicated that impairment in a component may feed forward to contribute to impairments in other components. *Solid lines* indicate that input from intact components. *Down arrow* (↓) indicates evidence for reduced function of this neural mechanism. *Up arrow* (↑) indicates evidence for increased or dysregulated function of this neural mechanism. *Left arrow Right arrow* (→←) indicates interactions between neural mechanisms. *BG* Basal ganglia, *DA*, Dopamine, *DLPFC* Dorsolateral prefrontal cortex

individuals with depression show consistent evidence of *reduced* in-the-moment hedonic experience and *impaired* implicit reinforcement learning, coupled with relatively *intact* explicit reinforcement learning, and less evidence for a contribution of cognitive control systems to altered reward processing and motivated behavior.

When integrated, these patterns (see Fig. 2) suggest that impaired incentive processing in schizophrenia may be more related to impaired goal representation and utilization mechanisms rather than to fundamental deficits in hedonic experience. Such impairments may reflect both altered DA function and altered activation of dorsal frontal–parietal cognitive control systems. Specifically, recent meta-analyses point to robust evidence for increased dopamine synthesis availability and some evidence for D2 receptor overexpression (Howes et al. 2012; Fusar-Poli and Meyer-Lindenberg 2013), as well as robust evidence for altered activity of cognitive control systems (Minzenberg et al. 2009). As described above, there is intriguing evidence from a murine model of the negative symptoms of schizophrenia (Simpson et al. 2011; Ward et al. 2012) that D2 overexpression occurring throughout development contributes to altered prefrontal function and DA sensitivity (Kellendonk et al. 2006; Li et al. 2011). Thus, even though individuals with schizophrenia can experience reward and pleasure from a variety of stimuli, they may have difficulty learning appropriate reward or salience representations (Howes and Kapur 2009) and difficulty representing and maintaining reward information over time so that information can drive further goal-directed behavior and action selection (Barch and Dowd 2010; Kring and Barch 2014).

In contrast, the data reviewed above suggest that in the context of depressive pathology (see Fig. 2), altered incentive processing may be more related to fundamental deficits in hedonic experience that propagate forward to result in impaired motivated behavior. Individuals with depression show consistent evidence for altered hedonic responses to a range of stimuli, with the severity of such deficits sometimes varying with self-reported levels of anhedonia. In turn, individuals with depression show impaired implicit reinforcement learning that is thought to be striatally mediated (and dependent on experiencing a stimulus as pleasurable) as well as impaired effort allocation for incentives. In contrast, there is less evidence that individuals with depression show impaired explicit reinforcement learning and less evidence that there are clear contributions impaired cognitive control systems to altered incentive processing and motivated behavior. As described above, the animal literature suggests that hedonic responsivity is associated with opioid and GABA-ergic function in the striatum, and this pattern is consistent with the hypothesis that altered opioid function may contribute to hedonic impairment in depression. This hypothesis is also consistent with a growing literature on opioid mechanisms in depression (Lalanne et al. 2014; Murphy 2015) and an emerging interest in modulation of the kappa opioid system as a treatment for depression (Connolly and Thase 2012), with a specific focus on anhedonia. At the same time, the results in depression could also suggest a role for altered DA function in the striatum. There is some evidence that depression may be associated with DA dysfunction, e.g., examining DA binding in the striatum (Cannon et al. 2009). However, in general, the literature on DA alterations is mixed and relatively small (Savitz and Drevets 2013; Camardese et al. 2014).

To illustrate how such differing patterns of impairments may lead to altered motivated behavior, consider the following scenarios. An individual with schizophrenia may report that they enjoy chocolate chip cookies and would find the experience of eating chocolate chip cookies quite pleasurable if you were to bring them a nice warm plateful (i.e., intact hedonics). However, they may have difficulty generating/initiating the behaviors necessary to obtain or make chocolate chip cookies on their own (Kring and Moran 2008; Barch and Dowd 2010). Planning, purchasing, preparing, or baking the cookies requires ongoing maintenance of contextual or cue information that trigger associations about the food's rewarding properties, which should drive the allocation of effort to obtain these outcomes (e.g., get dressed, leave the house, go to the store). These functions depend on the ability to associate relevant cues with rewarding outcomes, a process that is associated with striatal DA function, which is dysregulated in schizophrenia. In addition, these functions may depend on the intact ability to maintain appetitive cues or context over time—a process that may be reliant on cognitive control and working memory mechanisms, which are compromised in schizophrenia (Barch and Dowd 2010). In contrast, an individual with depression may report that they do not enjoy chocolate chip cookies and may not find the experience of eating chocolate chip cookies enjoyable even if you bring them fresh baked ones. Such hedonic impairments might reflect altered opioid signaling in the striatum. In turn, individuals with depression (at least those with anhedonia) may fail to learn about cues associated with delicious cookies and may be unwilling/unmotivated to allocate effort to making cookies since they do not anticipate them being particularly enjoyable.

These are of course simplified examples and do not reflect the vast complexity of motivated behaviors that we need to engage in every day. Further, this example does not capture all of the interactions among these systems or the heterogeneity that exists across individuals or even within individuals across time. However, these ideas may provide a heuristic framework for future research that attempts to understand the transdiagnostic or diagnosis-specific mechanisms contributing to altered motivated behavior in psychopathology. Studies that cut across diagnostic boundaries are clearly need now to explicitly test such hypotheses about common and distinction mechanisms leading to a motivation, as a means to develop more effective and targeted interventions that will hopefully lead to enhanced quality of life, reduced public health burden, and even preventative interventions that could preclude the development of psychopathology.

References

Abler B, Walter H et al (2006) Prediction error as a linear function of reward probability is coded in human nucleus accumbens. Neuroimage 31(2):790–795
Anderson IM, Shippen C et al (2011) State-dependent alteration in face emotion recognition in depression. Br J Psychiatry 198(4):302–308
Barch DM, Ceaser AE (2012) Cognition in schizophrenia: core psychological and neural mechanisms. Trends Cogn Sci 16:27–34
Barch DM, Dowd EC (2010) Goal representations and motivational drive in schizophrenia: the role of prefrontal-striatal interactions. Schizophr Bull 36(5):919–934
Barch DM, Sheline YI et al (2003) Working memory and prefrontal cortex dysfunction: specificity to schizophrenia compared with major depression. Biol Psychiatry 53(5):376–384
Barch DM, Treadway MT et al (2014) Effort, anhedonia, and function in schizophrenia: reduced effort allocation predicts amotivation and functional impairment. J Abnorm Psychol 123 (2):387–397
Beauregard M, Paquette V et al (2006) Dysfunction in the neural circuitry of emotional self-regulation in major depressive disorder. Neuroreport 17(8):843–846
Beck SM, Locke HS et al (2010) Primary and secondary rewards differentially modulate neural activity dynamics during working memory. PloS one 5(2):e9251
Beevers CG, Worthy DA et al (2013) Influence of depression symptoms on history-independent reward and punishment processing. Psychiatry Res 207(1–2):53–60
Beninger RJ, Wasserman J et al (2003) Typical and atypical antipsychotic medications differentially affect two nondeclarative memory tasks in schizophrenic patients: a double dissociation. Schizophr Res 61(2–3):281–292

Berenbaum H, Oltmanns TF (1992) Emotional experience and expression in schizophrenia and depression. J Abnorm Psychol 101(1):37–44

Berlin I, Givry-Steiner L et al (1998) Measures of anhedonia and hedonic responses to sucrose in depressive and schizophrenic patients in comparison with healthy subjects. Eur Psychiatry 13(6):303–309

Berridge KC (2004) Motivation concepts in behavioral neuroscience. Physiol Behav 81(2):179–209

Berridge KC, Robinson TE et al (2009) Dissecting components of reward: 'liking', 'wanting', and learning. Curr Opin Pharmacol 9(1):65–73

Blanchard JJ, Bellack AS et al (1994) Affective and social-behavioral correlates of physical and social anhedonia in schizophrenia. J Abnorm Psychol 103(4):719–728

Blanchard JJ, Horan WP et al (2001) Diagnostic differences in social anhedonia: a longitudinal study of schizophrenia and major depressive disorder. J Abnorm Psychol 110(3):363–371

Botvinick M, Braver T (2015) Motivation and cognitive control: from behavior to neural mechanism. Annu Rev Psychol 66:83–113

Botvinick MM, Huffstetler S et al (2009) Effort discounting in human nucleus accumbens. Cogn Affect Behav Neurosci 9(1):16–27

Bowie CR, Reichenberg A et al (2006) Determinants of real-world functional performance in schizophrenia subjects: correlations with cognition, functional capacity, and symptoms. Am J Psychiatry 163(3):418–425

Braver TS (2012) The variable nature of cognitive control: a dual mechanisms framework. Trends Cogn Sci 16(2):106–113

Braver TS, Cohen JD (1999) Dopamine, cognitive control, and schizophrenia: the gating model. Progr Brain Res 121:327–349

Bress JN, Smith E et al (2012) Neural response to reward and depressive symptoms in late childhood to early adolescence. Biol Psychol 89(1):156–162

Bress JN, Foti D et al (2013) Blunted neural response to rewards prospectively predicts depression in adolescent girls. Psychophysiology 50(1):74–81

Burbridge JA, Barch DM (2007) Anhedonia and the experience of emotion in individuals with schizophrenia. J Abnorm Psychol 116(1):30–42

Burgdorf J, Panksepp J (2006) The neurobiology of positive emotions. Neurosci Biobehav Rev 30(2):173–187

Bylsma LM, Morris BH et al (2008) A meta-analysis of emotional reactivity in major depressive disorder. Clin Psychol Rev 28(4):676–691

Camardese G, Di Giuda D et al (2014) Imaging studies on dopamine transporter and depression: a review of literature and suggestions for future research. J Psychiatr Res 51:7–18

Cannon DM, Klaver JM et al (2009) Dopamine type-1 receptor binding in major depressive disorder assessed using positron emission tomography and [11C]NNC-112. Neuropsychopharmacology 34(5):1277–1287

Carlson JM, Foti D et al (2011) Ventral striatal and medial prefrontal BOLD activation is correlated with reward-related electrocortical activity: a combined ERP and fMRI study. NeuroImage 57(4):1608–1616

Cavanagh JF, Bismark AJ et al (2011) Larger error signals in major depression are associated with better avoidance learning. Front Psychol 2:331

Ceaser AE, Goldberg TE et al (2008) Set-shifting ability and schizophrenia: a marker of clinical illness or an intermediate phenotype? Biol Psychiatry 64(9):782–788

Cella M, Dymond S et al (2010) Impaired flexible decision-making in major depressive disorder. J Affect Disord 124(1–2):207–210

Chapman LJ, Chapman JP et al (1994) Putatively psychosis-prone subjects 10 years later. J Abnorm Psychol 103(2):171–183

Chase HW, Frank MJ et al (2010) Approach and avoidance learning in patients with major depression and healthy controls: relation to anhedonia. Psychol Med 40(3):433–440

Chase HW, Nusslock R et al (2013) Dissociable patterns of abnormal frontal cortical activation during anticipation of an uncertain reward or loss in bipolar versus major depression. Bipolar Disord 15(8):839–854

Cheng GL, Tang JC et al (2012) Schizophrenia and risk-taking: impaired reward but preserved punishment processing. Schizophr Res 136(1–3):122–127

Chentsova-Dutton Y, Hanley K (2010) The effects of anhedonia and depression on hedonic responses. Psychiatry Res 179(2):176–180

Cicero DC, Martin EA et al (2014) Reinforcement learning deficits in people with schizophrenia persist after extended trials. Psychiatry Res 220(3):760–764

Clepce M, Gossler A et al (2010) The relation between depression, anhedonia and olfactory hedonic estimates–a pilot study in major depression. Neurosci Lett 471(3):139–143

Clery-Melin ML, Schmidt L et al (2011) Why don't you try harder? An investigation of effort production in major depression. PLoS One 6(8):e23178

Cohen AS, Minor KS (2010) Emotional experience in patients with schizophrenia revisited: meta-analysis of laboratory studies. Schizophr Bull 36(1):143–150

Collins AG, Brown JK et al (2014) Working memory contributions to reinforcement learning impairments in schizophrenia. J Neurosci 34(41):13747–13756

Connolly KR, Thase ME (2012) Emerging drugs for major depressive disorder. Expert Opin Emerg Drugs 17(1):105–126

Cools R, Clark L et al (2002) Defining the neural mechanisms of probabilistic reversal learning using event-related functional magnetic resonance imaging. J Neurosci 22(11):4563–4567

Cools R, Lewis SJ et al (2007) L-DOPA disrupts activity in the nucleus accumbens during reversal learning in Parkinson's disease. Neuropsychopharmacology 32(1):180–189

Crespo-Facorro B, Paradiso S et al (2001) Neural mechanisms of anhedonia in schizophrenia. J Am Med Assoc 286(4):427–435

Croxson PL, Walton ME et al (2009) Effort-based cost-benefit valuation and the human brain. J Neurosci 29(14):4531–4541

Croxson PL, Walton ME et al (2014) Unilateral medial frontal cortex lesions cause a cognitive decision-making deficit in rats. Eur J Neurosci 40(12):3757–3765

Curtis CE, Lebow B et al (1999) Acoustic startle reflex in schizophrenic patients and their first-degree relatives: evidence of normal emotional modulation. Psychophysiology 36:469–475

Culbreth AJ, Gold JM et al (in submission). Impaired activation in cognitive control regions predicts reversal learning in schizophrenia.

Cuthbert BN, Insel TR (2010) Toward new approaches to psychotic disorders: the NIMH research domain criteria project. Schizophr Bull 36(6):1061–1062

Cuthbert BN, Kozak MJ (2013) Constructing constructs for psychopathology: the NIMH research domain criteria. J Abnorm Psychol 122(3):928–937

Daw ND, Gershman SJ et al (2011) Model-based influences on humans' choices and striatal prediction errors. Neuron 69(6):1204–1215

Dayan P, Balleine BW (2002) Reward, motivation, and reinforcement learning. Neuron 36:285–298

Delawalla Z, Barch DM et al (2006) Factors mediating cognitive deficits and psychopathology among siblings of individuals with schizophrenia. Schizophr Bull 32:525–537

Dias R, Robbins TW et al (1996) Dissociation in prefrontal cortex of affective and attentional shifts. Nature 380:69–72

Dichter GS, Felder JN et al (2009) The effects of psychotherapy on neural responses to rewards in major depression. Biol Psychiatry 66(9):886–897

Dichter GS, Smoski MJ et al (2010) Unipolar depression does not moderate responses to the sweet taste test. Depress Anxiety 27(9):859–863

Dichter GS, Kozink RV et al (2012) Remitted major depression is characterized by reward network hyperactivation during reward anticipation and hypoactivation during reward outcomes. J Affect Disord 136(3):1126–1134

Docx L, de la Asuncion J et al (2015) Effort discounting and its association with negative symptoms in schizophrenia. Cogn Neuropsychiatry 20(2):172–185

Doll BB, Simon DA et al (2012) The ubiquity of model-based reinforcement learning. Curr Opin Neurobiol 22(6):1075–1081

Dowd E, Barch DM (2010) Subjective emotional experience in schizophrenia: neural and behavioral markers. Biol Psychiatry 67(10):902–911

Dowd EC, Barch DM (2012) Pavlovian reward prediction and receipt in schizophrenia: relationship to anhedonia. PloS one 7(5):e35622

Downar J, Geraci J et al (2014) Anhedonia and reward-circuit connectivity distinguish nonresponders from responders to dorsomedial prefrontal repetitive transcranial magnetic stimulation in major depression. Biol Psychiatry 76(3):176–185

Elliott R, McKenna PJ et al (1995) Neuropsychological evidence for frontostriatal dysfunction in schizophrenia. Psychol Med 25(3):619–630

Erk S, Mikschl A et al (2010) Acute and sustained effects of cognitive emotion regulation in major depression. J Neurosci 30(47):15726–15734

Esslinger C, Englisch S et al (2012) Ventral striatal activation during attribution of stimulus saliency and reward anticipation is correlated in unmedicated first episode schizophrenia patients. Schizophr Res 140(1–3):114–121

Evans CE, Bowman CH et al (2005) Subjective awareness on the Iowa gambling task: the key role of emotional experience in schizophrenia. J Clin Exp Neuropsychol 27(6):656–664

Farrar AM, Segovia KN et al (2010) Nucleus accumbens and effort-related functions: behavioral and neural markers of the interactions between adenosine A2A and dopamine D2 receptors. Neuroscience 166(4):1056–1067

Fellows LK, Farah MJ (2003) Ventromedial frontal cortex mediates affective shifting in humans: evidence from a reversal learning paradigm. Brain 126(Pt 8):1830–1837

Fellows LK, Farah MJ (2005) Different underlying impairments in decision-making following ventromedial and dorsolateral frontal lobe damage in humans. Cereb Cortex 15(1):58–63

Fenton WS, McGlashan TH (1991) Natural history of schizophrenia subtypes. II. Positive and negative symptoms and long-term course. Arch Gen Psychiatry 48(11):978–986

Fervaha G, Agid O et al (2013a) Impairments in both reward and punishment guided reinforcement learning in schizophrenia. Schizophr Res 150(2–3):592–593

Fervaha G, Graff-Guerrero A et al (2013b) Incentive motivation deficits in schizophrenia reflect effort computation impairments during cost-benefit decision-making. J Psychiatr Res 47(11):1590–1596

Forbes EE, Hariri AR et al (2009) Altered striatal activation predicting real-world positive affect in adolescent major depressive disorder. Am J Psychiatry 166(1):64–73

Fornito A, Yoon J et al (2011) General and specific functional connectivity disturbances in first-episode schizophrenia during cognitive control performance. Biol Psychiatry 70(1):64–72

Foti D, Hajcak G (2009) Depression and reduced sensitivity to non-rewards versus rewards: evidence from event-related potentials. Biol Psychol 81(1):1–8

Foti D, Kotov R et al (2011a) Abnormal neural sensitivity to monetary gains versus losses among adolescents at risk for depression. J Abnorm Child Psychol 39(7):913–924

Foti D, Weinberg A et al (2011b) Event-related potential activity in the basal ganglia differentiates rewards from nonrewards: response to commentary. Hum Brain Mapp 32(12):2267–2269

Foti D, Weinberg A et al (2011c) Event-related potential activity in the basal ganglia differentiates rewards from nonrewards: temporospatial principal components analysis and source localization of the feedback negativity. Hum Brain Mapp 32(12):2207–2216

Foti D, Carlson JM et al (2014) Reward dysfunction in major depression: multimodal neuroimaging evidence for refining the melancholic phenotype. Neuroimage 101:50–58

Frank MJ, O'Reilly RC (2006) A mechanistic account of striatal dopamine function in human cognition: psychopharmacological studies with cabergoline and haloperidol. Behav Neurosci 120(3): 497–517

Frank MJ, Loughry B et al (2001) Interactions between frontal cortex and basal ganglia in working memory: a computational model. Cogn, Affect Behav Neurosci 1(2):137–160

Frank MJ, Seeberger LC et al (2004) By carrot or by stick: cognitive reinforcement learning in parkinsonism. Science 306(5703):1940–1943

Fusar-Poli P, Meyer-Lindenberg A (2013) Striatal presynaptic dopamine in schizophrenia, part II: meta-analysis of [(18)F/(11)C]-DOPA PET studies. Schizophr Bull 39(1):33–42

Gard DE, Germans M et al (2006) Anticipatory and consummatory components of the experience of pleasure: a scale development study. J Pers Res 40:1086–1102

Gard DE, Kring AM et al (2007) Anhedonia in schizophrenia: distinctions between anticipatory and consummatory pleasure. Schizophr Res 93(1–3):253–260

Gard DE, Sanchez AH et al (2014) Do people with schizophrenia have difficulty anticipating pleasure, engaging in effortful behavior, or both? J Abnorm Psychol 123(4):771–782

Gilleen J, Shergill SS et al (2014). Impaired subjective well-being in schizophrenia is associated with reduced anterior cingulate activity during reward processing. Psychol Med: 1–12.

Glahn DC, Ragland JD et al (2005) Beyond hypofrontality: A quantitative meta-analysis of functional neuroimaging studies of working memory in schizophrenia. Hum Brain Mapp 25(1):60–69

Glahn DC, Laird AR et al (2008) Meta-analysis of gray matter anomalies in schizophrenia: application of anatomic likelihood estimation and network analysis. Biol Psychiatry 64(9):774–781

Glascher J, Daw N et al (2010) States versus rewards: dissociable neural prediction error signals underlying model-based and model-free reinforcement learning. Neuron 66(4):585–595

Glatt SJ, Stone WS et al (2006) Psychopathology, personality traits and social development of young first-degree relatives of patients with schizophrenia. Br J Psychiatry 189:337–345

Gold JM, Strauss GP et al (2013) Negative symptoms of schizophrenia are associated with abnormal effort-cost computations. Biol Psychiatry

Gold JM, Kool W et al (2014) Cognitive effort avoidance and detection in people with schizophrenia. Cogn Affect Behav Neurosci:1–10

Gold JM, Waltz JA et al (2008) Reward processing in schizophrenia: a deficit in the representation of value. Schizophr Bull 34(5):835–847

Gold JM, Waltz JA et al (2012) Negative symptoms and the failure to represent the expected reward value of actions: behavioral and computational modeling evidence. Arch Gen Psychiatry 69(2):129–138

Gooding DC, Tallent KA et al (2005) Clinical status of at-risk individuals 5 years later: further validation of the psychometric high-risk strategy. J Abnorm Psychol 114(1):170–175

Gorka SM, Huggins AA et al (2014) Neural response to reward anticipation in those with depression with and without panic disorder. J Affect Disord 164:50–56

Gotlib IH, Hamilton JP et al (2010) Neural processing of reward and loss in girls at risk for major depression. Arch Gen Psychiatry 67(4):380–387

Gradin VB, Kumar P et al (2011) Expected value and prediction error abnormalities in depression and schizophrenia. Brain: J Neurol 134(Pt 6):1751–1764

Green MF, Satz P et al (1992) Wisconsin card sorting test performance in schizophrenia: remediation of a stubborn deficit. Am J Psychiatry 149(1):62–67

Grimm O, Vollstadt-Klein S et al (2012) Reduced striatal activation during reward anticipation due to appetite-provoking cues in chronic schizophrenia: a fMRI study. Schizophr Res 134(2–3):151–157

Grimm O, Heinz A et al (2014) Striatal response to reward anticipation: evidence for a systems-level intermediate phenotype for schizophrenia. JAMA Psychiatry 71(5):531–539

Halari R, Simic M et al (2009) Reduced activation in lateral prefrontal cortex and anterior cingulate during attention and cognitive control functions in medication-naive adolescents with depression compared to controls. J Child Psychol Psychiatry 50(3):307–316

Hall J, Harris JM et al (2007) Emotional memory in schizophrenia. Neuropsychologia 45(6):1152–1159

Hall GB, Milne AM et al (2014) An fMRI study of reward circuitry in patients with minimal or extensive history of major depression. Eur Arch Psychiatry Clin Neurosci 264(3):187–198

Han G, Klimes-Dougan B et al (2012) Selective neurocognitive impairments in adolescents with major depressive disorder. J Adolesc 35(1):11–20

Hardin MG, Schroth E et al (2007) Incentive-related modulation of cognitive control in healthy, anxious, and depressed adolescents: development and psychopathology related differences. J Child Psychol Psychiatry 48(5):446–454

Hartmann MN, Hager OM et al (2014) Apathy but not diminished expression in schizophrenia is associated with discounting of monetary rewards by physical effort. Schizophr Bull

Hazy TE, Frank MJ et al (2007) Towards an executive without a homunculus: computational models of the prefrontal cortex/basal ganglia system. Philos Trans R Soc Lond B Biol Sci 362(1485):1601–1613

Heerey EA, Gold JM (2007) Patients with schizophrenia demonstrate dissociation between affective experience and motivated behavior. J Abnorm Psychol 116(2):268–278

Heerey EA, Bell-Warren KR et al (2008) Decision-making impairments in the context of intact reward sensitivity in schizophrenia. Biol Psychiatry 64(1):62–69

Hellman SG, Kern RS et al (1998) Monetary reinforcement and Wisconsin card sorting performance in schizophrenia: why show me the money? Schizophr Res 34(1–2):67–75

Henriques JB, Glowacki JM et al (1994) Reward fails to alter response bias in depression. J Abnorm Psychol 103:460–466

Herbener ES, Harrow M et al (2005) Change in the relationship between anhedonia and functional deficits over a 20-year period in individuals with schizophrenia. Schizophr Res 75(1):97–105

Herbener ES, Rosen C et al (2007) Failure of positive but not negative emotional valence to enhance memory in schizophrenia. J Abnorm Psychol 116(1):43–55

Herzallah MM, Moustafa AA et al (2010) Depression impairs learning whereas anticholinergics impair transfer generalization in Parkinson patients tested on dopaminergic medications. Cogn Behav Neurol 23(2):98–105

Herzallah MM, Moustafa AA et al (2013a) Learning from negative feedback in patients with major depressive disorder is attenuated by SSRI antidepressants. Front Integr Neurosci 7:67

Herzallah MM, Moustafa AA et al (2013b) Depression impairs learning, whereas the selective serotonin reuptake inhibitor, paroxetine, impairs generalization in patients with major depressive disorder. J Affect Disord 151(2):484–492

Hillman KL, Bilkey DK (2010) Neurons in the rat anterior cingulate cortex dynamically encode cost-benefit in a spatial decision-making task. J Neurosci 30(22):7705–7713

Hillman KL, Bilkey DK (2012) Neural encoding of competitive effort in the anterior cingulate cortex. Nat Neurosci 15(9):1290–1297

Horan WP, Foti D et al (2011) Impaired neural response to internal but not external feedback in schizophrenia. Psychol Med: 1–11

Horan WP, Green MF et al (2006) Does anhedonia in schizophrenia reflect faulty memory for subjectively experienced emotions? J Abnorm Psychol 115(3):496–508

Hornak J, O'Doherty J et al (2004) Reward-related reversal learning after surgical excisions in orbito-frontal or dorsolateral prefrontal cortex in humans. J Cogn Neurosci 16(3):463–478

Hosking JG, Cocker PJ et al (2014a) Dissociable contributions of anterior cingulate cortex and basolateral amygdala on a rodent cost/benefit decision-making task of cognitive effort. Neuropsychopharmacology 39(7):1558–1567

Hosking JG, Floresco SB et al (2014b) Dopamine antagonism decreases willingness to expend physical, but not cognitive, effort: a comparison of two rodent cost/benefit decision-making tasks. Neuropsychopharmacology

Hosking JG, Cocker PJ et al (2015) Prefrontal cortical inactivations decrease willingness to expend cognitive effort on a rodent cost/benefit decision-making task. Cereb Cortex

Howes OD, Kapur S (2009) The dopamine hypothesis of schizophrenia: version III–the final common pathway. Schizophr Bull 35(3):549–562

Howes OD, Kambeitz J et al (2012) The nature of dopamine dysfunction in schizophrenia and what this means for treatment. Arch Gen Psychiatry 69(8):776–786

Hutton SB, Puri BK et al (1998) Executive function in first-episode schizophrenia. Psychol Med 28(2):463–473

Insel T, Cuthbert B et al (2010) Research domain criteria (RDoC): toward a new classification framework for research on mental disorders. Am J Psychiatry 167(7):748–751

Insel C, Reinen J et al (2014) Antipsychotic dose modulates behavioral and neural responses to feedback during reinforcement learning in schizophrenia. Cogn Affect Behav Neurosci 14 (1):189–201

Jazbec S, McClure E et al (2005) Cognitive control under contingencies in anxious and depressed adolescents: an antisaccade task. Biol Psychiatry 58(8):632–639

Jazbec S, Pantelis C et al (2007) Intra-dimensional/extra-dimensional set-shifting performance in schizophrenia: impact of distractors. Schizophr Res 89(1–3):339–349

Jimura K, Locke HS et al (2010) Prefrontal cortex mediation of cognitive enhancement in rewarding motivational contexts. Proc Natl Acad Sci United States Am 107(19):8871–8876

Johnstone T, van Reekum CM et al (2007) Failure to regulate: counterproductive recruitment of top-down prefrontal-subcortical circuitry in major depression. J Neurosci 27(33):8877–8884

Joyce E, Hutton S et al (2002) Executive dysfunction in first-episode schizophrenia and relationship to duration of untreated psychosis: the West London Study. Br J Psychiatry Suppl 43:s38–s44

Juckel G, Schlagenhauf F et al (2006a) Dysfunction of ventral striatal reward prediction in schizophrenic patients treated with typical, not atypical, neuroleptics. Psychopharmacology (Berl) 187(2):222–228

Juckel G, Schlagenhauf F et al (2006b) Dysfunction of ventral striatal reward prediction in schizophrenia. Neuroimage 29(2):409–416

Juckel G, Friedel E et al (2012) Ventral striatal activation during reward processing in subjects with ultra-high risk for schizophrenia. Neuropsychobiology 66(1):50–56

Kanske P, Heissler J et al (2012) Neural correlates of emotion regulation deficits in remitted depression: the influence of regulation strategy, habitual regulation use, and emotional valence. Neuroimage 61(3):686–693

Keedwell PA, Andrew C et al (2005) The neural correlates of anhedonia in major depressive disorder. Biol Psychiatry 58(11):843–853

Kellendonk C, Simpson EH et al (2006) Transient and selective overexpression of dopamine D2 receptors in the striatum causes persistent abnormalities in prefrontal cortex functioning. Neuron 49(4):603–615

Keri S, Kelemen O et al (2000) Schizophrenics know more than they can tell: probabilistic classification learning in schizophrenia. Psychol Med 30(1):149–155

Keri S, Juhasz A et al (2005a) Habit learning and the genetics of the dopamine D3 receptor: evidence from patients with schizophrenia and healthy controls. Behav Neurosci 119(3):687–693

Keri S, Nagy O et al (2005b) Dissociation between medial temporal lobe and basal ganglia memory systems in schizophrenia. Schizophr Res 77(2–3):321–328

Kern RS, Green MF et al (1995) Modification of performance on the span of apprehension, a putative marker of vulnerability to schizophrenia. J Abnorm Psychol 104(2):385–389

Kester HM, Sevy S et al (2006) Decision-making impairments in adolescents with early-onset schizophrenia. Schizophr Res 85(1–3):113–123

Kim YT, Lee KU et al (2009) Deficit in decision-making in chronic, stable schizophrenia: from a reward and punishment perspective. Psychiatry Investig 6(1):26–33

Kirsch P, Ronshausen S et al (2007) The influence of antipsychotic treatment on brain reward system reactivity in schizophrenia patients. Pharmacopsychiatry 40(5):196–198

Knutson B, Westdorp A et al (2000) FMRI visualization of brain activity during a monetary incentive delay task. NeuroImage 12:20–27

Knutson B, Fong GW et al (2001) Dissociation of reward anticipation and outcome with event-related fMRI. Neuroreport 12(17):3683–3687

Knutson B, Bhanji JP et al (2008) Neural responses to monetary incentives in major depression. Biol Psychiatry 63(7):686–692

Kobayashi S, Nomoto K et al (2006) Influences of rewarding and aversive outcomes on activity in macaque lateral prefrontal cortex. Neuron 51(6):861–870

Koch K, Schachtzabel C et al (2009) Altered activation in association with reward-related trial-and-error learning in patients with schizophrenia. Neuroimage 50(1):223–232

Krawczyk DC, Gazzaley A et al (2007) Reward modulation of prefrontal and visual association cortex during an incentive working memory task. Brain Res 1141:168–177

Kring AM, Barch DM (2014) The motivation and pleasure dimension of negative symptoms: neural substrates and behavioral outputs. Eur Neuropsychopharmacol 24(5):725–736

Kring AM, Moran EK (2008) Emotional response deficits in schizophrenia: insights from affective science. Schizophr Bull 34(5):819–834

Kring AM, Kerr SL et al (1993) Flat affect in schizophrenia does not reflect diminished subjective experience of emotion. J Abnorm Psychol 102:507–517

Kring AM, Germans Gard M et al (2011) Emotion deficits in schizophrenia: timing matters. J Abnorm Psychol 120(1):79–87

Kujawa A, Proudfit GH et al (2014) Neural reactivity to rewards and losses in offspring of mothers and fathers with histories of depressive and anxiety disorders. J Abnorm Psychol 123(2):287–297

Kumar P, Waiter G et al (2008) Abnormal temporal difference reward-learning signals in major depression. Brain: J Neurol 131(Pt 8):2084–2093

Kunisato Y, Okamoto Y et al (2012) Effects of depression on reward-based decision making and variability of action in probabilistic learning. J Behav Ther Exp Psychiatry 43(4):1088–1094

Kurtz MM (2012) Cognitive remediation for schizophrenia: current status, biological correlates and predictors of response. Expert Rev Neurother 12(7):813–821

Kwapil TR, Miller MB et al (1997) Magical ideation and social anhedonia as predictors of psychosis proneness: a partial replication. J Abnorm Psychol 106(3):491–495

Lalanne L, Ayranci G et al (2014) The kappa opioid receptor: from addiction to depression, and back. Front Psychiatry 5:170

Lee Y, Kim YT et al (2007) Dissociation of emotional decision-making from cognitive decision-making in chronic schizophrenia. Psychiatry Res 152(2–3):113–120

Lee SW, Shimojo S et al (2014) Neural computations underlying arbitration between model-based and model-free learning. Neuron 81(3):687–699

Lesh TA, Westphal AJ et al (2013) Proactive and reactive cognitive control and dorsolateral prefrontal cortex dysfunction in first episode schizophrenia. Neuroimage Clin 2:590–599

Li YC, Kellendonk C et al (2011) D2 receptor overexpression in the striatum leads to a deficit in inhibitory transmission and dopamine sensitivity in mouse prefrontal cortex. Proc Natl Acad Sci U S A 108(29):12107–12112

Liu WH, Wang LZ et al (2014) The influence of anhedonia on feedback negativity in major depressive disorder. Neuropsychologia 53:213–220

Liverant GI, Sloan DM et al (2014) Associations among smoking, anhedonia, and reward learning in depression. Behav Ther 45(5):651–663

Macoveanu J, Knorr U et al (2014) Altered reward processing in the orbitofrontal cortex and hippocampus in healthy first-degree relatives of patients with depression. Psychol Med 44(6):1183–1195

Maddox WT, Gorlick MA et al (2012) Depressive symptoms enhance loss-minimization, but attenuate gain-maximization in history-dependent decision-making. Cognition 125(1):118–124

Manes F, Sahakian B et al (2002) Decision-making processes following damage to the prefrontal cortex. Brain 125(Pt 3):624–639

Mann CL, Footer O et al (2013) Spared and impaired aspects of motivated cognitive control in schizophrenia. J Abnorm Psychol 122(3):745–755

Martino DJ, Bucay D et al (2007) Neuropsychological frontal impairments and negative symptoms in schizophrenia. Psychiatry Res 152(2–3):121–128

Mathews JR, Barch DM (2004) Episodic memory for emotional and nonemotional words in schizophrenia. Cogn Emot 18(6):721–740

McCabe C, Cowen PJ et al (2009) Neural representation of reward in recovered depressed patients. Psychopharmacology (Berl) 205(4):667–677

McCabe C, Woffindale C et al (2012) Neural processing of reward and punishment in young people at increased familial risk of depression. Biol Psychiatry 72(7):588–594

McClure SM, Berns GS et al (2003) Temporal prediction errors in a passive learning task activate human striatum. Neuron 38(2):339–346

McFarland BR, Klein DN (2009) Emotional reactivity in depression: diminished responsiveness to anticipated reward but not to anticipated punishment or to nonreward or avoidance. Depression Anxiety 26(2):117–122

McGuire JT, Botvinick MM (2010) Prefrontal cortex, cognitive control, and the registration of decision costs. Proc Natl Acad Sci USA 107(17):7922–7926

Miller EK, Cohen JD (2001) An integrative theory of prefrontal cortex function. Ann Rev Neurosci 21:167–202

Minzenberg MJ, Laird AR et al (2009) Meta-analysis of 41 functional neuroimaging studies of executive function in schizophrenia. Arch Gen Psychiatry 66(8):811–822

Montague PR, Sejnowski TJ (1994) The predictive brain: temporal coincidence and temporal order in synaptic learning mechanisms. Learn Memory 1:1–33

Montague PR, Dayan P et al (1996) A framework for mesencephalic dopamine systems based on predictive Hebbian learning. J Neurosci 16:1936–1947

Moran EK, Mehta N et al (2012) Emotional responding in depression: distinctions in the time course of emotion. Cogn Emot 26(7):1153–1175

Morgan JK, Olino TM et al (2013) Neural response to reward as a predictor of increases in depressive symptoms in adolescence. Neurobiol Dis 52:66–74

Morris RW, Vercammen A et al (2012) Disambiguating ventral striatum fMRI-related BOLD signal during reward prediction in schizophrenia. Mol Psychiatry 17(3): 235:280–239.

Morris SE, Heerey EA et al (2008) Learning-related changes in brain activity following errors and performance feedback in schizophrenia. Schizophr Res 99(1–3):274–285

Morris SE, Holroyd CB et al (2011) Dissociation of response and feedback negativity in schizophrenia: electrophysiological and computational evidence for a deficit in the representation of value. Front Hum Neurosci 5:123

Mote J, Minzenberg MJ et al (2014) Deficits in anticipatory but not consummatory pleasure in people with recent-onset schizophrenia spectrum disorders. Schizophr Res 159(1):76–79

Mucci A, Dima D et al (2015) Is avolition in schizophrenia associated with a deficit of dorsal caudate activity? A functional magnetic resonance imaging study during reward anticipation and feedback. Psychol Med: 1–14

Murphy NP (2015) Dynamic measurement of extracellular opioid activity: status quo, challenges, and significance in rewarded behaviors. ACS Chem Neurosci 6(1):94–107

Murphy FC, Michael A et al (2003) Neuropsychological impairment in patients with major depressive disorder: the effects of feedback on task performance. Psychol Med 33(3):455–467

Murray GK, Corlett PR et al (2008) Substantia nigra/ventral tegmental reward prediction error disruption in psychosis. Mol Psychiatry 13(3):267–276

Must A, Szabo Z et al (2006) Sensitivity to reward and punishment and the prefrontal cortex in major depression. J Affect Disord 90(2–3):209–215

Must A, Horvath S et al (2013) The Iowa gambling task in depression—what have we learned about sub-optimal decision-making strategies? Front Psychol 4:732

Nelson BD, McGowan SK et al (2013) Biomarkers of threat and reward sensitivity demonstrate unique associations with risk for psychopathology. J Abnorm Psychol 122(3):662–671

Nelson BD, Shankman SA et al (2014) Intolerance of uncertainty mediates reduced reward anticipation in major depressive disorder. J Affect Disord 158:108–113

Nielsen MO, Rostrup E et al (2012a). Improvement of brain reward abnormalities by antipsychotic monotherapy in schizophrenia. Arch Gen Psychiatry: 1–10

Nielsen MO, Rostrup E et al (2012b) Alterations of the brain reward system in antipsychotic naive schizophrenia patients. Biol Psychiatry 71(10):898–905

Oades RD (1997) Stimulus dimension shifts in patients with schizophrenia, with and without paranoid hallucinatory symptoms, or obsessive compulsive disorder: strategies, blocking and monoamine status. Behav Brain Res 88(1):115–131

O'Doherty JP (2007) Lights, camembert, action! The role of human orbitofrontal cortex in encoding stimuli, rewards and choices. Ann NY Acad Sci 1121: 254–272

O'Doherty J, Critchley H et al (2003) Dissociating valence of outcome from behavioral control in human orbital and ventral prefrontal cortices. J Neurosci 23(21):7931–7939

Olino TM, McMakin DL et al (2013) Reduced reward anticipation in youth at high-risk for unipolar depression: a preliminary study. Dev Cogn Neurosci

Olino TM, McMakin DL et al (2011) I won, but I'm not getting my hopes up: depression moderates the relationship of outcomes and reward anticipation. Psychiatry Res 194(3):393–395

Olino TM, McMakin DL et al (2014) Reduced reward anticipation in youth at high-risk for unipolar depression: a preliminary study. Dev Cogn Neurosci 8:55–64

Olvet DM, Hajcak G (2008) The error-related negativity (ERN) and psychopathology: toward an endophenotype. Clin Psychol Rev 28(8):1343–1354

Otto AR, Skatova A et al (2015) Cognitive control predicts use of model-based reinforcement learning. J Cogn Neurosci 27(2):319–333

Padoa-Schioppa C (2007) Orbitofrontal cortex and the computation of economic value. Ann NY Acad Sci 441(7090):223–226

Padoa-Schioppa C, Assad JA (2006) Neurons in the orbitofrontal cortex encode economic value. Nature 441(7090):223–226

Padoa-Schioppa C, Cai X (2011) The orbitofrontal cortex and the computation of subjective value: consolidated concepts and new perspectives. Ann New York Acad Sci 1239:130–137

Pantelis C, Barber FZ et al (1999) Comparison of set-shifting ability in patients with chronic schizophrenia and frontal lobe damage. Schizophr Res 37(3):251–270

Paradiso S, Andreasen NC et al (2003) Emotions in unmedicated patients with schizophrenia during evaluation with positron emission tomography. Am J Psychiatry 160(10):1775–1783

Pechtel P, Dutra SJ et al (2013) Blunted reward responsiveness in remitted depression. J Psychiatric Res 47(12):1864–1869

Pecina S, Smith KS et al (2006) Hedonic hot spots in the brain. Neuroscientist 12(6):500–511

Penn DL, Combs D (2000) Modification of affect perception deficits in schizophrenia. Schizophr Res 46(2–3):217–229

Perlman G, Simmons AN et al (2012) Amygdala response and functional connectivity during emotion regulation: a study of 14 depressed adolescents. J Affect Disord 139(1):75–84

Pizzagalli DA, Iosifescu D et al (2008) Reduced hedonic capacity in major depressive disorder: evidence from a probabilistic reward task. J Psychiatric Res 43(1):76–87

Pizzagalli DA, Holmes AJ et al (2009) Reduced caudate and nucleus accumbens response to rewards in unmedicated individuals with major depressive disorder. Am J Psychiatry 166 (6):702–710

Plailly J, d'Amato T et al (2006) Left temporo-limbic and orbital dysfunction in schizophrenia during odor familiarity and hedonicity judgments. Neuroimage 29(1):302–313

Premkumar P, Fannon D et al (2008) Emotional decision-making and its dissociable components in schizophrenia and schizoaffective disorder: a behavioural and MRI investigation. Neuropsychologia 46(7):2002–2012

Prevost C, Pessiglione M et al (2010) Separate valuation subsystems for delay and effort decision costs. J Neurosci 30(42):14080–14090

Rassovsky Y, Green MF et al (2005) Modulation of attention during visual masking in schizophrenia. Am J Psychiatry 162(8):1533–1535

Reinen J, Smith EE et al (2014) Patients with schizophrenia are impaired when learning in the context of pursuing rewards. Schizophr Res 152(1):309–310

Remijnse PL, Nielen MM et al (2009) Differential frontal-striatal and paralimbic activity during reversal learning in major depressive disorder and obsessive-compulsive disorder. Psychol Med 39(9):1503–1518

Richardson DK, Reynolds SM et al (2005) Endogenous opioids are necessary for benzodiazepine palatability enhancement: naltrexone blocks diazepam-induced increase of sucrose-'liking'. Pharmacol Biochem Behav 81(3):657–663

Robinson OJ, Cools R et al (2012) Ventral striatum response during reward and punishment reversal learning in unmedicated major depressive disorder. Am J Psychiatry 169(2):152–159

Rock PL, Roiser JP et al (2013) Cognitive impairment in depression: a systematic review and meta-analysis. Psychol Med 1–12

Rodriguez-Sanchez JM, Crespo-Facorro B et al (2005) Prefrontal cognitive functions in stabilized first-episode patients with schizophrenia spectrum disorders: a dissociation between dorsolateral and orbitofrontal functioning. Schizophr Res 77(2–3):279–288

Roesch MR, Olson CR (2005) Neuronal activity in primate orbitofrontal cortex reflects the value of time. J Neurophysiol 94(4):2457–2471

Roiser JP, Stephan KE et al (2009) Do patients with schizophrenia exhibit aberrant salience? Psychol Med 39(2):199–209

Rolls ET, Sienkiewicz ZJ et al (1989) Hunger modulates the responses to gustatory stimuli of single neurons in the caudolateral orbitofrontal cortex of the Macaque monkey. Eur J Neurosci 1(1):53–60

Rottenberg J, Wilhelm FH et al (2002) Respiratory sinus arrhythmia as a predictor of outcome in major depressive disorder. J Affect Disord 71(1–3):265–272

Rudebeck PH, Murray EA (2011) Dissociable effects of subtotal lesions within the macaque orbital prefrontal cortex on reward-guided behavior. J Neurosci: Official J Soc Neurosci 31(29):10569–10578

Rudebeck PH, Buckley MJ et al (2006a) A role for the macaque anterior cingulate gyrus in social valuation. Science 313(5791):1310–1312

Rudebeck PH, Walton ME et al (2006b) Separate neural pathways process different decision costs. Nat Neurosci 9(9):1161–1168

Rudebeck PH, Walton ME et al (2007) Distinct contributions of frontal areas to emotion and social behaviour in the rat. Eur J Neurosci 26(8):2315–2326

Rushworth MF, Behrens TE et al (2007) Contrasting roles for cingulate and orbitofrontal cortex in decisions and social behaviour. Trends Cogn Sci 11(4):168–176

Sakagami M, Watanabe M (2007) Integration of cognitive and motivational information in the primate lateral prefrontal cortex. Ann NY Acad Sci 1104:89–107

Salamone JD (2007) Functions of mesolimbic dopamine: changing concepts and shifting paradigms. Psychopharmacology (Berl) 191(3):389

Salamone JD, Correa M (2012) The mysterious motivational functions of mesolimbic dopamine. Neuron 76(3):470–485

Salamone JD, Correa M et al (2007) Effort-related functions of nucleus accumbens dopamine and associated forebrain circuits. Psychopharmacology (Berl) 191(3):461–482

Salamone JD, Correa M et al (2009) Dopamine, behavioral economics, and effort. Front Behav Neurosci 3:13

Salamone JD, Correa M et al (2012) The behavioral pharmacology of effort-related choice behavior: dopamine, adenosine and beyond. J Exp Anal Behav 97(1):125–146

Savine AC, Braver TS (2010) Motivated cognitive control: reward incentives modulate preparatory neural activity during task-switching. J Neurosci: Official J Soc Neurosci 30(31):10294–10305

Savitz JB, Drevets WC (2013) Neuroreceptor imaging in depression. Neurobiol Dis 52:49–65

Schlagenhauf F, Juckel G et al (2008) Reward system activation in schizophrenic patients switched from typical neuroleptics to olanzapine. Psychopharmacology (Berl) 196(4):673–684

Schlagenhauf F, Sterzer P et al (2009) Reward feedback alterations in unmedicated schizophrenia patients: relevance for delusions. Biol Psychiatry 65(12):1032–1039

Schlagenhauf F, Huys QJ et al (2014) Striatal dysfunction during reversal learning in unmedicated schizophrenia patients. Neuroimage 89:171–180

Schlenker R, Cohen R et al (1995) Affective modulation of the startle reflex in schizophrenic patients. Euro Arch Psychiatry Clin Neurosci 245:309–318

Schlosser DA, Fisher M et al (2014) Motivational deficits in individuals at-risk for psychosis and across the course of schizophrenia. Schizophr Res 158(1–3):52–57

Schmidt L, Lebreton M et al (2012) Neural mechanisms underlying motivation of mental versus physical effort. PLoS Biol 10(2):e1001266

Schneider F, Habel U et al (2007) Neural substrates of olfactory processing in schizophrenia patients and their healthy relatives. Psychiatry Res 155(2):103–112

Schoning S, Zwitserlood P et al (2009) Working-memory fMRI reveals cingulate hyperactivation in euthymic major depression. Hum Brain Mapp 30(9):2746–2756

Schultz W (1992) Activity of dopamine neurons in the behaving primate. Semin Neurosci 4:129–138

Schultz W (2004) Neural coding of basic reward terms of animal learning theory, game theory, microeconomics, and behavioral ecology. Curr Opin Neurobiol 14:139–147

Schultz W (2007) Multiple dopamine functions at different time courses. Annu Rev Neurosci 30:259–288

Schultz W, Apicella P et al (1993) Responses of monkey dopamine neurons to reward and conditioned stimuli during successive steps of learning a delayed response task. J Neurosci 13(3):900–913

Schultz W, Dayan P et al (1997) A neural substrate of prediction and reward. Science 275:1593–1599

Scinska A, Sienkiewicz-Jarosz H et al (2004) Depressive symptoms and taste reactivity in humans. Physiol Behav 82(5):899–904

Sevy S, Burdick KE et al (2007) Iowa gambling task in schizophrenia: a review and new data in patients with schizophrenia and co-occurring cannabis use disorders. Schizophr Res 92(1–3):74–84

Shankman SA, Klein DN et al (2007) Reward sensitivity in depression: a biobehavioral study. J Abnorm Psychol 116(1):95–104

Shankman SA, Nelson BD et al (2013) A psychophysiological investigation of threat and reward sensitivity in individuals with panic disorder and/or major depressive disorder. J Abnorm Psychol 122(2):322–338

Sharp C, Kim S et al (2014) Major depression in mothers predicts reduced ventral striatum activation in adolescent female offspring with and without depression. J Abnorm Psychol 123(2):298–309

Sheline YI, Barch DM et al (2009) The default mode network and self-referential processes in depression. Proc Natl Acad Sci USA

Shenhav A, Botvinick MM et al (2013) The expected value of control: an integrative theory of anterior cingulate cortex function. Neuron 79(2):217–240

Sherdell L, Waugh CE et al (2012) Anticipatory pleasure predicts motivation for reward in major depression. J Abnorm Psychol 121(1):51–60

Shurman B, Horan WP et al (2005) Schizophrenia patients demonstrate a distinctive pattern of decision-making impairment on the Iowa gambling task. Schizophr Res 72(2–3):215–224

Siegle GJ, Thompson W et al (2007) Increased amygdala and decreased dorsolateral prefrontal BOLD responses in unipolar depression: related and independent features. Biol Psychiatry 61(2):198–209

Simon JJ, Biller A et al (2009) Neural correlates of reward processing in schizophrenia—relationship to apathy and depression. Schizophr Res 118(1–3):154–161

Simpson EH, Kellendonk C et al (2011) Pharmacologic rescue of motivational deficit in an animal model of the negative symptoms of schizophrenia. Biol Psychiatry 69(10):928–935

Smith KS, Berridge KC (2007) Opioid limbic circuit for reward: interaction between hedonic hotspots of nucleus accumbens and ventral pallidum. J Neurosci 27(7):1594–1605

Smoski MJ, Lynch TR et al (2008) Decision-making and risk aversion among depressive adults. J Behav Ther Exp Psychiatry 39(4):567–576

Smoski MJ, Felder J et al (2009) fMRI of alterations in reward selection, anticipation, and feedback in major depressive disorder. J Affect Disord 118(1–3):69–78

Somlai Z, Moustafa AA et al (2011) General functioning predicts reward and punishment learning in schizophrenia. Schizophr Res 127(1–3):131–136

Stoy M, Schlagenhauf F et al (2012) Hyporeactivity of ventral striatum towards incentive stimuli in unmedicated depressed patients normalizes after treatment with escitalopram. J Psychopharmacol 26(5):677–688

Taylor SF, Phan KL et al (2005) Neural response to emotional salience in schizophrenia. Neuropsychopharmacology 30(5):984–995

Treadway MT, Buckholtz JW et al (2009) Worth the 'EEfRT'? The effort expenditure for rewards task as an objective measure of motivation and anhedonia. PloS one 4(8):e6598

Treadway MT, Bossaller NA et al (2012a) Effort-based decision-making in major depressive disorder: a translational model of motivational anhedonia. J Abnorm Psychol 121(3):553–558

Treadway MT, Buckholtz JW et al (2012b) Dopaminergic mechanisms of individual differences in human effort-based decision-making. J Neurosci: Official J Soc Neurosci 32(18):6170–6176

Treadway MT, Peterman JS et al (2015) Impaired effort allocation in patients with schizophrenia. Schizophr Res 161(2–3):382–385

Tremeau F, Antonius D et al (2010) Anticipated, on-line and remembered positive experience in schizophrenia. Schizophr Res 122(1–3):199–205

Tremeau F, Antonius D et al (2014) Immediate affective motivation is not impaired in schizophrenia. Schizophr Res 159(1):157–163

Trifilieff P, Feng B et al (2013) Increasing dopamine D2 receptor expression in the adult nucleus accumbens enhances motivation. Mol Psychiatry 18(9):1025–1033

Tsujimoto S, Sawaguchi T (2005) Context-dependent representation of response-outcome in monkey prefrontal neurons. Cereb Cortex 15(7):888–898

Turnbull OH, Evans CE et al (2006) A novel set-shifting modification of the iowa gambling task: flexible emotion-based learning in schizophrenia. Neuropsychology 20(3):290–298

Turner DC, Clark L et al (2004) Modafinil improves cognition and attentional set shifting in patients with chronic schizophrenia. Neuropsychopharmacology 29(7):1363–1373

Tyson PJ, Laws KR et al (2004) Stability of set-shifting and planning abilities in patients with schizophrenia. Psychiatry Res 129(3):229–239

Ubl B, Kuehner C et al (2015) Altered neural reward and loss processing and prediction error signalling in depression. Soc Cogn Affect Neurosci

Vaidyanathan U, Nelson LD et al (2012) Clarifying domains of internalizing psychopathology using neurophysiology. Psychol Med 42(3):447–459

Velthorst E, Nieman DH et al (2009) Baseline differences in clinical symptomatology between ultra high risk subjects with and without a transition to psychosis. Schizophr Res 109(1–3):60–65

Ventura J, Hellemann GS et al (2009) Symptoms as mediators of the relationship between neurocognition and functional outcome in schizophrenia: a meta-analysis. Schizophr Res 113(2–3):189–199

Vollema MG, Geurtsen GJ et al (1995) Durable improvements in Wisconsin card sorting test performance in schizophrenic patients. Schizophr Res 16(3):209–215

Volz M, Hamm AO et al (2003) Temporal course of emotional startle modulation in schizophrenia patients. Int J Psychophysiol 49(2):123–137

Vrieze E, Pizzagalli DA et al (2013) Reduced reward learning predicts outcome in major depressive disorder. Biol Psychiatry 73(7):639–645

Wallis JD (2007) Orbitofrontal cortex and its contribution to decision-making. Ann Rev Neurosci 30:31–56

Walter H, Vasic N et al (2007) Working memory dysfunction in schizophrenia compared to healthy controls and patients with depression: evidence from event-related fMRI. Neuroimage 35(4):1551–1561

Walter H, Kammerer H et al (2009) Altered reward functions in patients on atypical antipsychotic medication in line with the revised dopamine hypothesis of schizophrenia. Psychopharmacology (Berl) 206(1):121–132

Walter H, Heckers S et al (2010) Further evidence for aberrant prefrontal salience coding in schizophrenia. Front Behav Neurosci 3:62

Walton ME, Bannerman DM et al (2003) Functional specialization within medial frontal cortex of the anterior cingulate for evaluating effort-related decisions. J Neurosci 23(16):6475–6479

Walton ME, Rudebeck PH et al (2007) Calculating the cost of acting in frontal cortex. Ann NY Acad Sci 1104:340–356

Waltz JA, Gold JM (2007) Probabilistic reversal learning impairments in schizophrenia: further evidence of orbitofrontal dysfunction. Schizophr Res 93(1–3):296–303

Waltz JA, Frank MJ et al (2007) Selective reinforcement learning deficits in schizophrenia support predictions from computational models of striatal-cortical dysfunction. Biol Psychiatry 62(7):756–764

Waltz JA, Schweitzer JB et al (2009) Patients with schizophrenia have a reduced neural response to both unpredictable and predictable primary reinforcers. Neuropsychopharmacology 34(6):1567–1577

Waltz JA, Schweitzer JB et al (2010) Abnormal responses to monetary outcomes in cortex, but not in the basal ganglia, in schizophrenia. Neuropsychopharmacology 35(12):2427–2439

Waltz JA, Kasanova Z et al (2013) The roles of reward, default, and executive control networks in set-shifting impairments in schizophrenia. PLoS One 8(2):e57257

Ward RD, Simpson EH et al (2012) Dissociation of hedonic reaction to reward and incentive motivation in an animal model of the negative symptoms of schizophrenia. Neuropsychopharmacology 37(7):1699–1707

Watanabe M (1996) Reward expectancy in primate prefrontal neurons. Nature 382:629–632

Weickert TW, Terrazas A et al (2002) Habit and skill learning in schizophrenia: evidence of normal striatal processing with abnormal cortical input. Learn Mem 9(6):430–442

Weickert TW, Goldberg TE et al (2009) Neural correlates of probabilistic category learning in patients with schizophrenia. J Neurosci 29(4):1244–1254

Weiler JA, Bellebaum C et al (2009) Impairment of probabilistic reward-based learning in schizophrenia. Neuropsychology 23(5):571–580

Westheide J, Wagner M et al (2007) Neuropsychological performance in partly remitted unipolar depressive patients: focus on executive functioning. Eur Arch Psychiatry Clin Neurosci 257(7):389–395

Whitmer AJ, Frank MJ et al (2012) Sensitivity to reward and punishment in major depressive disorder: effects of rumination and of single versus multiple experiences. Cogn Emot 26(8):1475–1485

Wilder KE, Weinberger DR et al (1998) Operant conditioning and the orbitofrontal cortex in schizophrenic patients: unexpected evidence for intact functioning. Schizophr Res 30(2):169–174

Wolf DH, Satterthwaite TD et al (2014) Amotivation in schizophrenia: integrated assessment with behavioral, clinical, and imaging measures. Schizophr Bull 40(6):1328–1337

Wynn JK, Horan WP et al (2010) Impaired anticipatory event-related potentials in schizophrenia. Int J Psychophysiol 77(2):141–149

Yang XH, Huang J et al (2014) Motivational deficits in effort-based decision making in individuals with subsyndromal depression, first-episode and remitted depression patients. Psychiatry Res 220(3):874–882

Yilmaz A, Simsek F et al (2012) Reduced reward-related probability learning in schizophrenia patients. Neuropsychiatr Dis Treat 8:27–34

Yip SW, Sacco KA et al (2009) Risk/reward decision-making in schizophrenia: a preliminary examination of the influence of tobacco smoking and relationship to Wisconsin card sorting task performance. Schizophr Res 110(1–3):156–164

Zalla T, Plassiart C et al (2001) Action planning in a virtual context after prefrontal cortex damage. Neuropsychologia 39(8):759–770

Zhang WN, Chang SH et al (2013) The neural correlates of reward-related processing in major depressive disorder: a meta-analysis of functional magnetic resonance imaging studies. J Affect Disord 151(2):531–539

Methods for Dissecting Motivation and Related Psychological Processes in Rodents

Ryan D. Ward

Abstract Motivational impairments are increasingly recognized as being critical to functional deficits and decreased quality of life in patients diagnosed with psychiatric disease. Accordingly, much preclinical research has focused on identifying psychological and neurobiological processes which underlie motivation. Inferring motivation from changes in overt behavioural responding in animal models, however, is complicated, and care must be taken to ensure that the observed change is accurately characterized as a change in motivation, and not due to some other, task-related process. This chapter discusses current methods for assessing motivation and related psychological processes in rodents. Using an example from work characterizing the motivational impairments in an animal model of the negative symptoms of schizophrenia, we highlight the importance of careful and rigorous experimental dissection of motivation and the related psychological processes when characterizing motivational deficits in rodent models. We suggest that such work is critical to the successful translation of preclinical findings to therapeutic benefits for patients.

Keyword Motivation · Rodent models · Anhedonia and avolition · Preference assessment · Taste-reactivity testing · Goal-directed behaviour · Effort-related choice · Outcome representation · D2R-OE mice

Contents

1	Aspects of Motivation Impacted in Psychiatric Disease	454
2	Assessing Anhedonia	455
	2.1 Preference Assessments	455
	2.2 Taste-Reactivity Testing	455
3	Assessing Avolition	456
	3.1 Fixed- and Progressive-Ratio Schedules	456
	3.2 Effort-Related Choice Tasks	457

R.D. Ward (✉)
Department of Psychology, University of Otago, PO Box 56, Dunedin 9054, New Zealand
e-mail: rward@psy.otago.ac.nz

© Springer International Publishing Switzerland 2015
Curr Topics Behav Neurosci (2016) 27: 451–470
DOI 10.1007/7854_2015_380
Published Online: 14 August 2015

4	Assessing Outcome Representation and Flexible Use of Represented Outcomes in Behaviour..	458
5	Separating Goal-Directed from Arousal Components of Motivation............................	459
6	Dissecting Psychological Components of Motivation: An Example Using an Animal Model of the Negative Symptoms of Schizophrenia..........................	461
7	Conclusions..	466
References..		467

Motivation comprises many related psychological processes, but can be defined generally as the energizing of behaviour in pursuit of a desired goal. There are a number of functional impairments in psychiatric disease which fall under the rubric of "motivational" deficits. For example, one of the critical diagnostic symptoms in major depressive episode is "marked diminished interest or pleasure in all, or almost all, activities most of the day, nearly every day", while the diagnostic criteria for schizophrenia explicitly include avolition as a negative symptom (American Psychiatric Association 2013). Uncovering the biological basis of these impairments requires the use of animal models. Rodents are particularly useful in this regard because of the many molecular manipulations that have been optimized for use in rats and mice.

Modelling motivational deficits in rodents presents a particular challenge. This is because motivational deficits by nature have to do with the absence of overt behavioural responding. This makes it all the more difficult to identify the crucial motivational components that are impacted by any experimental manipulations. There are any number of reasons why an animal might not engage in some behaviour, and not all of these have to do with the psychological or neurobiological processes researchers are interested in. If an animal responds less in a task, it could be because it is less motivated. However, it could also be that the animal is satiated, fatigued, has motor impairments, among other reasons. Conversely, if an animal responds more in a task, it could be due to increased motivation, but it could also be due to increased overall hyperactivity. Thus, without a careful experimental analysis, it is impossible to separate specific effects on goal-directed motivation from effects on other processes.

Aside from the difficulty of interpreting increases or decreases in behavioural output as specific changes in motivation, another reason why studying motivation is so difficult is that any behaviour that we study and infer motivation from is not caused by a monolithic process, but is itself the result of a combination of multiple psychological processes. Consider all of the psychological processes involved in the simple act of a rat pressing a lever for access to a food reward. For one, there is the physical effort required to press the lever. There is also a hedonic component to the behaviour. If the reward does not produce a positive hedonic reaction, that is, if the rat does not "like" the reward, it may not press the lever. The rat must also be able to represent not only the anticipated cost of the effort it will expend to press the lever (along with other costs, such as the time lost which could be engaged in other foraging activities due to engaging in the work requirement), but also the anticipated benefit of the reward at the end of the work. These expectations must then be compared via some type of cost/benefit calculation. If the outcome of the effort outweighs the associated costs, the rat will choose to expend the effort of the lever presses.

In addition, in experiments which assess the willingness of animals to work for many trials (and many rewards), there is the question of how quickly the animal satiates. If they are not hungry, they will not work for food. In cases where the work requirement takes some amount of time to complete, we must consider the extent to which the animal is tolerant of the delay to the reward. Related to this is how accurately the animal is able to represent the time it takes to complete the work requirement. If an animal overestimates or underestimates the elapsed time, it will have an impact on the willingness to expend the effort. All things being equal, an animal that is less tolerant of delays to reward, or has an exaggerated representation of the time it will take to fulfil the work requirement, will work less for a reward that is temporally distant.

We can see that interpreting deficits in performance on some behavioural task becomes all the more challenging when we parse apart the many psychological processes involved in these performances. The challenge is to isolate these specific psychological processes via a careful behavioural analysis and identify the neurobiology of these separable aspects of motivation (see Berridge and Robinson 2003, and also Robinson et al., in this volume for related discussion). Because of the many processes involved in motivated performance, and the potential for confounding influences, careful and rigorous behavioural analysis of motivation in animal models is as much about identifying experimental factors and psychological processes that *are not* related to a behavioural performance of interest, as it is about identifying factors and processes that are (Ward et al. 2011). We give an example of this type of an analysis below.

This careful behavioural analysis becomes even more critical given the pervasive use of genetically modified animal models of various diseases in contemporary biomedical research (Ellgood and Crawley 2015; Nestler and Hyman 2010; O'Tuathaigh et al. 2014). With the relative ease of manipulating specific genes, mouse (and increasingly, rat) models of various genetic risk factors are easily tractable. The impact of genetic deletion or overexpression of specific genetic products is difficult to predict or understand; however, it is most certainly the case that these manipulations impact much more than the circumscribed psychological or disease process of interest (Berridge and Robinson 2003; O'Tuathaigh et al. 2010, 2013).

Fortunately, there are a wealth of behavioural procedures, which have been developed throughout the history of experimental psychology with which to dissect motivational processes. Furthermore, researchers have employed these procedures to understand the neurobiology of various components of motivation and have revealed mostly separable neural substrates. In this chapter, I discuss some of these methods, with a particular focus on methods for dissecting specific psychological aspects of motivation, and ruling out alternative interpretations for observed behavioural deficits.

1 Aspects of Motivation Impacted in Psychiatric Disease

Diagnostic criteria for a range of psychiatric diseases include deficits in motivation. Although these types of deficits vary across diagnoses and patients, they generally fall into two categories: anhedonia and avolition. Anhedonia refers to the inability to experience pleasure from normally pleasurable activities (Ribot 1897). While it has been studied in a range of psychiatric conditions, it is historically most strongly associated with major depression (Gorwood 2008) and schizophrenia, although current research in patients suggests that the prevalence of anhedonia in schizophrenia may be exaggerated due to the format of clinical intake interviews, and may instead reflect decreased anticipatory motivation or deficits in other cognitive processes (Burbridge and Barch 2007; Gard et al. 2007; Strauss and Gold 2012). More information on the clinical concepts and methods of assessment of motivational deficits in patients with schizophrenia is provided by Reddy et al., in this volume.

Avolition refers to a general lack of motivation or drive to engage in goal-directed activities. It is considered to be a critical component of the functional deficits in schizophrenia (Foussias and Remington 2010) and is resistant to current therapeutic strategies. In recent years, there has been a renewed focus on avolition and motivational deficits in schizophrenia. An in-depth review of studies aimed at dissecting the behavioural components of deficits in motivation in patients with Schizophrenia is provided by Waltz and Gold et al., in this volume. The recent increase in such research is due to the critical nature of these deficits and their likely interaction with other symptom clusters in the disease given the aberrant function of prefrontal-striatal circuitry (critical to both motivational and cognitive processes) in patients (Barch and Dowd 2010; Barch 2005). Such an interaction may contribute to the lack of functional improvement in patients following cognitive-remediation therapy, even when such therapy results in significant improvement in specific domains of cognitive functioning (Hogarty et al. 2004; Medalia and Choi 2009; Velligan et al. 2006). Results like these suggest that successfully treating an underlying motivational impairment may be critical to achieve meaningful cognitive and functional gains. For an in-depth discussion on this subject, please see Saperstein and Medalia in this volume. In addition, symptoms in other psychiatric diseases, which have historically been considered separately from goal-directed motivation (apathy, emotional and mood disturbances), are increasingly recognized as having a core deficit in avolition. Thus, impairments in motivation likely play a critical role in producing functional deficits in depression (Chen et al. 2015) and Parkinson's disease (Pagonabarraga et al. 2015), among others. For a detailed discussion on the similarities and differences in the mechanisms underlying motivational deficits in depression and schizophrenia, please see Barch et al., in this volume.

2 Assessing Anhedonia

2.1 Preference Assessments

The most common way to assess anhedonia in rodents is to present them with a palatable substance and assess some behavioural measure of preference, pleasure or hedonic reaction. There are a number of ways to assay these types of reactions in rodents. The simplest is the sucrose preference test (Muscat and Willner 1989). This assay takes advantage of the fact that when given a free choice between water and a sucrose solution, rodents will generally choose to consume more of the sucrose solution (Muscat and Willner 1989). The development of this preference is thought to be indicative of a pleasurable hedonic response to the palatable sucrose solution. Thus, preference for the sucrose solution and overall sucrose consumed over some time period is thought to be a measure of hedonic capacity in rodents.

Interpreting increased sucrose preference and consumption as a measure of hedonic reaction to reward is complicated, however, by other factors that contribute to flavour preference. For example, preference for a specific flavour has been shown to be related to not only the sweetness of the solution, but also the caloric benefit the animal obtains from consuming the solution (Bolles et al. 1981; Mehiel and Bolles 1984). Thus, the hedonic contribution of increased preference for and consumption of sucrose over the course of an experiment is difficult to disentangle from other factors. Giving animals a brief, single exposure and measuring consumption is less likely to suffer from this confound, and this method is widely used as a quick general assay of hedonic functioning.

Another method for assessing taste preference is to measure the number of licks a rodent makes to a taste stimulus during an experimental session. In these assays, a number of flavours or concentrations can be delivered during the same session and the number of licks to each solution is measured (see Glendinning et al. 2002). These experiments are readily implemented with the use of commercially available gustometers. In addition, the within-subject nature of these assessments means that all animals experience all solutions or concentrations of the test solution, meaning that any differences are not due to differences in post-ingestional consequences.

2.2 Taste-Reactivity Testing

Another method for assaying hedonic reaction to reward, the taste-reactivity test (Grill and Norgren 1978), takes advantage of the evolutionary conservation of the hedonic response evoked by certain taste stimuli. In a number of species, from humans to mice, exposure to a sweet solution, such as sucrose, elicits a response consisting of rhythmic tongue and mouth movements and tongue protrusions (Berridge 2000). This characteristic response is quite different from that observed when a bitter solution (for example, quinine) is presented (mouth gapes, head shaking), and is thought to reflect a positive hedonic reaction to reward (Berridge and Robinson 2003; Pecina et al. 1997, 2003; Pecina and Berridge 2005). By exposing animals to varying solutions and videotaping, coding and scoring their facial expressions, researchers can assess differences in both positive and negative hedonic reactions and the impact of manipulating various neurobiological circuits on the hedonic response (see Robinson et al., in this volume for more detail).

3 Assessing Avolition

Avolition is generally assessed in rodents by determining how willing an animal is to make a particular response in order to procure some type of reward. This type of behaviour is thought to be goal-directed, meaning that the animal is motivated to engage in the behaviour in anticipation of earning the reward. Goal-directed behaviour is thought to involve a cost/benefit computation in which the anticipated benefit of the to-be-earned reward is weighed against the anticipated effort of the task at hand (Salamone et al. 2007; and see Redish et al., in this volume for further discussion). By manipulating aspects of the required effort and/or the characteristics of reward, combined with measuring and manipulating neural activity in putatively involved brain areas, researchers can interrogate different aspects of the cost/benefit computation, thus gaining insight into impaired motivation.

3.1 Fixed- and Progressive-Ratio Schedules

One of the most simple experimental arrangements to test avolition is a fixed-ratio (FR) schedule of reinforcement. In this schedule, once the animal is initially trained to lever press, it is rewarded after it makes a specified number of lever presses (e.g. FR1, FR5, FR25). Over the course of several sessions, the number of presses required for reward can be increased, and the extent to which animals are willing to keep pressing in the face of an increasing work requirement gives an index of how motivated an animal is to obtain the reward in the face of an increasing effort requirement and allows researchers to assess the effects of targeted manipulations (Salamone et al. 2003).

In a progressive-ratio (PR) schedule (Hodos 1961), a variant of an FR schedule, animals are rewarded after a certain number of lever presses are emitted, but the work requirement for each reward increases following reward delivery. With each subsequent reward, the effort becomes more and more difficult, until at some point the animal refuses to complete the work requirement. The work requirement at which the animal no longer responds is taken as an index of motivation. Animals are generally trained to first respond on a low requirement FR schedule. Once responding is established, the PR schedule is implemented. During this schedule the response requirement increases according to some mathematical rule following each reward. A number of progression steps have been used, including exponential, geometric and arithmetic progressions with various step sizes (Killeen et al. 2009; Richardson and Roberts 1996). The requirement continues to increase until the animal ceases responding for a set period of time (e.g. 3 min) or until the session has continued for a predetermined amount of time (e.g. 2 h).

An important consideration to be noted here is that choosing the right progression is a key to identifying differences in motivation using this schedule. Depending on the progression step used, the response requirement and required effort increases either slowly, or very rapidly. Choosing too difficult a progression will result in the animal quitting very soon, providing limited data to analyse. This can be particularly problematic if a difference in motivation between groups is subtle. Similarly, choosing too easy a progression may also obscure differences. Thus, it can be helpful to test animals using several different progressions in a parametric design (Simpson et al. 2011).

Several measures of performance on the PR schedule can be used to assay motivation. The most common is the break point, which is the last ratio completed before the animal quits responding or the session ends. This measure gives an indication of the amount of effort an animal is willing to expend for the reward. Other commonly used measures are the total number of lever presses and the number of obtained rewards. Another way to analyse PR data is to present a log-survivor plot, which shows the proportion of animals still working on the schedule as a function of elapsed time or response requirement (e.g. Drew et al. 2007; Simpson et al. 2011). These plots are particularly informative because the data from all subjects are shown and considered, whereas the average measures may obscure individual differences or be unduly influenced by extreme cases.

3.2 Effort-Related Choice Tasks

Another task that has been widely used to study the neurobiology of reward-motivated behaviour is the effort-based choice task first used by Salamone et al. (1991; for review, see Salamone et al. 2007). This task assesses the trade-off between expending effort to obtain a more-preferred reward and consuming a freely available less-preferred reward. Animals are trained on an FR schedule, which is gradually increased until it is moderately demanding. Working on this schedule is rewarded with a highly preferred reward. Concurrently with the FR schedule, a less-preferred reward (usually home-cage chow) is freely available to the animal, usually in a dish on the floor of the testing chamber. Thus, throughout the session, the animal can make a choice between working for the more-preferred reward or consuming the less-preferred home-cage chow. This task is thought to specifically assay reward-motivated behaviour. By including the choice between the lever and the freely available chow, it can also dissociate goal-directed motivated behaviour from motivation to consume reward. Usually, animals will work more for the preferred reward, and consume less of the freely available chow. By increasing the

FR requirement (or manipulating neural functioning in some way), researchers can determine how various conditions or manipulations affect the willingness to expend effort for a preferred reward (Salamone et al. 2007 and see Salamone et al., in this volume).

In addition to the operant effort-based procedure, a T-maze variant of this task has also been used (Salamone et al. 1994; Izquierdo and Belcher 2012). In this procedure, animals are trained in a T-maze and given forced choice trials where they experience both highly preferred and less-preferred rewards at the end of different arms of the maze. On free-choice trials, the rats will generally choose the arm with the highly preferred reward. The effort requirement for this reward can be increased by placing a short barrier (15 cm) in the highly preferred arm. Over trials, the height of the barrier is increased until the rat changes its preference to the less-preferred arm. This task, along with the operant version described above, has been successfully used to clarify the role of dopamine transmission in reward-motivated behaviour (see Salamone et al. 1991, 1994, 2003, 2007; and Salamone et al., in this volume).

4 Assessing Outcome Representation and Flexible Use of Represented Outcomes in Behaviour

As discussed above, motivation is a multifaceted concept, and research has demonstrated the critical involvement of a number of psychological processes in the overt manifestation of a reward-motivated response. Chief among these are processes that allow for the anticipation of rewards. These include the computation of the value of expected outcomes as well as the ability to update this representation with changing contingencies or to flexibly adapt behaviour in different situations. Two methods for assessing the extent to which animals can flexibly use represented outcomes in adaptive behaviour are devaluation and Pavlovian to instrumental transfer (PIT) tasks. A number of variations of the devaluation procedure have been employed, but the basic premise is the same across variations. In devaluation procedures, in separate sessions animals learn to associate a particular response with a particular outcome. For example, pressing on one lever during a morning session may result in a chocolate flavoured food pellet, while pressing on an opposite different lever in an afternoon session produces a banana-flavoured pellet. In these sessions, animals learn the association between a particular response (pressing a specific lever) and a particular outcome (pellet flavour). Prior to the test session, one of the rewards is devalued. This devaluation can be accomplished by pairing the reward with gastrointestinal malaise via an injection of lithium chloride to induce a conditioned taste aversion. Alternatively, the animals can be given free access to as much of one type of outcome (usually counterbalanced across test-session days) as they will consume. Directly after devaluation, animals are given access to both response options at the same time for the first time under extinction conditions, where making either response will not be reinforced with any outcome. If the animal has learned the association between a particular response and outcome, and if they update the value of the outcome and use the current value to guide behaviour, during this extinction session they will make fewer responses on the response option that normally produces the devalued outcome (Balleine and Dickinson 1998).

In devaluation procedures, an animal learns that different responses produce unique outcomes, and the extent to which these unique outcomes are valued dictates the animal's willingness to engage in the response during the devaluation test. Representations of unique outcomes can also impact performance in discrimination procedures. In a differential outcomes procedure, animals are trained to make different responses in the presence of different stimuli (Traphold 1970). Under these conditions, animals learn the discrimination much more quickly if correct responses produce different outcomes than if they produce the same outcome. This differential outcomes effect (DOE) provides more evidence that animals can use information about anticipated rewards to guide their behaviour.

Another way to assess the flexibility with which animals use information about the learned value of expected outcomes is the PIT task. In PIT, an animal is first trained in a simple Pavlovian conditioning protocol, in which a stimulus (e.g. a tone or light) predicts the delivery of a reward. Under these conditions, animals will readily learn to make an anticipatory response during the conditioned stimulus. Once this response is established, the animals are trained to make an instrumental lever-press response for the same reward, which they easily learn. The key test takes place in sessions in which the previously learned conditioned stimulus is presented to the animal while it is currently engaged in the lever-press task. Under these conditions, an animal that has learned the association between the conditioned stimulus and the reward will elevate its lever-press responding in the presence of this cue, indicating anticipation of the reward previously predicted by the cue. This procedure assays the ability of a cue learned in one context (Pavlovian conditioning) to transfer its value and impact behaviour in another context (instrumental lever-press responding). Furthermore, by including stimuli and outcomes that are not explicitly paired with instrumental responding, one can separate the general excitatory effects of reward predicting stimuli (stimuli associated with reward in the initial training phase but not explicitly paired with instrumental responding) from outcome-specific motivational activation (Corbit and Balleine 2005, 2011; Corbit et al. 2007).

5 Separating Goal-Directed from Arousal Components of Motivation

One of the most difficult aspects of studying and modelling motivation in rodents is separating goal-directed action from general arousal. For example, it is well-documented that amphetamine and other stimulants increase performance on

PR schedules (Mayorga et al. 2000; Olausson et al. 2006) and these results are often interpreted as evidence that these substances increase motivation. This interpretation is complicated, however, by the fact that these drugs also reliably increase measures of arousal, including locomotor activity (Hall et al. 2008; McNamara et al. 1993). Thus, pharmacological, and increasingly, genetic manipulations which may appear to enhance motivation and goal-directed behaviour may involve an enhancement in arousal and hyperactivity instead, or in addition to, an increase in motivation.

In tasks such as the FR or PR schedule, long considered the gold standard in terms of assaying motivation, it is impossible to differentiate the contribution of these two separate components to motivated performance. These shortcomings may also be relevant to the treatment of motivational deficits in patients. Motivational impairments produce a significant functional burden in patients, and there are currently no effective treatments. Deficits in goal-directed behaviour are seen as a critical aspect of the impairment in functioning. However, given the failure of most methodologies for studying motivation to separate this aspect of behaviour from general arousal, targeting specific aspects of impaired motivation in preclinical models and translation of efficacious treatment strategies to patients is difficult. Methods for separating these components of motivation will therefore be useful for the development of successful treatment strategies.

In a recent series of elegant experiments, Bailey and colleagues developed and validated a novel behavioural task with the express purpose of dissociating goal-directed from arousal components of motivated behaviour (Bailey et al. 2015). In their method, called the progressive hold down (PHD) task, mice are required to hold a response lever down for a progressively longer period of time after each reward. Bailey and colleagues show that this schedule is sensitive to manipulation of variables that are known to impact motivation, such as food deprivation and increasing sucrose concentration. Most importantly, they characterized the effects of methamphetamine on both PR and PHD performance. In the PR schedule, they replicated the results of numerous prior studies, showing that methamphetamine increased overall lever presses, rewards earned, and time spent working in the session. These results have sometimes been interpreted previously as indicating increased motivation.

When they tested the effects of methamphetamine on performance of the PHD task, they found that it also increased the number of lever presses, but because of the nature of the schedule, this increase in lever presses came at the expense of fewer earned rewards. More careful analysis showed that methamphetamine specifically increased bursts of short-duration presses, which the authors interpreted as indicating a general increase in arousal (hyperactivity), rather than an increase in goal-directed motivation. Thus, using their method, they were able to separate methamphetamine's effects on goal-directed motivation from its effects on general arousal. Future research using procedures such as this is critical to further elucidation and understanding of the neurobiology of motivation.

6 Dissecting Psychological Components of Motivation: An Example Using an Animal Model of the Negative Symptoms of Schizophrenia

Over the past several years, we have undertaken a behavioural characterization of the motivational deficits in a transgenic mouse model of the negative symptoms of schizophrenia. This mouse models the well-replicated finding of increased striatal dopamine D2 receptor activity in patients with schizophrenia (Abi-Dargham et al. 2000). In this model, transgenic expression of the human D2 receptor is directed via the camkIIα promotor, resulting in an increase in D2 receptor expression that is limited to the striatum and olfactory tubercle of D2R-OE mice (Kellendonk et al. 2006). Fortuitously, this model has a 15% increase in overall striatal D2 receptor expression, thus recapitulating the degree of increased D2 activity in patients. Importantly, by employing the tetracycline controlled gene expression system in this model (Mayford et al. 1996), Kellendonk et al. were able to control transgene expression in a temporal manner, thus allowing for excess D2 receptors to be "turned off" by feeding the mice doxycycline. This allows for separation of the acute, or reversible, effects of D2 overexpression from more permanent dysfunction that results from developmental changes.

The initial evidence of a motivational deficit in D2R-OE mice came from an experiment designed to assess interval timing in these mice (Drew et al. 2007). Interval timing deficits have been documented in schizophrenia (Carroll et al. 2009; Gomez et al. 2015), and we wished to determine if striatal D2R overexpression contributed to such timing deficits. We tested mice in the peak procedure (Roberts 1981). In this procedure, on some trials, called fixed interval (FI) trials, animals earn a reward for making a response after a specified interval of time since the beginning of the trial. Interval timing can then be assessed on probe, or peak, trials. During these trials, the reward is omitted and the trial continues on for much longer than the usual FI trial (three to four times as long). While the rate of responding as a function of time within the trial provides an index of the subject's timing accuracy (lever pressing should peak at the expected time of reward), the average rate of responding over the course of the entire trial provides a measure of the subject's motivation. When we tested D2R-OE mice on the peak procedure (see Fig. 1a), we found that D2R-OE mice were indeed impaired in timing accuracy (their peak rate of responding was shifted later than the reinforced interval). In addition, they displayed a dramatic reduction in overall response rates, indicative of decreased motivation compared to controls (Drew et al. 2007; Ward et al. 2009). Strikingly, turning-off the transgene rescued the motivation, as indexed by response rates, and this partially rescued their timing (Fig. 1a; Drew et al. 2007).

To more systematically assess motivation in D2R-OE mice, we first tested them on the PR paradigm (Fig. 1b). We used a schedule in which the work requirement doubled following each reward. We found that D2R-OE mice earned fewer rewards, had lower break points and quit working sooner than controls on this

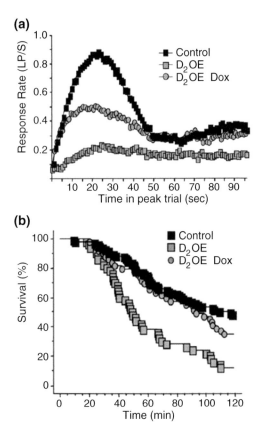

Fig. 1 **a** Performance of control and D2R-OE mice on the peak-interval procedure (see text for details). The figure shows response rate as a function of time in the peak trials. Data from Drew et al. (2007). **b** Performance of control and D2R-OE mice on the progressive-ratio schedule. The percentage of mice who were still responding on the schedule as a function of time in the session is shown. Also shown are the data from a group of D2R-OE mice in which the transgene was turned off by feeding mice doxycycline. Data from Drew et al. (2007)

schedule. Thus, D2R-OE mice displayed a motivational deficit. This deficit was rescued when the transgene was turned off (Fig. 1b).

The specific nature of the motivational deficit in D2R-OE mice is not clear from the PR performance. As discussed above, there are any number of psychological processes that combine to produce the outward behavioural outcome of persistence in pressing the lever (e.g. Ward et al. 2011). For example, perhaps D2R-OE mice do not find the reward hedonically satisfying (do not "like" the reward). Data collected from rewards retrieved suggested this was not the case, as D2R-OE mice and controls retrieved and consumed all of their earned rewards (Drew et al. 2007). Furthermore, D2R-OE mice displayed equivalent preference for sucrose as controls in a sucrose preference test (Ward et al. 2012). When we tested D2R-OE mice using the taste-reactivity test described above, we found no difference between D2R-OE mice and controls in several measures of hedonic reactivity to reward (Fig. 2), including positive hedonic reactions to increasing sucrose concentration (Fig. 2a) and increased lick rates as a function of increasing sucrose concentration in a gustometer test (Fig. 2b; Ward et al. 2012).

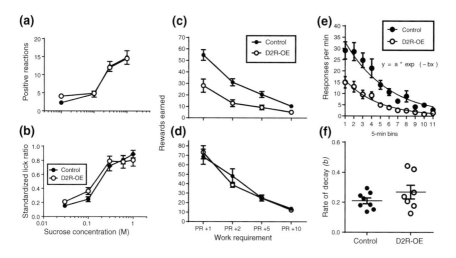

Fig. 2 **a** Positive hedonic reactions as a function of increasing sucrose concentration in control and D2R-OE mice. Data from Ward et al. (2012). **b** Standardized lick ratio (a measure of hedonic reaction) as a function of increasing sucrose concentration for control and D2R-OE mice. Data from Ward et al. (2012). **c** Number of rewards earned as a function of increases in the progressive-ratio work requirement for control and D2R-OE mice. Data from Simpson et al. (2011). **d** Number of rewards earned as a function of increases in the progressive-ratio work requirement for a separate group of control and D2R-OE mice in which transgene expression had been turned off by feeding the mice doxycycline. Data from Simpson et al. (2011). **e** Response rate as a function of time in extinction for control and D2R-OE mice. *Lines* through the data are the best fits of the negative exponential equation. **f** The value of the decay parameter (rate of extinction) from the negative exponential equation for all control and D2R-OE mice. In all figures, *error bars* represent one standard error above and below the mean

Another possible reason why D2R-OE mice could quit working sooner on the PR schedule is that they may satiate or fatigue more easily than controls, and are therefore unwilling or unable to produce the same behavioural output. We addressed this by parametrically manipulating the work requirement in the PR schedule (Simpson et al. 2011). We tested D2R-OE and control mice on PR +1, +2, +5 and +10 schedules (counterbalanced for order). We found that at all schedule requirements, D2R-OE mice pressed less than controls (Fig. 2c). This performance deficit was rescued by turning-off the transgene (Fig. 2d). The key data to address the question of satiety can be found in a comparison of rewards earned by the different genotypes during the PR + 1 and PR + 2 schedules. As shown in Fig. 2c, D2R-OE mice earned the same number of rewards under the PR + 1 schedule as control mice did under the PR + 2 schedule. Furthermore, response effort of D2R-OE mice, although lower at all PR schedule values than that of controls, was still modulated by work requirement, indicating that performance deficits were not the result of a ceiling on lever presses or reward consumption. These data indicate that the deficit in PR performance displayed by D2R-OE mice is not caused by satiety or fatigue, because they are capable under the right conditions of working just as hard and consuming just as many rewards as control mice.

In addition, as the work requirement increases in the PR schedule, so too does the amount of time that an animal must wait to obtain a reward. Perhaps D2R-OE mice are less tolerant of delays to reward? We tested D2R-OE mice on a schedule which was modelled after a PR schedule, except that instead of the response requirement increasing after each reward, the inter-reward interval doubled after each reward, and the mice received a reward if at least one response was made during this interval. Performance of D2R-OE mice was identical to that of controls under these conditions (Simpson et al. 2011).

Also related to the delay between response initiation and reward receipt is the question of whether D2R-OE mice's behaviour extinguishes (decreases when responses do not produce reward) more quickly than control animals. To test this, we trained D2R-OE mice on a variable ratio 25 schedule (VR25), in which rewards were given for completing a fixed number of lever presses. The number of lever presses required for reward varied from trial to trial with an average requirement of 25 presses. Following stable performance on this schedule, we then exposed them to one session of extinction in which no rewards were delivered. As shown in Fig. 2e, the overall rate of responding was lower over the course of the entire extinction session for D2R-OE mice than for controls. To assess the rate of extinction, we fit negative exponential equations to the extinction curves from individual mice. These fits yielded a parameter for the decay rate of the functions, a measure of the rate of extinction (Fig. 2f). This analysis showed that notwithstanding the overall lower rates displayed by D2R-OE mice, their rate of extinction was not different from controls (see Simpson et al. 2011, for similar analysis and results). This result supports the idea that D2R-OE mice are not quitting the PR task early because they are more sensitive to extinction, but are perhaps quitting due to their decreased willingness to work.

Performance on the PR schedule also involves the appreciation of the relationship between the animal's behaviour and the delivery of reward, that is, the contingency between responding and reward. In an unpublished experiment, we tested D2R-OE mice in a contingency-degradation protocol (Hammond 1980; Barker et al. 2014) in which we trained them to respond on a lever for rewards on a VR 25 schedule. This simple schedule was overlaid with a separate, independent schedule which presented free rewards with some specified frequency. Under these conditions, as the frequency of free rewards increases and the contingency between response and reward is degraded, animals usually decrease their rate of responding (Hammond 1980). As the frequency of free rewards increased, rate of responding decreased in both D2R-OE and control mice, but there was no difference in the degree of the decrease, indicating intact contingency appreciation in D2R-OE mice.

It should be noted that none of these results definitively determine the specific nature of the motivational deficit in D2R-OE mice. Thus, far, our analyses have only eliminated alternative explanations. To further specify the nature of this deficit, we tested D2R-OE mice on the effort-related choice paradigm (Ward et al. 2012). Similar to the results from the PR schedule, D2R-OE mice responded significantly less for a preferred reward (Fig. 3a). However, they consumed significantly more of the freely available home-cage chow (Fig. 3b). This shift in choice reversed and became matched to controls when the transgene was turned off (Fig. 3a, b).

Thus, the deficit in D2R-OE mice seems to be due to a decreased willingness to work for reward rather than a general decreased motivation for reward. We suggest that this decreased willingness to work is the result of distortions in the cost/benefit computation required for motivated behaviour. This could occur in at least two ways. First, D2R-OE mice could be impaired in their ability to represent the value

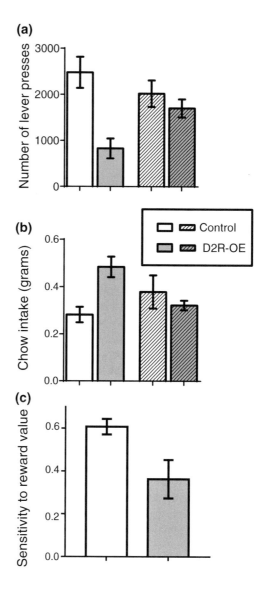

Fig. 3 Performance of D2R-OE mice on the effort-based choice task and sensitivity to differences in reward value. **a** Number of lever presses emitted for the preferred reward by control and D2R-OE mice. **b** Grams of freely available chow consumed by control and D2R-OE mice. *Scored bars* represent data from mice in which the transgene had been turned off by feeding the mice doxycycline. *Error bars* represent one standard error above and below the mean. **c** Sensitivity to differences in the distribution of rewards across response alternatives during concurrent-schedule testing. Data from Ward et al. (2012)

of the outcome of their work, leading to decreased motivation to engage effort. Second, D2R-OE mice could be impaired in their ability to accurately represent the anticipated cost of the effort required to procure reward, leading to an exaggerated anticipated cost.

Given the critical nature of outcome representation in motivated behaviour, we assessed the ability to which D2R-OE mice are able to represent the value of future outcomes (Ward et al. 2012). We exposed control and D2R-OE mice to a concurrent schedule in which we varied the frequency of rewards obtained for responding on two different levers. Mice received on average five rewards per minute (variable interval 20 s schedule) for pressing one lever, and 0.5 rewards per minute (variable interval 120 s schedule) for responding on the other lever. If mice are sensitive to the distribution of rewards across response options in this procedure, their ratio of lever presses will approximately match the ratio of rewards obtained from the two response options (Herrnstein 1961). We found that D2R-OE mice were less sensitive to the distribution of rewards across response options in the concurrent-schedule procedure (Fig. 3c), indicating that D2R-OE mice are less sensitive to the value of response options associated with different reward frequencies than controls.

7 Conclusions

In summary, the results of our experiments with the D2R-OE mice indicate that these mice have a deficit in goal-directed motivation. This impairment is not due to decreased hedonic reaction to reward, increased fatigue or satiation, decreased tolerance for delay, decreased sensitivity to contingency or increased susceptibility to extinction. The deficit appears to be due to a decreased willingness to expend effort, which results from a compromised cost/benefit computation, likely due to a deficit in either representing outcomes or using this information to guide behaviour. In fact, recent results indicate that although D2R-OE mice have relatively intact representations of different outcome types, these representations learned in one specific situation are not able to be used flexibly to contribute to adaptive behaviour in a new situation (Mezias et al., unpublished results).

The work described above with the D2R-OE mice is an example of the way that experimental methods can be leveraged to elucidate psychological mechanisms of motivational deficits in rodent models. By taking advantage of the reversible nature of the genetic manipulation in D2R-OE mice, we were able to gain much ground in uncovering the specific psychological processes that produced the performance deficits in D2R-OE mice. These types of methods can be used in concert with the ever-increasingly sophisticated molecular manipulations developed by researchers to make unprecedented gains in characterization of rodent models of psychiatric disease. This approach can also be used to investigate novel targets for therapies to enhance motivation in psychiatric disease (Simpson et al. 2011).

In conclusion, as noted above, motivational impairments are critical to functional deficits in a number of psychiatric diseases, and the need for effective treatments is great. Motivated behaviour involves a number of factors and psychological processes, including hedonic functioning, physical effort, satiety, tolerance of delay, ability to represent and adaptively use information about the value of represented outcomes, and ability to represent effort or time required to procure reward. All of these factors enter into a cost/benefit computation; the outcome of which will determine whether an animal will expend the time and effort required to obtain the reward. By utilizing the wealth of experimental methods available for dissecting component psychological processes involved in motivation, and continuing to invent new assays, researchers can better assess motivational impairments in animal models of psychiatric disease. Partitioning of motivated behaviour into its component psychological processes will deepen and enrich understanding of the separate and sometimes dissociable (as in the case of hedonic reaction and goal-directed motivation; Berridge and Robinson 2003) neurobiological underpinnings of this behaviour, and will lead to more specific and targeted treatment strategies. These improvements will in turn promote and foster more translatable outcomes from preclinical research to clinical populations.

References

Abi-Dargham A, Rodenhiser J, Printz D, Zea-Ponce Y, Gil R, Kegeles LS et al (2000) Increased baseline occupancy of D2 receptors by dopamine in schizophrenia. Proc Natl Acad Sci USA 97:8104–8109

American Psychiatric Association (2013) Diagnostic and statistical manual of mental disorders. American Psychiatric Publishing, Arlington, VA, USA

Bailey MR, Jensen G, Taylor K, Mezias C, Williamson C, Silver R, Simpson EH, Balsam PD (2015) A novel strategy for dissecting goal-directed action and arousal components of motivated behavior with a progressive hold-down task. Behav Neurosci 129:269–280

Balleine BW, Dickinson A (1998) The role of incentive learning in instrumental revaluation by specific satiety. Anim Learn Behav 26:46–59

Barch DM (2005) The relationships among cognition, motivation, and emotion in schizophrenia: how much and how little we know. Schizophr Bull 31:875–881

Barch DM, Dowd EC (2010) Goal representations and motivational drive in schizophrenia: the role of prefrontal-striatal interactions. Schizophr Bull 36:919–934

Barker JM, Zhang H, Villafane JJ, Wang TL, Torregrossa MM, Taylor JR (2014) Epigenetic and pharmacological regulation of 5HT3 receptors controls compulsive ethanol seeking in mice. Eur J Neurosci 39:999–1008

Berridge KC (2000) Measuring hedonic impact in animals and infants: microstructure of affective taste reactivity patterns. Neurosci Biobehav Rev 24:173–198

Berridge KC, Robinson TE (2003) Parsing reward. Trends Neurosci 26:507–513

Bolles RC, Hayward L, Crandall C (1981) Conditioned taste preferences based on caloric density. J Exp Psychol Anim Behav Process 7:59–69

Burbridge JA, Barch DM (2007) Anhedonia and the experience of emotion in individuals with schizophrenia. J Abnorm Psychol 116:30–42

Carroll CA, O'Donnell BF, Shekhar A, Hetrick WP (2009) Timing dysfunctions in schizophrenia span from millisecond to several-second durations. Brain Cogn 70:181–190

Chen C, Takahashi T, Nakagawa S, Inoue T, Kusumi I (2015) Reinforcement learning in depression: a review of computational research. Neurosci Biobehav Rev 55:247–267

Corbit LH, Balleine BW (2005) Double dissociation of basolateral and central amygdala lesions on the general and outcome-specific forms of pavlovian-instrumental transfer. J Neurosci 25:962–970

Corbit LH, Janak PH, Balleine BW (2007) General and outcome-specific forms of Pavlovian-instrumental transfer: the effect of shifts in motivational state and inactivation of the ventral tegmental area. Eur J Neurosci 26:3141–3149

Corbit LH, Balleine BW (2011) The general and outcome-specific forms of Pavlovian-instrumental transfer are differentially mediated by the nucleus accumbens core and shell. J Neurosci 31:11786–11794

Drew MR, Simpson EH, Kellendonk C, Herzberg WG, Lipatova O, Fairhurst S et al (2007) Transient overexpression of striatal D2 receptors impairs operant motivation and interval timing. J Neurosci 27:7731–7739

Ellgood J, Crawley JN (2015) Behavioral and neuroanatomical phenotypes in mouse models of autism. Neurotherapuetics. doi:10.1007/s13311-015-0360-z

Foussias G, Remington G (2010) Negative symptoms in schizophrenia: avolition and Occam's razor. Schizophr Bull 36:359–369

Gard DE, Kring AM, Gard MG, Horan WP, Green MF (2007) Anhedonia in schizophrenia: distinctions between anticipatory and consummatory pleasure. Schizophr Res 93:253–260

Glendinning JL, Gresack J, Spector AC (2002) A high-throughput screening procedure for identifying mice with aberrant taste and oromotor function. Chem Senses 27:461–474

Gomez J, Marin-Mendez J, Molero P, Atakan Z, Ortuno F (2015) Time perception networks and cognition in schizophrenia: a review and a proposal. Psychiatry Res 220:737–744

Gorwood P (2008) Neurobiological mechanisms of anhedonia. Dialogues Clin Neurosci 10:291–299

Grill HJ, Norgren R (1978) The taste reactivity test. I. Mimetic responses to gustatory stimuli in neurologically normal rats. Brain Res 143:263–279

Hall DA, Stanis JJ, Marquez-Avila H, Gulley JM (2008) A comparison of amphetamine- and methamphetamine-induced locomotor activity in rats: evidence for qualitative differences in behavior. Psychopharmacology 195:469–478

Hammond LJ (1980) The effect of contingency upon the appetitive conditioning of free-operant behavior. J Exp Anal Behav 34:297–304

Herrnstein RJ (1961) Relative and absolute strength of response as a function of frequency of reinforcement. J Exp Anal Behav 4:267–272

Hodos W (1961) Progressive ratio as a measure of reward strength. Science 134:943–944

Hogarty GE, Flesher S, Ulrich R, Carter M, Greenwald D, Pogue-Geile M et al (2004) Cognitive enhancement therapy for schizophrenia: effects of a 2-year randomized trial on cognition and behavior. Arch Gen Psychiat 61:866–876

Izquierdo A, Belcher AM (2012) Rodent models of adaptive decision making. Methods Mol Biol 829:85–101

Kellendonk C, Simpson EH, Polan HJ, Malleret G, Vronskaya S, Winiger V et al (2006) Transient and selective overexpression of dopamine D2 receptors in the striatum causes persistent abnormalities in prefrontal cortex functioning. Neuron 49:603–615

Killeen PR, Posadas-Sanchez D, Johansen EB, Thrailkill EA (2009) Progressive ratio schedules of reinforcement. J Exp Psychol Anim Behav Process 35:35–50

Mayford M, Bach ME, Huang YY, Wang L, Hawkins RD, Kandel ER (1996) Control of memory formation through regulated expression of a camkII transgene. Science 274:1678–1683

Mayorga AJ, Popke EJ, Fogle CM, Paule MG (2000) Similar effects of amphetamine and methylphenidate on the performance of complex operant tasks in rats. Behav Brain Res 109:59–68

McNamara CG, Davidson ES, Schenk S (1993) A comparison of the motor-activating effects of acute and chronic exposure to amphetamine and methylphenidate. Pharmacol Biochem Behav 45:729–732

Medalia A, Choi J (2009) Cognitive remediation in schizophrenia. Neuropsychol Rev 19:353–364
Mehiel R, Bolles RC (1984) Learned flavor preferences based on caloric outcome. Anim Learn Behav 12:421–427
Muscat R, Willner P (1989) Effects of dopamine receptor antagonists on sucrose consumption and preference. Psychopharmacology 99:98–102
Nestler EJ, Hyman SE (2010) Animal models of neuropsychiatric disorders. Nat Neurosci 13:1161–1169
Olausson P, Jentsch JD, Tronson N, Neve RL, Nestler EJ, Taylor JR (2006) DeltaFosB in the nucleus accumbens regulates food-reinforced instrumental behavior and motivation. J Neurosci 26:9196–9204
O'Tuathaigh CMP, Desbonnet L, Waddington JL (2014) Genetically modified mice related to schizophrenia and other psychoses: seeking phenotypic insights into the pathobiology and treatment of negative symptoms. Eur Neuropsychopharmacol 24:800–821
O'Tuathaigh CMP, Moran PM, Waddington JL (2013) Genetic models of schizophrenia and related psychotic disorders: progress and pitfalls across the methodological "minefield". Cell Tissue Res 354:247–257
O'Tuathaigh CMP, Kirby BP, Moran PM, Waddington JL (2010) Mutant mouse models: Genotype-phenotype relationships to negative symptoms in schizophrenia. Schizophr Bull 36:271–288
Pagonabarraga J, Kulisevsky J, Strafella AP, Krack P (2015) Apathy in Parkinson's disease: clinical features, neural substrates, diagnosis, and treatment. Lancet Neurol 14:518–531
Pecina S, Berridge KC (2005) Hedonic hot spot in nucleus accumbens shell: where do u-opioids cause increased hedonic impact of sweetness. J Neurosci 25:11777–11786
Pecina S, Cagniard B, Berridge KC, Aldridge JW, Zhuang X (2003) Hyperdopaminergic mutant mice have higher "wanting" but not "liking" for sweet rewards. J Neurosci 23:9395–9402
Pecina S, Berridge KC, Parker LA (1997) Pimozide does not shift palatability: separation of anhedonia from sensorimotor suppression by taste reactivity. Pharmacol Biochem Behav 58:801–811
Ribot T (1897) The psychology of emotions. W. Scott, London, UK
Richardson NR, Roberts DCS (1996) Progressive ratio schedule of drug self-administration studies in rats: a method to evaluate reinforcing efficacy. J Neurosci Methods 66:1–11
Roberts S (1981) Isolation of an internal clock. J Exp Psychol Anim Behav Process 7:242–268
Salamone JD, Steinpreis RE, McCullough LD, Smith P, Grebel D, Mahan K (1991) Haloperidol and nucleus accumbens dopamine depletion suppress lever pressing for food but increase free food consumption in a novel food choice procedure. Psychopharmacology 104:515–521
Salamone JD, Cousins MS, Bucher S (1994) Anhedonia or anergia? Effects of haloperidol and nucleus accumbens dopamine depletion on instrumental response selection in a T-maze cost/benefit procedure. Behav Brain Res 65:221–229
Salamone JD, Correa M, Mingote S, Weber SM (2003) Nucleus accumbens dopamine and the regulation of effort in food-seeking behavior: implications for studies of natural motivation, psychiatry, and drug abuse. J Pharmacol Exp Ther 305:1–8
Salamone JD, Correa M, Farrar A, Mingote SM (2007) Effort-related functions of nucleus accumbens dopamine and associated forebrain circuits. Psychopharmacology 191:461–482
Simpson EH, Kellendonk C, Ward RD, Richards V, Lipatova O, Fairhurst S et al (2011) Pharmacologic rescue of motivational deficit in an animal model of the negative symptoms of schizophrenia. Biol Psychiatry 69:928–935
Strauss GP, Gold JM (2012) A new perspective on anhedonia in schizophrenia. Am J Psychiatry 169:364–373
Trapold MA (1970) Are expectancies based upon different positive reinforcing events discriminably different? Learn Motivation 1:129–140
Velligan DI, Kern RS, Gold JM (2006) Cognitive rehabilitation for schizophrenia and the putative role of motivation and expectancies. Schizophr Bull 32:474–485

Ward R, Kellendonk C, Simpson EH, Lipatova O, Drew MR, Fairhurst S, Kandel ER, Balsam PD (2009) Impaired timing precision produced by striatal D2 receptor overexpression is mediated by cognitive and motivational deficits. Behav Neurosci 123:720–730

Ward RD, Simpson EH, Kandel ER, Balsam PD (2011) Modeling motivational deficits in mouse models of schizophrenia: Behavior analysis as a guide for neuroscience. Behav Processes 87:149–156

Ward RD, Simpson EH, Richards VL, Deo G, Taylor K, Glendinning JL, Kandel ER, Balsam PD (2012) Dissociation of hedonic reaction to reward and incentive motivation in an animal model of the negative symptoms of schizophrenia. Neuropsychopharmacology 37:1699–1707

Part IV
Addiction and the Pathological Misdirection of Motivated Behaviour

Motivational Processes Underlying Substance Abuse Disorder

Paul J. Meyer, Christopher P. King and Carrie R. Ferrario

Abstract Drug addiction is a syndrome of dysregulated motivation, evidenced by intense drug craving and compulsive drug-seeking behavior. In the search for common neurobiological substrates of addiction to different classes of drugs, behavioral neuroscientists have attempted to determine the neural basis for a number of motivational concepts and describe how they are changed by repeated drug use. Here, we describe these concepts and summarize previous work describing three major neural systems that play distinct roles in different conceptual aspects of motivation: (1) a nigrostriatal system that is involved in two forms of instrumental learning, (2) a ventral striatal system that is involved in Pavlovian incentive motivation and negative reinforcement, and (3) frontal cortical areas that regulate decision making and motivational processes. Within striatal systems, drug addiction can involve a transition from goal-oriented, incentive processes to automatic, habit-based responding. In the cortex, weak inhibitory control is a predisposing factor to, as well as a consequence of, repeated drug intake. However, these transitions are not absolute, and addiction can occur without a transition to habit-based responding, occurring as a result of the overvaluation of drug outcomes and hypersensitivity to incentive properties of drug-associated cues. Finally, we point out that addiction is not monolithic and can depend not only on individual differences between addicts, but also on the neurochemical action of specific drug classes.

Keywords Drug Addiction · Motivation · Neural Systems · Incentive Sensitization · Allostatsis · Habit · Drug Learning

P.J. Meyer (✉) · C.P. King
Behavioral Neuroscience Program, Department of Psychology, University at Buffalo, Park Hall B72, Buffalo, NY 14260, USA
e-mail: pmeyer@buffalo.edu

C.R. Ferrario
Department of Pharmacology, University of Michigan Medical School, A220B MSRB III, 1150 West Medical Center Drive, Ann Arbor, MI 48109-5632, USA
e-mail: ferrario@umich.edu

Contents

1 Introduction 474
   1.1 Rodent Models of Addiction 475
2 Overview of Motivational Processes and Their Neurobiological Substrates 476
3 Hedonic Allostasis 479
4 Incentive Sensitization 482
5 Addiction as a Disorder of Learning and Decision Making 485
   5.1 Goal- Versus Habit-Based Responding 486
   5.2 Neural Circuits Governing Goal- and Habit-Based Decision Making 487
   5.3 Pavlovian Cues: Reward Prediction, Incentive Salience,
        and Transfer to Instrumental Responding 489
6 Impulsive Action and Impaired Executive Function 489
7 Models that Integrate Motivational Concepts 490
   7.1 Dual-Process Models 490
   7.2 Transitions from Drug-Taking to Addiction 491
   7.3 Drug Addiction: A Compendium of Vulnerabilities 492
8 Conclusion 493
References 496

1 Introduction

Drug addiction is characterized by intense drug craving and compulsive drug use, with an excessive amount of time and effort spent procuring drug[1] (Jaffe 1975; Koob and Le Moal 2008; Tiffany 1999; Skinner and Aubin 2010; Gass et al. 2014; American Psychiatric Association 2013). In many cases, the drug has no direct benefit to the subject's survival, and drug-taking occurs despite adverse consequences to one's health, livelihood, and family. Smoking cigarettes is a classic example of this, and becoming addicted to prescription opioids highlights the complex balance between the benefits and harms of many drugs. So what drives drug-taking in the first place, and what shifts motivational systems into "overdrive," resulting in addiction? The obvious starting point is that the individual must first try the drug. This initiation depends on a number of factors, including individual personality traits, cultural norms, social context, and interactions with comorbid mental illness (Bucholz 1999; Compton et al. 2007; Homberg et al. 2014).

---

[1]Note that "substance use disorder" in humans is diagnosed on a continuum from mild to severe based on 11 behavioral criteria laid out in the DSM V, which does not mention "addiction." This choice is in part intended to help identify and treat overuse of drugs (e.g., alcohol and nicotine which can cause or worsen a number of other conditions) and also to avoid social stigmas associated with the term "addiction," which may result from the lack of clear biological markers that identify the addicted state. We avoid this symptomatological approach to describing addiction in favor of focusing on the core motivational processes involved in drug-taking, in a manner analogous to the National Institute of Mental Health's Research Domain Criteria (RDoC) initiative (Cuthbert 2014; Litten et al. 2015; NIMH 2015).

Following this initiation, a drug is taken repeatedly to obtain some perceived positive effect, be it physiological or psychological (e.g., euphoria or social acceptance), or to relieve a perceived negative state (e.g., to reduce anxiety). The experienced outcome of drug use reinforces the drug-taking behavior; that is, the experience of a positive outcome or the removal of a negative outcome increases the likelihood that the drug will be used again. These forms of positive and negative reinforcement, respectively (Khantzian 1997; Wise and Bozarth 1987), help maintain drug-taking behaviors, particularly in early, often experimental stages.

The transition from initial controlled drug use to uncontrolled use and addiction is not well understood, in part because this transition is gradual, and is not defined by a precise behavioral or biological tipping point (Ahmed and Koob 1998). However, it is clear that most drug users begin by experimenting with a variety of drugs, that multiple drug experiences are needed, and that not all drug-taking experiences will result in addiction (even if use is prolonged; think of four years on an American college campus). Below, we describe several key psychological processes that underlie initial drug-taking, how they are altered in the addicted state, and the neurobiological substrates thought to underlie these processes. Because these ideas have been presented in many reviews (Robinson et al. 2014; Belin et al. 2009; Wise and Koob 2014; Hogarth et al. 2013; Wiers et al. 2007; Kalivas and Volkow 2005; Redish et al. 2008), we will focus on core motivational processes involved in addiction and their key neurobiological substrates. In addition, we highlight how different motivational processes may be involved in different aspects of addiction depending on the individual, the drug class, and the duration of drug-taking behavior.

1.1 Rodent Models of Addiction

Much of what is known about the psychological and neurobiological substrates of addictive behavior has been learned using animal models (Lesscher and Vanderschuren 2012; Deroche-Gamonet et al. 2004). Here, we review a few key approaches to orient the reader to studies referred to in subsequent sections. Early studies used repeated administration of potentially addictive drugs to an animal (usually rats or mice) and measured the consequences on behavior and brain function. This approach was, and still is, very useful for understanding how drug exposure changes the brain and influences behavior, but lacks an aspect of voluntary drug-taking (Wolf and Ferrario 2010; Vezina and Leyton 2009; Schmidt and Pierce 2010). Ultimately, self-administration approaches became the gold standard not only for studying drug addiction, but also for assessing how likely it is that a drug will be abused (Schuster and Thompson 1969; Weeks 1962). More recently, self-administration models that attempt to differentiate "casual" drug-taking from "addiction-like" overconsumption have been developed. These models typically use extended access procedures in which drug is freely available over prolonged periods of time. These procedures induce escalated drug self-administration and incorporate the idea of continued pursuit of the drug despite adverse consequences (e.g., crossing an electrified grid to obtain the drug). Indeed, a number of studies have found that extended access self-administration procedures produce "addiction-like" behaviors and neuroadaptations that are distinct from those associated with limited access "drug-taking" behaviors (Loweth et al. 2014; Wolf and Ferrario 2010; Deroche-Gamonet et al. 2004; Lesscher and Vanderschuren 2012; Ahmed 2010). In addition to modeling different aspects of addiction, self-administration models have been informed by concepts from learning theory and psychology in order to ask questions not only about drug-taking and drug-seeking behaviors, but also to determine how these behaviors can be influenced and modified by non-drug experiences (e.g., stressors) and stimuli in the environment that are associated with drug (i.e., Pavlovian-conditioned cues; referred to as drug cues throughout). For example, neuroadaptations induced by drug self-administration have long-term behavioral consequences that can be measured using rodent models of relapse or "reinstatement." These relapse tests involve "extinguishing" drug self-administration by omitting drug delivery for several sessions and then presenting the rat with stressors, drug cues, or the drug itself. The degree to which these stimuli "reinstate" drug-seeking behavior is a model of relapse severity. Not surprisingly, there are different neurobiological systems underlying these different forms of reinstatement, which are themselves distinct from those involved in drug-taking (Torregrossa and Kalivas 2008; Shalev et al. 2002).

2 Overview of Motivational Processes and Their Neurobiological Substrates

Shifts in behavioral and psychological responses to drugs and drug cues underlie a transition from controlled or causal drug use to addiction. Experimental evidence from human and animal studies shows that drugs produce alterations in mesocorticolimbic circuits that underlie drug-seeking and drug-taking behaviors, particularly within basal ganglia circuits (Fig. 1; Nestler 2005; Koob and Nestler 1997; Belin et al. 2009; Fields et al. 2007; Sesack and Grace 2010; Vanderschuren and Kalivas 2000; Pierce and Kalivas 1997; Wolf 1998; Steketee 2003). The basal ganglia can be divided into a *ventral subsystem* in which the nucleus accumbens (NAcc) receives dopaminergic input from the ventral tegmental area (VTA) and glutamatergic input from the prefrontal cortex and amygdala, and a *dorsal subsystem* (comprised of dorsolateral and dorsomedial subregions) which receives input from the substantia nigra. The NAcc and dorsal striatum project to the ventral pallidum and globus pallidus, respectively, and both have reciprocal connections with frontal cortical areas, including the prefrontal, anterior cingulate, and orbitofrontal cortices. These reciprocal connections form parallel cortical–striatal–pallidal–cortical loops (Humphries and Prescott 2010; Alexander et al. 1990; Haber 2003) that are intimately involved in the initiation and maintenance of drug-taking and drug-seeking behaviors (as well as other motivated behaviors). This system, particularly the ventral

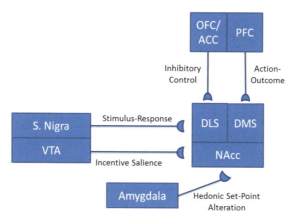

Fig. 1 An oversimplified diagram of key neural projections underlying motivational aspects of addiction. Note that this diagram does not reflect that these projections participate in multiple processes. For example, the amygdala is also involved in the hedonic response to drug rewards and in incentive–salience attribution to reward cues. Also, reciprocal connections between the cortex and midbrain areas are involved in all the concepts listed in the figure

subsystem, projects to and receives input from limbic areas including the amygdala and hippocampus (Schmidt et al. 2005; Berendse et al. 1992; Brog et al. 1993; Klitenick et al. 1992). Because of its connectivity with these emotion and learning centers, as well as the hypothalamus, the ventral subsystem of the basal ganglia is often referred to as a "functional interface" between motivation and action (Mogenson et al. 1980; Kelley et al. 2005). In comparison, the dorsal subsystem has been more tightly linked to the processing of goal-directed behavior and habit, which is discussed below (Belin et al. 2009; Yin et al. 2004; Faure et al. 2005). For further discussions of the role of these circuits in motivation, see chapters by Learning and Motivational Processes Contributing to Pavlovian–Instrumental Transfer and Their Neural Bases: Dopamine and Beyond, Multiple Systems for the Motivational Control of Behavior and Associated Neural Substrates in Humans, The Computational Complexity of Valuation and Motivational Forces in Decision-Making Processes, and Neurophysiology of Reward-Guided Behavior: Correlates Related to Predictions, Value, Motivation, Errors, Attention, and Action, in this volume.

Addiction is mediated by several different motivational processes that can be dissociated experimentally and neurobiologically. For example, the maintenance of drug-taking by negative reinforcement processes results from the removal of a negative drug withdrawal state by the drug (Koob and Le Moal 2001). This *hedonic allostasis* model also emphasizes that drugs produce a reward deficit in which previously pleasurable events now fail to produce the same hedonic effect. Thus, addiction is maintained by a need both to increase the amount of drug used in order to reach the new hedonic threshold and to overcome anhedonia (i.e., a generalized inability to feel pleasure) induced by repeated drug exposure. From a biological standpoint, this new set point is mediated by reduced activity within dopamine, opioid, and GABA neurotransmitter systems, while anhedonia is mediated by alterations in the function of the amygdala and corticotropin-releasing hormone (CRH) systems (Koob and Le Moal 2008; George et al. 2012) (Fig. 1). In comparison, the *incentive–sensitization* model of addiction (Robinson and Berridge 1993) proposes that in the addicted state, drugs and drug cues have a heightened ability to direct and influence behavior, ultimately focusing behavior toward drug-seeking and drug-taking. This concept also stresses that in the addicted state, drugs and drug cues are better able to direct behavior, not because they are pleasurable or "liked," but because they are highly noticeable, attractive, and "wanted" (for more detailed information on the distinct roles of wanting and liking in motivating behavior, please see Robinson et al., in this volume). Neurobiologically, the enhanced influence of drug cues over behavior (i.e., incentive sensitization) is linked to *increased* responsivity of mesolimbic dopamine systems, including inputs into this system from cortical areas (Fig. 1). Thus, alterations in dopamine- and glutamate-mediated transmission, particularly in the NAcc, drive incentive sensitization that underlies cue-triggered drug-seeking behaviors.

These first two models of addiction propose bivalent changes in brain reward systems to increase motivation: one in which the function of mesolimbic systems and hedonic responses are reduced and drug is needed to return to a "normal" state and one in which heightened responsivity of mesocorticolimbic systems to drug cues drives drug-seeking and drug-taking. These ideas describe different aspects of motivation that may be involved in different phases of addiction.[2] For example, hedonic allostasis may maintain drug-taking behaviors during early withdrawal or attempts to reduce drug use, whereas incentive sensitization may promote relapse even after physiological and emotional withdrawal symptoms have subsided.

Additional concepts of addiction focus on the relationship between behavioral actions (e.g., lighting a cigarette) and an outcome (e.g., smoking a cigarette). Within this framework, initial drug-taking involves learning that an action (drug-seeking) results in a given outcome (euphoria, anxiety reduction, etc.), but the nature of the relationship between the action and outcome changes as a consequence of drug use to produce addiction. Goal-directed (i.e., action–outcome) models posit that the expected drug outcome is improperly *overvalued* such that behaviors directed toward obtaining drug override all others. Similarly, drug-induced anhedonia may also contribute to this improper valuation through the underestimating of the value of alternative goals. On the other hand, habit-based models of addiction (i.e., stimulus–response) posit that after initial learning and repeated drug exposure, the outcome no longer drives drug-seeking behavior, but instead, drug-seeking is an *automated* response triggered by drug cues, emotional states, stress, etc. Like hedonic allostasis and incentive sensitization, goal-directed and habit models of addiction also propose that changes in mesolimbic function underlie these

---

[2]These models also disagree regarding the relationship between dopamine and hedonia, whereas an altered hedonic set point is proposed to be mediated in part by changes in dopamine; the "liking" reaction to rewards is dopamine independent (Berridge and Robinson 1998; Koob and Le Moal 2008). This discrepancy may lie in the definition of hedonia, which may be referred to generally as affect or pleasure, or operationalized into a specific subjective experience or behavioral measure.

behaviors. However, habit-based ideas emphasize a transition from behaviors mediated by the ventral subsystem of the basal ganglia to behaviors mediated by the dorsal subsystem (Fig. 1). In addition, goal-directed, habit-based, and incentive sensitization all incorporate the idea that the relationship between drug-seeking behavior and consumption of the drug itself changes as addiction develops, though they differ in the nature of the relationship that is thought to change. Below, we provide more specific details of the processes involved in each of these models described above as well as evidence in support of and contrary to them.

3 Hedonic Allostasis

The hedonic allostasis model of addiction (Koob and Le Moal 2001) is based on Solomon's opponent process theory, but has been modified to account for the chronic, pathological nature of drug-taking and relapse (Solomon and Corbit 1974). In Solomon's original work, he proposed that drug-taking consists of two dissociable processes: a short-lived *a-process*, which might include elevated heart rate, a state of well-being or euphoria (i.e., a hedonic affective state), or motoric effects (e.g., punding or chewing that are common with psychostimulants). At the same time, the drug also produces a *b-process* that has two key features: (1) It is in opposition to the *a-process* (e.g., reduced heart rate) and (2) it has a slower onset and offset relative to the *a-process*. The *b-process* can thus counteract aspects of the *a-process*. Solomon argued that the *b-process*, because it is longer lasting than the *a-process*, could explain the phenomenon of drug withdrawal. Further, he suggested that the *b-process* becomes larger after repeated drug administration. This change in the *b-process* has two effects: first, to further counteract the *a-process* (resulting in drug tolerance and a need to increase drug intake to compensate) and second, to worsen the withdrawal syndrome.

The hedonic allostasis model adds to this idea by positing that a portion of the *b-process* involves a reduction in the hedonic affective state (to counter euphoria induced by drug) that is maintained after drug use is discontinued. This maintained anhedonic *b-process* changes the "set point" for activation of the *a-process* such that the person remains in a relatively anhedonic state in the absence of drug. Koob and colleagues call this "allostatic state ... a new equilibrium, a state of chronic deviation [from the] normal (homeostatic) operating level to a pathological (allostatic) operating level" (George et al. 2012).[3] This anhedonic deficit creates a motivational drive to take drugs in order to reverse anhedonia (through negative reinforcement). However, because the set point for the desired *a-process* has changed, people must now take larger doses of the drug to activate the *a-process*

---

[3]Note that this is a deviation from McEwen's definition of allostasis, which states that allostasis consists of the adaptive physiological changes that are evoked in order to return a system to homeostasis. Thus, this Koob's allostatic state is better referred to as allostatic load, which McEwen defines as the long-term cost of allostasis that accumulates over time (McEwen 1998).

once more and overcome the enhanced *b-process*. In this manner, a shift from positive to negative reinforcement underlies drug addiction. In addition, the imbalance between *a-* and *b-processes* produces an unending "downward anhedonic spiral" where the *b-process* (anhedonia) continually grows stronger due to drug use and thus does not produce the desired *a-process* effect (euphoria to remove anhedonia). The motivational drive to relieve this hedonic deficit becomes ever stronger and leads to more drug-taking (Ahmed and Koob 1998; Koob and Le Moal 2001). Thus, drug withdrawal induces both physiological withdrawal (e.g., shakes in alcoholism or temperature dysregulation in the case of opioids) and "hedonic" withdrawal (i.e., anhedonia and malaise).

Drug-induced alterations in *a-* and *b-processes* are mediated by within-systems adaptations, in which changes occur in systems responsible for initial drug effects, and between-systems adaptations, which involve the recruitment of additional systems not initially engaged by the drug (Koob and Le Moal 2008). Neurobiologically, within-systems changes mediating a reduction in the set point *a-process* are due to drug-induced reductions in mesolimbic dopamine. For example, acute administration of alcohol stimulates VTA dopamine input into the NAcc (Brodie et al. 1999; Meyer et al. 2009; Nimitvilai et al. 2012; Gonzales et al. 2004). However, repeated alcohol exposure reduces the activity of VTA neurons (Diana et al. 1993; Shen 2003; Shen et al. 2007), including during withdrawal from chronic alcohol drinking (Bailey et al. 1998). This is consistent with the idea that increased alcohol intake is needed to restore reduced function of the mesolimbic system to prealcohol exposure levels. In another study, Barak et al. (2014) used a chronic intermittent alcohol procedure to induce escalated intake; this escalation was associated with a decrease in the firing rate of VTA dopamine neurons. When administered glial-derived neurotrophic factor (GDNF), the escalation of alcohol intake and the decrease in neural firing rates were blocked. These data suggest that normalizing an alcohol-induced deficit in VTA neuron firing (and by extension NAcc dopamine levels) resulted in a reversal of escalated alcohol intake. Additional evidence for *a-process* set point alterations comes from animal studies of intracranial self-stimulation (ICSS), in which lever pressing is reinforced by electrical stimulation of the medial forebrain bundle (Edmonds and Gallistel 1974; Miliaressis et al. 1986). ICSS reward thresholds are defined as the intensity of stimulation needed to support self-stimulation; *increases* in ICSS thresholds are interpreted as a need for stronger stimulation to achieve the same level or reinforcement (Diana 2011). Indeed, chronic exposure to many drugs increases ICSS thresholds (Kenny et al. 2006; Epping-Jordan et al. 1998; Kenny et al. 2003; but see Kenny and Markou 2006). Importantly, increased ICSS thresholds are long-lasting and persist even during protracted abstinence (Ahmed et al. 2002; Kenny et al. 2003). ICSS activates dopamine neurons (Yeomans 1989), and maintenance of ICSS requires activation of the hypothalamus (Kempadoo et al. 2013; Olds 1962). Thus, these experiments match electrophysiological data from VTA neurons supporting the hypothesis that a reduction in the activity of dopamine neurons drives the escalation of drug intake, but also engage additional hypothalamic circuits and transmitter systems.

Between-systems *b-process* adaptations that lead to enhanced anhedonia after repeated drug use may involve CRH within the extended amygdala as well as the dynorphin–κ opioid system (George et al. 2012). CRH receptor antagonists block withdrawal-induced anxiety (Basso et al. 1999; Logrip et al. 2011), withdrawal-induced increases in nicotine and alcohol self-administration (George et al. 2007; Finn et al. 2007) and escalated cocaine and heroin self-administration during extended access tests (Specio et al. 2008; Greenwell et al. 2009). Further, blockade of κ receptors reduces negative hedonic aspects of drug withdrawal and decreases drug-seeking (Koob 2013; Koob et al. 2014; Chartoff et al. 2012). After extended drug intake, input from CRH and dynorphin systems into the striatum is strengthened (e.g., see Turchan et al. 1997), leading to a further increase in the anhedonic *b-process*. Ultimately, these changes in *b-processes* are mediated by between-systems adaptations in CRH that lead to a further masking of the *a-processes* mediated by within-systems adaptations in mesolimbic dopamine systems and the persistence of drug-taking behavior (Aston-Jones et al. 1999; Nestler 2001; Koob 2003; George et al. 2012).

The negative reinforcement process of the hedonic allostasis model can account for drug-taking during early and late phases of addiction (especially in the case of alcoholics who can experience shakes and life-threatening seizures if alcohol use is stopped abruptly) and addresses why drug users may continually increase the amount of drug used. While the physiological, flu-like symptoms of withdrawal for most commonly abused drugs (shakes, nausea, malaise) subside within two to three weeks after drug use is discontinued, the hedonic effects, including mood alterations that manifest as depression and anxiety, may last longer (Soto et al. 1985; Ashton 1991; Rothwell et al. 2012; Hughes 2007; Satel et al. 1993). This may explain why many recovering addicts relapse weeks to months after drug use has stopped and when physical withdrawal symptoms are absent. However, it is important to note that drug-taking behavior can occur even in the absence of subjective, hedonic effects (Fischman and Foltin 1992; Fischman 1989; Lamb et al. 1991; Robinson and Berridge 2001), suggesting that the reinforcing properties of drugs are not mediated by their hedonic impact alone (Robinson and Berridge 2001). In addition, drug-seeking can often be context specific (e.g., only occurring in particular places) and triggered by drug cues (like hearing the click of a lighter when trying to quit smoking). An increase in hedonic set point is intrinsic and thus should not be modified by drug cues and drug contexts (Crombag et al. 2000, 2001; Badiani 2013). Thus, the hedonic allostasis model does not explain cue-induced reinstatement, context-dependent self-administration behaviors, or craving reported by recovering addicts when faced with a drug cue (Fox et al. 2005; Witteman et al. 2015; Celentano et al. 2009; Bossert et al. 2013). Further, a decrease in the *a-process* would be expected to generalize to other types of rewards, because there is a partial overlap in the neural circuits underlying the responses to "natural" rewards, such as food and sex, and drugs. However, it is not clear whether addicts are impaired in their ability to experience pleasure from non-drug sources (DiLeone et al. 2012; Hone-Blanchet and Fecteau 2014; Ziauddeen and Fletcher 2013; Volkow et al. 2012). Thus, the idea that drug history produces a generalized, long-lasting anhedonia cannot explain all instances of addiction, and is unlikely to drive relapse after long-term abstinence.

4 Incentive Sensitization

The incentive–sensitization concept of addiction proposes that addiction is due to adaptations in brain regions within the mesocorticolimbic systems that mediate a specific aspect of "incentive motivation": the process of "incentive–salience attribution" (Robinson and Berridge 1993, 2001; also see Robinson et al., this volume). Drug-induced neuroadaptations result in the sensitization of incentive–motivational processes such that drugs and drug-related stimuli acquire enhanced (sensitized) salience. Ultimately, it is sensitization of the incentive–motivational properties of drugs and drug-related stimuli that govern compulsive pursuit of drug.

Incentive motivation refers to a multi-component processes by which behavior is directed toward "incentives" such as food, a potential mate, drugs, and stimuli associated with these incentives through Pavlovian learning. During early drug use, the activation of neural systems that mediate pleasure triggers incentive–motivational processes that become progressively stronger across repeated drug use. An association between the act, event, or cues in the environment in which the pleasure occurs is then formed through Pavlovian learning processes. These Pavlovian cues can themselves gain motivational properties through the separate process of *incentive–salience attribution*. In this process, *after* the initial association has been formed, salience is then attributed to the mental representation of the reward/drug-associated cue. Incentive sensitization posits that the process of incentive–salience attribution occurs in a neural system that is separable from the system responsible for the pleasurable effects of the initial incentive (liking) and separable from the neural systems responsible for the initial Pavlovian associative learning processes. Thus, cues that have incentive salience are able to direct behavior, not because they are pleasurable or "liked," but because they are highly noticeable, attractive, and "wanted." Within this conceptual framework, addiction develops and persists due to drug-induced neuroadaptations in the brain systems that mediate incentive salience (wanting), making these systems hypersensitive (sensitized) to drugs and drug cues. Further, sensitization of brain circuits that mediate incentive salience, a component of incentive motivation, results in drug-seeking behavior. The sensitization within these neural circuits persists even after long periods without drug use, making addicts susceptible to relapse long after drug use has ceased. Furthermore, progressive alterations in incentive salience through associative learning mechanisms engaged by repeated drug-taking experiences enhance the overall incentive–motivational properties of drug cues leading to increasingly compulsive patterns of drug-seeking and drug-taking. Rodent studies have shown that "liking" and "wanting" processes can be dissociated neurobiologically (Berridge and Kringelbach 2015). This is discussed further by Robinson et al., in this volume.

Neurobiologically, incentive sensitization has traditionally been attributed to progressive enhancement (i.e., sensitization) of mesolimbic responses to drugs and drug cues (Berridge and Robinson 2003; Di Chiara et al. 1999; Wise 1987, 1988). Many studies have focused on the role of dopamine projections from the VTA to the NAcc in this process (Vanderschuren and Kalivas 2000; Flagel et al. 2008; Vezina 2004; Vezina and Leyton 2009), as well as the contribution of enhanced glutamatergic transmission particularly within the NAcc (Wolf and Ferrario 2010; Loweth et al. 2014; Kalivas and Nakamura 1999; Vanderschuren and Kalivas 2000; Cardinal et al. 2002). The focus on dopamine systems arose because many commonly abused drugs share the ability to enhance dopaminergic neurotransmission within the mesocorticolimbic system (Vanderschuren and Kalivas 2000; Vezina 2004; Wolf 2002) and to produce psychomotor sensitization (Babbini and Davis 1972; Joyce and Iversen 1979; Kita et al. 1992; Benwell and Balfour 1992; Littleton 1998; Nestby et al. 1997), an enduring hypersensitivity to the psychomotor activating effects of drugs that includes locomotion and stereotyped sniffing and rearing behaviors (Robinson and Becker 1986; Segal 1975).

The development of psychomotor sensitization is a progressive, incremental process, much like the development of addiction itself. Furthermore, psychomotor sensitization can persist for months to years after drug use has stopped (Paulson et al. 1991; Robinson and Becker 1986). Thus, incentive sensitization underlying addiction is thought to be mediated by sensitization-related adaptations in brain systems that mediate the incentive–salience and incentive–motivational properties of drug and drug cues.[4] Consistent with this, drug pretreatments that produce psychomotor sensitization facilitate the acquisition of drug self-administration in the monkey, rat, and mouse (Woolverton et al. 1984; Horger et al. 1990; Lessov et al. 2001), increase motivation for drug assessed by breakpoint on a progressive ratio schedule (Mendrek et al. 1998; Lorrain et al. 2000; Deroche et al. 1999), and enhance the escalation of drug intake in extended access models (Ferrario and Robinson 2007). Thus, psychomotor sensitization is linked to enhanced motivation for drugs and with addiction-like behaviors (Ferrario et al. 2005; Morgan and Roberts 2004 though see also Ahmed and Koob 1998). The incentive–sensitization model additionally proposes that drug-induced increases in incentive salience contribute to aberrant motivation underlying addiction. In support of this, drug cues themselves can support conditioned approach behavior (Meyer et al. 2012b; Uslaner et al. 2006; Yager and Robinson 2013), and pretreatment with a sensitizing regimen of amphetamine enhances the subsequent acquisition of Pavlovian-conditioned approach behavior (Harmer and Phillips 1998).

---

[4]Note that neuroadaptations accompanying *psychomotor* sensitization are thought to overlap with the neuroadaptations underlying the *incentive* sensitization that drives drug-seeking and drug-taking behavior. However, psychomotor sensitization and incentive sensitization are neurobiologically dissociable processes that mediate different aspects of behavior. Thus, the presence of psychomotor sensitization is indicative of changes underlying incentive sensitization, but they are not one and the same.

Although VTA-NAcc circuitry is heavily involved in psychomotor and incentive sensitization, alterations in glutamatergic transmission within the NAcc mediated by both cortical and amygdalar inputs play important roles (Ma et al. 2014; MacAskill et al. 2014; Everitt and Wolf 2002). For example, cocaine self-administration increases glutamatergic transmission in the NAcc, drug exposure increases the number of excitatory inputs from the basolateral amygdala to the NAcc shell, and remodeling of glutamatergic synapses from the PFC to NAcc contributes to cue-triggered drug-seeking behavior (MacAskill et al. 2014; Wolf 2003; Cornish et al. 1999; Wolf and Ferrario 2010; Robinson and Kolb 2004; Ma et al. 2014). Furthermore, the central nucleus of the amygdala (CeN) is necessary for Pavlovian-conditioned approach to reward predictive incentive stimuli (Everitt et al. 2003), and activation of dopaminergic input into the CeN is sufficient to elicit approach to drug cues (Cardinal et al. 2002).

The induction and expression of psychomotor sensitization is modulated by learning and the circumstances surrounding drug administration (Robinson et al. 1998). This concept is particularly important to the incentive–sensitization model of addiction because it posits that associative learning processes focus "wanting" onto drug cues enabling the expression of sensitization-related neuroadaptations to be influenced by factors that contribute to associative learning (Robinson and Berridge 1993). For example, even when neural sensitization is induced by repeated drug exposure, the expression of sensitization can be context specific. That is, if the drug is given in an environment that is different from where initial drug treatments were received, the sensitized response is not evoked (Post et al. 1981; Anagnostaras and Robinson 1996; Terelli and Terry 1999). However, these same animals will express a sensitized response when given drug in the initial treatment environment (Anagnostaras and Robinson 1996; Post et al. 1981). Thus, although neural sensitization occurred, the behavioral expression of sensitization is context dependent, i.e., it is influenced by conditioned stimuli paired with drug administration (Anagnostaras and Robinson 1996). Conversely, animals placed in a drug-paired environment will show conditioned locomotor hyperactivity in the absence of any drug (see Stewart 1992 for review). Enhanced incentive salience of drug cues and increased incentive–motivational power of these cues in rodent models are consistent with behavioral studies in human drug addicts, as well as self-reports of enhanced cue-triggered craving during abstinence (Childress et al. 1993; Gawin and Kleber 1986; for review, see Veilleux and Skinner 2015). In addition, incentive sensitization accounts for the context dependence of drug craving. More recent studies of incentive sensitization have focused on individual differences in the attribution of incentive salience in order to address why some individuals are able to maintain controlled drug use, while others develop addiction (Robinson et al. 2014; Meyer et al. 2012a).

In summary, at the heart of incentive sensitization is the concept that drugs produce neuroadaptations in systems that mediate the attribution of incentive salience to drug cues. Through these sensitization-related processes, Pavlovian cues associated with drug-taking become increasingly able to direct and focus behaviors, inducing cue-triggered motivational states that drive pathological drug "wanting" and addiction. Several other concepts of addiction presented below incorporate the idea that drug cues influence drug-seeking behaviors. However, they differ from incentive sensitization in that not all concepts include the induction of a motivational state by Pavlovian cues.

5 Addiction as a Disorder of Learning and Decision Making

In the above sections, we discussed ideas about addiction that focused on hedonic responses to drug- and emotion-driven states, versus conditioned incentive–motivational states elicited by drug cues. Addiction can also be described in the context of aberrant decision-making processes in which the drug becomes irresistible and leads to persistent drug-taking even in the face of negative consequences. Decision-making conceptualizations of motivated behavior are based on the idea that there is some value, either real or expected, to the outcome of a particular action. These actions have potential costs and rewards, and individuals vary in their ability to estimate these costs and rewards, just as they differ in how much they value these costs and rewards in the first place. Thus, making a decision to engage in some action depends on the individual's ability to weigh out the costs, benefits, and consequences of their actions. There are at least two strategies in forming this appreciation. In the *forward-search* strategy, the outcome of each possible action is independently considered every time the person makes a decision. In *caching*, the person learns essentially through trial and error that a particular action in a given situation is likely to produce a certain outcome; then, the appropriate action is chosen when that particular situation is encountered. Thus, the forward-search strategy involves planning, is flexible, and is sensitive to changes in the action–outcome relationship, while the cache strategy is inflexible, is fast, and leads to habitual responding (Redish et al. 2008; also see Redish et al., in this volume). In many instances, these strategies occur sequentially; the initial phase of motivated decision making involves forward-search planning and evolves later into a caching strategy. These forward-search and caching strategies are analogous to reflective (goal-directed) control and reflexive (habitual) control of decision-making processes (Balleine and Dickinson 1998; Dolan and Dayan 2013; Doll et al. 2012).

Many studies of drug addiction consider the transition from forward-search goal-directed behavior to caching or habitual behaviors a hallmark of addiction (Corbit et al. 2012; Zapata et al. 2010; Everitt and Robbins 2015; Redish et al. 2008). Below, we describe general processes involved in goal-directed and habit-based responding, and how alterations in these processes may contribute to addiction. The main distinction between these two ideas is that for goal-directed behavior, the outcome (i.e., the subjective or physiological drug experience) maintains addiction, whereas for habit-based responding, the outcome itself is no longer relevant; drug-taking behavior is instead maintained by automated processes elicited by drug cues, stress, etc.

5.1 Goal- Versus Habit-Based Responding

During initial drug-taking experiences, an individual learns that an action (drug-seeking) produces some outcome (drug experience). This action–outcome relationship, or *A-O* relationship, can be dependent on the presence of environmental stimuli (or, as the notation goes, *S:A-O*). In other words, the individual learns that drug-seeking actions produce drug-taking outcomes only in the presence of certain stimuli such as drug cues. As addiction develops, the drug-taking outcome, either actual or expected, becomes *overvalued*. This overvaluation can be a result of drug withdrawal (Hutcheson et al. 2001), or an inflated *expectation* of the euphoric effects of the drug, even when the euphoric event is actually decreased in magnitude (Kennett et al. 2013). Overvaluation of the drug-taking outcome (*O*) can lead to pathological drug-seeking (*A*), especially when triggered by drug cues (*S*). In this manner, goal-directed responding can lead to addiction when the expected drug outcome is improperly overvalued such that behaviors directed toward obtaining drug override all others.

In contrast, habit-based ideas of addiction suggest that a drug-seeking action is an automated response (*R*) elicited by a drug cue (*S*). This stimulus–response, or *S-R*, learning process leads to habitual drug-seeking and drug-taking, despite changes in the outcome of the drug-seeking action. Further, habitual, *S-R* drug-taking behaviors are automatically elicited in the absence of craving or urges, and occur independently of antecedent motivational states or the actual outcome of drug use itself (Hogarth et al. 2013; Belin et al. 2013; Ostlund and Balleine 2008; Everitt et al. 2001; Tiffany 1990). As frequency of use increases, more automated and routine actions are prompted that can further perpetuate drug-taking. These automatic drug-taking behaviors are initiated by drug cues and have been conceptualized as "incentive habits," or habit processes prompted by incentive stimuli (Pavlovian cues) that have undergone incentive sensitization (Belin et al. 2013). Importantly, incentive habits differ from incentive motivation because cue-driven *S-R* responses occur in the absence of an explicit motivational state, whereas Pavlovian stimuli with incentive motivational properties induce motivational states that produce drug-taking.

S-R habits are further characterized by the blunting of executive function such that conscious decision making (i.e., ability to weigh costs, benefits, and potential consequences) becomes increasingly compromised. A decline in function of executive control processes further enhances habit-based behaviors and reduces outcome-based decision-making processes (Jentsch and Taylor 1999). These features of habitual use develop over prolonged drug use, further drive drug-taking behavior, and likely occur in parallel with the other features of addiction. Therefore, through learning routine drug-taking patterns, automatic drug-use behaviors crystallize, becoming highly efficient and repetitive behaviors that are resilient to intervention.

In humans, many aspects of drug-taking behavior can be categorized as habitual, such as the tapping and packing of a pack of cigarettes, or the preparation of a needle, which can occur in an automated fashion analogous to driving the same route to work each day. The automation of complex behaviors can be adaptive, but these processes can also be usurped by drugs. Thus, habitual behaviors in the addicted state are aberrant not because these behaviors have become automated, but because these behaviors are fixed in a habitual state and do not move back into a goal-directed system (Everitt and Robbins 2005; Dickinson et al. 2002; Miles et al. 2003; Robbins and Everitt 1999). However, some aspects of human addiction do not fit as well with habit-based ideas of drug-taking. For example, addicts often engage in novel and even creative behaviors in order to maintain their drug use (e.g., a prescription opioid pill abuser that switches from oral administration to intravenous street drug use). This transition requires an entirely different set of actions that were never previously used to obtain drug and thus do not fit well with an automated, habitual behavioral response.

Another layer can be added to these concepts of goal-directed and habitual behaviors; that is the ability of Pavlovian-conditioned motivational states[5] (as described above in the Incentive Sensitization section) to invoke the retrieval of specific outcomes, such as drug-specific effects. Thus, while addiction can transition from *A-O* to *S-R*, addiction may also be marked by a transition from the ability of Pavlovian stimuli to elicit the representations of specific drug outcomes to elicit general motivational states (Hogarth 2012; Hogarth et al. 2013). As discussed next, these instrumental and Pavlovian processes are subserved by different neurobiological systems and thus may interact to produce craving and drug addiction.

5.2 *Neural Circuits Governing Goal- and Habit-Based Decision Making*

Studies of food reinforcement show that prefrontal cortical inputs to the dorsomedial section of the striatum are critical for the formation of associations between

---

[5]The notation for action–outcome (*A-O*) and stimulus–response (*S-R*) responding can be confusing because the drug-seeking action is denoted as either *A* or *R* in each of these conceptualizations. Indeed, notations such as $\xrightarrow{a} O$ (action–outcome), $S \xrightarrow{a} O$ (action–outcome in the presence of environmental stimuli), and $S \xrightarrow{a}$ (stimulus–response) are sometimes used to avoid this confusion (Redish et al. 2008). The notation for the activation of a conditioned motivational state by Pavlovian cues is $[SO] \xrightarrow{a}$, where *O* is the conditioned motivational or affective state elicited by drug stimuli *S* and *a* is the behavioral response elicited by this state. However, we avoid this notation because *O* has quite different meanings depending on whether it is the outcome of an action (as in *A-O*) or the motivational state elicited by a Pavlovian stimulus (as in is $[SO] \xrightarrow{a}$).

actions and outcomes (Shiflett and Balleine 2011; Balleine 2005). The neurochemical basis for goal-directed responding may be related to dopamine signals, where drug-induced dopamine release maladaptively increases the value of the outcome (i.e., drug intake; Redish 2004; Naude et al. 2014; Belin et al. 2009). Because the likelihood of engaging in a drug-seeking action is a function of the perceived value of the outcome (O), this "hijacking" of the dopamine system would increase drug-seeking progressively. A change in the valuation of an outcome as a result of drug exposure involves alterations in the amygdala and hippocampus (Wassum et al. 2011; Mahler and Berridge 2012), and the retrieval of this now altered outcome likely involves a circuit connecting areas of the prefrontal cortex (including the orbitofrontal cortex) and the NAcc (Shiflett and Balleine 2010; Jentsch and Taylor 1999; Puumala and Sirvio 1998; Simon et al. 2013; Winstanley et al. 2004; Johnson et al. 2007; Schoenbaum et al. 2006; Kalivas and Volkow 2005).

As described above, initial drug-seeking responses are mediated by dopamine transmission in the ventral striatum. However, as goal-directed drug-taking transitions to habitual drug-taking, prolonged activation of dopamine neurotransmission triggers a transition to behavior mediated by the dorsolateral striatum (Balleine et al. 2014; Yin et al. 2006). This transition from ventral to dorsal striatal activation reduces goal-directed behaviors and supports habit learning such that behavior is driven by drug cues, but no longer requires a representation of or motivational state driven by the outcome (Corbit et al. 2012; Vanderschuren et al. 2005; Willuhn et al. 2012). Dorsal and ventral regions of the striatum are connected through a complex "spiraling" or "looping" of reciprocal connections to the VTA and substantia nigra that allow one striatal subregion to regulate dopaminergic tone in adjacent regions. For example, inputs from the NAcc shell to the VTA can inhibit dopamine release within the NAcc core (Belin and Everitt 2008). Similarly, the NAcc core regulates dopaminergic tone within the dorsal striatum through the substantia nigra. The dorsal striatum is divided into multiple domains as well, and dopaminergic input is regulated through reciprocal nigrostriatal projections that are organized anatomically from ventral to dorsal. Each domain regulates the dopaminergic tone of the adjacent domain and, through their spatial location, produces a "spiral" of reciprocal connections that can regulate dopaminergic tone throughout the striatum. This anatomical arrangement allows for the transition from goal-directed behaviors mediated by the NAcc to habit-based behaviors mediated by the dorsal striatum. Specifically, initial exposure to a drug would induce dopaminergic responses and neuroadaptations in the ventral striatum, but because of the spiraling reciprocal connections, these responses and neuroadaptations would engage and alter dorsal striatal projections during extended drug-taking (e.g., Porrino et al. 2004; Letchworth et al. 2001; Ito et al. 2000, 2002; Belin and Everitt 2008). Consistent with the idea of a transition between goal-directed and habit-based processes, an intact connection between NAcc core and the dorsal striatum is required to elicit habit behaviors (Belin and Everitt 2008). The involvement of the dorsal striatum in habit learning is also dynamic throughout prolonged drug use, with activity shifting from the dorsal–medial striatum (DMS) to the dorsal–lateral striatum (DLS) in driving habit-based, stimulus–response behaviors (Murray et al. 2012). Consistent with this, inactivation of the DLS after prolonged cocaine exposure renders rats sensitive to changes in goal value and hence restores goal-directed behavior while attenuating automated stimulus–response processes (Zapata et al. 2010). Thus, reciprocal connections between these circuits allow for transitions between habitual and goal-directed behaviors as needed, whereas drug-induced alterations within these circuits perturb this flexibility. This is further compounded by drug-induced cortical dysregulation that compromises executive function and planning.

5.3 Pavlovian Cues: Reward Prediction, Incentive Salience, and Transfer to Instrumental Responding

Pavlovian cues play an important role in these processes by triggering sequences of actions that can be goal-directed and habitual, or by inducing a conditioned motivational state (incentive motivation). The ability of cues to guide goal-directed behavior likely involves the dopamine system as well, in which changes in dopamine neuron activity signal *prediction* error—outcomes that have a perceived value that was either greater or less than expected based on the presence of reward-predicting cues (Schultz et al. 1997). Therefore, increases in dopamine neuron activity, and the resultant increases in dopamine (e.g., Hart et al. 2014), would lead to more rapid associations between drugs and their cues. However, dopamine is also involved in the attribution of incentive salience to drug cues (Flagel et al. 2011; Saddoris et al. 2015). In other words, drug-associated cues acquire incentive salience as a consequence of drug-induced increases in dopamine. The incentive motivational state induced by these cues can then transfer to instrumental drug-taking, thereby increasing the drug-seeking and drug-taking behavior. These Pavlovian transfer effects during food self-administration are dependent on the orbitofrontal cortex (Ostlund and Balleine 2007a, b), while the devaluation of an instrumental outcome during A-O learning is dependent on the prelimbic cortex (Corbit and Balleine 2003). However, whether Pavlovian transfer effects during drug-taking also involve these brain areas is not known.

6 Impulsive Action and Impaired Executive Function

Layered on top of the concept of enhanced goal-directed behavior described above is the contribution of impulsivity and impulsive action. One consequence of the overvaluation of a drug is impulsive responding in which individuals (1) value immediate drug delivery over more adaptive, but long-term, outcomes and/or (2) value less probable drug availability over more certain, alternative outcomes. These patterns of responding could be due to the overvaluation of the drug itself and/or to altered discounting of the temporal delay between the action and receipt of the outcome, the probability of drug availability, or the negative consequences of drug-taking. This aberrant outcome valuation, in the form of impulsivity, may explain the correlation between impulsive behavior and drug-taking (Moeller and Dougherty 2002; de Wit and Richards 2004; Perry and Carroll 2008). However, impulsivity may also reflect a loss of inhibitory control over drug urges, including those elicited by drug cues (Jentsch and Taylor 1999; Belin et al. 2013). In this case, in contrast to overvaluing drug outcomes, individuals are simply unable to make choices based on long-term outcomes (Bechara 2001, 2005).

The crystallization of habit appears to be further exacerbated by cortical dysregulation, particularly in the prefrontal areas involved in planning (Volkow et al. 2003). Under normal circumstances, habitual behaviors can be inhibited by cortical systems. However, repeated drug use also produces neural alterations in prefrontal cortical systems that exacerbate habitual behaviors. These adaptations in human addicts occur within the orbitofrontal cortex and anterior cingulate cortex (Lubman et al. 2004; van Huijstee and Mansvelder 2014). In addition, dysregulations in these regions are implicated in other behavioral control disorders involving impulsivity, evaluation and planning, and inhibitory control (Verdejo-Garcia et al. 2006, 2007; Sakagami et al. 2006). Dampened cortical activation and reduced executive function after repeated drug exposure may explain the addict's failure to control habitual behaviors as well as the inflexibility in response to changing outcomes (Jentsch and Taylor 1999). Together, the loss of prefrontal inhibitory control and consolidation of dopaminergic signaling in the dorsal striatum are thought to perpetuate responses to drug cues by bypassing goal-directed behaviors entirely, and prompting drug use in a stimulus-dependent manner. These changes may also explain the ineffectiveness of cognitive control approaches to reducing drug-taking behaviors.

7 Models that Integrate Motivational Concepts

7.1 Dual-Process Models

Dual-process models of drug addiction describe drug-taking behavior as the output of two competing forces: an automatic, impulse-generating system that is approach-oriented and generates drug-taking behavior, and a controlled, reflective system that uses cognitive control to regulate the impulse-generating system. For example, in the case of alcoholism, Wiers et al. (2007) argues that alcohol-use disorders occur from an imbalance between automatic and controlled processes. In adolescence, where much of the evidence for this model is derived, this controlled process is weak—as any parent can attest. Upon repeated alcohol exposure, the appetitive/approach system is sensitized, and recreational users are attracted to and influenced by drug-associated cues (Duka and Townshend 2004; Field et al. 2005).

In contrast, the controlled process degrades as a result of chronic alcohol use (White et al. 2000). Thus, adolescents with already weak control systems are particularly prone to the further sensitization and weakening of approach and control systems, respectively.

While these ideas may seem analogous to the habit-based and goal-directed response systems described above, they should not be conflated. Instead, incentive motivation, action–outcome systems, and habit-based systems together produce drug-seeking behaviors. However, because each process can be localized to some extent to different neural circuits, they can be independently modified by drug exposure. Specifically, cue-triggered behaviors rely heavily on the ventral and dorsal striatum, as well as prelimbic and orbitofrontal cortices (Daw et al. 2011; de Wit et al. 2009, 2012). These systems are in turn under negative control of a regulatory gating system mediated by cortical regions. This cortical regulatory system can then influence whether forward-search or cached response patterns (discussed earlier in this section) are employed. While such an arbiter of response strategy selection has not been identified, it may involve the infralimbic and dorsolateral components of the prefrontal cortex, as manipulations of these regions can induce switches from one strategy to the other (McClure and Bickel 2014; Jentsch and Taylor 1999; Coutureau and Killcross 2003; Killcross and Coutureau 2003; Daw et al. 2005).

7.2 Transitions from Drug-Taking to Addiction

Given the neural overlap and interactions between the different motivational concepts described above, it seems likely that these concepts are explaining different aspects or stages of addiction (Piazza and Deroche-Gamonet 2013). For example, several have suggested that transition into addiction is characterized by a slow yet categorical change in the motivational processes underlying drug-taking. Koob and colleagues have argued that addiction transitions from positive reinforcement, to incentive sensitization, to allostasis (Koob et al. 2014). Goldstein and Volkow (2002) propose a model, based on neuroimaging data, where the intoxicating effects of drugs initially involve the frontal cortex, particularly the anterior cingulate and prefrontal cortices. As addiction develops, the degree of incentive motivation attributed to drugs (as measured by drug craving) is paralleled by activity in the orbitofrontal and anterior cingulate cortices.

7.3 Drug Addiction: A Compendium of Vulnerabilities

Redish (2004) and Redish et al. (2008) have utilized the goal-directed and habit distinction to describe how "vulnerabilities" in these decision-making processes can lead to addiction (Redish et al. 2008). Redish describes 10 such vulnerabilities or "failure points" in which the decision-making process can become maladaptive. These failure points can be grouped into general categories, including *negative reinforcement*, *positive reinforcement*, *learning processes (including reward prediction and action–outcome learning)*, *response inhibition and discounting*, and *habit*. However, it should be mentioned here that even though we have grouped these vulnerabilities together for discussion purposes, they are to an extent neurobiologically dissociable and thus represent separate psychological processes.

Each of these failure points can lead to addiction. For example, a person may form incorrect action–outcome relationships, or improperly categorize situations in which action–outcome relationships occur. They would thus develop an "illusion of control" in which they view the success in achieving a positive outcome as result of their own actions, or because of changing situations, even if this is not the case. In this case, they would be less likely to select alternatives to drug-taking. For habit-based system vulnerabilities, behavior becomes inflexible, and insensitive to changes in the consequences of behavioral actions. While a non-addict may be able to switch from habitual to goal-directed responding without difficulty, an addict that suffers from improper response inhibition vulnerability will have difficulty doing so.

These vulnerabilities can be applied to positive and negative reinforcement as well. For negative reinforcement, the increase in "need" that happens as a result of withdrawal increases the probability that a person will take drugs to satisfy the need, and the allostatic load that accumulates (strengthened opponent process) is a result of the neuroadaptive response to repeated drug exposure. For positive reinforcement, the euphoric effect of drugs creates an intense reward signal that is stored in memory and becomes a goal to be sought after. Stimuli that induce recall of this euphoric event induce craving. Action–outcome vulnerabilities are also related to the reward signal. For example, incentive sensitization leads to a sought-after event, but the expected reward is much greater than the actual reward, and thus the distinction between "wanting" and "liking" arises, as discussed earlier. Impulsivity, then, is a similar improper valuation of future events in favor of choices that result in immediate outcomes. In this sense, Redish's conceptualization is unique in that there is no single "culprit," such as the habit system, that is responsible for the transition into addiction (Ahmed 2008; see chapters by Redish et al., and Cornwell et al., in this volume for further discussions of multiple vulnerabilities in motivational systems). Each of the predominant models of addiction explains one or more, but not all, of vulnerabilities to addiction. The advantage of describing multiple vulnerabilities is that it acknowledges that individuals likely become addicts for different reasons and that the biological underpinnings of drug-taking may depend on which vulnerability is being measured. Further, different drugs may be influenced by these vulnerabilities to different degrees. For example, opioid drugs may exploit the opponent process and positive reinforcement (reward-mimicking) vulnerabilities more than psychostimulant drugs, which may act more on the incentive salience and habit vulnerabilities. Nicotine, which facilitates learning by enhancing the reinforcing properties of cues (Caggiula et al. 2009), may drive drug-taking and relapse by enhancing initial learning about these cues.

8 Conclusion

There has been a great deal of focus on how personality traits or other pre-existing factors can influence drug-taking behavior, including sensation-seeking, impulsivity, and incentive–salience attribution (Flagel et al. 2009; Bickel et al. 2012; Bardo et al. 1996; Everitt 2014). These studies demonstrate that the predisposition to take drugs is not due to a single factor. Similarly, in this chapter, we have summarized some basic motivational concepts underlying drug-taking and how these processes are dysregulated as addiction develops. Behavioral neuroscience views largely ignore, or even avoid, the heterogeneity of motivational processes in addiction and are limited by studies in which only a single concept is tested. This point is illustrated in part by Fig. 2, in which we graphed the collaboration network of the references cited in this chapter. Each researcher cited is a node, and two researchers are connected if they have coauthored a paper together. A few prominent groups emerge that correspond to the main concepts discussed in the chapter. What is apparent is that there are very few connections between these groups and only a few researchers bridge these concepts. Of course, this is not a complete representation of the entire addiction field. However, the absence of cross-concept collaborations and integration is a major impediment to progress. Human studies strongly support the idea that different addicts take different drugs for different reasons and that these reasons may change throughout the course of the addict's lifetime. Thus, ideas including hedonic allostasis, incentive sensitization, decision making and learning, impulsivity, and dual-process theory are applicable in some but not all situations. For example, much of what is known about negative reinforcement and allostasis comes from studies on opiates and alcohol, while incentive sensitization stems primarily from studies of psychostimulants (though see Robinson and Berridge 2000). Therefore, it is perhaps not surprising that none of these theories explain all aspects of addiction. Furthermore, it is likely that the transition into addiction is not driven by a single process, and people may transition to compulsive drug-taking for different reasons, both behavioral and neural. We believe that a unified framework of addiction is not a red herring. Instead, a framework that is flexible and adaptable to different drug classes, personality types, and environmental situations will be enormously useful in the development of specialized treatment for the wide variety of addicted individuals in the population.

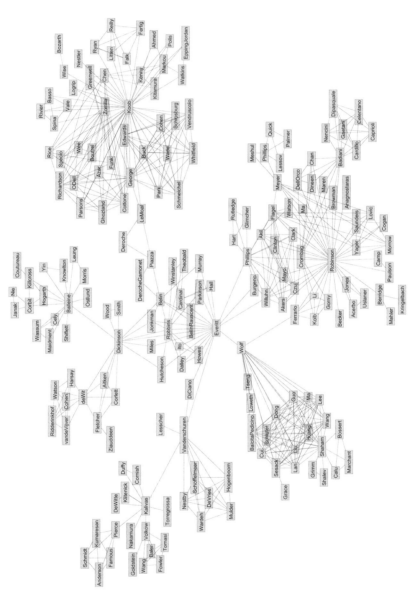

Fig. 2 Collaboration network of the references cited in this chapter. Each researcher cited is a node, and two researchers are connected if they have coauthored a paper together

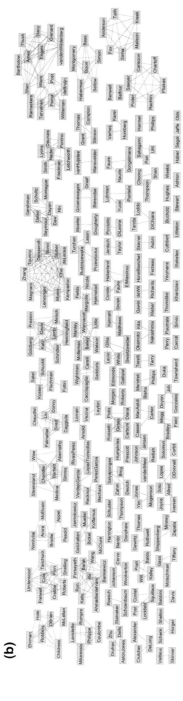

Fig. 2 (continued)

Acknowledgements The authors thank MEJ Newman for making the collaboration network depicted in Fig. 2.

References

Ahmed SH (2008) The origin of addictions by means of unnatural decision. Behav Brain Sci 31 (4):437–438

Ahmed SH (2010) Validation crisis in animal models of drug addiction: beyond non-disordered drug use toward drug addiction. Neurosci Biobehav Rev 35(2):172–184

Ahmed SH, Kenny PJ, Koob GF, Markou A (2002) Neurobiological evidence for hedonic allostasis associated with escalating cocaine use. Nat Neurosci 5(7):625–626

Ahmed SH, Koob GF (1998) Transition from moderate to excessive drug intake: change in hedonic set point. Science 282(5387):298–300

Alexander GE, Crutcher MD, DeLong MR (1990) Basal ganglia-thalamocortical circuits: parallel substrates for motor, oculomotor, "prefrontal" and "limbic" functions. Prog Brain Res 85:119–146

American Psychiatric Association (ed) (2013) Diagnostic and statistical manual of mental disorders: DSM-5

Anagnostaras SG, Robinson TE (1996) Sensitization to the psychomotor activating stimulant effects of amphetamine: modulation by associative learning. Behav Neurosci 110(6):1397–1414

Ashton H (1991) Protracted withdrawal syndromes from benzodiazepines. J Subst Abuse Treat 8 (1–2):19–28

Aston-Jones G, Delfs JM, Druhan J, Zhu Y (1999) The bed nucleus of the stria terminalis. A target site for noradrenergic actions in opiate withdrawal. Ann N Y Acad Sci 877:486–498

Babbini M, Davis WM (1972) Time-dose relationships for locomotor activity effects of morphine after acute or repeated treatment. Br J Pharmacol 46(2):213–224

Badiani A (2013) Substance-specific environmental influences on drug use and drug preference in animals and humans. Curr Opin Neurobiol 23(4):588–596

Bailey CP, Manley SJ, Watson WP, Wonnacott S, Molleman A, Little HJ (1998) Chronic ethanol administration alters activity in ventral tegmental area neurons after cessation of withdrawal hyperexcitability. Brain Res 803(1–2):144–152

Balleine BW (2005) Neural bases of food-seeking: affect, arousal and reward in corticostriatolimbic circuits. Physiol Behav 86(5):717–730

Balleine BW, Dickinson A (1998) Goal-directed instrumental action: contingency and incentive learning and their cortical substrates. Neuropharmacology 37(4–5):407–419

Balleine BW, Morris RW, Leung BK (2014) Thalamocortical integration of instrumental learning and performance and their disintegration in addiction. Brain Res. http://dx.doi.org/10.1016/j.brainres.2014.12.023

Barak S, Wang J, Ahmadiantehrani S, Ben Hamida S, Kells AP, Forsayeth J, Bankiewicz KS, Ron D (2014) Glial cell line-derived neurotrophic factor (GDNF) is an endogenous protector in the mesolimbic system against excessive alcohol consumption and relapse. Addict Biol 20:629

Bardo MT, Donohew RL, Harrington NG (1996) Psychobiology of novelty seeking and drug seeking behavior. Behav Brain Res 77(1–2):23–43

Basso AM, Spina M, Koob GF, Rivier J, Vale W (1999) Corticotropin-releasing factor antagonist attenuates the 'anxiogenic-like' effect in the defensive burying paradigm but not in the elevated plus-maze following chronic cocaine in rats. Psychopharmacology 145(1):21–30

Bechara A (2001) Neurobiology of decision-making: risk and reward. Semin Clin Neuropsychiatry 6(3):205–216

Bechara A (2005) Decision making, impulse control and loss of willpower to resist drugs: a neurocognitive perspective. Nat Neurosci 8(11):1458–1463

Belin D, Belin-Rauscent A, Murray JE, Everitt BJ (2013) Addiction: failure of control over maladaptive incentive habits. Curr Opin Neurobiol 23(4):564–572

Belin D, Everitt BJ (2008) Cocaine seeking habits depend upon dopamine-dependent serial connectivity linking the ventral with the dorsal striatum. Neuron 57(3):432–441

Belin D, Jonkman S, Dickinson A, Robbins TW, Everitt BJ (2009) Parallel and interactive learning processes within the basal ganglia: relevance for the understanding of addiction. Behav Brain Res 199(1):89–102

Benwell ME, Balfour DJ (1992) The effects of acute and repeated nicotine treatment on nucleus accumbens dopamine and locomotor activity. Br J Pharmacol 105(4):849–856

Berendse HW, Graaf YG-D, Groenewegen HJ (1992) Topographical organization and relationship with ventral striatal compartments of prefrontal corticostriatal projections in the rat. J Comp Neurol 316(3):314–347

Berridge KC, Robinson TE (1998) What is the role of dopamine in reward: hedonic impact, reward learning, or incentive salience?. Brain Res Brain Res Rev 28(3):309–369

Berridge KC, Kringelbach ML (2015) Pleasure systems in the Brain. Neuron 86(3):646–664

Berridge KC, Robinson TE (2003) Parsing reward. Trends Neurosci 26(9):507–513

Bickel WK, Jarmolowicz DP, Mueller ET, Koffarnus MN, Gatchalian KM (2012) Excessive discounting of delayed reinforcers as a trans-disease process contributing to addiction and other disease-related vulnerabilities: emerging evidence. Pharmacol Ther 134(3):287–297

Bossert JM, Marchant NJ, Calu DJ, Shaham Y (2013) The reinstatement model of drug relapse: recent neurobiological findings, emerging research topics, and translational research. Psychopharmacology 229(3):453–476

Brodie MS, Pesold C, Appel SB (1999) Ethanol directly excites dopaminergic ventral tegmental area reward neurons. Alcohol Clin Exp Res 23(11):1848–1852

Brog JS, Salyapongse A, Deutch AY, Zahm DS (1993) The patterns of afferent innervation of the core and shell in the "accumbens" part of the rat ventral striatum: immunohistochemical detection of retrogradely transported fluoro-gold. J Comp Neurol 338(2):255–278

Bucholz KK (1999) Nosology and epidemiology of addictive disorders and their comorbidity. Psychiatr Clin N Am 22(2):221–240

Caggiula AR, Donny EC, Palmatier MI, Liu X, Chaudhri N, Sved AF (2009) The role of nicotine in smoking: a dual-reinforcement model. Nebr Symp Motiv 55:91–109

Cardinal RN, Parkinson JA, Hall J, Everitt BJ (2002) Emotion and motivation: the role of the amygdala, ventral striatum, and prefrontal cortex. Neurosci Biobehav Rev 26(3):321–352

Celentano M, Caprioli D, Dipasquale P, Cardillo V, Nencini P, Gaetani S, Badiani A (2009) Drug context differently regulates cocaine versus heroin self-administration and cocaine-versus heroin-induced Fos mRNA expression in the rat. Psychopharmacology 204(2):349–360

Chartoff E, Sawyer A, Rachlin A, Potter D, Pliakas A, Carlezon WA (2012) Blockade of kappa opioid receptors attenuates the development of depressive-like behaviors induced by cocaine withdrawal in rats. Neuropharmacology 62(1):167–176

Childress AR, Hole AV, Ehrman RN, Robbins SJ, McLellan AT, O'Brien CP (1993) Cue reactivity and cue reactivity interventions in drug dependence. NIDA Res Monogr 137:73–95

Compton WM, Thomas YF, Stinson FS, Grant BF (2007) Prevalence, correlates, disability, and comorbidity of DSM-IV drug abuse and dependence in the United States: results from the national epidemiologic survey on alcohol and related conditions. Arch Gen Psychiatry 64 (5):566–576

Corbit LH, Balleine BW (2003) Instrumental and Pavlovian incentive processes have dissociable effects on components of a heterogeneous instrumental chain. J Exp Psychol Anim Behav Process 29(2):99–106

Corbit LH, Nie H, Janak PH (2012) Habitual alcohol seeking: time course and the contribution of subregions of the dorsal striatum. Biol Psychiatry 72(5):389–395

Cornish JL, Duffy P, Kalivas PW (1999) A role for nucleus accumbens glutamate transmission in the relapse to cocaine-seeking behavior. Neuroscience 93(4):1359–1367

Coutureau E, Killcross S (2003) Inactivation of the infralimbic prefrontal cortex reinstates goal-directed responding in overtrained rats. Behav Brain Res 146(1–2):167–174

Crombag HS, Badiani A, Chan J, Dell'Orco J, Dineen SP, Robinson TE (2001) The ability of environmental context to facilitate psychomotor sensitization to amphetamine can be dissociated from its effect on acute drug responsiveness and on conditioned responding. Neuropsychopharmacology 24(6):680–690

Crombag HS, Badiani A, Maren S, Robinson TE (2000) The role of contextual versus discrete drug-associated cues in promoting the induction of psychomotor sensitization to intravenous amphetamine. Behav Brain Res 116(1):1–22

Cuthbert BN (2014) The RDoC framework: facilitating transition from ICD/DSM to dimensional approaches that integrate neuroscience and psychopathology. World Psychiatry 13(1):28–35

Daw ND, Gershman SJ, Seymour B, Dayan P, Dolan RJ (2011) Model-based influences on humans' choices and striatal prediction errors. Neuron 69(6):1204–1215

Daw ND, Niv Y, Dayan P (2005) Uncertainty-based competition between prefrontal and dorsolateral striatal systems for behavioral control. Nat Neurosci 8(12):1704–1711

de Wit H, Richards JB (2004) Dual determinants of drug use in humans: reward and impulsivity. Nebr Sym Motiv 50:19–55

de Wit S, Corlett PR, Aitken MR, Dickinson A, Fletcher PC (2009) Differential engagement of the ventromedial prefrontal cortex by goal-directed and habitual behavior toward food pictures in humans. J Neurosci 29(36):11330–11338

de Wit S, Watson P, Harsay HA, Cohen MX, van de Vijver I, Ridderinkhof KR (2012) Corticostriatal connectivity underlies individual differences in the balance between habitual and goal-directed action control. J Neurosci 32(35):12066–12075

Deroche-Gamonet V, Belin D, Piazza PV (2004) Evidence for addiction-like behavior in the rat. Science 305(5686):1014–1017

Deroche V, Le Moal M, Piazza PV (1999) Cocaine self-administration increases the incentive motivational properties of the drug in rats. Eur J Neurosci 11(8):2731–2736

Di Chiara G, Loddo P, Tanda G (1999) Reciprocal changes in prefrontal and limbic dopamine responsiveness to aversive and rewarding stimuli after chronic mild stress: implications for the psychobiology of depression. Biol Psychiatry 46(12):1624–1633

Diana M (2011) The dopamine hypothesis of drug addiction and its potential therapeutic value. Front Psychiatry 2:64

Diana M, Pistis M, Carboni S, Gessa GL, Rossetti ZL (1993) Profound decrement of mesolimbic dopaminergic neuronal activity during ethanol withdrawal syndrome in rats: electrophysiological and biochemical evidence. Proc Natl Acad Sci 90(17):7966–7969

Dickinson A, Wood N, Smith JW (2002) Alcohol seeking by rats: action or habit? Q J Exp Psychol Sect B 55(4):331

DiLeone RJ, Taylor JR, Picciotto MR (2012) The drive to eat: comparisons and distinctions between mechanisms of food reward and drug addiction. Nat Neurosci 15(10):1330–1335

Dolan RJ, Dayan P (2013) Goals and habits in the brain. Neuron 80(2):312–325

Doll BB, Simon DA, Daw ND (2012) The ubiquity of model-based reinforcement learning. Curr Opin Neurobiol 22(6):1075–1081

Duka T, Townshend JM (2004) The priming effect of alcohol pre-load on attentional bias to alcohol-related stimuli. Psychopharmacology 176(3–4):353–361

Edmonds DE, Gallistel CR (1974) Parametric analysis of brain stimulation reward in the rat: III. Effect of performance variables on the reward summation function. J Comp Physiol Psychol 87(5):876–883

Epping-Jordan MP, Watkins SS, Koob GF, Markou A (1998) Dramatic decreases in brain reward function during nicotine withdrawal. Nature 393(6680):76–79

Everitt BJ (2014) Neural and psychological mechanisms underlying compulsive drug seeking habits and drug memories—indications for novel treatments of addiction. Eur J Neurosci 40:2163

Everitt BJ, Cardinal RN, Parkinson JA, Robbins TW (2003) Appetitive behavior: impact of amygdala-dependent mechanisms of emotional learning. Ann N Y Acad Sci 985:233–250

Everitt BJ, Dickinson A, Robbins TW (2001) The neuropsychological basis of addictive behaviour. Brain Res Rev 36(2–3):129–138

Everitt BJ, Robbins TW (2005) Neural systems of reinforcement for drug addiction: from actions to habits to compulsion. Nat Neurosci 8(11):1481–1489

Everitt BJ, Robbins TW (2015) Drug addiction: updating actions to habits to compulsions ten years on. Annu Rev Psychol 67:8.1–8.28

Everitt BJ, Wolf ME (2002) Psychomotor stimulant addiction: a neural systems perspective. J Neurosci 22(9):3312–3320

Faure A, Haberland U, Conde F, El Massioui N (2005) Lesion to the nigrostriatal dopamine system disrupts stimulus-response habit formation. J Neurosci 25(11):2771–2780

Ferrario CR, Gorny G, Crombag HS, Li Y, Kolb B, Robinson TE (2005) Neural and behavioral plasticity associated with the transition from controlled to escalated cocaine use. Biol Psychiatry 58(9):751–759

Ferrario CR, Robinson TE (2007) Amphetamine pretreatment accelerates the subsequent escalation of cocaine self-administration behavior. Eur Neuropsychopharmacol 17(5):352–357

Field M, Mogg K, Bradley BP (2005) Craving and cognitive biases for alcohol cues in social drinkers. Alcohol Alcohol 40(6):504–510

Fields HL, Hjelmstad GO, Margolis EB, Nicola SM (2007) Ventral tegmental area neurons in learned appetitive behavior and positive reinforcement. Annu Rev Neurosci 30:289–316

Finn DA, Snelling C, Fretwell AM, Tanchuck MA, Underwood L, Cole M, Crabbe JC, Roberts AJ (2007) Increased drinking during withdrawal from intermittent ethanol exposure is blocked by the CRF receptor antagonist D-Phe-CRF(12-41). Alcohol Clin Exp Res 31(6):939–949

Fischman MW (1989) Relationship between self-reported drug effects and their reinforcing effects: studies with stimulant drugs. NIDA Res Monogr 92:211–230

Fischman MW, Foltin RW (1992) Self-administration of cocaine by humans: a laboratory perspective. In: Bock GR, Whelan J (eds) Cocaine: scientific and social dimensions, CIBA Foundation Symposium, vol 166, pp 165–180

Flagel SB, Akil H, Robinson TE (2009) Individual differences in the attribution of incentive salience to reward-related cues: implications for addiction. Neuropharmacology 56(Suppl 1):139–148

Flagel SB, Clark JJ, Robinson TE, Mayo L, Czuj A, Willuhn I, Akers CA, Clinton SM, Phillips PE, Akil H (2011) A selective role for dopamine in stimulus-reward learning. Nature 469(7328):53–57

Flagel SB, Watson SJ, Akil H, Robinson TE (2008) Individual differences in the attribution of incentive salience to a reward-related cue: influence on cocaine sensitization. Behav Brain Res 186(1):48–56

Fox HC, Talih M, Malison R, Anderson GM, Kreek MJ, Sinha R (2005) Frequency of recent cocaine and alcohol use affects drug craving and associated responses to stress and drug-related cues. Psychoneuroendocrinology 30(9):880–891

Gass JC, Motschman CA, Tiffany ST (2014) The relationship between craving and tobacco use behavior in laboratory studies: a meta-analysis. Psychol Addict Behav 28(4):1162–1176

Gawin FH, Kleber HD (1986) Abstinence symptomatology and psychiatric diagnosis in cocaine abusers. Clinical observations. Arch Gen Psychiatry 43(2):107–113

George O, Ghozland S, Azar MR, Cottone P, Zorrilla EP, Parsons LH, O'Dell LE, Richardson HN, Koob GF (2007) CRF–CRF1 system activation mediates withdrawal-induced increases in nicotine self-administration in nicotine-dependent rats. Proc Natl Acad Sci 104(43):17198–17203

George O, Le Moal M, Koob GF (2012) Allostasis and addiction: role of the dopamine and corticotropin-releasing factor systems. Physiol Behav 106(1):58–64

Goldstein RZ, Volkow ND (2002) Drug addiction and its underlying neurobiological basis: neuroimaging evidence for the involvement of the frontal cortex. Am J Psychiatry 159(10):1642–1652

Gonzales RA, Job MO, Doyon WM (2004) The role of mesolimbic dopamine in the development and maintenance of ethanol reinforcement. Pharmacol Ther 103(2):121–146

Greenwell TN, Funk CK, Cottone P, Richardson HN, Chen SA, Rice KC, Zorrilla EP, Koob GF (2009) Corticotropin-releasing factor-1 receptor antagonists decrease heroin self-administration in long-but not short-access rats. Addict Biol 14(2):130–143

Haber SN (2003) The primate basal ganglia: parallel and integrative networks. J Chem Neuroanat 26(4):317–330

Harmer CJ, Phillips GD (1998) Enhanced appetitive conditioning following repeated pretreatment with d-amphetamine. Behav Pharmacol 9(4):299–308

Hart AS, Rutledge RB, Glimcher PW, Phillips PE (2014) Phasic dopamine release in the rat nucleus accumbens symmetrically encodes a reward prediction error term. J Neurosci 34(3):698–704

Hogarth L (2012) Goal-directed and transfer-cue-elicited drug-seeking are dissociated by pharmacotherapy: Evidence for independent additive controllers. J Exp Psychol Anim Behav Process 38(3):266–278

Hogarth L, Balleine BW, Corbit LH, Killcross S (2013) Associative learning mechanisms underpinning the transition from recreational drug use to addiction. Ann N Y Acad Sci 1282:12–24

Homberg JR, Karel P, Verheij MM (2014) Individual differences in cocaine addiction: maladaptive behavioural traits. Addict Biol 19(4):517–528

Hone-Blanchet A, Fecteau S (2014) Overlap of food addiction and substance use disorders definitions: analysis of animal and human studies. Neuropharmacology 85:81–90

Horger BA, Shelton K, Schenk S (1990) Preexposure sensitizes rats to the rewarding effects of cocaine. Pharmacol Biochem Behav 37(4):707–711

Hughes JR (2007) Effects of abstinence from tobacco: valid symptoms and time course. Nicotine Tob Res 9(3):315–327

Humphries MD, Prescott TJ (2010) The ventral basal ganglia, a selection mechanism at the crossroads of space, strategy, and reward. Prog Neurobiol 90(4):385–417

Hutcheson DM, Everitt BJ, Robbins TW, Dickinson A (2001) The role of withdrawal in heroin addiction: enhances reward or promotes avoidance? Nat Neurosci 4(9):943–947

Ito R, Dalley JW, Howes SR, Robbins TW, Everitt BJ (2000) Dissociation in conditioned dopamine release in the nucleus accumbens core and shell in response to cocaine cues and during cocaine-seeking behavior in rats. J Neurosci 20(19):7489–7495

Ito R, Dalley JW, Robbins TW, Everitt BJ (2002) Dopamine release in the dorsal striatum during cocaine-seeking behavior under the control of a drug-associated cue. J Neurosci 22(14):6247–6253

Jaffe JH (1975) Drug addiction and drug abuse. In: Goodman LS, Gilman A (eds) The pharmacological basis of therapeutics. MacMillan, New York, pp 284–324

Jentsch JD, Taylor JR (1999) Impulsivity resulting from frontostriatal dysfunction in drug abuse: implications for the control of behavior by reward-related stimuli. Psychopharmacology 146(4):373–390

Johnson A, van der Meer MA, Redish AD (2007) Integrating hippocampus and striatum in decision-making. Curr Opin Neurobiol 17(6):692–697

Joyce EM, Iversen SD (1979) The effect of morphine applied locally to mesencephalic dopamine cell bodies on spontaneous motor activity in the rat. Neurosci Lett 14(2–3):207–212

Kalivas PW, Nakamura M (1999) Neural systems for behavioral activation and reward. Curr Opin Neurobiol 9(2):223–227

Kalivas PW, Volkow ND (2005) The neural basis of addiction: a pathology of motivation and choice. Am J Psychiatry 162(8):1403–1413

Kelley AE, Baldo BA, Pratt WE, Will MJ (2005) Corticostriatal-hypothalamic circuitry and food motivation: integration of energy, action and reward. Physiol Behav 86(5):773–795

Kempadoo KA, Tourino C, Cho SL, Magnani F, Leinninger GM, Stuber GD, Zhang F, Myers MG, Deisseroth K, de Lecea L, Bonci A (2013) Hypothalamic neurotensin projections promote reward by enhancing glutamate transmission in the VTA. J Neurosci 33(18):7618–7626

Kennett J, Matthews S, Snoek A (2013) Pleasure and addiction. Front Syst Neurosci 4:117

Kenny PJ, Chen SA, Kitamura O, Markou A, Koob GF (2006) Conditioned withdrawal drives heroin consumption and decreases reward sensitivity. J Neurosci 26(22):5894–5900

Kenny PJ, Markou A (2006) Nicotine self-administration acutely activates brain reward systems and induces a long-lasting increase in reward sensitivity. Neuropsychopharmacology 31(6):1203–1211

Kenny PJ, Polis I, Koob GF, Markou A (2003) Low dose cocaine self-administration transiently increases but high dose cocaine persistently decreases brain reward function in rats. Eur J Neurosci 17(1):191–195

Khantzian EJ (1997) The self-medication hypothesis of substance use disorders: a reconsideration and recent applications. Harvard Rev Psychiatry 4(5):231–244

Killcross S, Coutureau E (2003) Coordination of actions and habits in the medial prefrontal cortex of rats. Cereb Cortex 13(4):400–408

Kita T, Okamoto M, Nakashima T (1992) Nicotine-induced sensitization to ambulatory stimulant effect produced by daily administration into the ventral tegmental area and the nucleus accumbens in rats. Life Sci 50(8):583–590

Klitenick MA, DeWitte P, Kalivas PW (1992) Regulation of somatodendritic dopamine release in the ventral tegmental area by opioids and GABA: an in vivo microdialysis study. J Neurosci 12(7):2623–2632

Koob GF (2003) Alcoholism: allostasis and beyond. Alcohol Clin Exp Res 27(2):232–243

Koob GF (2013) Addiction is a reward deficit and stress surfeit disorder. Front Psychiatry 4:72

Koob GF, Buck CL, Cohen A, Edwards S, Park PE, Schlosburg JE, Schmeichel B, Vendruscolo LF, Wade CL, Whitfield TW Jr, George O (2014) Addiction as a stress surfeit disorder. Neuropharmacology 76 Pt B:370–382

Koob GF, Le Moal M (2001) Drug addiction, dysregulation of reward, and allostasis. Neuropsychopharmacology 24(2):97–129

Koob GF, Le Moal M (2008) Addiction and the brain antireward system. Annu Rev Psychol 59:29–53

Koob GF, Nestler EJ (1997) The neurobiology of drug addiction. J Neuropsychiatry Clin Neurosci 9(3):482–497

Lamb RJ, Preston KL, Schindler CW, Meisch RA, Davis F, Katz JL, Henningfield JE, Goldberg SR (1991) The reinforcing and subjective effects of morphine in post-addicts: a dose-response study. J Pharmacol Exp Ther 259(3):1165–1173

Lesscher HM, Vanderschuren LJ (2012) Compulsive drug use and its neural substrates. Rev Neurosci 23(5–6):731–745

Lessov CN, Palmer AA, Quick EA, Phillips TJ (2001) Voluntary ethanol drinking in C57BL/6 J and DBA/2 J mice before and after sensitization to the locomotor stimulant effects of ethanol. Psychopharmacology 155(1):91–99

Letchworth SR, Nader MA, Smith HR, Friedman DP, Porrino LJ (2001) Progression of changes in dopamine transporter binding site density as a result of cocaine self-administration in rhesus monkeys. J Neurosci 21(8):2799–2807

Litten RZ, Ryan ML, Falk DE, Reilly M, Fertig JB, Koob GF (2015) Heterogeneity of alcohol use disorder: understanding mechanisms to advance personalized treatment. Alcohol Clin Exp Res 39(4):579–584

Littleton J (1998) Neurochemical mechanisms underlying alcohol withdrawal. Alcohol Health Res World 22(1):13–24

Logrip ML, Koob GF, Zorrilla EP (2011) Role of corticotropin-releasing factor in drug addiction: potential for pharmacological intervention. CNS Drugs 25(4):271–287

Lorrain DS, Arnold GM, Vezina P (2000) Previous exposure to amphetamine increases incentive to obtain the drug: long-lasting effects revealed by the progressive ratio schedule. Behav Brain Res 107(1–2):9–19

Loweth JA, Tseng KY, Wolf ME (2014) Adaptations in AMPA receptor transmission in the nucleus accumbens contributing to incubation of cocaine craving. Neuropharmacology 76 Pt B:287–300

Lubman DI, Yucel M, Pantelis C (2004) Addiction, a condition of compulsive behaviour? Neuroimaging and neuropsychological evidence of inhibitory dysregulation. Addiction (Abingdon, England) 99 (12):1491–1502

Ma YY, Lee BR, Wang X, Guo C, Liu L, Cui R, Lan Y, Balcita-Pedicino JJ, Wolf ME, Sesack SR, Shaham Y, Schluter OM, Huang YH, Dong Y (2014) Bidirectional modulation of incubation of cocaine craving by silent synapse-based remodeling of prefrontal cortex to accumbens projections. Neuron 83(6):1453–1467

MacAskill AF, Cassel JM, Carter AG (2014) Cocaine exposure reorganizes cell type-and input-specific connectivity in the nucleus accumbens. Nat Neurosci 17(9):1198–1207

Mahler SV, Berridge KC (2012) What and when to "want"? Amygdala-based focusing of incentive salience upon sugar and sex. Psychopharmacology 221(3):407–426

McClure SM, Bickel WK (2014) A dual-systems perspective on addiction: contributions from neuroimaging and cognitive training. Ann N Y Acad Sci 1327:62–78

McEwen BS (1998) Stress, Adaptation, and Disease: Allostasis and Allostatic Load. Ann NY Acad Sci 840(1):33–44

Mendrek A, Blaha CD, Phillips AG (1998) Pre-exposure of rats to amphetamine sensitizes self-administration of this drug under a progressive ratio schedule. Psychopharmacology 135(4):416–422

Meyer PJ, Lovic V, Saunders BT, Yager LM, Flagel SB, Morrow JD, Robinson TE (2012a) Quantifying individual variation in the propensity to attribute incentive salience to reward cues. PLoS ONE 7(6):e38987

Meyer PJ, Ma ST, Robinson TE (2012b) A cocaine cue is more preferred and evokes more frequency-modulated 50-kHz ultrasonic vocalizations in rats prone to attribute incentive salience to a food cue. Psychopharmacology 219(4):999–1009

Meyer PJ, Meshul CK, Phillips TJ (2009) Ethanol- and cocaine-induced locomotion are genetically related to increases in accumbal dopamine. Genes Brain Behav 8(3):346–355

Miles FJ, Everitt BJ, Dickinson A (2003) Oral cocaine seeking by rats: action or habit? Behav Neurosci 117(5):927–938

Miliaressis E, Rompre PP, Laviolette P, Philippe L, Coulombe D (1986) The curve-shift paradigm in self-stimulation. Physiol Behav 37(1):85–91

Moeller FG, Dougherty DM (2002) Impulsivity and substance abuse: what is the connection? Addict Disord Treat 1(1):3–10

Mogenson GJ, Jones DL, Yim CY (1980) From motivation to action: functional interface between the limbic system and the motor system. Prog Neurobiol 14(2–3):69–97

Morgan D, Roberts DC (2004) Sensitization to the reinforcing effects of cocaine following binge-abstinent self-administration. Neurosci Biobehav Rev 27(8):803–812

Murray JE, Belin D, Everitt BJ (2012) Double dissociation of the dorsomedial and dorsolateral striatal control over the acquisition and performance of cocaine seeking. Neuropsychopharmacology 37(11):2456–2466

Naude J, Dongelmans M, Faure P (2014) Nicotinic alteration of decision-making. Neuropharmacology 96:244

Nestby P, Vanderschuren LJ, De Vries TJ, Hogenboom F, Wardeh G, Mulder AH, Schoffelmeer AN (1997) Ethanol, like psychostimulants and morphine, causes long-lasting hyperreactivity of dopamine and acetylcholine neurons of rat nucleus accumbens: possible role in behavioural sensitization. Psychopharmacology 133(1):69–76

Nestler EJ (2001) Molecular basis of long-term plasticity underlying addiction. Nat Rev Neurosci 2(2):119–128

Nestler EJ (2005) Is there a common molecular pathway for addiction? Nat Neurosci 8(11):1445–1449

NIMH (2015) Research domain criteria (RDoC). http://www.nimh.nih.gov/research-priorities/rdoc/index.shtml

Nimitvilai S, Arora DS, McElvain MA, Brodie MS (2012) Ethanol blocks the reversal of prolonged dopamine inhibition of dopaminergic neurons of the ventral tegmental area. Alcohol Clin Exp Res 36(11):1913–1921

Olds J (1962) Hypothalamic substrates of reward. Physiol Rev 42:554–604
Ostlund SB, Balleine BW (2007a) The contribution of orbitofrontal cortex to action selection. Ann N Y Acad Sci 1121:174–192
Ostlund SB, Balleine BW (2007b) Orbitofrontal cortex mediates outcome encoding in Pavlovian but not instrumental conditioning. J Neurosci 27(18):4819–4825
Ostlund SB, Balleine BW (2008) On habits and addiction: an associative analysis of compulsive drug seeking. Drug Discov Today Dis Models 5(4):235–245
Paulson PE, Camp DM, Robinson TE (1991) The time coures of transient behavioral depression and persistent behavioral sensitization in relation to regional brain monoamine concentrations during amphetamine withdrawal in rats. Psychopharmacology 103:480–492
Perry JL, Carroll ME (2008) The role of impulsive behavior in drug abuse. Psychopharmacology 200(1):1–26
Piazza PV, Deroche-Gamonet V (2013) A multistep general theory of transition to addiction. Psychopharmacology 229(3):387–413
Pierce RC, Kalivas PW (1997) Repeated cocaine modifies the mechanism by which amphetamine releases dopamine. J Neurosci 17(9):3254–3261
Porrino LJ, Lyons D, Smith HR, Daunais JB, Nader MA (2004) Cocaine self-administration produces a progressive involvement of limbic, association, and sensorimotor striatal domains. J Neurosci 24(14):3554–3562
Post RM, Lockfeld A, Squillace KM, Contel NR (1981) Drug-environment interaction: context dependency of cocaine-induced behavioral sensitization. Life Sci 28(7):755–760
Puumala T, Sirvio J (1998) Changes in activities of dopamine and serotonin systems in the frontal cortex underlie poor choice accuracy and impulsivity of rats in an attention task. Neuroscience 83(2):489–499
Redish AD (2004) Addiction as a computational process gone awry. Science 306(5703):1944–1947
Redish AD, Jensen S, Johnson A (2008) A unified framework for addiction: vulnerabilities in the decision process. Behav Brain Sci 31(4):415–437 (discussion 437–487)
Robbins TW, Everitt BJ (1999) Drug addiction: bad habits add up. Nature 398(6728):567–570
Robinson TE, Becker JB (1986) Enduring changes in brain and behavior produced by chronic amphetamine administration: a review and evaluation of animal models of amphetamine psychosis. Brain Res 396(2):157–198
Robinson TE, Berridge KC (1993) The neural basis of drug craving: an incentive-sensitization theory of addiction. Brain Res Rev 18(3):247–291
Robinson TE, Berridge KC (2000) The psychology and neurobiology of addiction: an incentive-sensitization view. Addiction 95(Suppl 2):S91–117
Robinson TE, Berridge KC (2001) Incentive-sensitization and addiction. Addiction 96(1):103–114
Robinson TE, Browman KE, Crombag HS, Badiani A (1998) Modulation of the induction or expression of psychostimulant sensitization by the circumstances surrounding drug administration. Neurosci Biobehav Rev 22:347–354
Robinson TE, Kolb B (2004) Structural plasticity associated with exposure to drugs of abuse. Neuropharmacology 47(Suppl 1):33–46
Robinson TE, Yager LM, Cogan ES, Saunders BT (2014) On the motivational properties of reward cues: Individual differences. Neuropharmacology 76(Part B (0)):450–459
Rothwell PE, Thomas MJ, Gewirtz JC (2012) Protracted manifestations of acute dependence after a single morphine exposure. Psychopharmacology 219(4):991–998
Saddoris MP, Cacciapaglia F, Wightman RM, Carelli RM (2015) Differential dopamine release dynamics in the nucleus accumbens core and shell reveal complementary signals for error prediction and incentive motivation. J Neurosci 35(33):11572–11582
Sakagami M, Pan X, Uttl B (2006) Behavioral inhibition and prefrontal cortex in decision-making. Neural Networks Official J Int Neural Network Soc 19(8):1255–1265
Satel SL, Kosten TR, Schuckit MA, Fischman MW (1993) Should protracted withdrawal from drugs be included in DSM-IV? Am J Psychiatry 150(5):695–704

Schmidt HD, Anderson SM, Famous KR, Kumaresan V, Pierce RC (2005) Anatomy and pharmacology of cocaine priming-induced reinstatement of drug seeking. Eur J Pharmacol 526 (1–3):65–76

Schmidt HD, Pierce RC (2010) Cocaine-induced neuroadaptations in glutamate transmission: potential therapeutic targets for craving and addiction. Ann N Y Acad Sci 1187:35–75

Schoenbaum G, Roesch MR, Stalnaker TA (2006) Orbitofrontal cortex, decision-making and drug addiction. Trends Neurosci 29(2):116–124

Schultz W, Dayan P, Montague PR (1997) A neural substrate of prediction and reward. Science 275(5306):1593–1599

Schuster CR, Thompson T (1969) Self administration of and behavioral dependence on drugs. Annu Rev Pharmacol 9:483–502

Segal DS (1975) Behavioral and neurochemical correlates of repeated d-amphetamine administration. Adv Biochem Psychopharmacol 13:247–262

Sesack SR, Grace AA (2010) Cortico-Basal Ganglia reward network: microcircuitry. Neuropsychopharmacology 35(1):27–47

Shalev U, Grimm JW, Shaham Y (2002) Neurobiology of relapse to heroin and cocaine seeking: a review. Pharmacol Rev 54(1):1–42

Shen RY (2003) Ethanol withdrawal reduces the number of spontaneously active ventral tegmental area dopamine neurons in conscious animals. J Pharmacol Exp Ther 307(2):566–572

Shen RY, Choong KC, Thompson AC (2007) Long-term reduction in ventral tegmental area dopamine neuron population activity following repeated stimulant or ethanol treatment. Biol Psychiatry 61(1):93–100

Shiflett MW, Balleine BW (2010) At the limbic-motor interface: disconnection of basolateral amygdala from nucleus accumbens core and shell reveals dissociable components of incentive motivation. Eur J Neurosci 32(10):1735–1743

Shiflett MW, Balleine BW (2011) Molecular substrates of action control in cortico-striatal circuits. Prog Neurobiol 95(1):1–13

Simon NW, Beas BS, Montgomery KS, Haberman RP, Bizon JL, Setlow B (2013) Prefrontal cortical-striatal dopamine receptor mRNA expression predicts distinct forms of impulsivity. Eur J Neurosci 37:1779

Skinner MD, Aubin HJ (2010) Craving's place in addiction theory: contributions of the major models. Neurosci Biobehav Rev 34(4):606–623

Solomon RL, Corbit JD (1974) An opponent-process theory of motivation. I. Temporal dynamics of affect. Psychol Rev 81(2):119–145

Soto CBD, O'Donnell WE, Allred LJ, Lopes CE (1985) Symptomatology in alcoholics at various stages of abstinence. Alcohol Clin Exp Res 9(6):505–512

Specio SE, Wee S, O'Dell LE, Boutrel B, Zorrilla EP, Koob GF (2008) CRF(1) receptor antagonists attenuate escalated cocaine self-administration in rats. Psychopharmacology 196 (3):473–482

Steketee JD (2003) Neurotransmitter systems of the medial prefrontal cortex: potential role in sensitization to psychostimulants. Brain Res Brain Res Rev 41(2–3):203–228

Stewart J (1992) Conditioned stimulus control of the expression of sensitization of the behavioral activating effects of opiate and stimulant drugs. In: Gormezano I, Wasserman EA (eds) Learning and memory: the behavioral and biological substrates. Erlbaum, Hillsdale, NJ, pp 917–923

Terelli E, Terry P (1999) Amphetamine induced conditioned activity and sensitization: the role of habituation to the test context and the involvement of Pavlovian processes. Behav Pharmacol 9:409–419

Tiffany ST (1990) A cognitive model of drug urges and drug-use behavior: Role of automatic and nonautomatic processes. Psychol Rev 97(2):147–168

Tiffany ST (1999) Cognitive concepts of craving. Alcohol Res Health 23(3):215–224

Torregrossa MM, Kalivas PW (2008) Neurotensin in the ventral pallidum increases extracellular gamma-aminobutyric acid and differentially affects cue- and cocaine-primed reinstatement. J Pharmacol Exp Ther 325(2):556–566

Turchan J, Lason W, Budziszewska B, Przewlocka B (1997) Effects of single and repeated morphine administration on the prodynorphin, proenkephalin and dopamine D2 receptor gene expression in the mouse brain. Neuropeptides 31(1):24–28

Uslaner JM, Acerbo MJ, Jones SA, Robinson TE (2006) The attribution of incentive salience to a stimulus that signals an intravenous injection of cocaine. Behav Brain Res 169(2):320–324

van Huijstee AN, Mansvelder HD (2014) Glutamatergic synaptic plasticity in the mesocorticolimbic system in addiction. Front Cell Neurosci 8:466

Vanderschuren LJ, Di Ciano P, Everitt BJ (2005) Involvement of the dorsal striatum in cue-controlled cocaine seeking. J Neurosci 25(38):8665–8670

Vanderschuren LJ, Kalivas PW (2000) Alterations in dopaminergic and glutamatergic transmission in the induction and expression of behavioral sensitization: a critical review of preclinical studies. Psychopharmacology 151(2–3):99–120

Veilleux JC, Skinner KD (2015) Smoking, food, and alcohol cues on subsequent behavior: a qualitative systematic review. Clin Psychol Rev 36:13–27

Verdejo-Garcia A, Bechara A, Recknor EC, Perez-Garcia M (2007) Negative emotion-driven impulsivity predicts substance dependence problems. Drug Alcohol Depend 91(2–3):213–219

Verdejo-Garcia A, Rivas-Perez C, Lopez-Torrecillas F, Perez-Garcia M (2006) Differential impact of severity of drug use on frontal behavioral symptoms. Addict Behav 31(8):1373–1382

Vezina P (2004) Sensitization of midbrain dopamine neuron reactivity and the self-administration of psychomotor stimulant drugs. Neurosci Biobehav Rev 27(8):827–839

Vezina P, Leyton M (2009) Conditioned cues and the expression of stimulant sensitization in animals and humans. Neuropharmacology 56(Suppl 1):160–168

Volkow ND, Fowler JS, Wang G-J (2003) The addicted human brain: insights from imaging studies. J Clin Invest 111(10):1444–1451

Volkow ND, Wang GJ, Fowler JS, Tomasi D, Baler R (2012) Food and drug reward: overlapping circuits in human obesity and addiction. Curr Top Behav Neurosci 11:1–24

Wassum KM, Cely IC, Balleine BW, Maidment NT (2011) Micro-opioid receptor activation in the basolateral amygdala mediates the learning of increases but not decreases in the incentive value of a food reward. J Neurosci 31(5):1591–1599

Weeks JR (1962) Experimental morphine addiction: method for automatic intravenous injections in unrestrained rats. Science 138(3537):143–144

White AM, Ghia AJ, Levin ED, Swartzwelder HS (2000) Binge pattern ethanol exposure in adolescent and adult rats: differential impact on subsequent responsiveness to ethanol. Alcohol Clin Exp Res 24(8):1251–1256

Wiers RW, Bartholow BD, van den Wildenberg E, Thush C, Engels RC, Sher KJ, Grenard J, Ames SL, Stacy AW (2007) Automatic and controlled processes and the development of addictive behaviors in adolescents: a review and a model. Pharmacol Biochem Behav 86(2):263–283

Willuhn I, Burgeno LM, Everitt BJ, Phillips PE (2012) Hierarchical recruitment of phasic dopamine signaling in the striatum during the progression of cocaine use. Proc Natl Acad Sci USA 109(50):20703–20708

Winstanley CA, Theobald DE, Cardinal RN, Robbins TW (2004) Contrasting roles of basolateral amygdala and orbitofrontal cortex in impulsive choice. J Neurosci 24(20):4718–4722

Wise RA (1987) The role of reward pathways in the development of drug dependence. Pharmacol Ther 35(1–2):227–263

Wise RA (1988) The neurobiology of craving: Implications for the understanding and treatment of addiction. J Abnorm Psychol 97(2):118–132

Wise RA, Bozarth MA (1987) A psychomotor stimulant theory of addiction. Psychol Rev 94(4):469–492

Wise RA, Koob GF (2014) The development and maintenance of drug addiction. Neuropsychopharmacology 39(2):254–262

Witteman J, Post H, Tarvainen M, de Bruijn A, Perna ES, Ramaekers JG, Wiers RW (2015) Cue reactivity and its relation to craving and relapse in alcohol dependence: a combined laboratory and field study. Psychopharmacology (Berl) 232(20):3685–3696

Wolf ME (1998) The role of excitatory amino acids in behavioral sensitization to psychomotor stimulants. Prog Neurobiol 54(6):679–720

Wolf ME (2002) Addiction: making the connection between behavioral changes and neuronal plasticity in specific pathways. Mol Interventions 2:146–157

Wolf ME (2003) LTP may trigger addiction. Mol Interventions 3(5):248–252

Wolf ME, Ferrario CR (2010) AMPA receptor plasticity in the nucleus accumbens after repeated exposure to cocaine. Neurosci Biobehav Rev 35(2):185–211

Woolverton WL, Cervo L, Johanson CE (1984) Effects of repeated methamphetamine administration on methamphetamine self-administration in rhesus monkeys. Pharmacol Biochem Behav 21:737–741

Yager LM, Robinson TE (2013) A classically conditioned cocaine cue acquires greater control over motivated behavior in rats prone to attribute incentive salience to a food cue. Psychopharmacology 226(2):217–228

Yeomans JS (1989) Two substrates for medial forebrain bundle self-stimulation: myelinated axons and dopamine axons. Neurosci Biobehav Rev 13(2–3):91–98

Yin HH, Knowlton BJ, Balleine BW (2004) Lesions of dorsolateral striatum preserve outcome expectancy but disrupt habit formation in instrumental learning. Eur J Neurosci 19(1):181–189

Yin HH, Knowlton BJ, Balleine BW (2006) Inactivation of dorsolateral striatum enhances sensitivity to changes in the action-outcome contingency in instrumental conditioning. Behav Brain Res 166(2):189–196

Zapata A, Minney VL, Shippenberg TS (2010) Shift from goal-directed to habitual cocaine seeking after prolonged experience in rats. J Neurosci Official J Soc Neurosci 30(46):15457–15463

Ziauddeen H, Fletcher PC (2013) Is food addiction a valid and useful concept? Obes Rev 14(1):19–28

Skewed by Cues? The Motivational Role of Audiovisual Stimuli in Modelling Substance Use and Gambling Disorders

Michael M. Barrus, Mariya Cherkasova and Catharine A. Winstanley

Abstract The similarity between gambling disorder (GD) and drug addiction has recently been recognized at the diagnostic level. Understanding the core cognitive processes involved in these addiction disorders, and in turn their neurobiological mechanisms, remains a research priority due to the enormous benefits such knowledge would have in enabling effective treatment design. Animal models can be highly informative in this regard. Although numerous rodent behavioural paradigms that capture different facets of gambling-like behaviour have recently been developed, the motivational power of cues in biasing individuals towards risky choice has so far received little attention despite the central role played by drug-paired cues in successful laboratory models of chemical dependency. Here, we review some of the comparatively simple paradigms in which reward-paired cues are known to modulate behaviour in rodents, such as sign-tracking, Pavlovian-to-instrumental transfer and conditioned reinforcement. Such processes are thought to play an important role in mediating responding for drug reward, and the need for future studies to address whether similar processes contribute to cue-driven risky choice is highlighted.

Keywords Decision making · Cues · Dopamine · Gambling · Animal models

Contents

1	Introduction	508
2	The Impact of Conditioned Cues in Models of Drug Addiction	511
	2.1 Pavlovian Conditioning and Drug Addiction	511
	2.2 The Incentive Sensitization Theory of Drug Addiction	512
	2.3 Attentional Bias in Drug Addiction	513
3	The Role of Cues in Gambling	514
	3.1 Animal Models of the Influence CS Exert over Behaviour	516
	3.2 Sign-Tracking	516
	3.3 Pavlovian-to-Instrumental Transfer	518
	3.4 Conditioned Reinforcement	519
	3.5 Interim Summary	520
	3.6 The Addition of Cues to Decision-Making Tasks Fundamentally Alters Neurobiological Regulation of Choice	521
	3.7 Concluding Remarks	523
	3.8 Financial Disclosures	523
References		524

1 Introduction

Gambling is a common recreational pastime that can lead to debilitating and compulsive behaviour for some users. While most individuals are able to gamble within reasonable limits, some 12.5 % of the general public demonstrates subclinical problem gambling, and 2.5 % meet the criteria for gambling disorder (GD), a Diagnostic and Statistical Manual, Fifth Edition (DSM-V) recognized behavioural addiction characterized by a loss of control over gambling (Cunningham-Williams et al. 2005). Despite GD's prevalence and social and individual costs, the neurobiology of gambling behaviour is not well understood. This lack of insight has thus far limited treatment of the disorder (Williams et al. 2008). Laboratory-based models of gambling behaviour are thus extremely useful in that they allow researchers to isolate the cognitive and neurobiological processes implicated in gambling. Analogues of these paradigms with strong face, construct and predictive validity can then be designed for use with non-human laboratory animals, thereby enabling the causative nature of particular brain areas or neurotransmitter systems in maladaptive gambling behaviour to be determined (see Potenza 2009; Cocker and Winstanley 2015 for discussion). Establishing such robust models has the potential to catalyse the development of pharmacological treatments for GD, as well as inform our understanding of the very nature of GD and therefore remains an important research priority in the field.

Perhaps the most widely used cognitive task that assesses decision-making processes similar to those recruited during gambling behaviour is the Iowa Gambling Task (IGT) which provides a reliable measure of preference for risky (disadvantageous) over conservative (advantageous) options (Bechara et al. 1994). Although ostensibly designed to capture "real-world" decision-making in which all options could lead to both gains and losses according to initially ambiguous odds, it has been used as a proxy for gambling largely due to its strong superficial resemblance to the act of gambling. In the IGT, human participants must choose between decks of cards, each of which is associated with different schedules of risk and reward, in order to maximize the amount of money or points earned. Two of the decks (decks A and B) are associated with sizeable wins but also disproportionately larger losses, leading to a net loss over time. The remaining two decks (decks C and D) are associated with smaller wins but also smaller losses, and exclusive choice of these decks leads to a net gain over time. Subjects must learn to resist choosing the superficially tempting options (A and B) in order to succeed at the task, and work with the IGT has demonstrated impairment in a number of clinical populations including pathological gamblers (Goudriaan et al. 2005; Shurman et al. 2005; Verdejo-Garcia et al. 2007a, b). While there are numerous aspects of problematic gambling behaviour that are not captured by this task (see Cocker and Winstanley 2015 for discussion), there is no doubt that work with the IGT has made a significant contribution to our understanding of decision-making under conditions of risk and ambiguity. Understandably, developing rodent analogues of the IGT was considered by many researchers a logical first-step in generating a model of gambling behaviour that would hopefully prove useful in capturing elements of disordered gambling and identifying viable pharmacotherapeutic targets (de Visser et al. 2011).

One such model is the rat Gambling Task (rGT), in which animals are allowed to choose between four options, signalled by illumination of four response apertures, loosely analogous to the four decks of cards used in the IGT in that each is associated with unique schedules of food reward or "timeout" punishment (Fig. 1; Zeeb et al. 2009). As is true of the IGT, the best strategy on this task is to favour options associated with smaller rewards but also smaller punishments—this more conservative approach leads to the steady accumulation of the greatest amount of reward over time. In contrast, a preference for these tempting "high-risk high-reward" outcomes is ultimately disadvantageous: although such options can yield greater rewards per trial, the disproportionately larger punishments result in considerably less benefit during the course of a session. Critically, this task incorporates loss, a central component of naturalistic gambling paradigms, through the use of punishing timeout periods. Given the limited length of each session, time is a resource animals are at risk of losing if their wager is unsuccessful. In essence, the disadvantageous options and their longer timeout periods require animals to balance the desire for larger rewards with the risk of the loss of future earning potential. Most rats acquire the optimal strategy readily, and such decision-making appears to depend on similar neural circuitry as is implicated in performance of the IGT (Zeeb and Winstanley 2011, 2013; Paine et al. 2013; Zeeb et al. 2015).

While this task, and others, provides valuable insight into gambling-like behaviours, there are elements of real-world gambling that have not yet been addressed by animal models. To our knowledge, little work has been done evaluating the role of salient cues in modulating decision-making. This is a potentially rich area of research; real-world gambling is rife with salient cues, and their influence on

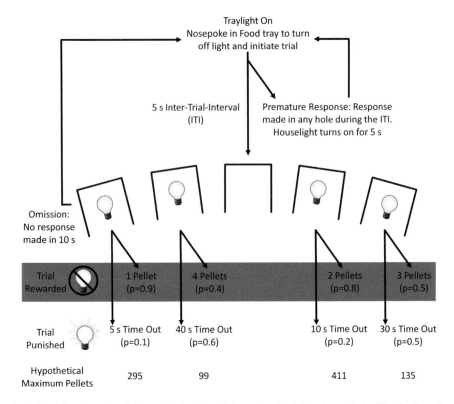

Fig. 1 Task schematic of the rat Gambling Task (*rGT*). Each trial begins with the illumination of the tray light. A nosepoke in the tray extinguishes the tray light and initiates a 5-s inter-trial interval (*ITI*), during which all lights in the chamber are off. Following the ITI, stimulus lights are illuminated in apertures 1, 2, 4 and 5, each of which has a different schedule of reward/punishment associated with it. If the animal nosepokes one of the apertures within 10 s, the animal is rewarded or punished according to the schedule associated with that aperture. The size of reward and duration of punishment for each option are indicated on the schematic; the *p*-value in brackets beneath each of those indicates the probability of a win or loss on any given trial. On a rewarded trial, the tray light is illuminated and the requisite pellets dispensed. A response at the tray then initiates a new trial. On a punished trial, the light in the chosen aperture flashes at a frequency of 0.5 Hz for the duration of the timeout period; all other lights are extinguished. At the end of the timeout, the tray light is once again illuminated and the animal can initiate a new trial. A nosepoke at an aperture during the ITI is scored as a premature response and initiates a 5-s timeout period during which the houselight is illuminated. Failure to make a response at an aperture within 10 s of the stimulus lights being illuminated is scored as an omission; the stimulus lights are extinguished, the tray light once again illuminated, and the animal is able to initiate a new trial. Adapted from Zeeb et al. (2009)

gambling behaviour may be significant. Experimental work with human subjects has demonstrated that manipulating the gambling environment can affect gambling behaviours (Brevers et al. 2015), and some have proposed that attentional biases towards salient cues may underlie the transition from recreational to problem gambling (Grant and Bowling 2014). The influence of salient cues on decision-making is not limited to gambling; the ability of drug-related stimuli to promote craving and relapse is well documented and represents one of the most destructive forms of cue-biased behaviour (Childress et al. 1992; Grimm et al. 2001; Shaham et al. 2003). Being able to demonstrate cue-induced maladaptive decision-making in animal models would be of value to both gambling and substance abuse research and could more generally aid in the characterization of how salient cues exert their effects on decision-making. We will first consider the ways in which the study of conditioned cues has influenced models of addiction.

2 The Impact of Conditioned Cues in Models of Drug Addiction

2.1 Pavlovian Conditioning and Drug Addiction

It is necessary to define the term "cue" as we will be using it before embarking on a discussion of the cues' significance and contributions to decision-making. In the light of the focus of this review on the motivational impact of cues, any stimuli that have come to be associated with reinforcement satisfy the definition. As such, our discussion of cues must essentially begin with classical conditioning, famously laid out by Ivan Pavlov following his discovery of the motivational power of a bell. Originally intending to study the role of salivation in digestion, Pavlov noticed that his canine subjects began to salivate upon exposure to the experimenter who regularly distributed meat powder (Pavlov 1927). Pavlov then paired a ringing bell with the distribution of meat powder and found, with time, that the bell alone was sufficient to evoke salivation in his animals. The bell therefore became what is termed a conditioned stimulus (CS) capable of eliciting a conditioned response (CR) as if it were the primary reward itself.

The real-world examples of this interaction are myriad, and research on the subject has placed particular emphasis on understanding the prominent role of drug-related cues in addiction and substance abuse (Childress et al. 1993). Drug-related cues can be anything the user associates with the drug-taking experience, be that individuals with whom the user takes drugs, locations in which the user commonly takes drugs or drug-associated paraphernalia such as pipes or syringes. After repeated pairings of these people, places and things with the drug-taking experience, these formerly neutral stimuli come to predict the delivery of reward and may even take on the motivational properties of the reward, promoting drug-seeking behaviour and CRs. Drug-associated cues such as paraphernalia and location can induce powerful craving and arousal states (Childress et al. 1993); exposure to smoking-related cues increases subjective craving for cigarettes (Carter and Tiffany 1999b), while exposure to alcohol-related cues increases subjective craving for alcohol (Schulze and Jones 1999). The degree of attentional bias towards these cues can distinguish between abusers and non-abusers/non-users, and among users, substance-related attentional bias tends to be positively related to the quantity and frequency of use, though this relationship has been less consistent for smokers (Robbins and Ehrman 2004; Cox et al. 2006).

While Pavlovian associations between drug-related stimuli and consumption are believed to contribute to compulsive drug use, this simple form of associative learning is unlikely to be the sole mechanism mediating behaviour. Were these simple stimulus–stimulus pairings, the CSs should produce effects that mimic either the appetitive, intoxication-like effects the substance produces or the aversive, withdrawal-like effects associated with physical withdrawal from the substance. While drug-paired cues can promote withdrawal-like experiences in some circumstances (Carter and Tiffany 1999a), these effects do not appear to be consistent. Instead, cues more readily produce increases in subjective craving and physiological arousal as described previously (Carter and Tiffany 1999b). Given that these are not the unconditioned responses evoked by the substance but instead appetitive behaviours directed towards the substance, it suggests that the affective properties of cues are more complex than a simple Pavlovian model can account for. Several compelling theories have been proposed to explain these effects.

2.2 *The Incentive Sensitization Theory of Drug Addiction*

One of the most prominent of these is the theory of incentive sensitization (Robinson and Berridge 1993), which proposes an elegant mechanistic explanation of the neurobiological mechanisms underlying the ability of cues to guide behaviour. Reward-paired cues can acquire incentive salience, i.e. motivational significance, through mesocorticolimbic dopamine signalling (Berridge and Robinson 1998). Repeated use of drugs of abuse can lead to the sensitization of dopaminergic systems related to reward, motivation and salience attribution. This "incentive sensitization" results in heightened sensitivity to drug-related stimuli, which increases subjective motivation (or "wanting") for drugs of abuse. Continued substance use can result in long-lasting "hypersensitivity to the incentive motivational effects of drugs and drug-associated stimuli" (Robinson and Berridge 1993), with dopamine mediating the "wanting" component. Reward-related cues themselves become "wanted" or motivationally salient and capable of driving behaviour to a greater extent than reward alone could (Heinz et al. 2004). The misattributions of salience to drug-related cues can lead to significant behavioural changes that long outlast physical dependence, while also providing a better explanation for complex patterns of drug-seeking behaviour seen in substance dependence than simple Pavlovian paradigms of stimulus–stimulus learning. In essence, the CS becomes an "incentive stimulus", capable of influencing action selection and goal-directed behaviour (Saunders and Robinson 2010; Yager and Robinson 2013).

Although the theoretical basis for the incentive salience model of addiction was elucidated using laboratory animals, considerable evidence points to the motivational significance of cues in human-addicted populations. Several studies have reported increases in self-reported liking for drug-paired locations following conditioning (Childs and de Wit 2009, 2013), and smokers preferred to listen to a smoking-paired cue over a control cue (Mucha et al. 1998). Furthermore, the interoceptive cues triggered by smoke inhalation have been found to significantly contribute to the desire to smoke, beyond the simple administration of the addictive chemical nicotine (Naqvi and Bechara 2005, 2006). An instrumentally conditioned cue that resulted in the opportunity to smoke produced greater attentional bias than a control cue did (Hogarth et al. 2003). Furthermore, previously cocaine-paired cues sustained responding in cocaine-dependent subjects, even though self-reports indicated that subjects were aware they were no longer receiving cocaine (Panlilio et al. 2005). Collectively, these results suggest that drug-paired cues become motivationally "wanted" following conditioning, consistent with the incentive sensitization model. In addition, both behavioural and dopamine drug sensitizations have now been demonstrated in humans (Boileau et al. 2006; O'Daly et al. 2011). Though the findings have not been entirely consistent, this could be in part explained in by the presence versus absence of drug-paired contextual cues, which appear to be critical for expression of sensitization (Leyton and Vezina 2013, 2014).

2.3 Attentional Bias in Drug Addiction

Other theories have expanded upon the contingencies necessary for cues to exert their effects. Field and Cox suggested that existing theories were incomplete, specifically arguing the motivational power of drug-related cues is contingent on the availability of the drug (Field and Cox 2008). In this model, drug-related cues come to gain significance not simply because of recurrent pairing with the substance but because these cues signify drug availability. It is this expectancy of drug availability then that elicits CRs such as subjective craving and attentional bias towards drug-paired cues. Therefore, cognitive appraisal of the availability of the substance is an important mediator of the ability of salient cues to promote conditioned responding. The difference between this and incentive salience is subtle, but it has some support in research demonstrating that smokers report greater cravings for cigarettes when there is some possibility of smoking as compared to no possibility to do so (Bailey et al. 2010). Furthermore, smoking-paired CSs have been shown to evoke craving only when subjects have an imminent opportunity to smoke (Dar et al. 2010). However, to the best of our knowledge, these effects have been difficult to replicate with non-tobacco substances such as alcohol (Davidson et al. 2003; MacKillop and Lisman 2005), suggesting the theory is imperfect. Nonetheless, it presents a compelling demonstration that at least in some cases the motivational force of CSs may be contingent on a variety of complex environmental factors.

3 The Role of Cues in Gambling

While the above theories are framed within the context of substance abuse, the powerful motivational effects of cues are not limited to drug-taking and extend to behavioural addictions and gambling in particular. The effects of gambling-associated cues in problem gamblers are comparable to the effects of drug cues in problem users in at least some ways. Exposure to gambling cues can induce craving in problem and frequent gamblers (Kushner et al. 2008; McGrath et al. 2013). Problem gamblers also appear to be more sensitive to such cues than non-problem gamblers. Adolescent pathological gamblers reported being more attracted by music, lights and noises produced by slot machines than non-pathological adolescent gamblers (Griffiths 1990). Removing sound from video lottery terminals and decreasing speed of play decreased ratings of enjoyment, excitement and tension-relief more in pathological than in non-pathological gamblers (Loba et al. 2001); pathological gamblers also experienced more difficulty stopping play in the presence of sound cues and at higher play speeds. Though it is not possible here to disentangle the effect of sound alone on gambling behaviour from the concurrent changes in rate of play, it at least suggests that sound can modulate the experience of gambling for individuals who exhibit disordered gambling. Furthermore, like problem substance users, problem gamblers show attentional biases towards gambling-related stimuli across different paradigms, including gambling Stroop, dual tasks, flicker and attentional blink tasks, as well as eye fixation and ERP reactivity measures though the findings have not been entirely consistent (for review, see Honsi et al. 2013). Attentional bias towards gambling cues has been suggested to play a critical role in the transition from recreational to problem gambling (van Holst et al. 2012; Grant and Bowling 2014).

Despite these similarities, there may also be important differences in the roles that cues play in substance versus gambling contexts. Similar to drug cues, gambling cues are associated with rewards (in this case, monetary), or the possibility of rewards. However, in the case of gambling cues are linked with rewards at multiple levels. Broad contextual cues, such as red lights, casino sounds and appearance of gambling tables and machines, are not specifically associated with outcomes, yet signal the possibility of a reward if gambling is initiated. These seem phenomenologically most similar to drug cues. Anticipatory cues, such as reel spins and accompanying music, signal the possibility of an imminent reward in a given play. Outcome-specific cues, such as flashing lights and sounds of tumbling coins of the slot machine when a win occurs, are concurrent with and symbolic of monetary rewards and hence might themselves help reinforce and maintain gambling once it has already been initiated. Whereas other research has posited that sound serves as an occasion setter or discriminative stimuli that essentially sets the stage for other stimuli to modulate gambling behaviour (Griffiths and Parke 2005), some have suggested that win-associated cues are second-order conditioned stimuli, which become rewarding in their own right (Dixon et al. 2014) (see below for discussion of conditioned reinforcement). This distinction is subtle but important. Again, describing salient cues such as win-related lights and sounds as mere occasion-setters relegates them to a supporting role in maintaining disadvantageous behaviour, rather than a driving force with direct influence on decision-making. They have frequently been described as the former; lights and sounds of fruit machines have been characterized as "psycho-structural... characteristics" that serve as "gambling inducers" (Griffiths 1993), serving to "create an atmosphere which is probably conducive to gambling" (Caldwell 1974). In contrast, Dixon et al.'s work regards gambling-related stimuli as having a function similar to that of drug-related CSs, in that sound is capably of directly influencing disadvantageous gambling behaviour.

Different types of gambling cues (contextual, anticipatory, outcome-specific) may influence the gambler's experiences and behaviour in different ways—a possibility that has not yet been comprehensively studied, but appears to be supported by at least some evidence. Though this research is in its infancy, the handful of existing studies suggest that contextual gambling cues affect subjective experiences and energize behaviour of the player, whereas outcome-specific (win-associated) cues additionally affect and distort gambling-related cognitions. Thus, ambient cues (red lights, casino sounds) that were not specifically associated with outcomes on the IGT had a positive effect on mood and speeded up reaction times to make choices following losses, but had no effect on choice behaviour (Brevers et al. 2015). Higher tempo of background music increased the speed of betting in a virtual roulette game, especially when combined with ambient red light, but did not affect bet size or the amount spent (Dixon et al. 2007). Though the effects of anticipatory gambling cues remain mostly unstudied, one experiment found that that sequential presentation of symbols on the different reels may be more reinforcing to the players than simultaneous presentation of the symbols on all the reels, as sequential presentation increased the number of games played (Ladouceur and Sevigny 2002); however, varying the duration of the reel spin did not affect any aspect of gambling behaviour (Sharpe et al. 2005).

Unlike contextual cues, outcome-specific cues appear to affect play-related cognitions. The presence versus absence of specifically win-associated auditory cues —jingles varying in length and intensity as a function of win size—not only resulted in increased arousal (measured via galvanic skin responses and self-report) and higher preference ratings for the cued version of the task, but also led the subjects to overestimate their frequency of winning (Dixon et al. 2014). Other evidence comes from studies of win-associated audiovisual cues that slot machines commonly present during "wins" that actually fall short of the amount wagered—in other words "losses disguised as wins" (LDW) (Dixon et al. 2010, 2015). Such audiovisual "disguise" proves compelling: LDWs resulted in indices of physiological arousal that were more similar to those produced by genuine wins than those produced by frank losses. Sounds accompanying LDWs, in their own right, had a significant impact on subjects' impression of winning or losing: when LDWs were accompanied by winning sounds, players miscategorized the majority of these trials as wins and overestimated their overall frequency of winning; when LDWs were accompanied by losing sounds, both categorization and recall of winning frequency were considerably more accurate. Gambling-related cognitive distortions are believed to play an important role in driving pathological gambling (Clark 2010). Therefore, the demonstrated effects of outcome-specific cues on cognitive variables raise the possibility that these cues could thereby help drive disadvantageous gambling-related choices and behaviour. To the best of our knowledge, this possibility has not yet been tested in humans, and the effects of outcome-associated cues on human choice behaviour remain unstudied. This area deserves more attention. Research with both human and animal models with careful manipulation of cues at every level would provide valuable insight that could ultimately inform prevention and treatment of disordered gambling. Further, given the sophistication of cues in gambling and gaming, systematic study of these cues and their effects could produce new insights regarding the role of cues in addiction more generally, which may have escaped recognition with the focus on the apparently simpler drug cues.

3.1 Animal Models of the Influence CS Exert over Behaviour

While the value of human gambling research is self-apparent, the use of animal models provides insight that complements and expands on the human literature. Examining the behavioural influence of cues in rodent models provides more explicit neurobiological information and allows for manipulations that are not possible in human subjects. While the research into gambling-specific effects of cues is more limited (if not non-existent) in animal models, several established animal paradigms do investigate the ability of CSs to affect behaviour.

3.2 Sign-Tracking

Pavlov's seminal research demonstrated that some animals began to treat the stimuli predictive of reward as though it were the reward itself (Fig. 2). He wrote "…the animal may lick the electric lamp (that is predictive of food), or appear to take the air into its mouth, or to eat the sound, licking his lips and making the noise of chewing with his teeth as though it were a matter of having the food itself" (Pavlov 1927). Approach to and engagement with the cue suggested that it had taken on motivational properties of its own, and was not merely predictive of reward for some animals but rewarding in and of itself. This sort of engagement with the CS has been well documented in the literature; pigeons will peck at a cue light that predicts reward delivery, even though food delivery is not contingent on any instrumental response (Brown and Jenkins 1968), while raccoons trained to deposit a token to receive a food reward treat the token as though it were food itself, washing it and gnawing on it for extended periods of time despite the fact that these behaviours prevent the acquisition of the food itself (Breland and Breland 1961). In each of these

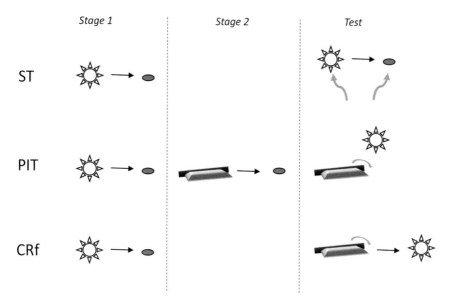

Fig. 2 Simplified illustration of the training stages need for sign-tracking (*ST*), Pavlovian-to-instrumental transfer (*PIT*) and conditioned reinforcement (*CRf*). *Black arrows* represent contingencies in place. *Grey arrows* represent the animal's behaviour. The first training stage for all three processes is identical: the animal learns that a conditioned stimulus (*CS*) is associated with an unconditioned stimulus (*US*), such as reward delivery, through classical conditioning. The CS is represented here as a visual stimulus for ease, but can theoretically be a cue of any modality. The US is depicted as a sugar pellet, but can likewise be any US. During the sign-tracking procedure, either in a designated test session or throughout acquisition, the experimenter then measures the number of times the animal approaches or interacts with the CS (sign-tracking), or instead approaches or interacts with the site of reward delivery (goal-tracking). During PIT, the animal learns that an operant response, such as depressing a lever as shown here, leads to delivery of reward. In this depiction, the reward is identical to that used in the classical conditioning session as for outcome-specific PIT, but could also be a reward of similar valence but different quality as in general PIT. In the critical test session, the animal makes the operant response in extinction, and the CS is presented non-contingently. Successful PIT is indicated by elevated responding during the CS. To test for CRf, the degree to which the animal must make a novel operant response, reinforced solely by delivery of the CS, is determined. For all three paradigms, an additional CS is typically also included, presentation of which does not have any consequences, to control for non-specific responding for cues (omitted from the figure for clarity)

examples, it appears that the reward-predictive cue acquired great incentive value of their own, sufficient to distract at least some of the animals from the US.

Engagement with reward-predictive cues has come to be known as sign-tracking (as opposed to goal-tracking, or engagement with the reward itself), and the study of this phenomenon has provided some of the most robust evidence for the incentive sensitization model of addiction described above. In one well-documented model of sign-tracking (ST, Robinson and Flagel 2009) (see chapters by Robinson et al. and Meyer et al., in this volume), a lever with an illuminated light above it is presented for a brief period of time. It is retracted, and a reward pellet is delivered to a food magazine immediately proximate to the lever. Importantly, the delivery of the reward pellet is not contingent on any instrumental response from the subject. Animals trained on this task can be divided into three groups based on their behavioural response to the illuminated lever. One group (the goal-trackers) orients towards the food magazine when the lever is illuminated, while another group (sign-trackers) engages with the lever itself. A third group spends approximately the same time with both the illuminated lever and the food magazine. Researchers have posited that the inclination to approach the reward-associated cue or "sign" over the reward itself represents a misattribution of salience that may be a marker for vulnerability to a host of behavioural disorders, including addiction (Robinson and Flagel 2009). In essence, the task provides a measure of the ability of salient reward-related cues to gain control over behaviour, roughly analogous to processes seen in the maintenance and reinstatement of drug addiction. Work with the task has demonstrated, among other findings, increased sensitivity to cocaine-induced plasticity in sign-trackers (Flagel et al. 2008), distinct alterations in the dopamine system in sign-trackers and goal-trackers (Flagel et al. 2007) and elevated corticosterone in sign-trackers relative to other groups (Tomie et al. 2000, 2004). ST also seems to be associated with other traits thought to confer vulnerability to addiction, including high reactivity to a novel environment (as measured by locomotor activity) (Flagel et al. 2010) and increased reinstatement of drug-seeking following extinction of cocaine self-administration (Saunders and Robinson 2010). ST thus has advantages in examining specific addiction-related behavioural profiles, and work with the model is providing valuable insights into individual attributions of incentive salience and the "misbehaviour of organisms", to borrow Breland's phrasing (Breland and Breland 1961).

3.3 Pavlovian-to-Instrumental Transfer

Pavlovian-to-Instrumental Transfer (PIT) examines the ability of CSs associated with an outcome to invigorate instrumental responding for either the same outcome (outcome-specific PIT), or one of a similar valence (appetitive or aversive), even when there is no formal association between CS and instrumental responding (see chapter by Corbit and Balleine in this volume). In outcome-specific PIT procedures, subjects learn two distinct contingencies. The first is a simple classical conditioning procedure in which the non-contingent delivery of a reinforcer is paired with a stimulus. Importantly, reinforcement is not dependent on any response from the subject. The second is an instrumental responding procedure where the subject must execute some behaviour in order to receive the same reinforcer (i.e. there is a causal relationship between the animal's behaviour and the delivery of reward). The testing period then takes place in extinction conditions, where the instrumental response is not rewarded, in order to see whether presentation of the reward-paired stimuli invigorates engagement in the previously reward-paired action.

The ability of CSs to encourage responding for an US in PIT paradigms probably reflects the CSs' ability to increase motivation (either generally or more specifically) for the reinforcer used. An alternative interpretation that deserves consideration is whether PIT instead reflects any crossover between the instrumental response and some behaviour evoked by the CS. For example, a CS can prompt subjects to approach the location of reward (Brown and Jenkins 1968), and if the instrumental response must be made proximally to the site of reinforcement, the effects of the CS on instrumental responding might merely be an interaction between responses to the CS and the particular instrumental behaviour. While there may well be something to this theory (Karpicke et al. 1977), PIT is not reducible to this mere interference effect. Lovibond conditioned a jaw-movement response to a CS in rabbits by pairing it with the administration of a sucrose solution, and then separately trained these same subjects to press a lever for the sucrose solution (Lovibond 1983). When the CS was presented while the subjects were working for reward, it invigorated lever-pressing. This suggests the CS evoked a general increase in motivation, as the jaw-movement CR did not promote anything resembling lever-pressing and in fact reduced lever-pressing when evoked by sucrose administration. Furthermore, the expression of PIT critically depends on the motivational state of the animal; in order for food-paired cues to evoke PIT, the animal must be hungry during the test phase (see Cardinal et al. 2002 for further discussion). PIT therefore appears to provide reliable evidence that CSs can produce a general increase in motivation for desirable USs, suggesting one method by which cues can come to influence behaviour. This phenomenon has also recently been described in human subjects, and stronger PIT observed in individuals that exhibit greater sign-tracking to a reward-paired cue (Garofalo and di Pellegrino 2015).

3.4 Conditioned Reinforcement

Tests of conditioned reinforcement (CRf) may look methodologically quite similar to PIT- again, the CS is first classically conditioned to reward delivery. However, the subsequent CRf test then determines the degree to which rats will perform a novel response, such as lever-pressing, that is reinforced solely by the CS. Thus, in contrast to PIT, presentation of the CS is *entirely contingent* on the animals' behaviour (Robbins 1978; Williams 1994). The process of CRf is thought to underlie second-order schedules of reinforcement of drug self-administration that are typically used to assess drug-seeking rather independent of drug-taking (Arroyo et al. 1998; Di Ciano and Everitt 2005). In such paradigms, animals initially make a single response to receive an infusion of an addictive substance, such as cocaine, paired with an audiovisual CS, such as a light or tone. Over successive iterations, an association is therefore formed between experience of the drug and the CS. The power of this association is so strong that this CS is then capable of supporting operant behaviour independent of drug delivery, as demonstrated in subsequent training sessions in which the response requirements are progressively increased such that animals must respond numerous times to receive presentation of the CS, and numerous CSs prior to receipt of a single drug infusion. Similar findings have been reported in humans (Panlilio et al. 2005). Such second-order schedules allow for the extensive study of the neurobiology underlying responding for drug in the absence of any confounding behavioural effects caused by drug delivery.

3.5 Interim Summary

All three processes, ST, PIT, and CRf, can be considered somewhat hierarchically in that the property of cues which they measure increases in behavioural significance, from attracting interest (ST), to influencing ongoing goal-directed behaviour (PIT), and finally to becoming the goal itself (CRf). All of these cue-driven processes have also been implicated in addiction, but in subtly different ways. As discussed above, ST is thought to reflect the degree to which cues paired with addictive drugs can induce the desire to use (Flagel et al. 2009). PIT taps into the process by which ongoing goal-directed behaviour can be influenced by encountering reward-paired cues and thus may reflect how cue-induced craving translates into active drug-seeking (Tiffany and Drobes 1990; Gawin 1991; O'Brien et al. 1998; Tomie et al. 2008). Evidence also suggests that the cues associated with drug-taking become CRfs and represent autonomous subgoals in their own right that are valued independently from the drug themselves (Williams 1994). This powerful observation helps explain why drug substitution therapy can combat the physiological symptoms associated with drug withdrawal but does not necessarily reduce craving and the desire to use; the addict still yearns for the sensory experience triggered by the drug-paired cues (Rose and Levin 1991; Naqvi and Bechara 2005, 2006). The degree to which individuals vary in their willingness to work for CRfs may therefore have a direct relationship to relapse vulnerability, particularly at timepoints distal to cessation of use, long after physiological withdrawal has passed. Interestingly, responding for CRfs is higher in rats during adolescence, a developmental period associated with higher vulnerability to addiction (Burton et al. 2011).

While PIT, CRf and sign-tracking tasks all provide valuable information into the ways in which cues modulate behaviour, they are somewhat removed from the specific type of decision-making that is recruited in the context of gambling and even relapse to addiction. Furthermore, although ST, PIT and CRf may look superficially quite similar, they depend on somewhat distinct neural and neurochemical systems that nevertheless overlap with those involved in addiction and affective decision-making within limbic corticostriatal loops (Cardinal et al. 2002). Given that very similar-looking cue-dependent behaviours can depend on dissociable neurobiological substrates, the question then remains as to whether the influence of cues in more complex cognitive processes, such as the kinds of cost/benefit decision-making involved in gambling behaviour, is subject to similar or distinct regulatory mechanisms.

3.6 The Addition of Cues to Decision-Making Tasks Fundamentally Alters Neurobiological Regulation of Choice

Although there are few reports of animal models in which the presence or absence of cues on decision-making has been explicitly studied, one exception is the delay-discounting model of impulsive choice. In this paradigm, animals choose between smaller-sooner versus larger-later rewards, thereby modelling the degree to which delay to gratification affects the subjective appraisal of reward value (see Mazur 1997 for review). If a cue light is illuminated during the delay, rats become less impulsive (Cardinal et al. 2000; Zeeb et al. 2010). Interestingly, whereas lesions or inactivations of the orbitofrontal cortex (OFC) decrease impulsive choice in the absence of the cue, the opposite pattern of results is observed if the delay is cued, and this increase in impulsive choice is most prominent in animals showing lower levels of impulsivity at baseline, i.e. those that were arguably using the cue to mitigate the negative impact of the delay (Zeeb et al. 2010). The role of the OFC in decision-making therefore appears particularly sensitive to the presence of cues, but whether this can be attributed to CRf mechanisms is currently unknown. Given that the cue is presented only after a response, it seems unlikely that ST or PIT would be acutely involved on a trial-by-trial basis. However, acquisition of complex operant tasks likely involves the formation of numerous associations, not all of which are immediately obvious to, or intended by, the experimenter. ST has been associated with impulsive choice even on an uncued delay-discounting paradigms (Flagel et al. 2010), implying there may be some neurobiological or phenomenological overlap between these processes.

With respect to addiction, one important consideration is the way that drugs of abuse boost the power that reward-paired cues have on behaviour due to hyper-stimulation of the dopamine (DA) system. Although not the only neurochemical system implicated, DA's influence is certainly the most well-established, and the nucleus accumbens (NAC) the neural target of most intensive study. Natural rewards simulate the firing of DA neurons in the mesolimbic pathway, but if those rewards are reliably predicted by a CS, this firing switches to presentation of the cue (Schultz et al. 1997; Schultz 1998; Clark et al. 2013). Increasing DA release actively recruits the NAC into the process of responding for CRf- under baseline conditions, lesions to accumbal regions have no effect on this behaviour (Taylor and Robbins 1984, 1986; Cador et al. 1991; Parkinson et al. 1999; Cardinal et al. 2002). Similarly, PIT can be enhanced by intra-NAC amphetamine and abolished by DA antagonists (Dickinson et al. 2000), and ST is likewise sensitive to DAergic manipulations of the NAC (Wyvell and Berridge 2000; Di Ciano et al. 2001; Dalley et al. 2002, 2005). Clearly changing DA signalling enhances the role of the NAC in cue-sensitive behaviours, but the addition of cues can likewise make behaviour DA-dependent. Administration of DA antagonists directly into the OFC only decreased impulsive choice in the cued version of the delay-discounting task, theoretically by reducing the ability of the cue to promote choice of the larger delayed reward (Zeeb et al. 2010).

Although systemic administration of DA receptor-type 2/3 $D_{2/3}$ antagonist moderately improved choice on the rGT, neither chronic nor acute administration of $D_{2/3}$ agonists impacted behaviour (Zeeb et al. 2009; Tremblay et al. 2013). The findings are in stark contrast to the ability of such drug regimens to enhance risky choice on a simpler test of preference for uncertainty, in which both cues and striatal $D_{2/3}$ receptors play a prominent role (Cocker et al. 2012; Tremblay et al. 2013). Furthermore, administration of the selective DA reuptake inhibitor GBR 12909 did not affect decision-making, although co-administration of this agent with the selective noradrenaline reuptake blocker did mimic the deleterious effects of amphetamine (Baarendse et al. 2012). In addition, while both D1 and D2-family antagonists can attenuate impulsive responses caused by amphetamine, neither of these compounds could attenuate amphetamine-induced impairments in choice (Zeeb et al. 2013). In sum, choice behaviour on the rGT does not seem to be predominantly driven by the DA system. While this may not alter the utility of the task in modelling decision-making under uncertainty, it may impact the ability of the task to accurately approximate certain aspects of pathological engagement in risky decision-making. The evidence reviewed above indicates that the addition of reward-paired cues may improve not only the face validity of the rGT, but also the construct and predictive validity of this paradigm.

In order to explore this hypothesis, we therefore disproportionately cued wins on the rGT's disadvantageous options to see whether these cues can shift animals' decision-making preferences (Barrus and Winstanley 2014, 2015). The pairing of salient cues to disadvantageously risky options is similar to human gambling paradigms in which large, often risky wins are more saliently cued than small wins or losses. The structure of the cued rGT was identical to that of the traditional rGT, save the introduction of salient cues to winning trials. On the cued rGT, a loss on any option was identical to a loss on that same option on the traditional rGT. However, while a win on the rGT was marked by the allocation of sucrose pellets and the solid illumination of the tray light, a win on the cued rGT was additionally marked by a combination of tones and flashing light. Although all wins, large or small, were accompanied by an audiovisual cue of equal length and intensity (i.e. brightness and loudness), the cues associated with the larger rewards were more complex and variable. Just as in a human gambling paradigm, the salience of the win-associated cues therefore increased with the size of the win.

Results to date indicate that animals performing the cued rGT adopt a riskier, more disadvantageous choice strategy than those on the uncued task (Barrus and Winstanley 2014). These results demonstrate that salient, audiovisual win-paired cues are sufficient to enhance choice of riskier, more disadvantageous options, thereby modelling the negative impact such cues may have on human choice. Furthermore, the presence of such cues alters the way in which certain dopaminergic ligands, namely those acting at subtypes of the D_2 receptor family, impact decision-making. While D_2- and D_4-selective agents were without effect on either version, choice on the cued task appears uniquely sensitive to modulation by DA

D_3 receptor ligands; D_3 agonism increased choice of the high-risk option associated with maximal uncertainty with respect to the delivery of reward or punishment, whereas D_3 antagonism had the opposite effect (Barrus and Winstanley 2015submitted). These compounds did not affect choice in the uncued paradigm (Di Ciano et al. 2015; Barrus and Winstanley 2015 submitted). Numerous studies specifically implicate D3 receptors in mediating the maladaptive influence of cues in substance use disorder, and recent data posit a critical role for this receptor subtype in GD (Le Foll et al. 2014; Lobo et al. 2015). The cued rGT may therefore provide a novel and relatively unique method to empirically determine the degree to which cue-sensitivity can promote poor choice in a cost/benefit model in a manner central to the addiction process.

3.7 Concluding Remarks

Associative learning is one of the fundamental building blocks of advanced cognitive processes. The degree to which associations are formed between cues and outcomes clearly shapes behaviour in both relatively simple ways, as in basic classical conditioning procedures, but also in more complex paradigms. The ability of drug-paired cues to influence drug-seeking and ongoing goal-directed behaviour lies at the heart of current theories of chemical dependency and has been investigated in tightly controlled animal experiments. The study of gambling in humans indicates that the numerous, salient, audiovisual cues used in commercial gambling scenarios can invigorate behaviour, but it is unclear whether these cues have as fundamental role to play in GD as they are theorized to have in drug addiction. Although much less is known about the importance of cues in processes relevant to the development of GD, recent data indicate that the presence of win-paired cues can bias animals towards risky choices and alter dopaminergic regulation of decision-making behaviour. Whether this cue-induced risky choice behaviour is driven by the same kinds of cue-driven behaviours as implicated in addiction (ST, PIT, CRf) remains to be experimentally determined. Understanding the similarities and differences in the motivational influence exerted by cues in chemical and behavioural addictions could further elucidate the degree to which these conditions can be considered homogeneous and therefore responsive to similar pharmacological and behavioural treatment interventions.

3.8 Financial Disclosures

This work was supported by open operating grants awarded to CAW from the Canadian Institutes for Health Research (CIHR). CAW also received salary support through the Michael Smith Foundation for Health Research (MSFHR) and a CIHR New Investigator Award program. In the past three years, CAW has consulted for

Shire on an unrelated matter. MC is supported by postdoctoral salary awards from both MSFHR and CIHR. No authors have any other conflicts of interest or financial disclosures to make.

References

Arroyo M, Markou A, Robbins TW, Everitt BJ (1998) Acquisition, maintenance and reinstatement of intravenous cocaine self-administration under second-order schedule of reinforcement in rats: effects of conditioned cues and continuous to cocaine. Psychopharmacology 140:331–344

Baarendse PJ, Winstanley CA, Vanderschuren LJ (2012) Simultaneous blockade of dopamine and noradrenaline reuptake promotes disadvantageous decision making in a rat gambling task. Psychopharmacology 225:719–731

Bailey SR, Goedeker KC, Tiffany ST (2010) The impact of cigarette deprivation and cigarette availability on cue-reactivity in smokers. Addiction 105:364–372

Barrus MM, Winstanley CA (2014) Win-cued lights and tones drive risky decision making in a rodent gambling task. IBNS June 2014

Barrus MM, Winstanley CA (2015) Dopamine D3 receptors modulate the ability of win-paired cues to increase risky choice in a rat gambling task. J Neurosci. (Submitted)

Bechara A, Damasio AR, Damasio H, Anderson SW (1994) Insensitivity to future consequences following damage to human prefrontal cortex. Cognition 50:7–15

Berridge KC, Robinson TE (1998) What is the role of dopamine in reward: hedonic impact, reward learning, or incentive salience? Brain Res Brain Res Rev 28:309–369

Boileau I, Dagher A, Leyton M, Gunn RN, Baker GB, Diksic M, Benkelfat C (2006) Modeling sensitization to stimulants in humans: an [11C] raclopride/positron emission tomography study in healthy men. Arch Gen Psychiatry 63:1386–1395

Breland K, Breland M (1961) The misbehavior of organisms. Am Psychol 16:681–684

Brevers D, Noel X, Bechara A, Vanavermaete N, Verbanck P, Kornreich C (2015) Effect of casino-related sound, red light and pairs on decision-making during the Iowa gambling task. J Gambl Stud 31:409–421

Brown PL, Jenkins HM (1968) Auto-shaping of the pigeon's keypeck. J Exp Anal Behav 11:1–8

Burton CL, Noble K, Fletcher PJ (2011) Enhanced incentive motivation for sucrose-paired cues in adolescent rats: possible roles for dopamine and opioid systems. Neuropsychopharmacology 36:1631–1643

Cador M, Taylor JR, Robbins TW (1991) Potentiation of the effects of reward-related stimuli by dopaminergic-dependent mechanisms in the nucleus-accumbens. Psychopharmacology 104:377–385

Caldwell G (1974) The gambling Australian. In: Edgar DE (ed) Social change in Australia. Cheshire, Melbourne, pp 13–28

Cardinal RN, Robbins TW, Everitt BJ (2000) The effects of d-amphetamine, chlordiazepoxide, alpha- flupenthixol and behavioural manipulations on choice of signalled and unsignalled delayed reinforcement in rats. Psychopharmacology 152:362–375

Cardinal RN, Parkinson JA, Hall J, Everitt BJ (2002) Emotion and motivation: the role of the amygdala, ventral striatum, and prefrontal cortex. Neurosci Biobehav Rev 26:321–352

Carter BL, Tiffany ST (1999a) Meta-analysis of cue-reactivity in addiction research. Addiction 94:327–340

Carter BL, Tiffany ST (1999b) Cue-reactivity and the future of addiction research. Addiction 94:349–351

Childress AR, Ehrman R, Roohsenow DJ, Robbins SJ, O'Brien CP (1992) Classically conditioned factors in drug-dependence. In: Lowinson W, Luiz P, Millman RB, Langard JG (eds) Substance abuse: a comprehensive textbook. Williams and Wilkins, Baltimore

Childress AR, Hole AV, Ehrman RN, Robbins SJ, McLellan AT, O'Brien CP (1993) Cue reactivity and cue reactivity interventions in drug dependence. NIDA Res Monogr 137:73–95

Childs E, de Wit H (2009) Amphetamine-induced place preference in humans. Biol Psychiatry 65:900–904

Childs E, de Wit H (2013) Contextual conditioning enhances the psychostimulant and incentive properties of d-amphetamine in humans. Addict Biol 18:985–992

Clark JJ, Collins AL, Sanford CA, Phillips PE (2013) Dopamine encoding of Pavlovian incentive stimuli diminishes with extended training. J Neurosci 33:3526–3532

Clark L (2010) Decision-making during gambling: an integration of cognitive and psychobiological approaches. Philos Trans R Soc Lond B Biol Sci 365:319–330

Cocker PJ, Winstanley CA (2015) Irrational beliefs, biases and gambling: exploring the role of animal models in elucidating vulnerabilities for the development of pathological gambling. Behav Brain Res 279:259–273

Cocker PJ, Dinelle K, Kornelson R, Sossi V, Winstanley CA (2012) Irrational choice under uncertainty correlates with lower striatal D(2/3) receptor binding in rats. J Neurosci 32:15450–15457

Cox WM, Fadardi JS, Pothos EM (2006) The addiction-stroop test: theoretical considerations and procedural recommendations. Psychol Bull 132:443–476

Cunningham-Williams RM, Grucza RA, Cottler LB, Womack SB, Books SJ, Przybeck TR, Spitznagel EL, Cloninger CR (2005) Prevalence and predictors of pathological gambling: results from the St. Louis personality, health and lifestyle (SLPHL) study. J Psychiatr Res 39:377–390

Dalley JW, Chudasama Y, Theobald DE, Pettifer CL, Fletcher CM, Robbins TW (2002) Nucleus accumbens dopamine and discriminated approach learning: interactive effects of 6-hydroxydopamine lesions and systemic apomorphine administration. Psychopharmacology 161:425–433

Dalley JW, Laane K, Theobald DE, Armstrong HC, Corlett PR, Chudasama Y, Robbins TW (2005) Time-limited modulation of appetitive Pavlovian memory by D1 and NMDA receptors in the nucleus accumbens. Proc Natl Acad Sci USA 102:6189–6194

Dar R, Rosen-Korakin N, Shapira O, Gottlieb Y, Frenk H (2010) The craving to smoke in flight attendants: relations with smoking deprivation, anticipation of smoking, and actual smoking. J Abnorm Psychol 119:248–253

Davidson D, Tiffany ST, Johnston W, Flury L, Li TK (2003) Using the cue-availability paradigm to assess cue reactivity. Alcohol Clin Exp Res 27:1251–1256

de Visser L, Homberg JR, Mitsogiannis M, Zeeb FD, Rivalan M, Fitoussi A, Galhardo V, van den Bos R, Winstanley CA, Dellu-Hagedorn F (2011) Rodent versions of the iowa gambling task: opportunities and challenges for the understanding of decision-making. Front Neurosci 5:109

Di Ciano P, Everitt BJ (2005) Neuropsychopharmacology of drug seeking: Insights from studies with second-order schedules of drug reinforcement. Eur J Pharmacol 526:186–198

Di Ciano P, Cardinal RN, Cowell RA, Little SJ, Everitt BJ (2001) Differential involvement of NMDA, AMPA/kainate, and dopamine receptors in the nucleus accumbens core in the acquisition and performance of pavlovian approach behavior. J Neurosci 21:9471–9477

Di Ciano P, Pushparaj A, Kim AS, Hatch J, Masood T, Khaled MA, Boileau I, Winstanley CA, Le Foll B (2015) The impact of selective dopamine D2, D3 and D4 ligands on the rat gambling task. PLoS ONE EPub ahead of print

Dickinson A, Smith J, Mirenowicz J (2000) Dissociation of Pavlovian and instrumental incentive learning under dopamine antagonists. Behav Neurosci 114:468–483

Dixon L, Trigg R, Griffiths M (2007) An empirical investigation of music and gambling behaviour. Int Gambl Stud 7:315–326

Dixon MJ, Harrigan KA, Sandhu R, Collins K, Fugelsang JA (2010) Losses disguised as wins in modern multi-line video slot machines. Addiction 105:1819–1824

Dixon MJ, Collins K, Harrigan KA, Graydon C, Fugelsang JA (2015) Using sound to unmask losses disguised as wins in multiline slot machines. J Gambl Stud 31:183–196

Dixon MJ, Harrigan KA, Santesso DL, Graydon C, Fugelsang JA, Collins K (2014) The impact of sound in modern multiline video slot machine play. J Gambl Stud 30:913–929

Field M, Cox WM (2008) Attentional bias in addictive behaviors: a review of its development, causes, and consequences. Drug Alcohol Depend 97:1–20

Flagel SB, Akil H, Robinson TE (2009) Individual differences in the attribution of incentive salience to reward-related cues: Implications for addiction. Neuropharmacology 56(Suppl 1):139–148

Flagel SB, Watson SJ, Robinson TE, Akil H (2007) Individual differences in the propensity to approach signals vs goals promote different adaptations in the dopamine system of rats. Psychopharmacology 191:599–607

Flagel SB, Watson SJ, Akil H, Robinson TE (2008) Individual differences in the attribution of incentive salience to a reward-related cue: influence on cocaine sensitization. Behav Brain Res 186:48–56

Flagel SB, Robinson TE, Clark JJ, Clinton SM, Watson SJ, Seeman P, Phillips PE, Akil H (2010) An animal model of genetic vulnerability to behavioral disinhibition and responsiveness to reward-related cues: implications for addiction. Neuropsychopharmacology 35:388–400

Garofalo S, di Pellegrino G (2015) Individual differences in the influence of task-irrelevant Pavlovian cues on human behavior. Front Behav Neurosci 9:163

Gawin FH (1991) Cocaine addiction: psychology and neurophysiology. Science 250:1580–1586

Goudriaan AE, Oosterlaan J, de Beurs E, van den Brink W (2005) Decision making in pathological gambling: a comparison between pathological gamblers, alcohol dependents, persons with Tourette syndrome, and normal controls. Brain Res Cogn Brain Res 23:137–151

Grant LD, Bowling AC (2014) Gambling attitudes and beliefs predict attentional bias in non-problem gamblers. J Gambl Stud

Griffiths M (1993) Fruit machine gambling: the importance of structural characteristics. J Gambl Stud 9:101–120

Griffiths MD (1990) The acquisition, development, and maintenance of fruit machine gambling in adolescents. J Gambl Stud 6:193–204

Griffiths MD, Parke J (2005) The psychology of music in gambling environments: An observational research note. J Gambl Issues 13–18

Grimm JW, Hope BT, Wise RA, Shaham Y (2001) Neuroadaptation. Incubation of cocaine craving after withdrawal. Nature 412:141–142

Heinz A, Siessmeier T, Wrase J, Hermann D, Klein S, Grusser SM, Flor H, Braus DF, Buchholz HG, Grunder G, Schreckenberger M, Smolka MN, Rosch F, Mann K, Bartenstein P (2004) Correlation between dopamine D(2) receptors in the ventral striatum and central processing of alcohol cues and craving. Am J Psychiatry 161:1783–1789

Hogarth L, Dickinson A, Duka T (2003) Discriminative stimuli that control instrumental tobacco-seeking by human smokers also command selective attention. Psychopharmacology 168:435–445

Honsi A, Mentzoni RA, Molde H, Pallesen S (2013) Attentional bias in problem gambling: a systematic review. J Gambl Stud 29:359–375

Karpicke J, Christoph G, Peterson G, Hearst E (1977) Signal location and positive versus negative conditioned suppression in the rat. J Exp Psychol-Anim Behav Process 3:105–118

Kushner M, Thurus P, Sletten S, Frye B, Abrams K, Adson D, Demark JV, Maurer E, Donahue C (2008) Urge to gamble in a simulated gambling environment. J Gambl Stud 24:219–227

Ladouceur R, Sevigny S (2002) Symbols presentation modality as a determinant of gambling behavior. J psychol 136:443–448

Le Foll B, Collo G, Rabiner EA, Boileau I, Merlo Pich E, Sokoloff P (2014) Dopamine D3 receptor ligands for drug addiction treatment: update on recent findings. Prog Brain Res 211:255–275

Leyton M, Vezina P (2013) Striatal ups and downs: their roles in vulnerability to addictions in humans. Neurosci Biobehav Rev 37:1999–2014

Leyton M, Vezina P (2014) Dopamine ups and downs in vulnerability to addictions: a neurodevelopmental model. Trends Pharmacol Sci 35:268–276

Loba P, Stewart SH, Klein RM, Blackburn JR (2001) Manipulations of the features of standard video lottery terminal (VLT) games: effects in pathological and non-pathological gamblers. J Gambl Stud 17:297–320

Lobo DS, Aleksandrova L, Knight J, Casey DM, El-Guebaly N, Nobrega JN, Kennedy JL (2015) Addiction-related genes in gambling disorders: new insights from parallel human and pre-clinical models. Mol Psychiatry 20:1002–1010

Lovibond PF (1983) Facilitation of instrumental behavior by a Pavlovian appetitive conditioned stimulus. J Exp Psychol Anim Behav Process 9:225–247

MacKillop J, Lisman SA (2005) Reactivity to alcohol cues: isolating the role of perceived availability. Exp Clin Psychopharmacol 13:229–237

Mazur J (1997) Choice, delay, probability and conditioned reinforcement. Anim Learn Behav 25:131–147

McGrath DS, Dorbeck A, Barrett SP (2013) The influence of acutely administered nicotine on cue-induced craving for gambling in at-risk video lottery terminal gamblers who smoke. Behav Pharmacol 24:124–132

Mucha RF, Pauli P, Angrilli A (1998) Conditioned responses elicited by experimentally produced cues for smoking. Can J Physiol Pharmacol 76:259–268

Naqvi NH, Bechara A (2005) The airway sensory impact of nicotine contributes to the conditioned reinforcing effects of individual puffs from cigarettes. Pharmacol Biochem Behav 81:821–829

Naqvi NH, Bechara A (2006) Skin conductance responses are elicited by the airway sensory effects of puffs from cigarettes. Int J Psychophysiol 61:77–86

O'Brien CP, Childress AR, Ehrman R, Robbins SJ (1998) Conditioning factors in drug abuse: can they explain compulsion? J Psychopharmacol 12:15–22

O'Daly OG, Joyce D, Stephan KE, Murray RM, Shergill SS (2011) Functional magnetic resonance imaging investigation of the amphetamine sensitization model of schizophrenia in healthy male volunteers. Arch Gen Psychiatry 68:545–554

Paine TA, Asinof SK, Diehl GW, Frackman A, Leffler J (2013) Medial prefrontal cortex lesions impair decision-making on a rodent gambling task: reversal by D1 receptor antagonist administration. Behav Brain Res 243:247–254

Panlilio LV, Yasar S, Nemeth-Coslett R, Katz JL, Henningfield JE, Solinas M, Heishman SJ, Schindler CW, Goldberg SR (2005) Human cocaine-seeking behavior and its control by drug-associated stimuli in the laboratory. Neuropsychopharmacology 30:433–443

Parkinson JA, Olmstead MC, Burns LH, Robbins TW, Everitt BJ (1999) Dissociation in effects of lesions of the nucleus accumbens core and shell on appetitive pavlovian approach behavior and the potentiation of conditioned reinforcement and locomotor activity by d-amphetamine. J Neurosci 19:2401–2411

Pavlov IR (1927) Conditioned reflexes. Oxford University Press, Oxford

Potenza MN (2009) The importance of animal models of decision making, gambling, and related behaviors: implications for translational research in addiction. Neuropsychopharmacology 34:2623–2624

Robbins SJ, Ehrman RN (2004) The role of attentional bias in substance abuse. Behav Cogn Neurosci Rev 3:243–260

Robbins TW (1978) The acquisition of responding with conditioned reinforcement: effects of pipradrol, methylphenidate, d-amphetamine, and nomifensine. Psychopharmacology 58:79–87

Robinson TE, Berridge KC (1993) The neural basis of drug craving: an incentive-sensitisation theory of addiction. Brain Res Rev 18:247–291

Robinson TE, Flagel SB (2009) Dissociating the predictive and incentive motivational properties of reward-related cues through the study of individual differences. Biol Psychiatry 65:869–873

Rose JE, Levin ED (1991) Inter-relationships between conditioned and primary reinforcement in the maintenance of cigarette smoking. Br J Addict 86:605–609

Saunders BT, Robinson TE (2010) A cocaine cue acts as an incentive stimulus in some but not others: implications for addiction. Biol Psychiatry 67:730–736

Schultz W (1998) Predictive reward signal of dopamine neurons. J Neurophysiol 80:1–27

Schultz W, Dayan P, Montague PR (1997) A neural substrate of prediction and reward. Science 275:1593–1599

Schulze D, Jones BT (1999) The effects of alcohol cues and an alcohol priming dose on a multi-factorial measure of subjective cue reactivity in social drinkers. Psychopharmacology 145:452–454

Shaham Y, Shalev U, Lu L, De Wit H, Stewart J (2003) The reinstatement model of drug relapse: history, methodology and major findings. Psychopharmacology 168:3–20

Sharpe L, Walker M, Coughlan MJ, Enersen K, Blaszczynski A (2005) Structural changes to electronic gaming machines as effective harm minimization strategies for non-problem and problem gamblers. J Gambl Stud 21:503–520

Shurman B, Horan WP, Nuechterlein KH (2005) Schizophrenia patients demonstrate a distinctive pattern of decision-making impairment on the Iowa Gambling Task. Schizophr Res 72:215–224

Taylor JR, Robbins TW (1984) Enhanced behavioral control by conditioned reinforcers following microinjections of D-amphetamine into the nucleus accumbens. Psychopharmacology 84:405–412

Taylor JR, Robbins TW (1986) 6-Hydroxydopamine lesions of the nucleus accumbens, but not of the caudate nucleus, attenuate enhanced responding with reward- related stimuli produced by intraaccumbens D-amphetamine. Psychopharmacology 90:390–397

Tiffany ST, Drobes DJ (1990) Imagery and smoking urges: the manipulation of affective content. Addict Behav 15:531–539

Tomie A, Grimes KL, Pohorecky LA (2008) Behavioral characteristics and neurobiological substrates shared by Pavlovian sign-tracking and drug abuse. Brain Res Rev 58:121–135

Tomie A, Aguado AS, Pohorecky LA, Benjamin D (2000) Individual differences in pavlovian autoshaping of lever pressing in rats predict stress-induced corticosterone release and mesolimbic levels of monoamines. Pharmacol Biochem Behav 65:509–517

Tomie A, Tirado AD, Yu L, Pohorecky LA (2004) Pavlovian autoshaping procedures increase plasma corticosterone and levels of norepinephrine and serotonin in prefrontal cortex in rats. Behav Brain Res 153:97–105

Tremblay M, Hosking JG, Winstanley CA (2013) Effects of chronic D-2/3 agonist ropinirole medication on rodent models of gambling behaviour. Movement Disord 28:S218–S218

van Holst RJ, Lemmens JS, Valkenburg PM, Peter J, Veltman DJ, Goudriaan AE (2012) Attentional bias and disinhibition toward gaming cues are related to problem gaming in male adolescents. J Adolesc Health 50:541–546

Verdejo-Garcia A, Benbrook A, Funderburk F, David P, Cadet JL, Bolla KI (2007a) The differential relationship between cocaine use and marijuana use on decision-making performance over repeat testing with the Iowa Gambling Task. Drug Alcohol Depend 90:2–11

Verdejo-Garcia AJ, Perales JC, Perez-Garcia M (2007b) Cognitive impulsivity in cocaine and heroin polysubstance abusers. Addict Behav 32:950–966

Williams BA (1994) Conditioned reinforcement: neglected or outmoded explanatory construct? Psychon Bull Rev 1:457–475

Williams WA, Grant JE, Winstanley CA, Potenza MN (2008) Current concepts in the classification, treatment and modeling of pathological gambling and other impulse control disorders. In: McArthur RA, Borsini F (eds) Reward deficit disorders. Elsevier, Burlington MA, pp 317–357

Wyvell CL, Berridge KC (2000) Intra-accumbens amphetamine increases the conditioned incentive salience of sucrose reward: enhancement of reward "wanting" without enhanced "liking" or response reinforcement. J Neurosci 20:8122–8130

Yager LM, Robinson TE (2013) A classically conditioned cocaine cue acquires greater control over motivated behavior in rats prone to attribute incentive salience to a food cue. Psychopharmacology 226:217–228

Zeeb FD, Winstanley CA (2011) Lesions of the basolateral amygdala and orbitofrontal cortex differentially affect acquisition and performance of a rodent gambling task. J Neurosci 31:2197–2204

Zeeb FD, Winstanley CA (2013) Functional disconnection of the orbitofrontal cortex and basolateral amygdala impairs acquisition of a rat gambling task and disrupts animals' ability to alter decision-making behavior after reinforcer devaluation. J Neurosci 33:6434–6443

Zeeb FD, Robbins TW, Winstanley CA (2009) Serotonergic and dopaminergic modulation of gambling behavior as assessed using a novel rat gambling task. Neuropsychopharmacology 34:2329–2343

Zeeb FD, Floresco SB, Winstanley CA (2010) Contributions of the orbitofrontal cortex to impulsive choice: interactions with basal levels of impulsivity, dopamine signalling, and reward-related cues. Psychopharmacology 211:87–98

Zeeb FD, Wong AC, Winstanley CA (2013) Differential effects of environmental enrichment, social-housing, and isolation-rearing on a rat gambling task: dissociations between impulsive action and risky decision-making. Psychopharmacology 225:381–395

Zeeb FD, Baarendse PJ, Vanderschuren LJM, Winstanley CA (2015) Inactivation of the prelimbic or infralimbic cortex impairs decision-making in the rat gambling task. Psychopharmacology (Berl)

Part V
Developments in Treatments for Motivation Pathologies

The Role of Motivation in Cognitive Remediation for People with Schizophrenia

Alice M. Saperstein and Alice Medalia

Abstract Motivation impairment is an often prominent component of schizophrenia symptomatology that impacts treatment engagement and reduces the functional benefit from psychosocial interventions. Intrinsic motivation in particular has been shown to be impaired in schizophrenia. Nowhere is the role of intrinsic motivation impairment more evident than in cognitive remediation for schizophrenia. This chapter describes the theoretical determinants of motivation to learn and illustrates how those determinants have been translated into therapeutic techniques that enhance intrinsic motivation in a clinical context. We review the extant research that indicates how motivation enhancing techniques yield treatment-related improvements within cognitive remediation therapy and, more broadly, in other behavioral skills-based interventions for schizophrenia.

Keywords Cognitive remediation · Intrinsic motivation · Self-determination theory · Expectancy-value theory

Contents

1	Introduction	534
2	The Role of Motivation in Learning	535
3	Determinants of Intrinsic Motivation	536
4	Intrinsic Motivation to Learn in Schizophrenia	537
	4.1 Do I Expect Success?	537
	4.2 Do I Value the Task?	538
	4.3 Are My Needs for Autonomy, Competence, and Relatedness Met When I Engage in This Learning Activity?	539

A.M. Saperstein (✉) · A. Medalia
Department of Psychiatry, Columbia University Medical Center,
1051 Riverside Drive, Mailbox 104, New York, USA
e-mail: saperst@nyspi.columbia.edu

A. Medalia
e-mail: am2938@columbia.edu

© Springer International Publishing Switzerland 2015
Curr Topics Behav Neurosci (2016) 27: 533–546
DOI 10.1007/7854_2015_373
Published Online: 15 March 2015

5	Translating Motivation Theory to Cognitive Remediation Practice	540
	5.1 The Learning Environment	540
	5.2 Instructional Techniques	541
6	Conclusion	544
References		545

1 Introduction

The earliest conceptualizations of schizophrenia included motivation deficits as a key feature of the disorder (Bleuler 1911; Krapelin 1919). Symptoms such as amotivation and apathy are considered core negative symptom features that are distinctly associated with substantial functional impairment (Strauss et al. 2013) when measured cross-sectionally (Foussias et al. 2009; Gard et al. 2009) as well as longitudinally (Foussias et al. 2011). Recently, there has been an increase in focus on understanding motivational impairment—the component processes, underlying neurocognitive mechanisms, observable behaviors, and functional consequences—so as to better target this major source of disability in people with schizophrenia (Barch 2008; Fervaha et al. 2013; Medalia and Saperstein 2011).

To better understand the "why" of motivated or unmotivated behavior in schizophrenia, researchers have examined goal-directed behavior in reference to intrinsic versus extrinsic rewards. *Intrinsic motivation* describes actions and behaviors that are accomplished in the absence of an external consequence-behaviors performed because they provide inherent, intrinsic satisfaction. *Extrinsic motivation*, in contrast, describes actions and behaviors that are a means to some end, to either achieve some external gain, or to avoid a prospective loss or punishment (Ryan and Deci 2000a). In most environments, both intrinsic and extrinsic motivation are operative. For example, motivation to work may be driven by the opportunity for financial gain (extrinsic reward), but being employed may also support a sense of fulfillment and enhance self-esteem (intrinsic reward). In this chapter, we will consider the roles of intrinsic and extrinsic motivation in the treatment of people with schizophrenia.

Recovery-oriented therapeutic programs for people with schizophrenia center around skills training, so that individuals can learn the social, cognitive, employment, and educational skills necessary to navigate daily challenges and progress toward meaningful goals. Cognitive remediation is one evidence-based skills-training intervention that focuses on improving those deficient cognitive skills that impede daily functioning and goal attainment. It is a learning activity through which people learn how to better pay attention, to process information quickly, to remember better, and to problem-solve through a combination of cognitive exercise and the learning of strategies for carrying out cognitive tasks in everyday life. In this therapeutic context, cognitive and motivational processes interact to either enhance or detract from the development of cognitive skills. There is a wealth of research that has delineated the relative contributions of intrinsic and extrinsic motivation in facilitating learning in healthy people, and an accumulating literature on the role of motivation in cognitive learning in people with schizophrenia.

2 The Role of Motivation in Learning

Research conducted in educational settings with non-psychiatric samples provides an empirically based and informative framework for understanding how intrinsic and extrinsic motivation influence learning as it may occur in skills-based interventions, such as cognitive remediation, employed with people with schizophrenia. On the whole, intrinsic motivation is appreciated as a central force in learning. Research indicates that when intrinsic motivation for learning is high, there is greater engagement in learning activities, greater creativity, learning, and greater persistence of learning over time (Vansteenkiste et al. 2004). Extrinsic motivators play some role in learning, although evidence indicates that when intrinsic motivation effectively drives learning behaviors in the absence of extensive external rewards, those learning behaviors are likely to be maintained. There is some research to suggest that when extrinsic rewards are offered, or contingencies are introduced, when intrinsic motivation is already strong, there is a negative impact on the amount of learning that takes place, in part by undermining intrinsic motivation (Deci et al. 1999; Dweck 1986).

Consistent with the core symptomatology of the disorder, people with schizophrenia tend to exhibit low intrinsic motivation, whether it is assessed in reference to daily activities, interest and curiosity, or goal-directed behavior in everyday life (Nakagami et al. 2008). Empirical studies of motivation in schizophrenia also indicate low intrinsic motivation in reference to treatment-related task performance and skill learning (Choi et al. 2010b). Findings of low motivation for skill learning have serious implications for the effectiveness of psychosocial skills interventions. The impact of baseline intrinsic motivation on treatment-related learning was examined in a study, which reported dramatic differences in effect sizes for learning outcomes when people with schizophrenia participating in a cognitive remediation program were divided into those exhibiting high versus low intrinsic motivation for treatment. In this study, participants were attending a community-based psychosocial program where attendance is not mandated, but is dependent upon each individual's own volition. Given the lack of external contingencies for participation, the frequency of voluntary attendance in cognitive remediation was the measure of intrinsic motivation. A large effect size was found for improvement on an untrained vocational task requiring processing speed for those who demonstrated greater baseline intrinsic motivation. By contrast, the participants who were not intrinsically motivated achieved a very small effect size on this outcome measure (Choi and Medalia 2005). These data suggest that when people with schizophrenia are intrinsically motivated for treatment, treatment engagement and outcomes may be enhanced. Knowing that people with schizophrenia tend to have intrinsic motivation deficits, it is important to understand whether intrinsic motivation in schizophrenia is inherently stable or dynamic. If the latter, the next step is to discern the person-related or environmental variables associated with shifts in intrinsic motivation.

Recent research with schizophrenia patients suggests that intrinsic motivation can indeed change over time. In a prospective study, Nakagami et al. (2010) examined intrinsic motivation among 130 individuals with schizophrenia or schizoaffective disorder attending community-based psychosocial rehabilitation programs. Clinician ratings of intrinsic motivation (i.e., ratings of sense of purpose, motivation, and curiosity) indicated a significant change over 6 and 12 months following the initial assessment. Importantly, positive changes in intrinsic motivation were strongly associated with positive changes in psychosocial functioning, thereby underscoring the significance of intrinsic motivation in the context of treatment (Nakagami et al. 2010).

In a separate study of 57 outpatients with schizophrenia, Choi and Medalia used the intrinsic motivation inventory for schizophrenia research (IMI-SR) before and after a 4-week cognitive training intervention. Again, results supported the dynamic nature of intrinsic motivation; intrinsic motivation to participate in the training program increased in a subgroup that received a motivationally engaging version of the intervention relative to the subgroup who did not receive motivational enhancements in the training program (Choi and Medalia 2010; for a description of interventions see below). Taken together, these results indicate that intrinsic motivation is dynamic and, even when impaired, it is still sensitive to manipulation in a therapeutic context. The next section will discuss theories of motivation that have aided our understanding of how to treat motivation deficits in schizophrenia.

3 Determinants of Intrinsic Motivation

Two theories of motivation have figured prominently in schizophrenia research—self-determination theory (SDT) and expectancy-value theory (EVT) by providing a basis for the empirical study of therapeutic techniques that increase motivated behavior and learning outcomes.

SDT posits that behavior is driven by the need to gratify psychological needs for autonomy, competency, and relatedness (Deci and Ryan 1985). The extent to which these basic needs are met will impact motivation to learn and thus learning behavior. First, the need for *autonomy* is supported when an individual experiences an internal perceived locus of control or causality. When an individual feels, their behavior is self-determined, as opposed to being controlled by external factors, when there are opportunities for choice and self-direction, intrinsic motivation may be enhanced (Deci and Ryan 1985). Second, the need for *competency* is supported when the learning environment promotes feelings of mastery, where a sense of confidence in achieving success is supported (Ryan and Deci 2000b). SDT postulates an interactive role of autonomy and competency; for intrinsic motivation to be optimally supported, feelings of competence must be accompanied by the attribution of success to one's own self-determined behavior. Third, the need for *relatedness* is the need to have meaningful connections with others. When a sense of security in the learning environment is supported, intrinsic motivation and learning outcomes will be supported too (Ryan and Deci 2000b).

EVT (Eccles and Wigfield 2002; Wigfield and Eccles 2002) posits that expectations for success as a learner and the perceived value of the task are interacting variables that drive motivation to learn. Expectations for success are influenced by one's assessment of past performance, ability beliefs, and one's appraisal of current task properties such as perceived task difficulty, the clarity of the task goals, and the temporal proximity of goal attainment. Similar to what is posited in SDT, an expectation to succeed yields a sense of competency and thus weighs heavily on one's drive to pursue a learning goal. In the learning context, those who feel competent are more likely to choose challenging learning tasks and will be more willing to try new ones (Schunk and Zimmerman 2008). Subjective value appraisals also impact the initiation and maintenance of learning behavior. There is nuance to the way a learning activity is valued, with respect to how interesting and enjoyable the task is, how useful it is for meeting goals, and whether attainment of the task goals is held with importance and esteem. These positive appraisals of task value are offset by perceived costs such as time, effort, stress, financial burden, or lost opportunity for other pursuits. Research indicates that the expectancy-value model predicts many learning outcomes. Expectancy for success predicts effort, engagement, and achievement in a learning context. Value appraisals predict behavioral choices to initiate a learning activity and to persist (Schunk et al. 2014). Thus when the value of engaging in a learning task is salient to the individual, and the individual expects to be successful as a learner, behavior that is oriented toward learning will occur.

4 Intrinsic Motivation to Learn in Schizophrenia

The theoretical perspectives on intrinsic motivation yield three basic questions that can guide research on motivation to learn in patients with schizophrenia. From the patient's point of view, these questions are as follows: "Do I expect success if I engage in the learning task?" "Do I value the task?" "Are my needs for autonomy, competence, and relatedness met when I engage in this learning activity?" Each question will be addressed in turn, using theories of motivation and empirical studies to describe intrinsic motivation to learn in schizophrenia.

4.1 Do I Expect Success?

Expectation of success is related to one's perceived ability, past performance, as well as present task difficulty (Schunk et al. 2014). In the setting of cognitive remediation, people with schizophrenia often have varied histories of successes and struggles in school or work, and therefore come to cognitive remediation with different expectations. Frequently, as the illness progresses, success in learning activities declines and leads to expectations of further failure. There is considerable data demonstrating the impact of perceived competency on learning in non-psychiatric research samples (Jones 2009). In schizophrenia samples, data indicate that perceived competency plays an instrumental role in determining intrinsic motivation for current task performance (Choi et al. 2012) and accumulating data indicating a large role for perceived competency in learning outcomes.

Choi and Medalia (2010) examined the degree to which perceptions of competency for a difficult cognitive task contributed to the amount of learning in 57 adults with schizophrenia-spectrum disorders. Higher perceived competency on an arithmetic task prior to training predicted greater learning. These findings were corroborated in a follow-up study using an independent sample of 70 schizophrenia outpatients which reported that greater expectation of success—the perception of competence on a learning task—was the most important factor in explaining how much was learned during the training intervention and how much was retained at 3-month follow-up (Choi et al. 2010a).

Expectations of success and competency also play a significant role in translating the skills one has learned into behavior. This is illustrated in a randomized controlled trial in which 97 adults with schizophrenia were enrolled in one of two psychosocial skills-training interventions for improving functioning. The study focused on assessing daily functioning, functional capacity, and self-efficacy for performing everyday functional tasks. Data showed that only when self-efficacy was high, functional capacity was significantly associated with actual real-world daily functioning. These data suggest that without confidence in their ability to perform the skills they have learned, individuals who otherwise have the capacity to do so will maintain relatively high levels of disability (Cardenas et al. 2013). Therefore, when behavioral interventions target those skills deficits that impede functioning, motivational processes such as self-efficacy need to be targeted as well, so as to support the translation of skill capacity to skill performance in everyday life.

4.2 Do I Value the Task?

The second question "do I value the task?" is important because people are more likely to engage in a learning activity if the value is perceived. Choi et al. (2010a) demonstrated that task value is an important factor in supporting intrinsic motivation and learning in schizophrenia. In fact, individuals who ascribed greater value to the learning task also reported greater expectations for success for the learning program, thereby illustrating how value and competency beliefs may interact to impact motivation and learning. Several factors contribute to why learning experiences, such as cognitive remediation, may be valued. First, *intrinsic value* is ascribed to a task if it is interesting or enjoyable. The potential intrinsic value of cognitive exercises is important to consider when implementing cognitive remediation for people with schizophrenia. Second, a learning activity may be perceived as having *utility value* if it is seen as instrumental in meeting one of the individual's long- or short-range goals. In cognitive remediation, the utility value of cognitive learning tasks may be enhanced by making explicit links between cognitive improvement and meaningful goals, like returning to school, work, or living independently. Third, a learning activity may have *attainment value* if it facilitates achievement of a desired self-image. For example, personal importance ascribed to remembering names, faces and details may promote motivation for engaging in memory exercises during cognitive remediation. Finally, the *cost* of engaging in the learning activity impacts the value appraisal. If cost is too high in terms of time, stress, or finances, then the value is diminished.

4.3 Are My Needs for Autonomy, Competence, and Relatedness Met When I Engage in This Learning Activity?

According to SDT, individuals will be intrinsically motivated to engage in learning tasks if they believe they have choice in the matter (autonomy), if they feel able to master the goals of the task (competence), and experience the accompanying social interactions as positive (relatedness) (Ryan and Deci 2000b). There is ample evidence to support this from studies conducted with students in educational settings (Jones 2009). Several studies with schizophrenia samples in learning settings suggest that a qualitatively similar motivational process is operative. Autonomy, as measured by perceived control, contributes to overall levels of intrinsic motivation in schizophrenia (Choi et al. 2010b) and when enhanced is associated with greater learning outcomes (Choi and Medalia 2010; Tas et al. 2012). Perceived competency was found to mediate the relationship between trait approach motivation and in-the-moment, state intrinsic motivation for learning in schizophrenia (Choi et al. 2012). Nakagami et al. (2010) and Silverstein (2010) describe the importance of creating therapeutic contexts where there is a collaborative-supportive, as opposed to controlling-hierarchical relationship between the therapist and client with schizophrenia. This is because controlling contexts, which rely on extrinsic incentives, pressure, or punishment to influence learning behavior and use directive or judgmental language in the process of learning, reduce self-determination and instead result in passivity, decreased persistence in learning, and poor learning overall (Grolnick and Ryan 1987; Vansteenkiste et al. 2004). Conversely, learning contexts where supportive language is used during instruction, and learning is personalized, foster intrinsic motivation and elicit positive learning behaviors and outcomes (Vansteenkiste et al. 2004). Taken together, studies of intrinsic motivation support the premise of SDT that gratification of needs for autonomy, competence, and relatedness are pertinent to learning outcomes in people with schizophrenia. Additionally, the factors postulated by SDT to impact intrinsic motivation for learning appear to overlap or interact with those put forth by EVT, thus providing a comprehensive framework for translating motivation theory to cognitive remediation practice.

5 Translating Motivation Theory to Cognitive Remediation Practice

We have reviewed the data that indicate that when intrinsic motivation in a learning context is high or is enhanced, so too are learning outcomes. Providing a context in which intrinsic motivation for cognitive learning is enhanced requires the translation of theoretical principles, grounded in research, to clinical practice. In an educational setting, the variables that impact intrinsic motivation to learn are manifested in the interpersonal context and in the instructional techniques employed. In cognitive remediation, basic psychological needs can be supported through the structure of the learning environment and the therapist-client relationship.

5.1 The Learning Environment

Within the learning environment, a cognitive remediation therapist can support autonomy by encouraging personal goal setting, and by guiding clients through the use of learning activities that suit their interests, learning needs and goals so that the enjoyment and utility value of cognitive exercise are emphasized. Acknowledging each client's unique learning needs and personal goals supports autonomy and may also fulfill the need for relatedness through generating an atmosphere of personal respect and support for each client's perspective. There is emerging empirical evidence that intrinsic motivation to learn is enhanced in an autonomy-supportive cognitive remediation environment where people with schizophrenia are allowed to exercise some control over their learning experience, where the value of the activity is salient and when opportunities for demonstrating competency exist (Choi and Medalia 2010).

Cognitive remediation is often conducted in a group setting, given that it is not only an efficient form of service delivery, but more so because of the positive impact of social learning on cognitive performance. Bandura's social cognitive theory (2006) posits that modeling can instill self-efficacy when, for example, a peer demonstrates effective skill learning. Thus, even if cognitive remediation clients are working individually on cognitive exercises, sharing in the learning experience with others supports a positive and supportive learning environment (Bandura 2006). In addition, when cognitive remediation group members share common goals of learning, the social supports can help to fulfill the need for relatedness and can thus boost intrinsic motivation for participating in the learning process. Some cognitive remediation programs include a verbal discussion component, where individuals have the opportunity to relate cognitive skill learning to everyday life, and to provide feedback and support to other group members. By providing a supportive environment in which skill learning is reinforced, the experiential context of cognitive remediation can bolster intrinsic motivation for learning and facilitate the successful application of skills and strategies for recovery goal attainment.

5.2 Instructional Techniques

Instructional techniques are another social contextual determinant of intrinsic motivation to learn. Instructional variables such as personalization, choice, and contextualization (Cordova and Lepper 1996) can be embedded into specific learning activities. *Personalization* refers to the tailoring of a learning activity to coincide with topics of interest or utility value, given the individuals goals for recovery. Some computer-based learning programs personalize the learning experience by allowing the learner to enter into the program as an identifiable and independent agent, for example, by signing in by name or taking on a role within a learning task that simulates a real-world activity. *Choice* in cognitive remediation means that clients can participate in planning the course of learning activities, based on their interests, cognitive needs and learning style, and to exercise choice within cognitive exercises with regard to the difficulty level or incidental features of the program. Finally, *contextualization* means that rather than presenting material in the abstract, where the relevance for everyday life is less salient, it is put into a meaningful context, whereby the practical application to activities of daily life or personal goals is made more apparent. Some cognitive programs, for example, utilize real-world scenarios such as a coffee shop or restaurant to engage learners in exercises to improve attention and memory. By designing cognitive exercises as games, where practicing memory and attention are presented in the context of a restaurant simulation, the interest, enjoyment, and utility value of the learning task may help to drive continued learning behavior.

Empirical data support the use of these techniques in people with schizophrenia. A previously described randomized controlled trial conducted by Choi and Medalia (2010) directly compared the impact of instructional techniques on motivation and learning in people with schizophrenia. Participants were either exposed to an enhanced learning program that contextualized the cognitive task into a meaningful game-like context, created a personalized character for each participant, and provided opportunities for choice within the learning activity, or were exposed to a contrasting program that provided generic instruction. Following the course of intervention, those who were assigned to the motivationally enhanced learning program demonstrated greater intrinsic motivation for the learning experience, acquired more cognitive skill, and reported greater feelings of competency after treatment. These findings underscore the malleability of the motivational system in people with schizophrenia and provide further credence to the applicability of motivation theory to the practice of cognitive remediation for schizophrenia, and likely for other patient populations as well.

Instructional techniques in cognitive remediation can also harness the impact of enhancing self-efficacy. As discussed above, perceived competency is a predictor of learning achievement, and level of task difficulty is a predictor of perceived competency. Instructors can carefully titrate the complexity level of cognitive exercises so that an 80 % or greater success rate is achieved, thus promoting perceived competency while also maintaining a level of challenge that supports cognitive growth. Another instructional technique that may enhance perceived competency to learn during cognitive remediation is the careful titration of the goal properties of the learning task. Tasks which have distal goals may be perceived as too difficult, or the end-goal may be perceived as too vague, thereby lessening a sense of competency for successful task completion. In contrast, tasks that have proximal goals that are clearly defined may be viewed as more attainable. Selection of learning activities should therefore consider both task complexity and goal properties, so as to best account for individuals' cognitive capacity and ability to persist, and thereby promote self-efficacy and motivation to initiate and complete a learning task.

A third instructional technique to promote self-efficacy and motivation is through provision of feedback during learning activities. *Attributional feedback*, provided by the therapist, verbalizes the link between the individual's effort or ability and their success on the task, thereby promoting a sense of agency as well as self-efficacy during the process of learning. *Performance feedback* provides information about progress in learning. Some software programs depict level of cognitive performance and improvements after each learning trial. Cognitive remediation clients can also track their own performance to encourage self-awareness of learning progress. Research suggests that people with schizophrenia who utilize feedback to self-monitor progress in learning have been shown to reap greater benefit in cognitive remediation (Wykes et al. 2007; Choi and Kurtz 2009).

A fourth instructional technique to promote expectation of learning competence is to emphasize that cognition is malleable, because of the human capacity for neuroplasticity. This contrasts with a perception of cognition as a fixed entity, which Dweck (1999) found to negatively impact learning outcomes in educational settings. Vinogradov et al. (2012) reported that schizophrenia participants who believe that intelligence is malleable show better cognitive outcomes after computerized training than those who believe it is a fixed entity, even after controlling for baseline cognition and number of hours of training. For people with schizophrenia who may have experienced repeated failures across multiple domains of functioning, emphasizing the process of learning and malleability of cognition, rather than the outcome, may best support feelings of competency. This will support active engagement in, rather than avoidance of, the learning environment and create the potential for learning to occur.

The Neuropsychological Educational Approach to Remediation (NEAR; Medalia et al. 2009) is a holistic approach to cognitive remediation for schizophrenia that incorporates theoretical principles and practices from educational psychology, learning theory, rehabilitation psychology, and neuropsychology. To optimize learning, motivation enhancing principles are incorporated into the NEAR model in a variety of ways.

First, the NEAR approach begins with an assessment of baseline cognitive abilities, which provides an understanding of the individual's cognitive strengths and weaknesses, and an understanding of how cognitive deficits interfere with functioning. A joint client–therapist process of goal setting builds an atmosphere of respect for the individual's learning needs and a mutual understanding of the client's personal goals for recovery. Second, intrinsic motivation and task engagement are promoted through incorporating learning activities that contextualize cognitive exercise in real-world situations, through providing opportunities for personal control of the non-essential aspects of the learning environment, using multisensory presentation of cognitive learning tasks, and through personalization of learning material. These features allow for the interest, enjoyment, and utility value of cognitive tasks to be salient. Furthermore, task difficulty and goal properties of tasks are tailored to suit the needs of each individual so that opportunities for success are maximized. As clients improve, task difficulty is carefully modulated within and between different exercises, to keep rates of successful performance high while continuing to challenge and teach new skills to promote learning.

Third, the overall structure of NEAR sessions is designed to enhance intrinsic motivation. Learning occurs in a group setting to support social learning, yet there is an individualized treatment plan for each client. Treatment plans are tailored to the individual's cognitive and learning needs so that clients will readily appreciate the relevance of the tasks for their particular situation. Further, linking the cognitive remediation program to each person's overall rehabilitation goals such as work, socialization, or independent living clarifies the relevance and utility of participation in the program. Fourth, the therapist is responsible for creating an autonomy-supportive learning environment. Provision of choice of learning activities provides opportunities for self-direction and fosters autonomy. The therapist avoids use of controlling language and provides frequent positive reinforcement for adaptive learning behavior such as persistence, effective strategy use, and effort allocation. In addition, the therapist helps clients track their learning progress on cognitive exercises which promotes self-reflection on cognitive progress and demonstrates competency.

To date, the NEAR approach has been successfully implemented in outpatient and inpatient settings which serve diverse psychiatric populations (Medalia et al. 2000, 2001, 2009; Hodge et al. 2010). This is but one of many holistic approaches to treating cognitive impairment in psychiatric disorders, and treatment techniques that systematically address motivation impairment to improve treatment engagement and outcomes are increasingly being studied in a broader range of interventions for schizophrenia (Grant et al. 2012; Silverstein 2010). Given ongoing investigations of the cognitive and neural mechanisms associated with motivational impairments in schizophrenia (see Waltz and Gold in this volume), therapeutic strategies may continue to evolve. We have illustrated that taking into account the social contextual variables that impact intrinsic motivation can create the optimal therapeutic conditions in which effective skill learning and functional goal attainment can be achieved.

6 Conclusion

Motivation is a key determinant of psychosocial treatment outcome and is thus an important target for behavioral intervention. When it comes to learning, motivation theories posit the central role of intrinsic motivation. Research indicates that the motivational factors that affect cognitive skills learning in people with schizophrenia are similar to those in non-psychiatric populations. Motivation theory and research have thus served as useful guideposts for the development of strategies to target low levels of intrinsic motivation during learning to augment treatment outcomes.

Although intrinsic motivation may be lowered in schizophrenia, importantly for therapeutic opportunity, it is malleable. Specific environmental conditions are able to support the expression, maintenance, and enhancement of this motivational capacity. When learning environments are able to gratify psychological needs of autonomy, competence, and relatedness, and foster expectations of success and the value of learning, intrinsic motivation is supported.

Empirical studies have demonstrated that in people with schizophrenia, there is a natural variability in the extent to which learning tasks are perceived as having interest, enjoyment, and utility value. There is also variability in the extent to which people with schizophrenia feel competent when engaging in learning tasks. We have described data from schizophrenia samples that show that these experiences, conceptualized as underlying intrinsic motivation, not only predict volitional learning behavior in the absence of extrinsic rewards, but also account for the magnitude and sustainability of learning. Importantly, data show that when instructional techniques support intrinsic motivation to learn, there are indeed measurable changes in intrinsic motivation, accompanied by greater participation in learning activities, and greater learning outcomes. It is important to note that the enhanced learning outcomes evidenced by pre- and post-treatment assessment in these studies were not due to baseline level of ability or the intensity with which learning sessions occurred. Rather, the data indicate that by providing the supportive conditions, intrinsic motivation for learning is indeed enhanced, and learning outcomes are improved.

We have also shown that perceived competency during learning is a powerful determinant of learning in and of itself. Perceived competency appears to be an important link between motivational drive measured more broadly and intrinsic motivation within a specific learning context. Data also suggest that perceived competency is an important mechanism through which motivational enhancements in cognitive remediation result in significant changes in cognitive ability. In accordance with motivation theory, people with schizophrenia must believe that their actions can lead to positive outcomes or else they may have little incentive to take on challenging treatment tasks. This has been demonstrated not only in the context of cognitive remediation, but also in the context of utilizing learned psychosocial skills to carry out daily life tasks important for real-world functioning.

How do the social and contextual determinants of intrinsic motivation translate into treatment-related improvements? To date, the data suggest that when people with schizophrenia value learning, experience successes, and attribute their success to their self-directed effort, subsequent engagement in the learning activities is deepened, effort on learning activities persists, and learning outcomes are enhanced. While these data have demonstrably influenced the design of cognitive remediation interventions, there are clear implications for the integration of these principles to support engagement in any psychosocial skills-training intervention for people with schizophrenia and other psychiatric disorders. A motivationally supportive therapeutic context may not only enhance learning within the psychosocial treatment milieu, but also set the stage for the deployment of learned skills in real-world contexts, thereby promoting recovery goal attainment and functional improvement.

References

Bandura A (2006) Toward a psychology of human agency. Perspect Psychol Sci 1:164–180
Barch DM (2008) Emotion, motivation and reward processing in schizophrenia spectrum disorders: what we know and where we need to go. Schizophr Bull 34:816–818
Bleuler E (1911) Dementia praecox or the group of schizophrenias. International Universities Press, Oxford
Cardenas V, Abel S, Bowie CR et al (2013) When functional capacity and real-world functioning converge: the role of self-efficacy. Schizophr Bull 39:908–916
Choi J, Kurtz MM (2009) A comparison of remediation techniques on the Wisconsin card sorting test in schizophrenia. Schizophr Res 107:76–82
Choi J, Medalia A (2005) Factors associated with a positive response to cognitive remediation in a community psychiatric sample. Psychiatr Serv 56:602–604
Choi J, Medalia A (2010) Intrinsic motivation and learning in a schizophrenia-spectrum sample. Schizophr Res 118:12–19
Choi J, Fiszdon JM, Medalia A (2010a) Expectancy-value theory in persistence of learning effects in schizophrenia: role of task value and perceived competency. Schizophr Bull 36:957–965
Choi J, Mogami T, Medalia A (2010b) Intrinsic motivation inventory (IMI): an adapted scale for schizophrenia research. Schizophr Bull 36:966–976
Choi KH, Saperstein AM, Medalia A (2012) The relationship of trait to state motivation: the role of self-competency beliefs. Schizophr Res 139:73–77
Cordova DI, Lepper MR (1996) Intrinsic motivation and the process of learning: beneficial effects of contextualization, personalization and choice. J Educ Psychol 88:715–730
Deci E, Ryan R (1985) Intrinsic motivation and self-determination in human behavior. Springer, New York
Deci EL, Koestner R, Ryan RM (1999) A meta-analytic review of experiments examining the effects of extrinsic rewards on intrinsic motivation. Psychol Bull 125:627–668
Dweck CS (1986) Motivational processes affecting learning. Am Psychol 41:1040–1048
Dweck CS (1999) Self-theories: their role in motivation, personality, and development. Psychology Press, Philadelphia
Eccles JS, Wigfield A (2002) Motivational beliefs, values, and goals. Annu Rev Psychol 53:109–132
Fervaha G, Foussias G, Agid O et al (2013) Amotivation and functional outcomes in early schizophrenia. Psychiatry Res 210:665–668
Foussias G, Mann S, Zakzanis KK et al (2009) Motivational deficits as the central link to functioning in schizophrenia: a pilot study. Schizophr Res 115:333–337
Foussias G, Mann S, Zakzanis KK et al (2011) Prediction of longitudinal functional outcomes in schizophrenia: the impact of baseline motivational deficits. Schizophr Res 132:24–27

Gard DE, Fisher M, Garrett C et al (2009) Motivation and its relationship to neurocognition, social cognition, and functional outcome in schizophrenia. Schizophr Res 115:74–78

Grant PM, Huh GA, Perivoliotis D et al (2012) Randomized trial to evaluate the efficacy of cognitive therapy for low-functioning patients with schizophrenia. Arch Gen Psychiatry 69:121–127

Grolnick WS, Ryan RM (1987) Autonomy in children's learning: an experimental and individual difference investigation. J Pers Soc Psycho 81:143–154

Hodge MA, Siciliano D, Withey P et al (2010) A randomized controlled trial of cognitive remediation in schizophrenia. Schizophr Bull 36:419–427

Jones BD (2009) Motivating students to engage in learning: the MUSIC model of academic motivation. Int J Teach Learn High Edu 21:272–285

Krapelin E (1919) Dementia praecox and paraphrenia. E & S Livingstone, Edinburgh

Medalia A, Saperstein A (2011) The role of motivation for treatment success. Schizophr Bull 37: S122–S128

Medalia A, Dorn H, Watras-Gans S (2000) Treating problem-solving deficits on an acute care psychiatric inpatient unit. Psychiatry Res 97(1):79–88

Medalia A, Revheim N, Casey M (2001) The remediation of problem-solving skills in schizophrenia. Schizophr Bull 27:259–267

Medalia A, Revheim N, Herlands T (2009) Cognitive remediation for psychological disorders: therapist guide. Oxford University Press, New York

Nakagami E, Xie B, Hoe M et al (2008) Intrinsic motivation, neurocognition, and psychosocial functioning in schizophrenia: testing mediator and moderator effects. Schizophr Res 105: 95–104

Nakagami E, Hoe M, Brekke JS (2010) The prospective relationships among intrinsic motivation and psychosocial functioning in schizophrenia. Schizophr Bull 36:935–948

Ryan RM, Deci EL (2000a) Intrinsic and extrinsic motivations: classic definitions and new directions. Contemp Edu Psychol 25:54–67

Ryan RM, Deci EL (2000b) Self-Determination theory and the facilitation of intrinsic motivation, social development, and well-being. Am Psychol 55:68–78

Schunk DH, Zimmerman BJ (eds) (2008) Motivation and self-regulated learning: theory, research, and applications. Lawrence Erlbaum Associates, New York

Schunk DH, Meece JL, Pintrich PR (2014) Motivation in education: theory, research, and applications, 4th edn. Pearson, Boston

Silverstein SM (2010) Bridging the gap between extrinsic and intrinsic motivation in the cognitive remediation of schizophrenia. Schizophr Bull 36:949–956

Strauss GP, Horan WP, Kirkpatrick B et al (2013) Deconstructing negative symptoms of schizophrenia: avolition-apathy and diminished expression clusters predict clinical presentation and functional outcome. J Psychiatr Res 47:783–790

Tas C, Brown EC, Esen-Danaci A et al (2012) Intrinsic motivation and metacognition as predictors of learning potential in patients with remitted schizophrenia. J Psychiatr Res 46:1086–1092

Vansteenkiste M, Simons J, Lends W et al (2004) Motivating learning, performance, and persistence: the synergistic effects of intrinsic goal contents and autonomy supportive contexts. J Pers Soc Psychol 87:246–260

Vinogradov S, Fisher M, de Villers-Sidani E (2012) Cognitive training for impaired neural systems in neuropsychiatric illness. Neuropsychopharmacol Rev 37:43–76

Wigfield A, Eccles JS (2002) Development of achievement motivation. Academic Press, San Diego

Wykes T, Reeder C, Landau S et al (2007) Cognitive remediation therapy in schizophrenia: randomized controlled trial. Br J Psychiatry 190:421–427

Distress from Motivational *Dis*-integration: When Fundamental Motives Are *Too* Weak or *Too* Strong

James F.M. Cornwell, Becca Franks and E. Tory Higgins

Abstract Past research has shown that satisfying different kinds of fundamental motives contributes to well-being. More recently, advances in motivational theory have shown that z is also tied to the *integration* of different motives. In other words, well-being depends not only on maximizing effectiveness in satisfying specific motives, but also on ensuring that motives *work together* such that no individual motive is too weak or too strong. In this chapter, we review existing research to show that specific forms of psychological distress can be linked to specific types of motivational imbalance or *dis*-integration. Such disintegration can arise from either excessive weakness of a specific motive or the excessive strength and/or dominance of a specific motive, thereby inhibiting other motives. Possible neural correlates and avenues of intervention are discussed.

Keywords Regulatory focus · Regulatory mode · Well-being · Depression · Anxiety · Parkinson's · Narcissism

---

J.F.M. Cornwell (✉)
Department of Behavioral Sciences and Leadership, United States Military Academy, 281 Thayer Hall, West Point, New York, NY 10996, USA
e-mail: jamesfcornwell@gmail.com

B. Franks
Animal Welfare Program, University of British Columbia, 2357 Main Mall, Vancouver, BC V6T 1Z4, Canada
e-mail: beccafranks@gmail.com

E.T. Higgins
Department of Psychology, Columbia University, 406 Schermerhorn Hall, 1190 Amsterdam Ave. MC 5501, New York, NY 10027, USA
e-mail: tory@psych.columbia.edu

© Springer International Publishing Switzerland 2015
Curr Topics Behav Neurosci (2016) 27: 547–568
DOI 10.1007/7854_2015_389
Published Online: 30 September 2015

Contents

1	Four Distinct Motivational Concerns	549
2	Effective Motivational Pursuit	552
3	Dis-integration from Motives Being Too Weak	553
	3.1 Depression	553
	3.2 Parkinson's Disease	555
	3.3 Discussion: Dis-integration from Motives Being Too Weak	556
4	Dis-integration from Motives Being Too Strong	557
	4.1 Anxiety	557
	4.2 Narcissism	559
	4.3 Discussion: Dis-integration from Motives Being Too Strong	561
	4.4 An Important Caveat	562
5	Possibilities for Improvement: Micro-interventions	562
6	Motivational Integration as a Source of Well-Being	564
References		564

An inescapable characteristic of human beings is that we are *motivated* animals. Research on motivation is as old as psychology itself, and, beginning at least as early as the work of Freud (2011/1927), motivation has been explicitly tied to the well-being of individuals. Recent theoretical advances have distilled and organized past research on motivation, identifying three fundamental motives of humans and other animals (see Franks and Higgins 2012; Higgins 2012): value (wanting to have desired results), truth (wanting to establish what is real), and control (wanting to manage what happens). Value can be further divided (Higgins 1997, 1998) into ensuring better results (promotion) and ensuring against worse results (prevention). These general motives represent fundamental and distinct ways that humans and other animals want to be effective.

The link between motivation and effectiveness is crucial for our discussion of well-being. First, it is important to distinguish among the four different motives in order to understand how they each contribute separately to being effective, because being effective in satisfying a particular motive is associated with improved well-being. Extensive review of the behavioral research confirms this perspective, showing that success in value, truth, and control motives are each associated with well-being in both humans and other animals (Franks and Higgins 2012). While most research already assumes that well-being can result from subjective experience of satisfying desired life outcomes (e.g., Diener et al. 1985), research has also shown that individuals report greater well-being when they have a greater sense of meaning in their lives and a greater sense of engagement with the world around them (Peterson et al. 2005). Having a meaningful life and being engaged in life contribute to truth and control effectiveness, respectively. In non-human animals, research has shown that animals are motivated to understand how things work and to learn, even when there is no material outcome to be gained (e.g., Harlow 1950; Langbein et al. 2009). In addition, there is evidence that non-human animals want to engage in activities rather than simply acquiring the outcomes of that activity (e.g., Spinka and Wemelsfelder 2011), and they are willing to pay a cost for the opportunity to explore novel environments (Franks et al. 2013; Sherwin 2004). In addition, non-human animals distinguish between the pursuit of promotion-focused gains and prevention-focused safety (e.g., Franks et al. 2012, 2014).

Second, beyond being effective in satisfying each motive, overall well-being also depends on these concerns working together effectively—an effective organization or "fit" among them (Cornwell et al. 2014; Higgins et al. 2014). For example, if individuals are so concerned with understanding and making sense of experiences (a strong "truth" motive) that they pay little attention to either effectively managing their surroundings (a weak "control" motive) or ending up with desired end-states (a weak "value" motive), these individuals can become "lost in thought." An example of such a condition is rumination, which has been linked to deficits of well-being (Nolen-Hoeksema 2000). Interestingly, this well-being deficit takes three paths: increased consideration of negative emotions and outcomes (i.e., value ineffectiveness); inhibition of instrumental behaviors that would provide a sense of control (i.e., control ineffectiveness); and interference of problem-solving skills (i.e., truth ineffectiveness; Nolen-Hoeksema 1998). This last case highlights the important point that "strength" is not the same thing as "effectiveness." A strong motive, which in this case is strong truth, can actually be ineffective at satisfying its own concerns when it is not adequately complemented by the other fundamental motives. In the following sections, we unpack these ideas in more detail.

1 Four Distinct Motivational Concerns

In this chapter, we consider the different ways of being effective and their relation to well-being by drawing upon the motivational systems proposed by regulatory focus theory (Higgins 1997, 1998) and by regulatory mode theory (Kruglanski et al. 2000; Higgins et al. 2003). Regulatory focus theory proposes that motivation for value (i.e., having desired results) takes place in two independent systems, each with its particular focus (Higgins 1997, 1998). The *promotion* focus is primarily concerned with advancement and making progress, with attaining something better than the current status quo; it is a motivational focus on *ideals* (hopes, aspirations). The *prevention* focus is primarily concerned with safety and security, with maintaining a current satisfactory status quo against something worse or restoring a satisfactory state if the current state is unsatisfactory (e.g., restoring safety if you are currently in danger); it is a motivational focus on *oughts* (duties, responsibilities).

Thus, two ways of being effective in having desired end-states are the achievement of advancement from current states to better states (i.e., promotion value effectiveness) and the achievement of securing a satisfactory state against worse states (i.e., prevention value effectiveness). Being effective occurs with reference to some status quo because the status quo determines the value of one's current condition and the choices one makes. For prevention, being at the status quo "0" (a satisfactory state) is experienced as a positive non-loss, whereas being below it at "−1" is experienced as a negative loss. For promotion, on the other hand, being at the status quo "0" is experienced as a negative non-gain, and exceeding the status quo at "+1" is experienced as a positive gain (see Higgins 2014). This difference in the value of the status quo impacts choices. For example, when individuals with a strong prevention focus are below the status quo (e.g., in danger), they want to restore the status quo (e.g., safety) and are willing to choose a risky option if that is what is needed to make it happen. In contrast, those with a strong promotion focus are not motivated to choose a risky option if it simply restores the status quo because for them, the final state must exceed the status quo to count as a positive gain (Scholer et al. 2010). In contrast, when at the status quo, those with a strong promotion focus are willing to choose a risky option that can move them to a "+1" clearly better state, whereas those with a strong prevention focus are not willing to choose that risky option because they are satisfied with the status quo "0" and are not willing to risk moving to "−1" (Zou et al. 2014).

Complementing these value motives are the motivational systems proposed by regulatory mode theory. According to regulatory mode theory, the process of goal pursuit itself takes place through the exercise of two independent self-regulatory modes (Kruglanski et al. 2000; Higgins et al. 2003). The *locomotion* mode is primarily concerned with the initiation and maintenance of movement from state to state in the service of effecting change. It takes *control* as its reference point. Locomotion mode does not refer to motor movement per se or even a desire for motor movement. It refers to *psychological movement*, as originally discussed by Lewin (1951). Psychological movement includes activities such as making progress toward coming to a decision (Avnet and Higgins 2003) or transitioning from a state of conflict to a state of resolution (Webb et al. 2014). Moreover, locomotion mode is not necessarily concerned with movement toward any particular end-state, but only with initiating or sustaining movement *away* from the current state. At times, people in locomotion mode may do something just to experience having an effect on something, managing to make something happen. When in a locomotion state, people prefer to do *anything* than nothing at all—"just move on."

The *assessment* mode concerns a different function of self-regulation—evaluating potential options for change rather than effecting change. It is primarily concerned with the comparison and critical evaluation of alternative options in the service of finding the correct or right choice. It takes *truth* as its reference point rather than control. Like the locomotion mode, the assessment mode is orthogonal to the achievement of valued outcomes per se. The assessment mode, for example, may motivate an individual to continue seeking additional information even after a satisfactory state has been achieved. It may also motivate an individual to leave behind a safe status quo in order to explore alternatives for comparison. It is the motivation to acquire as much true knowledge as possible, by any means necessary, at whatever cost in the service of making the right choices. It is the "look" before the "leap."

The two ways of being effective here involve the achievement of control in the face of change—either in one's surrounding environment or in one's own shifting from state to state (i.e., locomotion control effectiveness)—and the integration of information into a coherent "picture" of reality in the face of multiple forms and sources of such information (i.e., assessment truth effectiveness). It should be noted that the achievement of locomotion control and assessment truth can co-occur (e.g., someone who learns by doing). In addition, they can work together effectively as motivational partners. Indeed, there is evidence that task achievement is highest when locomotion and assessment are both high (Kruglanski et al. 2000; Pierro et al., in press). Nonetheless, locomotion and assessment also have distinct relations to psychological outcomes. As one example (see Higgins et al. 2003), individuals with a strong assessment motive are more sensitive to social criticism because to arrive at the truth or the correct course of action, one must pay close attention to the information input and feedback from any and all sources, regardless of whether it is critical; that is, strong assessors take criticism to heart. In contrast, individuals with a strong locomotion motive are less sensitive to social criticism because effectively managing a change involves maintaining a plan or course of action in spite of potential obstacles, which would include social criticism; that is, strong locomotors do not let criticism throw them off course.

These four motivational systems—promotion, prevention, locomotion, and assessment—with their distinct concerns can be thought of as each pointing in a particular motivational direction. The motivational systems of regulatory focus theory represent the vertical motivational dimension, with the promotion focus motivating individuals to reach a better or "higher" state (ideals) and the prevention focus motivating individuals to maintain the status quo (oughts) and ensure against a worse state. The motivational systems of regulatory mode theory are depicted as the horizontal motivational dimension. The locomotion mode motivates individuals to effectively plan an unobstructed path forward (i.e., away from the current state)

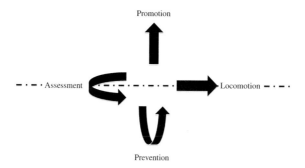

Fig. 1 A visualization of the motivations represented by regulatory focus and regulatory mode theory. The *promotion* focus is pointing upward, because it is primarily concerned with achieving states exceeding the status quo. The *prevention* focus is oriented below the current state, but is also pointing upward, because it is primarily concerned with maintaining the status quo or restoring it after dropping below it. The *locomotion* mode is pointing forward, because it is primarily concerned with movement from state to state, regardless of whether those states involve improvements or dangers. The *assessment* mode is oriented behind the current state, but is also pointing forward, because it is primarily concerned with the integration of information from a number of sources in order to make the best choices or decisions in goal pursuit. The *dashed line* represents the psychological "status quo" with respect to valued outcomes

and initiate the movement necessary for the execution of that plan (control). It is orthogonal to the value dimension because the change from moving away from the status quo could be to either a better or worse state given that the movement need not be in the service of satisfying a promotion or prevention concern. The assessment mode motivates individuals to synthesize present experience with past experience, weigh alternative choice options, and construct a sense of how things really are regarding each choice option (truth). It is also orthogonal to the value dimension because it is not committed to any choice until it knows the truth about all of the options (in contrast to promotion's commitment to advancement and prevention's commitment to maintenance). A visual representation of these four motives and their associated psychological "directions" is shown in Fig. 1.

2 Effective Motivational Pursuit

These motives, representing as they do different psychological "directions," can exist in tension with one another and can potentially constrain one another, but they can also support and sustain one another (Higgins 2012). Because each motive has both potential benefits and potential costs as a function of how strongly it functions, both support and constraint from other motives can contribute to a motive's effectiveness. Well-being would be enhanced when the different motives *work together*, i.e., interact, in a manner to support the benefits of one another and constrain the costs. Indeed, there is some evidence that motives can work together in a way that enhances well-being (for reviews, see Cornwell et al. 2014; Higgins et al. 2014). What we have found in our own work (Cornwell et al. 2015) is that not only do individuals benefit from strong motives in an additive sense, but overall effectiveness is dependent upon their *integration* as well. When an individual's value, truth, and control motives are all effective, and the variance among these three kinds of effectiveness is low (suggesting that integration has occurred), the experience of this harmony among strong motives is associated with experiencing one's life as being happier and more meaningful, along with other standard measures of well-being (Cornwell et al. 2015). When an individual's motives are not well integrated, motives may become disproportionately weak *or* disproportionately strong relative to other motives, and the experience of this ineffective organization, or lack of harmony, can diminish well-being.

What follows from this perspective is that when motivations are *dis*-integrated—that is, when either a motivation is so weak so that it does not function properly or so strong that it dominates the other motives—well-being is reduced and people can experience psychological distress. The remainder of this chapter will illustrate how disintegration is associated with impaired functioning. We will also suggest ways that a motivational perspective emphasizing integration can work to improve overall well-being. We begin by reviewing established cases of motives being too weak to function properly and the kinds of psychological distress that result from this. Later, we will review cases of psychological dysfunction from motives being too strong and dominating other motives.

3 Dis-integration from Motives Being Too Weak

When any particular motive is too weak, then the result is motivational disintegration that may lead to psychological distress for at least two reasons. The first reason is that each form of motivation is aiming to satisfy some concern and thus failure to do so produces a negative experience, such as feeling confused from failure to satisfy truth concerns or feeling frustrated from failure to satisfy control concerns. The second reason is that a weak motive will fail to act as a check or constraint on another motive, such as too weak locomotion allowing assessment to dominate. Because every motive has potential downsides as well as benefits, a motive unconstrained by others can become maladaptive, such as a dominating assessment motive leading a person to become "lost in thought" and fail to take necessary action.

In this section, we will give an example of each—the consequences of a motive being too weak in absolute terms (resulting in failure experiences) and relative terms (resulting in other motivations becoming too strong). We will discuss how weak promotion motivation can lead to depression from a failure to advance toward promotion goals. We will also discuss how symptoms of Parkinson's disease may be related to both absolute and relative weakness in assessment.

3.1 Depression

The area of disintegration that has received the most attention is in the value domain. We will discuss the prevention focus and its relation to anxiety in the following section, but here, we consider depression, which has been related to failures to be effective in promotion, especially when individuals are predominantly promotion focused. In this section, we discuss how promotion focus failures produce a specific type of emotional distress. Specifically, research has shown that in both clinical and non-clinical populations, promotion focus failures are associated with suffering from dejection-related emotions (e.g., feeling sad and discouraged) and depression (Higgins 1987; Moretti and Higgins 1990; Strauman and Higgins 1987, 1988; Strauman et al. 2006).

For example, a series of studies conducted by Strauman and Higgins (1988, 1999) found a positive relation between promotion failure and depressed mood. Promotion failure is measured by the amount of discrepancy between participants' report of their actual self (the kind of person they currently are) and their report of the kind of person they aspire or wish or hope to be (their *ideal* self)—the magnitude of their actual–ideal discrepancy. Participants who reported failing to attain their promotion goals—those with a high actual–ideal discrepancy—also showed higher rates of dysphoric and depressed mood (compared to those with low actual–ideal discrepancies). Moretti and Higgins (1990) extended these findings to show that promotion failure (i.e., high actual–ideal discrepancy) also predicts low self-esteem. Importantly, there is also evidence that the strength of the association between having a high actual–ideal discrepancy and suffering dejection-related emotions depends on the strength of individuals' promotion focus. That is, it is the interaction between magnitude of actual–ideal discrepancy and strength of promotion focus that predicts suffering dejection-related emotions (see Higgins et al. 1997).

Taken as a group, these studies suggest that when individuals are chronically unable to attain their hopes and aspirations, they experience depressed moods, lose their confidence, and self-esteem, and may thereby begin to anticipate failure and disengage from situations that could bring positive results. Such tendencies are associated with a number of conditions related to depression, including learned helplessness (Abramson et al. 1978) and anhedonia (Snaith 1993).

The theoretical and behavioral association between promotion failure and depressive symptomology is further bolstered by recent neuroscience studies that show considerable overlap between activation of the promotion system and the experience of depression (Strauman et al. 2013). For example, fMRI research has shown that the activation of promotion goals is associated with greater activation of the left prefrontal cortex (Eddington et al. 2007). Specifically, researchers found that incidental priming of individuals with their own promotion ideal goals resulted in greater activation of the left orbital PFC. In addition, activation in this area was also stronger for those who scored higher on a questionnaire measure of promotion strength. Relatedly, research on the neuroscience of depression has found that diminished activation in the prefrontal cortex, particularly the left PFC, is associated with depression (Davidson et al. 1999). This appears to be due to the region's relation to emotion regulation, which is a diminished capacity among those experiencing depression (George et al. 1994). It is important to note that as neuroimaging methods and research have become more sophisticated, the brain systems associated with depression are proving to be more complex, but hypo-activity in portions of the left prefrontal cortex does continue to be associated with depressive symptoms (specifically, in scans of patients with major depressive disorder relative to healthy subjects; Grimm et al. 2008).

Thus, we see that having a weak promotion focus is associated with diminished well-being and the presentation of symptoms, including dejection-related emotions, that are associated with depression. The brain region associated with the functioning of the promotion system also shows decreased blood flow in individuals who suffer from depressive conditions. This pattern of results supports the idea that depression is related to a weak promotion focus.

3.2 Parkinson's Disease

Not only can a weakened motivation lead to negative psychological outcomes in its own right, but, as noted above, also weakened motivations are also poor at acting as checks or constraints on other motivations. Thus, if a motivational constraint is diminished (because it is weakened), other motivations can become too strong, leading to maladaptive behaviors. For example, research has shown that those who attain the highest levels of achievement are those who have *both* strong assessment motives *and* strong locomotion motives—having only one of these motives be strong is insufficient for optimal performance (Kruglanski et al. 2000; Pierro et al., in press). For example, individuals with a strong locomotion mode and strong assessment mode have higher cumulative GPAs. Specifically, the relation between high locomotion strength and high GPA is only significant among those with high assessment strength—a significant interaction of locomotion strength and assessment strength when predicting GPA (Kruglanski et al. 2000). This same locomotion X assessment interaction has also been found when predicting work performance in companies (Pierro et al., in press).

Similarly, research has shown that teams in which strong assessors and strong locomotors are equally represented complete tasks faster and more accurately than teams consisting of either all strong locomotors or all strong assessors (Mauro et al. 2009). The theoretical grounding for this empirical outcome is that *effective* locomotion requires the synthesis of information in order to get to the correct course of action, setting aside incorrect courses of action (in other words, effective assessment). Similarly, *effective* assessment requires that evaluations and comparisons eventually come to an end in the form of a judgment that is then carried out (in other words, effective locomotion). Thus, without locomotion, assessors can get "lost in thought," and without assessment, locomotors can "leap before they look" or act in ways that make things worse rather than better. Thus, we see in the examples discussed above that within individuals and within groups, having high and balanced locomotion and assessment strength is related to achieving the best performance outcomes.

Recent research on Parkinson's disease suggests that some of the illness's symptoms may be due, in part, to a weakening of the assessment mode. Compared to age-matched control individuals (Foerde et al. 2014), people with Parkinson's disease score lower on assessment strength as measured by the Regulatory Mode Questionnaire (RMQ; Kruglanski et al. 2000). With respect to locomotion, however, there are no differences between the groups. The RMQ measures the strength of individuals' locomotion mode to initiate and maintain movement (e.g., "I am a go-getter"; "When I decide to do something, I can't wait to get started"), as well as the strength of their assessment mode to make critical evaluations and comparisons of multiple options and targets (e.g., "I spend a great deal of time taking inventory of my positive and negative characteristics"; "I like evaluating other people's plans").

Past research has shown that those with Parkinson's disease suffer from a loss of dopaminergic innervation in the striatum (Aarsland et al. 2012), and past research has also shown that learning based on immediate feedback is associated with activity in that region, whereas learning based on delayed feedback is processed in the hippocampus (Foerde et al. 2013). Based on these two sets of findings, more recent research showed, as expected, that those with Parkinson's disease showed greater learning deficits associated with immediate feedback processed in the striatum compared to age-matched matched controls, but no deficit for delayed feedback (Foerde et al. 2014). Interestingly, low assessment strength in this research also was related to greater learning deficits with immediate feedback but not delayed feedback, and, in addition, low assessment strength significantly mediated the relation between disease status (i.e., Parkinson's patient vs. age-matched control) and performance on the immediate (vs. delayed) feedback task (Foerde et al. 2014). This research suggests that there is an important link between weak assessment mode and the dopaminergic functioning of the striatum. Interestingly, one central problem in Parkinson's disease is translating internal motivation into selecting the appropriate action (e.g., Balleine and O'Doherty 2010), and there is evidence that striatal activity is involved in goal-directed action selection (Balleine et al. 2007; Tai et al. 2012). See chapters in this volume by Corbitt et al. for reviews.

Is there other evidence regarding Parkinson's deficits that suggests that Parkinson patients are low in assessment but normal in locomotion? Notably, if assessment is low and locomotion is normal, it means that locomotion is predominant. The question, then, can be restated as, Is there other evidence regarding Parkinson's deficits that suggests that Parkinson patients are locomotion predominant? Yes, there is. Some Parkinson's patients, for example, talk so quickly that they are difficult to understand and their handwriting movements are also fast, producing writing that is difficult to read. In addition, some Parkinson's patients turn quickly by crossing one leg over the other rather than taking an extra step, which throws them off balance. Each of these cases involves a speed/accuracy trade-off in which accuracy is being sacrificed for speed. This is precisely what is found when locomotion dominates because assessment is too weak—a preference for speed over accuracy (see Kruglanski et al. 2000).

3.3 Discussion: Dis-integration from Motives Being Too Weak

In both of the above examples, the diminished strength of different systems of motivation (promotion; assessment) can be associated with diminished functioning. In addition, each system of motivation appears to be linked with particular structures and functions in the brain. These relations are intriguing and need to be studied further. What we do not yet know is whether an intervention at one level of analysis will have effects at another level of analysis. Would increasing dopamine in the striatum strengthen assessment motivation? Would dopamine in the striatum be increased by using a standard experimental induction to strengthen assessment—using questions from the RMQ to have participants recall times in their past when they behaved like high assessors ("Write down a time in the past when you enjoyed evaluating another person's plans"; see Avnet and Higgins 2003; Cornwell and Higgins 2014)? It is notable in this regard that Foerde et al. (2014) found no difference in assessment strength between those Parkinson's patients who currently were on versus off dopaminergic medication, suggesting that, while connected, the precise relationship between Parkinson's disease and assessment strength is still unclear.

4 Dis-integration from Motives Being Too Strong

In the previous section, we detailed the ways in which motivational dis-integration resulting from particularly weak motives is related to the experience of different forms of psychological distress. In this section, we look at dis-integration from the opposite perspective—when a motive's excessive strength, both absolutely and relatively, can cause problems. Notably, psychological distress in these circumstances can occur both because a motive is so strong that it dominates other motives leaving its own downsides unchecked, and its dominance of other motives makes them too weak to provide their benefits to self-regulation. In this section, we will provide examples of these problems. We will consider how an unchecked strong prevention focus can lead to ineffective functioning associated with anxiety and then discuss the ways in which a disproportionately strong assessment motive can develop into a problematic personality disorder (narcissism).

4.1 Anxiety

What are the negative consequences of a motivation that is too strong? To date, we have the most evidence about the negative consequences of a too strong prevention focus that intensifies vigilance to the point of becoming hypervigilance. When prevention motivation is extremely strong, it is associated with maladaptive responses that actually make it difficult for a person to achieve even their prevention goals. For example, when action (fight or flight) is needed to prevent harm, but a person is paralyzed by fear, or when attention needs to be directed widely to prevent harm, but hypervigilance is narrowing attention. Moreover, an overly predominant prevention motivation could also interfere with the ability of other motives to satisfy their respective concerns, such as inhibiting what is needed to satisfy promotion focus concerns. For example, a hypervigilant prevention focus would create a strong concrete, here-and-now focus (Förster and Higgins 2005; Semin et al. 2005), which would then interfere with the need of a promotion focus to think about the future and advance toward an ideal end-state. In this section, we discuss the sorts of behaviors that are associated with unchecked prevention, and how an excessively strong prevention focus actually fails to be effective in spite of its strength.

If individuals repeatedly fail to achieve their prevention goals—leading them to experience a self-discrepancy between their actual self and their ought self (i.e., the kind of person they believe it is their duty and responsibility to be)—they experience agitation-related emotions and even possibly generalized anxiety (Strauman and Higgins 1987, 1988). The experience that one is failing to maintain a status quo can lead one to adopt a strong prevention focus and become highly vigilant (e.g., Scholer et al. 2010). Hypervigilance can lead to negative mental, behavioral, and physiological states that are related to general anxiety, including intense worry (Borkovec and Inz 1990) and hypertension (Barger and Sydeman 2005). The prevention focus has two elements that in combination make a prevention-focused individual particularly prone to anxiety: an emphasis on perseverance and repetition (Friedman and Förster 2001), and a focus on ensuring against negative outcomes and thus vigilantly seeking out negative stimuli (Franks et al. 2013). Both of these factors can contribute to perseverant processing of negative stimuli in a manner similar to rumination when left unchecked by other motives.

The brain region most strongly associated with the experience of anxiety and fear emotions is most certainly the amygdala (e.g., Phan et al. 2006), but less is known about how motivation relates to brain regions that are associated with the successful or unsuccessful regulation of such fear experiences. Some research supports a model linking an overly strong prevention focus and the activation of brain regions associated with anxiety. For example, the active priming of prevention goals was related to greater activation in the anterior cingulate cortex in an exploratory fMRI study (Eddington et al. 2007). This same region (along with the dorsal medial prefrontal cortex) has shown more persistent activation among patients diagnosed with general anxiety disorder (GAD) compared to controls when presented with sentences intended to induce worry (Paulesu et al. 2010). Specifically, researchers found that all individuals showed the increased activation during the presentation of the worry-inducing stimuli—the attendant emotional experience of vigilance (Higgins 2001)—but only GAD patients showed persistent activation even after the stimuli were no longer present during resting state scans—which, since it was no longer serving a motivational function, we might call hypervigilance.

As we noted earlier, we do not equate "strong" with "effective." Stronger motivation can improve performance, but it does not always do so. Moreover, our emphasis is on the case of a motivation being *too* strong, similar to Atkinson's notion of "overmotivation" (Atkinson 1976). A motive that is too strong not only does not ensure successful completion of one's goals, but, as we suggested earlier, can also actually inhibit one's ability to pursue those goals effectively. There is some additional neuroscience evidence supporting this distinction. Research has examined the effects on brain activation associated with presentation of individuals' ought and ideal goals compared to yoked control words and nonwords. Researchers found that when participants were presented with ought goals, the more success they reported for their prevention goals, the stronger was the activation of a cluster that included the precuneus cortex (along with the lateral occipital cortex, angular gyrus, and superior parietal lobule; see Strauman et al. 2013); that is, the less prevention success, the weaker was the activation of this cluster. Interestingly, reduced functional connectivity between the precuneus and the amygdala has been linked to greater experience of state anxiety among a combined sample of individuals with social phobia and control individuals (Hahn et al. 2011). It is important not to overstate the connection here since prevention success shows associations with a number of brain areas and not succeeding is not always the same as failing. However, it is notable that one of the brain areas whose connectivity with the amygdala was negatively associated with state anxiety also appears to be associated with prevention effectiveness where prevention involves effective vigilance rather than overly strong and ineffective hypervigilance.

Not only does hypervigilance have negative consequences in and of itself, it can also crowd out other motives, leading to motivational ineffectiveness in multiple domains. For example, there is a well-known comorbidity between general anxiety and depression (Kessler et al. 1996), but the onset is not equally balanced across the two conditions: Comorbidity runs more from anxiety to depression than the reverse, with age of onset for anxiety disorders typically being earlier than for depression disorders (Kessler et al. 2005). According to our model, hypervigilance against potential negative outcomes—overly predominant prevention motivation—leads individuals to be much less likely to pursue situations that carry some risk or challenge—the kind of challenge that may be needed for promotion growth and advancement (Franks et al. 2015). This would reduce the likelihood of satisfying promotion concerns. In other words, an excessively strong prevention focus can also lead to less expression of the promotion focus, with the attendant forms of psychological distress noted above, including depression if it occurs chronically. In the next section, we examine how this kind of self-regulatory dominance of a particular motivation over other motives can also eventually lead to the development of a personality disorder.

4.2 Narcissism

As noted above, the assessment mode is associated with engaging in comparing and weighing the costs and benefits of different options, and the critical evaluation of the self and others. When done effectively, strong assessment can lead to the establishment of what is real or correct about oneself, one's environment, and others. Research has shown that this truth motive, when effectively pursued, leads to greater well-being in both humans and non-human animals (Higgins 2012; Franks and Higgins 2012). However, as we have argued, strong motives only translate into effective motives when they are constrained and supported by other motivations. When they are pursued in isolation from other motives, they can be ineffective in spite of and, in many cases, because of their strength. For assessment, unconstrained strength leads to ruminative experiences. Illustrating how "strong" is not the same as "effective," rumination has been linked to impaired performance on a variety of tasks, among them problem-solving (Watkins and Moulds 2005)—despite the fact that the assessment mode is the regulatory mode associated with getting at the truth (Higgins 2012).

Not only can disproportionately strong motivations present continual psychological difficulties, but, over time, they can develop into particular personality disorders. This is because the strength of a motivation is related to its chronic accessibility (Higgins et al. 1997), with "stronger" motives having higher chronic accessibility and therefore monopolizing the psychological resources of an individual. The dominant motivation becomes *the* relevant motivation, which then determines what comes to mind and what you pay attention to (Eitam and Higgins 2010). This monopolization, in addition to causing the disproportionately strong motive to be ineffective in its own right, can lead to the further weakening of alternative motives. Over time, there will be motivational dis-integration that constitutes a dysfunctional personality. In this section, we will look at one such personality disorder associated with a disproportionately strong assessment motivation: narcissism (Boldero et al. 2015).

Narcissistic personality typically presents two correlated dimensions: grandiosity and vulnerability (Luchner et al. 2011). Narcissists are insecure about their self and their status, particularly in relation to others. This leads to both a disproportionate view of themselves as being uniquely good—grandiosity—but also being especially reactive to negative assessments and personal criticisms that call attention to their deficiencies—vulnerability. Research in personality has found that, with respect to self-regulatory focus and mode, there is a relation between the assessment mode and both aspects of narcissistic personality (Boldero et al. 2015). Those with strong assessment are particularly critical of both themselves and others, which, in addition to being tied to negative views of the self, can also lead to very low opinions of others (Boldero et al. 2015; Higgins 2008; Kruglanski et al. 2000). Grandiosity itself has been related to engaging in social comparisons (Krizan and Bushman 2011), which is a central feature of strong assessment.

For our purposes, we see narcissistic personalities as a result of disproportionate assessment motives, which inhibit the effective functioning of motives associated with locomotion, promotion, and prevention, by either disengaging them or engaging them too much in the service of assessment goals. For example, those with narcissistic personalities tend to present both depressive and anxiety-related symptoms (Miller et al. 2007), which, as noted above, have been associated with an overly weak promotion focus and an overly strong prevention focus, respectively. Furthermore, research on psychological resources has found that narcissism is negatively associated with attentional control (Claes et al. 2009) and with self-control more generally (Raskin and Terry 1988; Wink and Gough 1990), suggesting that narcissism also involves weak locomotion given that locomotion is associated with the kind of planning for smooth, uninterrupted maintenance of movement that is part of being conscientious and efficient (Higgins 2008, 2012). All of this research presents a picture of narcissism in which the assessment mode dominates the other motivations, leading to their dysfunction in the service of assessment and, therefore, the poor functioning of narcissistic individuals.

There is also some convergent evidence of this hypothesis from neuroscience. Neuroscience research on narcissism is much more sparse than research on the other well-being deficiencies discussed above, but some research does exist. An exploratory fMRI study conducted by Fan et al. (2011) has demonstrated a link between narcissistic personality and reduced deactivation in the right anterior insula. In the experimental trials, participants were presented with emotional faces and instructed to empathize with them. During control trials, participants were presented with no emotional faces and instead presented with "smoothed" faces. Both high and low narcissistic personality participants showed similar activation in the right anterior insula during the empathy trials, but those scoring highly on the narcissistic personality inventory showed *reduced de*activation of this region during the control trials. This research is consistent with a link between narcissism and assessment because the right anterior insula has been particularly implicated in the monitoring or assessment of task performance (in this case, a word recognition task and an Eriksen Flanker Task; Eriksen and Eriksen 1974), as well as the selection of strategies in response to the performance assessment (Eckert et al. 2009). Furthermore, researchers also found that increased activation in the anterior insula was associated with poorer performance on the tasks (Eckert et al. 2009). The authors speculate that this is because there is an optimal level of arousal for ideal task performance (consistent with the Yerkes–Dodson law) and that overstimulation of the sympathetic nervous system (which is associated with the right anterior insula; see Abboud et al. 2006 or Critchley et al. 2002) can actually lead to poor performance. Thus, similar to what we said earlier about how disproportionately strong assessment can lead to the dominance and ineffective functioning of locomotion motivation (control concerns) and promotion and prevention motivation (value concerns), we see in this study that greater activation in the anterior insula is associated both with longer reaction times and with poorer performance.

4.3 Discussion: Dis-integration from Motives Being Too Strong

Not only is it problematic from the standpoint of well-being for a motive to be too weak, motives that are too strong can cause psychological difficulties as well. In one of the "weakness" cases, we saw that having an assessment mode that was too weak resulted in a drop in monitoring and attendant symptoms associated with Parkinson's disease. Here in one of the "strength" cases, we see that having an assessment mode that is too *strong* results in excessive monitoring and symptoms associated with rumination (in the short term) and the eventual development of a narcissistic personality that can be considered a kind of chronic hypermonitoring.

In each of these cases, we noted how neuroscience research of these forms of psychological distress is consistent with our perspective of dis-integration. Each was associated with the disproportionate activation of particular brain systems that are normally important elements of effective functioning but, through the diminished activation in areas associated with attention and self-control and increased activation associated with negative outcome experiences, became dysfunctional given the patterns of motivational strength and weakness. Unfortunately, unlike the research on Parkinson's disease and depression and anxiety described earlier, we do not have direct experimental links between the patterns of neural functioning in narcissism and the various motivational orientations. The patterns are merely suggestive. Future research will need to be conducted in order to determine whether the suggested associations hold true.

4.4 An Important Caveat

The above four examples of motivational dysfunction that we have presented are intended to illustrate motivational *dis*-integration. Our exploration of this issue is currently restricted to the quite limited available research on the subject. We expect that future research will find other combinations of strong and weak motives that are associated with other kinds of psychological distress. What we have presented, then, is just the beginning of what we believe will be an exciting story.

We are also restricted in what neuroscience evidence is currently available on these particular motivational concerns (promotion, prevention, locomotion, assessment). This limits what we can say about the relations between these motivations and neuroscience. Just as there are likely to be several kinds of motivational dis-integration, the neural correlates of these motivations are likely to be very complex. Therefore, the above discussions of associations between motivations and neuroscience should not be taken as authoritative in a final way. They are intriguing, but there is much to be learned. Moreover, and importantly, our proposals regarding the underpinnings of the psychological distress and dysfunctions produced by too weak or too strong motivations stand apart from any associations with neuroscience per se. Their validity or usefulness does not depend on what we ultimately learn about their connections to neuroscience. Such connections are interesting and can provide insights, but each level of analysis or "language" must be evaluated in its own terms and with its own evidence.

5 Possibilities for Improvement: Micro-interventions

Before closing, we want to consider, if only briefly, how the proposed relations between specific forms of motivational dis-integration to particular kinds of psychological distress suggest the possibility for new interventions to enhance well-being. The study of motivation has included the development of tools or "micro-interventions" that can be used to strengthen or weaken targeted motivations. As noted above, the strength of any particular motivational orientation is related to its chronic accessibility, and simple techniques have been designed to increase or decrease this accessibility, with important implications for well-being.

Most of the existing research on this issue has been about regulatory focus (though the same principles have been applied to some regulatory mode research; see Avnet and Higgins 2003). The research suggesting a connection between promotion and depressive symptoms and prevention and anxiety-related symptoms was discussed earlier, but there is new research making use of micro-interventions that show promise for ameliorating these forms of distress through the strengthening of promotion (too weak when depressed) and weakening of prevention (too strong when anxious). For example, the Regulatory Focus Questionnaire (a measure of promotion focus and prevention focus) contains the item "Compared to most people, are you typically unable to get what you want out of life?" Asking participants to recall a time when "compared to most people" they *were* "able to get what they wanted out of life" strengthens their promotion focus (Higgins et al. 2001). Intriguingly, similar motivational interventions have been shown to improve well-being; for example, focusing on the attainment of promotion goals in therapy (termed "self-system therapy") can help ameliorate the dysphoric mood associated with promotion goal failure (Strauman et al. 2006).

Most recently, Strauman et al. (in press) found that those suffering from dysphoric or anxious mood could have these elements of distress ameliorated by targeting micro-interventions to strengthen or weaken, respectively, their levels of engagement with their everyday problems. Specifically, during a 10-min interaction with the experimenter, participants were asked to discuss their problems as either "obstacles to be overcome" (which would increase promotion engagement strength) or as "nuisances to be coped with" (which would decrease prevention engagement strength). The research found that interventions involving the "overcoming obstacles" intervention were more effective at reducing dysphoric mood and that those involving the "coping with a nuisance" intervention were more effective at reducing anxious mood (Strauman et al., in press).

Though these approaches offer useful techniques in combatting different forms of psychological distress, they were designed to modify each motivation in isolation from the other motivations. Throughout this chapter, we have noted that not only is it important that individuals achieve effectiveness in each form of motivation to maximize well-being, but that the motivations need to "work together" in order to achieve the greatest amount of well-being. Therefore, future research should explore not only the possibility of strengthening or weakening individual motivation strength, but also adjusting overall motivational organization in order to achieve integration. The model we have sketched here offers a potential starting point for the development of this novel form of intervention.

6 Motivational Integration as a Source of Well-Being

In this chapter, we have argued that certain forms of psychological distress have their root in the "dis-integration" of different kinds of motivational orientation. Specifically, we have shown that when motives are *too weak* (either through their absolute weakness or their weakness relative to other motives), particular forms of psychological distress are likely to occur. We have also shown that particular forms of psychological distress are likely to occur when motives are *too strong*—again, either through their absolute strength or through their dominating strength relative to other motives. In each case, we have highlighted evidence from neuroscience that is consistent with this model of well-being and provided an example of a potential aid for ameliorating these forms of psychological distress based on this motivational model.

What we have tried to make clear is that *both* the *absolute* strengths and weaknesses of promotion, prevention, locomotion, and assessment and their *relative* strength (related to their integration or dis-integration) have important implications for well-being. Certain patterns of strength and weakness among these motives can produce motivational dysfunction and psychological distress. We propose that to help ameliorate the problems associated with such motivational dysfunctions, it is important to consider what would best contribute to an effective *integration* among these motives—an organization of motives whereby each motive supports the benefits of each other motive while also constraining its potential downsides. To make this happen, we will need to examine these motives at multiple levels of analysis, from the social to the neural. Each language of analysis can provide unique insights into how an effective integration of motives can be created.

References

Aarsland D, Påhlhagen S, Ballard CG, Ehrt U, Svenningsson P (2012) Depression in Parkinson disease—epidemiology, mechanisms and management. Nat Rev Neurol 8(1):35–47

Abboud H, Berroir S, Labreuche J, Orjuela K, Amarenco P (2006) Insular involvement in brain infarction increases risk for cardiac arrhythmia and death. Ann Neurol 59(4):691–699

Abramson LY, Seligman ME, Teasdale JD (1978) Learned helplessness in humans: critique and reformulation. J Abnorm Psychol 87(1):49–74

Atkinson JW (1976) Resistance and overmotivation in achievement-oriented activity. In: Serban G (ed) Psychopathology of human adaptation. Springer, New York, pp 193–209

Avnet T, Higgins ET (2003) Locomotion, assessment, and regulatory fit: value transfer from "how" to "what". J Exp Soc Psychol 39(5):525–530

Balleine BW, O'Doherty JP (2010) Human and rodent homologies in action control: corticostriatal determinants of goal-directed and habitual action. Neuropsychopharmacology 35(1):48–69

Balleine BW, Delgado MR, Hikosaka O (2007) The role of the dorsal striatum in reward and decision-making. J Neurosci 27(31):8161–8165

Barger SD, Sydeman SJ (2005) Does generalized anxiety disorder predict coronary heart disease risk factors independently of major depressive disorder? J Affect Disord 88(1):87–91

Boldero JM, Higgins ET, Hulbert CA (2015) Self-regulatory and narcissistic grandiosity and vulnerability: common and discriminant relations. Pers Individ Differ 76:171–176

Borkovec TD, Inz J (1990) The nature of worry in generalized anxiety disorder: a predominance of thought activity. Behav Res Ther 28(2):153–158

Claes L, Vertommen S, Smits D, Bijttebier P (2009) Emotional reactivity and self-regulation in relation to personality disorders. Pers Individ Differ 47(8):948–953

Cornwell JF, Higgins ET (2014) Locomotion concerns with moral usefulness: when liberals endorse conservative binding moral foundations. J Exp Soc Psychol 50:109–117

Cornwell JFM, Franks B, Higgins ET (2014) Truth, control, and value motivations: the "what," "how," and "why" of approach and avoidance. Front Syst Neurosci 8. doi:10.3389/fnsys.2014.00194

Cornwell JFM, Franks B, Higgins ET (2015) Effectiveness of motive organization: the character of integrity in "the good life." Unpublished manuscript, Columbia University

Critchley HD, Melmed RN, Featherstone E, Mathias CJ, Dolan RJ (2002) Volitional control of autonomic arousal: a functional magnetic resonance study. Neuroimage 16(4):909–919

Davidson RJ, Abercrombie HC, Nitschke JB, Putnam KM (1999) Regional brain function, emotion and disorders of emotion. Curr Opin Neurobiol 9:228–234

Diener E, Emmons RA, Larsen RJ, Griffin S (1985) The satisfaction with life scale. J Pers Assess 49:71–75

Eckert MA, Menon V, Walczak A, Ahlstrom J, Denslow S, Horwitz A, Dubno JR (2009) At the heart of the ventral attention system: the right anterior insula. Hum Brain Mapp 30(8):2530–2541

Eddington KM, Dolcos F, Cabeza R, Krishnan KRR, Strauman TJ (2007) Neural correlates of promotion and prevention goal activation: an fMRI study using an idiographic approach. J Cogn Neurosci 19(7):1152–1162

Eitam B, Higgins ET (2010) Motivation in mental accessibility: relevance of a representation (ROAR) as a new framework. Soc Pers Psychol Compass 4(10):951–967

Eriksen BA, Eriksen CW (1974) Effects of noise letters upon the identification of a target letter in a nonsearch task. Percept Psychophys 16(1):143–149

Fan Y, Wonneberger C, Enzi B, De Greck M, Ulrich C, Tempelmann C, Bogerts B, Doering S, Northoff G (2011) The narcissistic self and its psychological and neural correlates: an exploratory fMRI study. Psychol Med 41(8):1641–1650

Foerde K, Braun EK, Shohamy D (2013) A trade-off between feedback-based learning and episodic memory for feedback events: evidence from Parkinson's disease. Neuro-degener Dis 11(2):93–101. doi:10.1159/000342000

Foerde K, Braun EK, Higgins ET, Shohamy D (2014) Motivational modes and learning in Parkinson's disease. Soc Cogn Affect Neurosci. doi:10.1093/scan/nsu152

Förster J, Higgins ET (2005) How global versus local perception fits regulatory focus. Psychol Sci 16:631–636

Franks B, Higgins ET (2012) Effectiveness in humans and other animals: a common basis for well-being and welfare. In: Olson JM, Zanna MP (eds) Advances in experimental social psychology, vol 46. Academic Press, Burlington, pp 285–346

Franks B, Higgins ET, Champagne FA (2012) Evidence for individual differences in regulatory focus in rats, *Rattus norvegicus*. J Comp Psychol 126(4):347–354

Franks B, Champagne FA, Higgins ET (2013) How enrichment affects exploration trade-offs in rats: implications for welfare and well-being. PLoS ONE 8(12):e83578. doi:10.1371/journal.pone.0083578

Franks B, Higgins ET, Champagne FA (2014) A theoretically based model of rat personality with implications for welfare. PLoS ONE 9(4):e95135. doi:10.1371/journal.pone.0095135

Franks B, Chen C, Manley K, Higgins ET (2015) Effective challenge regulation coincides with promotion focus-related success and emotional well-being. J Happiness Stud. doi:10.1007/s10902-015-9627-7

Freud S (2011) The ego and the id—first edition text. Martino Fine Books, Eastford (Original work published in 1927)

Friedman RS, Förster J (2001) The effects of promotion and prevention cues on creativity. J Pers Soc Psychol 81(6):1001–1013

George MS, Ketter TA, Post RM (1994) Prefrontal cortex dysfunction in clinical depression. Depression 2(2):59–72

Grimm S, Beck J, Schuepbach D, Hell D, Boesiger P, Bermpohl F, Niehaus L, Boeker H, Northoff G (2008) Imbalance between left and right dorsolateral prefrontal cortex in major depression is linked to negative emotional judgment: an fMRI study in severe major depressive disorder. Biol Psychiatry 63(4):369–376

Hahn A, Stein P, Windischberger C, Weissenbacher A, Spindelegger C, Moser E, Kasper S, Lanzenberger R (2011) Reduced resting-state functional connectivity between amygdala and orbitofrontal cortex in social anxiety disorder. Neuroimage 56(3):881–889

Harlow HF (1950) Learning and satiation of response in intrinsically motivated complex puzzle solving performance by monkeys. J Comp Physiol Psychol 43(4):289–294

Higgins ET (1987) Self-discrepancy: a theory relating self and affect. Psychol Rev 94:319–340

Higgins ET (1997) Beyond pleasure and pain. Am Psychol 52:1280–1300

Higgins ET (1998) Promotion and prevention: regulatory focus as a motivational principle. Adv Exp Soc Psychol 30:1–46

Higgins ET (2001) Promotion and prevention experiences: relating emotions to nonemotional motivational states. In: Forgas JP (ed) Handbook of affect and social cognition. Lawrence Erlbaum, Mahwah, pp 186–211

Higgins ET (2008) Culture and personality: variability across universal motives as the missing link. Soc Pers Psychol Compass 2(2):608–634

Higgins ET (2012) Beyond pleasure and pain: how motivation works. Oxford University Press, New York

Higgins ET (2014) Promotion and prevention: how "0" can create dual motivational forces. In: Sherman JW, Gawronski B, Trope Y (eds) Dual-process theories of the social mind. Guilford Press, New York, pp 423–435

Higgins ET, Shah J, Friedman R (1997) Emotional responses to goal attainment: strength of regulatory focus as moderator. J Pers Soc Psychol 72(3):515–525

Higgins ET, Friedman RS, Harlow RE, Idson LC, Ayduk ON, Taylor A (2001) Achievement orientations from subjective histories of success: promotion pride versus prevention pride. Eur J Soc Psychol 31:3–23

Higgins ET, Kruglanski AW, Pierro A (2003) Regulatory mode: locomotion and assessment as distinct orientations. Adv Exp Soc Psychol 35:293–344

Higgins ET, Cornwell JFM, Franks B (2014) "Happiness" and "the good life" as motives working together effectively. In Elliot A (ed) Advances in motivation science, vol 1, pp 135–179

Kessler RC, Nelson CB, McGonagle KA, Liu J (1996) Comorbidity of DSM-III—R major depressive disorder in the general population: results from the US National Comorbidity Survey. Br J Psychiatry 17–30

Kessler RC, Berglund P, Demler O, Jin R, Merikangas KR, Walters EE (2005) Lifetime prevalence and age-of-onset distributions of DSM-IV disorders in the National Comorbidity Survey Replication. Arch Gen Psychiatry 62(6):593–602

Krizan Z, Bushman BJ (2011) Better than my loved ones: social comparison tendencies among narcissists. Pers Individ Differ 50(2):212–216

Kruglanski AW, Orehek E, Higgins ET, Pierro A, Shalev I (2000) Modes of self-regulation: assessment and locomotion as independent determinants in goal-pursuit. In: Hoyle R (ed) Handbook of personality and self-regulation. Blackwell, Boston, pp 374–402

Langbein J, Siebert K, Nürnberg G (2009) On the use of an automated learning device by group-housed dwarf goats: do goats seek cognitive challenges? Appl Anim Behav Sci 120(3–4):150–158

Lewin K (1951) Field theory in social science. Harper, New York

Luchner AF, Houston JM, Walker C, Houston MA (2011) Exploring the relationship between two forms of narcissism and competitiveness. Pers Individ Differ 51(6):779–782

Mauro R, Pierro A, Mannetti L, Higgins ET, Kruglanski AW (2009) The perfect mix: regulatory complementarity and the speed-accuracy balance in group performance. Psychol Sci 20:681–685

Miller JD, Campbell WK, Pilkonis PA (2007) Narcissistic personality disorder: relations with distress and functional impairment. Compr Psychiatry 48(2):170–177

Moretti MM, Higgins ET (1990) Relating self-discrepancy to self-esteem: the contribution of discrepancy beyond actual-self ratings. J Exp Soc Psychol 26:108–123

Nolen-Hoeksema S (1998) Ruminative coping with depression. In: Heckhausen J, Dweck CS (eds) Motivation and self-regulation across the lifespan. Cambridge University Press, New York, pp 237–256

Nolen-Hoeksema S (2000) The role of rumination in depressive disorders and mixed anxiety/depressive symptoms. J Abnorm Psychol 109(3):504–511

Paulesu E, Sambugaro E, Torti T, Danelli L, Ferri F, Scialfa G, Ruggiero GM, Bottini G, Sassaroli S (2010) Neural correlates of worry in generalized anxiety disorder and in normal controls: a functional MRI study. Psychol Med 40(1):117–124

Peterson C, Park N, Seligman ME (2005) Orientations to happiness and life satisfaction: the full life versus the empty life. J Happiness Stud 6(1):25–41

Phan KL, Fitzgerald DA, Nathan PJ, Tancer ME (2006) Association between amygdala hyperactivity to harsh faces and severity of social anxiety in generalized social phobia. Biol Psychiatry 59(5):424–429

Pierro A, Pica G, Mauro R, Kruglanski AW, Higgins ET (in press) How regulatory modes work together: locomotion-assessment complementarity in work performance. TPM Test Psychom Methodol Appl Psychol

Raskin R, Terry H (1988) A principal-components analysis of the narcissistic personality inventory and further evidence of its construct validity. J Pers Soc Psychol 54:890–902

Scholer AA, Zou X, Fujita K, Stroessner SJ, Higgins ET (2010) When risk seeking becomes a motivational necessity. J Pers Soc Psychol 99(2):215–231

Semin GR, Higgins ET, Gil de Montes L, Estourget Y, Valencia JF (2005) Linguistic signatures of regulatory focus: how abstraction fits promotion more than prevention. J Pers Soc Psychol 89:36–45

Sherwin CM (2004) The motivation of group-housed laboratory mice, Mus musculus, for additional space. Anim Behav 67:711–717. doi:10.1016/j.anbehav.2003.08.018

Snaith P (1993) Anhedonia: a neglected symptom of psychopathology. Psychol Med 23(4):957–966

Spinka M, Wemelsfelder F (2011) Environmental challenge and animal agency. In: Appleby MC, Mench JA, Olsson IAS, Hughes BO (eds) Animal welfare, 2nd edn. CAB, Cambridge, pp 27–43

Strauman TJ, Higgins ET (1987) Automatic activation of self-discrepancies and emotional syndromes: when cognitive structures influence affect. J Pers Soc Psychol 53:1004–1014

Strauman TJ, Higgins ET (1988) Self-discrepancies as predictors of vulnerability to distinct syndromes of chronic emotional distress. J Pers 56:685–707

Strauman TJ, Vieth AZ, Merrill KA, Kolden GG, Woods TE, Klein MH, Papadakis AA, Schneider KL, Kwapil L (2006) Self-system therapy as an intervention for self-regulatory dysfunction in depression: a randomized comparison with cognitive therapy. J Consult Clin Psychol 74(2):367–376

Strauman TJ, Detloff AM, Sestokas R, Smith DV, Goetz EL, Rivera C, Kwapil L (2013) What shall I be, what must I be: neural correlates of personal goal activation. Front Integr Neurosci 6

Strauman T, Socolar Y, Kwapili L, Cornwell JFM, Franks B, Sehnert S, Higgins ET (in press) Microinterventions targeting regulatory focus and fit selectively reduce dysphoric and anxious mood. Behav Res Ther

Tai L-H, Lee AM, Benavidez N, Bonci A, Wilbrecht L (2012) Transient stimulation of distinct subpopulations of striatal neurons mimics changes in action value. Nat Neurosci 15:1281–1289

Watkins E, Moulds M (2005) Distinct modes of ruminative self-focus: impact of abstract versus concrete rumination on problem solving in depression. Emotion 5(3):319–328

Webb CE, Franks B, Romero T, Higgins ET, de Waal FBM (2014) Individual differences in chimpanzee reconciliation relate to social switching behaviour. Anim Behav 90:57–63. doi:10.1016/j.anbehav.2014.01.014

Wink P, Gough HG (1990) New narcissism scales for the California psychological inventory and the MMPI. J Pers Assess 54:446–462

Zou X, Scholer AA, Higgins ET (2014) In pursuit of progress: promotion motivation and risk preference in the domain of gains. J Pers Soc Psychol 106(2):183–201

Motivation and Contingency Management Treatments for Substance Use Disorders

Kimberly N. Walter and Nancy M. Petry

Abstract Contingency management (CM) is a highly efficacious psychosocial treatment for substance use disorders based on the principles of behavioral analysis. CM involves delivering a tangible positive reinforcer following objective evidence of submission of a drug-negative urine sample. Although CM interventions primarily involve applying extrinsic rewards, a patient's intrinsic motivation to change substance use behavior may also be impacted by CM. This chapter provides an introduction to CM interventions for substance use disorders and examines the impact of CM on intrinsic motivation. It also addresses applications of this intervention to other conditions and patient populations.

Keywords Psychosocial treatments · Substance use disorders · Contingency management · Intrinsic motivation

Contents

1 Preface 570
2 Overview of Contingency Management 570
   2.1 Fundamentals of Contingency Management 571
   2.2 Research Evidence 572
3 Patient Motivation and Contingency Management 574
   3.1 External Rewards and Intrinsic Motivation 574
   3.2 Contingency Management and Intrinsic Motivation 576
4 Summary 578
References 578

K.N. Walter · N.M. Petry (✉)
Calhoun Cardiology Center and Department of Medicine, University of Connecticut School of Medicine (MC 3944), 263 Farmington Ave., Farmington, CT 06030-3944, USA
e-mail: npetry@uchc.edu

K.N. Walter
e-mail: kwalter@uchc.edu

© Springer International Publishing Switzerland 2015
Curr Topics Behav Neurosci (2016) 27: 569–582
DOI 10.1007/7854_2015_374
Published Online: 12 March 2015

1 Preface

Contingency management (CM) is a highly efficacious psychosocial treatment for substance use disorders that applies extrinsic motivators to change patients' substance use behaviors (Petry 2012). Based on the principles of behavioral analysis, CM involves delivering a tangible positive reinforcer following objective evidence of submission of a drug-negative urine sample (Petry 2012). The hope is that patients will be more motivated to obtain the therapeutic reinforcer than the positive effects derived from drug use. Although CM interventions primarily involve applying extrinsic rewards, a patient's intrinsic motivation to change substance use behavior may also be impacted by CM.

The aim of this chapter was twofold: first, to provide an introduction to CM interventions for substance use disorders and second, to examine the impact of CM on intrinsic motivation. The first section presents the fundamentals of CM interventions and an overview of the empirical evidence demonstrating the efficacy of CM interventions for substance use disorders in psychosocial treatment settings. The second section examines the differing perspectives on the influence of external rewards on intrinsic motivation and then discusses research on the effects of CM on substance use disorder patients' intrinsic motivation. The chapter concludes by suggesting future research directions on CM and intrinsic motivation.

2 Overview of Contingency Management

There is a wealth of empirical evidence demonstrating the efficacy of CM for improving substance use disorder treatment outcomes (Lussier et al. 2006; Prendergast et al. 2006). It is included in the National Institute for Health and Clinical Excellence (NICE 2007) guidelines in the UK and is being implemented nationwide throughout the Veterans Administration in the USA (Petry et al. 2014). However, for CM to be effective in changing substance use behaviors, it is critical that it is appropriately implemented with careful attention to behavioral principles (Petry 2012). The three key principles of CM interventions are (1) monitoring behavior frequently, (2) providing tangible positive reinforcers immediately following the behavior, and (3) withholding the positive reinforcer if the behavior does not occur. This section will provide an overview of the fundamentals of CM interventions for substance use disorders followed by a discussion of the empirical evidence of its efficacy in substance use disorder populations. A more comprehensive explanation of each component of CM and specific information about designing and implementing CM interventions can be found in Petry (2012).

2.1 Fundamentals of Contingency Management

2.1.1 Monitoring of Behavior

CM interventions for substance use disorders most commonly reinforce abstinence. During the monitoring and reinforcement period, it is critical that abstinence, typically verified by a drug-negative sample, be monitored regularly, reinforced as immediately as possible, and objectively quantified (Petry 2012). Frequent monitoring increases the chance that each period of abstinence is reinforced and also helps the patient to learn the connection between abstinence and the reinforcer. Most drug urine tests can assess use over 48–72 h, and in these cases, samples should be collected and tested every two to three days. The reinforcer also should be delivered immediately following the drug test. For example, if the patient submits a drug-negative sample, he/she should receive the reinforcer as soon as the test reads negative. If the patient submits a drug-positive sample, reinforcement should be withheld until the next negative sample is submitted. Empirical evidence has shown that immediate reinforcement is associated with better treatment outcomes than delayed reinforcement, e.g., the reinforcer is not provided the same day as the drug test (Lussier et al. 2006).

Monitoring of drug use should involve objective assessments that provide immediate results rather than patients' self-reports of drug use. The monitoring schedule should be based on the frequency of drug use and the test's ability to detect the drug. On-site drug tests should be utilized such as urine toxicology kits (e.g., OnTrak TesTstiks), alcohol breathalyzers (e.g., Intoximeter Breathalyze), or exhaled breath carbon monoxide monitors (e.g., Bedfont Somkerlyzer). On-site drug tests are important because they allow for immediate reinforcement, which is critical because the behavior is less likely to be altered if there is a delay between the behavior and the reinforcer (Rowan-Szal et al. 1994; Roll et al. 2000). The monitoring and reinforcement schedule should be set up according to test's ability to detect drug use. For example, the monitoring schedule for cocaine testing should be three days a week because on-site urine toxicology kits can detect cocaine use over the past two to three days, while the monitoring and reinforcement schedule when using exhaled breath carbon monoxide monitors for cigarette smoking should be several times a day because exhaled carbon monoxide monitors can detect smoking only over the past few hours.

2.1.2 Types of Reinforcers

CM interventions for substance use disorders typically use either vouchers exchangeable for goods and services (Higgins et al. 2008) or chances to win prizes of varying magnitudes (Petry 2012) to reinforce abstinence. Vouchers are similar to monetary incentives, but money is not directly given to patients. Instead, the vouchers are worth a specific monetary amount, and each time a patient provides a drug-negative sample he/she is given a voucher that is deposited in a "clinic bank account." When the patient earns enough vouchers, he/she can exchange them for an item he/she desires such as restaurant gift certificates, clothing, bus tokens, electronic equipment, or movie tickets.

The prize reinforcement system involves earning draws from a prize bowl following submission of a drug-negative sample. The prize bowl contains slips of paper labeled with prizes of various magnitudes (e.g., small prizes worth $1, large prizes worth $20, and jumbo prizes worth $100). About half of the slips do not result in a prize and instead say "Good job!" Most of the prize slips are for small prizes worth $1 and only one strip is labeled with a jumbo prize. Thus, the prize system is less costly than the voucher system because not every draw results in a prize, and the most frequently won prizes are inexpensive.

Both the voucher and prize systems typically use an escalating reinforcement schedule, which promotes longer periods of abstinence during treatment (Roll et al. 1996; Roll and Higgins 2000). Increasing the duration of abstinence is important because longer durations of abstinence are associated with long-term abstinence after the reinforcers are removed, i.e., after the CM intervention ends (e.g., Higgins et al. 2000a, 2002; Petry et al. 2005a, 2007). Escalating reinforcement schedules involve increasing the voucher amounts or number of prize draws as the number of consecutive negative urine samples increase. For example, patients earn one dollar or one draw for their first negative sample, two dollars or two draws for their second consecutive negative sample, and so forth. Additionally, voucher amounts and prize draws are reset back to the lowest value when the patient provides a drug-positive sample or fails to submit a scheduled sample.

2.2 Research Evidence

Multiple clinical trials have demonstrated the efficacy of voucher-based and prized-based CM interventions in reducing drug use in a variety of research- and community-based settings (Higgins et al. 1994; Petry et al. 2005b; see also Lussier et al. 2006; Prendergast et al. 2006). Although a majority of the studies have focused on decreasing cocaine use, studies have also found that CM is effective in reducing cigarette smoking (Roll et al. 1996; Hunt et al. 2010; Alessi and Petry 2014), alcohol (Petry et al. 2000), opioids (Silverman et al. 1996; Petry and Martin 2002), marijuana (Budney et al. 2006; Kadden et al. 2007), and benzodiazepines (Stitzer et al. 1992). Results of two meta-analyses comparing CM interventions to control conditions found CM to be efficacious in decreasing drug use (Lussier et al. 2006; Prendergast et al. 2006). Furthermore, a meta-analysis of all psychosocial treatments for substance use disorders found that CM interventions had the largest effect size, $d = 0.58$, while the next largest effect size was for relapse prevention interventions, $d = 0.32$ (Dutra et al. 2008).

As previously mentioned, the CM literature is extensive and this chapter will only discuss a few examples of CM interventions for substance use disorders conducted in psychosocial treatment settings, the most common and generalizable settings. In one of the earliest well-designed studies, Higgins et al. (1994) randomized 42 cocaine-dependent patients to either standard treatment or standard treatment plus voucher-based CM. The standard treatment in this study was based on the community reinforcement approach, which involved relationship counseling, employment counseling, behavioral skills training, relapse prevention, and social and recreational skills counseling. Results showed that 55 % of patients in the CM condition achieved at least 10 weeks of continuous abstinence, but only 15 % of the patients in the standard treatment condition achieved at least 10 continuous weeks of abstinence. Furthermore, the rates of treatment completion at 3 and 6 months for patients in the CM condition were 90 and 75 %, compared to 65 and 40 % for patients in the standard treatment condition.

In the largest CM study conducted in the National Drug Abuse Treatment Clinical Trials Network, Petry et al. (2005b) randomized over 400 patients with cocaine or methamphetamine use disorders to either standard treatment at the patients' clinics, consisting primarily of group counseling, or standard treatment plus prize-based CM. Compared to patients in the standard treatment condition, patients in the CM condition achieved longer durations of continuous abstinence (4.4 vs. 2.6 weeks, respectively) and stayed in treatment longer (19.2 vs. 8.0 weeks, respectively).

Most studies of CM in psychosocial treatment settings reinforce abstinence explicitly, but these studies also find that CM improves treatment retention along with abstinence outcomes, such as staying in treatment longer and attending more therapy sessions (Higgins et al. 1994, 2000b; Petry et al. 2000, 2005b). Improving treatment retention is important because attrition from substance abuse treatment programs is extraordinarily high and length of time in treatment is a stable predictor of treatment outcomes (Simpson and Sells 1982; Hubbard 1989; Hubbard et al. 1997). Overall, when CM is added to standard care, patients remain in treatment longer (e.g., Higgins et al. 1994; Petry et al. 2000, 2005b). Figure 1 shows the percentage of patients completing treatment in the standard treatment and CM conditions from three representative studies: Higgins et al. (1994), Petry et al. (2000, 2005b).

It is clear that CM interventions targeting substance use are effective at improving treatment outcomes, especially during the treatment period. However, there is inconsistent evidence on the long-term efficacy of CM. Some studies have found CM maintains statistically significant post-treatment benefits on abstinence (e.g., Higgins et al. 1995, 2000b) while others have not (Rawson et al. 2006; Sigmon and Higgins 2006). However, in no study have patients who received CM demonstrated worse outcomes compared to their non-CM counterparts. Thus, applying extrinsic reinforcers to substance use treatment patients does not ever reduce their long-term likelihood of abstinence. The underlying mechanisms involved in the post-treatment maintenance effects of CM, as well as other psychotherapies are not clear and multiple factors likely play a role in long-term outcomes beyond the treatment period. Intrinsic motivation to remain abstinent is one possible factor that may contribute to the long-term effects of psychotherapies, and the next section will discuss the influence of CM on patient motivation.

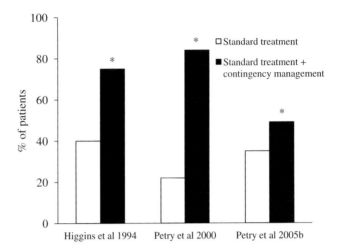

Fig. 1 Percentage of patients completing treatment in the standard treatment and contingency management conditions from Higgins et al. (1994) and Petry et al. (2000, 2005b). The *asterisks* indicate statistically significant differences between groups, $p < 0.05$

3 Patient Motivation and Contingency Management

Because both intrinsic and extrinsic motivation appears to play important roles in substance use disorder treatment and recovery (Miller 1985; DiClemente 1999), it is important to consider how treatments targeting extrinsic motivational processes, such as CM, also affect intrinsic motivation to change behavior and remain abstinent. This section provides an overview of the differing perspectives on the influence of external rewards on intrinsic motivation to change behavior and then reviews research on the effects of CM interventions targeting substance use behavior on intrinsic motivation.

3.1 External Rewards and Intrinsic Motivation

Substance use behavior is influenced by intrinsic and extrinsic motivation. Intrinsic motivation is the internal desire to do something because it is self-fulfilling and is influenced by feelings of autonomy, self-determination, and competence; extrinsic motivation involves doing something for reasons external to the individual, such as to receive rewards or avoid punishment (DiClemente 1999; Ryan and Deci 2000). There are conflicting perspectives regarding the effects of rewards on intrinsic motivation (Deci et al. 1999; Cameron et al. 2001; Promberger and Marteau 2013). Some argue that there is clear evidence that rewards undermine intrinsic motivation (Deci et al. 1999), while others suggest that the evidence is inconclusive and in certain situations, external rewards can actually enhance intrinsic motivation (Cameron et al. 2001; Promberger and Marteau 2013).

Much of the research in psychological literature on the effects of rewards on intrinsic motivation is grounded in the cognitive evaluation theory, which proposes that intrinsic motivation is necessary for sustained behavior change and that external rewards undermine intrinsic motivation (Deci and Ryan 1985; Ryan and Deci 2000). According to the cognitive evaluation theory, feelings of competence, autonomy, and self-determination are essential to behavior change and factors that increase these feelings enhance intrinsic motivation, while factors that decrease these feelings decrease intrinsic motivation (Deci and Ryan 1985; Ryan and Deci 2000). Cognitive evaluation theorists argue that external rewards reduce feelings of competence, autonomy, and self-determination because they shift the locus of causality from factors internal to the individual to factors external to the individual (Deci and Ryan 1985; Ryan and Deci 2000). Furthermore, they reason that although rewards can change behavior initially, once they are removed, behavior will return to baseline levels because individuals are not intrinsically motivated to maintain the behavior change (Deci et al. 1999).

Numerous studies in the cognitive evaluation theory literature have investigated the effects of rewards on behavior changes in non-clinical settings. Results of a meta-analysis revealed that external rewards undermine intrinsic motivation (Deci et al. 1999). However, in a similar meta-analysis, Cameron et al. (2001) concluded that rewards do not always undermine intrinsic motivation and actually may enhance intrinsic motivation in certain situations, such as when initially the behavior rarely occurs. One explanation for the disparate findings is that the two meta-analyses included different studies. Deci et al. (1999) only included studies with high-interest tasks, such as playing a game, and excluded studies with low-interest tasks, for example proofreading a paper. They only investigated high-interest tasks because the cognitive evaluation theory field was primarily concerned with the effects of rewards on intrinsic motivation for interesting activities (Deci et al. 1999). Cameron et al. (2001) believed a comprehensive assessment of rewards' effects on intrinsic motivation should include studies with high-interest and low-interest tasks.

Another difference between the two meta-analyses was the procedure used to categorize studies by reward contingency. Deci et al. (1999) classified studies as task non-contingent, engagement-contingent, completion-contingent, or performance-contingent. A potential issue with this classification system was that the categories were too broad, and studies with different procedures were included in the same category (Cameron et al. 2001). For example, the performance-contingent category included studies that provided rewards for doing well, each problem solved, achieving a certain score, or exceeding a norm. Cameron et al. (2001) used a more specific classification system and created separate categories for providing rewards for doing well, doing a task, finishing or completing a task, each unit solved, surpassing a score, and exceeding a norm.

The results from these meta-analyses, however, are not easily generalizable to health-related behaviors such as substance use in treatment seeking populations, e.g., substance abusers who have at least some motivation to initiate treatment. In general, the early studies included in the meta-analyses tested the effects of rewards on time spent in simple activities such as completing puzzles or drawing pictures, and participants in these studies typically were college students or children (Cameron et al. 2001; Deci et al. 1999), not adults with serious physical and mental health problems who were receiving extrinsic reinforcers for health-related behavior changes.

Few studies have specifically examined the effects of rewards on intrinsic motivation to change health-related behavior. Promberger and Marteau (2013) reviewed studies examining the effects of rewards on health behaviors and concluded that there is little evidence that supports the hypothesis that rewards undermine intrinsic motivation in these contexts. In fact, for health-related behaviors that depend on self-control, rewards may actually increase intrinsic motivation because they enhance feelings of competence (Promberger and Marteau 2013). However, most of the studies did not explicitly assess motivation to change health-related behaviors. These studies found that that patients receiving the rewards were more likely to change their behavior compared to patients in the non-reward group, but they did not assess internal motivation to change explicitly (e.g., Paul-Ebhohimhen and Avenell 2008; Volpp et al. 2008, 2009).

The results from previous studies indicate that rewards used in CM interventions do not necessarily undermine intrinsic motivation. This might occur for several reasons. First, CM interventions for substance use disorders typically reinforce abstinence, which initially occurs at low levels, and according to the cognitive evaluation theory literature, rewards are less likely to be associated with reductions in intrinsic motivation when the initial levels of the behavior are low. Second, the escalating reinforcement schedule typically used in CM interventions may actually enhance intrinsic motivation by increasing perceived self-determination and competence. The following section will further discuss the impact of CM on intrinsic motivation.

3.2 Contingency Management and Intrinsic Motivation

Only two CM intervention studies conducted in psychosocial treatment settings have specifically assessed the influence of CM on substance use disorder patients' intrinsic motivation to change substance use, as measured by the University of Rhode Island Change Assessment (URICA) (Budney et al. 2000; Ledgerwood and Petry 2006). The URICA is a self-report measure of intrinsic motivation that assesses readiness to change substance use behavior (Prochaska et al. 1992; DiClemente et al. 2004). It contains four subscales that coincide with the URICA's stages of change: precontemplation, contemplation, action, and maintenance (Diclemente et al. 2004). Patients respond to each item using a five-point Likert-type scale ranging from 1 (strongly disagree) to 5 (strongly agree), and higher scores are suggestive of higher perceived readiness to change and intrinsic motivation.

Budney et al. (2000) conducted a randomized trial comparing three psychosocial treatments for cannabis dependence: Motivational enhancement therapy that involved motivational interviewing techniques to promote changes in marijuana use; motivational enhancement plus behavioral coping-skills therapy that included additional sessions focusing on coping skills related to maintaining abstinence; and motivational enhancement plus behavior coping-skills therapy and voucher-based CM. Results indicated that CM had no effect relative to other conditions on impacting readiness to change substance use, as measured by the URICA. Ledgerwood and Petry (2006) assessed intrinsic motivation to change substance use in drug-dependent patients who were enrolled in a randomized clinical trial of prize-based CM. They also did not find that CM interventions positively or negatively affected patients' readiness to change substance use assessed by the URICA relative to standard care.

A possible limitation of these studies is that the measure of intrinsic motivation, the URICA, may not be a valid measure of intrinsic motivation in patients who are engaged in or recently completed CM treatment. Although the URICA is one of the most commonly used measures of motivation to change substance use behavior, studies assessing its psychometric properties have produced mixed results (DiClemente and Hughes 1990; Willoughby and Edens 1996; El-Bassel et al. 1998; Edens and Willoughby 2000; Blanchard et al. 2003; Field et al. 2009). Further and perhaps most importantly in terms of assessing the impact of CM on motivation to change, changes in URICA scores in the context of treatment do not reliably parallel changes in substance use behavior (Callaghan et al. 2008; Field et al. 2009). Most psychometric validation studies of the URICA were either cross-sectional studies evaluating construct and concurrent validity (e.g., DiClemente and Hughes 1990; El-Bassel et al. 1998; Siegal et al. 2001) or longitudinal studies assessing the predictive validity of pre-treatment URICA scores on subsequent treatment outcomes (e.g., Willoughby and Edens 1996; Edens and Willoughby 2000; Pantalon et al. 2002; Blanchard et al. 2003). Cross-sectional and predictive validity studies do not provide information about the validity of using the URICA to assess the effects of a treatment on intrinsic motivation, i.e., change in intrinsic motivation throughout treatment.

Several items on the maintenance and action subscales of the URICA may be problematic and confusing for patients who have experienced sustained periods of abstinence. Many of the items are more relevant to a patient's feelings about substance use before beginning treatment such as these items from the maintenance subscale: "It worries me that I might slip back on a problem I have already changed, so I am here to seek help" and "I'm not following through with what I already changed as well as I had hoped, and I'm here to prevent a relapse of the problem." Patients who have successfully completed treatment cannot respond appropriately to these items. Problematic action subscale items include: "Even though I'm not always successful in changing, I am at least working on my problem," and "I have started working on my problems, but I would like help." Hypothetically, patients experiencing sustained periods of abstinence may respond that they "strongly disagree" or "disagree" to these items because they have experienced some degree of treatment success and are not seeking additional help.

Taken together, the findings from the two previous studies do not indicate that CM affects patients' intrinsic motivation, but because of potential issues with the URICA it is difficult to determine if the results from previous studies investigating the effects of CM or other treatments on intrinsic motivation are valid. Future research should develop measures of intrinsic motivation that are more appropriate for patients in treatment or who have successfully completed it and include more items that relate to maintaining behavior change with less emphasis on pre-contemplation and contemplation issues. Ideally, a comprehensive instrument that can address a range of behaviors and motivation related to them would be useful, especially because CM interventions are now being applied to a multitude of health behavior issues, including enhancing weight loss efforts (Volpp et al. 2008; Petry et al. 2011), increasing exercise (Petry et al. 2013; Andrade et al. 2014), and improving medication adherence (Petry et al. 2012, 2015).

4 Summary

Intrinsic motivators, such as feelings of competence and self-determination, and extrinsic motivators, such as financial incentives and legal pressures, influence patients' desire to change substance use behavior and maintain abstinence. Intrinsic motivation to remain abstinent is a possible factor that may contribute to the long-term effects of CM as well as other psychotherapies, but there has been little research on this topic. Studies that have examined CM's impact on intrinsic motivation did not find evidence that CM affects patients' intrinsic motivation, but these studies may not have used an instrument that sensitively or accurately assesses intrinsic motivation in patients who have maintained periods of sustained abstinence. Clearly, additional research is needed before definitive conclusions can be made regarding CM's positive or negative effects on intrinsic motivation. Future research should focus on developing measures of intrinsic motivation that are better able to assess intrinsic motivation to maintain abstinence for patients engaged in treatment and who have already experienced sustained periods of behavior change. Greater understanding of mechanisms involved in the efficacy of CM and its impact on motivation to change ultimately may assist in improving treatments for substance use disorders as well as other health care conditions.

References

Alessi SM, Petry NM (2014) Smoking reductions and increased self-efficacy in a randomized controlled trial of smoking abstinence-contingent incentives in residential substance abuse treatment patients. Nicotine Tob Res Off J Soc Res Nicotine Tob 16:1436–1445. doi:10.1093/ntr/ntu095

Andrade LF, Barry D, Litt MD, Petry NM (2014) Maintaining high activity levels in sedentary adults with a reinforcement-thinning schedule. J Appl Behav Anal 47:523–536. doi:10.1002/jaba.147

Blanchard KA, Morgenstern J, Morgan TJ et al (2003) Motivational subtypes and continuous measures of readiness for change: concurrent and predictive validity. Psychol Addict Behav 17:56–65. doi:10.1037/0893-164X.17.1.56

Budney AJ, Higgins ST, Radonovich KJ, Novy PL (2000) Adding voucher-based incentives to coping skills and motivational enhancement improves outcomes during treatment for marijuana dependence. J Consult Clin Psychol 68:1051–1061

Budney AJ, Moore BA, Rocha HL, Higgins ST (2006) Clinical trial of abstinence-based vouchers and cognitive-behavioral therapy for cannabis dependence. J Consult Clin Psychol 74:307–316. doi:10.1037/0022-006X.4.2.307

Callaghan RC, Taylor L, Moore BA et al (2008) Recovery and URICA stage-of-change scores in three marijuana treatment studies. J Subst Abuse Treat 35:419–426. doi:10.1016/j.jsat.2008.03.004

Cameron J, Banko KM, Pierce WD (2001) Pervasive negative effects of rewards on intrinsic motivation: the myth continues. Behav Anal MABA 24:1–44

Deci EL, Ryan RM (1985) Intrinsic motivation and self-determination in human behavior. Plenum, New York

Deci EL, Koestner R, Ryan RM (1999) A meta-analytic review of experiments examining the effects of extrinsic rewards on intrinsic motivation. Psychol Bull 125:627–668. doi:10.1037/0033-2909.125.6.627

DiClemente CC (1999) Motivation for change: implications for substance abuse treatment. Psychol Sci 10:209–213. doi:10.1111/1467-9280.00137

DiClemente CC, Hughes SO (1990) Stages of change profiles in outpatient alcoholism treatment. J Subst Abuse 2:217–235

DiClemente CC, Schlundt D, Gemmell L (2004) Readiness and stages of change in addiction treatment. Am J Addict 13:103–119. doi:10.1080/10550490490435777

Dutra L, Stathopoulou G, Basden SL et al (2008) A meta-analytic review of psychosocial interventions for substance use disorders. Am J Psychiatry 165:179–187. doi:10.1176/appi.ajp.2007.06111851

Edens JF, Willoughby FW (2000) Motivational patterns of alcohol dependent patients: a replication. Psychol Addict Behav J Soc Psychol Addict Behav 14:397–400

El-Bassel N, Schilling RF, Ivanoff A et al (1998) Stages of change profiles among incarcerated drug-using women. Addict Behav 23:389–394. doi:10.1016/S0306-4603(97)00036-1

Field CA, Adinoff B, Harris TR et al (2009) Construct, concurrent and predictive validity of the URICA: data from two multi-site clinical trials. Drug Alcohol Depend 101:115–123. doi:10.1016/j.drugalcdep.2008.12.003

Higgins ST, Budney AJ, Bickel WK et al (1994) Incentives improve outcome in outpatient behavioral treatment of cocaine dependence. Arch Gen Psychiatry 51:568–576

Higgins ST, Budney AJ, Bickel WK et al (1995) Outpatient behavioral treatment for cocaine dependence: one-year outcome. Exp Clin Psychopharmacol 3:205–212. doi:10.1037/1064-1297.3.2.205

Higgins ST, Badger GJ, Budney AJ (2000a) Initial abstinence and success in achieving longer term cocaine abstinence. Exp Clin Psychopharmacol 8:377–386

Higgins ST, Wong CJ, Badger GJ et al (2000b) Contingent reinforcement increases cocaine abstinence during outpatient treatment and 1 year of follow-up. J Consult Clin Psychol 68:64–72. doi:10.1037/0022-006X.68.1.64

Higgins ST, Alessi SM, Dantona RL (2002) Voucher-based incentives: a substance abuse treatment innovation. Addict Behav 27:887–910. doi:10.1016/S0306-4603(02)00297-6

Higgins ST, Silverman K, Heil SH (2008) Contingency management in substance abuse treatment. Guilford Press, New York

Hubbard RL (1989) Drug abuse treatment: a national study of effectiveness. University of North Carolina Press, Chapel Hill

Hubbard RL, Gail S, Flynn PM et al (1997) Overview of 1-year follow-up outcomes in the drug abuse treatment outcome study (DATOS). Psychol Addict Behav 11:261–278. doi:10.1037/0893-164X.11.4.261

Hunt YM, Rash CJ, Burke RS, Parker JD (2010) Smoking cessation in recovery: comparing 2 different cognitive behavioral treatments. Addict Disord Their Treat 9:64–74. doi:10.1097/ADT.0b013e3181bf0310

Kadden RM, Litt MD, Kabela-Cormier E, Petry NM (2007) Abstinence rates following behavioral treatments for marijuana dependence. Addict Behav 32:1220–1236. doi:10.1016/j.addbeh.2006.08.009

Ledgerwood DM, Petry NM (2006) Does contingency management affect motivation to change substance use? Drug Alcohol Depend 83:65–72. doi:10.1016/j.drugalcdep.2005.10.012

Lussier JP, Heil SH, Mongeon JA et al (2006) A meta-analysis of voucher-based reinforcement therapy for substance use disorders. Addiction 101:192–203. doi:10.1111/j.1360-0443.2006.01311.x

Miller WR (1985) Motivation for treatment: a review with special emphasis on alcoholism. Psychol Bull 98:84–107. doi:10.1037/0033-2909.98.1.84

National Institute for Health and Clinical Excellence (2007) NICE clinical guideline 51: drug misuse: psychosocial interventions. 2007. http://www.nice.org.uk/guidance/cg51. Accessed 29 Dec 2014

Pantalon MV, Nich C, Franckforter T, Carroll KM (2002) The URICA as a measure of motivation to change among treatment-seeking individuals with concurrent alcohol and cocaine problems. Psychol Addict Behav 16:299–307. doi:10.1037/0893-164X.16.4.299

Paul-Ebhohimhen V, Avenell A (2008) Systematic review of the use of financial incentives in treatments for obesity and overweight. Obes Rev Off J Int Assoc Study Obes 9:355–367. doi:10.1111/j.1467-789X.2007.00409.x

Petry NM (2012) Contingency management for substance abuse treatment: a guide to implementing this evidence-based practice. Routledge, New York

Petry NM, Martin B (2002) Low-cost contingency management for treating cocaine- and opioid-abusing methadone patients. J Consult Clin Psychol 70:398–405

Petry NM, Martin B, Cooney JL, Kranzler HR (2000) Give them prizes, and they will come: contingency management for treatment of alcohol dependence. J Consult Clin Psychol 68:250–257

Petry NM, Martin B, Simcic F (2005a) Prize reinforcement contingency management for cocaine dependence: integration with group therapy in a methadone clinic. J Consult Clin Psychol 73:354–359. doi:10.1037/0022-006X.73.2.354

Petry NM, Peirce JM, Stitzer ML et al (2005b) Effect of prize-based incentives on outcomes in stimulant abusers in outpatient psychosocial treatment programs: a national drug abuse treatment clinical trials network study. Arch Gen Psychiatry 62:1148–1156. doi:10.1001/archpsyc.62.10.1148

Petry NM, Alessi SM, Hanson T, Sierra S (2007) Randomized trial of contingent prizes versus vouchers in cocaine-using methadone patients. J Consult Clin Psychol 75:983–991. doi:10.1037/0022-006X.75.6.983

Petry NM, Barry D, Pescatello L, White WB (2011) A low-cost reinforcement procedure improves short-term weight loss outcomes. Am J Med 124:1082–1085. doi:10.1016/j.amjmed.2011.04.016

Petry NM, Rash CJ, Byrne S et al (2012) Financial reinforcers for improving medication adherence: findings from a meta-analysis. Am J Med 125:888–896. doi:10.1016/j.amjmed.2012.01.003

Petry NM, Andrade LF, Barry D, Byrne S (2013) A randomized study of reinforcing ambulatory exercise in older adults. Psychol Aging 28:1164–1173. doi:10.1037/a0032563

Petry NM, DePhilippis D, Rash CJ et al (2014) Nationwide dissemination of contingency management: the veterans administration initiative. Am J Addict Am Acad Psychiatr Alcohol Addict 23:205–210. doi:10.1111/j.1521-0391.2014.12092.x

Petry NM, Alessi SM, Byrne S, White WB (2015) Reinforcing adherence to antihypertensive medications. J Clin Hypertens Greenwich Conn 17:33–38. doi:10.1111/jch.12441

Prendergast M, Podus D, Finney J et al (2006) Contingency management for treatment of substance use disorders: a meta-analysis. Addiction 101:1546–1560. doi:10.1111/j.1360-0443.2006.01581.x

Prochaska JO, DiClemente CC, Norcross JC (1992) In search of how people change: applications to addictive behaviors. Am Psychol 47:1102–1114. doi:10.1037/0003-066X.47.9.1102

Promberger M, Marteau TM (2013) When do financial incentives reduce intrinsic motivation? Comparing behaviors studied in psychological and economic literatures. Health Psychol 32:950–957. doi:10.1037/a0032727

Rawson RA, McCann MJ, Flammino F et al (2006) A comparison of contingency management and cognitive-behavioral approaches for stimulant-dependent individuals. Addict Abingdon Engl 101:267–274. doi:10.1111/j.1360-0443.2006.01312.x

Roll JM, Higgins ST (2000) A within-subject comparison of three different schedules of reinforcement of drug abstinence using cigarette smoking as an exemplar. Drug Alcohol Depend 58:103–109

Roll JM, Higgins ST, Badger GJ (1996) An experimental comparison of three different schedules of reinforcement of drug abstinence using cigarette smoking as an exemplar. J Appl Behav Anal 29:495–504; quiz 504–505. doi: 10.1901/jaba.1996.29-495

Roll JM, Reilly MP, Johanson C-E (2000) The influence of exchange delays on cigarette versus money choice: a laboratory analog of voucher-based reinforcement therapy. Exp Clin Psychopharmacol 8:366–370. doi:10.1037/1064-1297.8.3.366

Rowan-Szal G, Joe GW, Chatham LR, Simpson DD (1994) A simple reinforcement system for methadone clients in a community-based treatment program. J Subst Abuse Treat 11:217–223

Ryan RM, Deci EL (2000) Self-determination theory and the facilitation of intrinsic motivation, social development, and well-being. Am Psychol 55:68–78. doi:10.1037/0003-066X.55.1.68

Siegal HA, Li L, Rapp RC, Saha P (2001) Measuring readiness for change among crack cocaine users: a descriptive analysis. Subst Use Misuse 36:687–700

Sigmon SC, Higgins ST (2006) Voucher-based contingent reinforcement of marijuana abstinence among individuals with serious mental illness. J Subst Abuse Treat 30:291–295. doi:10.1016/j.jsat.2006.02.001

Silverman K, Wong CJ, Higgins ST et al (1996) Increasing opiate abstinence through voucher-based reinforcement therapy. Drug Alcohol Depend 41:157–165

Simpson DD, Sells SB (1982) Effectiveness of treatment for drug abuse. Adv Alcohol Subst Abuse 2:7–29. doi:10.1300/J251v02n01_02

Stitzer ML, Iguchi MY, Felch LJ (1992) Contingent take-home incentive: effects on drug use of methadone maintenance patients. J Consult Clin Psychol 60:927–934

Volpp KG, John LK, Troxel AB et al (2008) Financial incentive-based approaches for weight loss: a randomized trial. J Am Med Assoc (JAMA) 300:2631–2637. doi:10.1001/jama.2008.804

Volpp KG, Troxel AB, Pauly MV et al (2009) A randomized, controlled trial of financial incentives for smoking cessation. N Engl J Med 360:699–709. doi:10.1056/NEJMsa0806819

Willoughby FW, Edens JF (1996) Construct validity and predictive utility of the stages of change scale for alcoholics. J Subst Abuse 8:275–291

Index

A
Accumbens, 232–236, 240, 241, 243–245
Accumbens DA, 232–237, 239–241
Accumbens dopamine, 231
Action–outcome association, 261, 268, 276
Activity/rest rhythms, 145
Actor-critic, 380, 401
Addiction, 114–116, 119, 123, 127, 327
　alcohol, 281, 282
　craving, 115
　drug euphoria, 118
　drug habits, 118
　relapse, 114, 117, 119
　withdrawal, 114, 115, 118, 119
Addicts, 118, 125
Adenosine, 147
Adipose, 23
Adiposity, 24, 28
Adiposity-signaling, 25
Adiposity signals, 22, 23
Adolescence, 148
Adrenal steroids, 39
　glucocorticoid receptors (GRs), 39
　mineralocorticoid receptors (MRs), 39
Affiliative, 52
Affiliative behaviors, 53, 55
Aggression, 53–56, 63, 65–75, 79, 81
Aging, 145
Agouti-related protein, 25
AgRP, 24, 25
Alcohol, 160
Allostasis, 23, 477–479, 481, 491, 493
α-melanocyte concentrating hormone, 24
αMSH, 24, 25
Alzheimer's, 145
Amplitude, 141, 142, 144

Amygdala, 27, 28, 271, 277, 283, 320, 384, 392, 397, 398
Anergia, 242
Anhedonia, 159, 338, 375, 378, 385, 389–392, 395, 396, 398, 399, 402, 454, 455
Animal models, 516, 521
　conditioned reinforcement, 519
　rGT, 509, 522, 523
　sign-tracking, 517
Anorexia nervosa, 29
Anterior cingulate cortex (ACC), 391, 393
Anterior insula (AI), 383, 399, 397
Anticipate, 151, 156
Anticipation, 139, 146, 158, 160
Antipsychotic, 377
ARC, 24, 25
Arcuate nucleus, 24
Arousal, 139, 141, 146
Artificial sweeteners, 29
Association
　affiliative, 55
　affiliative behavior, 52, 53, 62, 63, 65, 66, 69, 74
　aggression, 55, 70–72
Attention, 199, 208, 220
Attention deficit hyperactivity disorder, 159
Avolition, 376, 377, 385, 387, 389–391, 395, 397, 398, 402, 403, 452, 454, 456

B
Balloon Analog Risk Task (BART), 388
Basal ganglia (BG), 379, 380, 387, 389
Behavior, 146
Behavioral activation, 231, 232, 237, 239, 243, 245, 246
Binge eating, 123, 126

Bipolar affective disorder, 388
Blood–brain barrier, 23
Blood glucose, 16
Body fat, 18, 24
Body weight, 24–26, 28

C
Caffeine, 147
Cancers, 145
Cardiovascular disease, 145
Cephalic insulin, 20
Cephalic responses, 20
Cerebral cortex, 152
Cholecystokinin (CCK), 17, 19, 25, 26
Circadian, 139, 146
Clock, 143
Cognition, 28
Cognitive, 21, 26, 29
Cognitive performance, 144
Cognitive remediation, 534
Comfort foods, 28
Common currency, 314
Conditioned, 220
Contingency management, 327, 570, 571, 574, 576
Corticosterone, 157
Craving, 117, 122, 123, 326
Cross-sensitization, 116, 117, 121
Cue, 511, 514–523, 520
    audiovisual, 515
    light, 514–517, 521
    sound, 514, 515
    tone, 519, 522

D
D2R-OE mice, 461–466
DA (Dopamine),*see* Dopamine, 74, 75, 78, 79, 83, 231, 233–237, 239–241, 243–246
Dating, 139
Decision-making, 199, 205, 208, 219, 220, 314, 508, 509, 511, 515, 520–523
Decision making (DM), 379, 383–385, 389, 391
Decision-making systems, 316
    approach and avoidance, 320
    deliberation, 317
    deliberative, 321
    intransitivity, 318
    pavlovian, 321
    pavlovian action-selection systems, 317
    procedural, 319, 321
    procedural action chains, 317

    revealed-preference, 318
    willingness-to-pay, 318, 319
Deficit syndrome, 377, 387
Delay discounting (DD), 386, 389, 391
Deliberation, 318, 319
Depression, 140, 145, 159, 231, 242, 243, 245, 246
Desynchrony, 142
Devaluation, 267, 268, 281, 284, 386, 389, 393
Dopamine, 22, 27, 53, 111, 113, 115, 116, 121, 124, 232, 340, 512, 513, 518, 521
    D_2, 522
    D_3, 523
    D_4, 522
Dopaminergic, 27, 28, 158, 161
Dorsal striatum (DS), 152, 387, 399
    dorsolateral striatum, 277
    ventral striatum, *see* nucleus accumbens
Dorsolateral prefrontal cortex (DLPFC), 393, 398
Dorsomedial hypothalamus, 147
Drinking, 146
Drug, 127
Drug addiction, 112, 115, 118, 473, 475, 480, 485, 492
    relapse, 112
    withdrawal, 113
Drug learning, 478, 482, 484, 486, 493

E
Eating, 139, 142, 150
Effort, 375, 386
Effort-cost, 386, 385
Effort-related choice, 457, 464
Elasticity, 327
Endowment effect, 325
Episodic future thinking, 317
Estrogens, 37
    chromatin, 37
    estrogen receptor-alpha, 37
    response elements, 37
Estrus, 155
Exercise, 144, 149, 160
Expectancy-value, 536
Expected value (EV), 375, 377, 379–386, 388, 389, 391–395, 398, 400–403
Extinction, 263, 264, 266, 270, 275
Extinction learning, 390, 393

F
Fatigue, 242, 245
Fatty acid, 17–19

Fear-conditioning, 390, 393
Feedback, 148, 149, 157
Feeding, 143, 144, 146, 153
fMRI, 296, 298, 302, 381, 382, 391, 397, 399
Fronto-striatal, 380, 394, 395, 397

G

Gambling, 119–122, 127, 508–511, 514–516, 522, 523
    Iowa Gambling Task (IGT), 508, 509, 515
    pathological gamblers, 121
    slot machine, 514, 515
Gastrointestinal tract, 151
Ghrelin, 20, 148, 153, 157, 160
Glucose, 17–20, 23, 27
Goal-directed, 293, 296–300, 302, 306–308
Goal-directed behaviour, 460
Goal-tracking, 323

H

Habit, 262, 277, 477–479, 486, 491, 492
Habitual, 298–300, 302, 306–308
Hedonic, 21, 22, 27, 29
Hedonically, 26, 28
Hedonics, 26, 29
High-fructose corn syrup, 29
Hippocampus, 152, 317, 393
Histamine, 147
Homeostasis, 17, 23
Homeostatic, 16, 17, 21–23, 25–29, 142, 146, 147, 160
Homeostatic adiposity signals, 25
Homeostatic signals, 21
Hormones, 146
Hypocretin, 147
Hypothalamic, 25, 28
Hypothalamus, 22–25, 27

I

Imagination, 319
Incentive, 18
    instrumental, 262
    pavlovian, 262
Incentive motivation, 482
Incentive salience, 106, 107, 112, 116, 124, 125
Incentive sensitization, 112, 114, 115, 118, 119, 127, 478, 482–486, 491–493
    craving, 112
Inhibition, 280
Insomnia, 160

Instrumental learning, 262, 276, 277, 284
Insulin, 17, 20, 22–25
International Affective Pictures Set (IAPS), 392, 396
Intrinsic motivation, 534, 569, 570, 573, 574–578
Iowa Gambling Task (IGT), 387, 388

J

Jet lag, 142, 145, 154, 158
Jet lag or shift work, 148
Junk food, 124

L

Lateral hypothalamus (LH), 23, 24, 27, 28
Leptin, 22–26, 148
LH orexinergic, 28
Light, 144
Liking, 27, 106, 107, 109, 111–119, 123, 125, 127
    euphoria, 113, 114
    hot spots, 109
    pleasure, 112, 114
    taste reactivity, 109
    tolerance, 118
Limbic areas, 27
Limbic brain, 23
Lnsulin, 23
Locus coeruleus, 147
Lordosis, 36
    estrogens, 36
    progesterone, 36
Luteinizing hormone, 155

M

Maternal, 146
Maternal aggression, 72
Maternal behavior, 65, 157
Mating, 146, 156
Mesolimbic, 113, 122
Mesolimbic dopamine, 114, 123
Mesolimbic tract, 27
Midbrain (MB), 382, 391, 392, 399
Monetary Incentive Delay (MID) task, 381
Morality, 324
Motivation, 36, 113, 199, 202, 207, 214, 231–233, 237, 242, 246, 390, 403, 451–454, 456–461, 465–467, 477, 478, 482, 483, 489, 491
    estradiol, 36
    female sex behavior, 36

Motivation (*cont.*)
  hunger, 262, 264–266
  satiety, 266, 282
  thirst, 262, 264
Motivational, 451–454, 459–462, 464, 466, 467
Motivational deficits, 362, 363, 365, 368
MRI, 381, 391, 393, 397, 399

N
Narcolepsy, 147
Negative symptom, 357–368, 375, 377, 378, 386, 387, 389, 390, 393, 394, 398, 400–402
Neophobic, 20
Neural systems, 482
Neuroimaging, 382, 383, 391, 393–395, 397, 398, 402
Neuropeptide Y (NPY), 24, 25
Neurotransmitters
  acetylcholine, 274, 276
  dopamine, 272–274
  opiates, 274
  opioid, 275
Night eating syndrome, 153
Non-homeostatic, 17, 21–23, 26–29
NPY/AgRP, 25
Nucleus, 27, 147
Nucleus accumbens, 23, 27, 28, 152, 259, 270, 272, 278, 317, 321
Nucleus of the solitary (NTS), 21

O
Obese, 124
Obesity, 26, 29, 123, 125, 126, 145, 282
Obsessive-compulsive disorder, 323
Operant learning, 378, 385
Optimal foraging theory, 314, 326
Orbitofrontal, 318, 321
Orbitofrontal cortex (OFC), 380, 382–384, 386, 388, 401
Orexin, 147
Orexin-A, 28
Oscillation, 144
Outcome Representation, 458, 466
Ovulation, 155

P
Pair bonds
  affiliative behaviors, 54
Parasympathetic nervous systems, 25
Paraventricular nuclei (PVN), 23–25
Paraventricular nucleus of the thalamus (PVT), 21, 28

Parenting, 139, 146
Parkinson's, 145
Patch-foraging, 326
Pavlovian, 292, 300–304, 306–308
Pavlovian action-selection, 320, 321
Pavlovian conditioning, 391
Pavlovian–instrumental transfer (PIT), 393, 316, 325
Pavlovian learning, 284
Pavlovian learning system, 320
Peptide proopiomelanocortin, 24
Periaqueductal gray, 320
Period, 143
Peripheral oscillators, 153
Piriform cortex, 397
Pituitary gland, 25
Positron emission tomography (PET), 397
Posterior cingulate gyrus (PCG), 393
Prediction error (PE), 199, 208, 220, 379, 380, 398
Preference assessments, 455
Prefrontal cortex (PFC), 317, 379, 380, 384, 385, 392, 397, 399, 401, 402
  prelimbic cortex, 277
Primates, 54, 61, 62, 68
Proopiomelanocortin (POMC), 24, 25
Prospect Theory, 383
Psychosis, 392, 400

Q
Q-learning, 380, 401

R
Raphe, 147
Reinforcement, 375, 377, 379, 382, 383, 387, 400–402
Reinforcement learning (RL), 314, 376–380, 386, 387, 389, 392, 398–401
Relapse, 327
Reproductive behavior, 40
  follicular stimulating hormone (FSH), 41
  gonadotropin-releasing hormone (GnRH), 40
  luteinizing hormone (LH), 40
Response-outcome association, 277
Restricted feeding, 152
Reward, 199–201, 220, 233, 234, 240, 242, 260, 264, 268, 270, 272, 275, 282, 284, 376, 378, 380, 382, 384, 385, 388–391, 393, 394, 396–399, 401, 402
Reward circuitry, 26
Rewarding, 28, 29
Reward prediction error (RPE), 376
Rewards, 235, 246, 265, 283

Reward system of the brain, 160
Reward value, 28
Rhythmicity, 144
Rodent models, 451, 466
Rodents, 55, 65, 68

S
Salience, 376, 382–385, 387, 392, 395, 399
Satiation, 17
Satiation signals, 18, 19, 25, 26, 28
Satiety center, 23
Schizophrenia, 140, 159, 357, 358, 368, 375–378, 381, 382, 385, 386, 390
Schizophrenia, 391, 393, 395–398, 400, 402, 403
Scizophrenia, 359–366
Self-determination theory (SDT), 536
Sensitization, 112, 115–117, 120, 123
Sensory-specific satiety (SSS), 390
Set points, 23
Sexual Behavior, 154
Sexually dimorphic, 154
Shift work, 143, 145, 158
Sign-tracking, 323
Sleep, 141, 146, 147
Sleep deprivation, 142
Sleep disorders, 145
Sleep–wake, 146
Sleeping, 139, 142, 146
Social behavior neural network (SBNN), 56, 57, 58, 61, 64, 67, 72–74, 83
Social cognition, 76–78, 82, 83
Social memory, 63, 65
Social recognition memory, 56, 64
Stimulus–outcome associations, 277, 279
Stimulus-response association, 262
Stress, 29, 39, 42, 283
    adrenocorticotropin hormone (ACTH), 39
    arginine vasopressin (AVP), 39
    corticosterone, 39
    corticotropin-releasing hormone (CRF), 39
    hypothalamic–pituitary–adrenal axis (HPA), 39
    kisspeptin inhibition, 42
    LH suppression, 42
    lordosis inhibition, 42, 43
Stress-induced eating, 28

Striatum, 382, 383, 391, 398, 399, 402
Subparaventricular zone, 147
Substance use disorders, 569–572, 576
Substantia nigra, 277
Suprachiasmatic nucleus (SCN), 143
Sympathetic, 25
Synchronization, 145
Synchrony, 144

T
Taste-reactivity testing, 455
Thalamus, 277, 278, 391, 393
Tourette's syndrome, 323
Tuberomammillary, 147

U
Uncertainty, 120, 122

V
Value, 199, 202, 203, 207, 220, 314, 315, 322, 325, 328
    approach/avoidance, 315, 316
    revealed preference, 315
    willingness to pay, 315
Ventral pallidum, 277, 278
Ventral striatum (VS), 317, 380–383, 392, 393, 397–399
Ventral tegmental area (VTA), 21–23, 27, 28, 277, 278, 280
Ventrolateral prefrontal cortex (VLPFC), 384, 397
Ventrolateral preoptic area, 147
Ventromedial nuclei, 23
Ventromedial prefrontal cortex, 318, 321

W
Wake, 146
Wanted, 113
Wanting, 27, 106–108, 110–112, 114–116, 118–124, 126, 127
    cravings, 112, 113
    mesolimbic, 110
    sign-tracking, 107
    stress, 126
Wheel running, 148

Printed in the United States
By Bookmasters